Fundamentals of Soil Dynamics and Earthquake Engineering

Bharat Bhushan Prasad

Professor and Head of Civil Engineering
Ajay Kumar Garg Engineering College, Ghaziabad

Formerly, Director
Department of Science and Technology, Government of Bihar
Patna

PHI Learning Private Limited

Delhi-110092
2026

In fond memory of ***Shri Asoke K. Ghosh*** *(October 1942 – February 2024), Founder Chairman and Managing Director of PHI Learning, whose vision endlessly inspires.*

The Legacy Continues....

Published by Pushpita Ghosh, PHI Learning Private Limited, Rimjhim House, 111, Patparganj Industrial Estate, Delhi-110092 and Printed by Syndicate Binders, A-20, Hosiery Complex, Noida, Phase-II Extension, Noida-201305 (N.C.R. Delhi).

₹1150.00

FUNDAMENTALS OF SOIL DYNAMICS AND EARTHQUAKE ENGINEERING
Bharat Bhushan Prasad

ISBN-978-81-203-2670-5 (Print Book)
ISBN-978-93-5443-026-8 (e-Book)

The export rights of the book are vested solely with the publisher.

In memory of
my wife **Dayamanti Devi**

To my grandsons
Akshay, Unraj, Shivam and **Sahil**

CONTENTS

PREFACE

This text essentially presents the fundamentals of soil dynamics and earthquake engineering for students, young faculty members and practising engineers and consultants. The book is the result of several long years spent in developing the text. The association of the author in the field of teaching, guiding research and providing consultancy services in geotechnical engineering during this period, has provided the opportunity to develop the text.

The text describes the fundamental features of soil dynamics and earthquake engineering—a new discipline of civil engineering which is also popularly known as geotechnical earthquake engineering. The text is a synthesis of various disciplines like geology, geophysics and engineering seismology, classical vibration theory together with probability and reliability analysis. The theme is universal and multidimensional, multidisciplinary where physical and geophysical principles, mathematical theorems and good engineering practice mingle. Earthquakes have been known for centuries and this text is another step in human endeavours to build earthquake resistant designs, which will ultimately minimize the loss of life and property.

This textbook is essentially meant for senior undergraduate students in civil engineering and architecture for a course in Soil-Structure Interaction Studies, and also covers the course in Soil Dynamics and Earthquake Engineering for postgraduate civil engineering students specializing in the area of Soil Dynamics. This book is also intended to provide valuable information to professional geotechnical consultants engaged in investigation, analysis and seismic design of earth retaining structures. The text will be extended to cover the requirements of foundation engineers. For teachers it is a useful reference guide too, for preparation of their lectures and for designing short courses in geotechnical earthquake engineering.

The ultimate goal of the author is to present the basics of soil dynamics and earthquake engineering as a first course to students who have no previous background of vibration theory or dynamics of elastic systems. In order to present soil-structure interactions in a sophisticated manner, a new demand for rigour in analysis has emerged. This text addresses itself by adopting an approach that is mathematically as rigorous as possible, while attempting to provide a large degree of physical insight into principles of soil dynamics and their application to earthquake engineering.

As this subject is developing very fast, an attempt has been made to exclude such analysis, conclusions and recommendations, which are not verified in practice or of dubious nature. Only those theories which are generally universally acceptable and supported by the IS code or other relevant codes and practices at the international level, have been included.

Chapter 1 introduces the basic parameters of soil dynamics and earthquake engineering. This chapter presents historical review of past earthquakes and its effect on structures leading to loss of life and property. In short, this chapter presents the challenges of seismic hazards in India as well as in the global context. Basic concepts and fundamentals of seismology have been presented in Chapter 2 to enable an overview of complete spectrum of earthquakes, their size, intensity and magnitude as well as damage potential. Assuming that the readers have no formal background of theory of vibrations or dynamics of elastic system, Chapters 3 and 4 present the basic principles of vibrations and their practical applications. Chapter 5 introduces the propagation of waves in soil media, propagation of strains, volume change in terms of compression and distortions. The detailed treatment of one-, two- and three-dimensional analysis of body wave propagation as well as surface wave propagation has been presented. Chapter 6 contains the dynamic soil properties and constitutive laws. The experimental aspects of soil dynamics are very important as dynamic soil is location specific which is very different from steel or concrete.

Chapter 7 presents the dynamic earth pressure theory. In classical theory of elasticity the analysis of such long retaining structures compared to the cross-section, presents a classical case of plane strain problem of elasticity. However, under dynamic conditions during earthquakes the retaining structures are subjected to dynamic motion and consequently owing to ground motion the dynamic earth pressure becomes very important. In Chapter 7 the evaluation of dynamic earth pressure and deformations (sliding and overturning) of retaining structures have been presented.

Chapter 8 describes the characteristics of strong ground motion and their measurements which are of major concern to the engineers. Proper earthquake-resistant design requires the estimation of the level of strong ground motion to which structures are subjected. This chapter describes the approach and methodology to measure strong ground motion. Chapter 9 presents seismic hazard analysis. Such analysis considers the uncertainty in design in terms of assessment of strong ground motion. Strong ground motions are primarily due to seismic occurrence, source process, propagation, and local site conditions. The seismic hazard analysis is presented in this chapter to facilitate mean evaluation of various properties of earthquake motion on a deterministic basis or on a probabilistic basis which are likely to occur over the specified period in the future.

Chapter 10 deals with liquefaction of soil. Earthquake liquefaction is a major contributor to urban seismic risk. The shaking causes increased pore water pressure which reduces the effectives stress, and therefore reduces the shear strength of the sand. Studies of liquefaction have been presented in detail, analytically as well as experimentally. The criteria for assessing liquefaction potential as well as recent advances in liquefaction studies have been included.

Chapter 11 introduces risk, uncertainty and reliability with reference to soil dynamics and earthquake engineering. In the consideration of various uncertainties, it is important to represent the properties of earthquake motion along with a "risk index", a parameter describing the possibility of their occurrence. Thus, earthquake-hazard analysis can also mean evaluation of various properties of earthquake motion likely to occur at a given point over the specified period in the future in terms of the risk index. The probability of earthquake occurrence in a year, or recurrence time, is frequently used as the risk index.

In preparation of this text the published works have been consulted and all efforts have been made to collate such references at the end of the book. These references may be used by the interested readers for further study of the subject matter.

The author owes special thanks to the management of PHI Learning, New Delhi, for undertaking the publication of the book and specially to Darshan Kumar, Senior Editor in processing the manuscript and in bringing it finally to its present compact form in the best possible manner. This is indeed gratefully acknowledged.

Finally, I profusely thank my daughter Mrs Jyoti for encouraging my pursuit of this book. More importantly, my love, gratitude and apologies to my grandson Akshay for bearing with me during my long periods of pre-occupation with this work.

It is possible that some errors might have crept in despite the best efforts to eliminate them. It will be appreciated if such errors are brought to the notice of the author or the publisher. Helpful suggestions and critical comments with a view to improving the text in the subsequent editions will be welcomed.

Bharat Bhushan Prasad

1

INTRODUCTION

1.1 GEOTECHNICAL ENGINEERING AND SOIL DYNAMICS

Dr. Karl von Terzaghi who is rightly recognized as the "father of soil mechanics" introduced this new discipline of civil engineering for the evaluation of engineering properties and behaviour of soil under various loadings. The birth of geotechnical engineering as a widely recognized discipline was perhaps the year 1925 and that was the year when Terzaghi published the first ever comprehensive book on the subject. Since the publication of this book entitled *Erd bau mechanik auf Bodenphysikalischer Grundlage* (German for the Mechanics of Earth Construction based on Soil Physics) in Vienna, there has been considerable contribution of knowledge and research in this area and various new aspects have been addressed too. Further by synthesis with engineering geology, geophysics, and theory of elasticity and above all engineering judgments, geotechnical engineering has emerged as a modern branch of civil engineering. Terzaghi provided the leadership at the right time and by the synthesis of theoretical analysis, practical and field observations with the necessary skills and engineering judgments he established geotechnical engineering as a rational and legitimate branch of civil engineering. As geotechnical engineering matured, it developed a personality trait of its own slightly different from other civil engineering disciplines. However, these personality traits by and large remained confined to static state only, so there is a need for studying response of soil under dynamic state as well, and thus emerged soil dynamics as an essential component of geotechnical engineering.

Soil dynamics is thus that offshoot of geotechnical engineering, which deals with material properties of soil under dynamic stress. Soil dynamics essentially consists of classical dynamics of elastic continuum and yet relies on dynamics of vibrations. Although the dynamic theories for evaluation of soil behaviour under dynamic state are the same as those of any other mechanical system, specific improvisation and adaptations are needed for soil as an engineering material. The soils or rocks are essentially natural materials. As such, their engineering properties are complex and can be only evaluated by field and laboratory tests, in contrast to material properties of, say, steel which can be easily obtained from a structural handbook. Although treated as elastic material, soil is very different from concrete or steel and hence there is a specific need for study of soil mechanics in general and soil dynamics in particular. The dynamics of earthquake motion are expressed in terms of acceleration–time trace, velocity–time trace and displacement–time trace.

The term soil has originated from the Latin word *solum* and this term has different meanings in different disciplines. Foundations of all structures have to be placed on mother's earth and that is why we all are concerned and interested in its engineering behaviour. Richard L. Handy wrote in ASCE (1995) on "The Day the House Fell" that virtually every structure is supported by soil or rock. Those that are not either fly, float, or fall over.

1.2 SOIL DYNAMICS AND STRUCTURAL DYNAMICS

The structural response to a dynamic load in terms of resulting deflection and stress is essentially time varying and is studied as structural dynamics. There are fundamentally two ways in structural dynamics for evaluating the structural response to types of dynamic loads depicted in Figure 1.1:

(a) Deterministic
(b) Non-deterministic

In case the time variation of loading is completely known, then it is termed *prescribed dynamic loading* and the analysis of the response to prescribed dynamic loading is defined as a deterministic analysis. In case the time variation is not completely known and is prescribed in statistical sense, the said loading is essentially *random dynamic loading* and the analysis of the response to random loading is defined as a non-deterministic analysis.

In general, the structural response to dynamic loading is essentially in terms of displacement of the structure. Thus, a deterministic analysis leads to a displacement–time history wherein the stresses, strains, internal forces, etc. are determined in the secondary phase of the analysis. On the other hand, a non-deterministic analysis provides only the statistical information about the displacements. As such, the time variation of displacements is not determined and consequently stresses, strains or internal forces, etc. are evaluated directly by an independent non-deterministic analysis rather than from the displacements.

The structural dynamics is largely associated with material properties of steel or concrete wherein the stress history of the material had no significant hangover. The phenomenon of loading, unloading and reloading is taken care of by assuming a linearly elastic behaviour of steel or concrete. But in case of soil, the stress history is very significant.

Thus, the exclusive dynamic properties unique to soils that are dominant in soil dynamics include classical dynamics of elastic continua and the classical theory of vibrations as prevalent in structural dynamics, but the special case and other adoptions are needed to fit in with geometry of practical problems involving subsoil regions. The relevant properties of soil have to be ascertained by dynamic tests whereas such determinations are not at all necessary for steel or concrete in structural dynamics. In special cases, exclusive dynamic properties are dominant in studies of liquefaction wherein the entire shear strength of soil is lost and in such cases the related theory of classical dynamics and the theory of vibrations are non-significant.

The dynamics of earthquake motion are expressed in terms of acceleration–time trace as shown in Figures 1.2(a) and 1.2(b), wherein the ground motions in the form of accelerograms are shown for Koyna earthquake (1967) and Port Hueneme earthquake (1957). In Figure 1.2(b), in addition to acceleration–time trace the corresponding velocity–time trace and displacement–time trace have also been shown.

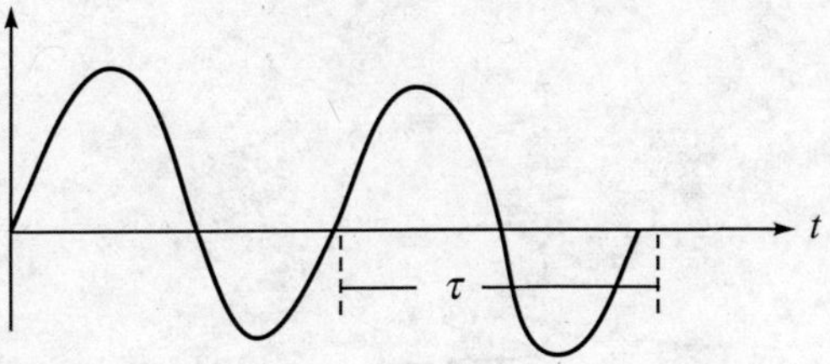

(a) Simple harmonic—machine induced motion

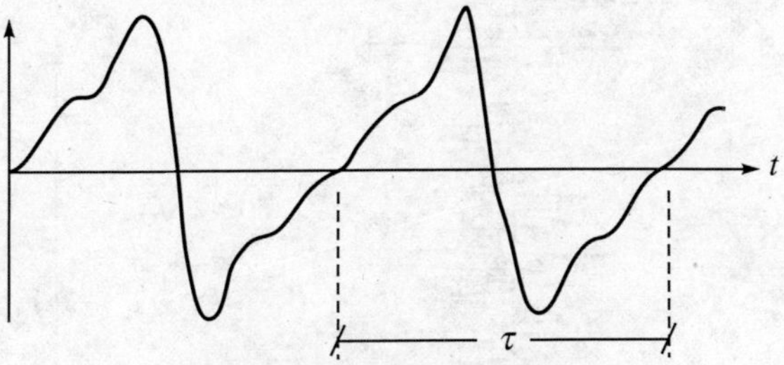

(b) Complex periodic—propeller forces

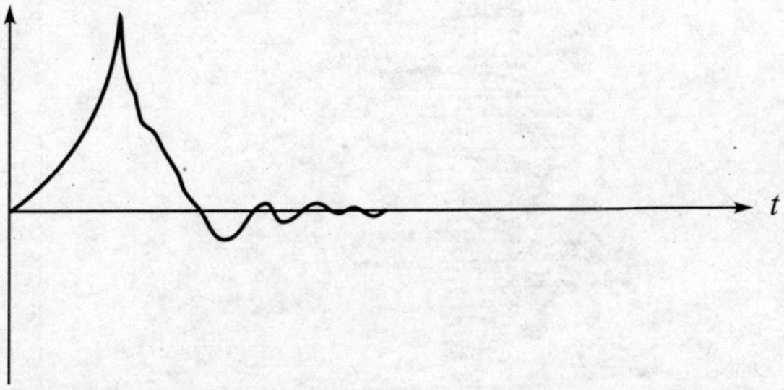

(c) Impulsive loading—bomb blast on building

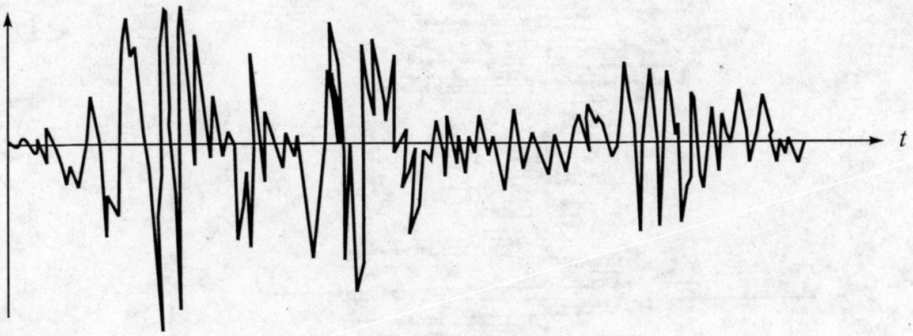

(d) Earthquake loading (acceleration–time trace)

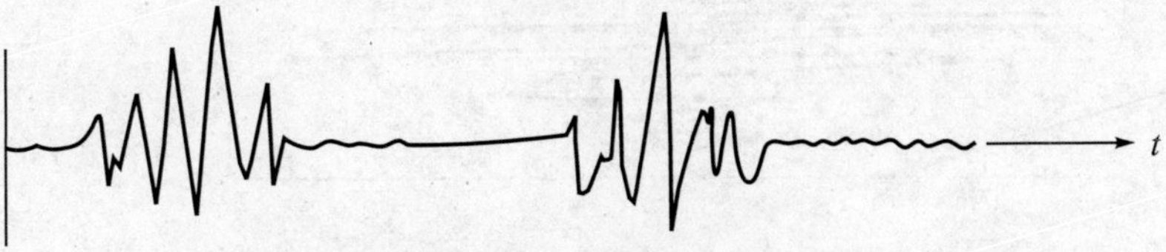

(e) Ground motion due to pile driving (acceleration–time trace)

Figure 1.1 Types of dynamic loadings.

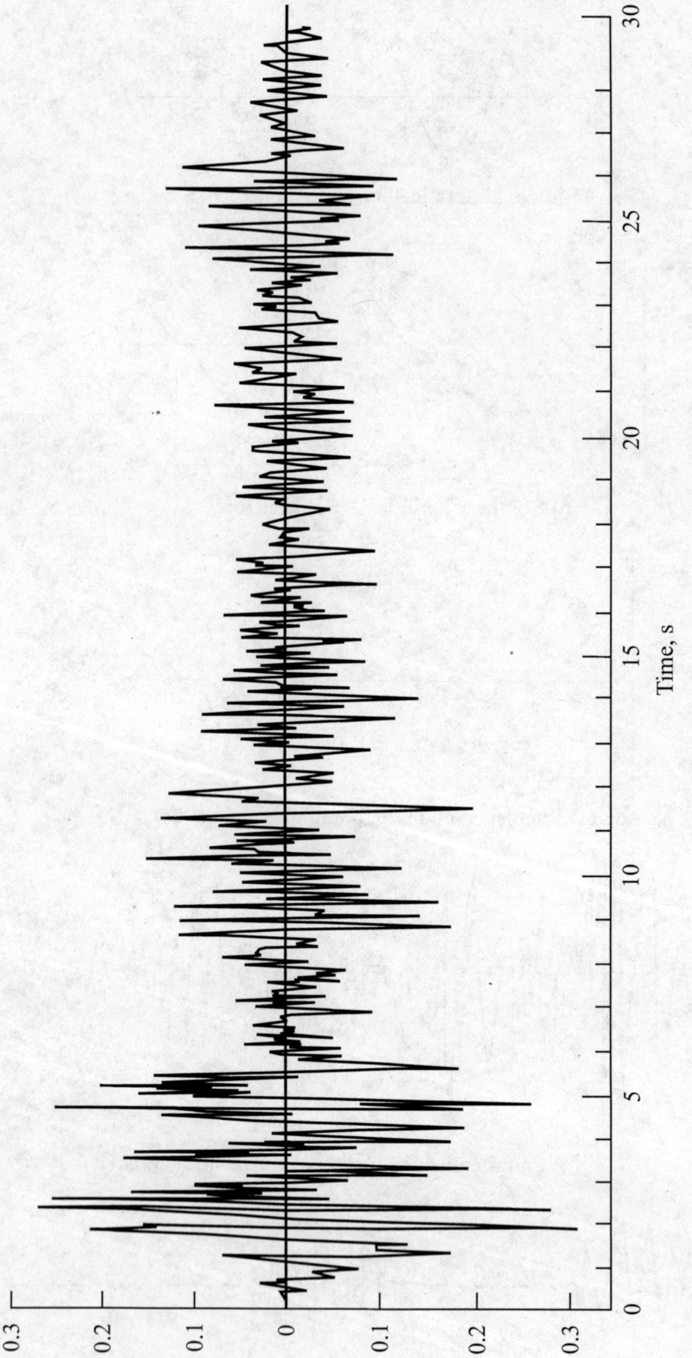

Figure 1.2(a) Accelerogram of Koyna earthquake (1967).

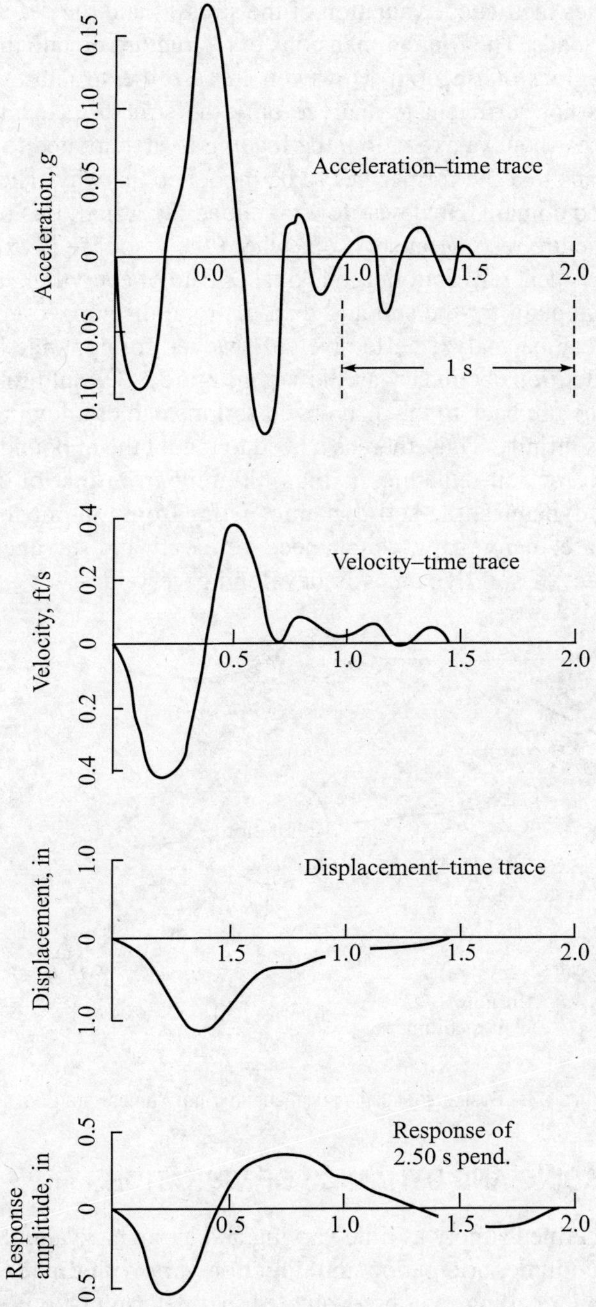

Figure 1.2(b) Acceleration–time trace, velocity–time trace, displacement–time trace and amplitude–time trace (N–S component) for Port Hueneme earthquake, March 18,1957.

Structural dynamics facilitates evaluation of the stresses and the deformations of a structure subjected to dynamic loads. The finite dimensions of a structure dictate the dynamic model with a finite number of degrees of freedom. However, in case the structure does interact with the surrounding soil, it is not sufficient to analyze only the structure. In many cases of dynamic loading, specially the earthquake excitation, the loading is first applied to the soil region around the structure; this means that the former has to be modelled anyway. The soil is a semi-infinite medium, an unbounded domain. However, for static loading, a fictitious boundary at a sufficient distance from the structure resting on soft soil, where the response is expected to die out from a practical standpoint, is generally introduced and takes care of everything as shown in Figure 1.3. This leads to a finite domain for the soil and then the total discretized system consisting of the structure and the soil can be analyzed effectively. However, for dynamic loading, this procedure cannot be used. The fictitious boundary as shown in Figure 1.3 would reflect waves originating from the vibrating structure back to the discretized soil instead of allowing them to pass through and propagate towards infinity. Thus, there is a need to model the unbounded foundation medium realistically. The study of soil dynamics is thus different from that of structural dynamics.

Unlike structural dynamics, the soil dynamics is far from a homogeneous body of knowledge wherein there are major gaps which need research and advancement of the subject. Nonetheless the subject of soil dynamics is developing very fast.

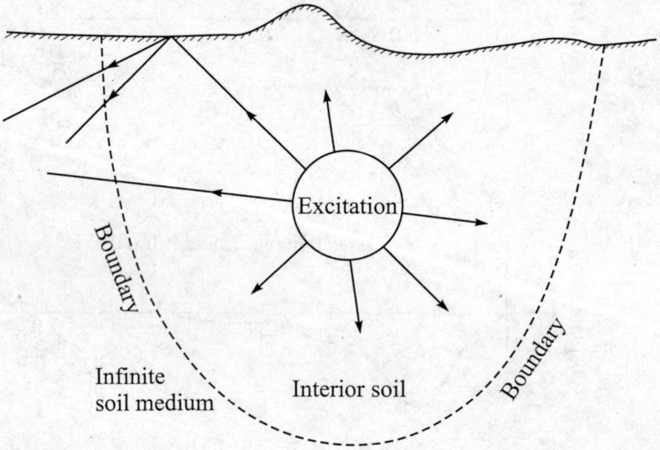

Figure 1.3 System for infinite soil medium [After Kameswara Rao, 1998]

1.3 DYNAMIC LOADING AND DYNAMICS OF VIBRATIONS

The term dynamic is defined simply as time varying and as we have already seen, a dynamic load is any load whose magnitude, orientation and direction vary with time. As stated in Section 1.2 the response to dynamic loading may be evaluated in a deterministic way or non-deterministic way depending upon whether the variation of loading is totally known or partially known. Further the deterministic loadings are of two types, namely, periodic and non-periodic. Figure 1.1(a), (b) and (c) represent periodic, non-periodic and random loading. Figure 1.1(d) shows the natural ground motion produced by earthquakes, whereas Figure 1.1(e) shows the ground motion produced by pile driving.

The periodic loadings are repetitive loads, which exhibit the same variation with time for a large number of cycles. The non-common and simple example is that of a sinusoidal variation as shown in Figure 1.1(a). Such loading is characteristic of unbalanced mass effects in rotating machinery or that caused by hydrodynamic pressure generated by a propeller at the stern of a ship or by inertial effects in reciprocating machinery.

Non-periodic loading is either a short-duration impulse loading or a long-duration general type of dynamic loading. An impact owing to explosion (Bomb blast on building) is typical source of impulsive loading as shown in Figure 1.1(c). Simplified forms of analysis are required to evaluate the dynamic response, whereas a long-duration loading which might result from an earthquake excitation may require a comprehensive dynamic analysis procedure.

1.4 STRESS CONDITIONS OF SOIL UNDER DYNAMIC LOADING

Stress conditions, shear deformations and strength characteristics of soil subjected to static loads depend on soil characteristics such as initial void ratio, relative density, initial static stress level and above all stress history. The stress deformations and strength characteristics of soils subjected to dynamic loads also depend upon initial static stress field, initial void ratio, pulsating stress level and the frequency of the loading. In this context various problems in geotechnical engineering require determination of the dynamic soil properties. In case of dynamic loading such problems are either of small strain amplitude response type or of large strain amplitude response type. Machine foundations subjected to dynamic loads can sustain small levels of strains while structural elements subjected to seismic forces or bomb blast loading must sustain large strain levels. Ishihara (1971) suggested the values of strain levels from various field and laboratory tests and the corresponding state of soil. The dynamic soil properties are strain level dependent.

The IS 5249 has recommended various field and laboratory tests for evaluating dynamic soil properties. As the dynamic properties of soils are strain level dependent, various laboratory and field tests have been developed to include a wide range of strain amplitudes. The large strain amplitude responses are of the order of 0.01% to 0.1%, whereas small strain amplitude responses are of the order of 0.0001% to 0.001%.

1.5 SOIL DYNAMICS AND EARTHQUAKE ENGINEERING

The soil dynamics and earthquake engineering are so interlinked that, in fact, the two should be termed a single subject, namely, geotechnical earthquake engineering. Normally earthquake engineering is treated as an application of structural engineering with regard to earthquake resistant design of superstructures. In earthquake prone areas the important problem that concerns structural engineers is the behaviour of the structures subjected to earthquake induced motion of the base of the structure. The displacement of the ground is, therefore, better studied in soil dynamics and its application in earthquake engineering.

As far as seismology and earthquake engineering are concerned, D. Oldham of Geological Society of India (GSI) was the founder of modern seismology whose systematic account of the great Assam earthquake (1897) is the first well-recorded earthquake of the world and Robert Mallet's (1862) contribution has been that of towards the early organization of knowledge about earthquake into science. No one man contributed more to the early organization of knowledge about earthquake into a science than Robert Mallet. He formed definite hypothesis of: what earthquakes are, how they are caused, and how they ought to be investigated. He reported the great Neapolitan earthquake of Italy in 1857. H.F. Reid, an American geologist presented the elastic rebound theory after classical observations of April 18, Great California earthquake of 1906. B. Guntenberg, a German, was the first to accurately determine the depth of the earth's core and developed many equations for size and occurrences of earthquakes. Gray, Miline, and Edwing were the first who developed effective seismographs in Japan in 1880.

In seismic zones, as and when motion originates not from forces acting on a superstructure but from the supporting soil, it is transmitted to the structure which then reacts in accordance with its own characteristics and those of the soil as well. Often the motion of the soil is caused by the earthquakes. Either the ground motions are taken care of in a deterministic way or else they are postulated by probabilities methods or random processes. As all structures on earth are bound by ground realities the problems of dynamic loading of soils and foundations have existed ever since the art came into existence. Earthquakes produce damage, deformation and rupture of earth mass, and so while tackling them in a seismic and technical way both the soil dynamics and earthquake engineering are in use simultaneously. The concepts of random process, probability theory reliability analysis providing positive definite confidence level in analysis and design are methods of the present time to ensure earthquake resistant design and construction.

The earthquake resistant design of structures taking into account the seismic data from studies of past earthquakes has become very essential, particularly in view of the heavy non-structural programme at present all over the country and, in general, all over the globe. With the availability of additional seismic data and further use of knowledge and experience, there is always a value addition to analysis and earthquake resistant design.

1.6 LITHOLOGICAL AND SEISMOTECTONICS PROFILE OF INDIA

It is interesting to compare the map of various soil deposits of India with the seismic zoning map of India, as shown in Figure 1.4 and Figures 1.5 and 1.6, respectively. The major soil deposits of principal lithological groups based on climate, topography and their origin of formation have been classified into the groups shown in Figure 1.4.

The foothills in the hilly terrain carry large boulders downstream. Such deposits are found in the sub-Himalayan regions of Himachal Pradesh, Uttaranchal and Uttar Pradesh. Marine deposits are mainly confined along a narrow belt near the coast. In the Southwest coast of India, there are thick layers of sand above deep deposits of soft marine clays, which are soft and plastic in nature. In North India, a large part is covered with alluvial deposits. The thickness of alluvium in the Indo-gangetic and Brahmaputra flood plains varies from a few centimetres to more than hundred of metres. Even in peninsular India alluvial deposits occur in some places. Black cotton soil is the Indian name given to expansive soil deposits and they are mostly located in the central part of India. They are widespread in Maharashtra, Madhya Pradesh, Karnataka, Andhra Pradesh, Tamil Nadu and Uttar Pradesh. Lateritic soils cover an area of about 100,000 sq. km and extend over Kerala, Orissa and West Bengal. The presence of iron oxide gives these soils the characteristic red or pink colour.

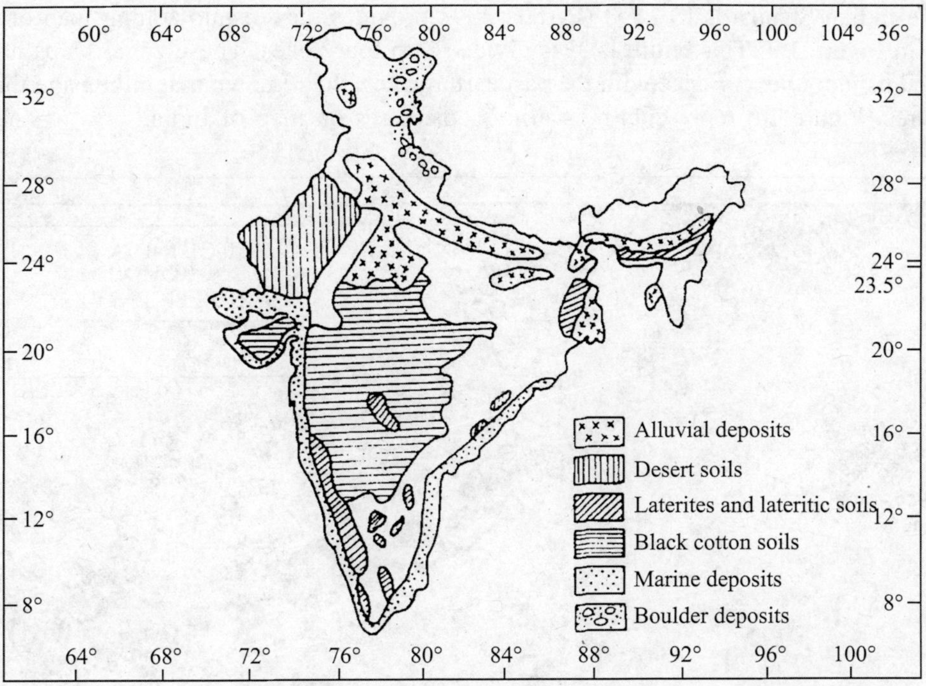

Figure 1.4 Map showing soil deposits of India

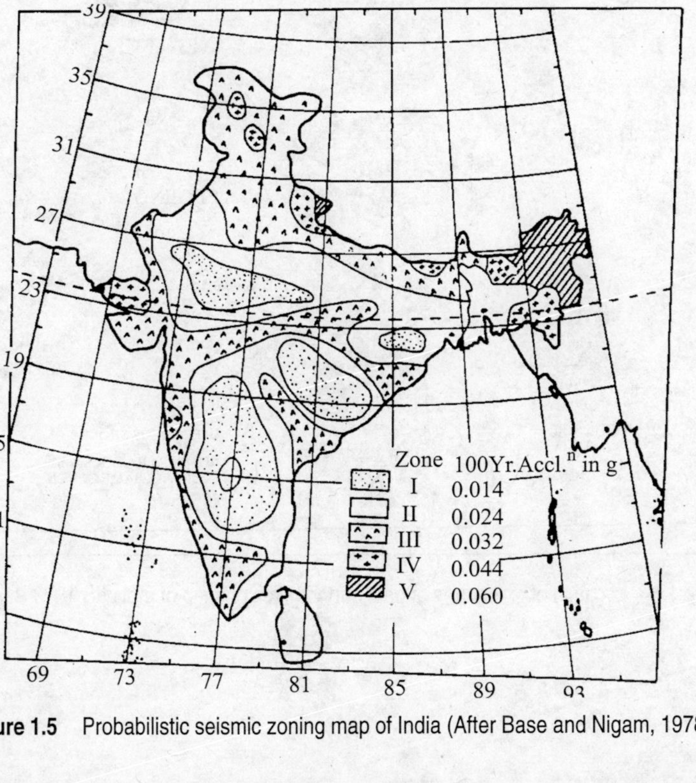

Figure 1.5 Probabilistic seismic zoning map of India (After Base and Nigam, 1978)

The Indian Standards IS 1893 (Part 1) 2002 provides the seismic zoning map of India as shown in Figure 1.7. This entire land is divided into four zones. These zonal maps have been prepared using our experience with the past earthquakes, their known magnitude and the known epicentres. Figure 1.6 represents the various epicentres on map of India.

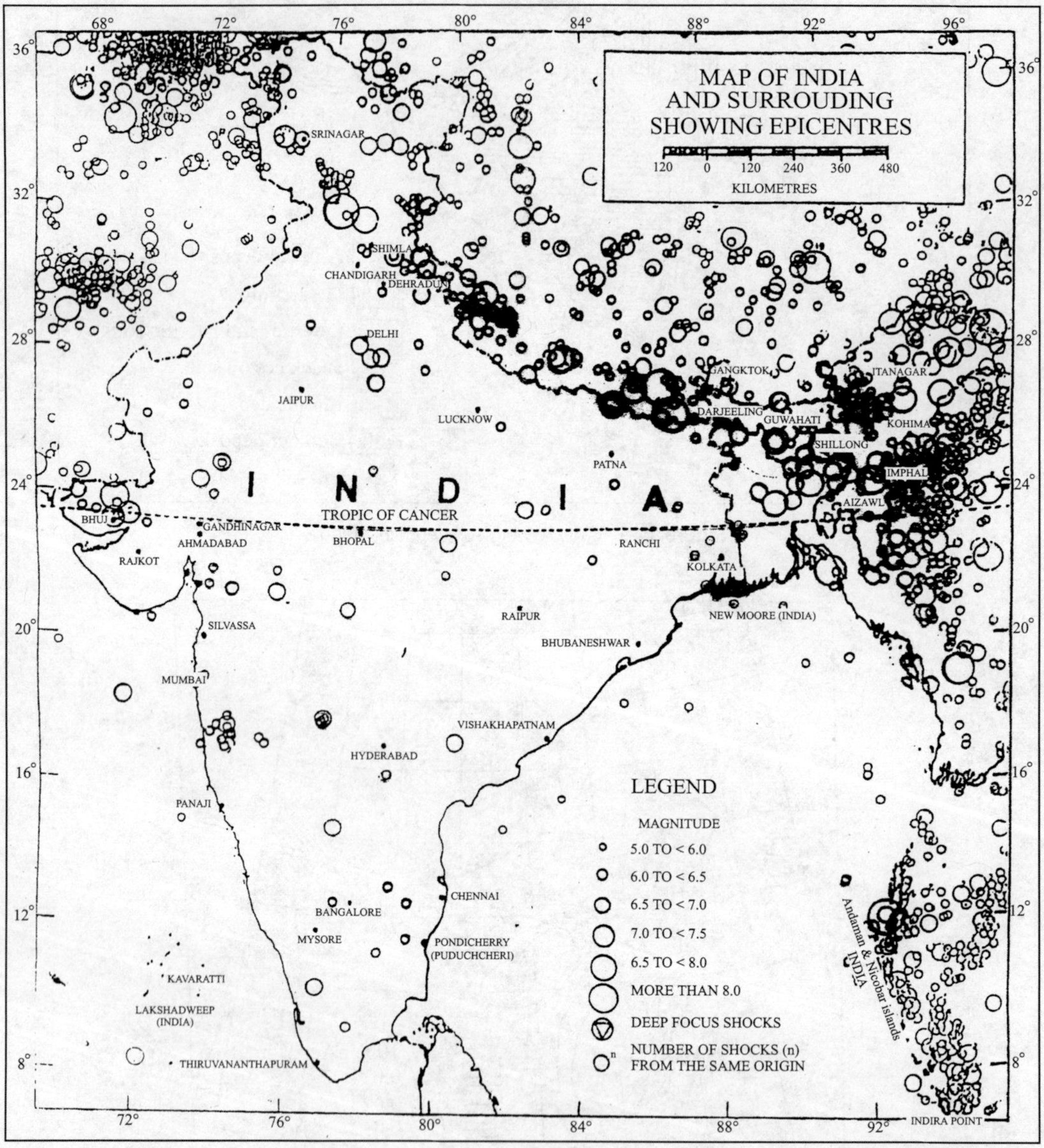

Figure 1.6 Location of epicentres of past earthquakes on map of India [IS 1893 Part 1-2002]

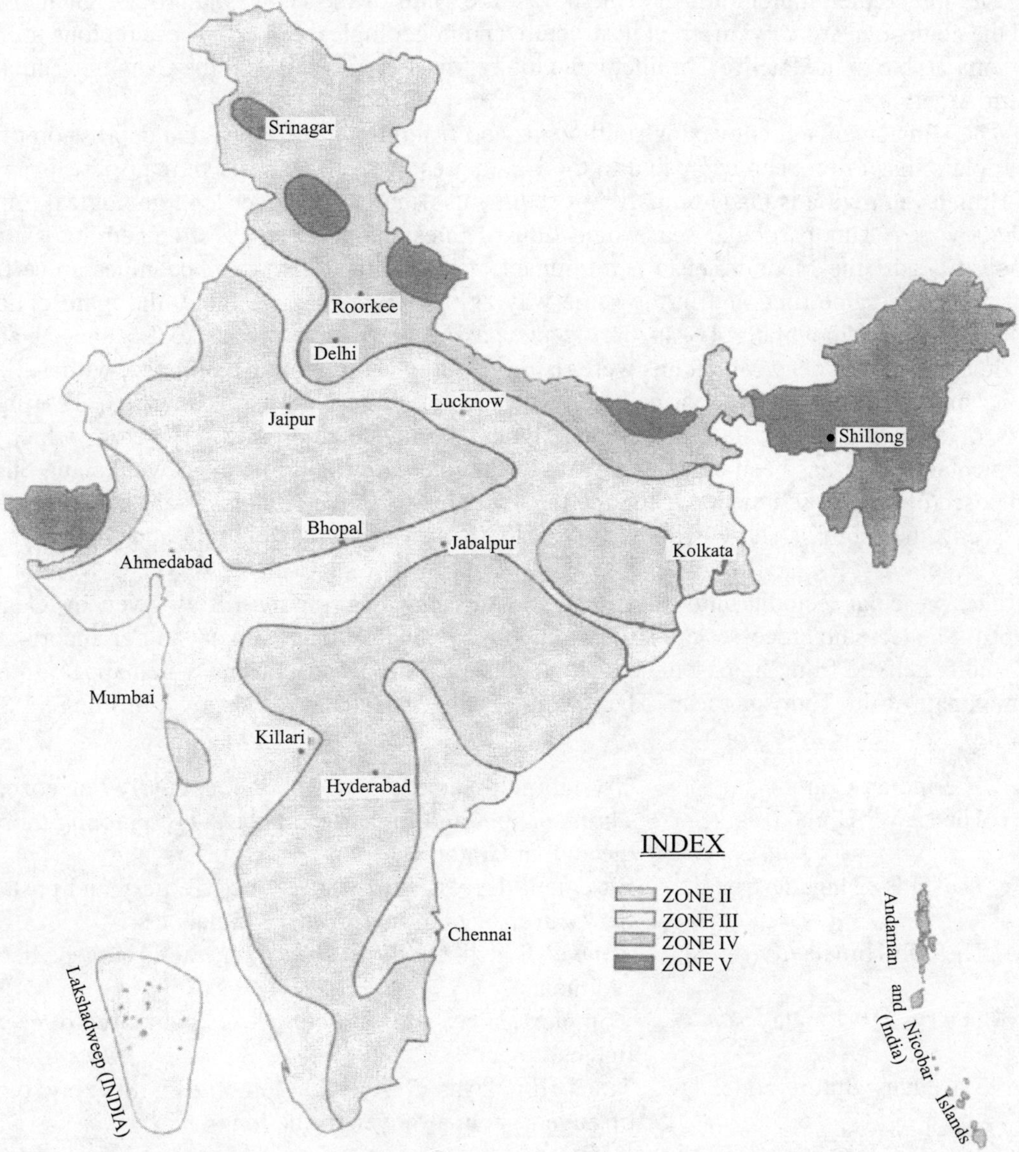

Figure 1.7 Seismic zoning map of India [After IS 1983 (Part 1) 2002]

The depth of alluvium in the Ganges plain is unknown, but it is certainly deep. Like an ocean, this great depression separates the Himalayan region from the peninsula, which is an ancient stable area, a continental old land. Archean rocks are exposed over more than half of the peninsula; much of the remainder portion is covered by the basaltic flows of the Deccan Trap, which were extruded in the Cretaceous-Eocene interval. The peninsula has no marine sediments of any consequence younger than the Cambrian, except near the coast and in one long narrow belt where shallow waters entered at the peak of Cretaceous floods.

The three chief subregions, the Himalayas, the plain of the Ganges and other great rivers, and the peninsula, are very different in structure and in geological history. These regions in India are comparable to the Pacific Cordillera, the lower Mississippi Plain and the Canadian Shield in North America.

The Himalayan arc, convexing southwards and fronting on the alluviated depression of the great plain, has often been compared to the island arcs of the Pacific. Like many great ranges, the Himalayan region is made primarily of sediments accumulated over long geological time in a shallow sea. This particular sea, which Eduard Suess named Tethys, stretched across what is now Eurasia; the Mediterranean is a remnant of it, and the Alps and Apennines arose from it at about the same time and in the same way as the Himalayas. In India, the main collapse and folding into mountains began during the passage from Cretaceous to Eocene, at about the time when the Rocky Mountains were rising. Folding and thrusting continued, with a climax in the mid-Tertiary; Eocene marine sediments are found as high as 20,000 feet. The higher parts of the present Himalayas consist of igneous and metamorphic rocks from which the sedimentary cover has been eroded. In front of the range are foothills, the Siwaliks and others, composed of tertiary sediments. Although the great thrusts of the Himalayas are now apparently quiescent, the foothills show evidence of geologically very recent faulting and thrusting on a large scale.

The principal tectonic units of Himalayas are shown in Figure 1.8 as given by Gansser (1966). The tectonic processes are still continuously going on. The Himalayan belt incorporates rock units derived from the basement and fills one or more marine basins which appear to have formed part of the Tethyan ocean. The tectonic zones of Himalayas as shown in Figure 1.8 are as follows:

- The Indian Craton: Crystalline basement, Precambrian and early Palaeozoic.
- The Lower Himalayas: Thrust nappes and folded-complexes—resembling those of the Indian Craton.
- The Indo-Gangetic trough: Basement-depressed beneath a thick Tertiary and post Tertiary cover of detritus from the Himalayas.
- The Sub-Himalayas: Zone of folded and thrust Palaeogene and Neogene detrital sediments
- The Higher Himalayas: Complex nappes and fold-complexes composed of crystalline basement
- The Indus Suture-zone: Trans-Himalayas Flysch and Ophiolites with exotic blocks of eugeosynclinal cover-formations

The terms used in describing the tectonic units of Himalayas and the Indian subcontinent like Permian, Cambrian, Archean, etc. are associated with geological events of earth's history. Geochronology provides a system of dating of events in the earth's history in a definite order (era, period, epoch, age). The geological time scale has been listed in Table 1.1.

The later history of the Indian peninsula was dominated by the rise of the massive Himalayan ranges in late Tertiary and Quaternary times. Material eroded from the rising mountains was swept down onto the craton and beyond it to the Indian Ocean, filling the *alluvial basins* of the northern Indian plains and constructing the huge deltas of the *Indus, the Ganges* and *the Brahmaputra*. The eruption of the *Deccan Traps*, unlike the comparable igneous episodes in other fragments of Gondwanaland, continued well into the Tertiary. A bodily migration of the

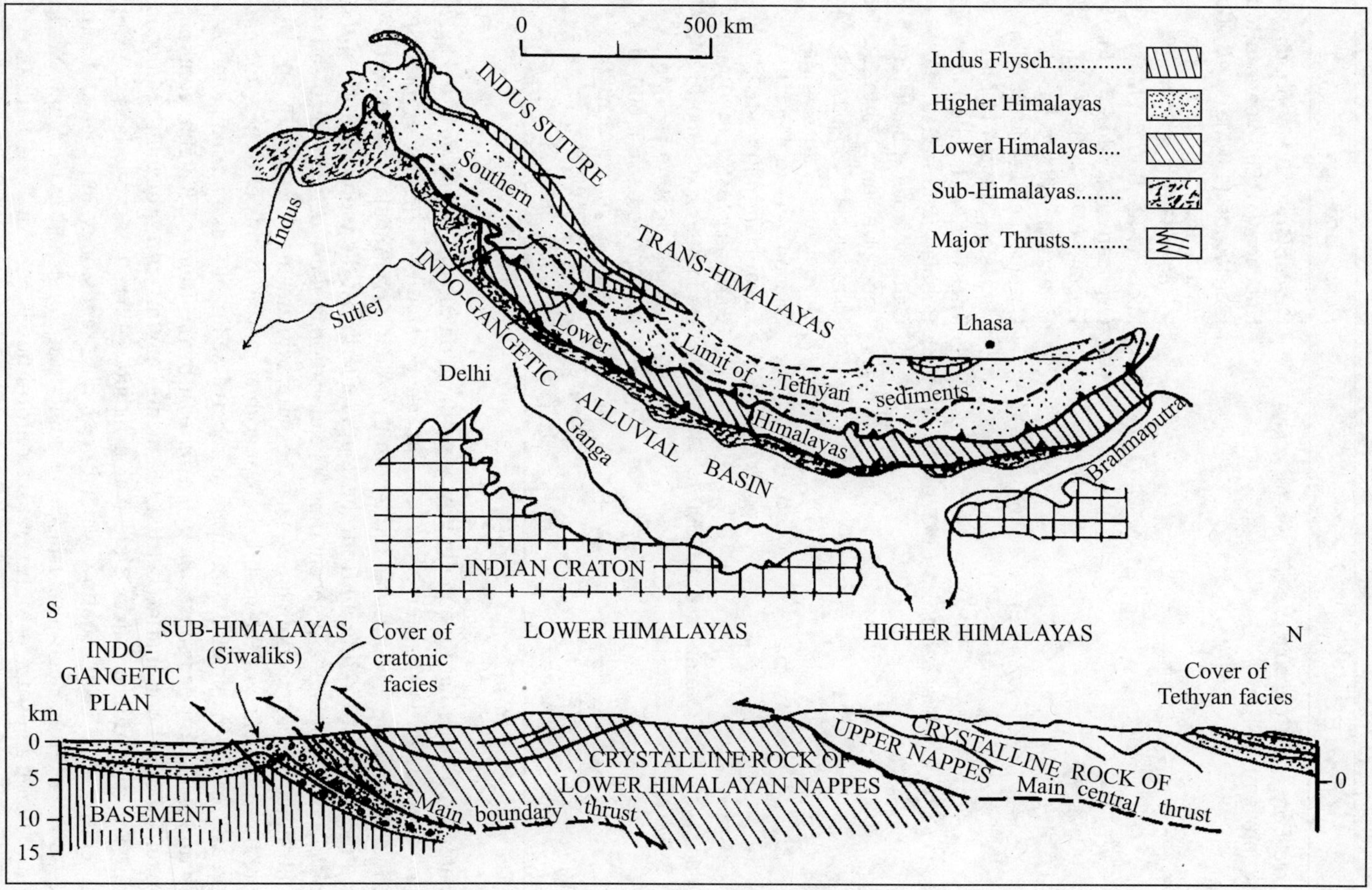

Figure 1.8 The principal tectonic units of India. (After Gansser, 1966)

Table 1.1 Geological Time Scale

Eon	*Era*	*Period and Age*
Phanerozoic Eon (543 Ma to present)	Cenozoic Era (65 Ma to today)	Quaternary (1.8 Ma to today)
		Hdocene (10,000 years to today)
		Pleistocene (1.8 Ma to 10,000 yrs)
		Tertiary (65 to 1.8 Ma)
		Pliocene (5.3 to 1.8 Ma)
		Miocene (23.8 to 5.3 Ma)
		Oligocene (33.7 to 23.8 Ma)
		Eocene (54.8 to 33.7 Ma)
		Palaeocene (85 to 54.8 Ma)
	Mesozoic Era (248 to 65 Ma)	Cretaceous (144 to 65 Ma)
		Jurassic (206 to 144 Ma)
		Triassic (248 to 206 Ma)
	Palaeozoic Era (543 to 248 Ma)	Permian (290 to 248 Ma)
		Carboniferous (354 to 290 Ma)
		Pennsylvanian (323 to 290 Ma)
		Mississippian (354 to 323 Ma)
		Devonian (417 to 354 Ma)
		Silurian (443 to 417 Ma)
		Ordovician (490 to 443 Ma)
		Cambrian (543 to 490 Ma)
Precambrian Time (4,500 to 543 Ma)	Proterozoic Era (2500 to 543 Ma)	Neoproterozoic (900 to 543 Ma)
		Vendian (650 to 543 Ma)
		Masoproterozoic (1600 to 900 Ma)
		Palaeoproterozoic (2500 to 1600 Ma)
	Archaen	(3800 to 2500 Ma)
	Hadean	(4500 to 3800 Ma*)

Ma* – million year ago (mya)

Indian craton on a scale hardly equalled by any other continental fragment is indicated by the changes of palaeolatitude registered by palaeomagnetic studies. The history of displacement suggests that the union of peninsular India with the Asiatic continent, as a result of which the raft of continental crust moving up from the south underthrust the mobile border of the Asiatic plate, took place at a geologically recent time. The elevation of the Tibetan plateau and Himalayas may be attributed in part to the consequent isostatic adjustment.

Before the emergence of the mobile belt, marginal marine basins at or near the eastern and western coasts of the peninsula continued to receive sediments as had happened during late Mesozoic times. The oil-bearing *Cambay basin*, east of the Rann of Kutchh, contains 2000 or 3000 m of marine and non-marine detrital sediments ranging from Eocene to Pliocene, resting on Deccan Traps. A thinner succession, which includes limestones, overlaps onto the craton north and east of this basin. Shallow-water limestones, sandstones and shales of early Cretaceous to Lower Miocene age also fringe the south-east coast and extend into Sri Lanka. A Tertiary succession interrupted by several unconformities is seen in West Bengal.

The Tertiary and post-Tertiary sediment-masses which flank the Himalayan mobile belt occupy an arcuate tract crossing northern India and Pakistan. Towards the northern side of the arc, syn-orogenic sediments are strongly folded, often thrust and incorporated in the Himalayan ranges. The successions of this *sub-Himalayan zone* reach more than 10 km in thickness and are almost entirely detrital. The Lower Tertiary members are partly marine, whereas the Upper Tertiary and Quaternary formations are non-marine. The incoming of the Upper Siwalik conglomerates reflects the vigorous stages of uplift and erosion to the north. On the plains, south of the mountain-front, the corresponding successions of the *Indo-Gangetic basin* consist mainly of Upper Tertiary and Quaternary fluviatile sediments whose latest units constitute the alluvium of the Ganges and other modern rivers. The youngest formations overlap southward to rest directly on the basement.

Perhaps even more remarkable for bulk are the deposits which underlie the lower reaches of the rivers draining the Himalayas and which form enormous deltas. Both the *Indus basin* on the west of the craton and the *Assam basin* on the east are underlain by late Mesozoic and Tertiary sequences locally reaching more than 10 km in thickness. These sequences thin rapidly into shelf-facies towards the peninsular craton. That of the Indus basin is gently folded, that of Assam is interrupted by several unconformities. Virtually all of the basin-fill consists of detrital material, with minor coals and limestones; the lower members are partly marine, but the later Miocene, Pliocene and Quaternary are almost entirely non-marine, laid down on advancing delta-plains. Recent surveys show that the sub-aerial deltas are fronted by abyssal cones channelled by many submarine canyons and passing into blankets of sediment which extend for at least 1000 km southwards from the mouths of the rivers.

1.7 SOME PAST INDIAN EARTHQUAKES

1.7.1 The Bhuj Earthquake 2001

A devastating earthquake struck the Bhuj area of Gujarat in the morning of January 26, 2001 while the entire country was celebrating the 51st anniversary of the Republic Day. Loss of human life in thousands and extensive damage to property was reported (see Figures 1.9 and 1.10).

Geological Survey of India's broadband Seismic Observatory at Jabalpur recorded the main shock of the devastating earthquake on 26.1.2001 at 08.46 hours (see Table 1.2). The aftershocks that took place were also recorded and analyzed. For measuring the intensity of aftershocks, three digital microseismic recorders were used in Ahmedabad; four digital and ten analogue recorders were also deployed.

Table 1.2 Seismic data of Bhuj earthquake

Date	*January 26, 2001*
Origin time (IST)	08:46:41.8
P-arrival time (IST)	08:48:47.16 eP'c'
S-arrival time (IST)	08:50:26.09
A-P duration (s)	98.93
Latitude (°)	23.31° N
Longitude(°)	70.41° E
Epicentral distance	968 km
Magnitude(Ml)	ND

(Contd.)

Table 1.2 Seismic data of Bhuj earthquake

Date	*January 26, 2001*
Magnitude(Ms)	7.6
Focal depth	ND
Geographical location	76 km east of Bhuj or 100 km NNE of Jamnagar, Gujarat

Figure 1.9 Structural damage during Bhuj earthquake January 26, 2001.

Figure 1.10 Total collapse of an RCC water tank at Manfera village (Bhuj earthquake January 26, 2001).

1.7.2 The Assam Earthquake 1897

Dr. Thomas Oldham, the first director of the geological survey of India is credited with laying the foundation of the scientific studies of earthquakes in India. His son R.D. Oldham also went on to become director of GSI (Geological Survey of India) and contributed very substantially to the earthquake studies.

The name of R.D. Oldham is associated with much pioneer work during the years when seismology was passing from the pre-instrumental period into the era of the seismograph. As head of the Geological Survey of India, he directed and personally carried out most of the investigation of the great earthquake of June 12, 1897. His monograph is one of the most valuable source books in seismology. Its contents fall principally into five categories: (1) determination of intensities and drawing of isoseismals; (2) estimation of displacement, velocity, and acceleration; (3) investigation of the meizoseismal area; (4) study of seismograms; (5) hypotheses as to the cause of the earthquake. The Assam earthquake of 1897 and the Bihar earthquake of 1934 can be compared as follows:

Table 1.3 Comparative study of Assam earthquake 1897 and Bihar earthquake 1934

Parameters of the earthquake	*1897 Assam earthquake*	*1934 Bihar earthquake*
Mean radius of area of perceptibility	900 miles	800 miles
Mean radius of area of serious damage	300 miles	200 miles
Longest dimension of meizoseismal area	160 miles	65 miles

These figures establish the 1897 event as of greater intensity than that of 1934.

Amplitudes and acceleration

Like Mallet, Oldham estimated amplitudes from cracks in the ground and in buildings; but he was dissatisfied with the results and searched for better data. His best evidence he considered to be that of a pair of damaged brick tombs at Cherrapunji, which had impinged against each other and against the walls of the depression in which they stood. He inferred an amplitude of 10 to 18 inches, probably near the mean of 14 inches. His observations were minutely carved and his reasoning ingenious as described by C.F. Richter (1957).

Seismograms of large earthquakes often indicate quite large amplitudes at short distances. Near the epicentre of a great earthquake the amplitudes of slow elastic wave motion may be comparable with the observed displacements which in the 1897 earthquake reached 35 feet.

Earthquake effects do not remain on the ground to long; many of them are erased by the weather or by human activity in a single season. The ground has to be gone over in a hurry and the investigation simply cannot be thorough. Unfortunately, there is little real chance to accumulate sound experience. Earthquakes differ and few workers have the opportunity to investage strong earthquakes in the field. The only source of help is to become acquainted with the literature so as to profit from what has already been written into the record.

1.7.3 The Bihar–Nepal Earthquake 1934

Turning from the Assam Earthquake, 1897 to the Bihar–Nepal earthquake of January 15, 1934, one sees the importance of the period in studying seismology. The investigators of 1934 were a well-trained team, all of whom made significant contributions. They were familiar with the progress of seismology up to 1934, including familiarity with Oldham's work. Seismograms at stations in India as well as in all distant parts of the world made it possible to locate the epicentre and to fix the magnitude of Bihar–Nepal earthquake as 8.4 on Richter's scale as mentioned by C.F. Richter (1957).

The extent of the isoseismals places this earthquake only a little below than that of 1897. Intensity X on the Mercalli scale was assigned to a belt about 80 miles long by 20 miles wide, and to two spots almost 100 miles distant from the main belt, at Monghyr to the south and in the Nepal Valley to the north. The isoseismal of intensity IX was drawn to include an area, which the authors of the report named "the slump belt", about 190 miles long and of irregular width exceeding 40 miles at some places. The main belt of intensity X lied entirely within the slump belt (Figure 1.11).

The known loss of life in India was given as 7253. In the Nepal Valley it was estimated as 3400. This is not high for so great an earthquake, especially in view of the widespread devastation. Fortunately the event occurred in the early winter afternoon, when most people were awake and many were outdoors. C.F. Richter in his book on *Elementary Seismology* has presented the description of seismic events with great excellence.

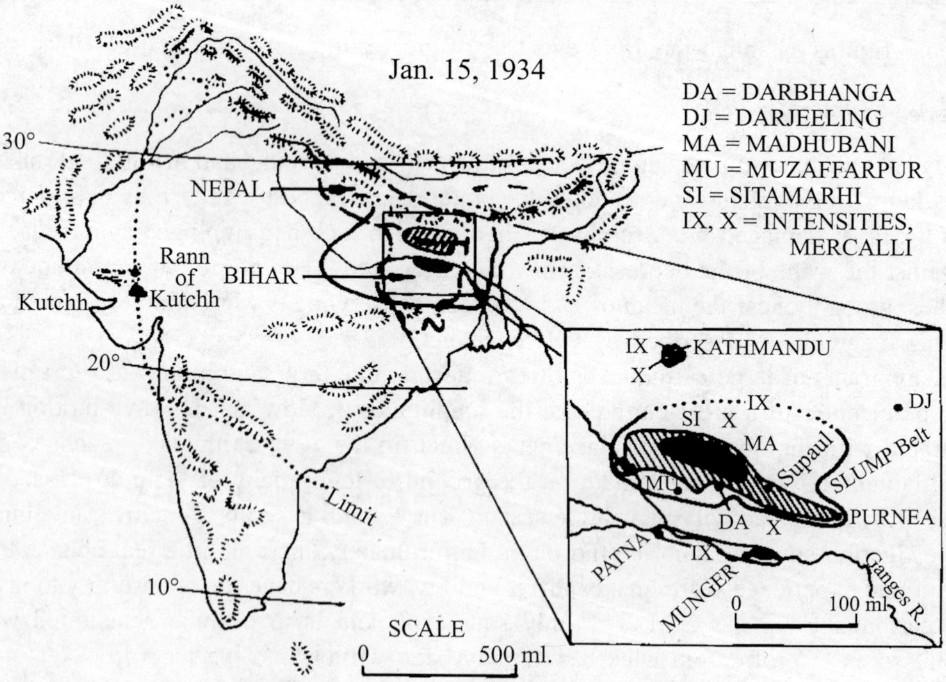

Figure 1.11 Bihar–Nepal Earthquake, Jan. 15, 1934.

1.8 OTHER EARTHQUAKES OF INDIA

In view of the size of India, great earthquakes are relatively no more frequent than those in California or in New Zealand. They are not nearly so frequent as in Japan. Moderate earthquakes, damaging a small area, appear to be relatively uncommon. Some of the historically important events are:

- 1819, June 16. Kutchh. This great earthquake provides the earliest well-documented instance of faulting during an earthquake.
- 1905, April 4. Kangra. The earliest large Indian earthquake for which a well-documented instrumental magnitude (8.6±) can be assigned. This was a great disaster; the loss of life is stated as 19,000. Instrumental data are not adequate to fix the epicentre. The meizoseismal area, including Kangra, was on the tertiary rocks of the foothills of the Himalaya. An isolated area of high intensity, lower than that at Kangra but not approached elsewhere, included Dehra Dun, also in the foothills; this was separated from the Kangra meizoseismal area by about 100 miles. The available evidence does not support the idea of two separate earthquakes; it is more likely that there was a great linear extent of faulting.
- 1935, May 30. Quetta. This earthquake devasted the city of Quetta, the capital of Baluchistan (now part of Pakistan), with a loss of about 30,000 lives. While its magnitude (7.6) was less than those of the others discussed in this chapter, the epicentre was close to the city, resulting in a relatively high intensity in that area.
- 1950, August 15. Assam and Tibet. The epicentre was near Rima. It is one of the few earthquakes to which the instrumentally determined magnitude, 8.7, is assigned. This shock caused more damage in Assam, in terms of property loss, than that caused during the earthquake of 1897. To the effects of shaking were added those of flood; the rivers rose high after the earthquake, bringing down sand, mud, trees, and all kinds of debris. Pilots flying over the meizoseismal area reported great changes in topography; this was largely due to enormous slides, some of which were photographed. The only available on-the-spot account is that of F. Kingdon-Ward, a botanical explorer who was at Rima. However, he had little opportunity for making observations; he confirms violent shaking at Rima, extensive slides, and the rise of the streams, but his attention was perforce directed to the difficulties of getting out and back to India. Aftershocks were numerous; many of them were of magnitude 6 and over and well enough recorded at distant stations for reasonably good epicentre location. From such data Dr. Tandon, of the Indian seismological service, established an enormous geographical spread of this activity, from about 90° to 97° east longitude, with the epicentre of the great earthquake being near the eastern margin. One of the more westerly aftershocks, a few days later, was felt more extensively in Assam than the main shock; this led certain journalists to the absurd conclusion that the later shock was "bigger" and must be the greatest earthquake of all time! This is a typical example of confusion between the essential concepts of magnitude and intensity. The extraordinary sounds heard by Kingdon-Ward and many others at the time of the main earthquake have been specially investigated.

1.8.1 Some Past Indian Earthquakes

Table 1.4 presents a brief description of some of the significant past earthquakes of India.

Table 1.4 Glimpses of some past Indian earthquakes

Date	*Event*	*Time*	*Magnitude*	*Max. intensity*	*Deaths*
16 June, 1819	Kutchh	11:00	8.3	IX	1500
12 June, 1897	Assam	17:00	8.7	XII	1500
8 February, 1900	Coimbatore	03:11	6.0	VII	Not known
4 April, 1905	Kangra	06:20	8.0	X	19,000
15 January, 1934	Bihar–Nepal	14:13	8.3	X	11,000
15 August, 1950	Assam	19:31	8.6	XII	1530
21 July, 1956	Anjar	21:02	6.1	IX	115
10 December, 1967	Koyna	04:30	6.5	VIII	200
23 March, 1970	Bharuch	20:56	5.7	VII	30
21 August, 1988	Bihar–Nepal	04:39	6.6	IX	1004
20 October, 1991	Uttarkashi	02:53	6.4	IX	768
30 September, 1993	Killari(Latur)	03:53	6.2	VIII	7928
22 May, 1997	Jabalpur	04:22	6.0	VIII	38
29 March, 1999	Chamoli	12:35	6.6	VIII	63
26 January, 2001	Bhuj	08:46	7.7	X	13,805

Further, Figure 1.12 shows the epicentres of earthquakes that have occurred in Asia recently. This reflects the seismic activities in various regions. The location of the epicentre is

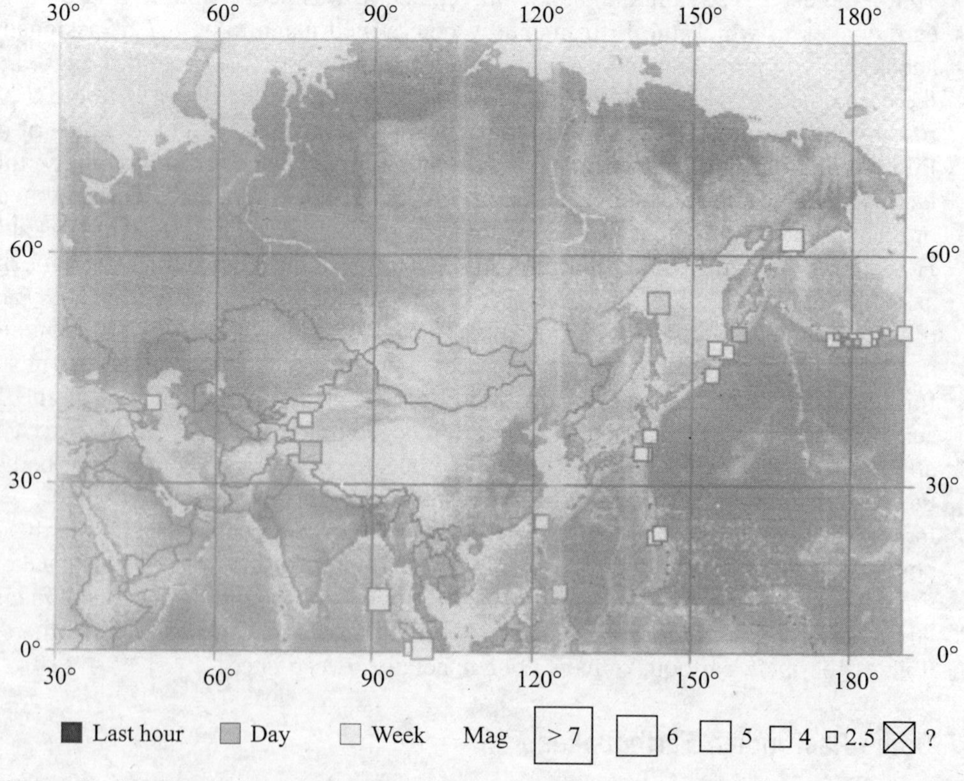

Figure 1.12 Location of epicentres of recent earthquakes in Asia (After http: // earthquake. usgs. Maps. /Asia)

marked by a rectangle and the size of the rectangle represents the magnitude of the earthquake on Richter's scale. Though the magnitudes of the different earthquakes are known to a reasonable accuracy, the intensities of the earthquakes so far, have been mostly estimated by the damage surveys.

1.9 GLOBAL INTERNATIONAL SEISMICITY—SEISMICITY OF THE EARTH

Seismic regions of the world have been identified. Seismologically speaking, the most important subdivisions of the earth's surface are:

- The Circum–Pacific belt
- The Alpide belt
- The Pamir–Baikal zone of Central Asia
- Rift zones of East Africa
- A wide triangular area between the Alpide belt are the Pamir–Baikal zone.
- The central basin of the northern Pacific ocean
- The stable shields of the continents
- The Atlantic–Arctic belt
- Non-seismic belts/regions

Epicentres occur chiefly in a few narrow belts or zones. Certain wider areas show fairly general moderate seismicity. In the chief seismic zones, shallow earthquakes occur in two different environments which may be termed conditions of arc and block tectonics. Arcuate structures are dominant in most of the Circum–Pacific and Alpide belts. In Alpide belts there are chiefly mountain areas like Himalaya. Block faulting is dominant in certain parts of the Circum-Pacific belt as in California and central New Zealand.

The Circum–Pacific belt

The Circum–Pacific belt is the principal seismic and tectonic feature of the globe. It is complex, with several main branches including arc structures, areas of block tectonics and having example of ridge and rift type structures. The longer-sector of the Circum–Pacific belt characterized by block tectonics to the exclusion of arc features extends from Southern Alaska to northern Mexico. Block faulting occurs in the interior of arc structures as in Peru and Japan.

The Alpide belt

The Alpide belt can be traced westwards as a series of arcs with generally southward front, in Burma, the Himalaya, Baluchistan, Iran and the eastern Mediterranean. Most of the Alpide belts are shallow. Intermediate shocks are fairly frequent in Burma and in the Hindu Kush, near 36.5° N, 70.5° E. In this region there has been a remarkable and persistent repetition of earthquakes.

The Pamir–Baikal zone

Most of the severe earthquakes outside the Pacific and Alpide belts occur in the zone from the Pamir plateau to lake Baikal, along the southern margin of the Asiatic stable mass. This is probably the broadest of known seismic regions: some of its earthquakes such as those of Kansu in north-western China are very large.

Non-seismic regions

It is now well-settled that no large area is permanently unaffected by earthquakes, but there are many to which no epicentres can yet be assigned. The least seismic regions are those of Pacific basin excluding Hawaiian islands and the stable continental shields. In the Atlantic region, seismicity is very low in basins east and west of the mid-Atlantic ridge. The same is apparently true for similarly placed areas in the Indian Ocean.

The world seismicity is reflected from the details of past earthquakes listed in Table 1.5. The seismic activities are constantly reported on the Internet. These reports are being updated regularly. The Geological Society of United States of America (USGS) and the Geological Society of India have their own websites. The other countries also have their own websites on which everyday seismic activities are presented, and every seventh day updatings are regularly carried out.

Table 1.5 Glimpses of some global eathquakes

Year	*Region*	*Magnitude*	*Fatalities*
780 B.C.	China	Not known	First reliable record
79 A.D.	Italy	Not known	Not known
856	Damghan, Iran	Not known	200,000
893	India	Not known	180,000
893	Iran, Ardabil	Not known	1,50,000
1138	Aleppo, Syria	Not known	2,30,000
1290	China, Chihli	Not known	100,000
1556	China	M 8.0	8,30,000
1619	Trullilo Peru	M 7.7	350
1668	Anatolia, Turkey	M 8.0	8,000
1687	Lima, Peru	M 8.5	600
1692	Jamaica	Not known	2,000
1693	Sicily, Italy	M 7.5	60,000
1737	India (Calcutta)	Not known	3,00,000
1755	Portugal	8.6	60,000
1795	Italy	Not known	50,000
1819	India	M 8.0	1500
1833	India	M 7.7	Several
1857	California, USA	M 8.3	Not known
1872	California, USA	M 8.5	27
1886	Southern California, USA	M 7.0	110

(*Contd.*)

Table 1.5 Glimpses of some global eathquakes

Year	*Region*	*Magnitude*	*Fatalities*
1906	California, USA	M 7.9	3000
1908	Italy	M 7.5	83,000
1923	Japan	M 7.9	1,43000
1960	Chile	M 9.5	5,000
1964	Alaska	M 9.2	131
1974	China	M 6.8	20,000
1976	Turkey-Iran border region	M 7.3	5,000
1976	Mindanao, Philippines	M 7.9	8,000
1976	Tangshan, China	M 8.0	255,000
1976	Papua, Indonesia	M 7.1	5,000
1981	Southern Iran (11 June)	M 7.3	1,600
1981	Southern Iran (18 July)	M 6.9	3,000
1998	Afghanisten-Tajikistan border region	M 6.6	4,000
1999	Izmit, Turkey	M 7.4	17,118
2000	Sumatra, Indonesia	M 7.9	103
2001	Bhuj, India	M 7.6	20,023
2002	Hindu Kush region, Afghanistan	M 7.4	166
2004	Sumatra-Andaman islands	M 9.0	1,106/Sumatra
2005	Northern Sumatra, Indonesia	M 8.6	1,313
2006	Pakistan	M 7.6	80,000
2006	Java, Indonesia	M 7.7	5749
2007	Near Coast of Central Peru	M 8.0	650/Central

A number of major earthquakes have been recorded that resulted in massive losses of human lives and destruction of thousands of buildings and structures. So, the Calcutta earthquake of 1737 destroyed 300,000 lives. Portugal, Spain and northern Morocco were subjected to three strong shocks in the forenoon of November 1, 1775. The Lisbon earthquake of 1775 literally devastated Lisbon, the loss of life was heavy. The disaster was colossus as the first shock was followed by a massive whirling wall of water sweeping out every object in its path. The major Skopje, Yugoslavia, earthquake is still in the memory of everyone.

The earthquake that literally devastated Tokyo and Yokohama on September 1, 1923, laid a heavy toll on human lives and property. Nearly, 11,000 buildings were ruined and 59,000 houses devastated in Yokohama as a result of the earthquake-induced fires. Throughout the affected area in Tokyo, the death toll was 100,000, while 43,000 remained missing. Over 300,000 houses were damaged. Nearly 45% of brick buildings and 10% of reinforced concrete buildings collapsed during that event. The 1950 Himalayas earthquake, one of the severest seismic events, recorded instrumentally, was equivalent to an energy released by explosions of 100,000 A-bombs.

An extremely severe earthquake which took place on December 4, 1956 in the Mongolian People's Republic and the adjacent regions of the USSR and China brought about vast devastations. A mountain peak was split into two. Part of a mountain, 400 m in height, collapsed and fell down. A depression, up to 18 km in length and 800 m in width, originated. Broad fissures, up to 20 m in width appeared on the ground surface. One of these fissures extended to a length of 250 km. The intensity of the earthquake approached force (XI).

The American scientists consider the 1964 Alaskan earthquake, the intensity of which was over (XI), as being the most severe of all known seismic events in the world's history. However, the most violent earthquake of the present century took place in 1960, Chile (see Figure 1.21). It affected an area of over 200,000 km^2 and caused numerous landslides.

During the last decades, several large-scale earthquakes have been recorded: the Yalta earthquake of 1927; the Ashkhabad earthquake of 1948; the 1966-1967 Tashkent earthquake and the 1976 Gazli, Uzbekistan earthquake.

Most earth tremors are very hard to detect and can only be recorded by sensitive instruments and seismographs. Yet as many as a hundred earthquakes per annum are destructive and at least one catastrophic. This suggests that destructive earthquakes are violent movements of the earth crust after a period of accumulation of stress.

It may be assumed that earthquakes are caused by major and sudden discontinuities of the crust, ruptures and faults as well as displacements of the crust. They are associated with the physico-chemical processes that are at work in the earth's bowels and are also associated with changes in the thermodynamic conditions in the inner reaches of the earth.

The pattern of ground surface vibrations during an earthquake can be inferred from Figure 1.2(a) which presents an accelerogram of vibrational translations (or displacements) as recorded by a seismograph at a recording station.

The seismic impulse and vibrational movement caused by an earthquake often lasts only a few seconds. However, during a major (or strong-motion) event, even this short-lived shock generally brings about catastrophic consequences. Seismic events are known to have caused continuous vibrations, as in the Alma-Ata earthquake of 1910 which lasted for 5 minutes. Earthquakes lasting 10–15 s or more occur very commonly.

The examination of any accelerogram will show that the seismic vibrations attain a maximum amplitude only after a weaker vibration has occured. Differently speaking, practically any earthquake has an initial stage. This stage is heralded by weaker seismic waves called *precursor waves.*

Seismic waves of an earthquake originate at a place in the earth crust some distance from the surface called the *focus* or *hypocentre*.

The foci of earthquakes have generally been found at depths not exceeding 20–50 km. However, we know of seismic events whose foci were located 500–600 km below the earth surface. This is a convincing proof of that tectonic processes which take place in the deep inner reaches of the earth.

The sites of the most frequent and intensive earthquakes are regions of folded mountains of recent origin. Thus, seismic events are closely linked to tectonic processes and particularly, to modern folded mountain-building. This is the reason why the severest earthquakes in this country take place in mountainous areas of young origin, such as Transcaucasia, mountainous regions of Turkmenistan (Ashkhabad), the Crimea, the Baikal region, the Far East, Kamchatka and the Kurile Isles. Rather severe (up to intensity 9) earthquakes typically occur in mountain regions of Middle Asia.

It should be made clear that the zones of strong-motion earthquakes almost invariably coincide with the zones of faults or folds of tectonic origin. Lowland regions representing less prone areas of the earth crust (continental platform) demonstrate inappreciable seismicity. These include the Europian regions of old USSR and Siberian lowland. Major fracture faults and displacements caused by an earthquake are characterized by dramatic relative deformations and shifts of the adjacent regions. Seismic faults often break for several kilometres. So, the 1891 earthquake in Japan caused fissures and crustal displacements over 100 km in extent and formed ledges that attained 20 m in depth. Lateral displacements of individual ground surface areas are common in an earthquake. This phenomenon, in particular, was caused by the major Californian earthquake of 1906 (see Figure 1.14) where the fault and shear zone broke for 500 km.

If the epicentre of a seismic event is located in the floor of a sea or an ocean, the seismic waves are called tsunami which propagate from the site of origin at velocities up to hundred or more than 1000 km/hour. The Chile earthquake of 1960 caused major deformations of relief covering an area of 200,000 km^2. The Alpine regions of the country were displaced 300 m for a length of 40 km. The San Francisco earthquake of 1906 (see Section 1.10.1) caused a downslope sliding of moist pastures by 800 m.

Earthquakes have repeatedly disturbed the stability of bridges and approach embankments. Such events were particularly numerous during the 1923 earthquakes in Japan.

It is now well-settled that the zones of strong-motion earthquakes almost invariably coincide with zones of faults or folds of tectonic region. The Crimean earthquakes are associated with tectonic disturbances at the floor of the Black Sea. Similar conditions prevailed during the Alaskan earthquake of 1964.

An earthquake is generally accompanied by subterranean roar, deafening thunder and involving fractures and crustal displacements. The events often cause depression in one area and crustal upheavals in another. For example, during the 1892 earthquakes a substantial portion of Port Royal, Jamaica, went thundering down to the sea.

1.9.1 Global Seismic Hazard Assessment

The Global Seismic Hazard Assessment Programme (GSHAP) was launched in 1992 by the International Lithosphere Program (ILP) with the support of the International Council of Scientific Unions (ICSU), and endorsed as a demonstration programme in the framework of the United Nations International Decade for Natural Disaster Reduction (UN/IDNDR). The GSHAP project terminated in year 1999. The findings revealed that the small earthquakes are much more abundant than the great ones. Their occurrences per year are listed below.

Type of earthquakes	*Magnitude*	*No. per year*
Great earthquakes	≥ 8	1.1
Major earthquakes	7–7.9	18
Destructive earthquakes	6–6.9	120
Damaging earthquakes	5–5.9	180
Minor earthquakes	4–4.9	6,200
Smallest generally felt	3–3.9	49,000
Sometimes felt	2–2.9	300,000

From the preceding table, it may be inferred that a great earthquake like the 1934 Bihar–Nepal earthquake occurs almost every year once somewhere in the world. Worldwide, each year there occur about 18 earthquakes of magnitude (M) 7.0 or larger. Actual annual numbers since 1968 ranged from lows of 6–7 events/year in 1986 and 1990 to highs of 20–23 events/year in 1970, 1971 and 1992. Although we are not able to predict individual earthquakes, the world's largest earthquakes do have a clear spatial pattern, and therefore, "forecasts" of the locations and magnitudes of some future large earthquakes can be made. It may never be possible to predict the exact time when a damaging earthquake would occur, because when enough strain has built up, a fault may become inherently unstable, and any small background earthquake may or may not continue rupturing and turning into a large earthquake. While it may eventually be possible to accurately diagnose the strain state of faults, the precise timing of large events may continue to elude us. In the Pacific north-west, earthquake hazards are well-known and future earthquake damage can be greatly reduced by identifying and improving or removing our most vulnerable and dangerous structures. Figure 1.13 shows the global seismic hazard map where in gray and dark (in depth) represents the earthquake prone area with high seismicity. Seismic hazard map represents basically the degree of earthquake shaking that can be expected in a given place during a given time. A global seismic hazard assessment may be evaluated using the probabilistic approach in conjunction with a modified means of evaluating the seismicity parameters. The earthquake occurrence rate function may be formulated for area source cells from recent instrumental earthquake catalogues. The seismic hazard at a particular site may be obtained by integrating the hazard contribution from influencing cells, and the results were combined with the Poisson distribution to obtain the seismic hazard in terms of the intensity at 10% probability of excellence for the next 50 years.

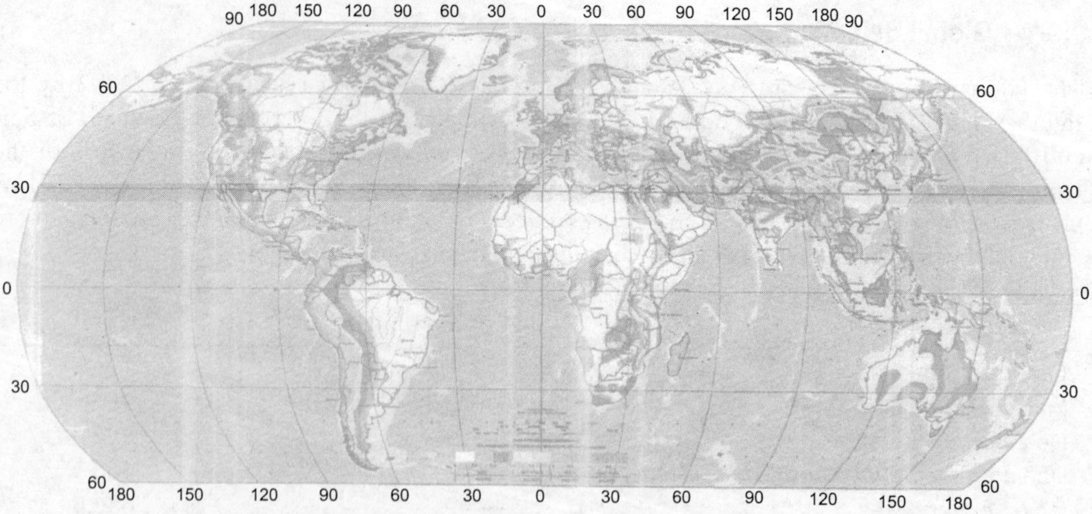

Figure 1.13 Global seismic hazard map I [After D. Giardini et al., GSHAP, 1999].

1.10 SIGNIFICANT CASE HISTORY OF SOME PAST EARTHQUAKES

1.10.1 San Francisco, California, Earthquake (April 18, 1906)

Figure 1.14 The 1906 San Francisco earthquake was one of the largest events (magnitude 7.9) to occur in the United States in the 20th century. Recent estimates indicate that as many as 3000 people lost their lives in the earthquake and ensuing fire. In terms of the year 1906 dollars, the total property damage amounted to about $ 24 million from the earthquake and $ 350 million from the fire. The fire destroyed 28,000 buildings in a 520-block area of San Francisco.

1.10.2 Loma Prieta Earthquake, Part 1

Figure 1.15 On October 17, 1989, a 7.1 magnitude earthquake occurred near Loma Prieta in the Santa Cruz mountains. Movement occurred along a 40-km segment of the San Andreas fault from south-west of Los Gatos to north of San Juan Bautista.

1.10.3 Loma Prieta Earthquake, Part 2

Figure 1.16 On October 17, 1989, a 7.1 magnitude earthquake occurred near Loma Prieta in the Santa Cruz mountains. This earthquake is also known as the "San Francisco World Series Earthquake."

1.10.4 San Fernando Valley California Earthquakes

Figure 1.17 This figure compares two earthquakes that were separated by a distance of 10 miles and a time of 23 years. Disproving the notion that once an earthquake has occurred, that area is safe from future earthquakes, these events affected much of the same area and even some of the same structures. These two events were the largest of 17 moderate-sized main shock/aftershock sequences that have occurred in the Los Angeles area since 1920. The 1971 shock is referred to in the scientific literature as the San Fernando earthquake. The 1994 shock (also in the San Fernando Valley) is called the Northridge earthquake.

1.10.5 Great Hanshin-Awaji (Kobe) Earthquake, January 17, 1995

Figure 1.18 On the morning of January 17, 1995, a major earthquake occurred near the City of Kobe, Japan. The greatest intensity of shaking for the 6.9 magnitude earthquake was in a narrow corridor of 2-4 km stretching 40 km along the coast of Osaka Bay. The worst destruction ran along the previously undetected fault on the coast, east of Kobe. Kobe's major businesses and port facilities, and residences are located on this strip. This earthquake caused 5480 deaths, and totally destroyed more than 192,000 houses and buildings.

1.10.6 Izmit (Kocaeli) Turkey Earthquake, August 17, 1999-Set 1, Coastal Effects

Figure 1.19 On August 17, 1999, at 3:02 a.m. local time a magnitude 7.4 earthquake occurred on the northern Anatolian fault. The epicentre was located very close to the south shore of the Bay of Izmit, an eastward extension of the Marmara Sea. The location of this earthquake and its proximity to populous region of the Bay of Izmit contributed greatly to its damaging effects. The total estimated loss for port facilities in the region was around $ 200 million (US). Subsidence and slumping caused much of the coastal damage, but a tsunami was generated that also caused coastal damage and deaths.

1.10.7 Duzce, Turkey Earthquake, November 12, 1999

Figure 1.20 The magnitude 7.2 quake occurred at 6.57 p.m. local time (16:57 GMT). Duzce lies on the eastern fringe of the region hit by the August 17 quake. Some areas experienced a one-two punch from the 1999 earthquakes. The death toll from the November quake was reported to be 260 people. More than 1282 were injured and at least 102 buildings were destroyed.

1.10.8 Great Chile Earthquake of May 22, 1960

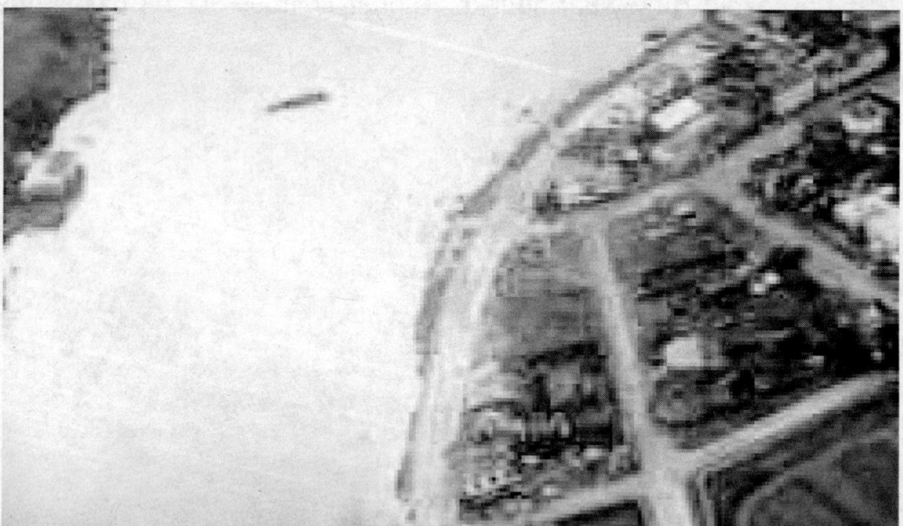

Figure 1.21 On May 22, 1960, a M_w 9.5 earthquake, the largest earthquake ever instrumentally recorded, occurred in southern Chile. The series of earthquakes that followed ravaged southern Chile and ruptured over a period of days a 1.0 km section of the fault, one of the longest ruptures ever reported.

1.11 UNCERTAINTY, HAZARD, RISK, RELIABILITY AND PROBABILITY OF EARTHQUAKES

1.11.1 Uncertainty and Hazard

Geology reveals the basic source for uncertainty in geotechnical engineering. As such the main objective is to identify potential hazards and also the other side of the coin, i.e., the favourable features of the geology. The knowledge of both sides is essential to estimate risk as part of the overall cost–benefit ratio of the scheme. Geophysical records often entail processing to enhance the signal and subsequent interpretation, which may involve several steps of reasoning. Unless these are clearly described and the level of confidence assessed and stated, the engineering judgment may accept one out of several possible representations without any appreciation of the degree of uncertainty. The risk may be defined as the probability that a particular adverse event like an earthquake occurs during the stated period of time or results from a particular challenge.

The term risk in seismic-prone zones is usually associated with the concept of danger to life or property. Earthquakes produce natural disaster, whereas bomb blasts are due to socio-political disruptions on account of terrorism, racial tension and nuclear explosions. Dynamic loads due to earthquakes, bomb blasts and vibrations of machines are very different and distinct. The nature of hazard is different in all these three cases.

In general, the assessment of risk is to a large extent a qualitative concept. To live in an active seismic zone, there is always risk involved with it. In engineering the use of risk analysis lies in evolving procedures to arrive at a quantitative measure of risk. It is usual to combine the probability of occurrence of earthquakes and the consequences of that event. Here the frequency of occurrence of earthquakes is not that serious; rather the magnitude of the event is far more serious. As loss of life and property is associated with the consequences of the earthquake, the possible definition of risk is:

Risk (consequence of earthquake/unit time) = frequency of occurrence (earthquake/unit time) × magnitude (earthquake as event)

The seismic hazard in this context is generally defined as the predicted level of ground acceleration which would be exceeded with ten per cent probability at the site under consideration due to the occurrence of an earthquake anywhere in the region in next fifty years. The assessment of seismic hazard will take care of the risk involved. In general terms, this means predicting the properties of an earthquake that is likely to take place in future at a given site. This can be done either with deterministic approach or with probabilistic approach. The seismic properties given in IS 1893 Code are based on deterministic approach. A probabilistic seismic hazard map of India is shown in Figure 1.22.

The failure of roads/buildings/soil retaining structures occurs due to:

(i) Mistakes in design, i.e., underestimating the effects of dynamics
(ii) Mistakes in construction
(iii) Poor materials used in construction
(iv) Seismic forces exceeding the design values
(v) Other environmental factors like water table, etc.

The evaluation of reliability is intimately associated with probability, the study of probability, and its associated aspects therefore assume importance.

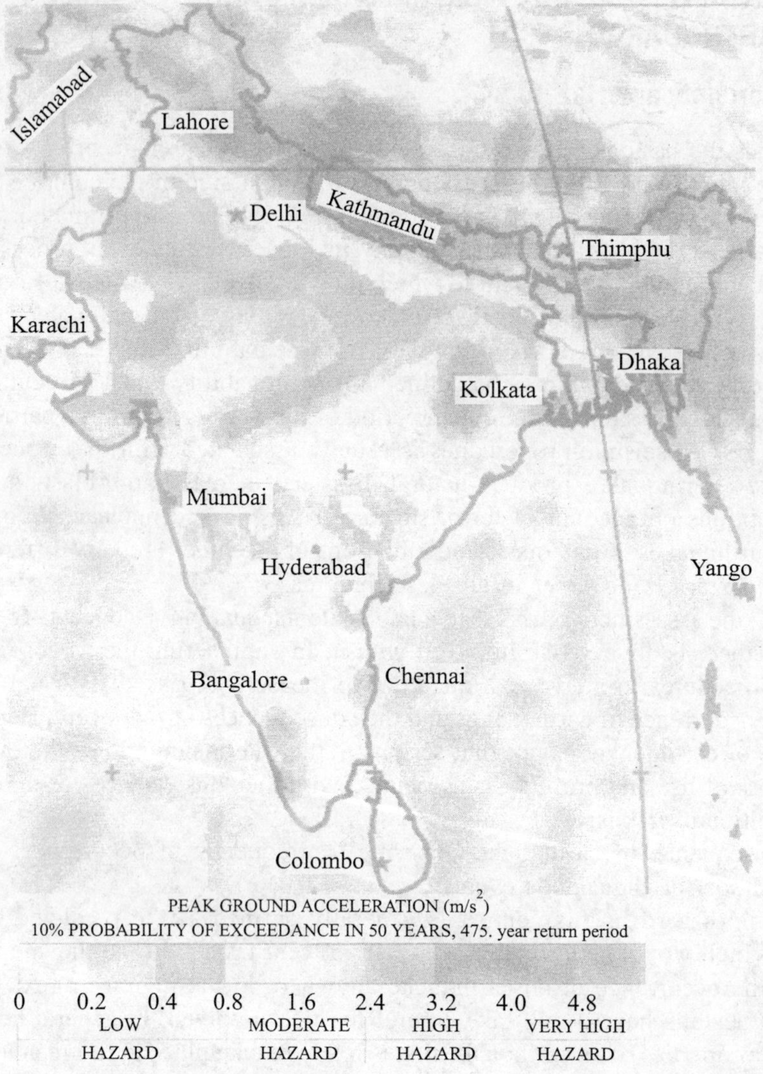

Figure 1.22 Seismic Hazard Map, India [After http: //geology. b1/maps/b1india. htm]

Probability is a number between 0 and 1 (both inclusive), which measures the uncertainty about the occurrence of a particular event or a set of events. The event may be an earthquake or a series of bomb blasts.

The uncertainty factor is more pronounced in soil dynamics than in structural dynamics. The confidence level is much higher in dealing with engineering materials like steel or concrete than dealing with soil. Earthquakes are uncertain in size, location and propagation, and therefore consequences are uncertain owing to inherent variability of soil as well as resistance of soil. Thus, when both loading and resistance are uncertain, the consequences are doubly uncertain. In this context, the need to increase the confidence level in design, construction and maintenance, the study of probability theory and the related reliability concepts have become very important.

The hazard projects a situation, which in particular circumstances, could lead to harm. Thus, earthquakes/bomb blasts represents a potential cause for apprehension, which relates to the likelihood and consequences of such an occurrence.

1.11.2 Risk, Reliability and Probability of Earthquakes

In the consideration of various uncertainties, it is important to represent the properties of earthquake motion along with a "risk index", a parameter describing the possibility of their occurrence. Thus, earthquake-hazard analysis can also mean evaluation of various properties of earthquake motion likely to occur at a given point within the specified period in the future in terms of the risk index.

The probability of earthquake occurrence in a year, or recurrence time, is frequently used as the risk index. For this purpose, earthquake occurrence or properties of earthquake motion are expressed in terms of a probability model. This does not mean that the earthquake phenomenon is a statistical (probability) phenomenon, but the element of uncertainty present in the quantitative evaluation of related parameters can be considered a model in the form of relative frequency (probability distribution). The problem can then be superimposed on the process of determining the risk index.

Earthquake-hazard analysis based on the probability model exhibits clarity in the steps involved and the result obtained. As such, it is very useful from the engineering point of view. Hence, we shall explain the methods of earthquake-hazard analysis based on the probability model.

The probability model for earthquake-hazard analysis naturally reflects the physical properties of the region concerned but due to lack of adequate data the model cannot be made highly rational. Calculation of model parameters is also difficult and quite often the reliability of the model itself is questionable.

1.12 EARTHQUAKE PREDICTION AND PREVENTION

Because of their devastating potential, there is a great deal of interest in predicting the location and time of large earthquakes. Although a great deal is known about where earthquakes are likely, there is currently no reliable way to predict the days or months when an event will occur in any specific location. Most large earthquakes occur on long fault zones around the margin of the Pacific Ocean. This is because the Atlantic Ocean is growing a few inches wider each year, and the Pacific is shrinking as the ocean floor is pushed beneath the Pacific Rim continents. Geologically, earthquakes around the Pacific Rim are normal and expected. This phenomenon will be explained in Chapter 2.

The long fault zones that ring the Pacific are subdivided by geologic irregularities into smaller fault segments which rupture individually. Earthquake magnitude and timing are controlled by the size of a fault segment, the stiffness of the rocks and the amount of accumulated stress. Where faults and plate motions are well-known, the fault segments most likely to break can be identified. If a fault segment is known to have broken in a past large earthquake, recurrence time and probable magnitude can be estimated based on fault segment size, rupture history and strain accumulation. This forecasting technique can only be used for well-understood faults, such as the San Andreas. No such forecasts can be made for poorly-understood faults, such as those that caused the 1994 Northridge, CA and 1995 Kobe, Japan quakes. Although there are clear seismic hazards in such area, Pacific north-west faults are complex and it is not yet possible to forecast when any particular fault segment in Washington or Oregon will break.

Along the San Andreas Fault, the segment considered most likely to rupture is near Parkfield CA. Earlier, it produced a series of identical earthquakes (about M 6.0) at fairly regular time intervals. USGS scientists are monitoring Parkfield for a wide variety of possible precursory effects. Using a set of assumptions about fault mechanics and the rate of stress accumulation, the seismologists are working hard to discover means of predicting earthquakes. These include the measurements of foreshocks, water depth in wells, tilting of the ground, magnetism, radon in wells and electrical conductivity. So far, successes are few. Earthquake prevention is even more difficult than earthquake prediction. Prevention is obviously not yet possible, and may never be possible. However, it is known that faults can be lubricated with water to cause slippage, and it has been suggested that major strains along a fault zone might be relieved in this manner. Proper building construction can reduce earthquake damage, but it is even better to delineate particularly hazardous areas and avoid constructing buildings in such areas.

One well-known successful earthquake prediction was for the Haicheng, China earthquake of 1975, when an evacuation warning was issued the day before a M 7.3 earthquake. In the preceding months, changes in land elevation and in ground water levels, widespread reports of peculiar animal behaviour, and many foreshocks had led to a lower-level warning. An increase in foreshock activity triggered the evacuation warning. Unfortunately, most earthquakes do not have such obvious precursors. In spite of their success in 1975, there was no warning of the 1976 Tangshan earthquake (Hebei Province) magnitude 7.6, which caused an estimated 250,000 fatalities.

Earthquake prediction is a popular pastime for psychics and pseudoscientists, and extravagant claims of past success are common. Predictions claimed as “successes” may rely on a restatement of well-understood long-term geologic earthquake hazards, or be so broad and vague that they are fulfilled by typical background seismic activity. Neither tidal forces nor unusual animal behaviour have been useful for predicting earthquakes. If an unscientific prediction is made, scientists cannot state that the predicted earthquake will not occur, because an event could possibly occur by chance on the predicted date, though there is no reason to think that the predicted date is more likely than any other day. Scientific earthquake predictions should state *where, when, how big* and *how probable* the predicted event is, and why the prediction is made. The national Earthquake Prediction Evaluation Council of USA reviews such predictions, but no generally useful method of predicting earthquakes has yet been found.

It may never be possible to predict the exact time when a damaging earthquake will occur, because when enough strain has built up, a fault may become inherently unstable, and any small background earthquake may or may not continue rupturing and turn into a large earthquake. While it may eventually be possible to accurately diagnose the strain state of faults, the precise timing of large events may continue to elude us. In the Pacific north-west, earthquake hazards are well-known and future earthquake damage can be greatly reduced by identifying and improving or removing our most vulnerable and dangerous structures.

The goal of earthquake prediction is to give warning of potentially damaging earthquakes early enough to allow appropriate response to the disaster, enabling people to minimize loss of life and property. The US Geological Survey conducts and supports research on the likelihood of future earthquakes. This research includes field, laboratory and theoretical investigations of earthquake mechanisms and fault zones. A primary goal of earthquake research is to increase the reliability of earthquake probability estimates. Ultimately, scientists would like to be able to specify a high probability for a specific earthquake on a particular fault within a particular year. Scientists estimate earthquakes probabilities in two ways: by studying the history of large earthquakes in a specific area and the rate at which strain accumulates in the rock.

Scientists study the past frequency of large earthquakes in order to determine the future likelihood of similar large shocks. For example, if a region has experienced four magnitude 7 or larger earthquakes during 200 years of recorded history, and if these shocks occurred randomly in time, then scientists would assign a 50 per cent probability (that is, just as likely to happen) to the occurrence of another magnitude 7 or larger quake in the region during the next 50 years.

But in many places, the assumption of random occurrence with time may not be true, because when strain is released along one part of the fault system, it may actually increase on another part. Four magnitude 6.8 or larger earthquakes and many magnitude 6–6.5 shocks occurred in the San Francisco Bay region during the 75 years between 1836 and 1911. For the next 68 years (until 1979), no earthquakes of magnitude 6 or larger occurred in the region. Beginning with a magnitude 6.0 shock in 1979, the earthquake activity in the region increased dramatically; between 1979 and 1989, there were four magnitude 6 or greater earthquakes, including the magnitude 7.1 Loma Prieta earthquake. This clustering of earthquakes leads scientists to estimate that the probability of a magnitude 6.8 or larger earthquake occurring during the next 30 years in the San Francisco Bay region is about 67 per cent.

Another way to estimate the likelihood of future earthquakes is to study how fast strain accumulates. When plate movements build the strain in rocks to a critical level, like pulling a rubber band too tight, the rocks will suddenly break and slip to a new position. Scientists measure how much strain accumulates along a fault segment each year, how much time has passed since the last earthquake along the segment, and how much strain was released in the last earthquake. This information is then used to calculate the time required for the accumulating strain to build to the level that results in an earthquake. This simple model is complicated by the fact that such detailed information about faults is rare. In the United States, only the San Andreas fault system has adequate records for using this prediction method.

Both of these methods, and a wide array of monitoring techniques, are being tested along part of the San Andreas fault. For the past 150 years, earthquakes of about magnitude 6 have occurred on an average of every 22 years on the San Andreas fault near Parkfield, California. The last major shock was in 1966. Because of the consistency and similarity of these earthquakes, scientists have started an experiment to "capture" the next Parkfield earthquake. A dense web of monitoring instruments was deployed in the region during the late 1980s. The main goals of the ongoing Parkfield Earthquake Prediction Experiment are to record the geophysical signals before and after the expected earthquake; to issue a short-term prediction; and to develop effective methods of communication between earthquake scientists and community officials responsible for disaster response and mitigation. This project has already made important contributions to both earth science and public policy.

Scientific understanding of earthquakes is of vital importance to any nation. As the population increases, expanding urban development and construction works encroach upon areas susceptible to earthquakes. With a greater understanding of the causes and effects of earthquakes, we may be able to reduce damage and loss of life from this destructive phenomenon.

On the basis of the research conducted since the 1989 Loma Prieta earthquake, US Geological Survey (USGS) and other scientists conclude that there is a 62% probability of at least one magnitude 6.7 or greater quake, capable of causing widespread damage, striking the San Francisco Bay region before 2032. Major quakes may occur in any part of this rapidly growing region. This emphasizes the urgency for all communities in the Bay region to continue preparing for earthquakes.

The mission of the National Earthquake Information Center (NEIC) is to rapidly determine location and size of all destructive earthquakes worldwide and to immediately disseminate this information to concerned national and international agencies, scientists, and the general public. As World Data Centre for Seismology, Denver, the NEIC compiles and maintains an extensive, global database on earthquake parameters and their effects that serves as a solid foundation for basic and applied earth science research.

Earthquake prediction for different seismic active regions of India are also going on in their own way. A sudden drop in atmospheric temperature and smaller earthquakes have convinced seismologists of the need to alert the Assam government of a major earthquake likely to strike in the near future. All district magistrates have been alerted for a possible earthquake of a strong intensity Meanwhile, the Geological Survey of India wrote back that such changes in temperature could take place and there was no need to jump into conclusion, said the official source. Assam has witnessed two major earthquakes in the past. One was on June 12, 1897 and another on Aug. 15, 1950. Both were higher than intensity 8 in the Richter scale and killed thousands of people.

Earthquake prevention is even more difficult than earthquake prediction. Prevention is obviously not yet possible, and may never be possible. There is no technology available till date to prevent the occurrance of earthquake. Thus the only course left is to design and construct earthquake proof structures and earthquake resistant structures.

PROBLEMS

1.1 What is geotechnical earthquake engineering? Explain its relationship with soil dynamics and structural dynamics.

1.2 Describe any three past earthquakes (major/great) of the world. What is the frequency of occurrence of great earthquakes with magnitude $M > 8.0$. Give your engineering comments and interpretations of two great earthquakes with magnitude $M > 8.0$.

1.3 Suppose that you are considering buying a house in Bhuj area in the state of Gujarat. The house is a well-designed frame structure resting on medium soft soil. For a repeat of 26 Jan. 2001 Bhuj earthquake, what type of damage would be expected for foundation and superstructure of the house?

1.4 Describe the seismicity of any three of the following regions:

1. Indian subcontinent
2. Iraq
3. Japan
4. Europe
5. New Zealand
6. USA

1.5 What is meant by Reservoir induced seismicity? Discuss the Koyna earthquake of 1967 in this context.

1.6 Discuss the nature of seismic forces and bomb blast loading. What is the basic difference between response mechanism of the structure to the earthquake excitation and bomb blast loading?

1.7 Discuss in detail the case history of Bhuj earthquake of January 26, 2001. How is it different from four other major earthquakes in India? Describe the awakening and awareness generated towards coping with earthquakes in India after this event.

1.8 Define risk, reliability and hazard for an earthquake prone site. How will you ascertain probability of earthquake occurrence or recurrence time as risk index?

2

SEISMOLOGY AND EARTHQUAKES

2.1 INTRODUCTION

The earthquake is a natural phenomenon occurring as a result of sudden rupture of the rocks, due to some reason or the other, which constitute the earth. The vibrations generated due to the occurrence of an earthquake are termed earthquake motion. The terms 'earthquake' and 'earthquake motion' are used interchangeably.

The rupture of rocks causing an earthquake extends over quite some distance but the point beneath the earth's surface at which the rupture is initiated is called the *hypocentre* (or focus). Its depth is called the *hypocentral* (focal) depth, while the point on the earth's surface straight above the hypocentre is called the *epicentre*. The distance from the epicentre or the hypocentre to any given point is called the epicentral or hypocentral distance, respectively, as shown in Figures 2.1 and 2.2. Information about the major earthquakes is displayed on the website of Geological Survey of America (USGS), giving such information as magnitude and intensity of seismic motion. Seismic waves originating at the hypocentre propagate in all directions and reach the earth's surface following different paths. Seismic waves vary in a complex manner depending upon the path of propagation, types of soil/rock media and their topography.

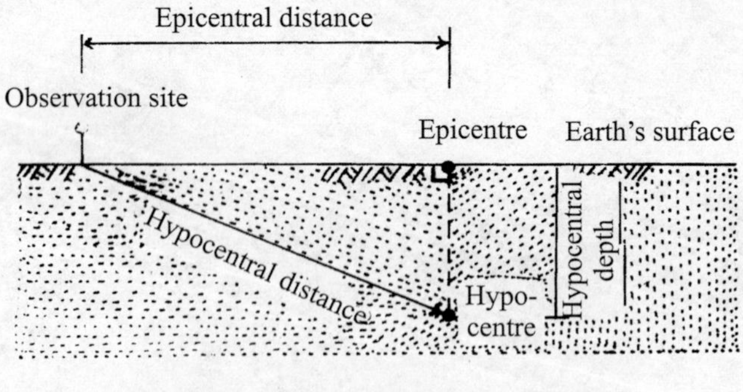

Figure 2.1 Schematic diagram of hypocentre.

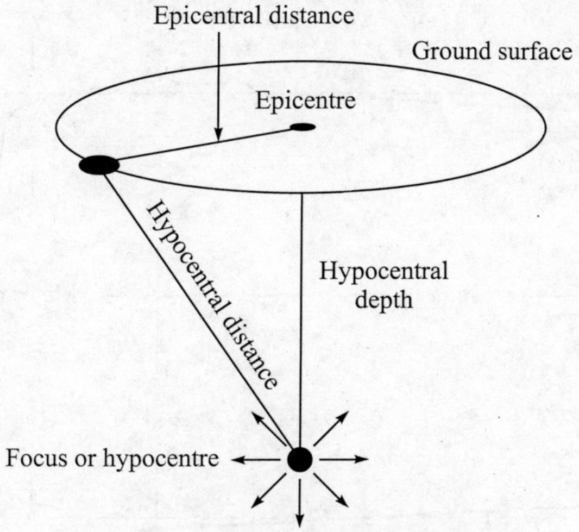

Figure 2.2 Schematic diagram of epicentre.

The two general type of waves produced by earthquake are *surface waves* which travel along the earth's surface and *body waves* which travel through the earth. Surface waves usually have the strongest ground motions and vibrations. The seismic waves are classified from the geotechnical earthquake engineering angle as described in the Figure 2.3. The surface waves cause most of the damage done during earthquakes. In short, body waves are designated as P-waves and S-waves. Both types pass through the earth's interior from the focus of an earthquake to distant points on the surface. As compressional waves travel at greater speeds as explained below in Eq. (2.1), they are called primary waves or simply P-waves. The shear waves do not travel as rapidly as P-waves as explained in Eq. (2.2) below, so they ordinarily reach the earth's surface later and are called secondary or S-waves.

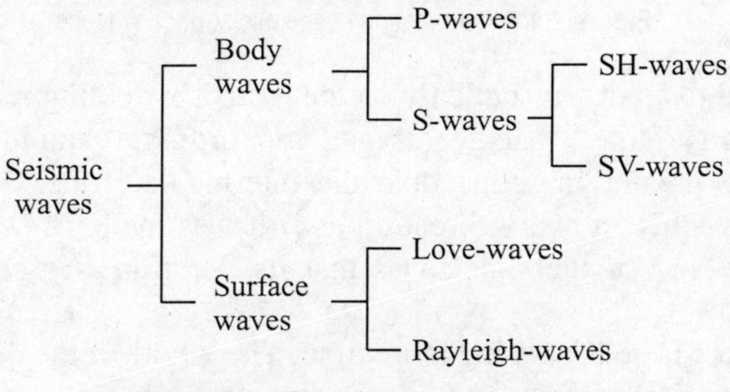

Figure 2.3 Seismic wave propagation.

The first physical indication of an earthquake is often a sharp thud, signalling the arrival of P-waves. This is followed by the shear waves and later the ground roll caused by the surface waves as shown in Figure 2.4. Oldham, R.D. as Director of Geological Survey at India (GSI), who was in Shillong on a morning walk during 1897 Assam earthquake described this sequence as:

> …a deep rumbling sound like thunder commenced … followed by the shock. The ground began to rock violently and in a few seconds it was impossible to stand

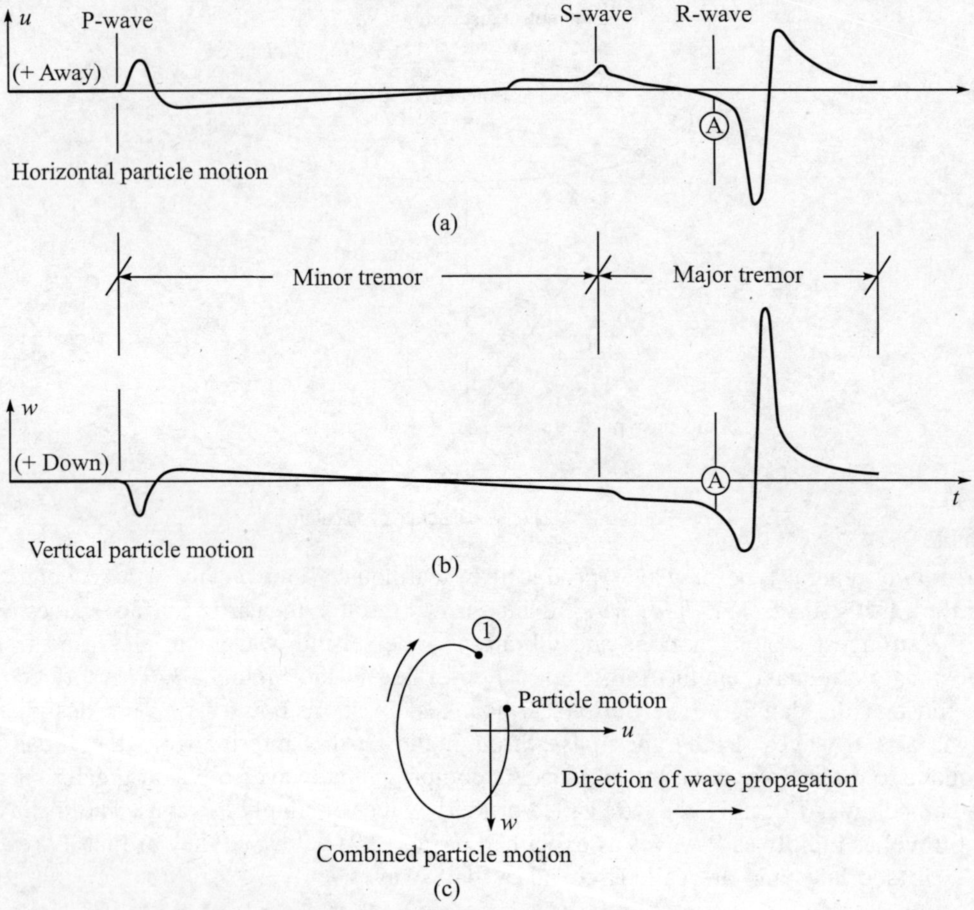

Figure 2.4 Propagation of seismic surface waves

> upright and ... had to sit down suddenly on the road. The feeling was as if the ground was being violently jerked, backwards and forwards very rapidly, every third or fourth jerk being greater in scope than the intermediate one. The surface at the ground vibrated visibly in every direction as if it was made of soft jelly. Oldham's impression at the end of the shock was that its duration was certain's under one minute....
>
> Subsequent tremors lasted for some more time. The whole of the damage was caused in the first 10 or 15 seconds...

The physical feel of the outset of an earthquake and the sequence of events have been described by a geologist as follows—who was at Valdez, Alaska during the 1964 earthquake.

> The first tremor was hard enough to stop a moving person and shock waves were immediately noticeable on the surface of the ground. The shocks continued with a rather long frequency, which gave the observer an impression of a rolling feeling rather than abrupt hard jolts. After about one minute, the amplitude or strength of the shock waves increased in intensity and failures in building as well as the frozen

ground surface began to occur.... After about 3½ minutes the severe shock waves ended and people began to react as could be expected.

The personal experience of the author during the 1988 Bihar–Nepal earthquake at Muzaffarpur during the early hours of Sunday 21st August (midnight of 20th August as per international date and time) was that of sleeping on a rolling and rocking bed ... at 4.39 am, awakening from sleep.... I quickly realized that it was the occurrence of an earthquake. There was a huge and loud sound, probably from deformations of the wooden chaukhats/frames of doors and windows. By the time I came out of the residential bungalow to the road, it was all over in about 20 seconds or so. The whole event of shaking lasted about a minute or so.

The severity of an earthquake can be expressed by intensity of the earthquake and magnitude of the earthquake. The intensity is a subjective measure that describes how strong a ground motion was felt as a particular site. The magnitude of an earthquake, usually expressed by the Richter's scale, is a measure of the amplitude of the seismic wave. The moment magnitude of an earthquake is a measure of the amount of the energy released.

The propagation velocities of P- and S-waves, i.e., V_P, V_S are determined by the modulus of elasticity of the propagation medium. The following relationships apply.

$$V_P = \sqrt{\frac{(1-\nu)E}{(1+\nu)(1-2\nu)\rho}} \qquad (2.1)$$

$$V_S = \sqrt{\frac{G}{\rho}} \qquad (2.2)$$

where,

E = Young's modulus of elasticity
G = shear modulus of elasticity
ν = Poisson ratio
ρ = density.

It is clear from Eqs. (2.1) and (2.2) that $V_P > V_S$. If the medium is purely elastic, both P- and S-waves can propagate at whatever depth. From this point of view, these waves are called body waves. However, since there are a number of discontinuities in the earth's crust, these body waves are subject to complex phenomena, such as reflection, refraction, diffraction, scattering, amplification, damping, etc. Reflection or refraction of P- and S-waves at these discontinuities follow Snell's law just as light rays do. The wave propagation will be presented in greater detail in Chapter 5.

The very term *earthquake,* when mentioned, generally creates a sense of panic and calamity in the minds of people, since many earthquakes have taken heavy tolls of life and property in the past, in many countries. Even now, with the prevailing advanced state of knowledge, earthquake occurrence still remains a mystery and is unpredictable.

Since earthquakes are capable of causing severe damage to any civil engineering structure, it is necessary to know what they are, why they occur, how they occur, what kind of harmful effects they will produce from the civil engineering point of view, what precautionary measures can be taken to minimize such harm, and other related factors.

Thus, an earthquake may be simply described as a sudden vibrating (or jerking or jolting or trembling or shivering) phenomenon of the earth's surface for some reason or the other. The intensity of this jolting phenomenon of the earth's surface may be insignificant at one extreme and highly catastrophic at the other extreme. From the physical geology point of view, an earthquake may be described as a natural force which originates below the earth's surface, works randomly and creates irregularities on the earth's surface. Therefore, it is an endogenous geological agent. The study of earthquakes is known as *seismology*. The records of earthquakes are known as *seismograms* and the recording instrument are known as *seismographs*. In Greek, *seismo* means shaking and *logy* stands for science, and so shaking of earth is studied in the branch of science known as seismology.

In other words, earthquakes are powerful manifestations of sudden release of strain energy accumulated over extensive time intervals. They radiate seismic waves of various types which propagate in all directions through the Earth's interior. The passage of seismic waves through rocks causes shaking which we feel as an earthquake.

Earthquake terminology

Before proceeding further, let us acquaint ourselves with the earthquake terminology frequently used in this book:

1. The place of origin of the earthquake in the interior of the earth, as already stated, is known as focus or origin or centre or hypocentre as shown in Figure 2.1.
2. The place on the earth's surface, which lies exactly above the centre of the earthquake, is known as the epicentre. For obvious reasons, the destruction caused by the earthquake at this place will always be maximum, and with increasing distance from this point, the intensity of destruction will decrease. The point on the earth's surface diametrically opposite to the epicentre is called the *anti-centre*.
3. The imaginary line which joins the hypocentre and the epicentre is called the *seismic vertical*, and this represents the minimum distance which the earthquake has to travel to reach the surface of the earth.
4. An imaginary line joining the points of same intensity of the earthquake is called *isoseismal*. In plan, the different isoseismals will appear more or less as a point. On the other hand, if the focus happens to be a linear tract, the isoseismals will occur elongated. Naturally, the areas or zones enclosed by any two successive isoseismals would have suffered the same extent of destruction.
5. An imaginary line which joins the points at which the earthquake waves have arrived at the earth's surface at the same time is called *coseismal*. In homogeneous grounds with plain surfaces, the isoseismals and coseismals coincide. Of course, in many cases due to surface and subsurface irregularities, such coincidences may not occur.
6. The enormous energy released from the hypocentre at the time of the earthquake is transmitted in all directions in the form of waves, known as *seismic waves*.
7. The earthquakes can produce long period sea-waves called *tsunamis* (soo-NAH-mees), however the earthquake-induced waves in enclosed bodies of water are called *seiches*. The word seiche originated in Switzerland; Forel introduced this word for general usage. Tsunami is a Japanese word represented by the characters 'tsu' and 'nami'. The character 'tsu' means harbour while the character 'nami' means wave.

8. The intensity as expressed by the Modified Mercalli Scale (MMS), is a subjective measure that describes how strong a shock was felt at a particular location.
9. The Richter scale, named after Dr. Charles F. Richter of the California Institute of Technology, is the best-known scale for measuring the magnitude of earthquakes. The scale is logarithmic so that a recording of 7, for example, indicates a disturbance with ground motion 10 times as large as recording of 6. Earthquakes with a Richter scale of 6 or more are commonly considered major; great earthquakes have magnitude of 8 or more (see Tables 1.4 and 1.5).

The chapter provides the brief structure of earth's interior, mechanics of continental drifts, theories explaining why, earthquakes occur and the various terminologies used to describe them. The study of geotechnical earthquake engineering should be based on the various seismic processes by which earthquakes occur and their effect on ground motion. Seismology is essentially the study of earthquakes and seismic waves. Although earthquakes and seismic waves propagations are highly complex phenomena, advances in seismology provide ways and means to understand and estimate the rates of occurrence of earthquakes in most seismically active areas of the world. The related field of strong motion seismology (which is of interest to engineers) uses waves from large earthquakes to study the earthquakes source in detail, predict the strength of future shaking, establish safer building codes and hopefully improve the seismic engineering design.

Seismic deformation

Earthquakes originate from spontaneous slippage along planes of weakness, i.e., faults after elastic strain accumulation over a long period of time. The faulting process may be modelled mathematically as a shear dislocation in an elastic medium which is equivalent to a double couple body force. The earthquake cycle progresses from under stress state to an overstressed state as the plate tectonics motion drives the weaker zones to rupture during an earthquake and a nearly-relaxed but deformed state is formed. Typically, a straight component in pre-rupture state takes the distorted shape as shown in Figure 2.5(a). This process of seismic deformation is also called elastic rebound.

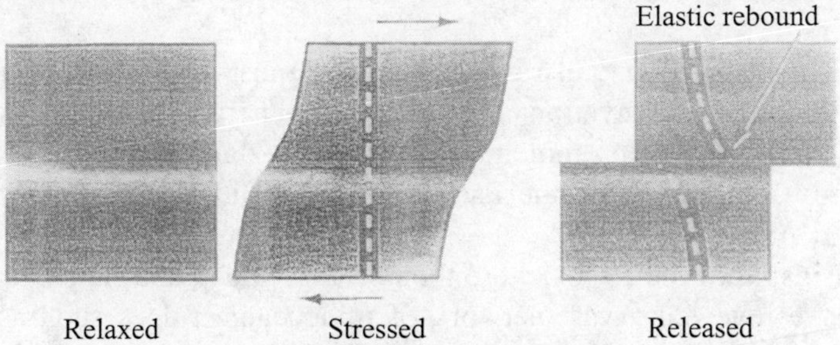

Figure 2.5(a) Seismic deformations.

Particle motion and seismic waves

The second type of deformation, i.e., dynamic motion is essentially comprised of waves radiated from the earthquake as the earth surface ruptures [see Figure 2.5(b)].

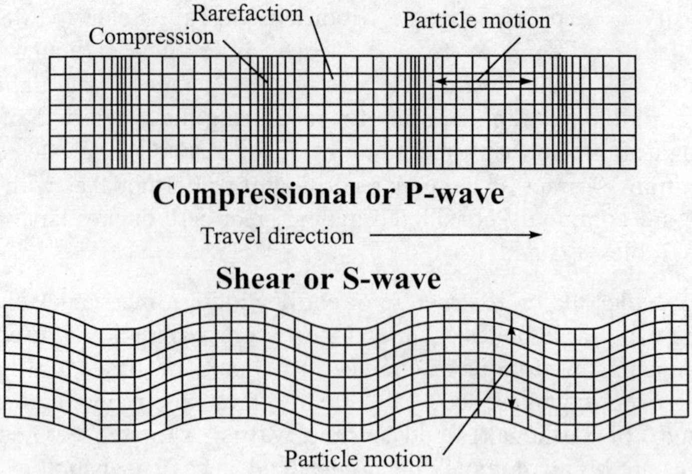

Figure 2.5(b) Particle motion produced by seismic surface waves.

2.2 STRUCTURE OF THE EARTH'S INTERIOR

The earth is a unique planet, with 71 per cent of its surface covered by water.

The equatorial diameter of the earth is 12,757 km (7927 miles) and the polar diameter of the earth is 12,714 km (7900 miles). This equatorial bulge is due to the earth's rotation. The higher diameter is caused by higher equatorial velocities. The nearly spherical earth consists of a very thin crust (8 to 35 km thick), a thick mantle (about 2900 km thick), a fluid outer core (2300 km thick), and a solid inner core (radius of about 1200 km). The crust and the mantle are made of rock material, and both parts of the core are largely made of iron. The earth weighs (5.4×10^{21} tons). The earth's overall density is 5.52; since crustal rocks have densities of about 2.6 (the granite rocks of the continents) to 3.0 (the basalts of the ocean basins), the earth must have a dense interior. The model of the earth's interior is shown in Figure 2.6.

The seven continents extend under the oceans, encompassing the continental shelves and the continental slopes. Thus, the shelves and the slopes are parts of the land masses rather than of the ocean basins.

On the continents, the mountain ranges form the most spectacular topographic features. Plateaus, generally of medium elevation, and the plains, generally of the lowest elevation and the lowest relief, are the other two prominent features on the land surface. The highest mountain ranges are generally located around the Pacific Ocean or lie along an east-west line between Africa, Europe and Asia.

The ocean basins, until the 1940s, were thought to be deep and rather featureless. Oceanographic studies have revealed a vast network of midoceanic ridges such as the Mid-Atlantic ridge, and the off-centre oceanic rises such as the East Pacific Rise, both of which, if on land, would be prominent mountain ranges. These are zones of active volcanism and faulting (breaking and moving) of the earth's crust or exterior layer.

The present state of knowledge of the structure of the interior of the earth is from seismological observations and modelling [see Figure 2.7(a)]. The majority of detailed information about the composition and structure of the earth's interior has come from the seismological

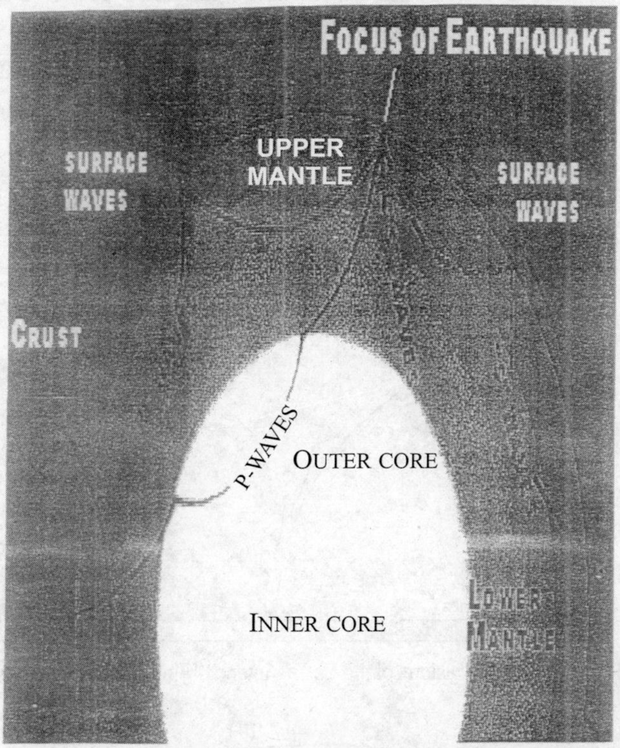

Figure 2.6 Seismic surface waves and model of Earth's interior.

observations as shown in Figures 2.7(b) and 2.7(c). There are other elements, too, that provide some basic information such as gravity observations and magnetic field studies.

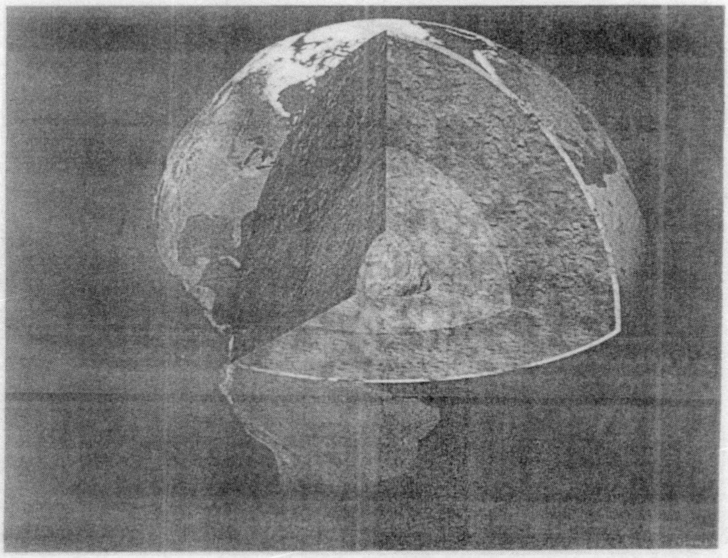

Figure 2.7(a) Model of structure of Earth's interior.

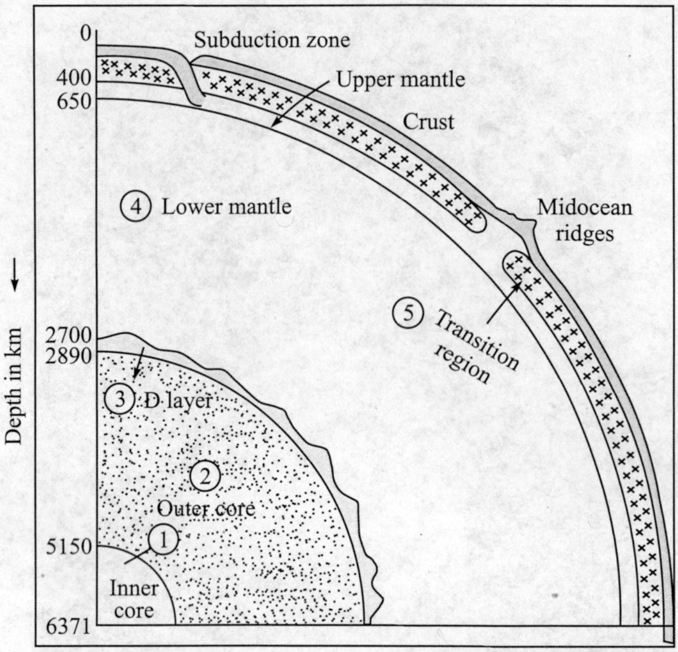

Figure 2.7(b) Divisions of the Earth's interior. [After Beatty et al. 1990]

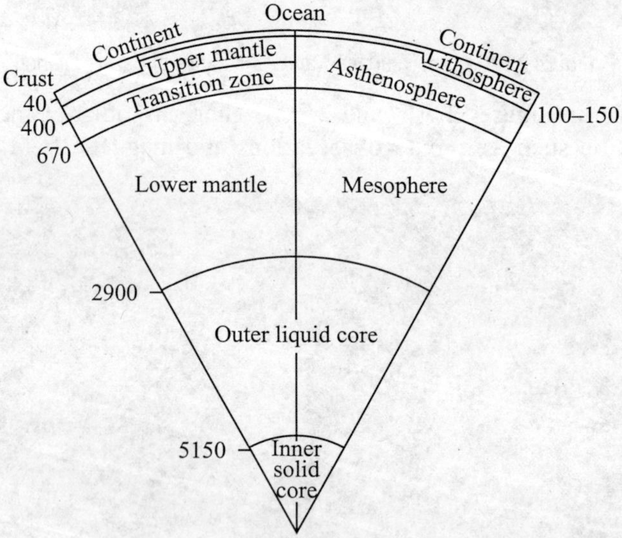

Figure 2.7(c) Compositional (left) and Rheological (right) divisions of the earth's interior (in km).

Gravity field observations on the surface of the earth, combined with the knowledge of the diameter of the earth, allow us to conclude that the average density of the earth must be about 5520 kg/m^3. We know from actual measurement, however, that the surface rocks are not denser than 3200 kg/m^3. Because the earth, like all other bodies of the solar system, accreted from the primordial dust of the solar nebula, we may conclude that deep inside the earth there must be some heavier material, which is most likely to be iron. The seismic velocities as in Eqs. (2.1) and (2.2) are governed by three parameters, namely, bulk modulus K, rigidity modulus G and the density ρ. The compressional wave velocity v_p and the shear wave velocity v_S can be measured from the observation of travel time of earthquake waves.

The other features of the structure of the earth's interior as follows:

Crust: The crust consists of the region from the surface to the Mohorovicic discontinuity, popularly known as Moho. The Moho occurs at a depth of about 6–12 km beneath the ocean and about 30–50 km beneath continents. The crust is further divided into two layers by the *Conrad* discontinuity across which P-wave velocity increases from about 5.6 km/s to 6.3 km/s. The first discontinuity was discovered by Andrija Mohorovicic, following a large earthquake in Croatia in 1909, from two arrivals separated in time. This discontinuity lies at a depth of about 35 km on continents and about 7 km beneath the oceanic crust. (Later on, it was found that the seismic P-wave velocity rapidly increases from ~6 km/s to more than about 8 km/s at this boundary.) Rai, S.N. et al., (2002) have presented detailed description of earth's interior and may be referred for further study.

Mantle: The mantle extends from the Moho to the Gutenberg discontinuity at a depth of about 2900 km, of which the velocity of P-waves decreases rapidly.

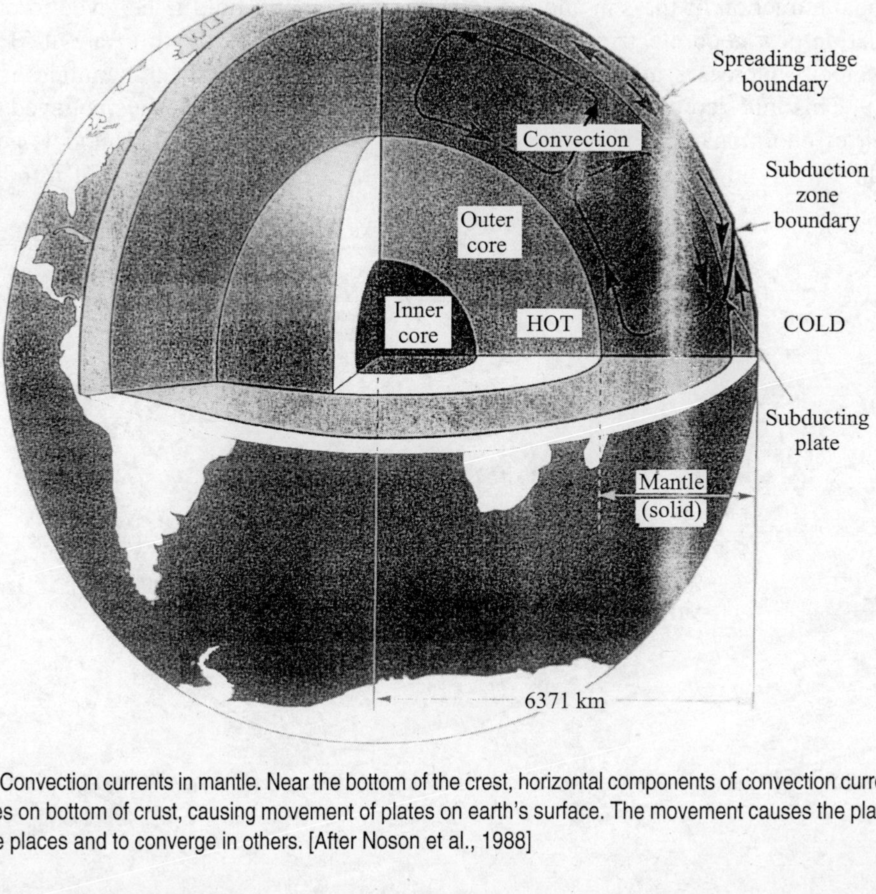

Figure 2.8 Convection currents in mantle. Near the bottom of the crest, horizontal components of convection currents impose shear stresses on bottom of crust, causing movement of plates on earth's surface. The movement causes the plates to move apart in some places and to converge in others. [After Noson et al., 1988]

Beno Gutenberg discovered another seismic discontinuity. He observed that P-waves died out about 105° around the globe from an earthquake, and reappeared 140° away, but about two minutes later than expected. This resulted in a shadow zone, about 35° wide, where P-waves were absent. He realized that this could be explained with a core having a lower velocity. Later it was discovered that S-waves were totally blocked by the core, producing a complete S-wave-shadow zone beyond 105°. This total blockage revealed that the outer part of the core must be a liquid that absorbed the S-wave. The Moho-to-the-Gutenberg discontinuity represents the boundary between the core and the mantle. Within the mantle, a transition zone exists at depths of 400–670 km. This zone is characterized by changes in mineralogy and structure of the silicates and separates the upper mantle from the lower mantle. The P-wave velocity in the upper mantle is about 8 km s^{-1} [After Noson et al., 1998]. The lower mantle extends from 670 km to the core mantle boundary. The lower mantle is characterized by the constant increase in velocity and density up to its lower boundary.

The mantle thus consists of rock through which sound waves move at a higher velocity than they do in crustal rocks. The upper mantle is called lithosphere as shown in Figure 2.7(c). Beneath this is a 100 m thick, low velocity plastic zone called the asthenosphere.

Core: The core occurs from depths of 2900 km to the centre of the earth. The outer core lies from 2900 km to a discontinuity at about 5150 km. It does not transmit the shear wave and is interpreted to be liquid. A fluid state is also indicated by the response of the earth to the gravitational attraction of the sun and moon. The geomagnetic field is believed to originate by the circulation of a good electrical conductor in this region. At a depth of about 5150 km, the P-wave velocity increases abruptly and the S-waves are once again transmitted as shown in Figure 2.9. This inner core from 5150 km to the centre of the earth is thus believed to be solid as a result of enormous confining pressure. Seismic wave velocities and models generated to explain the density of the earth, indicate that the core is mainly composed of iron and nickel.

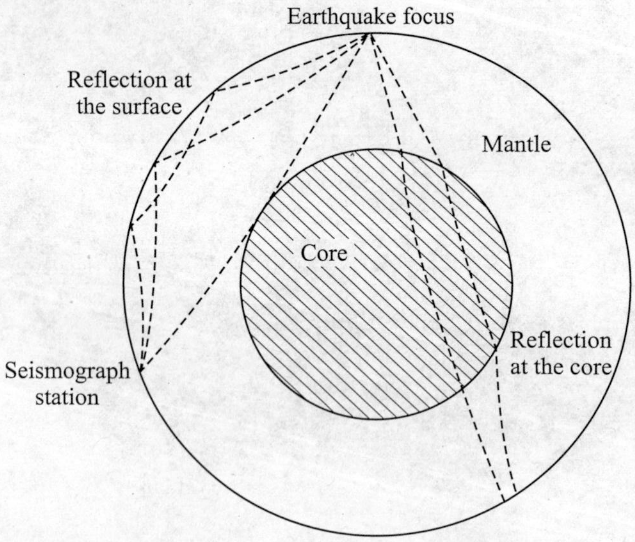

Figure 2.9 Propagation of P-waves and S-waves from focus of the earthquake by different layers of the earth.

2.2.1 Rheological Division of the Earth's Interior

Another type of subdivision of the earth's interior up to the mantle core boundary is based on rheological properties. The outermost strong layer, which is made of crust and the uppermost mantle and deform in an essentially elastic manner, is called *lithosphere*. Its thickness varies between 100 and 150 km beneath the continents and between 50 and 70 km beneath the oceans. The strength of the lithosphere is not uniform. The 20 to 40 km thick upper layer is brittle and responds to stress by elastic deformation. This is followed by a ductile zone, which deforms by plastic flow above a load of about 100 MPa. The lower part of the lithosphere is again brittle in nature. A much weaker layer that reacts to stresses in a fluid manner underlies the lithosphere. This layer is known as the *asthenosphere* and extends up to 670 km. A low velocity zone (LVZ) generally occurs at the top of the asthenosphere. Low seismic velocities, high seismic attenuation and probably high electrical conductivity characterize this zone. It is generally accepted that the lower seismic velocities arise because of the presence of molten material. From the base of this zone, seismic velocities increase slowly to a depth of 400 km, making the top boundary of the transition zone. The asthenosphere represents that location in the mantle where the melting point is most closely approached. This layer is certainly not completely molten because it transmits S-waves, but small amounts of melt may be present. The depth of the asthenosphere depends on the geothermal gradient and melting point of the mantle material and occurs at shallow depths beneath the oceanic ridge. The lithosphere is divided into approximately 12 major plates. These plates move relatively over the asthenosphere due to the dragging force of mantle convection exerted at the base of the lithospheric plates. These phenomena of the plate movement gave rise to the theory of plate tectonics. The mantle low-velocity zone is of major importance to plate tectonics as it represents a low-viscosity layer along which relative movements of lithospheric plates and the asthenosphere can be accommodated. The lower part of the mantle beneath the asthenosphere is of high strength. The seismic waves in this region do not suffer great attenuation. This zone is known as *mesosphere* (Kearey and Vine, 1990). The compositional and rheological divisions of the earth's interior are illustrated in Figure 2.7(c).

In the beginning, it was thought that the main force driving the plates arises from the viscous drag exerted on the base of the lithosphere by the underlying moving asthenosphere. If the velocity of the asthenosphere exceeds that of the plate, the resulting drag would help the plate motion, but if this velocity were lower, the drag would resist and impede the motion of the plate. The motion of the convection cells in the mantle would be rising under the oceanic ridges, and descending below the trenches, and be generally absent under the continental areas. This would require that the oceanic lithosphere be in a state of tension at the ridges and under compression at the trenches.

This mechanism has a number of difficulties. The region of contact between the convecting mantle and the lithosphere is the zone at the top of the asthenosphere, which has a low seismic velocity (the low-velocity zone LVZ) and a low viscosity. It is estimated that the asthenosphere must move at a rate of about 200 mm/year to move the lithosphere at a rate of 40 mm/year. This rate is rather too high to be reasonable, and the very small relative motion of the hotspots in the recent geological past indicates that this is not likely.

Another difficulty is that if this were currently the main mechanism, the major convection cells would have to have about half the width of the large oceans, with a pattern of motion that would have to be more or less constant over very large areas under the lithosphere. This would fail to explain the relative motion of plates with irregularly shaped margins at the Mid-Atlantic ridge and the Carlsberg ridge, and the motion of small plates such as the Caribbean and Philippine plates.

It has been argued by Ziegler (1993) that the mantle drag may have been a significant mechanism during the break-up of the supercontinent.

The other mechanism that has been gaining acceptance is one in which the plates move in response to forces applied to their edges, the role of the asthenosphere being largely passive. The ideas have been developed by Orowan (1965), Elasser (1969), Bott (1982), and others.

Four main edge forces have been considered. At the ocean ridges the ridge push force arises from the hot, buoyant, rising mantle material that results in an elevation of the ridge, pushing the newly created oceanic plates away from the ridge crest. At the subduction zones, the lithosphere is cooler and denser than the underlying material, and sinks as a result of this negative buoyancy. Part of this downward force is transmitted to the lithospheric plate as slab pull (the rest being taken up by the viscous drag resistance, and the friction due to the overriding plate, to the descent of the plate). The part of the overriding lithospheric plate being dragged into the subduction trench may be put into a state of tensile stress by the force designated as the trench suction force. These forces may have to work against the mantle drag, against the resistance of the subducting plate to bending, and against the frictional and viscous forces mentioned above.

The edge-force mechanism appears to be able to better explain the plate motions, the observed pattern of interpolate stresses, the observation that the plate velocities are independent of the plate areas, the more rapid movement of plates attached to the down-going slabs, and the slower movement of plates with a large area of continental crust. Most workers now accept the edge-force mechanism, the mantle drag being something that generally inhibits plate motion.

Drill holes have been penetrated only about 9000 m (9.0 km) into the earth's crust, and the deepest-mines are not that deep. Thus, humans have barely scratched the surface. The knowledge of the earth's interior is based on indirect observation from seismic waves. The study of the internal structure of the earth by wave propagation has been presented in Figure 2.10. The propagation of shear wave into first 900 km into the earth has been shown in Figure 2.11. The plate tectonics will be discussed in greater detail in Section 2.4.

Seismic discontinuities as discussed in this section aid in distinguishing between the various divisions of the earth into inner core, outer core, D layer, lower mantle, transition region, upper mantle and crust as shown in Figures 2.7(a) and 2.7(b). These divisions may described as follows:

Inner core: It constitutes 1.7% of the earth's mass, at depths between 5150 km and 6370 km. The inner core is solid and unattached to the mantle and is suspended in the molten core.

Outer core: It is 30.8% of the earth's mass, at depths between 2890 km and 5150 km. The outer core is hot, electrically conducting liquid within which convective motion occurs. This conductive layer combines with the earth's motion (rotation) to create a dynamo effect that maintains a system of electrical currents known as the earth's magnetic field. It is also responsible for the subtle jerking of earth's rotation.

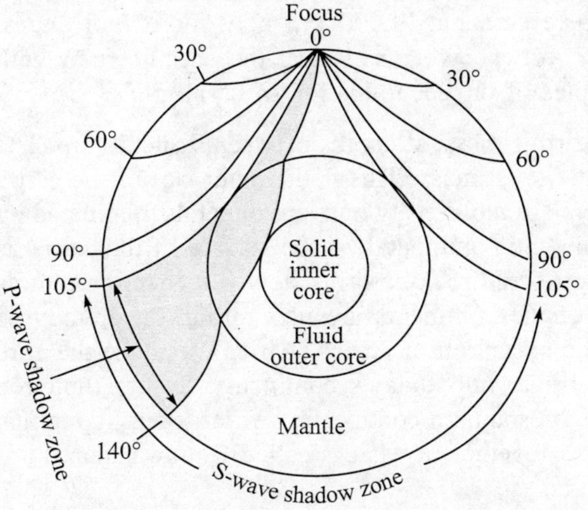

Figure 2.10 Study of the internal structure of the earth by wave propagation.

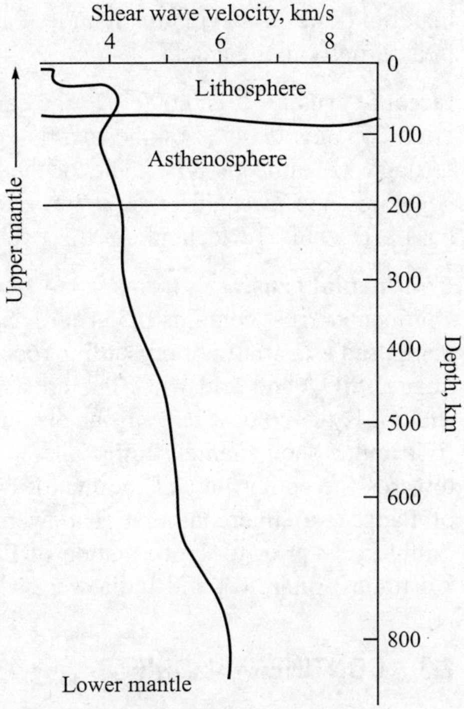

Figure 2.11 Propagation of shear wave into first 900 km into the earth.

D layer: It is 3% of the earth's mass, at depths between 2700 km and 2890 km. This layer is often identified as part of the lower mantle. Seismic discontinuities suggest that the D layer might differ chemically from the lower mantle lying above it.

Lower mantle: It forms 49.2% of the earth's mass, at depths between 650 km and 2890 km. The lower mantle contains 72.9% of the mantle-current mass and is probably composed of silicon, magnesium and oxygen. It probably also contains some iron, calcium and aluminium. Scientists make there deductions by assuming that the earth has a similar abundance and proportions of cosmic elements as found in the sun and primitive meteorites.

Transition region: It constitutes 7.5% of earth's mass at depths between 400 km and 650 km. The transition region or mesosphere (for middle mantle), sometimes called the fertile layer, contains 11.1% of the mantle-crust mass and is the source at basaltic magma. It also contains calcium, aluminium and garnet which is a complex aluminium–bearing silicate mineral. This layer is dense when cold because of the garnet. It is buoyant when hot because these materials melt easily to form basalt which can rise through the upper layers as magma.

Upper mantle: It is 10.3% of earth's mass; depth between 10 km and 400 km. The upper mantle contains 15.3% of the mantle-crust mass. Fragments have been excavated for observations by eroded mountain belts and volcanic eruptions. Olivine and pyroxene have been the primary minerals found in this way. These and the other materials are refractory and crystalline at high temperature; therefore, most settle out of rising magma, either forming new crustal material or never leaving the mantle. Part of the upper mantle called the asthenosphere might be in a partially molten state.

Oceanic crust: It is 0.099% of the earth's mass, at depths between 0 and 10 km. The oceanic crust contains 0.147% of the mantle-crust mass. The majority of the earth's crust was made through volcanic activity. The oceanic ridge system, a 40,000 km network of volcanoes, generates new oceanic crust at the rate at 17 km^3 per year, covering the oceanic floor with basalt. Hawaii and Iceland are the two examples of the accumulation of basalt piles.

Continental crust: It forms 0.374% of the earth's mass, at depths between 0 and 50 km. The continental crust contains 0.554% of the mantle-crust mass. This is the outer part of the earth, composed essentially of crystalline rocks. There are tuo-density buoyant minerals dominated by quartz (SiO_2) and feldspars. The crust (oceanic and continental) is the surface of the earth, as such, it is the coldest part of our planet. Because cold rock deforms slowly, it is referred to as rigid outer shell, namely, lithosphere. Most of the continents are now sitting on or moving towards the cooler part of the mantle, with the exception of Africa. Africa was once the core of Pangea, a supercontinent that eventually broke into today's continents. Several hundred million year prior to the formation of Pangea, the southern continents—Africa, South America, Australia, Antarctica and India were assembled together in what is called Gondwanaland.

2.3 CONTINENTAL DRIFTS

The observations of similarity between the coastlines of southern India and the eastern south America and the western Africa suggested the possibility of continental drift. The fit of the several continents has been shown in Figure 2.12. Although several early writers toyed with the possibility that the continents now separated were once united, the first serious attempt was made by Taylor (1910) and by Alfred Wegener of Germany (1912). Alfred Wegener was the first to use the phrase "continental drift" (in German "die Verschiebung der Kontinente") and formally publish the hypothesis that the continents had somehow "drifted" apart.

Watson, J.(1975) narrated that Wegener book, *Die Entestehung der Kontinents* appeared during the First World War. Wegener believed that the earth had only one continent Pangea (meaning all the earth) 200 m.y. (million years) ago. Pangea broke into pieces that slowly drifted into the present configurations of the continents. The drift theory had its proponents but, in general, its popularity died with Wegener in 1930. But the early 1960s saw the rousing rebirth of this concept. Thus, by 1960s the drift hypothesis had become an essential feature of geological thought in the most communities of earth' scientists. Further this led to a more complete concept discussed in the next section on Plate Tectonic theory. Tectonics became a fancy word for the study of the earth's major structural features and the process that formed them.

The earth is a unique planet with seventy-one per cent of its surface covered by water. The seven continents extend under the ocean. Since evolution of the earth, the continents have been moving relative to each other. This is shown in Table 2.1 wherein the relative motions are in cm per year. It has been observed that the continents are moving, colliding and sliding past one another for a very long time. The evidence also suggests that the ocean crust has been widened. The history of the oceanic crust, is essentially dominated by the phenomenon of sea-floor spreading. The record phases of sea-floor spreading are closely related to the process of continental drift. The generation of new crustal material in the ocean basins took place at

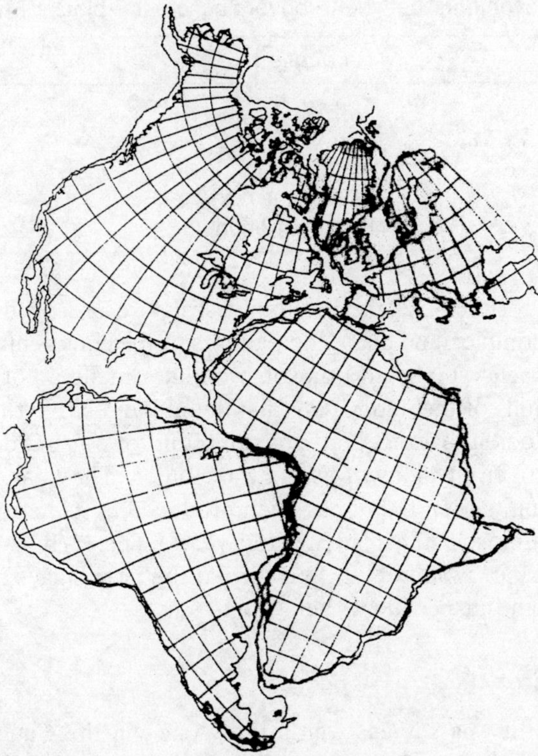

Figure 2.12 The fit of the American, African and European continents

approximately the rate required to fill the gaps opened up by separation of the continents as shown in Figure 2.12. The rates of convergence are controlled by the growth rate of the new ocean basins from which the advancing plates move, and have been calculated on this basis by Le Pichon (1968) at values of 5–10 cm per year as listed in Table 2.1.

Table 2.1 Differential movements between converging crustal blocks [After Le Pichon, 1968]

Plates	*Location*	*Rate (cm per year)*
E. Asian: Pacific	Kuriles trench	7.9
	Kuriles trench	8.5
	Japan trench	8.8
	Japan trench	9.0
	Mariana trench	9.0
	Mariana trench	8.9
Indian: Pacific	N. Tonga trench	9.1
	S. Kermadec trench	4.7
	S. New Zealand trench	1.7
	New Guinea	11.0
American: Pacific	E. Aleutian trench	5.3
	W. Aleutian trench	6.2
	W. Aleutian trench	6.3

(*Contd.*)

Table 2.1 Differential movements between converging crustal blocks [After Le Pichon, 1968]

Plates	*Location*	*Rate (cm per year)*
Indian: Eurasian	Turkey	4.3
	Iran	4.8
	Tibet	5.6
	W. Java trench	6.0
	E. Java trench	4.9

The ocean basins with their relatively simple structure are by this means adjusted with remarkable precision to complex and changing patterns of the continental masses. The integrated movements whereby the relatively rigid lithospheric plates made up of continental and oceanic crust, together with portions of the underlying mantle, move relative to one another in response to the processes of sea-floor spreading, and 'consumption' of crust at destructive plate margins, defines a style of crustal activity known as plate tectonics. The characteristic arrangement of continental and oceanic units seen today has been produced by plate movements whose effects can be traced back for almost 1000 m.y. (million years) The early stages were dominated by the evolution of a network of mobile belts and then at the later stages involved the break-up of the supercontinents and the dispersal of their fragments.

2.3.1 The Mobile Belt

The belt of Alpine mobility passes eastwards into Asia via the Caucasian Mountains of the U.S.S.R., the Anatolian peninsula, Cyprus, the Elburz and Zagros ranges of Iran and the desert mountains of Afghanistan as shown in Figure 2.13. As in the regions already considered, the components resolve themselves into northern and southern units with contrasting tectonic symmetry. The Caucasus has much in common with the Carpathians and continues the tract bordering the Eurasian craton. Their structures show a northward vergence and they are fronted towards the north by force deep basins. In the Taurus Mountains and adjacent parts of Anatolia, nape structures suggestive of southward transport have been described. The Zagros Mountains of Iran show a southward vergence, expressed by the overthrusting of rocks of an internal zone over those of a more southerly external zone along the Zagros lineament. In the complex region between these outward-facing components, some blocks in which Mesozoic and Tertiary rocks remain little disturbed, are seen in central Anatolia and possibly also in Iran.

From the Middle east the Alpine mobile tract swings north-eastwards to merge with the earth's greatest highland block. A vast complex at mountains and plateaux occupies central Asia, Mongolia, Tibet and the Himalayas. The geophysical observations suggest that the crust is almost twice its normal thickness, the Moho descending to 70–75 km beneath the high Himalayas and 65 km beneath the Pamirs. The Himalayas incorporate a portion of an uplifted marginal mobile belt, together with the thrust masses sliced from the frontal parts of the Indian craton.

2.3.2 The Gondwanaland Group

The Alpine–Himalayan belt which forms the zone of separation between continental masses derived from Laurasia and Gondwanaland has a history of mobility extending through the entire Phannerozoic eon (see Table 1.1). This belt began its evolution as a portion of the peripheral

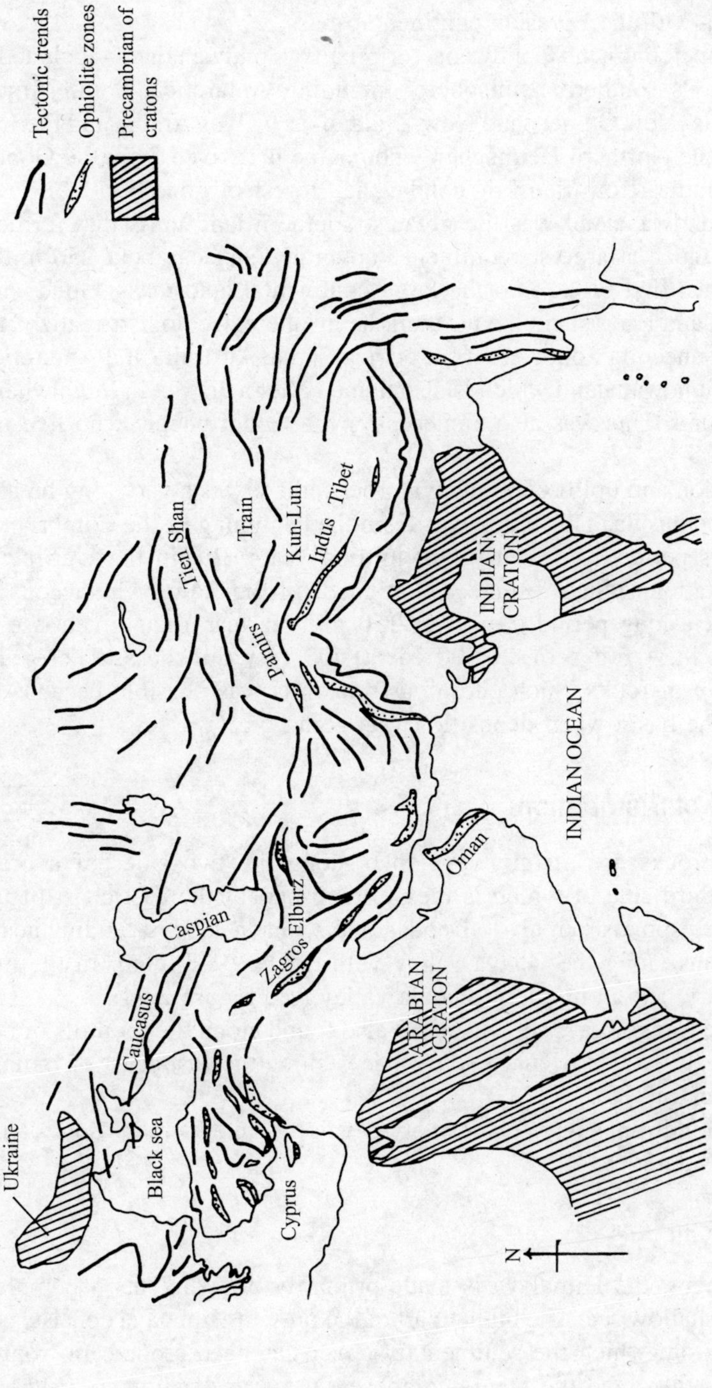

Figure 2.13 The mobile belt of the Middle East and Central Asia [After Gansser, 1966].

mobile system encircling the supercontinents. It was transformed to a largely intra-continental structure by and the collision of northward-moving fragments of Gondwanaland Africa, Arabia and Peninsular India–with the Eurasian continental mass.

The southern supercontinent Gondwana (originally Gondwanaland) included most of the land masses in today's southern hemisphere, including Antarctica, South America, Africa, Madagascar, Australia–New Guinea, and New Zealand, as well as Arabia and the Indian subcontinent, which are in the Northern Hemisphere. The name is derived from the Gondwana region of central northern India (from Sanskrit gondavana "forest of gond").

As described Gondwanaland was the greatest southern land mass that formed as a result of the division of a much a large subcontinent known as Pangaea about 250 million year ago (m.y.a or Ma) Pangaea first broke into the large continental land mass, Gondwanaland in the southern hemisphere and Laurasia in the northern hemisphere. Geologists realized that there had once a land bridge connecting South America, Africa, India, Australia and Antarctica. He named large land mass as Gondwanaland which is also named after a district in India where fossil plant Glossopteris was found. This was also named after the Gondi people who live in this part of India.

After the formation and uplift of Vindhayan rocks, there was a very long break in sedimentation in the Indian peninsula. This break was from the beginning of the Cambrian period to the Upper Carboniferous period, i.e., nearly 300 million years. But in the Upper Carboniferous period, the sedimentation, which resumed, continued till the Lower Cretaceous period. Such deposition, again for a long period (i.e., over 150 million years) has given rise to a massive sequence of sedimentary rocks of 20,000 to 30,000 feet thickness. These are called the *Gondwanaland group* of rocks. Such enormous thickness was possible because of the simultaneous sinking of the basin when deposition was going on.

2.3.3 Occurrence of Distribution

The Gondwanaland rocks are mainly developed along the two sides of a great (inverted) triangular area, the third side of which is the northern part of the east coast of the peninsula. The northern side corresponds roughly to Deodar, and Armada valleys, tending nearly east-west. The southern side runs along the Guava valley with the NW-SE trend. In the interior of this triangle is a subsidiary belt along the Maharaja valley (see Figure 2.14).

In addition to the foregoing, Gondwanas are also found along the foothills of the Himalayan Assam and Kashmir. They also occur at some places along the east coast of India, Rajah Hills, Madhya Pradesh, Gujarat, etc.

Outside India, Gondwanas have extensively developed in Australia, South America, South Africa and even in Antarctica.

2.3.4 The Himalayas

Like many great ranges, the Himalaya is made primarily of sediments accumulated our long geological time in a shallow sea. The higher part of the present Himalaya consists of igneous and metamorphic rocks from which the sedimentary cover has been eroded. In front of the range are foothills, the Siwaliks, and the others composed at tertiary sediments. Although the great thrusts of the Himalayas are now apparently quiescent, the foothills show an evidence of faulting

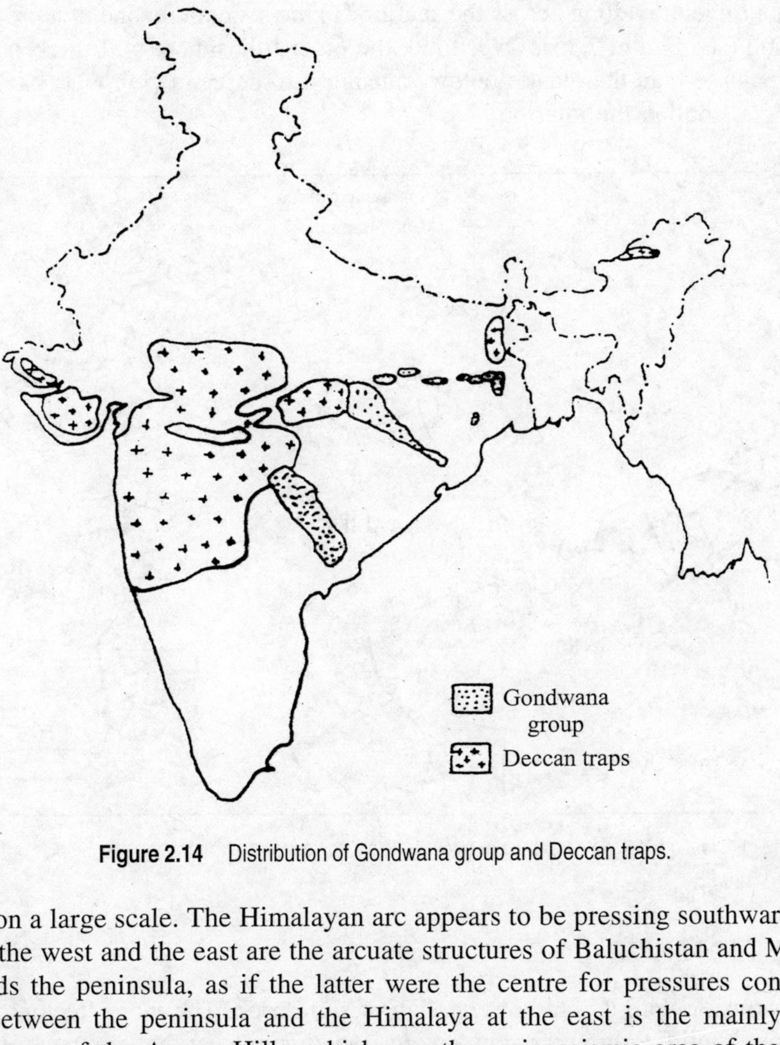

Figure 2.14 Distribution of Gondwana group and Deccan traps.

and thrusting on a large scale. The Himalayan arc appears to be pressing southward towards the peninsula. To the west and the east are the arcuate structures of Baluchistan and Myanmar, also convex towards the peninsula, as if the latter were the centre for pressures converging from three sides. Between the peninsula and the Himalaya at the east is the mainly igneous and metamorphic mass of the Assam Hills, which was the meizoseismic area of the great Assam earthquake of 1899. Since then three more Himalayan earthquakes having magnitude (M7.8) have occurred is 1905, 1934 and 1950. Presently, more than 60% of the Himalayas are overdue for a great earthquake.

Figure 2.15 shows the crustal plates and the arrows indicate the motions resulting from sea-floor spreading. Rapid sea-floor spreading which is accompanied by increased volume of oceanic ridges in effect displaces much water that it encroaches upon the continents. Therefore, rapid sea-floor spreading causes a rise in sea level. On the contrary, it shows that spreading causes drop in sea level.

The rigid, outermost layer of the earth comprising the crust and upper mantle is called lithosphere [see Figure 2.7(c)]. New oceanic lithosphere forms through volcanism in the form of fissures at mid-ocean ridges which are creaks that encircle the globe. Heat escapes from the interior as this new lithosphere emerges from below. It gradually cools, contracts and moves away from the ridge, travelling across the seafloor to the seduction zone in a process called sea-floor spreading (see Figure 2.15). Over time, the older lithosphere will thicken and eventually become more dense than the mantle below, causing it to descend (subduct) back into the earth at a steep angle, cooling the interior.

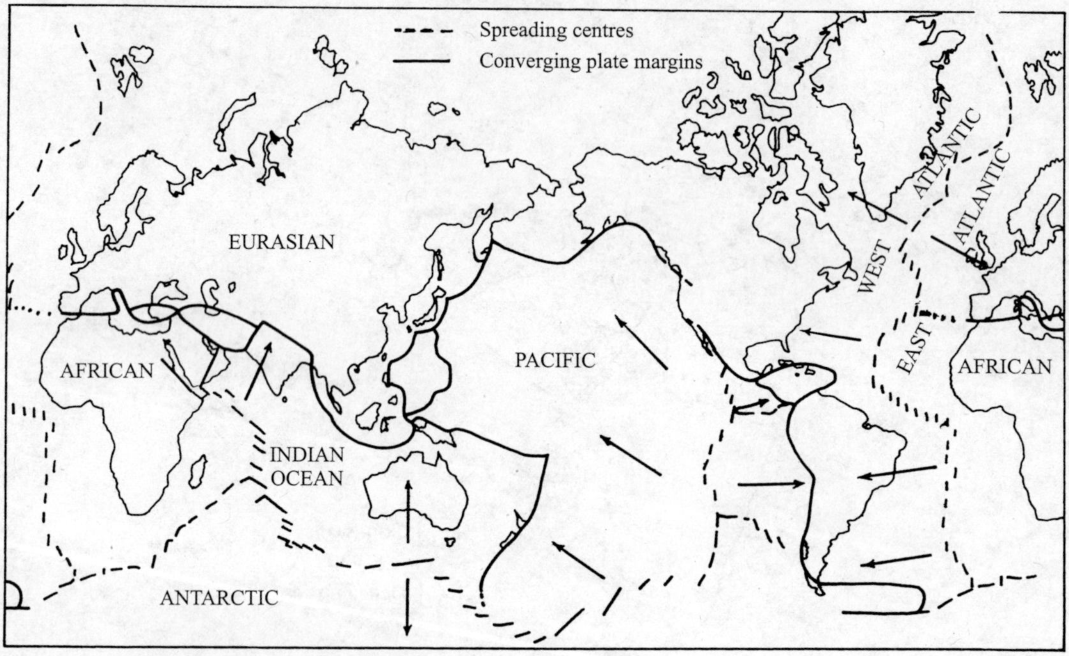

Figure 2.15 The principal crustal plates of the earth today. The arrows indicate the motions resulting from sea-floor spreading (After Le Pichon, 1968).

2.4 PLATE TECTONICS

In geologic terms, a plate is a large rigid slab of solid rock. The word 'tectonics' comes from the Greek root to build. Putting these two words together, we get the term tectonics which refers to how the earth surface is built of plates. The theory of plate tectonics states that the earth's outermost layer is fragmented into a dozen or more large and small plates that are moving relative to one another as they ride atop a hotter, more mobile material.

Plate tectonics has been used for the last several years to try to understand different aspects of the history of evolution of the earth's continents and oceans. Many observations on the surface of the earth have been very elegantly explained by this paradigm. That the 'plates' constituting the lithosphere have been moving apart, colliding and moving past each other over a long period of time is now accepted by all. There are, however, some questions that need clear answers. Many of these relate to the structure and dynamics of the deep interior of the earth, covering the entire mantle and the outer and the inner cores. Current research aims at answering these questions, and this involves the disciplines of geodesy, convection modeling, geomagnetism, seismology and mineral physics. The earth's rigid lithosphere—approximately the outer 100 km of the crust and the upper mantle that overlies the plastic atmosphere of the upper mantle—consists of six major plates and several smaller ones that are in motion relative to each other at slow rates measured at only a few centimetres per year (see Table 2.1).

Some plates are moving away from each other (divergent motion), some are moving towards each other (convergent motion), and others are moving sideways past each other (strike-slip or transform motion). If two plates are moving away from each other, with new crust formed in the zone between, doesn't this mean that either the earth is getting larger or that elsewhere on the surface, plates are being destroyed or consumed? Only a very few earth scientists support the idea of an expanding earth; most believe that there are zones of seduction where two converging plates meet, with one moving downwards beneath the other and melting at depth, and major folded and volcanic mountain ranges form along convergent boundaries.

Convergent plate boundaries can be classified into three types, based upon whether the leading edges of the converging plates consist of oceanic or continental crust. Ocean-ocean collisions, ocean-continent collisions and continent-continent collisions all occur.

It is becoming clear that there are vigorous movements of rock masses in not only the upper mantle, which are immediately responsible for the motion of the lithospheric plates, but at all levels down to the centre of the earth (Wysession, 1995). It turns out that the core mantle boundary (CMB) of the earth is quite complex, and it influences the processes that go on practically at all other levels. From the 1920s it was clear that solids can 'flow' in a manner similar to the flow of a liquid, if enough time is allowed. In fact, in some medieval cathedrals stained-glass windows have experienced 'flow' and in course of about a millennium, have actually thinned at the top. Although solid rocks can flow like this through the mechanisms of diffusion of atoms and dislocation of atomic bonds in mineral grains, the presence of fluids greatly facilitates the process of bulk deformation and also lowers the melting point.

It is agreed by all that the main mechanism that could drive convective motions inside the earth is the loss of heat from the interior, associated with changes in buoyancy created by one or more mechanisms, the simplest being thermal expansion or contraction.

The major milestones in the history of the development of plate tectonics are: [Rai et al., 2002]

1. In the fifteenth century, navigators noticed that continents might be approximately fitted like pieces of jigsaw puzzle.
2. In 1912, Alfred Wegener, a climatologist, noticed geologic features between the continents of Africa and South America, and concluded that continents had once been joined (see Figure 2.12). The mechanism proposed was the movement of rigid continental crust through weak oceanic crust (Press and Siever, 1994). But it failed to explain why mountain building occurred on the edge of continents.
3. In 1962, geologist Harry Hess proposed sea-floor spreading as a mechanism to explain the jigsaw-type fit among continents. He documented flat-topped undersea mountains gradually being submerged as they moved away from mid-ocean ridges. He proposed that the new ocean crust cooled and subsided and moved away from the mid-ocean ridge spreading centres. This new crust was compensated for by the subduction of old oceanic crust beneath continental or oceanic crust.
4. In 1965, Fred Vine and Drumond Mathews provided evidence supporting sea-floor spreading by documenting symmetric magnetic reversal patterns on either side of the

spreading ridges. New crust formed at mid-ocean ridges, recorded the polarity of the earth's magnetic field as it cooled and solidified. Thus, reversals in the field over time produced a striped pattern of alternating magnetic polarity, which was symmetrical about the ridge.

5. In 1968, Tuzo Wilson and Lyn Sykes provided further convincing evidence for sea-floor spreading with the discovery of transform faults that offset the mid-ocean ridges. These faults marked areas of the ocean crust that were moving past each other. This proved that spreading was really occurring along the offset portions of ridges.
6. In 1968, cores from the ocean crust collected by the Glomar Challenger Expedition provided independent evidence for sea-floor spreading by checking that the oceanic crust was progressively older at larger distances from the mid-ocean ridges.
7. During the 1970s, Dan Makenzie and Palmer coined the term plate tectonics to describe the global framework of horizontal motion of the continental and oceanic plates.

The plate boundaries as associated with worldwide seismicity have been shown in Figure 2.16.

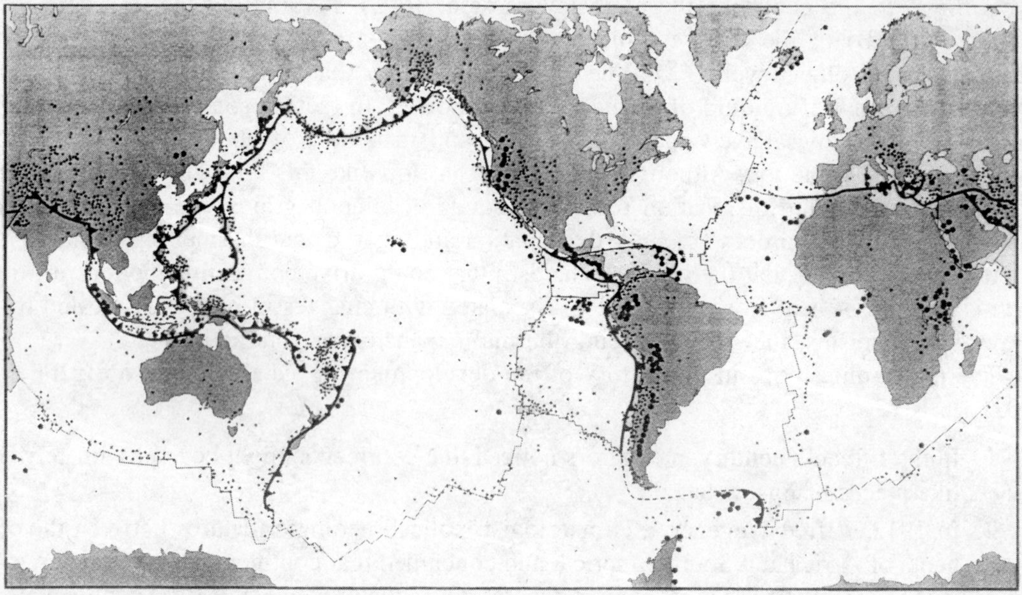

Figure 2.16 Worldwide seismicity and plate boundaries. The dots represent the epicentres of significant earthquakes (After Bolt, 1988).

This plate tectonics involves the formation, lateral movement, interaction and destruction of the lithospheric plates. Much of the earth's internal heat is relieved through this process and many of the earth's large structural and topographic features are consequently formed. Continental rift valleys and vast plateaus of basalt are created at plate build-up when magma ascends from the mantle to the ocean floor, forming new crust and separating mid-ocean ridges. Plates collide and are destroyed as they descend at subduction zones to produce deep ocean trenches, strings of volcanoes, extensive transform faults, broad linear rises and folded mountain belts. The earth's lithosphere is presently divided into eight large plates with about two dozen smaller ones that are drifting above the mantle at the rate of 5 to 10 cm per year. The eight large plates are the African, Antarctic, Eurasian, Indian, Australian, Nazca, North American, Pacific and South American plates. A few of the smaller plates are the Anatolian, Arabian, Caribbean, Cocos, Philippine and the Somali plates, as shown in Figure 2.17.

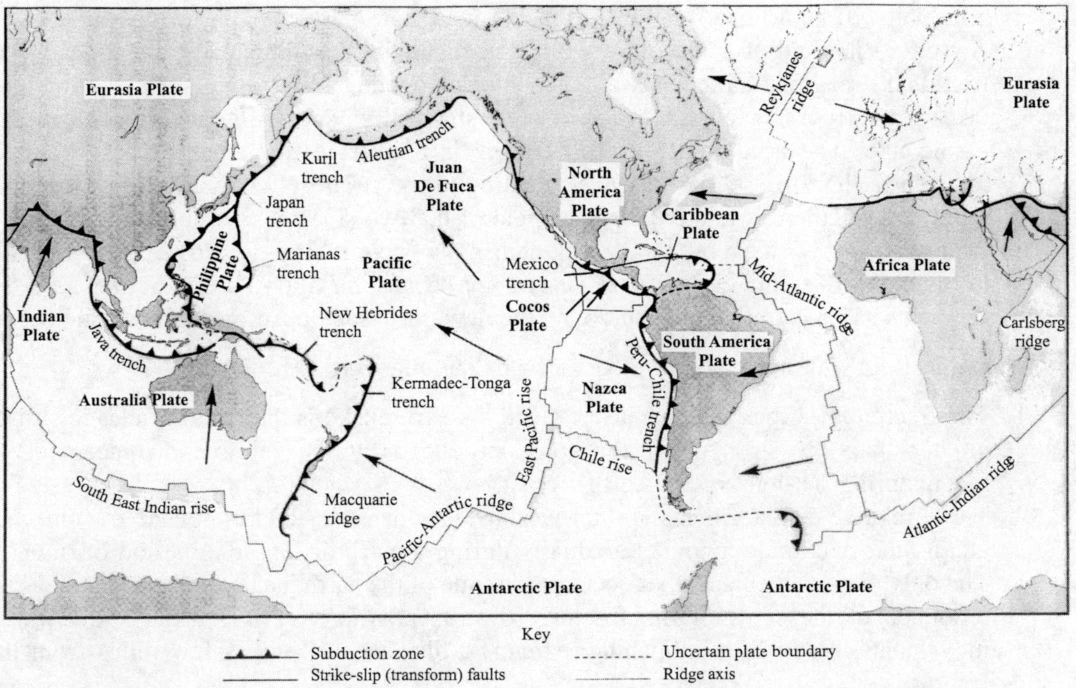

Figure 2.17 The major tectonic plates, mid-oceanic ridges, trenches and transform faults of the earth. Arrows indicate directions of plate movement (After Fowler, 1990).

2.5 ELASTIC REBOUND THEORY

The basic concepts of the elastic rebound theory are illustrated in Figure 2.18 and are outlined below in three stages A, B and C.

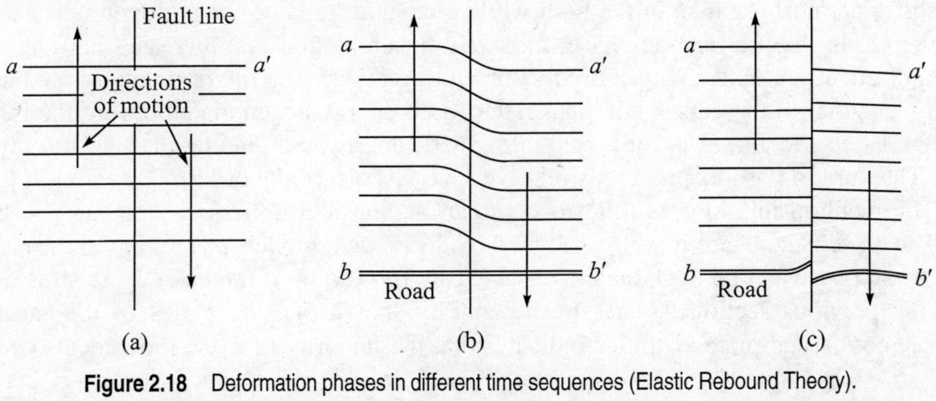

Figure 2.18 Deformation phases in different time sequences (Elastic Rebound Theory).

A. This stage represents conditions that are supposed to exist after an earthquake has completely relieved accumulated strains, leading to an unstrained state. Let *aa′* be a linear feature in this state perpendicular to the fault.
B. Crustal rocks accumulate an increasing amount of strain with time. Because the rocks are distorted, the linear feature *aa′* is deformed into a curve. This represents the strained state. Let *bb*' be a linear feature in this strained state perpendicular to the fault.
C. At some time and at some point along the fault, the accumulated strain exceeds the frictional strength holding the two crustal blocks together. Very rapid crustal motion then takes place (rocks snap back to the original shapes in a spring-like action), causing an earthquake. The accumulated strain energy is converted into kinetic energy and is radiated in the form of elastic waves, leading to the strain relief condition. During this rapid motion, there is relative displacement of the two sides of the fault. Consequently, the feature *aa′* is offset, but its two segments now become straight. However, the feature *bb′* is both offset and curved. It is curved not because of drag faulting but because the segments assume a position dictated by the new unstrained position of the crustal rocks.

Thus, the Reid's elastic rebound theory can be summarized as follows:

1. The fracture of a rock which causes a tectonic earthquake is the result of elastic strains greater than the strength of the rock, produced by the relative displacement of neighbouring portions of the earth's crust.
2. These relative displacements are not produced suddenly at the time of the fracture, but attain their maximum amounts gradually during a more or less long period of time.
3. The only mass movements that occur at the time of the earthquake are the sudden elastic rebounds of the sides of the fracture towards positions of no elastic strain; these movements, gradually diminishing, extend to distances of only a few miles from the fracture.
4. The energy liberated at the time of an earthquake is, immediately before the rupture, in the form of energy of elastic strain of the rock.

Scholz, C.H. (2002) has expanded the Reid concept of elastic rebound in terms of four phases of crustal deformation relative to earthquakes:

(a) Inter-seismic, (b) Pre-seismic, (c) Co-seismic, (d) Post-seismic.

The inter-seismic phase is the strain accumulation phase and is generally attributed to locking of the uppermost segment of the fault while a seismic slope on the fault continues at secular rate at depth. In the pre-seismic phase, the strain accumulation rate increases and the medium behaves elastically. Rapid changes of any sort during this period might be interpreted as earthquake precursors. In the co-seismic phase, the strain energy accumulated during the inter-seismic and pre-seismic phases is converted into kinetic energy and released in the form of seismic waves. The duration of this process is relatively very short, of the order of a few minutes at the most. The medium in which faulting occurs can be considered perfectly elastic for this time scale. The co-seismic phase is well-explained by dislocation models of faulting in the earth. The post-seismic phase can be explained as viscoelastic relaxation of the co-seismic stresses.

In the previous sections, it has been candidly shown that the plates of the earth are in constant motion and plate tectonics indicates that the majority of these movements occur near their boundaries. As a relative movement of the plate occurs, elastic strain energy is stored in the material near the boundary. Consequently, shear stresses also increase on fault planes that separate the plates. When the shear stress reaches the shear strength of the rock along the fault, the rock fails and the accumulated strain energy is released. The theory of elastic rebound as proposed by Reid (1910) describes this process of successive build-up and release of strain energy in the rock adjacent to faults.

The theory of elastic rebound implies that the occurrence at an earthquake will relieve stresses along the portion of a fault on which rupture occurs, and that a subsequent rupture will not occur on that segment until the stresses have had time to build up again. The chances of occurrence of an earthquake, therefore, become related to the time that has elapsed since the last earthquake. As earthquake relieve the strain energy that builds up on faults, they should be more likely to occur in areas where little or no seismic activity has been observed for some time. By plotting fault movement and historical earthquake activity along a fault, it is possible to identify gaps in seismic activity at certain locations along faults. A number of seismic gaps have been identified around the world. The use of seismic gaps offers promise for improvement in earthquake prediction capabilities and seismic risk evaluation.

2.6 RESERVOIR TRIGGERED SEISMICITY

In the previous section, it has been pointed out that the sustained and prolonged continual collection of strain energy inside the lithosphere due to interplay of body and surface traction leads to seismic failures. Such failures are more probable along weak zones such as pre-existing faults leading to the inheritable generation of earthquakes. This force system acts on the upper crust, which is prone to seismicity and as such the crust is almost on threshold of brittle failure and such failures can be easily indicated by small perturbations due to:

1. Plate tectonic forces
2. Local tectonic forces
3. Reservoir associated forces

During the last 100 years about 90 cases of Reservoir Triggered Seismicity (RTS) have been repeated. The important RTS in India is Koyna earthquake of 1967. The earthquake associated with a (Lake Mead) reservoir in Colarado, USA was reported by Carder (1945) way back in 1945. In such situations, the pore water reduces the normal stress, thereby leading to seismicity. In past, the epicentres of RTS earthquakes have been confined in both space and time.

2.6.1 Mechanism of RTS Earthquakes

Hubbert and Rubey (1959) were the first to describe the mechanism of triggering earthquakes by increasing the fluid pressure. The field experiments have shown that fluid pressure can control the occurrence of earthquakes by lowering or increasing the fluid pressure in the epicentre zone. Hubbert and Rubey (1959) showed that the force F exerted on a fluid-filled process medium is

$$F = -\iint_A u\,dA$$

where u is the pore pressure.

The coulomb shear failure criteria is

$$\tau = \sigma + (\sigma_n - u) \tan \phi \tag{2.3}$$

When, $\sigma_n - u = 0, \quad \tau = \tau_0,$ (2.4)

then, $\sigma_{crit} = \tau_0 + (\sigma_n - u) \tan \phi$ (2.5)

in which σ_{crit} is the critical value of shear stress at which failure or slippage occurs.

After initiation of slippage, $\tau_0 \to 0$, then

$$\tau_{crit} = (\sigma_n - u) \tan \phi$$
$$= \sigma' \tan \phi$$

If $u = \lambda\sigma_n$, where $\lambda = u/\sigma_n$ = pore pressure/normal stress, then

$$\tau_{crit} = (1 - \lambda)\sigma_n \tan \phi \tag{2.6}$$

The above equation suggests the worst condition when $\lambda \to 1$, i.e., when the fluid pressure approaches the normal stress. Hence, a large fault block can be pushed over a nearly horizontal surface to a few kilometres provided the pore pressure is sufficiently high, indicating the key role played by the pore pressure in the genesis of the earthquake.

Koyna earthquake

In several dams, which are situated in places of zero seismic activity, earthquakes were found to appear when their reservoirs were filled up [see Figure 2.19(a)]. The magnitude of the earthquakes increased when reservoirs became full. In all such cases, the epicentre was inside or around the reservoir.

The Koyna earthquake in Maharashtra is a typical example of this kind. The Koyna dam rests over stable rocks of very ancient times and the area was never active in terms of earthquake occurrences. As the reservoir commenced to take in water, seismic activity increased in the area. In 1967, a severe earthquake of magnitude 6.5 shook the region. Since then, earthquakes are not uncommon there. In all such cases, the epicentres lay within the reservoir area. [see Figure 2.19(b)]

Such cases clearly show that there is some sort of link between the location of reservoirs and the commencement of seismic activity there. Many scientists have carefully examined this aspect and some views expressed are as follows:

1. According to Cardar (1945), such places had very old inactive faults underground and when the reservoir was filled with water, the load of the water reactivated those faults and earthquakes followed.
2. According to Hubbert and Rubey (1959), these earthquakes related to reservoirs should be attributed to the increased pore pressure. They felt that the increase in pore pressure lowers the shearing strength of the rock formations and this results in releasing the tectonic strain in the form of earthquakes.

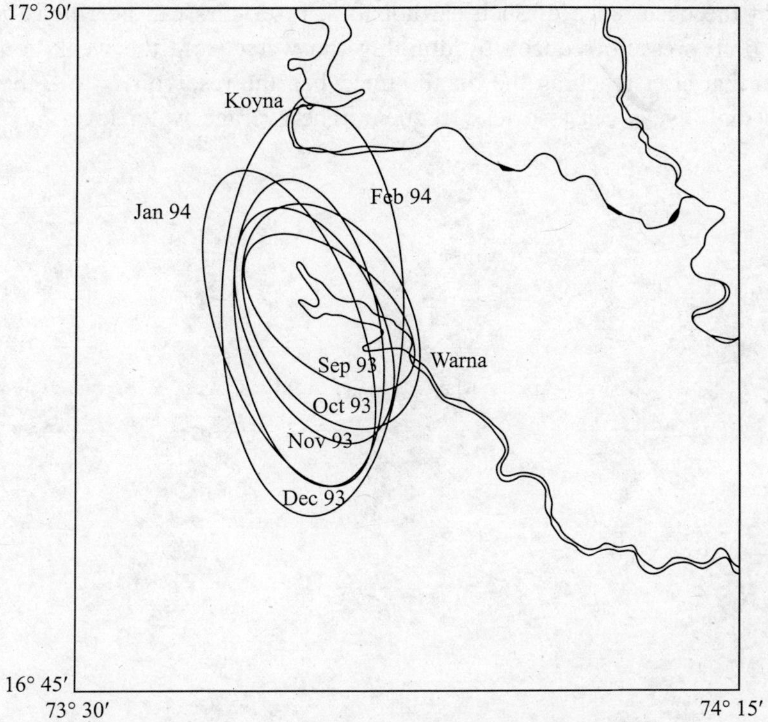

Figure 2.19(a) Epicentre growth in the Koyna-Warna zone during September 1993–February 1994. [After Talwani et al., 1996]

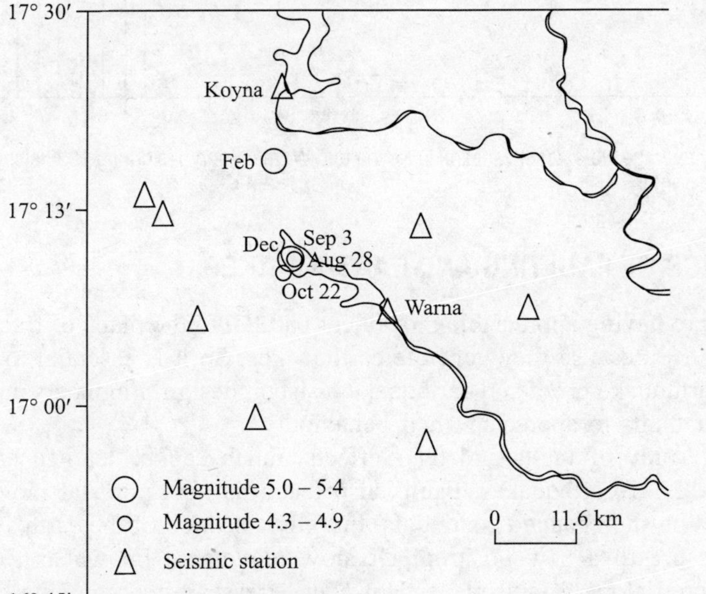

Figure 2.19(b) Epicentre locations for events with magnitude greater than 4.5 in the Koyna-Warna zone during September 1993–February 1994. (After Talwani et al., 1997)

To prevent the occurrence of such earthquakes, reservoirs can be filled to a limited safe level and the pore pressure reduced by draining the water from the weak zone. However, it is also noticed that after reaching the maximum, when the reservoir is full, these shocks tend to become not only weak but also less frequent. The average water level has been shown in Figure 2.20.

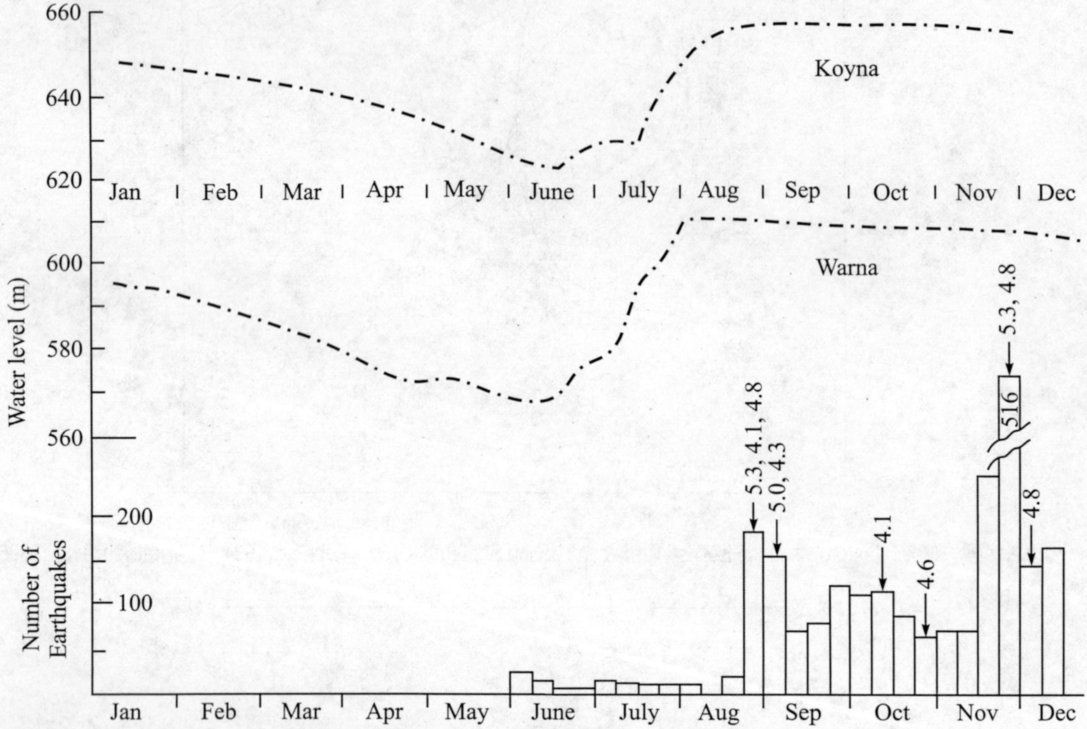

Figure 2.20 Weekly average water level variation in Koyna and Warna reservoirs during 1995 along with number of earthquakes (Rastogi et al., 1997)

2.7 MECHANICS OF FAULTING AND EARTHQUAKES

A fault is a fracture having appreciable movement parallel to the plane of fracture. Faults are of practical importance because they generate earthquakes. So it is essential to understand faults for facilitating earthquake proof and earthquake resistant design. Engineers must understand the basic anatomy of faults to appreciate their behaviour.

Through the study of faults and their effects, much can be learned about the size and recurrence intervals of earthquakes. Faults also teach us about crustal movements that have produced mountains and changed continents. Initially a section of the earth's crust may merely bend under pressure to a new position. Or slow movement known as seismic creep may continue unhindered along a fault plane. However, stresses often continue to build until they exceed the strength of the rock in that section of the crust. The rock then breaks, and an earthquake occurs, sometimes releasing massive amounts of energy. The ensuing earth displace-

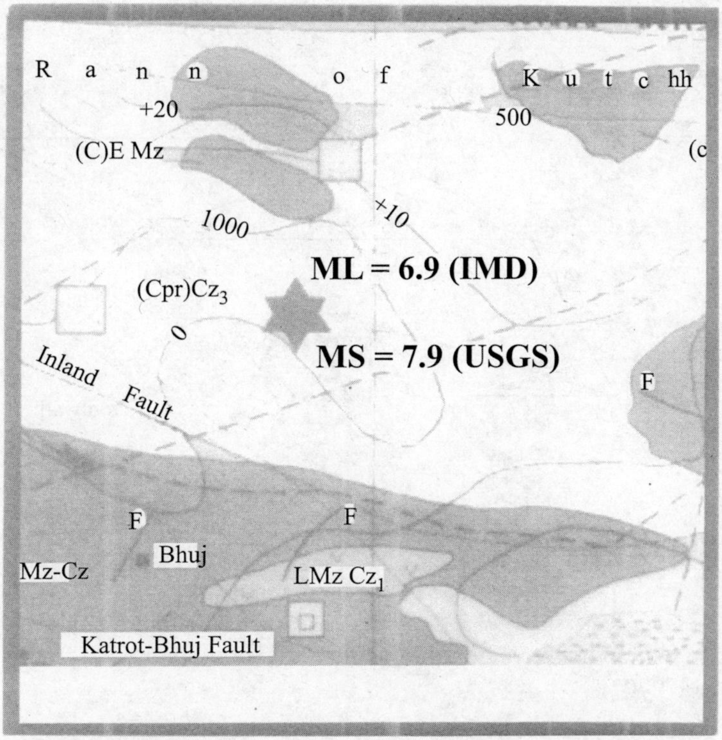

Figure 2.21 Fault movements during 2001 Bhuj earthquake.

ment is known as a fault. This slide set describes the mechanism and types of faulting. It illustrates a variety of fault expressions in natural and man-made features. Faults represent zones of crustal weakness. Seismic events will continue to be related to them. The mass of the rock below an inclined fault plane is known as the *footwall* and the mass of the rock above it as the *hanging wall*. Figure 2.22 shows the surface of the footwall. The line of intersection of the fault plane with the surface of the earth is known as the *strike direction* or the *strike of the fault*. Its orientation is expressed in terms of an angle λ ($0 \leq \lambda < 2\pi$), measured anticlockwise from the south, known as the strike angle. According to Ben-Menahem A. and Singh S.T. (1981), a line on the earth's surface perpendicular to the strike drawn in the direction in which the fault plane is dipping is known as the *dip direction*. We take the positive direction of the strike to the right of an observer facing the footwall.

In the case of a shear fault, the slip u_0 is parallel to the fault. The angle λ between u_0 and the strike of the fault is known as the slip angle ($0 \leq \lambda < 2\pi$). It is measured anticlockwise when viewed from the hanging wall side of the fault plane. The angle δ that the fault plane makes with the horizontal plane is known as the dip angle ($0 \leq \delta \leq \pi/2$).

Considering the direction of crustal block movements, there are two types of shear faults (Figure 2.23):

1. Strike-slip (also known as transcurrent wrench or lateral), in which the movement is parallel to the strike of the fault ($\lambda = 0°$ or $180°$).

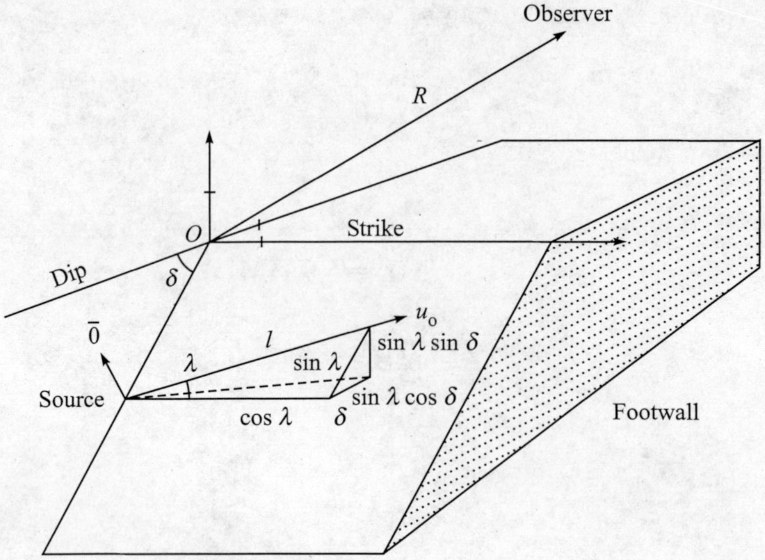

Figure 2.22 Geometry of a shear fault.

2. Dip-slip, in which the movement along the fault is perpendicular to the strike of the fault (λ = 90° or 270°).

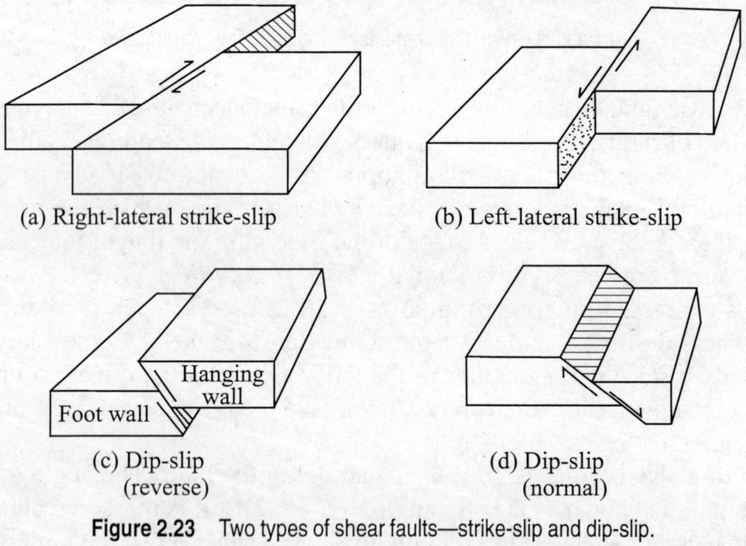

Figure 2.23 Two types of shear faults—strike-slip and dip-slip.

The strike-slip faults are of two types: right lateral and left lateral. In a right-lateral strike-slip fault, the direction of the relative displacement of the side of the fault opposite the observer who is facing the fault is to the right [Figure 2.23(a)]. In a left-lateral strike-slip fault, the direction of the relative displacement of the opposite side is to the left [Figure 2.23(b)]. The San Andreas and most other strike-slip faults in California have been associated with right-lateral displacements.

Dip-slip faults are also of two types: reverse and normal. In a reverse (or thrust) dip-slip fault, the hanging wall moves up relative to the footwall [Figure 2.23(c)]. In a normal dip-slip fault, the hanging wall moves down relative to the footwall [Figure 2.23(d)]. In reverse faults, the horizontal extent decreases, whereas in normal faults, the horizontal extent increases. A reverse fault occurs when the oceanic lithosphere is thrust under the adjacent continental lithosphere at a trench. Normal faulting occurs on the flanks of ocean ridges where a new lithosphere is being created.

By using the electrodynamics representation theorem, it can be shown that a shear fault, in point source approximation, is equivalent to a double couple of moment $M_0 = GUA$, where G is the rigidity of the surrounding medium, U is the magnitude of the relative displacement (slip), and A is the fault area. A shear fault is not a double couple but represents physically relative tangential movement across a break in the medium. The double couple serves merely as an alternative mathematical representation of the seismic source. As required by the principles of conservation of linear and angular moment, the double couple has zero net force and zero net moment.

A shear fault can be specified completely by two unit vectors, n and e, and a single scale, $M_0 = GUA$, where n is a unit vector normal to the fault (Figure 2.24) and describes the fault geometry. e is a unit vector parallel to the direction of the relative displacement, $u_0 = Ue$, of the two sides of the fault and describes the slip geometry. M_0 is called seismic moment and describes the size of the source. The directions of the forces of the equivalent double couple are parallel to the directions of the unit vectors e and n of the shear fault (Figure 2.24).

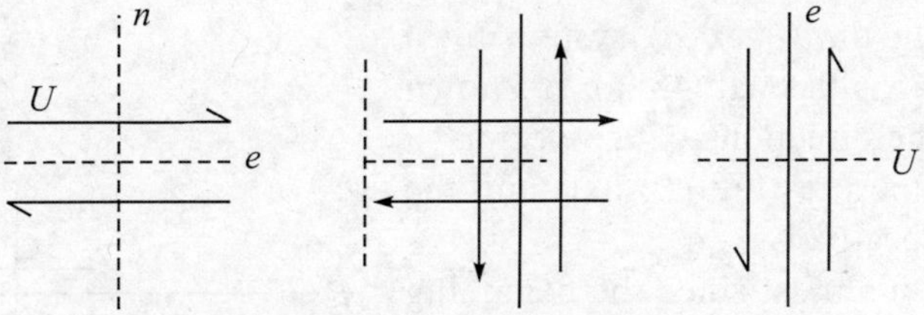

Figure 2.24 Double couple equivalent to two shear faults

The seismic moment determines the intensity of the emitted seismic radiation and is therefore a good measure of the size of an earthquake, at least as far as the elastic radiation is concerned. It is a more logical indication of the size of an event than is the earthquake magnitude, which is based on an arbitrary measure of the radiated elastic energy (either body waves or surfaces waves).

In 1776, Coulomb postulated that a brittle material under stress fractures along a plane of greatest tangential stress. Let τ_1, τ_2, τ_3, where $\tau_1 > \tau_2 > \tau_3$, be the principal stresses just before a fracture. From the theory of elasticity, the maximum shearing stress is equal to one-half the difference between the largest and the smallest principal stresses and acts on a plane that bisects the angle between the directions of these principal stresses. Therefore, from the Coulomb postulate, the plane of fracture passes through the direction of τ_2 and bisects the angle between the directions of τ_1 and τ_3, thus making an angle of $\pm 45°$ with τ_1; the magnitude of the greatest shear stress being $(\tau_1 - \tau_3)/2$. Since there is no tangential stress at a liquid–solid boundary, at the surface of the earth or at the ocean bottom, one of the principal stresses can be taken to be vertical. Then there is normal faulting if the vertical stress is the largest of the three principal stresses, reverse faulting if the vertical stress is the least and strike-slip faulting if the vertical stress is intermediate, as shown in Figure 2.25.

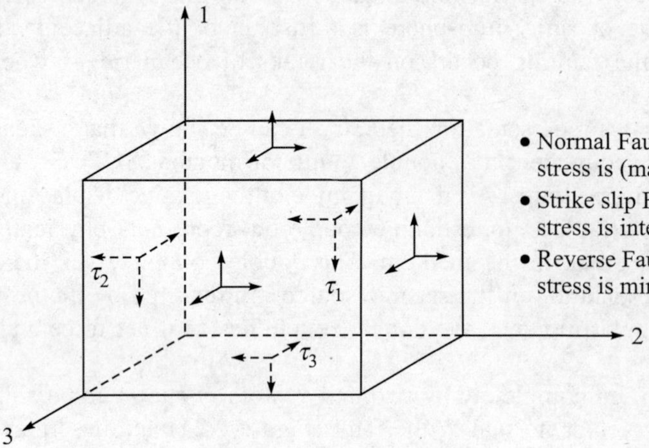

- Normal Faulting-vertical principal stress is (major)
- Strike slip Faulting-vertical principal stress is intermediate
- Reverse Faulting-vertical principal stress is minor

Figure 2.25 Definition of faulting based on principal stresses τ_1, τ_2, τ_3 ($\tau_1 > \tau_2 > \tau_3$) along axes 1, 2, 3 (axes 1 is vertical).

Let p be a unit vector in the direction of τ_1 and t be a unit vector in the direction of τ_3. Let b be the unit vector chosen in such a manner that (p, b, t) form a right-handed system. This system can be obtained from the (e, b, n) system by a rotation about the b-axis through 45°. b is known as the null vector. The directions of the vectors p and t are known as the pressure (P) axis and the tension (T) axis, respectively.

It has been known almost since the beginning of instrumental seismology that during an earthquake, certain stations record a P-wave impulse upwards and away from the epicentre (which is called a compression), whereas other stations record an impulse downwards and towards the epicentre (a dilatation or rarefaction). Further, the areas of compression and dilatation are arranged in a pattern. For a shear fault, the quadrant in the e–n plane in which t lies will yield compression and the quadrant in which p lies will yield dilatation. The opposite quadrants have similar patterns. In Figure 2.26, the first and the third quadrants will have compressions, whereas the second and the fourth quadrants will have dilatations. This property is of use in source mechanism studies using first-motion observations. The boundaries that separate those stations that record compressional impulses from those that record dilatational impulses are the conjugate planes. One of these planes is the fault plane; the other plane is known as the auxiliary plane. There is an ambiguity between the fault plane and the auxiliary plane. The

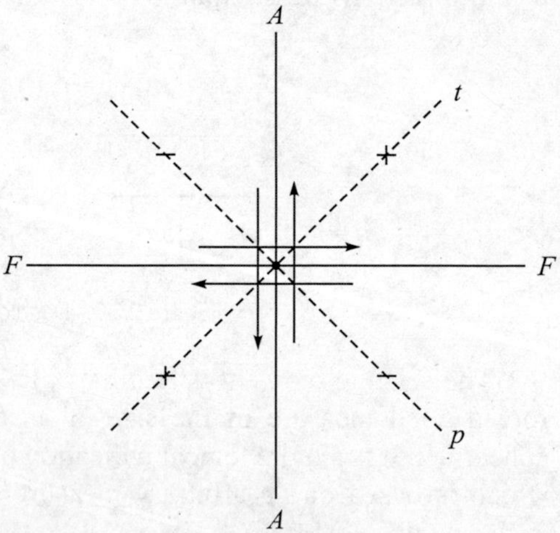

Figure 2.26 Compression (+); Dilatation (–); Pressure axis (p); Tension axis (t); Fault plane (FF); Auxiliary plane (AA) associated with a double couple.

resultant seismic motion at the observation point may be expressed (Heaton, T.H. and Hazrell, S.H., 1986) as

$$U(t) = \int_0^L \int_0^W \dot{D}(x, y, t) * G(x, y, t)\, dy dx$$

Where L and W represent the length and width of the fault, * represents a time convolution, $\dot{D}$ is the slip velocity, and G the green function (the double-couple impulse response of the medium). For more details, the reader may refer Berlin (1980), Gubbins (1990), Kasahara (1981) and Rai, S.N. et al. (2002).

2.8 SIZE OF EARTHQUAKE

Since an earthquake is a phenomenon resulting from a complex rupture of the earth's crust, it is extremely difficult to express its exact size. For an approximate estimation of the size of this complex physical phenomenon in a simple and numerical manner, a scale termed *magnitude* is used. But since the strength of seismic vibrations for earthquakes of the same size varies from place to place, the seismic-intensity scale is used to express the severity of vibrations at a given place. The seismic-intensity scale is not determined from the mechanical measurement of seismic vibrations, but rather from the damage caused as perceived by human beings and the behaviour of various objects or structures. The standard for such determination also varies from country to country. In Japan, the Japan Meteorological Agency's (JMA) seismic-intensity scale is used and in Europe and the USA the Modified-Mercalli (MM) scale is used. The former has a range of 0-7 divided into 8 steps while the latter has the same range divided into 12 steps. Since there is no one-to-one correspondence between these two scales, care must be exercised when comparing the seismic intensity of Japanese and Euro-American earthquakes in terms of the seismic-intensity scale.

2.8.1 Intensity of Earthquake

In the field of engineering, the intensity of earthquake force is expressed as a ratio of acceleration of earthquake motion α to gravitational acceleration g and is called the *seismic coefficient*. Care must be taken not to mistake the seismic-intensity coefficient used for engineering purposes for the seismic-intensity scale. The seismic-intensity scale is decided by human judgment based on various phenomena and the response of various structures. As such, there is no one-to-one correspondence with physical quantities of earthquake motion such as acceleration, velocity, displacement, etc. The correspondence between the JMA scale, the MMI scale, the RF scale and the MSK scale is shown in Figure 2.27.

It may be concluded that the seismic-intensity scale is not based on any quantitative measurement; it is still a useful parameter for indicating the overall response of the various structures to an earthquake and ought to be considered in earthquake-resistant analyses.

Intensity scales

In Mallet's day, it was generally known that the distribution of the macroseismic effects of earthquakes could be represented by the drawing of isoseismals, i.e., lines of equal apparent intensity of shaking.

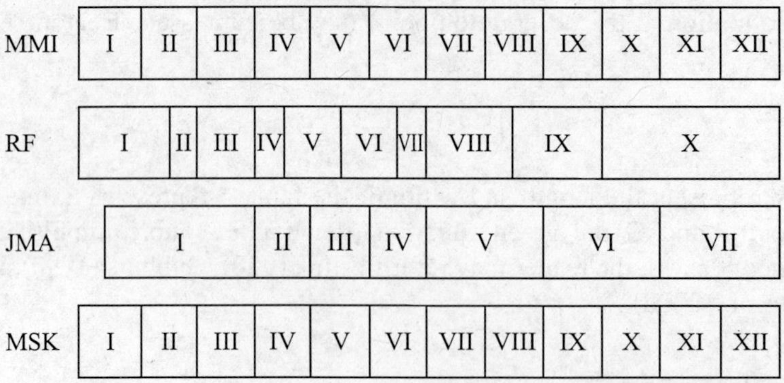

Figure 2.27 Comparison of intensity of MMI, RF, JMA and MSK scales.

Special scales

At first each earthquake was quite properly investigated independently; even at the present time this is considered a good practice. Especially when a large earthquake is being investigated and many observations are being correlated, it is scientifically preferable to begin by setting up isoseismals with reference to local conditions which sometimes almost force a special scale on the investigator. Thus, workers who took the field after the Turkish earthquake of 1939 found that conventional intensity scales failed to describe the damage to the earth construction common in that region, and they fell back on estimates of the percentage of damage in the various localities.

The Rossi–Forel scale

Intensity scales intended for general application developed gradually, as the comparison of individual investigations led towards a common pattern. De Rossi in Italy and Forel in Switzerland, who had been working in this direction more or less independently, joined forces in 1883 to set up the Rossi–Forel scale. It was widely adopted. In seismological and engineering literature, when no particular scale is specified, earthquake intensity is usually expressed in terms of this scale; it is commonly indicated by the abbreviation RF, followed by the Roman numeral of the scale degree (see Figure 2.27).

With the general advance of technology, the RF scale progressively went out of date. An enormous range of intensity was lumped together at its highest level, X. Moreover, the descriptions of effects both on construction and on natural objects proved to be too specifically European.

The Mercalli scale

The drawbacks as discussed were largely removed in an improved scale put forward by Mercalli in 1902 at first with ten grades of intensity, later with twelve following a suggestion by Cancani who attempted to express these grades in terms of acceleration. An elaboration of the Mercalli scale, including earthquake effects of many kinds and ostensibly correlated with Cancani's scheme, was published by Sieberg in 1923. This form was, in turn, used as the basis for the Modified Mercalli Scale of 1931 (commonly abbreviated MM) by H.O. Wood and Frank Neumann.

Modified Mercalli scale restated

The original publication gives the MM scale in two forms: one a lengthy statement modelled on that of Sieberg, with additions and modifications suggested by later experience; the other an abridgment meant for rough-and-ready use. The abridged form (see Table 2.2) was prepared chiefly by Richter and at a few points is in conflict with the main scale. At the risk of putting a third version into circulation, this chapter presents an expansion of the shorter form, including most of the items in the complete form. Some items are omitted for definite reasons and a few additional notes are included, with initials (CFR) to separate them from the scale properly. To eliminate many verbal repetitions in the original scale, the following convention has been adopted. Each effect is named at that level of intensity at which it first appears frequently and characteristically. Each effect may be found less strongly, or in fewer instances, at the next lower grade of intensity; more strongly or more often at the next higher grade. A few effects are named at two successive levels to indicate a more gradual increase.

To avoid ambiguity of language, the quality of masonry, brick or otherwise, is specified by the following lettering (which has no connection with the conventional Class A, B, C type of construction).

Masonry A: Good workmanship, mortar and design; reinforced, especially laterally and bound together by using steel, concrete, etc.; designed to resist lateral forces.

Masonry B: Good workmanship and mortar; reinforced, but not designed in detail to resist lateral forces.

Masonry C: Ordinary workmanship mortar; no extreme weaknesses like failing to tie in at corners, but neither reinforced nor designed against horizontal forces.

Masonry D: Weak materials such as adobe; poor mortar; low standards or workmanship; weak horizontally.

Intensity and Acceleration

Richter has participated in an attempt to correlate the degrees of the MM scale with ground acceleration in the manner attempted by Cancani. Many excellent seismograms written by the U.S. Coast and Geodetic Survey instruments in California and elsewhere are available for such study.

$$\log a = \frac{I}{3} - \frac{1}{2} \tag{2.7}$$

where a is the acceleration in cm/s^2 and I is the MM intensity. This is similar to Cancani's result, although it differs somewhat numerically. Here, of course, the intensity grades must be treated as true numerical quantities, which they are not. If one lets $I = 1\frac{1}{2}$ represent the limit of perceptibility between intensities I and II, $\log a = 0$ or $a = 1$ cm/s^2. Various lines of evidence point to this as the level of shaking ordinarily perceptible to persons. If one lets $I = 7\frac{1}{2}$, $\log a = 2$ or a 100 cm/s^2 = 0.1g approximately. This is the acceleration commonly accepted by engineers as that, which damages ordinary structures not designed to be earthquake resistant.

Mr. Frank Neumann engaged himself in an elaborate effort using the same data to correlate intensity with acceleration, and eventually to complete Cancani's project by redefining intensity in quantitative physical terms. The chief difficulties are:

1. Extreme variations introduced by differing types of ground
2. Effect of increasing magnitude in altering the proportion between the long-period and short-period vibrations, and consequently between the corresponding groups of phenomena.
3. Crudity of the non-instrumental data used to assign intensities, which often leads to legitimate debate as to their significance in relation to actual earth motion.

Table 2.2 Earthquake Intensity Scales: Modified Mercalli Intensity Scale (Abridged)

Class of earthquake	*Remarks (Reaction of observers and types of damage)*
I	Reactions : Not felt except by a very few persons under especially favourable circumstances. Damage : No damage.
II	Reaction : Felt only by a few persons at rest, especially on upper floors of buildings. Damage : No damage; delicately suspended objects may swing.
III	Reaction : Felt quite noticeably indoors, especially on upper floors of buildings but many people do not recognize it as an earthquake. Damage : No damage; standing motor cars may rock slightly; and vibrations may be felt like the passing of a truck.
IV	Reaction : During the day, felt indoors by many, outdoors by a few, at night some awakened. Damage : No damage; dishes, windows, doors disturbed; walls make creaking sound, sensation like heavy truck striking the building; standing motor cars rocked noticeably.
V	Reaction : Felt by nearly everyone; many awakened. Damage : Some dishes, windows, etc. broken; a few instances of cracked plaster; unstable objects overturned; disturbance of trees, poles and other tall objects noticed sometimes; and pendulum clocks may stop.
VI	Reaction : Felt by all, many frightened and run outdoors. Damage : Some heavy furniture moved; a few instances of fallen plaster or damaged chimneys; damage is slight.
VII	Reaction : Everybody runs outdoors, noticed by persons driving motor cars. Damage is negligible in buildings of good design and construction; slight to moderate damage in well-built ordinary structures; considerable damage in poorly-built or badly-designed structures; and some chimneys may get broken.
VIII	Reaction : Disturbs persons driving motorcars. Damage : Slight damage in especially designed structures; considerable damage in ordinary but substantial buildings with partial collapse; very heavy damage in poorly-built structures; panel walls may get thrown out of framed structures; falling of chimneys, factory stacks, columns, monuments, and walls; heavy furniture may get overturned, sand and mud ejected in small amounts; changes in well water.
IX	Damage : Considerable damage in especially designed structures; well-designed framed structures thrown; out of plumb; very heavy damage in substantial buildings with partial collapse; buildings shifted off foundations; ground cracked conspicuously; underground pipes broken.

X	Damage : Some well-built wooden structures destroyed; most masonry and framed structures with foundations destroyed; ground badly cracked; rails bent; considerable landslides from river banks and steep slopes; shifted sand and mud; water splashed over banks.
XI	Damage : Few, if any, masonry structures remain standing; bridges destroyed; broad fissures in ground, underground pipelines completely out of service; earth slumps and landslips in soft ground; rails get bent greatly.
XII	Reaction : Waves seen on ground surface; lines of sight and levels distorted; Damage : total damage with practically all works of construction greatly damaged or destroyed; objects are thrown upwards into the air.

Comprehensive Intensity Scale (CIS)

This scale was discussed generally at the inter-governmental meeting convened by UNESCO in April 1964. Though not finally approved, the scale is more comprehensive and describes the intensity of earthquake more precisely. The main definitions used are given in Table 2.3.

Table 2.3 Comprehensive intensity scales

(a) Type of structures (buildings)

Structure A	Buildings in fieldstone, rural structures, unburnt-brick houses, clay houses.
Structure B	Ordinary brick buildings, buildings of the large block and prefabricated type, half-timbered structures, buildings in natural hewn stone.
Structure C	Reinforced buildings, well-built wooden structures.

(b) Definition of quantity

Single; a few	About 5 per cent
Many	About 50 per cent
Most	About 75 per cent

(c) Classification of damage to buildings

Grade 1	Slight damage	Fine cracks in plaster; fall of small pieces of plaster.
Grade 2	Moderate damage	Small cracks in walls; fall of fairly large pieces of plaster, pantiles slip off; cracks in chimneys; parts of chimney fall down.
Grade 3	Heavy damage	Large and deep cracks in walls; fall of chimneys.
Grade 4	Destruction	Gaps in walls; parts of buildings may collapse; separate parts of the building lose their cohesion; inner walls collapse.
Grade 5	Total damage	Total collapse of buildings.

(d) Intensity scales

I. Not noticeable:
The intensity of the vibration is below the limit of sensibility; the tremor is detected and recorded by seismographs only.

II. Scarcely noticeable (very slight):

(*Contd.*)

Vibration is felt only by individual people at rest in houses, especially on upper floors of buildings.

III. Weak, partially observed only:

The earthquake is felt indoors by a few people, outdoors only in favourable circumstances. The vibration is like that due to the passing of a light truck. Attentive observers notice a slight swinging of hanging objects, somewhat more heavily on upper floors.

IV. Largely observed:

The earthquake is felt indoors by many people, outdoors by a few. Here and there people awake, but no one is frightened. The vibration is like that due to the passing of a heavily-loaded truck. Windows, doors and dishes rattle. Floors and walls crack. Furniture begins to shake. Hanging objects swing slightly. Liquids in open vessels are slightly disturbed. In standing motorcars the shock is noticeable.

V. Awakening:

(a) The earthquake is felt indoors by all, outdoors by many—many sleeping people awake. A few run outdoors. Animals become uneasy. Buildings tremble throughout. Hanging objects swing considerably. Pictures knock against walls or swing out of place. Occasionally, pendulum clocks stop. Unstable objects may be overturned or get shifted. Open doors and windows are thrust open and slam back again. Liquids spill in small amounts from well-filled open containers. The sensation of vibration is like that due to a heavy object falling inside the buildings.

(b) Slight damages to buildings of Type A are possible.

(c) Sometimes change in flow of springs.

VI. Frightening:

(a) Felt by most, both indoors and outdoors. Many people in buildings are frightened and run outdoors. A few persons lose their balance. Domestic animals run out of their stalls. In few instances, dishes and glassware may break, books may fall down. Heavy furniture may possibly move and small steeple bells may ring.

(b) Damage of Grade 1 is sustained in single buildings of Type B and in many of Type A. Damage in few buildings of Type A is of Grade 2.

(c) In a few cases, cracks up to widths of 1 cm possible in wet ground; in mountains occasional landslips; change in flow of springs and in level of well water are observed.

VII. Damage to buildings:

(a) Most people are frightened and run outdoors. Many find it difficult to stand. Persons driving motorcars notice the vibrations. Large bells ring.

(b) In many buildings of Type C, damage of Grade 1 is caused; in many buildings of Type B, damage is of Grade 2. Most buildings of Type A suffer damage of Grade 3, a few of Grade 4. In single instances, landslips of roadways on steep slopes; cracks appear in roads; seams of pipelines get damaged; cracks appear in stone walls.

VIII. Destruction of buildings:

(a) Fright and panic; also persons driving motorcars are disturbed. Some branches of trees break off. Even heavy furniture moves and partly overturns. Hanging lamps are damaged in part.

(b) Most buildings of Type C suffer damage of Grade 2, and a few of Grade 3. Most buildings of Type B suffer damage of Grade 3, and most buildings of Type A suffer damage of Grade 4. Many buildings of Type C suffer damage of Grade 4. Occasional breaking of pipe seams. Memorials and monuments move and twist. Tombstones overturn. Stone-walls collapse.

Table 2.3 Comprehensive Intensity Scales (Contd.)

(c) Small landslips in hollows and on banked roads on steep slopes; cracks in ground up to widths of several centimetres. Water in lakes becomes turbid. New reservoirs come into existence. Dry wells refill and existing wells become dry. In many cases change in flow and level of water is observed.

IX. General damage to buildings:

(a) General panic; considerable damage to furniture. Animals run to and fro in confusion and cry.

(b) Many buildings of Type C suffer damage of Grade 3, and a few of Grade 4. Many buildings of Type B show damage of Grade 4, and a few of Grade 5. Many buildings of Type A suffer damage of Grade 5. Monuments and columns fall. Considerable damage to reservoirs; underground pipes partly broken. In individual cases, railway lines are bent and roadways damaged.

(c) On flat land, overflow of water, sand and mud is often observed. Ground cracks to widths of up to 10 cm, on slopes and river banks more than 10 cm; furthermore a large number of slight cracks appear in ground; falls of rock, many landslides and earth flows; large waves in water. Dry wells renew their flow and existing wells dry up.

X. General destruction of buildings:

(a) Many buildings of Type C suffer damage of Grade 4 and a few of Grade 5. Many buildings of Type B show damage of Grade 5; most of Type A show destruction of Grade 5. Critical damage to dams and dykes and severe damage to bridges. Railway lines are bent slightly. Underground pipes are broken or bent. Road pavings and asphalt show waves.

(b) In ground, cracks up to widths of several centimetres, sometimes up to 1 metre. Parallel to watercourses, occur broad fissures. Loose ground slides from steep slopes. From riverbanks and steep coasts, considerable landslides are possible. In coastal areas, displacement of sand and mud; change of water level in wells; water from canals, lakes, rivers, etc. thrown on land. New lakes occur.

XI. Destruction:

(a) Severe damage even to well-built buildings, bridges, water dams and railway lines; highways become useless; underground pipes get destroyed.

(b) Ground considerably distorted by broad cracks and fissures as well as by movement in horizontal and vertical directions; numerous landslips and falls of rock occur. The intensity of the earthquake requires to be especially investigated.

XII. Landscape changes:

(a) Practically all structures above and below ground are greatly damaged or destroyed.

(b) The surface of the ground is radically changed. Considerable ground cracks with extensive vertical and horizontal movements are observed. Falls of rock and slumping of riverbanks over wide areas, lakes are dammed, waterfalls appear and rivers are deflected. The intensity of the earthquake requires to be especially investigated.

2.8.2 Magnitude of Earthquake

The intensity of ground motion varies considerably from place to place even for the same earthquake. Nevertheless, if we consider measuring points at the same epicentre distance and having the same soil properties, the amplitude of ground motion does increase with earthquake size. Hence, if we take into account the difference in epicentral distance or in soil properties, it is possible to estimate the magnitude of the earthquake indirectly from the instrumentally recorded amplitude of ground motion. The size of the earthquake is derived from such considerations.

Richter was the first to determine earthquake magnitude. He studied the relation between the maximum amplitude of ground motion, A, as measured by a specified seismograph and the epicentre distance Δ which was almost parallel irrespective of the size of the earthquake. He thus defined magnitude M in terms of the epicentral distance Δ (km) and the maximum recorded amplitude A (μm) in the following manner:

$$M = \log_{10} A + k \log_{10} \Delta + c \tag{2.8}$$

where k and c are constants.

The magnitude as determined by Richter assumes that the hypocentre of the earthquake is not too deep and the epicentral distance is $\Delta < 600$ km. This is called the Richter local magnitude M_L.

If different types of seismographs are used for measuring the earthquake magnitude, there may be considerable differences in the seismic waves recorded even at the same point of measurement. Thus, if the displacement seismograph is sensitive to waves of longer periods, it will mainly record surface waves of long periods; if it is sensitive to waves of short periods, it will mainly record body waves of shorter periods. So if we use different types of seismographs, different amplitudes of vibrations will get recorded. Thus, the magnitudes derived from the amplitudes recorded by two different records may not be identical for the same earthquake. Given this consideration, different magnitudes are now defined depending on the part of the seismic wave recorded. They are surface-wave magnitude M_s, body-wave magnitude M_b, and moment magnitude M_w. The relation between these magnitudes has been established empirically.

Surface wave magnitude scale M_s

The Richter local magnitude scale does not distinguish between different types of waves. So [Gutenberg and Richter (1956)] introduced a surface wave magnitude scale M_s. The surface wave magnitude scale is based on the amplitude of surface wave having a period of about 20 seconds. The surface wave magnitude scale M_s is defined as follows:

$$M_s = \log A' + 1.66 \log \Delta + 2.0$$

where,

M_s = surface wave magnitude scale
A' = maximum ground displacement, in μm
Δ = epicentral distance to seismograph measured in degree (360° corresponds to the circumference of earth)

The surface wave magnitude scale has an advantage over the local magnitude scale in the sense that it uses the maximum ground displacement, rather than the maximum trace amplitude from a standard Wood–Anderson seismograph. The magnitude is typically used for moderate to large earthquakes, having a shallow focal depth (less than 70 km) and the seismograph should be at least 1000 km from the epicentre.

Body wave magnitude M_b

As far as deep earthquakes (focal depth > 300 km) are concerned, surface waves are often too small to permit reliable evaluation of the surface wave magnitude. The body wave magnitude (Gutenberg, 1945) is a worldwide magnitude scale based on the amplitude of the first few cycles of P-waves which is not strongly influenced by the focal depth (Bolt, 1989).

The body wave magnitude M_b can be expressed as

$$M_b = \log A - \log T + 0.01\Delta + 5.9$$

where,

A = P-wave magnitude, in μm
T = period of P-wave (about one second)

The body-wave magnitude can be related to surface-wave magnitude M_s as (Darragh et al., 1994)

$$M_b = 2.5 + 0.63\ M_s$$

Moment magnitude scale M_w

The seismic moment can also be estimated, from the fault displacement as follows: (Idriss, 1985)

$$M_o = \mu\ A_f\ D$$

where,

M_o = seismic moment, in N-m
μ = shear modulus of material along fault plans, in N/m^2
D = average displacement of ruptured segment of fault, in m

The moment magnitude scale has become the more commonly used method for determining the magnitude of large earthquakes. This is due to the fact that it tends to take into account the entire size of the earthquake. The first step in the calculation of moment magnitude is to calculate the seismic moment M_o as given by the above equation.

The seismic moment is based on a concept different from the conventional one as known to engineers. The reason is because the seismic force and the moment are in the same direction. In engineering, a moment is calculated as the force times the moment arm, and the moment arm is always perpendicular to the force. Setting aside the problem with the moment arm, the seismic moment does consider the energy radiated from the entire fault, rather than the energy from an assumed point source. Thus, the seismic moment is a more useful measure of strength of an earthquake.

Kanamori (1977) and Hanks & Kanamori (1979) introduced the moment magnitude M_w scale in which the magnitude is calculated from the seismic moment by using the following equation.

$$M_w = -6.0 + 0.07\ \log\ M_o$$

where,

M_w = moment magnitude of earthquake
M_o = seismic moment of earthquake, in N-m

The same equation is also expressed as

$$M_W = \frac{\log M_o}{1.5} - 10.7$$

where M_o = seismic moment in dyne-cm.

In Japan, magnitude M_J is determined from the maximum amplitude of ground motion A as recorded by a Wiechert type seismograph or one of similar properties [recorded amplitude divided by magnification factor corresponding to that period, unit (μm)], and the epicentral distance Δ (km) using the following expression:

$$M_J = 1.73 \log_{10} \Delta + \log_{10} A - 0.83 \tag{2.9}$$

Magnitude M_J is the Japan Meteorological Agency's (JMA) magnitude and is applicable to earthquakes having hypocentral depths up to about 60 km only. For earthquakes with greater hypocentral depths, there is another M_J. From its approach of derivation, M_J may be considered similar to surface-wave magnitude M_s. M_J, M_s and M_L may be considered comparable.

At the occurrence of an earthquake, the Japan Meteorological Agency determines its magnitude M_J using Eq. (2.9) at each of the observation sites; the average of all sites is then declared as the magnitude of that earthquake. As such, there is some variation in the value of M_J. For example, for the 1978 earthquake of Miyagi prefecture, the maximum value of magnitude recorded was 8.25, minimum value 6.25, average 7.41 and standard deviation 0.3. These values are indicative of the accuracy of M_J as stated by JSCE (1997).

Magnitude is a simple parameter to express the size of an earthquake but not good enough to express its scale as a physical phenomenon. It is similar to expressing the height of a human body in terms of foot length. As will be seen later, with the development of a fault model, efforts have been made to express earthquake magnitude as a physical phenomenon; nevertheless one has to accept the fact that several simplifying assumptions are involved. On the other hand, although not so accurate the term magnitude, being an effective parameter from a practical point of view, has long been used to indicate the size of an earthquake and, furthermore, is quickly understood. Consequently, even in the field of earthquake engineering, various earthquake properties are described in terms of magnitude.

2.8.3 Energy Associated with Earthquake

Magnitude is evidently related to that energy, which is radiated from the earthquake source in the form of elastic waves. Part of the original potential energy of strain stored in the rock must go into mechanical work, as in raising crustal blocks against gravity, or as in crushing material in the fault zone; part must be dissipated as heat.

Reid estimated the work done in displacing crustal blocks during the California earthquake of 1906 as 1.75×10^{24} ergs. Energies of a number of earthquakes have been estimated from seismograms, for it is fairly well-established that there is relatively little absorption of seismic waves after they leave the vicinity of the hypocentre. Consequently the energy in the expanding wave front, which can be estimated from the recorded amplitudes and periods, represents most of the energy radiated. In this way, Jeffreys derived from the surface waves of the Pamir earthquake of 1911 (magnitude 7.6) and of the Montana earthquake of 1925 (magnitude 6.75) energies of about 10^{21} ergs.

Energy in an elastic wave of given period is proportional to the square of the amplitude. If seismograms of different earthquakes at a fixed distance actually differed only in amplitude, the periods would be unchanged, and may be expressed as

$$\log E = c + 2M \tag{2.10}$$

where c is a constant. Preliminary work using the results of Jeffreys and others gave $c = 8$, but this value gives incredibly small energies for the smallest recorded shocks. More elaborate calculations by Gutenberg and Richter led to $\log E = 11.3 + 1.8\, M$; introducing the overlooked factors, and a little further hypothesis, produced the formula

$$\log E = 12 + 1.8M \tag{2.11}$$

which was used in seismicity of the Earth. Especially for the larger shocks, energies given by this formula are too high. In the interim, di Filippo and Marcelli published a calculation, which led to

$$\log E = 9.15 + 2.15M \tag{2.12}$$

All these formulas depend on theoretical study of the radiation of energy at short distances, near the epicentre. In a recent revision, Gutenberg and Richter (1956) made extensive use of seismograms written by the strong motion instruments operated by the U.S. Coast and Geodetic Survey, including those for the Kern County earthquakes of 1952. The remaining uncertainties of this method have been a principal factor in Gutenberg's preference for the "unified magnitude" m, derived form body waves recorded at teleseismic distances. The relation of m to the radiated energy can be set up with less theoretical difficulty and a minimum of observational inaccuracy; it takes the form

$$\log E = 5.8 + 2.4m \tag{2.13}$$

Since $m = 2.5 + 0.63M$, this is equivalent to

$$\log E = 11.4 + 1.5M \tag{2.14}$$

In Eq. (2.14), M is at least an approximation to the magnitude determined from surface waves of shallow teleseisms.

Gutenberg has used every available means to relate m to the magnitude M_L derived in the original manner from local earthquake records in California. His preferred result is

$$m = 1.7 + 0.8M_L - 0.01M_L^2 \tag{2.15}$$

which leads to

$$\log E = 9.9 + 1.9M_L - 0.024M_L^2 \tag{2.16}$$

The terms in M_L^2 are highly empirical in nature and difficult to interpret satisfactorily in terms of physical dimensions. The relation between m and M_L, and consequently that between $\log E$ and M_L, will probably be modified soon by new data.

Putting $M = 8$ in the above equations, gives $\log E = 26.4$, 26.35, and 23.4; thus revision leads to greatly reduced values for the energies of the largest shocks. However, the values of M have generally been on increase, so that it would be better to put $M = 8.5$ in Eq. (2.12), giving $\log E = 24.15$. Since most of the energy of all earthquakes is in such shocks, the revision materially reduces the estimates of the annual total energy of seismic activity. On the earlier basis, this energy was given in publications as 1.2×10^{27} ergs per year. Since the energy of the annual flow of heat from the interior through the surface of the earth is roughly 8×10^{27} ergs, the two numbers were close enough to suggest various geophysical speculations. Revision for the seismic energy now gives a figure near to 9×10^{24} ergs per year, which is hardly more than one-thousandth of the heat energy.

A further point of chiefly journalistic interest relates to comparison between large earthquakes and atomic bombs. The official figure for the energy released by a normal atomic bomb of the Hiroshima type is 8×10^{20} ergs; a very large earthquake, on the old basis, might have an energy of 8×10^{26} ergs, hence comparable with a million atom bombs. On the new basis, the largest earthquakes are found to have an energy not much over 10^{25} ergs, roughly equivalent to 12,000 of such atom bombs.

2.9 LOCATING THE EARTHQUAKES

The principal use of a seismograph network is to locate earthquakes. Although it is possible to infer a general location for an event from the records of a single station, it is most accurate to use three or more stations. Locating the source of any earthquake is important, of course, in assessing the damage that the event may have caused, and in relating the earthquake to its geologic setting. The location of an earthquake essentially includes:

(a) The location of the epicentre
(b) The determination of the epicentral distance
(c) The location of hypocentre or focus
(d) The determination of the depth of the earthquake
(e) The determination of the local distance or hypocentral distance

2.9.1 Location of the Epicentre

As already described, the epicentre of an earthquake is the point on the earth's surface vertically above the focus (see Figure 2.1). This point can be easily located for any earthquake, taking advantage of the time lag noticed between P- and S-waves. When the earthquake waves are recorded at different stations, it will be observed that the time lag between the arrival of P- and S-waves increases gradually with the distance from the epicentre. Thus, this factor gives a measure of the distance between the epicentre and the seismic station. Therefore, if seismic recordings are made at three different well-spaced stations (say, A, B and C) and circles with measures of distance as radius are drawn, they (i.e., the circles) intersect at a common point. This point is the epicentre of the concerned earthquake as shown in Figure 2.28.

2.9.2 Determining the Depth of Focus of Earthquake

According to Oldham, the depth of the focus can be estimated by comparing the intensities at the epicentre with those at another station. Figure 2.29 illustrates this aspect.

where,

E = epicentre
G = a station where the intensity is known
m = intensity at the epicentre
n = intensity at G

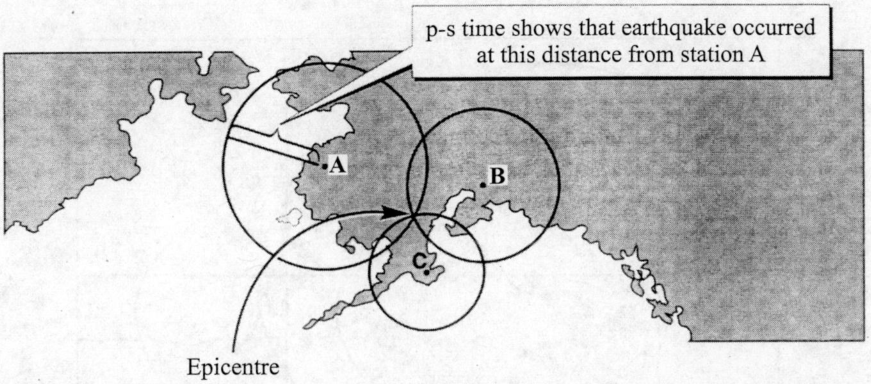

Figure 2.28 Preliminary location of epicentre from differential wave–arrival time measurements at seismographs A, B and C. Most likely epicentral location is at the intersection of the three circles. (After Foster, R.J., 1971).

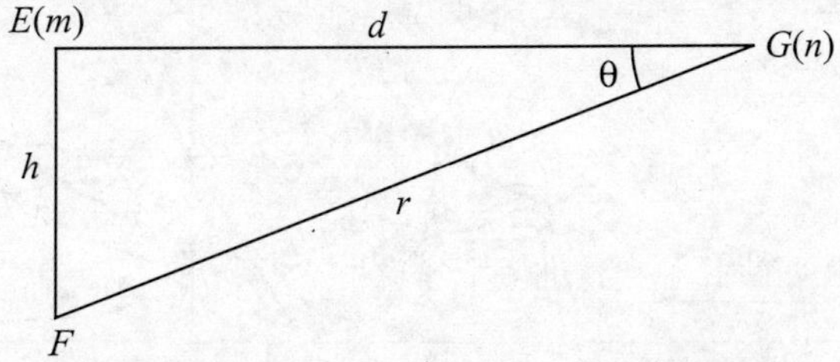

Figure 2.29 Determination of depth of earthquake origin.

d = distance between E and G
h = depth of focus

First, θ is calculated as follows:

$$\frac{n}{m} = \frac{h^2}{r^2} = \sin^2\theta$$

Based on the θ value, h is calculated from the relation: $h = d \tan \theta$.

2.9.3 Isoseismal Maps

A map of the earthquake affected area is usually prepared, and on which the intensity values assigned to various places are maked. Apart from intensity and magnitude of the earthquake, the extent of the area affected by the seismic ground motion is also a measure of the earthquake. After preparing the map of the affected area with the assigned values of intensity, the areas having the same intensity are then enclosed by contour lines. Such a map showing contours of same intensity of the earthquake is called an *isoseismal map*.

Figure 2.30(a) shows the isoseismal map of Agadir earthquake of 1967. Figure 2.30(b) shows the isoseismal map of Inangahua earthquake, New Zealand, 1968 whereas Figure 2.30(c) shows the map of the famous California earthquake of 1989. The reservoir triggered earthquake occurred in Koyna 1967 and the isoseismal map has been shown in Figure 2.30(d).

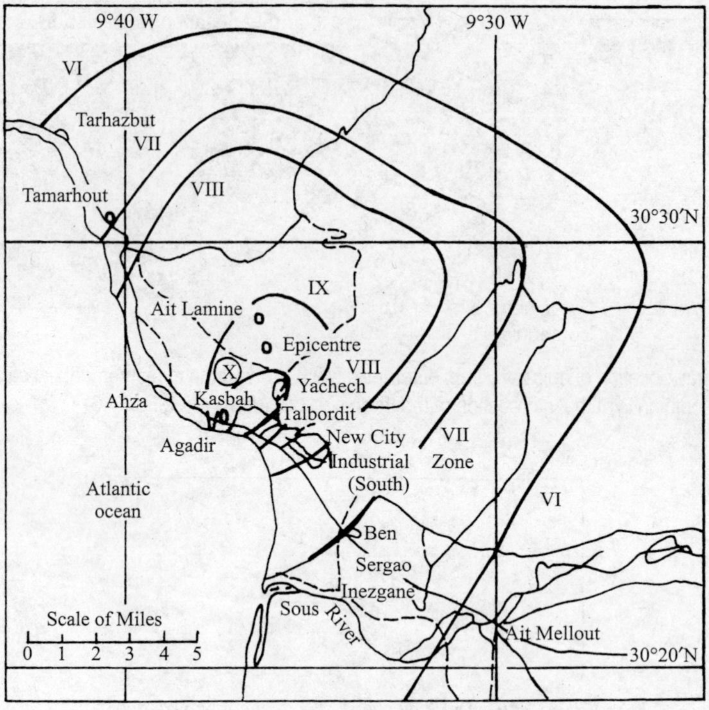

(a)

Figure 2.30(a) Isoseismal map of Agadir earthquake, 1960.

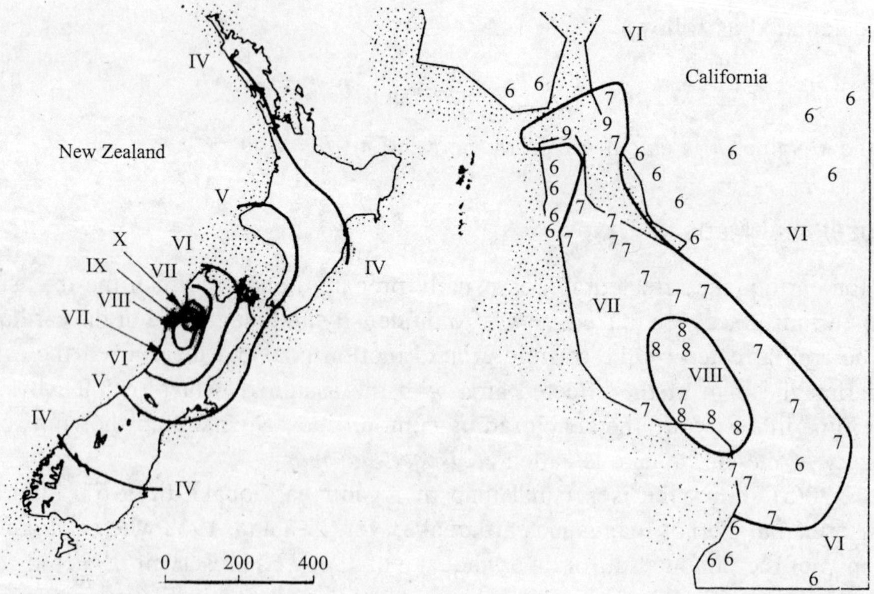

(b) (c)

Figure 2.30 (b) and (c) (b) Isoseismal map of Inangahua earthquake, New Zealand, 1968; and (c) Isoseismal map for California earthquake, 1989. [After Housner, 1990]

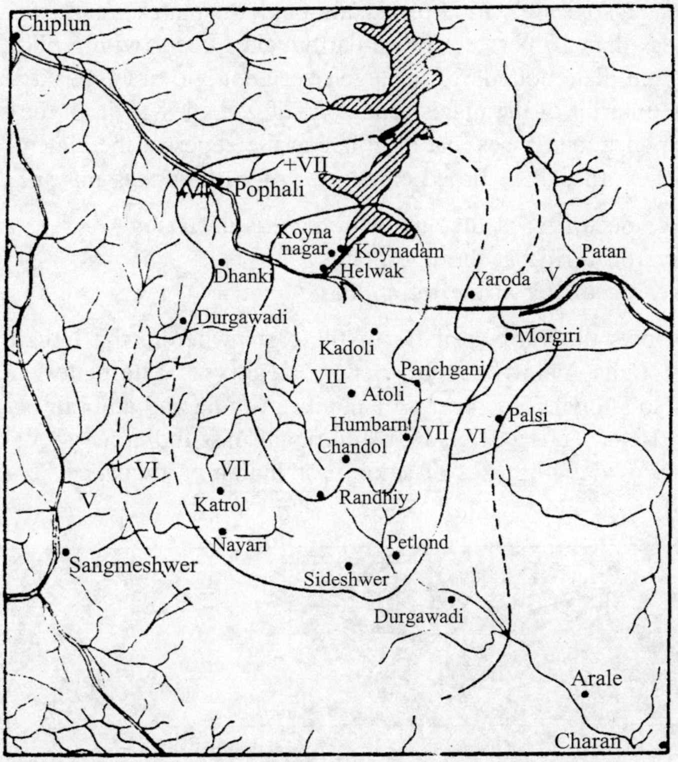

Figure 2.30(d) Isoseismal map of Koyna earthquake of Dec. 11, 1967.

2.10 PLATE TECTONICS, PLATE BOUNDARIES AND EARTHQUAKES IN INDIA

As explained in previous sections, the occurrence of the earthquake can be explained on the basis of mechanics of continental drift and plate tectonics. It is now almost well settled that the outer layer of the earth consists of about a dozen large irregularly-shaped plates that slide over, under and past each other on top of the partly molten magma (see Figure 2.31). Most earthquakes occur at the boundaries where the plates meet. On the contrary the locations of earthquakes and the kind of rupture they produce help seismologists define the plate boundaries. There are three types of plate boundaries—subduction zones, transform faults and spreading zones.

- Subduction zones are found where one plate overides or subducts another, pushing it downwards into the mantle where it may melt. An example of such plate boundary is along the N-W coast of the United States, southern Alaska and western Canada.
- Transform faults are found where plates slide past one another. An example of a transform-fault plate boundary is the San Andreas fault along the coast of California and north-western Mexico.
- At spreading zones, the molten rock mass rises, pushing two plates apart and adding new material at their edges. Most spreading zones are found in oceans.

Earthquakes can also occur within plates, although the plate-boundary earthquakes are much more common. Less than 10 per cent of all earthquakes occur within plate interiors. As plates continue to move and plate boundaries change over geologic time, weakened boundary regions become part of the interior of the plates. The zone of weakness, within the continents can cause earthquakes in response to stresses that originate at the edges of the plate or in the deeper crust.

Stated in other words, three broad categories of earthquakes may be recognized as

1. Earthquakes occurring at the subduction/collision zones
2. Earthquakes occurring at the interplates
3. Earthquakes occurring at the intraplates.

Figure 2.31 shows the various plates of the earth wherein the Indian plate may be seen. Seismic events in India mainly belong to the category of interplates, though a few events (intraplates) are also known. The recent seismicity in India and adjoining areas in Asia may be seen in Figure 2.31. In Section 1.7, some significant past Indian earthquakes have been mentioned. Table 1.4 provides glimpses of some past Indian earthquakes.

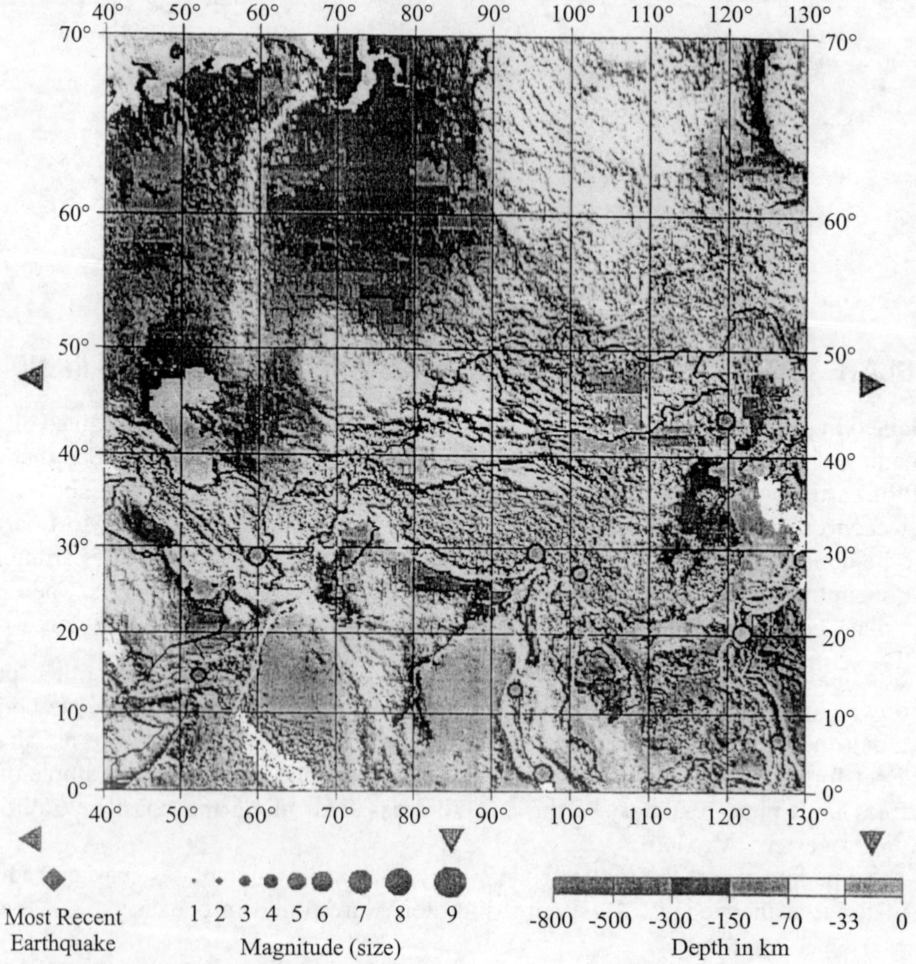

Figure 2.31 Earthquake activity in Indian subcontinent and plate boundaries (After USGS).

Figure 2.32 indicates the location, the year and the number of fatalities (in paranthesis) for earthquakes in India in the past 200 years. The earthquakes events in India are reported mainly from four regions, namely:

(a) Peninsular region
(b) Indo-Gangetic plain
(c) Andaman and Nicobar islands
(d) Himalayan mountain

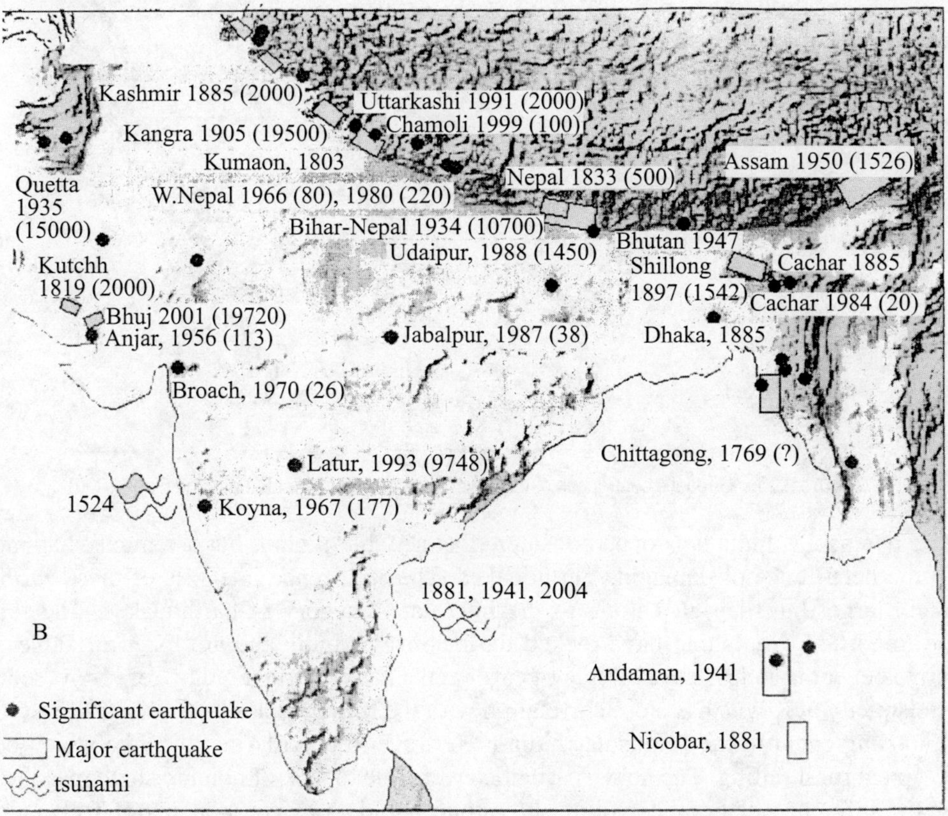

Figure 2.32 Location of Indian Earthquake in past 200 years.

2.10.1 Earthquakes in Peninsular India

The southern peninsula of India has deformed very little in the past 160 years. The derived rate indicates that points in southern and northern India converge by less than 3 ± 2 mm per year. The first exception is found on the Malabar coast of India where landing and tide gauge data indicate that the western coast of southern India is apparently sinking rapidly near Kochi.

Further, India contracts at 3 mm/year (with an uncertainty of 2 mm/year) in the direction A-A as shown in Figure 2.33. It may come as a surprise to many seismologists that India is moving southwards relative to the Earth's spin axis. This may be caused by the vanishing northern ice sheets and the subsequent mantle mass adjustment. The southward rate of motion of India at present is 4 cm/year whereas Tibet is moving south at 8 cm per year.

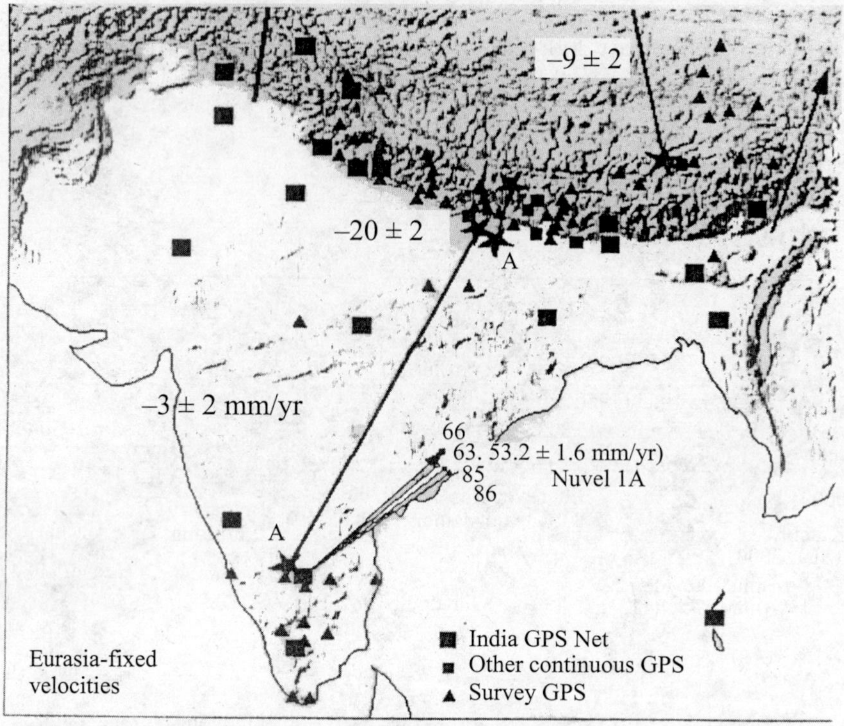

Figure 2.33 Geodefic contraction of the Indian subcontinent (After Bilham and Gaur, 2000).

The peninsular India was once considered as a stable region, but seismicity has increased due to the occurrence of damaging earthquakes. The recurrence intervals of these earthquakes are much larger but they all belong to the interplate category of earthquakes. The following are the important events that have rocked the peninsular India and are listed in Table 2.4.

Koyna event is a classical example of earthquake activity triggered by reservoir (see section 2.6). Seismicity at Koyna has close correlation with the filling cycles of the Koyna reservoir. The most puzzling event in the peninsular India is, however, the Killari earthquake which occurred in the typical rural setting. The heavy casualties were due to lack of bond in stone masonry walls. Under the influence of ground motion, big stone boulders as components of wall gained big momentum and their impact proved fatal. This event was least expected from the tectonic

Table 2.4 Earthquakes in Peninsular India

Place	*year*	*Magnitude*	*Casualty*
Kutchh	June 16, 1819	8.5	No record
Jabalpur	June 2, 1927	6.5	–
Indore	March 14, 1938	6.3	–
Bhadrachalam	April 14, 1969	6.0	–
Koyna	December 10, 1967	6.0	>200
Killari (Later)	September 30, 1993	6.3	>10,000
Jabalpur	May 22, 1997	6.0	>55

consideration, as it is located in the Deccan Trap covered stable Indian shield. There is no record of any historical earthquake in this region. This has been considered as SCR (Stable Continental Region) event in the world. Moreover, its spatial association with the Narmada Son Lineament has triggered a lot of interest from the seismotectonics point of view.

2.10.2 Earthquake in Himalayan Region

Subduction earthquakes in India occur in the Himalayan Frontal Arc (HFA). This arc is about 2500 km long from Kashmir in the west to Assam in the east (see Figure 2.35). It constitutes the central part of the Alpine seismic belt and is one of the most active regions in the world. The India plate came into existence after initial rifting of the Southern Gondwanaland in late Triassic period and subsequent drifting in mid-Jurrasic to late Cretaceous time. The force responsible for this drifting came from the spreading of the Arabian Sea on either side of the Carlsberg ridge (see Figure 2.34). It eventually collided with the Eurasian plate in Middle Eocene

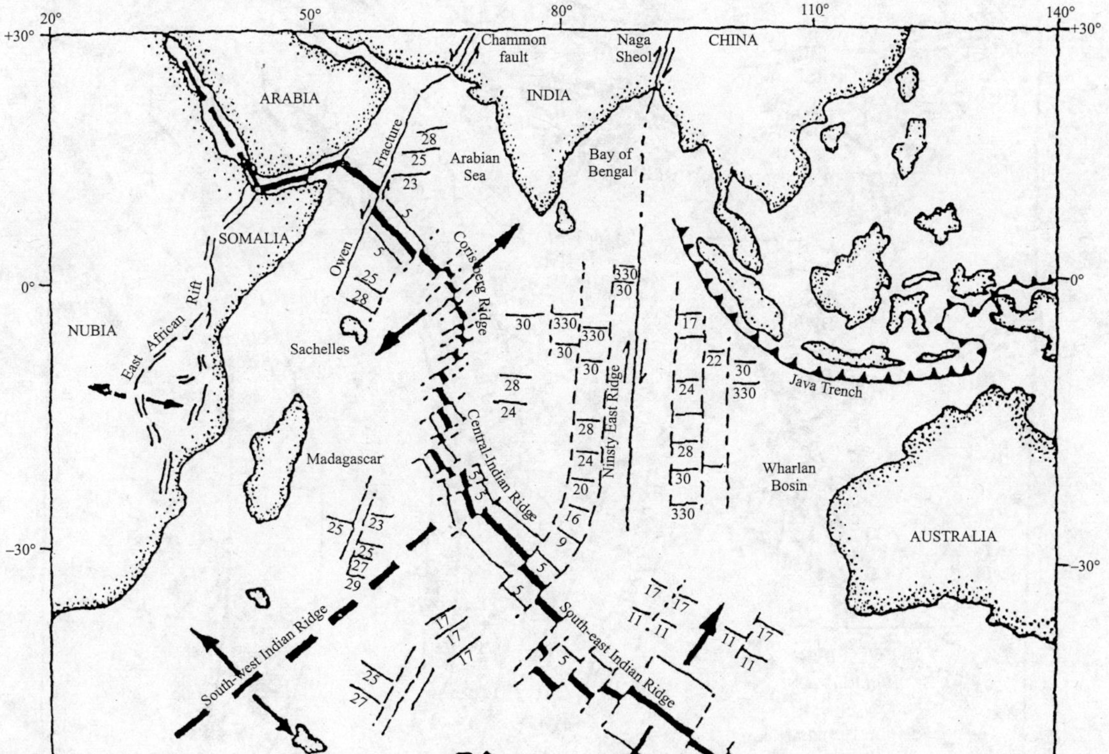

Figure 2.34 Major tectonic features of the Indian Ocean showing spreading of Arabian Sea on either side of Carlsberg Ridge (After Chatterjee, 2000)

after drifting along counter-clockwise path. The movement of the Indian plate caused continental collision with the rates of convergence varying from 44-66 mm per year. This led to the creation of Himalayan mountain range. The present seismicity of this region is due to continued collision between the Indian and the Eurasian plate. The plate boundaries between the two are of special significance. The important events that have rocked in this region and have visited the Himalayan Frontal Arc (HFA) are listed in Table 2.5.

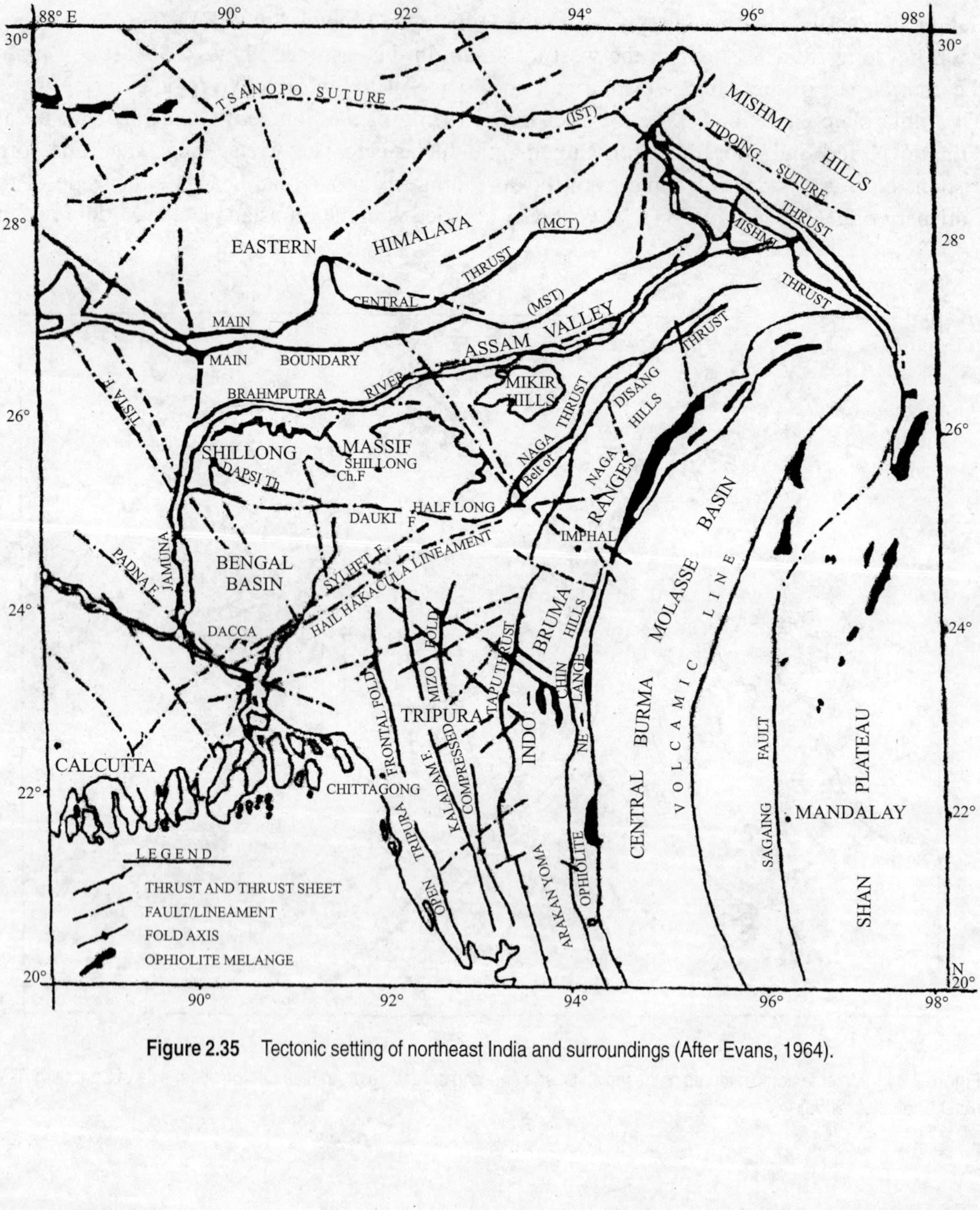

Figure 2.35 Tectonic setting of northeast India and surroundings (After Evans, 1964).

Table 2.5 Earthquakes in Himalayan region

Place	*Year*	*Magnitude*	*Casualty*
Kangra valley	April 14, 1905	8.6	>20,000
Bihar-Nepal border	Jan., 1934	8.4	>10,653
Quetta	May 30, 1935	7.6	>30,000
North Bihar	Aug., 1988	6.5	>1000
Uttarkashi	Oct. 20, 1991	6.6	>2000
Chamoli	March 29, 1999	6.8	>150
Hindukush	Nov. 11, 1999	6.2	None

2.10.3 Earthquakes in the North-Eastern Region

The north-eastern region of India lies at the junction of the Himalayan arc to the north and the Burmese arc to the east and is one of the six most seismically active regions of the world. The other five regions are Mexico, Japan, Taiwan, Turkey and California. Eighteen large earthquakes with magnitude (M>7) occurred in this region during the last hundred years (Kayal, 2001). High seismic activity in the north-east region may be attributed to the collision tectonics in the north (Himalayan arc) and subduction tectonics in the east (Burmese arc). The syntaxis zone (The Mishmi Hills Block) is the meeting place of the Himalayan and Burmese arcs and is another specific tectonic domain in the region. The main central thrust (MCT) and the Main Boundary Thrust (MBT) are two major crystal discontinuities in the Himalayan arc of the north-eastern region. In the Burmese arc, the structural trend of the Indo-Myanmar Ranges (IMR) swing from the NE-SW in the Naga Hills to N-S along the Arakan Yoma and Chin Hills. Naga thrust is the prominent discontinuity in the north. It connects the Tapu thrust to the south and Dauki fault to the east. This fold belt appears to be continuous with the Andaman–Nicobar ridge to the south. The Mishmi Thrust and the Lohit Thrust arc the major discontinuties in the Syntaxis zone (Kayal, 2001). The important earthquakes events in this region are listed in Table 2.6.

Table 2.6 Earthquakes in the north-eastern region

Place	*Year*	*Magnitude*	*Remarks*
Cachar	March 21, 1869	7.8	Earth fissures and sand crates
Shillong	June 12, 1897	8.7	First detailed scientific reporting in world by R.D. Oldham
Sibsagar	Aug. 31, 1906	7.0	Property damage
Srimangal	July 8, 1918	7.5	Property damage [4500 sq. km area]
SW Assam	September 9, 1923	7.1	Property damage
Dhubri	January 27, 1931	7.6	Railway line, culverts and bridges cracked
N-E Assam	Oct. 23, 1943	7.2	Destruction of property
Upper Assam	July 29, 1949	7.6	Severe damage
Upper Assam	Aug. 15, 1950	8.7	One of the largest known earthquakes of the history
Indo-Myanmar border	Aug. 6, 1988	7.5	No casualty reported

2.10.4 Earthquakes in Andaman and Nicobar Islands

The seismicity in Andaman and Nicobar islands is reflected by the location of epicentre of the past earthquakes. The tectonic and structural features of this region are essentially those of thrust. The principal lithological indicates sedimentary rocks in Port Blair and adjoining areas.

The Andaman and Nicobar Islands are located near the boundary of the Indian plate and Burmese Microplate. The Andaman trench marks the boundary and lies in the Bay of Bengal to the west of archipelago. Another prominent feature in the north-south west-Andaman fault which is strike-slip in nature and lies in the Andaman sea to the east of the island chain. The Andaman sea just like Atlantic ocean is presently being induced by a tectonic process called sea floor spreading. This is taking place along undersea ridges of the sea floor. The Indian plate is diving beneath the Burmese Microplate along the Andaman trench in a process known as subduction.

All of the Andaman and Nicobar islands lie in Zone V. The entire island chain is also susceptible to tsunamis both from large local earthquakes and also from massive distant shocks. No warning system is presently in place for any of the island in chain. A very great earthquake ($M_W = 9.1$) struck the North Indian Ocean and Bay of Bengal on 26 Dec 2004. The Nicobar island (and to lesser extent the Andaman islands) were hardest hit territory in India with as many as 4486 deaths.

Figure 2.36 shows the earthquake hazard map of Andaman Islands region by USGS.

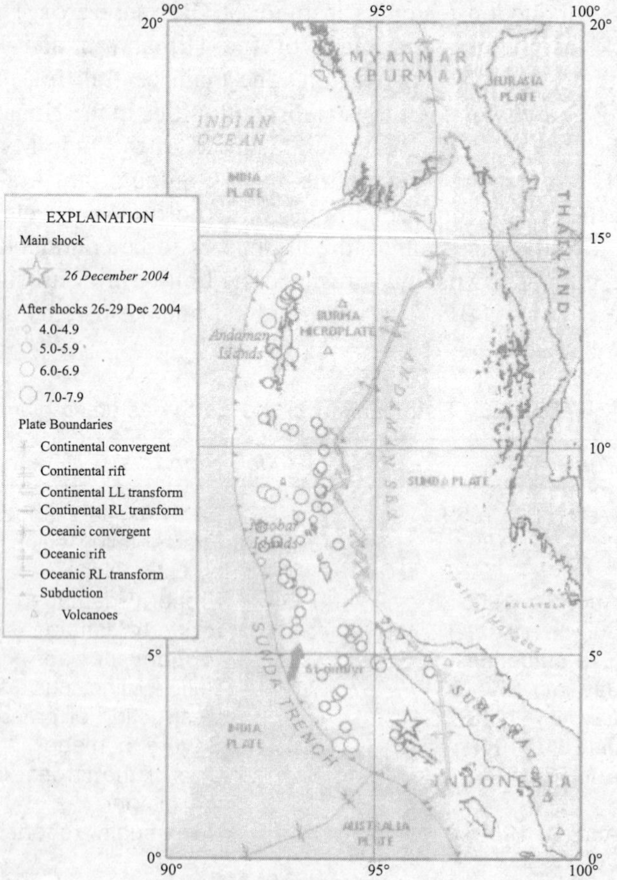

Figure 2.36 Seismic hazard map of Andaman Islands, India region. (After USGS).

2.11 MEASURING EARTHQUAKES

The vibrations and ground motion produced by the earthquakes are detected, recorded and measured by the instruments called seismographs as shown in Figure 2.37. The zigzag line (trace) made by a seismograph, called a seismogram, reflects the changing intensity of the vibrations by responding to the ground motion of the surface beneath the instrument. From the data expressed in seismograph, scientists can determine the time, the epicentre, the focal depth and the type of faulting of an earthquake and can estimate how much energy was released.

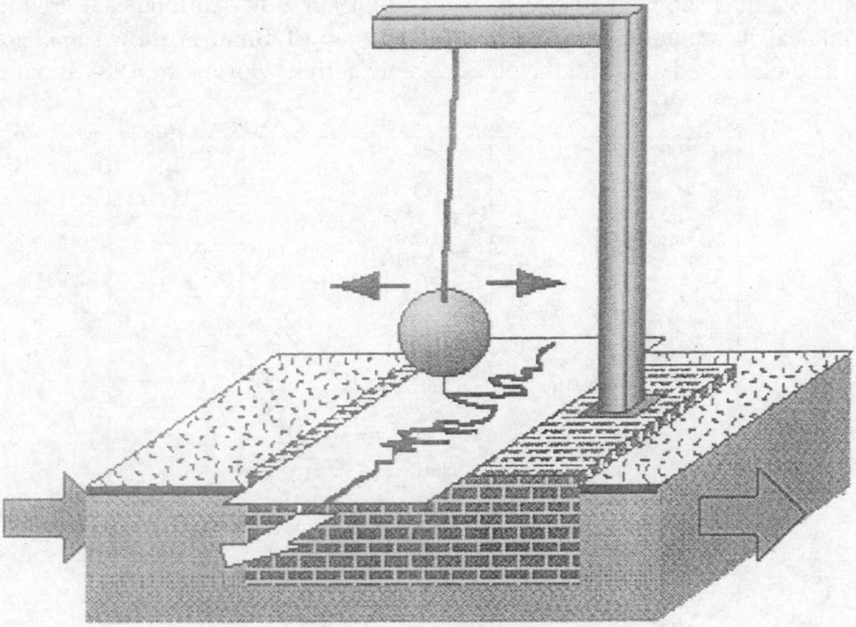

Figure 2.37 Seismograph for recording seismogram.

Sensitive seismographs are the principal tool of seismologists who study earthquakes. Thousands of seismograph stations are in operation through the world and the instruments have been transported to the Moon, Mars and Venus. Fundamentally a seismograph is a simple pendulum. When the ground shakes, the base and the frame of the instrument move with it but inertia keeps the pendulum bob in place. As it moves it records the pendulum displacements as they change with time, tracing out a record called a seismogram.

PROBLEMS

2.1 Describe the internal structure of the Earth. What do you mean by Moho discontinuity? Discuss the variation of shear wave velocity in different layers of the Earth.

2.2 What do you mean by Continental Drift? What are the primary cause of Continental Drift? Explain with suitable diagrams the mechanism of Continental Drift. Discuss the Plate Tectonic theory.

2.3 How will you measure the energy of the earthquake? If the energy released by atom bomb dropped on Hiroshima was 8×10^{20} ergs which is equivalent to energy associated with an earthquake of magnitude 6.33 on Richter scale, what was the energy associated with Bhuj earthquake of Jan. 2001, and how many such bombs will release that equivalent energy?

2.4 The seismograph records amplitude of 11.5 mm long on N-S direction. The distance of the recording station to the epicentre is 195 km. The distance correction may be taken as 3.4. Determine the magnitude of the earthquake assuming zero station correction.

2.5 Figure P2.5(a) shows the isoseismal map of 1934 Bihar earthquake. Describe the seismological information from this map. The map of Bihar is shown in Figure P2.5(b), locate the affected area and the other seismic activity during the 1988 Bihar earthquake.

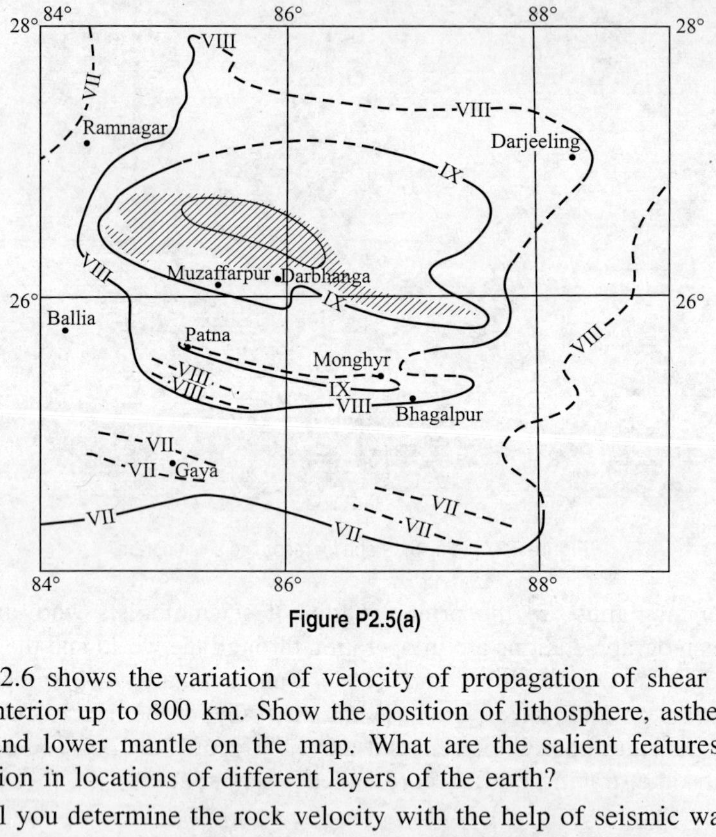

Figure P2.5(a)

2.6 Figure P2.6 shows the variation of velocity of propagation of shear wave inside the earth's interior up to 800 km. Show the position of lithosphere, asthenosphere, upper mantle and lower mantle on the map. What are the salient features of shear wave propagation in locations of different layers of the earth?

2.7 How will you determine the rock velocity with the help of seismic wave velocities?

2.8 What is the density of the earth? What are the densities of the rock at the earth's surface? How does density vary with the depth? What is the volume of the total earth? What are the percentage volumes of the crust, mantle and core with respect to the volume of the total earth.

2.9 Can the earthquakes be prevented? How can future earthquakes be minimized? How did the Chinese successfully predict an earthquake in 1976?

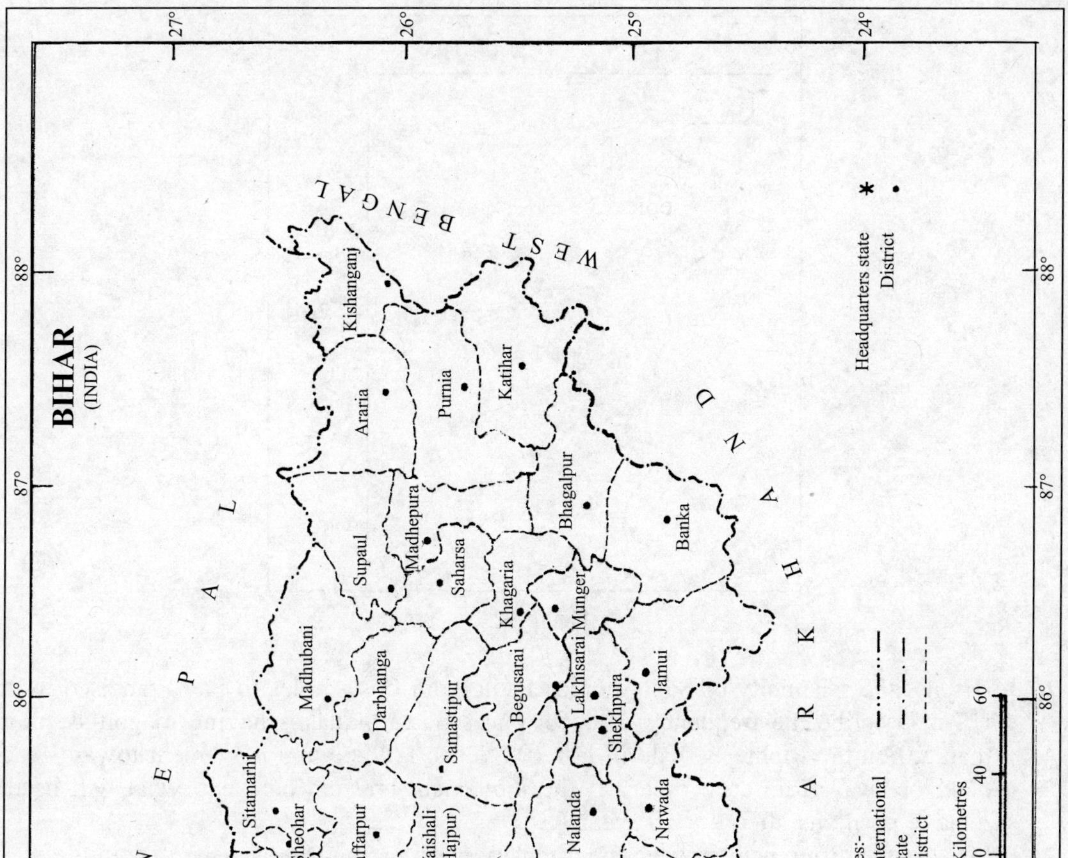
BIHAR
(INDIA)
NEPAL
WEST BENGAL
JHARKHAND
UTTAR PRADESH
Paschim Champaran
(Betiah)
Purba champaran
(Motihari)
Gopalganj
Siwan
Saran
(Chhapra)
Sitamarhi
Sheohar
Muzaffarpur
Vaishali
(Hajipur)
Madhubani
Darbhanga
Samastipur
Supaul
Madhepura
Saharsa
Araria
Kishanganj
Purnia
Katihar
Khagaria
Begusarai
Bhagalpur
Munger
Lakhisarai
Shekhpura
Jamui
Banka
Patna
Nalanda
Nawada
Jehanabad
Gaya
Aurangabad
Bhojpur
(Arah)
Buxar
Rohtas
(Sasaram)
Bhabhua
Boundaries:-
International
State
District
Kilometres
40 20 0 20 40 60
Headquarters state
District
27°
26°
25°
24°
84°
85°
86°
87°
88°

Figure P2.5(b)

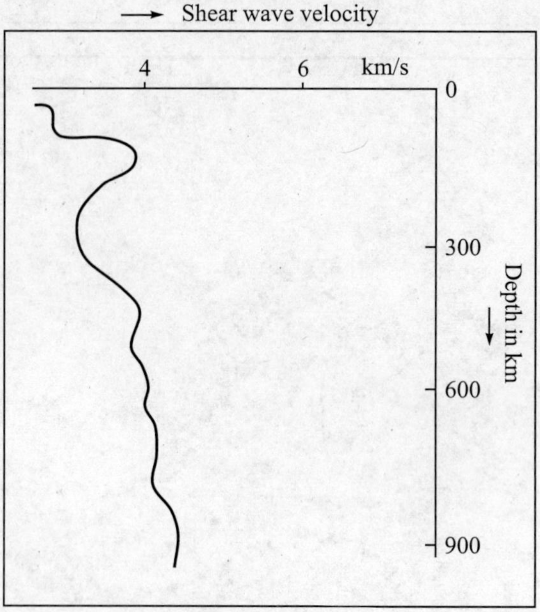

Figure P2.6

2.10 Discuss the seismicity of San Andreas Fault from Los Angles to San Francisco in the USA. Describe the frequency of occurrences of earthquakes having magnitude more than 5.0 on the Richter's scale in last 100 years. Is there any movement towards each other or away from each other? If the movement is 2 cm per year, what will be the relative positions after 3×10^7 years?

2.11 What is the difference between the intensity of an earthquake and the magnitude of an earthquake? What scales are MMI, RF, JMA and MSK? With the help of a suitable proportionate sketch, explain their comparative descriptions.

3

THEORY OF VIBRATIONS

3.1 INTRODUCTION

The dynamics of an elastic system include the study of mass of the system, its elastic properties, energy loss mechanism (dissipation of energy) in terms of damping, and the influence of the external loading or source of excitation. Vibrations are initiated when the energy is imparted to the elastic system by an external source. For example, vibration of a foundation or the supporting structure induced during an earthquake can lead to large stresses and may result in failure. The study of vibrations requires synthesis of basic engineering sciences and mathematics. It is rightly said that if the language of science is mathematics, then most of its prose and poetry is occupied by differential equations.

Vibrations are often classified in a number of ways depending upon various factors. If the external energy source is applied only to initiate the vibrations and then suddenly removed, the resulting vibrations are termed *free vibrations*. But if vibrations occur under the presence of an external energy source, the resulting vibrations are called *forced vibrations*.

The number of degrees of freedom of a system is the number of independent variables necessary to describe the motion of the system. The set of independent coordinates is called a set of generalized coordinates. For a system it may be one, two, three or, in general, n degrees of freedom, requiring n number of coordinates to describe entirely the motion of the system. A system with a finite number of degrees of freedom is called a *discrete system*. A continuous system has infinite number of degrees of freedom. If only one coordinate is needed to describe the entire motion of system, it is called a single degree of freedom (SDF) system. The essential physical properties of an elastic structural system subjected to a dynamic load include its mass, its elastic properties, its energy loss mechanism and the external source of excitation or loading. The entire mass m such a system is shown in Figure 3.1. The single coordinate z completely defines its position. The elastic resistance to displacement is provided by the weightless spring of stiffness k.

The energy loss mechanism is represented by the damper c. The external source of excitation of loading is represented by the time varying load. A system is defined linear if its motion

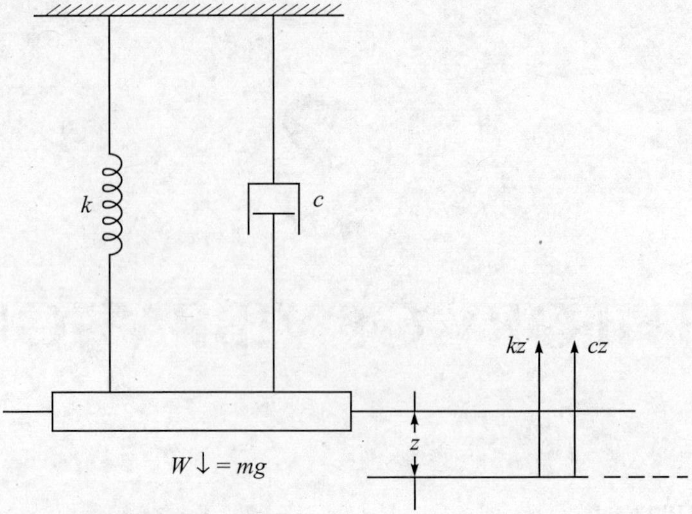

Figure 3.1 Single degree of freedom (SDF) system.

is governed by a linear set of differential equation. If the system is non-linear, its motion is governed by a non-linear system of equations. All systems are basically non-linear, however, simplifying assumptions are made for linear approximation. Linear systems are much easier to analyze than the non-linear systems. But for complex problems, often it is more realistic to analyze through non-linear techniques. The response of soils and rocks to dynamic loading is a highly complex phenomenon, and so more realistic models are needed for analytical analysis.

At any instant of time if the motion is described and the value of the excitation force is known, then the excitation is said to be deterministic. But if the excitation force is unknown, and in such case if the average or mean deviations are only known, then the excitation is random in nature and as such only statistical values of the response can be evaluated. Random vibration analysis is often used to analyze the earthquake (seismic) excitation of foundations and the supporting structures.

The modelling of a physical system results in the formation of a mathematical problem. Mathematical modelling of vibration problems leads to differential equations. Vibration of a single degree of freedom (SDF) discrete system is governed by a single ordinary differential equation. Vibrations of multi degree of freedom systems are governed by a system of ordinary differential equations. Vibrations of a continuous system having infinite degrees of freedom are governed by partial differential equations.

The energy loss mechanism is expressed in the form of damping. The damping is classified into the following categories:

(a) Viscous damping
(b) Coulomb damping
(c) Hysteretic damping
(d) Aerodynamic drag induced damping
(e) Other types of damping

The response of a system with any type of damping continues indefinitely with decaying amplitude. The energy dissipations for various types of damping are different from each other, and the resulting motion is termed *damped vibrations.*

3.2 PERIODIC MOTION

Oscillatory motion may repeat itself regularly or display considerable irregularities, as in earthquakes, which are represented in Figures 3.2 and 3.3, respectively. When the motion is repeated in equal intervals of time τ, it is called periodic motion. The repetition time τ is called the period of oscillation and its reciprocal $\omega = 1/\tau$ is called the frequency. The simplest form of periodic motion is harmonic motion, which is often represented as the projection on a straight line of a point, that is, moving on a circle at constant speed, as shown in Figure 3.2. With the angular velocity of the line OP designated by ω, the true displacement of the point P can be written as

$$z = A \sin \omega t \tag{3.1}$$

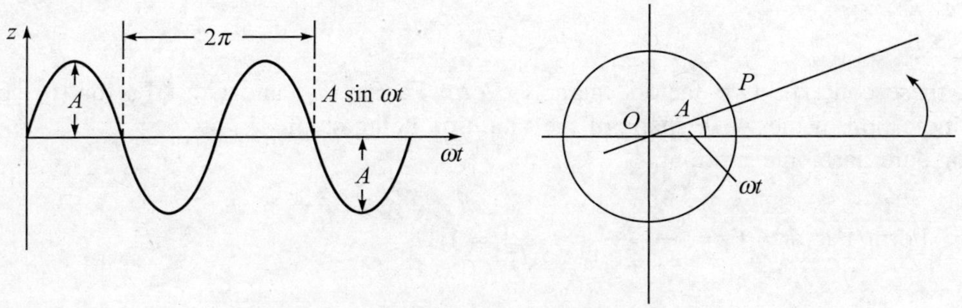

Figure 3.2 Representation of harmonic motion.

This quantity is generally measured in radians per second and is referred to as the circular frequency or simply frequency, since the motion repeats in 2π radians. So we have the relationship

$$\omega = \frac{2\pi}{\tau} = 2\pi f \tag{3.2}$$

where τ and f are the period and the frequency of the harmonic motion usually measured in seconds and cycles per second respectively whereas ω is the angular frequency measured in rad/s.

Further, another descriptive quantity, which takes the time history into account, is the average absolute value as shown in Figure 3.3.

$$Z_{\text{average}} = \frac{1}{T}\int_0^T |z|\,d\tau \tag{3.3}$$

However, a much more useful descriptive quantity which also takes the time history into account is the RMS (root mean square value) value as shown in Fig. 3.3.

$$Z_{\text{rms}} = \sqrt{\frac{1}{T}\int_0^T |z^2|\,d\tau} \tag{3.4}$$

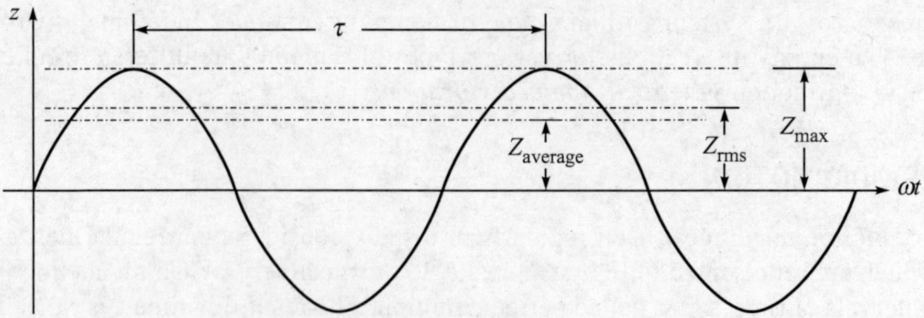

Figure 3.3 Variation of amplitude with time.

For pure harmonic motion,

$$Z_{\text{rms}} = \frac{\pi}{2\sqrt{2}} Z_{\text{average}}$$

or,

$$Z_{\text{rms}} = \frac{1}{\sqrt{2}} Z_{\text{max}} \tag{3.5}$$

In these contexts, two factors, namely, Form Factor (F_f) and Crest Factor (F_c) provide some indication of the wave shape of the vibration being studied.

For pure harmonic motion,

$$\text{Form Factor} = F_f = \frac{Z_{\text{rms}}}{Z_{\text{average}}} = \frac{\pi}{2\sqrt{2}} = 1.11 \tag{3.6}$$

$$\text{Crest Factor} = F_c = \frac{Z_{\text{max}}}{Z_{\text{rms}}} = \sqrt{2} = 1.414 \tag{3.7}$$

Exponential form: The trigonometric functions of sine and cosine are related to the exponential function by Euler's equation

$$e^{i\theta} = \cos\theta + i\sin\theta \tag{3.8}$$

where $i = \sqrt{-1}$.

Thus, Eq. (3.1) may be rewritten in exponential form as

$$z(t) = Ae^{i\omega t} \tag{3.9}$$

$$= A\cos\omega t + iA\sin\omega t \tag{3.10}$$

$$= A\cos\omega t + B\sin\omega t \tag{3.11}$$

where $B = iA$.

The vector of amplitude A rotating at constant angular speed ω can be represented as a complex quantity $z(t)$ as shown in the Argand diagram in Figure 3.4.

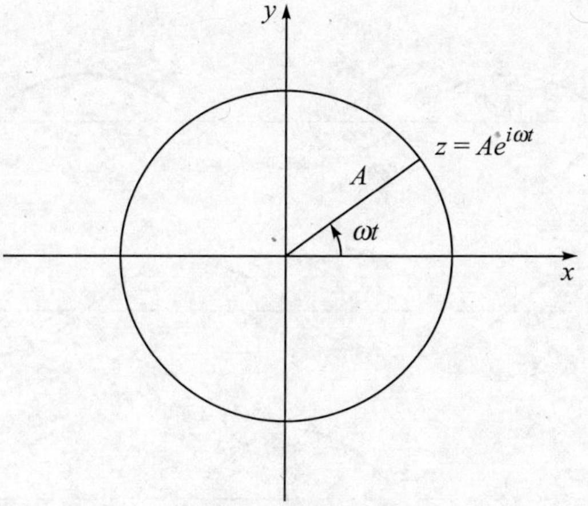

Figure 3.4 Argand diagram

Figure 3.4 is representation in graphical form of a pure translational oscillation along the z-axis only, then the instantaneous displacement $z(t)$ can be mathematically described as

$$z(t) = A \sin \omega t = Z_{\max} \sin \omega t \tag{3.12}$$

where $Z_{\max}$ is maximum displacement from the reference position. The velocity v as time rate change of the displacement may be expressed as

$$v(t) = \dot{z}(t) = \frac{dz(t)}{dt} = \omega Z_{\max} \cos \omega t = V_{\max} \sin(\omega t + \pi/2) \tag{3.13}$$

where,

$$V_{\max} = \omega Z_{\max} \tag{3.14}$$

Finally, the acceleration of the motion is the time rate of change of velocity and may be expressed as

$$a(t) = \frac{d^2 z(t)}{dt^2} = \ddot{z} = \frac{dv(t)}{dt} = -\omega^2 Z_{\max} \sin \omega t = A_{\max} \sin(\omega t + \pi) \tag{3.15}$$

Obviously from the above equations, the period of vibrations remain the same for displacement, velocity or acceleration. However, the velocity leads the displacement by a phase angle of 90° ($\pi/2$) and the acceleration again leads the velocity by a phase angle 90°($\pi/2$). (See Figure 3.5 and 3.6). The description in terms of maximum value or peak values are quite useful as peak values describe the vibration in terms of a quantity which depends only upon the instantaneous vibration magnitude regardless of the time history producing it.

3.2.1 Frequency Analysis

There are various types of vibrations, which are not pure harmonic motions even though many of them may be characterized as periodic. One of the most powerful descriptive methods in the

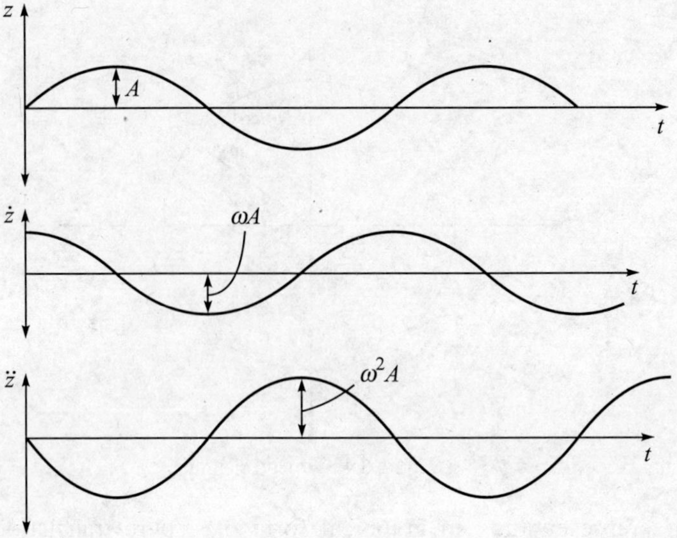

Figure 3.5 Variation (time history) of displacement, velocity and acceleration with time.

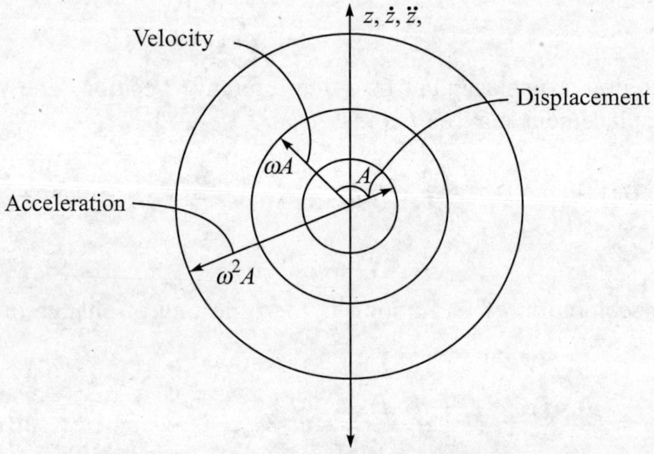

Figure 3.6 Rotating vector representation of displacement, velocity and acceleration.

method of frequency analysis is that due to J. Fourier. This is based in mathematical theorem first formulated by J. Fourier (1768–1830). According to this theorem, any periodic motion can be represented by a series of sine and cosine terms that are harmonically related. If $z(t)$ is a periodic function of the period τ, it is represented by the Fourier series as

$$F(t) = \frac{a_0}{2} + a_1 \cos \omega_1 t + a_2 \cos \omega_2 t + \cdots + b_1 \sin \omega_1 t + b_2 \sin \omega_2 t + \cdots$$

or, $$F(t) = \frac{a_0}{2} + \sum_{n=1}^{\infty} a_n \cos \omega_n t + b_n \sin \omega_n t \tag{3.16}$$

where, $\omega_1 = \dfrac{2\pi}{\tau} = 2\pi f_1$

$\omega_n = n\omega_1 = 2\pi f_n$

The coefficients are given by

$$a_0 = \frac{\omega_n}{\pi}\int_0^{\tau} z(t)\; dt \tag{3.17}$$

$$a_n = \frac{2}{\tau}\int_{-\tau/2}^{\tau/2} z(t)\cos\;\omega_n t\, dt \tag{3.18}$$

$$b_n = \frac{2}{\tau}\int_{-\tau/2}^{\tau/2} z(t)\sin\;\omega_n t\, dt \tag{3.19}$$

In exponential form, the Fourier series can be rewritten as

$$F(t) = \sum_{n=-\infty}^{\infty} C_n e^{i\omega_n t} \tag{3.20}$$

where, $$C_n = \frac{1}{\tau}\int_{-\tau/2}^{\tau/2} z(t)\; e^{i\omega_n t}\; dt \tag{3.21}$$

or, $$C_n = \frac{1}{2}(a_n - ib_n) \tag{3.22}$$

Often the coefficients of the Fourier series are plotted against frequency ω_n, the result is a series of discrete lines called the Fourier spectrum. With the rapid advancement in digital computer programming, a computer algorithm known as FFT (Fast Fourier Transform) is commonly used to minimize the computation time. In the Fourier series the number of terms may be infinite but in that case as the number of elements in the series is increased it becomes an increasingly better approximation to the original curve. Several elements constitute the vibration frequency spectrum.

Fourier series is an important useful tool in many branches of engineering and is also applicable to geotechnical earthquake engineering. The complex loading function imposed by seismic ground motion may be expressed as the sum of series of simple harmonic loading functions for linear systems.

3.3 CLASSICAL THEORY

In dynamic analysis, mass, stiffness and damping of the system come into play. These three together resist the applied loads, stiffness is linearly proportional to deformation and is independent of velocity and acceleration. The mass of the structure offers resistance proportional to the acceleration. Although the resistance offered by damping is quite complex, however, for linear variations, the viscous damping (Newtonian dashpot) offers resistance proportional to the velocity.

Newton's law

Mass is the measure of matter and is basically defined by the mass density. Newton's second law states that (Figure 3.7)

$$F(t) = m\frac{d^2z}{dt^2} = m\ddot{z} \tag{3.23}$$

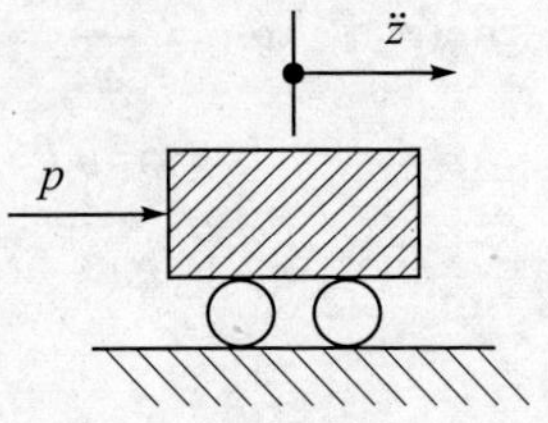

Figure 3.7 Equilibrium of mass (Newton's law).

D′ Alembert's principle

The equation of equilibrium treating $m\ddot{z}$ as a force is given by D′ Alembert's principle as

$$F(t) - m\ddot{z} = 0 \tag{3.24}$$

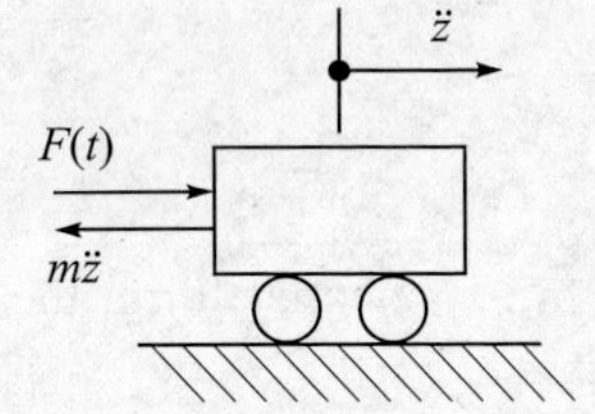

Figure 3.8 Equilibrium of mass (*D′* Alembert's principle)

The term $m\ddot{z}$ is called the inertial force or force of inertia. See Figure 3.8.

Equation (3.24) is taken as an equivalent static system wherein the applied dynamic load $F(t)$ and the inertia force $m\ddot{z}$ are in equilibrium at a given instant of time. The inertia force vector is taken to act in a direction opposite to that of the acceleration, hence the minus sign. Thus, conceptually the inertial force may be taken as one of the resisting forces similar to that offered by spring or the damper.

The following forces are taken into consideration in formulating the governing differential equation:

1. Resistance elastic force in the Hookean spring
2. Viscous force in the Newtonian dashpot.
3. Inertia force.
4. External dynamic force.

For dynamic equilibrium, the relationship between the forces as shown in Figure 3.1 may be expressed as

Inertia force $[F_i]$ + Viscous force in dashpot $[F_D]$ + Resistance force in the spring $[F_S]$ = External dynamic force $[F(t)]$.

Taking the spring to be linear elastic, the resistance force in the spring $F = kz$, where k is the stiffness of the weightless elastic spring.

Considering a viscous damping as shown by a Newtonian dashpot in Figure 3.1 wherein the force is proportional to the velocity, i.e.,

$$\text{Viscous force (damping)} = F_D = c\frac{dz}{dt} = c\dot{z}$$

wherein c is the coefficient of viscous damping. The linear variations of the resisting force in the spring, the viscous force in the dashpot and the inertia force have been shown in Figure 3.9.

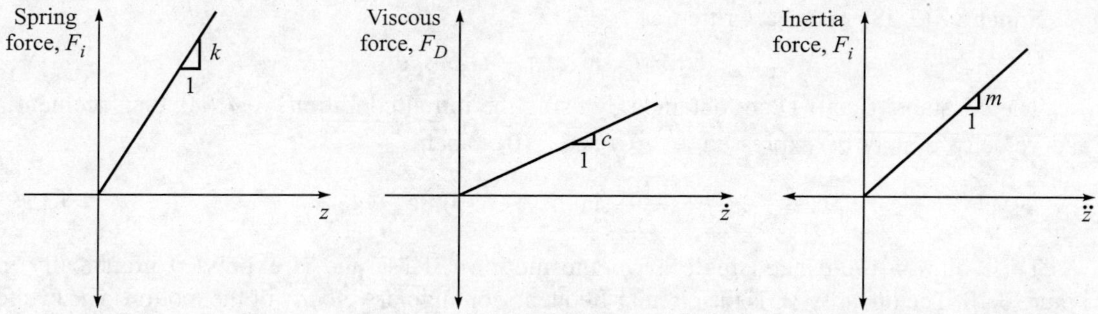

Figure 3.9 Linear variation of inertia, viscous force and spring force against acceleration, velocity and displacement, respectively.

The external dynamic force in general terms may be expressed as

$$\text{External dynamic force} = F(t)$$

Thus, the dynamic equilibrium is expressed as

$$m\ddot{z}(t) + c\dot{z}(t) + kz(t) = F(t) \tag{3.25}$$

for free and undamped vibration.

When, $F(t) = 0$ and $c\dot{z} = 0$, the equation of motion takes the form,

$$m\ddot{z}(t) + kz(t) = 0 \tag{3.26}$$

Dividing throughout by m and setting

$$\frac{k}{m} = \omega_n^2$$

the governing differential equation reduces to another form, which is known as the canonical form

$$\ddot{z}(t) + \omega_n^2 z(t) = 0 \tag{3.27}$$

This is a linear, second-order differential equation with constant coefficients. The auxiliary equation for solving the above differential equation (3.27) is

$$s^2 + \omega_n^2 = 0 \quad \text{i.e.,} \quad s = \pm\, i\omega_n$$

Otherwise if the solution of Eq. (3.27) is of the form

$$z(t) = A \cdot e^{st}$$

Substituting this solution in Eq. (3.27) yields

$$s^2 + \omega_n^2 = 0 \quad \text{i.e.,} \quad s = \pm\, i\omega_n$$

Thus, the response given by Eq. (3.27) is

$$z(t) = A_1\, e^{i\omega_n t} + A_2\, e^{-i\omega_n t} \tag{3.28}$$

in which the two values result from the two values of s and the constants A_1 and A_2 represent the arbitrary amplitudes of the motion. Equation (3.28) may be expressed in another form by introducing Euler's equation, i.e.,

$$e^{\pm i\omega_n t} = \cos\, \omega_n t \pm i \sin\, \omega_n t$$

Equation (3.28) may be written as

$$z(t) = C \sin \omega_n t + D \cos \omega_n t$$

The constants C and D are obtained by using the initial condition. At $t = 0$, displacement z and velocity $\dot{z}$ may be expressed as $z(0)$ and $\dot{z}(0)$. Then,

$$z(t) = \frac{\dot{z}(0)}{\omega_n} \sin \omega_n t + z(0) \cos \omega_n t \tag{3.29}$$

This solution represents single harmonic motion (SHM) and is expressed graphically in Figure 3.10. The quantity ω_n is the circular frequency or angular velocity of the motion. The cyclic frequency f_n, which is popularly known as frequency of the motion expressed in hertz is given by

$$f_n = \frac{\omega_n}{2\pi}$$

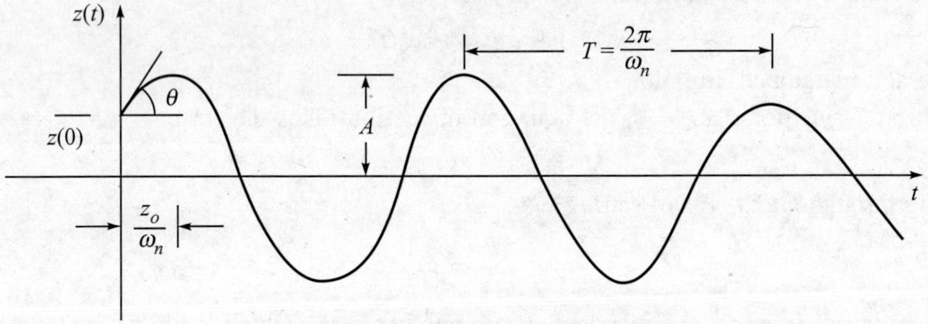

Figure 3.10 Undamped free vibrations.

and its reciprocal is called the period T in seconds, i.e.,

$$T = \frac{2\pi}{\omega} = \frac{1}{f_n}$$

The motion represented by Eq. (3.29) may also be expressed in the form

$$z(t) = A \sin(\omega_n t + \theta) \tag{3.30}$$

$$= A \cos \omega_n t \cdot \cos \theta + A \sin \omega_n t \cdot \sin \theta \tag{3.31}$$

and, $\quad \theta = \tan^{-1} \dfrac{\dot{z}(0)}{\omega_n z(0)} \quad$ and $\quad A = \left\{ z^2(0) + \left(\dfrac{\dot{z}(0)}{\omega_n} \right)^2 \right\}^{\frac{1}{2}}$

where A is the amplitude of the motion and θ is the phase angle. The graphic representation of Eq. (3.31) is shown in Figure 3.11.

The natural frequency ω_n is dependent on stiffness of the spring and the mass of the system. It is independent of the initial conditions, i.e., no matter how the system is set into motion, the frequency remains the same. The initial conditions determine the energy initially present in the system.

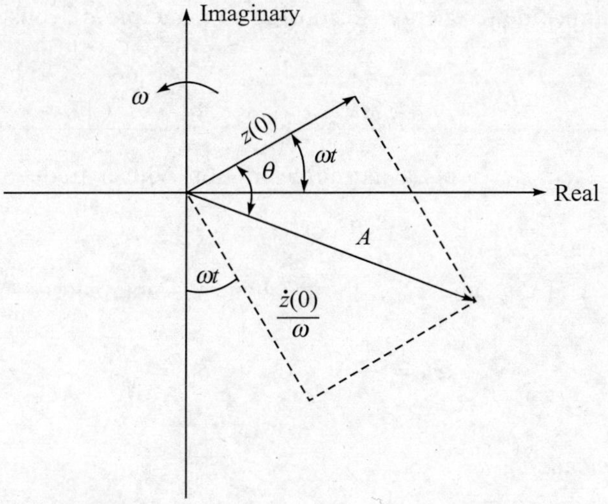

Figure 3.11 Rotating vector representation of free vibration

Energy method

The energy consideration of the single degree freedom system undergoing free vibrations may be expressed as

$$T + V = E \tag{3.32}$$

where, T = kinetic energy, V = potential energy and E = energy constant.

If the spring is displaced by an amount z, then

$$V = \text{potential energy stored}$$
$$= (1/2)\text{Force} \times z = (1/2) \times k \times z \times z = (1/2)kz^2$$

The kinetic energy of the mass m moving with velocity $\dot{z}$ is

$$T = \text{kinetic energy}$$
$$= (1/2)\text{m}\,\dot{z}^2$$

Substituting the values of T and V in Eq. (3.32),

$$(1/2)m\dot{z}^2 + (1/2)kz^2 = E \tag{3.33}$$

Since we are considering a closed system, where no energy can enter or leave the system, the time rate of change of energy must be zero, i.e.,

$$\frac{d}{dt}(T + V) = 0 \tag{3.34}$$

It can be observed that when the mass is at the extreme end of its stroke, the velocity is zero and all the energy possessed by the system is potential in nature and may be termed V_{max}, i.e., the maximum potential energy. When the mass passes through the equilibrium position, its displacement is zero and all the energy possessed by system is kinetic in nature and equals

T_{max}, i.e., the maximum kinetic energy. According to principle of conservation of energy,

$$T_{max} = V_{max}$$

or,
$$(1/2)m\dot{z}^2_{max} = (1/2)kz^2_{max}$$

Assuming that the system performs a harmonic motion with a frequency ω_n such that,

$$z = A \cos (\omega_n t - \alpha)$$

Substituting the values of z_{max} and $\dot{z}_{max}$, the frequency is obtained as

$$\omega_n^2 = \frac{k}{m} \qquad \text{or} \qquad \omega_n = \sqrt{\frac{k}{m}}$$

Natural frequency calculation

The calculation of natural frequency is an important aspect of vibration. The solution of various problems of soil dynamics and geotechnical earthquake engineering largely depends upon accurate knowledge of natural frequencies of the system. It is always desirable to avoid the condition of resonance at the design stage itself. Resonance occurs when the natural frequency of the system coincides with the excitation frequency, resulting in a large undesirable amplitude. Thus, the various methods, like the energy method, are very important for determining the natural frequencies of the foundation and the supporting structural systems.

Influence of gravitational forces

For all linear systems the static deflection of springs cancels with the gravity forces causing the static deflection when the governing differential equation is simplified. The SDF system with weight W acting along the gravity is shown in Figure 3.12,

where, $W = mg$ and static deflection $\delta_{st} = \dfrac{W}{k} = \dfrac{mg}{k} = d$ (say)

Then,
$$\omega_n = \sqrt{\frac{k}{m}} = \sqrt{\frac{g}{\delta_{st}}} = \sqrt{\frac{g}{d}}$$

However, if the total displacement z_1 is expressed as the sum of the static displacement δ_{st} caused by the weight W plus the additional dynamic displacement z, then

$$z_1 = \delta_{st} + z \tag{3.35}$$

$$\text{Total spring force} = k\delta_{st} + kz \tag{3.36}$$

Also,
$$k\delta_{st} = W$$

Then the equilibrium of forces gives

$$m\ddot{z} + k\delta_{st} + kz = F(t) + W$$

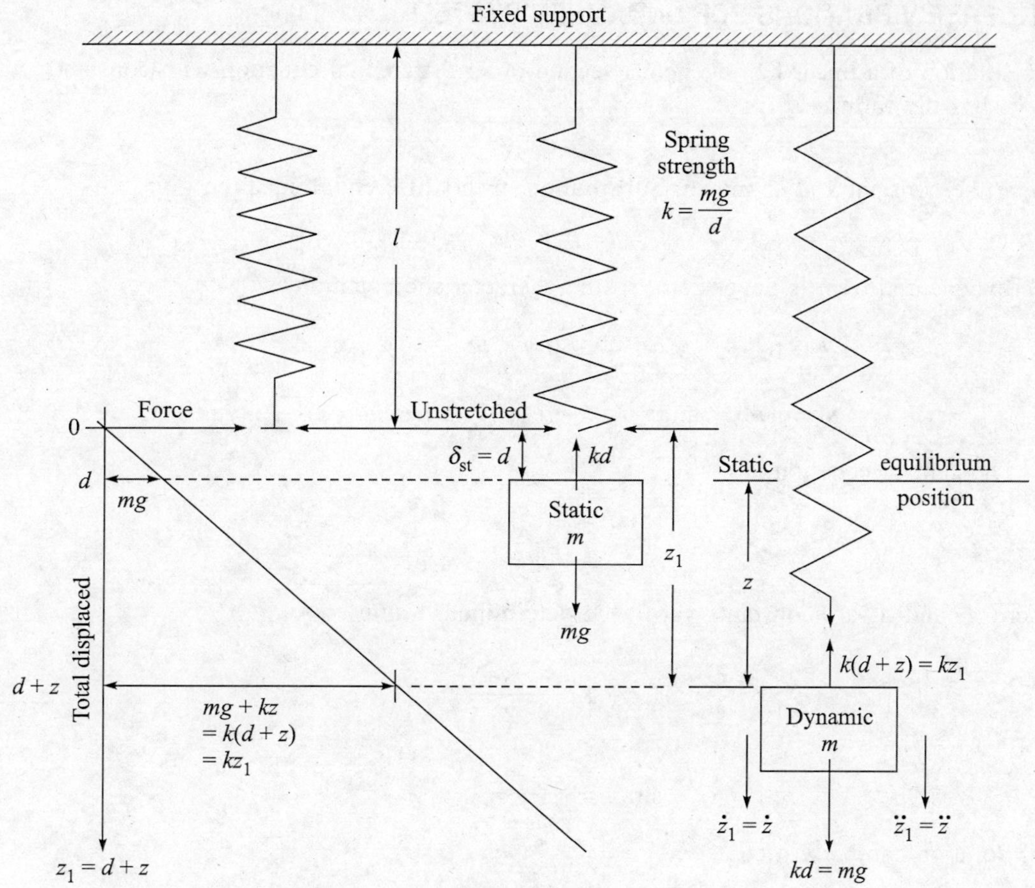

Figure 3.12 SDF system under the influence of gravity.

For free vibration, $F(t) = 0$ and $W = k\delta_{st}$. Thus,

$$m\ddot{z} + kz = 0$$

As δ_{st} does not vary with time, differentiating Eq. (3.35) gives

$$\ddot{z} = \ddot{z}_1 \tag{3.37}$$

Thus,

$$m\ddot{z}_1 + kz = 0 \quad \text{or} \quad m\frac{d^2z}{dt^2} + kz = 0 \tag{3.38}$$

Equation (3.38) demonstrates that the equation of motion expressed with reference to the static equilibrium portion of dynamic system is not affected by gravity forces. This is important to the extent that for obtaining the dynamic response the displacements are referenced from the static portion and therefore that deflection, stresses, etc. can be obtained by adding the appropriate static quantities to the result of dynamic analysis.

3.4 FREE VIBRATIONS SDF UNDAMPED SYSTEM

The solution of a linear homogeneous second-order differential equation with constant coefficients has the form

$$z = e^{st}$$

where the constant s is unknown. Substitution in the differential equation gives

$$(ms^2 + k)e^{st} = 0$$

The exponential term is never zero, so the characteristic equation

$$ms^2 + k = 0 \quad \text{or} \quad s^2 + \frac{k}{m} = 0 \quad \text{or} \quad s^2 + \omega_n^2 = 0$$

So, $\quad s_1 = +i\omega_n \quad \text{and} \quad s_2 = -i\omega_n, \qquad \text{where } i = \sqrt{-1}$

The general equation is

$$z(t) = A_1 e^{st} + A_2 e^{st}$$
$$= A_1 e^{i\omega_n t} + A_2 e^{-i\omega_n t}$$

where A_1 and A_2 are constants yet to the determined. Using

$$\cos x = \frac{e^{ix} + e^{-ix}}{2}$$

$$\sin x = \frac{e^{ix} - e^{-ix}}{2i}$$

The solution can be written as

$$z(t) = A \cos \omega_n t + B \sin \omega_n t \tag{3.39}$$

Using the initial condition,

$$t = 0,\ z(0) = z_0 \quad \text{and} \quad \dot{z}(0) = \dot{z}_0,$$

we have $\quad A = z(0) \text{ and } B = \dfrac{\dot{z}(0)}{\omega_n}$

So the solution takes the form

$$z(t) = z(0) \cos \omega_n t + \frac{\dot{z}(0)}{\omega_n} \sin \omega_n t$$

The same equation can be written as

$$z = C \cos(\omega_n t - \theta)$$

where C and θ are constants like A and B.

Using the trigonometric form, the above equation may be written as

$$z = C \cos \omega_n t \cdot \cos \theta + C \sin \omega_n t \cdot \sin \theta \tag{3.40}$$

Comparing Eq. (3.39) with Eq. (3.40) and using the trigonometric identity

$$A = C \cos \theta$$
$$B = C \sin \theta$$

So, $\tan \theta = B/A$ or $\theta = \tan^{-1}(B/A) = \tan^{-1} \dfrac{\dot{z}(0)}{\omega_n z(0)}$

and, $$B^2 + A^2 = C^2(\cos^2\theta + \sin^2\theta)$$

or, $$C = \sqrt{B^2 + A^2} = \sqrt{\left(\frac{\dot{z}(0)}{\omega_n}\right)^2 + z^2(0)}$$

Thus, $$z(t) = z(0) \cos \omega_n t + \frac{\dot{z}(0)}{\omega_n} \sin \omega_n t$$

$$= \sqrt{\left(\frac{\dot{z}(0)}{\omega_n}\right)^2 + z^2(0)} \cdot \cos \left\{\omega_n t - \tan^{-1} \frac{\dot{z}(0)}{\omega_n z(0)}\right\}$$

or else the solution can be written as

$$z(t) = D \sin (\omega_n t + \phi)$$

then, $$z(t) = D \cos \phi \sin \omega_n t + D \sin \phi \cos \omega_n t$$

as $$\sin(a + b) = \sin a \cos b + \cos a \sin b$$

Comparing the above two equations and using the trigonometric identity, see Fig. 3.13.

$$D \cos \phi = B$$
$$D \sin \phi = A$$
$$\tan \phi = \frac{A}{B},$$
$$\phi = \tan^{-1} \frac{\omega_n z(0)}{\dot{z}(0)}$$

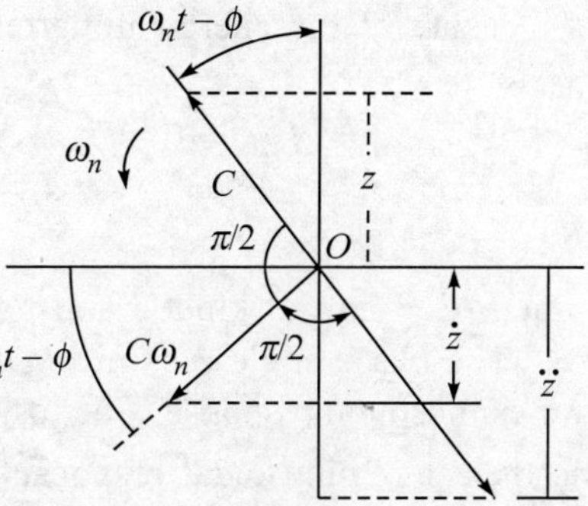

Figure 3.13

Thus, $$z(t) = \sqrt{z^2(0) + \left(\frac{\dot{z}(0)}{\omega_n}\right)^2} \cdot \sin(\omega_n t + \phi)$$

Quite often the amplitude in SHM is expressed by its average value D_{av}, the root mean square value D_{rms} or peak-to-peak value D_{pp}. The average and the root mean square amplitude are

$$D_{av} = \frac{1}{T^*} \int_0^{T^*} |z(t)|\, dt$$

$$D_{rms} = \left[\frac{1}{T^*} \int_0^{T^*} \{z(t)\}^2\, dt\right]^{1/2}$$

where T^* = average time.

For a sinusoidal motion,

$$\text{Average amplitude} = \frac{2}{\pi D}$$

$$\text{rms amplitude} = \frac{1}{\sqrt{2}} D$$

$$\text{Peak-to-peak amplitude} = 2D$$

Representation of free vibration in complex plane

Simple harmonic motion may be presented by a rotating phase in a complex plane (Figure 3.14). The solution of the Eq. (3.40) may be written as

$$z = A \cdot e^{i(\omega_n t - \phi)}$$

$$= A \cos(\omega_n t - \phi) + iA \sin(\omega_n t - \phi)$$

$$\dot{z} = \frac{dz}{dt} = i\omega_n \, A \cdot e^{i(\omega_n t - \phi)}$$

$$\ddot{z} = \frac{d^2 z}{dt^2} = -\omega_n^2 \, A \cdot e^{i(\omega_n t - \phi)}$$

It is readily observed from Figure 3.14 that velocity and acceleration phases lead the displacement phase by 90° and 180°, respectively. This has also been shown in Figure 3.11.

From Figure 3.15 the amplitude A is the maximum displacement from equilibrium. The amplitude is a function of system parameters and the initial conditions. The amplitude is a measure of the energy imparted to the system through the initial conditions. For the linear system

$$D = \sqrt{\frac{2E}{K}}$$

where E = sum of kinetic and potential energies.

The phase angle ϕ as in Figure 3.15 represents the lead or lag between the response and a pure sinusoidal response. The response is purely sinusoidal with $\phi = 0$ and if $z = 0$. The response leads a pure sinusoidal response by $\pi/2$ radian if $\dot{z} = 0$. The system takes a time of

$$t = \begin{cases} \dfrac{2\pi - \phi}{\omega_n} & \phi > 0 \\ -\phi/\omega_n & \phi \le 0 \end{cases}$$

to reach its equilibrium position from its initial position.

Phase plane representation

The phase plane representation method is a graphical method to solve the vibration problem. The motion of the single degree freedom system is decided completely by the displacement and

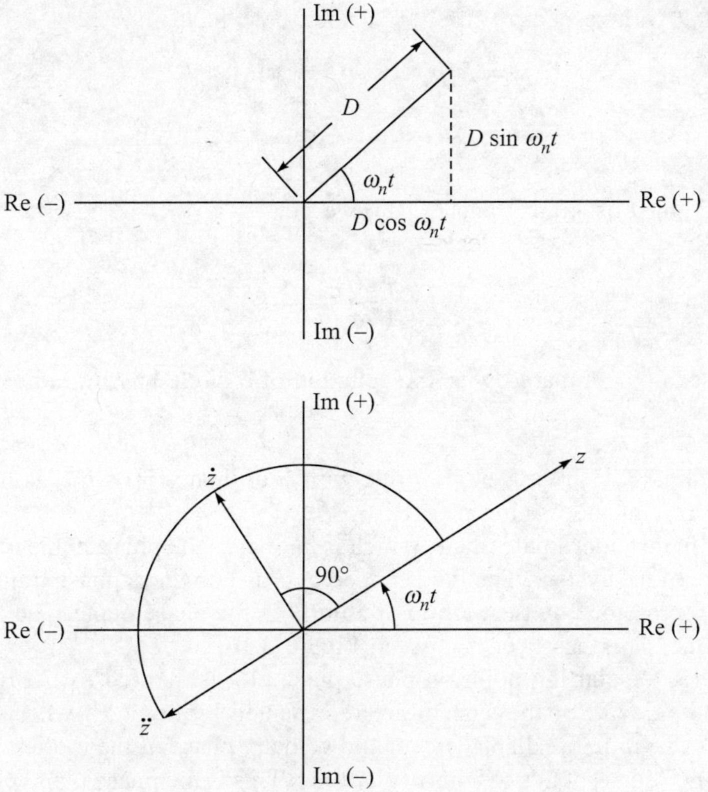

Figure 3.14 Representation of free vibration in complex plane.

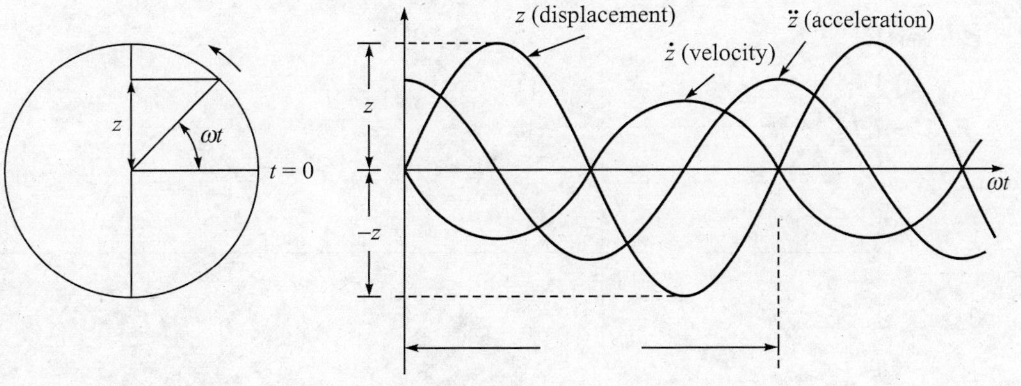

Figure 3.15 Representation of phases of z, $\dot{z}$, $\ddot{z}$

velocity. If the displacement and velocity are taken as coordinate axes, the resulting graphical representation is known as phase plane representation.

Considering the SHM represented by Eq. (3.40) as

$$z(t) = D \sin(\omega_n t + \phi) \tag{3.41}$$

and differentiating with respect to time for velocity, we have

$$\dot{z}(t) = D\omega_n \cos(\omega_n t + \phi)$$

or

$$\frac{\dot{z}(t)}{\omega_n} = D \cos(\omega_n t + \phi) \tag{3.42}$$

Squaring and adding Eqs. (3.41) and (3.42), we get

$$z^2(t) + \left(\frac{\dot{z}(t)}{\omega_n}\right)^2 = D^2 \tag{3.43}$$

Equation (3.43) can be compared with the equation of a circle having radius a in x–y plane as

$$x^2 + y^2 = a^2$$

Graphically, Eq. (3.43) represents a circle with coordinates $z(t)$ and regular fraction having radius D with centre at the origin.

Any point P in this coordinate plane, which is known as the phase plane of motion, indicates the dynamic state of the system. The locus traced by P is known as phase trajectory. The motion of the system is represented by the motion of point P in the phase plane. The state of the system depends upon time. This has been shown in Figure 3.16.

In Figure 3.15, the starting point on phase plane plot is marked P_1. At t_1 seconds later the displacement and velocity of the system are represented by point P_2 where $\angle P_1\ 0\ P_2 = \omega_n t_1$ radian. From this diagram, the displacement and velocity phase of the motion are available from a single point which corresponds to a particular time. This is the phase plane plot. The horizontal projections of the phase trajectory on a time base shows the displacement-time plot of the motion (Figure 3.15a).

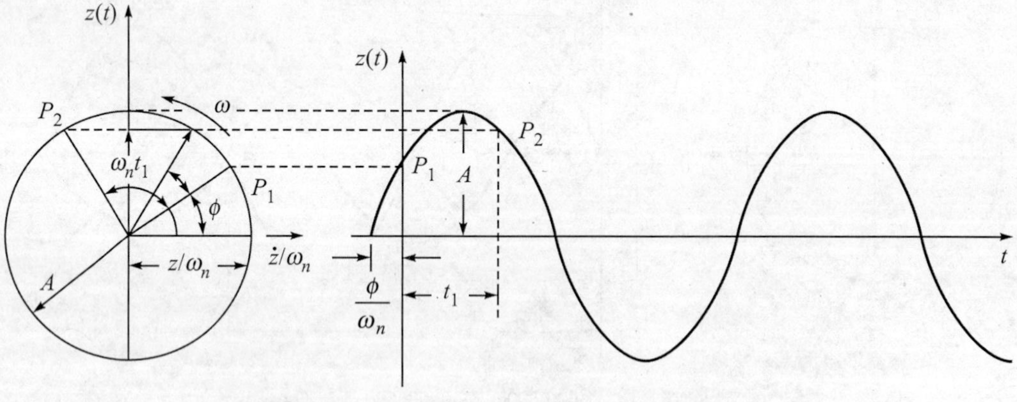

Figure 3.15a Phase plane representation for SDF system.

3.5 FREE VIBRATIONS SDF DAMPED SYSTEM

In general, physical systems are associated with some type of damping. When damped free vibrations take place, the amplitude of vibrations gradually becomes smaller and smaller and finally is completely lost. The energy loss of dissipation of energy mechanism controls the rate at which the amplitude decays. The dampings in a physical system are of several types. The most common type of damping which is known as viscous damping is described herein.

Viscous damping

This is the most important type of damping. The amount of resistance force due to damping will depend upon the relative velocity. For a particular system the damping resistance is always proportional to the relative velocity. One of the reasons for the importance of this type of damping is that the governing differential equation is linear. As such many system are often represented to include an equivalent damping even though the damping may not be truly viscous.

This type of damping occurs when the system vibrates in a viscous medium. A simple viscous damper may be represented by Newtonian dashpot or by piston moving in a cylinder filled with a viscous medium (Figure 3.16). If the instantaneous velocity equals $\dot{z}$,

Resisting force by virtue of damping, $F = c\dot{z}$

where c = damping coefficient.

And energy dissipated by the dashpot will be

$$E_d = c\,\dot{z}\,dz \tag{3.44}$$

It can be observed that $\dot{z} = \dot{z}_{max}$ at $z = 0$ and vice versa. The hysteresis loop for viscous damping is an ellipse. Thus, the integration of Eq. (3.44) gives

$$E_d = \pi\, c\, \omega\, z^2$$

The energy dissipated is a function of the amplitude and frequency.

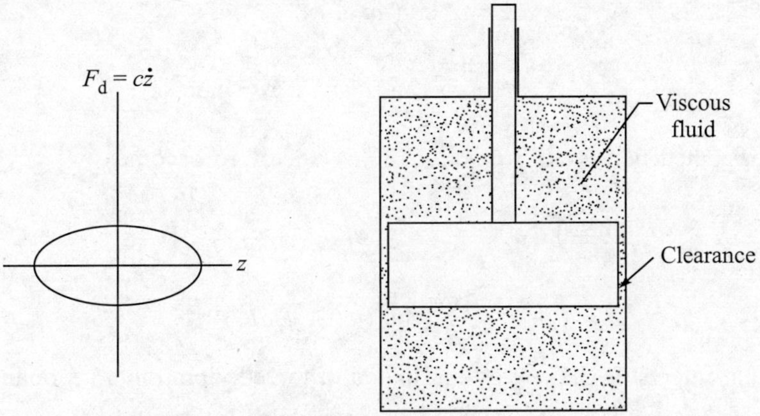

Figure 3.16 A simple viscous dashpot and its hysteresis loop.

Energy dissipated by damping

Damping is present in all oscillatory systems. Its effect is to remove energy from the system. Energy in a vibrating system is either dissipated into heat or radiated away. Simply bending a piece of metal back and forth a number of times can cause dissipation of energy into heat. We are all aware of the sound which is radiated from an object, given a sharp blow. When a buoy is made to bob up and down in the water, waves radiate out and away from it, thereby resulting in its loss of energy.

In vibration analysis, we are generally concerned with damping in terms of system response. The loss of energy from the oscillatory system results in the decay of amplitude of free vibration. In steady state forced vibration, the loss of energy is balanced by the energy which is supplied by the excitation.

A vibrating system may encounter many different types of damping forces, from internal molecular friction to sliding friction and fluid resistance. Generally, their mathematical description is quite complicated and not suitable for vibration analysis. Thus, simplified damping models have been developed which in many cases are found to be adequate in evaluating the system response. For example, we have already used the viscous damping model, designated by the dashpot, which leads to manageable mathematical solutions.

Energy dissipation is usually determined under conditions of cyclic oscillations. Depending on the type of damping present, the force–displacement relationships when plotted may differ greatly. In all cases, however, the force–displacement curve will enclose an area, referred to as the hysteresis loop, that is, proportional to the energy lost per cycle. The energy lost per cycle due to a damping force F_d is computed from the general equation

$$W_d = \int F_d \cdot dz \tag{3.45}$$

where W_d = energy lost per cycle.

We consider in this section the simplest case of energy dissipation, that of a spring–mass system with viscous damping. The damping force in this case is $F_d = c\,\dot{z}$. With the steady state displacement and velocity

$$z = Z \sin(\omega t - \phi)$$

$$\dot{z} = \omega Z \cos(\omega t - \phi), \text{ where } \dot{z} = \frac{dz}{dt}$$

The energy dissipated per cycle, from Eq. (3.45), therefore becomes

$$W_d = \int c\,\dot{z}\,dz = \int c\,\dot{z}^2\,dt \quad \text{as} \quad c\dot{z}dz = c\,\dot{z}\frac{dz}{dt}\cdot dt = c(\dot{z})^2 dt$$

$$= c\omega^2 Z^2 \int_0^{2\pi/\omega} \cos^2(\omega t - \phi)dt = \pi c \omega Z^2 \tag{3.46}$$

Of particular interest is the energy dissipated in forced vibration at resonance. Substituting $\omega_n = \sqrt{k/m}$ and $c = 2\xi\sqrt{km}$, the preceding equation at resonance becomes

$$W_d = 2\xi\pi k Z^2 \tag{3.47}$$

The energy dissipated in forced vibration can be represented graphically as in Figure 3.17. Writing the velocity in the form

$$\dot{z} = \omega Z \cos(\omega t - \phi) = \pm\, \omega Z\sqrt{1 - \sin^2(\omega T - \phi)}$$

$$= \pm\, \omega\sqrt{Z^2 - z^2}$$

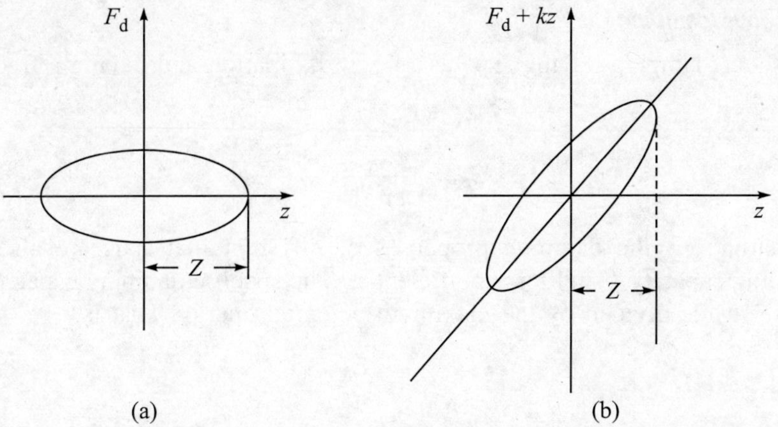

Figure 3.17 Energy dissipated by viscous damping.

The damping force becomes

$$F_d = c\dot{z} = \pm c\,\omega\sqrt{Z^2 - z^2} \tag{3.48}$$

Rearranging the above equation to

$$\left(\frac{F_d}{c\omega Z}\right)^2 + (z/Z)^2 = 1 \tag{3.49}$$

We recognize Eq. (3.49) as that of an ellipse with F_d and z plotted along the vertical and horizontal axes, as shown in Figure 3.17(a). The energy dissipated per cycle is then given by the area enclosed by the ellipse. If we add to F_d the force kz of the massless spring, the hysteresis loop is rotated as shown in Fig. 3.17(b). This representation then conforms to the Voigt model, which consists of a dashpot in parallel with a spring.

The damping properties of materials are listed in many different ways depending on the technical areas to which they are applied. Of these we list two relative energy units that have wide usage. The first of these is specific damping capacity, defined as the energy loss per cycle W_d divided by the peak potential energy U. That is,

$$\text{Specific damping capacity} = \frac{W_d}{U} \tag{3.50}$$

The second quantity is the loss coefficient, defined as the ratio of damping energy loss per radian, $W_d/2\pi$ divided by the peak potential or strain energy U. That is,

$$\text{Loss coefficient, } \eta = \frac{W_d}{2\pi U} \tag{3.51}$$

For the case of linear damping where the energy loss is proportional to the square of the strain or amplitude, the hysteresis curve is an ellipse. When the damping loss is not a quadratic function of the strain or amplitude, the hysteresis curve is no longer an ellipse.

Equivalent viscous damping

For a non-viscous damping having E_{dnv} as energy dissipation and having a frequency p, then

$$E_{dnv} = C_{eq}\,\pi\, p\, z^2$$

or,
$$C_{eq} = \frac{E_{dnv}}{\pi p z^2} \tag{3.52}$$

In such situations the damping properties of different materials are also expressed as specific damping capacity β and loss coefficient η. The specific damping capacity is defined as energy loss per cycle divided by the maximum potential energy. That is,

$$\beta = \frac{E_d}{V}$$

For viscous damping the maximum potential energy for the SDF system is in terms of strain energy and so

$$V = (1/2)(kz)(z) = (1/2)kz^2$$

then,
$$\beta = \frac{\pi c \omega z^2}{(1/2)kz^2}$$

$$= \frac{2\pi c\omega}{k}$$

The energy loss coefficient is defined as the ratio of energy loss per radian to the maximum potential energy. That is,

$$\eta = \frac{E_d}{2\pi V}$$

For a viscous system, $\eta = \dfrac{\omega}{k}$.

3.5.1 Free Vibrations of Viscously Damped System

The damped single degree freedom system as shown in Figure (3.1a) is governed by the differential equation

$$m\frac{dz^2}{dt^2} + c\frac{dz}{dt} + kz = 0 \tag{3.53}$$

or,
$$m\ddot{z} + c\dot{z} + kz = 0$$

The solution of the above equation may be taken as

$$z = e^{st}$$

Substituting in the governing differential equation,

$$ms^2e^{st} + cse^{st} + ke^{st} = 0$$

After cancellation of the common factors the above equation reduces to an equation called the characteristic equation of the system, namely

$$ms^2 + cs + k = 0$$

The above quadratic equation gives two roots for s. They are:

$$s_{1,2} = -\frac{c}{2m} \pm \sqrt{\left(\frac{c}{2m}\right)^2 - k/m}$$

The general solution is, therefore, given by

$$z = Ae^{ist} + Be^{-ist} \tag{3.54}$$

where A and B are constants to be evaluated from the initial conditions. It may be recalled that expressing the solution in the form of Eq. (3.54) is possible only where the differential equation is linear and so the principle of superposition holds good. Such superposition does not hold good in the case of non-linear differential equations.

Equation (3.54) substituted into Eq. (3.53) gives

$$z = e^{-(c/2m)t}\left(A \cdot e^{i\sqrt{(c/2m) - (k/m)t}} + B \cdot e^{-i\sqrt{(c/2m) - (k/m)t}}\right) \tag{3.55}$$

The mathematical form of the solution of Eq. (3.55) and the physical behaviour of the system depends upon the sign of the discriminant (quantity under the square root), i.e., whether the numerical value within the radical is positive, zero or negative. If the discriminant is positive, Eq. (3.30) has two real roots. If the discriminant is negative, there are two complex conjugate roots. If the discriminant is zero, then there are two equal real roots.

The physical nature of vibration is dependent upon the sign of the discriminant. In case of positive value of the quantity under the root, the system is overdamped, and if negative the system would be underdamped. As a special case when the discriminant is zero, the system will be critically damped. From Eq. (3.55), in that case

$$c = c_c = 2\sqrt{km} = 2m\,\omega_n \tag{3.56}$$

where c_c is the critical damping coefficient. The non-dimensional damping ratio ζ is defined as the ratio of the actual value of c to the critical damping coefficient. That is,

$$\zeta = \frac{c}{c_c} = \frac{c}{2\sqrt{k/m}} \tag{3.57}$$

This damping ratio is a property of the system parameters. Using Eqs. (3.24) and (3.25), the roots may be written as

$$s_{1,2} = -\zeta\omega_n \pm \omega_n\sqrt{\zeta^2 - 1}$$

The three cases of damping discussed here will depend upon whether ζ is greater than unity ($\zeta > 1$), less than unity ($\zeta < 1$) or equal to unity ($\zeta = 1$). Further more, the differential Eq. (3.53) may be expressed in terms of ζ and ω_n as

$$z + 2\zeta\omega_n \dot{z} + \omega_n^2 \ddot{z} = 0 \tag{3.58}$$

Figure 3.18 shows Eq. (3.58) plotted in a complex plane with ζ along the horizontal axis (real axis). If $\zeta = 0$, $s_{1,2} = \pm i$, so the roots on the imaginary axis correspond to the undamped case for ($0 \le \zeta \le 1$), i.e.,

$$s_{1,\,2} = \omega_n(-\zeta + i\sqrt{1-\zeta^2})$$

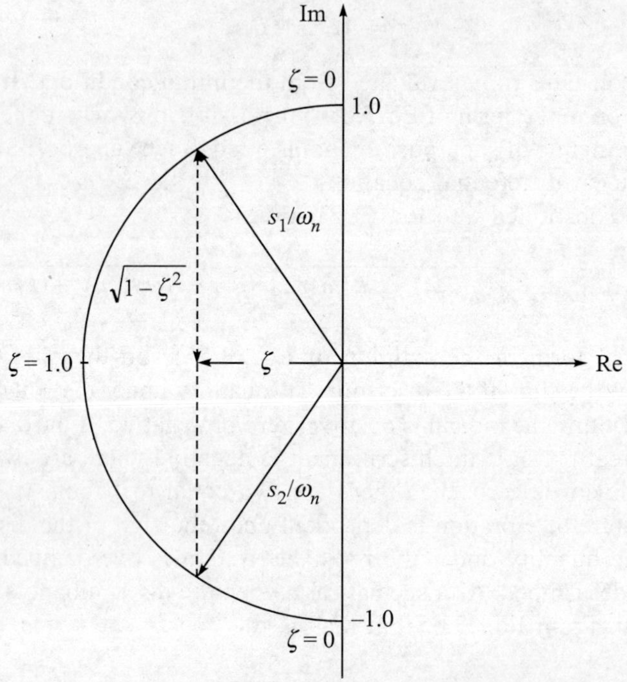

Figure 3.18 Representation of damping.

The roots s_1 and s_2 are then conjugate complex points on a circular arc converging at the point $s_{1,2}/\omega_n = \pm\, 1.0$. As ζ increases beyond unity, the roots separate along the horizontal axis and remain real numbers.

Critically damped motion (ζ = 1.0)

The double roots $s_1 = s_2 = -\,\omega_n$ and thus the terms combine to form a single term as the general solution is

$$z = (A + Bt)e^{-\omega_n t} \tag{3.59}$$

with initial condition at $t = 0$, $z = z(0)$ and $dz/dt = \dot{z}(0)$. Therefore,

$$z = e^{-\omega_n t}\,[z(0) + (\dot{z}(0) + \omega_n z(0)t]$$

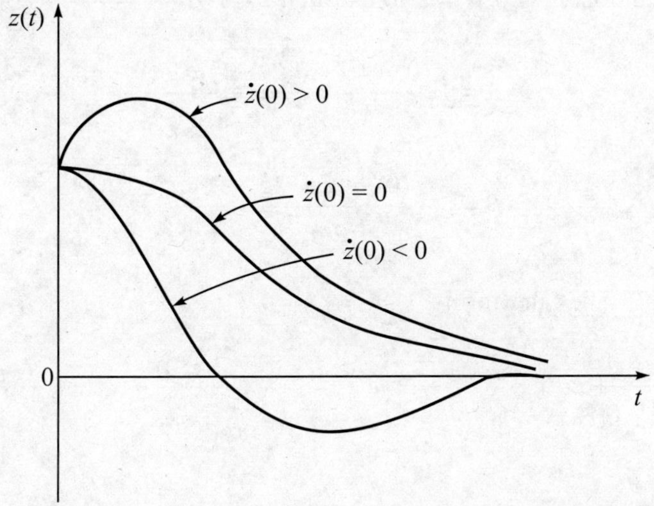

Figure 3.19 Critically damped motion ($\zeta = 1.0$)

The response of single degree of freedom system subjected to critical viscous damping is shown for different initial conditions in Figure 3.19. If the initial condition, is such as $\dot{z}(0) = 0$, the motion decays immediately. If both the initial conditions have the same sign or if $z(0) = 0$, the absolute value of z initially increases and reaches a maximum value of

$$z_{\max} = \exp\left(\frac{\dot{z}(0)}{\dot{z}(0) + \omega_0 z(0)}\right)\left(\dot{z}(0) + \frac{\dot{z}(0)}{\omega_n}\right)$$

at $$t = \dot{z}_0/\omega_n(\dot{z}(0) + \omega_n z(0))$$

If the signs of the initial conditions are opposite, i.e., $z(0)$ being positive and $\dot{z}(0)$ being < 0, the response overshoots the equilibrium position before eventually decaying. Thus, the damping force that leads to critical damping is sufficient to dissipate all of the system's initial energy before one cycle of motion is complete. A critically damped system can pass through equilibrium at most once before the motion decays. However, the total energy decays exponentially but never reaches zero. One useful definition of the critically damped condition is that it is the smallest amount of damping for which no oscillation occurs in the free vibrations. Thus, critically damped motion is an aperiodic motion.

Overdamped system **($\zeta > 1.0$)**

As ζ exceeds unity, the two roots remain on the real axis and separate, in which one is increasing with time and the other decreasing with time. The general solution for $\zeta > 1$ may be written as

$$z(t) = A\ \exp\left[\left(-\zeta + \sqrt{\zeta^2 - 1}\right)\omega_n t\right] + B\ \exp\left[\left(-\zeta - \sqrt{\zeta^2 - 1}\right)\omega_n t\right] \tag{3.60}$$

With the initial conditions, $t = 0$, $z = z(0)$ and $dz/dt = \dot{z}(0)$

$$A = \frac{\dot{z}(0) + (\zeta + \sqrt{\zeta^2 - 1})\omega_n z(0)}{2\omega_n\sqrt{\zeta^2 - 1}} \tag{3.61}$$

$$B = \frac{-\dot{z}(0) - (\zeta + \sqrt{\zeta^2 - 1})\omega_n z(0)}{2\omega_n\sqrt{\zeta^2 - 1}} \tag{3.62}$$

Thus, Eq. (3.60) takes the form

$$z(t) = \frac{e^{-\zeta\omega_n t}}{2\sqrt{\zeta^2 - 1}}\left[\frac{z(0)}{\omega_n} + x_0(\zeta + \sqrt{\zeta^2 - 1})\right]e^{\omega_n\sqrt{(\zeta 2 - 1)}t} + \left[-\frac{\dot{z}(0)}{\omega_n} + x_0(-\zeta + \sqrt{\zeta - 1})\right]e^{-\omega_n\sqrt{(\zeta 2 - 1)}t} \tag{3.63}$$

The resulting motion is an exponentially decreasing function of time as shown in Figure 3.20.

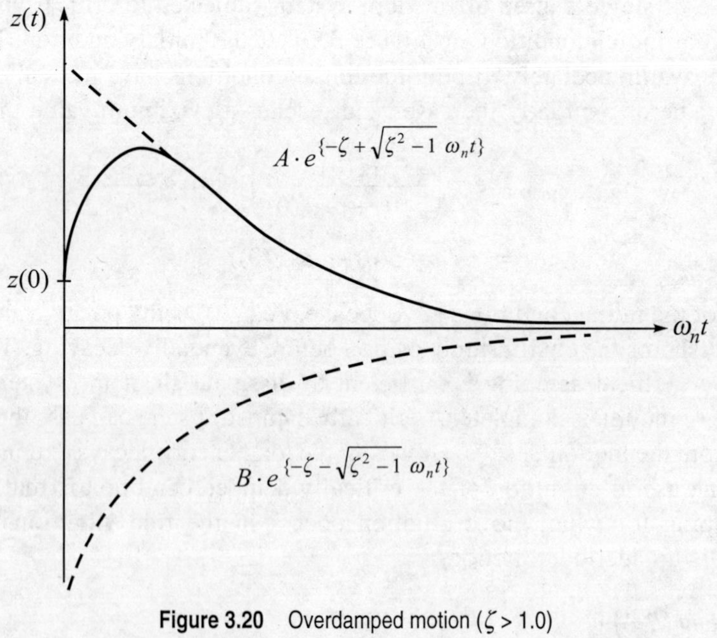

Figure 3.20 Overdamped motion ($\zeta > 1.0$)

The response of an overdamped system is also not periodic. It attains its maximum either at $t = 0$ or else expressed by $z(t)$ as

$$z(t) = -\frac{1}{2\omega_n\sqrt{\zeta^2-1}} \cdot \ln\left[\frac{\zeta-\sqrt{\zeta^2-1}}{\zeta+\sqrt{\zeta^2-1}} \cdot \frac{\dfrac{\omega(0)}{\omega_n}+z(0)(\zeta+\sqrt{\zeta^2-1})}{\dfrac{\omega(0)}{\omega_n}+z(0)(\zeta-\sqrt{\zeta^2-1})}\right] \tag{3.64}$$

For example, for ω_n = 30 rad/s and ζ = 2.0 with initial conditions $t = 0$, $z = z(0)$ and $dz/dt = 0$, then the motion is given by

$$z = z(0)[1.4045\ e^{-16.1t} - 0.4045\ e^{-55.9t}]$$

Underdamped system (ζ < 1.0)

If ζ is less than unity, the two roots are complex conjugate pair, i.e.

$$s_{1,2} = \omega_n(-\zeta \pm i\sqrt{1-\zeta^2})$$

Hence,

$$z(t) = Ae^{(-\zeta+\sqrt{1-\zeta^2})\omega_n t} + Be^{(-\zeta+\sqrt{1-\zeta^2})\omega_n t}$$

$$= e^{-\zeta\omega_n t}[Ae^{i\sqrt{1-\zeta^2}\omega_n t} + Be^{-i\sqrt{1-\zeta^2}\omega_n t}] \tag{3.65}$$

Using Euler identity to replace complex exponential as

$$e^{\pm ix} = \cos x \pm i \sin x$$

so,

$$z(t) = e^{-\zeta\omega_n t}[a \cos\sqrt{1-\zeta^2}\,\omega_n t + b \sin\sqrt{1-\zeta^2}\,\omega_n t]$$

$$= De^{-\zeta\omega_n t} \cos(\sqrt{1-\zeta^2}\,\omega_n t - \phi)$$

Putting the initial condition to evaluate the two constants a and b or D and ϕ for $t = 0$, $z = z(0)$ and $\dot{z} = v(0)$

$$z = e^{-\zeta\omega_n t}\left[z_{(0)}^2 + \frac{(v(0)+\zeta\omega_n z(0))}{\omega_n\sqrt{1-\zeta^2}}\right]^2 \cos\{\sqrt{1-\zeta^2}\,\omega_n t - \phi\}$$

where, $\phi = \tan^{-1}\left(\dfrac{v(0)+\zeta\omega_n z(0)}{\omega_n\sqrt{1-\zeta^2}}\right)$. The undamped motion is shown in Figure. 3.21(a).

The natural frequency of the damped system from the above equation is given by

$$\omega_d = \sqrt{1-\zeta^2}\,\omega_n \text{ rad/s}$$

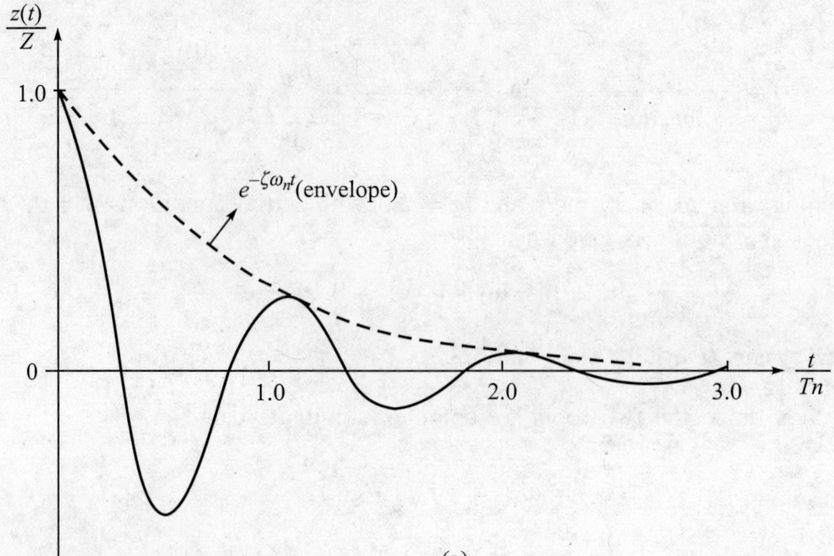

(a)

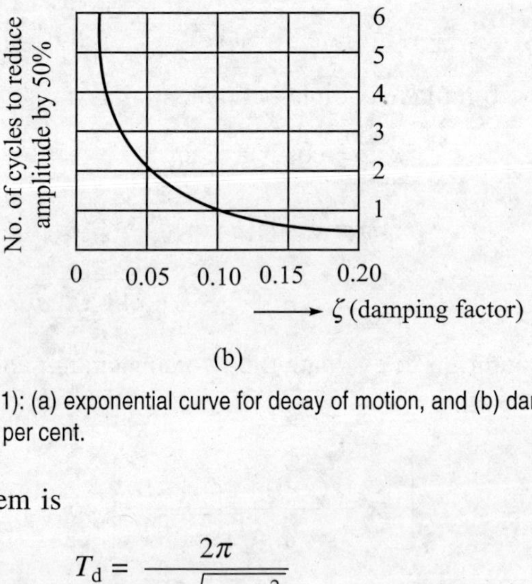

(b)

Figure 3.21 Underdamped motion ($\zeta < 1$): (a) exponential curve for decay of motion, and (b) damping ratio vs. number of cycles required to reduce amplitude by 50 per cent.

and time period of the system is

$$T_{\mathrm{d}} = \frac{2\pi}{\omega_n\sqrt{1-\zeta^2}}$$

Equation (3.65) may be written in a different form with another constant A and ϕ_d as

$$z(t) = Ae^{-\zeta\omega_n t}\,\sin(\omega_d t + \phi_d)$$

$$A = z(0)^2 + \left(\frac{v(0)+\zeta\omega_n z(0)}{\omega_d}\right)^2$$

$$\phi_d = \tan^{-1}\,\frac{\omega_d}{\dot{z}(0)+\zeta\omega_n z(0)}$$

$$\omega_d = \omega_n\sqrt{1-\zeta^2}$$

For example, taking $\zeta = 0.4$, $z(t) = Ae^{-\zeta\omega_n t}\sin\{\sqrt{(1-\zeta^2)}\,\omega_n t + \phi_d\}$

Thus, $$z(t) = Ae^{-\zeta\omega_n t}\sin(0.916\omega_n t + \phi_d)$$

With initial conditions at $t = 0$, $z(0) = 4$ and $\dot{z}(0) = 0$, we have

$$4 = A\sin\phi_d$$

and, $$0 = A\,[0.916\omega_n\cos\phi - 0.4\omega_n\sin\phi_d]$$

or, $$A = \frac{4}{\sin\phi_d} = 4.37$$

Therefore, $$z(t) = 4.37e^{-0.4\omega_n t}\cdot\sin[0.916\omega_n t + 66°\ 24']\tan\phi_d$$
$$= 2.29$$

and, $$\phi_d = 66°24'$$

Logarithmic decrement

Figure 3.22(a) shows a typical response curve of an underdamped SDF system. The envelope of the response is given by

$$z = z(0)\cdot e^{-\zeta\omega_n t}$$

The envelope of the response is shown in Figure 3.21(b).

From underdamped motion it is evident that there is reduction in amplitude with time. The relationship for decay can be conveniently presented graphically as shown in Figure 3.21(c).

As a ready reckoner, it is convenient to remember that for ten per cent of critical damping, the amptitude is reduced by 50 per cent in one cycle.

The logarithmic decrement δ is defined as the natural logarithm of the ratios of the amplitudes of vibration on successive cycles, that is

$$\delta = \log\frac{z(t)}{z(t+T_d)} \tag{3.66}$$

$$= \log\frac{e^{-\zeta\omega_n t}}{e^{-\zeta\omega_n(t+T)}}$$

$$= \zeta\omega_n T$$

$$= \frac{(\zeta\omega_n 2\pi)}{\omega_n\sqrt{1-\zeta^2}}$$

$$= \frac{2\pi\zeta}{\sqrt{1-\zeta^2}}$$

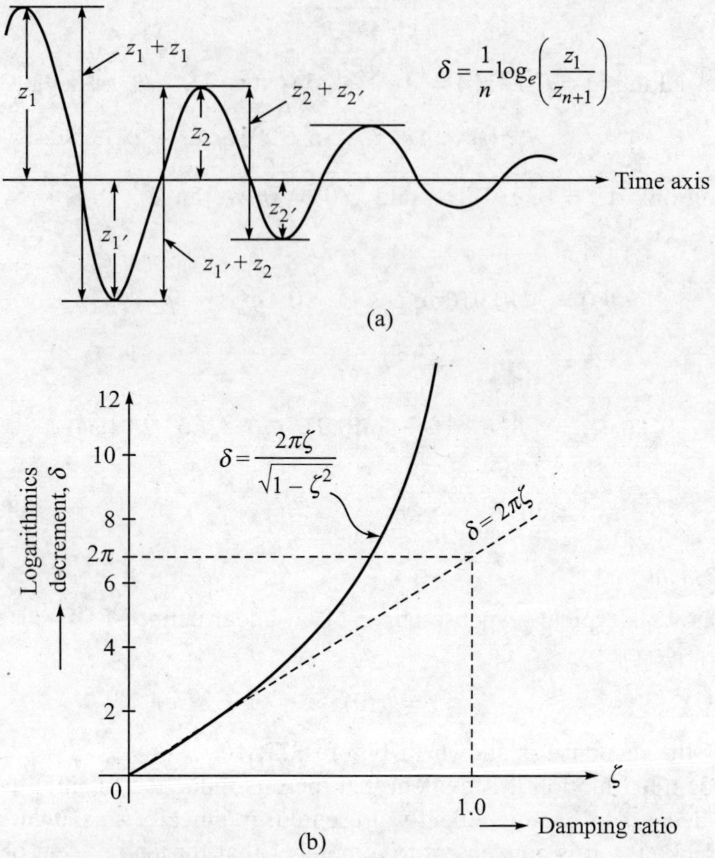

Figure 3.22 (a) Representation of underdamped motion (logarithmic decrement), and (b) variation of logarithmic decrement with damping ratio.

If the successive amplitudes are denoted by $z_1, z_2, \cdots, z_n$ then for a viscously damped system from Eq. (3.66)

$$\frac{z_1}{z_2} = \frac{z_2}{z_3} = \frac{z_3}{z_4} = \cdots = \frac{z_{n-1}}{z_n} = \delta$$

$$\frac{z_0}{z_n} = \frac{z_0}{z_1} = \frac{z_1}{z_2} = \frac{z_2}{z_3} = \cdots = \frac{z_{n-1}}{z_n} = (e\delta)^n$$

and hence,

$$\delta = \frac{1}{n}\log_e\left(\frac{z_1}{z_{n+1}}\right) \tag{3.67}$$

This method is very powerful for evaluating the damping of a soil by a suitable free vibration test. The details of the test and determination of damping of the soil shall be discussed in Chapter 6.

Example 3.1: In a free vibration test, the amplitude vs. time trace of a system is shown in Figure 3.23 where in 4 cycles the amplitude decreases from 5 mm to 0.10 mm. Find the damping ratio of the system.

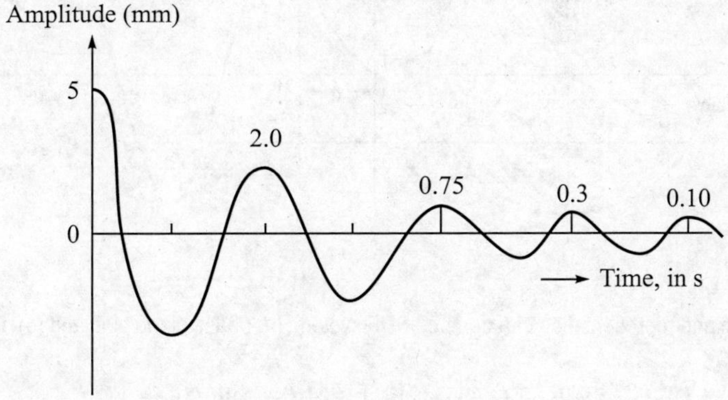

Figure 3.23 Example 3.1.

Solution: In 4 cycles the amplitude of 5 mm becomes 0.10 mm.

The logarithmic decrement $\delta = \frac{1}{4} \cdot \log_e \frac{5}{0.10} = 0.958$

Further $\delta = \frac{2\pi\zeta}{\sqrt{1-\zeta^2}} = 0.958$

Therefore, $\zeta = 0.15$

Figure 3.22(b) shows the variation of δ with damping ratio ζ.

Phase plane method

The response of a SDF system with viscous damping is given by

$$z = z\, e^{-\zeta\omega_n t} \cos\left(\sqrt{1-\zeta^2}\, \omega_n t - \phi\right)$$

or,

$$\frac{\dot{z}}{\omega_n} = -z e^{-\zeta\omega_n t} \sin\left(\sqrt{1-\zeta^2}\, \omega_n t - \phi + \sigma\right)$$

where $\sigma = \tan^{-1} \frac{\zeta}{\sqrt{1-\zeta^2}}$

If the squares of z and $\dot{z}/\omega_n$ quantities as expressed alone and added, it is not possible to get a simple circular trajectory as in the case of the undamped system [see Eq. (3.43)]. However, it is more convenient to represent on an oblique coordinate system with z on the vertical axis and $\dot{z}/\omega_n$ on an oblique coordinate making an angle ϕ with the horizontal axis as shown in Figure 3.24.

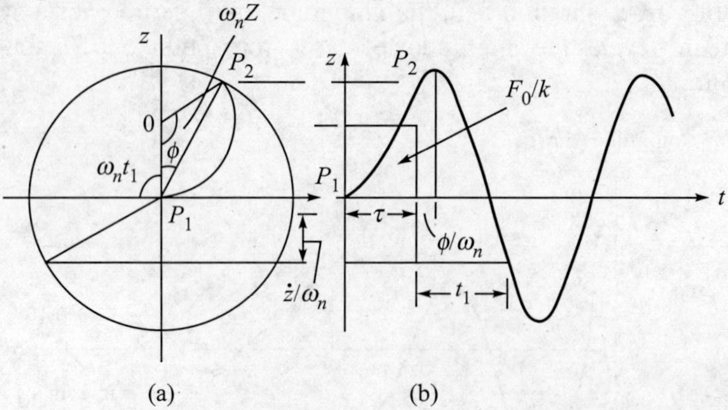

Figure 3.24 Phase plane representation of a damped SDF system: (a) phase plane plot, and (b) displacement–time trace.

Thus,
$$r^2 = [(\dot{z}/\omega_n \cos\ \phi]^2 + [z + (\dot{z}/\omega_n) \sin\ \phi]^2$$
$$= (\dot{z}/\omega_n)^2 \cos^2\ \phi + z^2 + 2z(\dot{z}/\omega_n) \sin\ \phi + (\dot{z}/\omega_n)^2 \sin^2\ \phi$$
$$= z^2 + (\dot{z}/\omega_n)^2 (\cos^2\ \phi + \sin^2\ \phi) + 2z(\dot{z}/\omega_n) \sin\ \phi$$
$$= z^2 + (\dot{z}/\omega_n)^2 + 2z(\dot{z}/\omega_n) \sin\ \phi$$

Thus, the trajectory of the radius vector may be written as

$$r = z \cos\ \phi\ e^{-\zeta\omega_n t} \tag{3.68}$$
$$= z_0\ e^{-\zeta\omega_n t} \tag{3.69}$$

The trajectory is, therefore, given by a simple exponential spiral in polar coordinates. Since the angular velocity of the trajectory is $\sqrt{1-\zeta^2}\ \omega_n$, the time, t for an angle θ on the oblique coordinate phase plane is

$$t = \frac{\theta}{\omega_d\sqrt{1-\zeta^2}}$$

Hence, Eq. (3.69) reduces to

$$r = X \cdot e^{-(\tan\phi)\theta}$$

The phase plane of a damped system is shown in Figure 3.24(c) which is extension of Figure 3.15 (undamped case).

For construction of the spiral geometry, selecting any convenient length $r_0 = 2$ units, taking $\zeta = 0.2$ and $\theta = \pi/2$, we have

$$r_1 = 2 \exp\left[-(\pi/2)\cdot 0.1/\sqrt{(1-0.2)^2}\right] = 1.70 \text{ units}$$

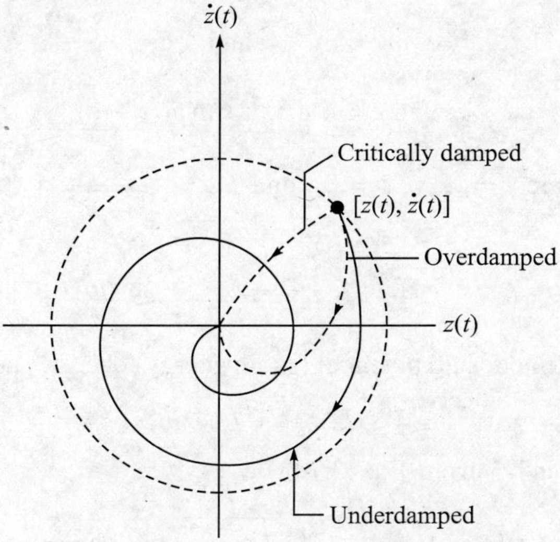

Figure 3.24(c) Phase plane of a damped system.

With $r_0 = 2$, $r_1 = 1.7$ and $\theta = \pi/2$, the phase plane by approximate spiral has been shown in Figure 3.25.

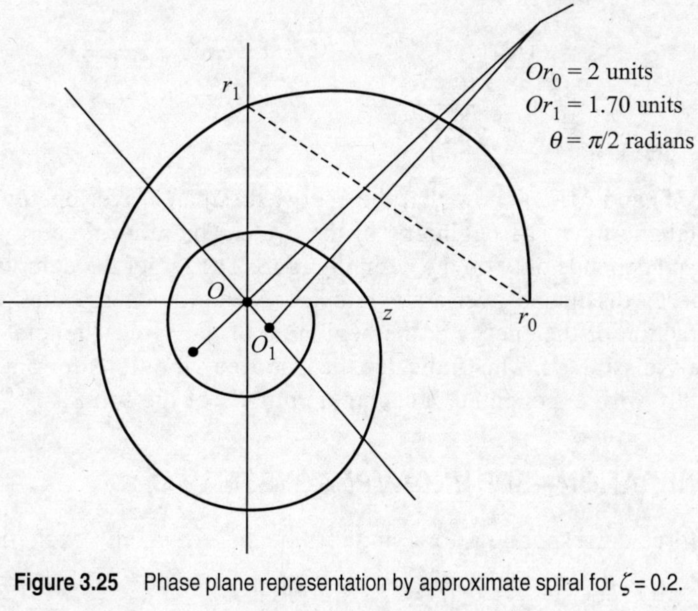

Figure 3.25 Phase plane representation by approximate spiral for $\zeta = 0.2$.

Energy dissipation during different cycles of viscous damping

For the underdamped system,

$$\omega_d = \sqrt{1 - \zeta^2}\, \omega_n \text{ rad/s}$$

Obviously, $\omega_d < \omega_n$ and $T_d = \dfrac{2\pi}{\omega_d}$

Let $z(t) = A\, e^{-\zeta\omega_n t} \sin(\omega_d t + \phi)$

If the initial conditions are $t = 0$, $\phi = 0$ and $A = \dfrac{z_0}{\sqrt{1-\zeta^2}}$

then,

$$z(t) = \frac{z_0}{\sqrt{1-\zeta^2}} e^{-\zeta\omega_n t} \sin \omega_d t \tag{3.70}$$

Let E be the sum of the kinetic and potential (strain energy) energies at any instant of time, then

$$E = (1/2)kz^2 + (1/2)m(\dot{z}^2)$$

Putting the values of z and $\dot{z}$ from Eq. (3.70),

$$E = (1/2)kz_0^2\, e^{-4\pi\zeta}\sqrt{1-\zeta^2} \text{ at } t = \frac{2\pi}{\omega_d}$$

But the energy dissipated over one cycle is given by

$$\Delta E_n = \frac{E 2\pi n}{\omega_d} - \frac{E 2(n+1)\pi}{\omega_d} \tag{3.71}$$

$$= (1/2)\, z^2\, e^{-4\pi n\zeta}\sqrt{1-\zeta 2}\, (1 - e^{-4\pi j}\sqrt{1-\zeta 2})$$

$$\frac{\Delta E_n}{\Delta E_{n+1}} = e^{-4\pi\zeta}\sqrt{1-\zeta 2} \tag{3.72}$$

Equations (3.71) and (3.72) show that the energy dissipated over one cycle of motion is a fraction of the total energy at the beginning of the cycle. The ratio of energy dissipated over a cycle is constant and depends only on the damping ratio. The larger the damping ratio, the larger the fraction of energy dissipated over a single cycle. Since the energy dissipated over a given cycle is a fixed fraction of the energy at the beginning of the cycle, the total remaining energy is never completely dissipated. This indicates that the free vibrations for an undamped system continue indefinitely with exponential decaying amplitude of the form $Ae^{-\zeta\omega_n t}$.

3.6 FORCED VIBRATION—SDF UNDAMPED SYSTEM

The generalized force corresponding to a single frequency harmonic excitation takes the form

$$F(t) = F_0 \sin(\omega t + \psi) \tag{3.73}$$

where,

F_0 = magnitude of excitation.

ω = forcing frequency.

ψ = phase differentiation between excitation and a pure sinusoidal excitation.

The differential equation for the undamped forced vibrations subjected to an excitation of the form as in is Eq. (3.73) is

$$\ddot{z} + \omega_n^2 z = \frac{F_0}{m}\sin(\omega t + \psi)$$

But firstly let us take the case when $\psi = 0$, i.e., for purely sinusoidal harmonic excitation, then

$$\ddot{z} + \omega_n^2 z = \frac{F_0}{m}\sin \omega t$$

The complementary function, i.e., solution for $\ddot{z} + \omega_n^2 z = 0$ may be taken as

$$z(t) = A \sin \omega_n t + B \cos \omega_n t$$

Particular integral

The general solution includes also the particular solution, i.e., the specific behaviour generated by the form of the dynamic loading. The response to the harmonic loading can be assumed to be harmonic and in phase with the loading, thus,

$$z(t) = D \sin \omega t$$

In which the amplitude D is to be evaluated. Substituting this in the governing differential equation, we have

$$-\omega^2 D \sin \omega t + \omega_n^2 D \sin \omega t = \frac{F_0}{m}\sin \omega t$$

Dividing throughout by sin ωt (which is non-zero in general),

$$(\omega_n^2 - \omega^2)D = \frac{F_0}{m}$$

Thus,

$$D = \frac{F_0}{k}\cdot\frac{1}{1-r^2}$$

where $r = \dfrac{\omega}{\omega_n}$ = frequency ratio.

General solution

The general solution to the harmonic excitation of the undamped SDF system is that given by the combination of the complementary solution and particular solution, therefore

$$z(t) = A \sin \omega_n t + B \cos \omega_n t + \frac{F_0}{k}\cdot\frac{1}{1-r^2}\sin \omega t \tag{3.74}$$

where the constants A and B are evaluated by placing the initial conditions as used. Taking the initial conditions as

$$t = 0,\ z(0) = 0 \quad \text{and} \quad \frac{dz}{dt} = \dot{z}(0) = 0$$

and solving, yields

$$A = -\frac{F_0}{k} \cdot \frac{1}{\sqrt{1-r^2}} \quad \text{and} \quad B = 0$$

Thus, the response of Eq. (3.74) becomes

$$z(t) = -\frac{F_0}{k} \cdot \frac{1}{1-r^2} (\sin \omega_n t - r \sin \omega t) \tag{3.75}$$

Magnification Factor,

$$M = \frac{|z_{\text{dynamic}}|}{|z_{\text{static}}|} = \frac{\frac{F_0}{k} . \frac{1}{1-r^2}}{\frac{F_0}{k}} = \frac{1}{1-r^2}$$

where $M = H(\omega)$ is called the magnification factor or dynamic factor or dynamic load factor (DLF). The variation of magnification factor with frequency ratio is shown in Figure 3.26.

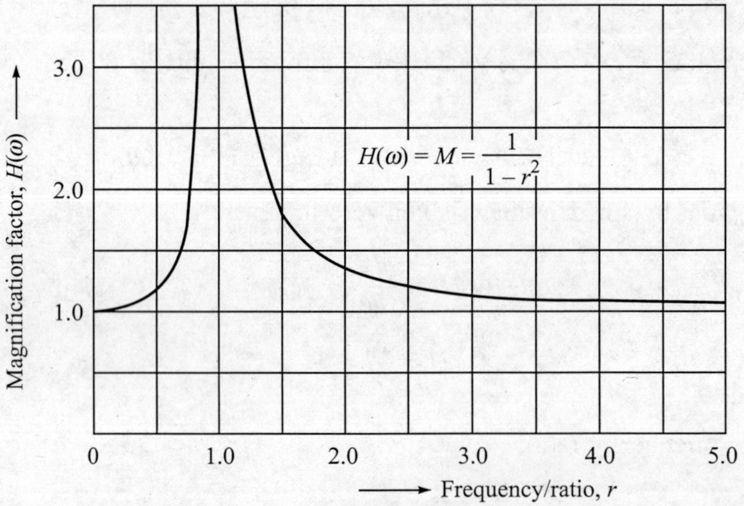

Figure 3.26 Variation of magnification factor with frequency ratio.

3.7 FORCED VIBRATION—SDF DAMPED SYSTEM

Consider a single degree freedom system (m, c, k) subjected to a harmonic dynamic external force $F \sin\omega t$, where ω is the forcing frequency. The equation of motion for the SDF system is

$$m\frac{d^2z}{dt^2} + c\frac{dz}{dt} + kz = F \sin \omega t \tag{3.76}$$

or, $$\ddot{z} + 2\zeta\omega_n \dot{z} + \omega_n^2 z = \frac{F}{m} \sin \omega t$$

The dynamic response will consist of transient response and steady state motion. The solution of the differential equation will be a particular integral in addition to the homogeneous solution. The transient vibration will eventually die out because of damping in the system. Since it is steady state vibration with the excitation frequency ω, it will persist as long as the force is acting. Naturally, the steady state responses are more important. For steady state vibration ignoring the transients, let the solution be

$$z = A \sin \omega t + B \cos \omega t \tag{3.77}$$

Substituting Eq. (3.77) in Eq. (3.76),

$$A = \frac{(k - m\omega^2) F}{(k - m\omega^2)^2 + (c\omega)^2}$$

$$B = \frac{-c\omega F}{(k - m\omega^2)^2 + (c\omega)^2}$$

Equation (3.77) may be written in another form

$$z = D \sin(\omega t - \phi)$$

where, $$D = \frac{F}{\sqrt{(k - m\omega^2)^2 + (c\omega)^2}} \quad \text{and} \quad \tan \phi = \frac{c\omega}{k - m\omega^2}$$

Thus, the dynamic response is given by

$$z(t) = \frac{F}{\sqrt{(k - m\omega^2)^2 + (c\omega)^2}} \sin(\omega t - \phi)$$

If F were a static load, then

$$z_{\text{static}} = \frac{F}{k}$$

and, with r as frequency ratio, $$r = \frac{\omega}{\omega_n} = \frac{\omega}{\sqrt{\frac{k}{m}}}$$

Then, $$z_{(t)\text{dynamic}}/z_{\text{static}} = M$$

$$= \frac{1}{\sqrt{(1 - r^2)^2 + (2\zeta r)^2}} = H(\omega) \tag{3.78}$$

where M or $H(\omega)$ is the magnification factor or dynamic factor or dynamic load factor (DLF). It is evident that the magnification factor is a function of frequency ratio for a given damped system. The response $z(t)$ lags behind the excitation force by an angle ϕ. This has been shown in Figure 3.27. Further, the phase relationships for different value of ζ have been shown separately in Figure 3.28.

Now for external excitation having phase difference ψ with the pure sinusoidal excitation, the governing differential equation may be expressed as

$$\ddot{z} + \omega_n^2 z = \frac{F_0}{m} \sin(\omega t + \psi) \tag{3.79}$$

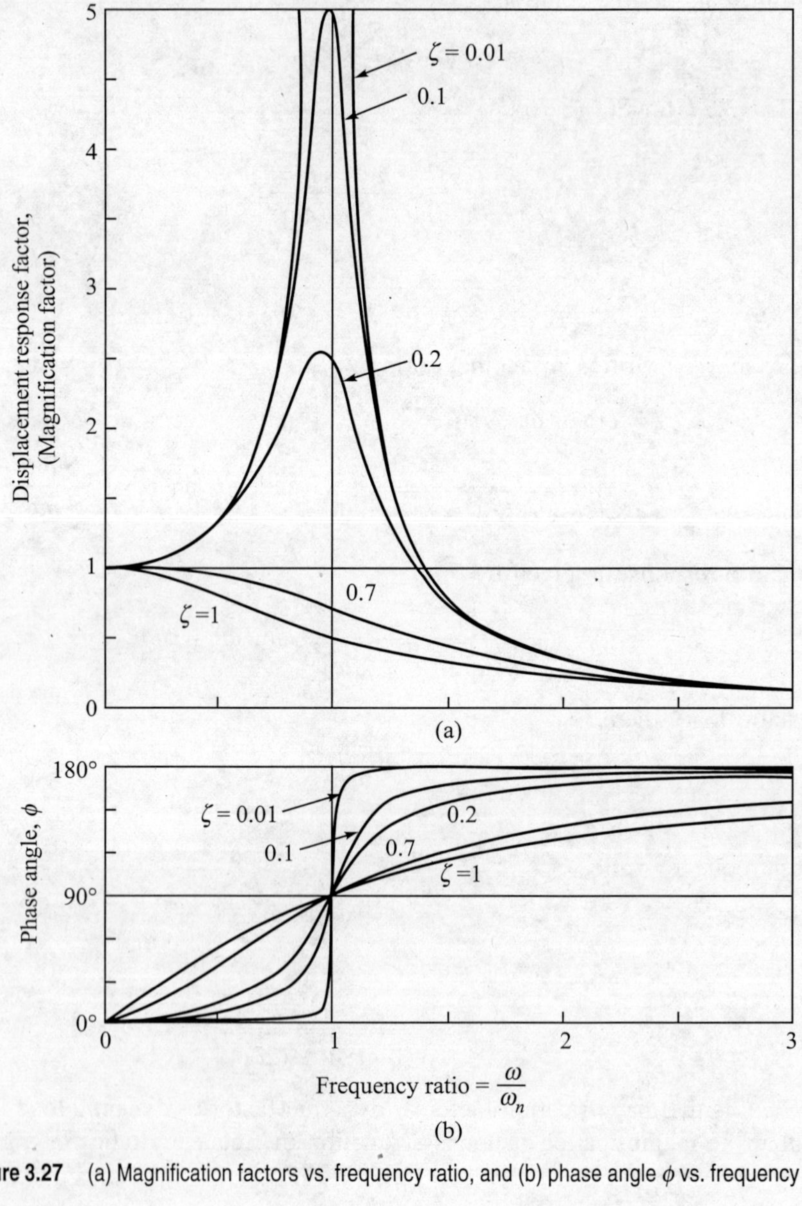

Figure 3.27 (a) Magnification factors vs. frequency ratio, and (b) phase angle ϕ vs. frequency ratio.

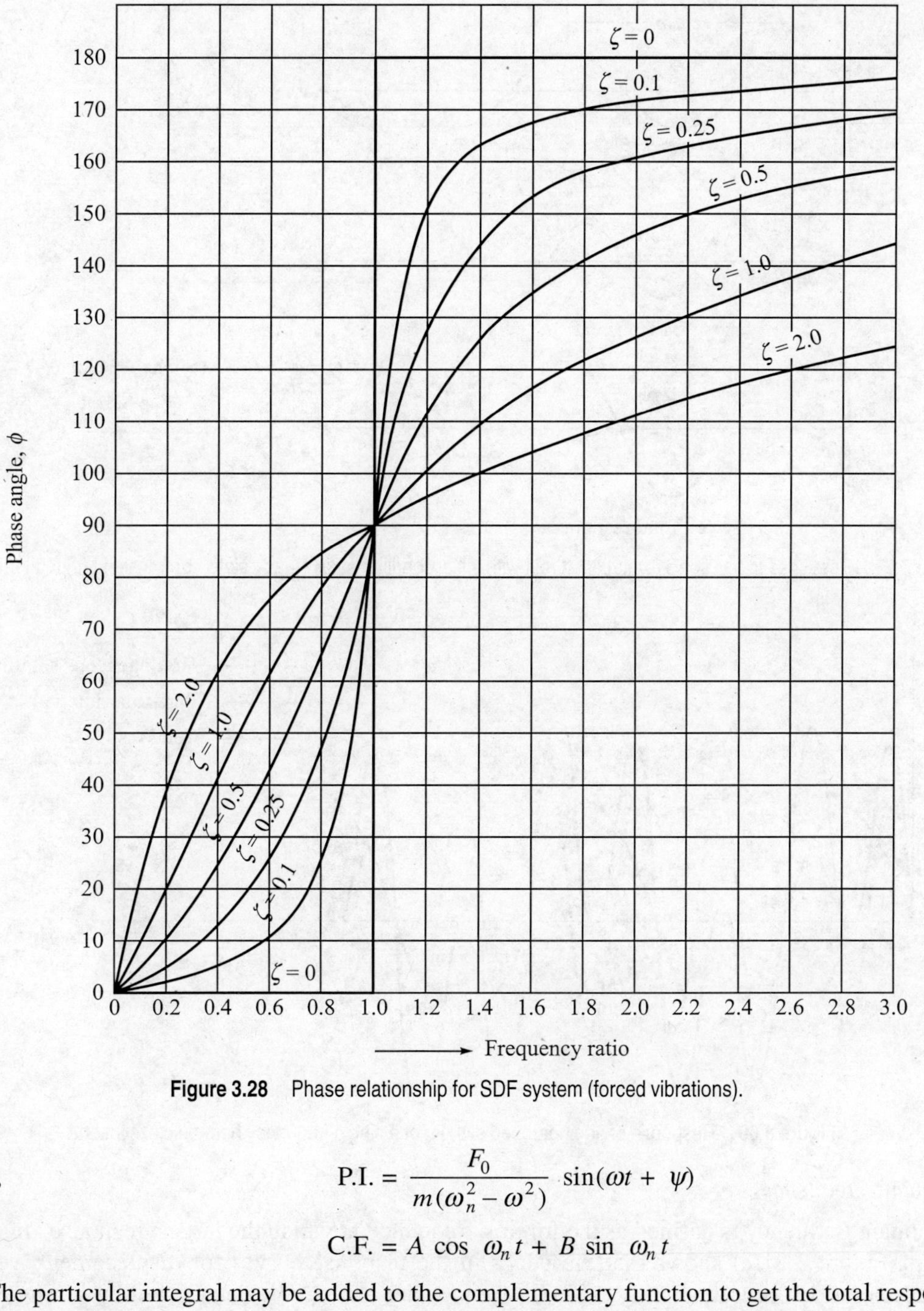

Figure 3.28 Phase relationship for SDF system (forced vibrations).

Thus,
$$\text{P.I.} = \frac{F_0}{m(\omega_n^2 - \omega^2)} \sin(\omega t + \psi)$$

and,
$$\text{C.F.} = A \cos \omega_n t + B \sin \omega_n t$$

The particular integral may be added to the complementary function to get the total response. The initial conditions will felicitate evaluation of the two constants. The response plotted in Figure 3.30 is the sum of two trigonometric terms of different frequencies.

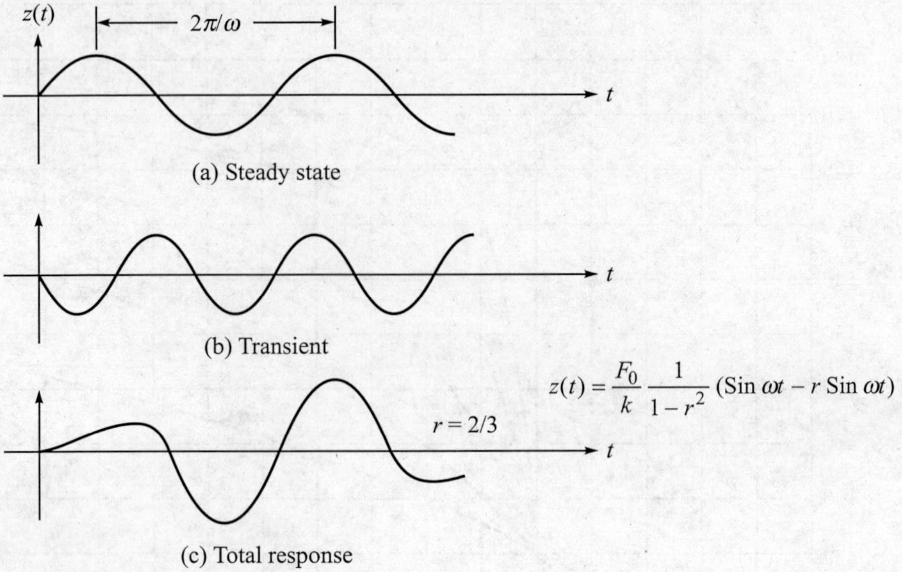

Figure 3.29 Response to harmonic excitation at rest with initial conditions: (a) steady state, (b) transient, (c) total response.

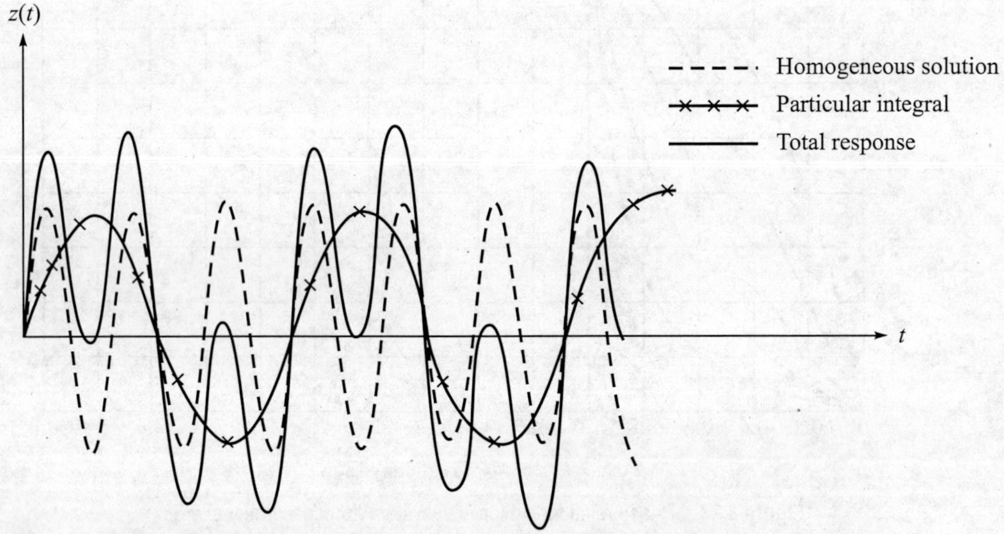

Figure 3.30 Response of an undamped system of a single frequency harmonic excitation.

Resonant Frequency

A resonant frequency is defined as the forcing frequency at which the largest response amplitude occurs. Figure 3.31(a) shows that the peaks in the response curves for displacement, velocity and acceleration occur at slightly different frequencies and can be determined by setting to zero the first derivatives of M_z, M_v, M_a with respect to frequency ratio ω/ω_n. For $\zeta < 1/\sqrt{2}$, they are:

$$\text{Displacement resonant frequency} = \omega_n \sqrt{1 - 2\zeta^2}$$

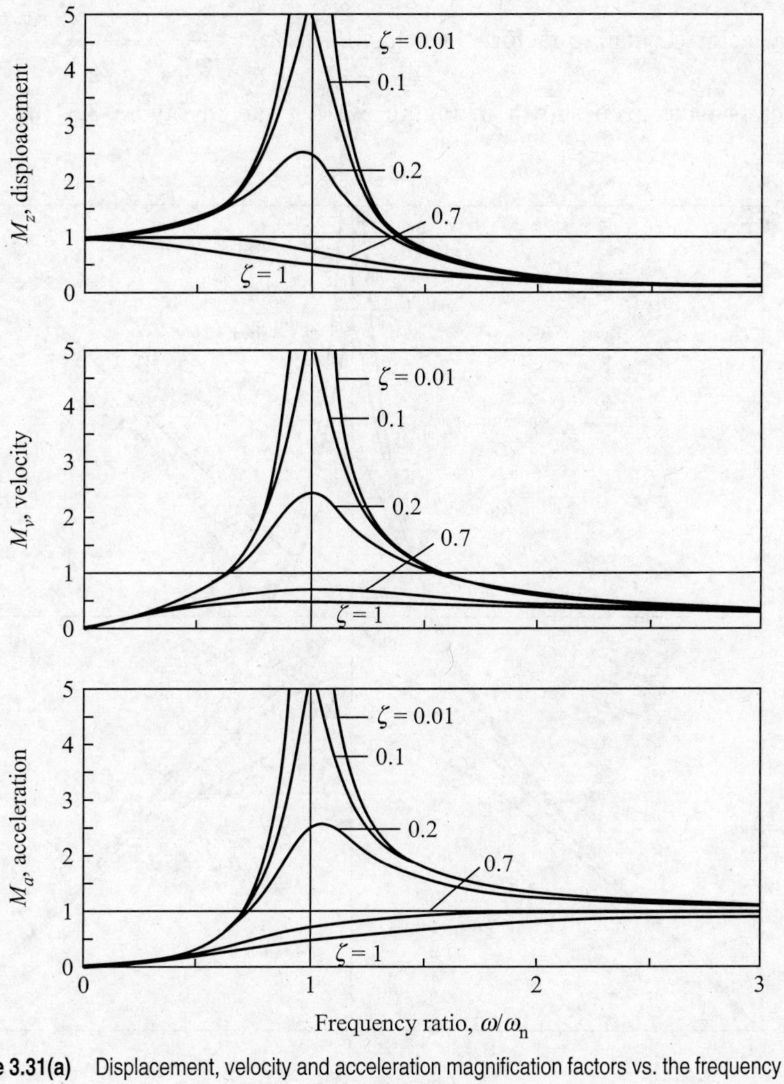

Figure 3.31(a) Displacement, velocity and acceleration magnification factors vs. the frequency ratio.

$$\text{Velocity resonant frequency} = \omega_n$$

$$\text{Acceleration resonant frequency} = \frac{\omega_n}{\sqrt{1-2\zeta^2}}$$

And correspondingly,

$$\text{Magnification factor (Dynamic factor)—displacement } (M_z) = \frac{1}{2\zeta\sqrt{1-2\zeta^2}}$$

$$\text{Magnification factor (Dynamic factor)—velocity } (M_v) = \frac{1}{2\zeta}$$

Magnification factor (Dynamic factor)—acceleration $(M_a) = \dfrac{1}{2\zeta\sqrt{1-2\zeta^2}}$

These factors have been shown in Figure 3.31(a) and in Figure 3.31(b) on a four-way logarithmic plot.

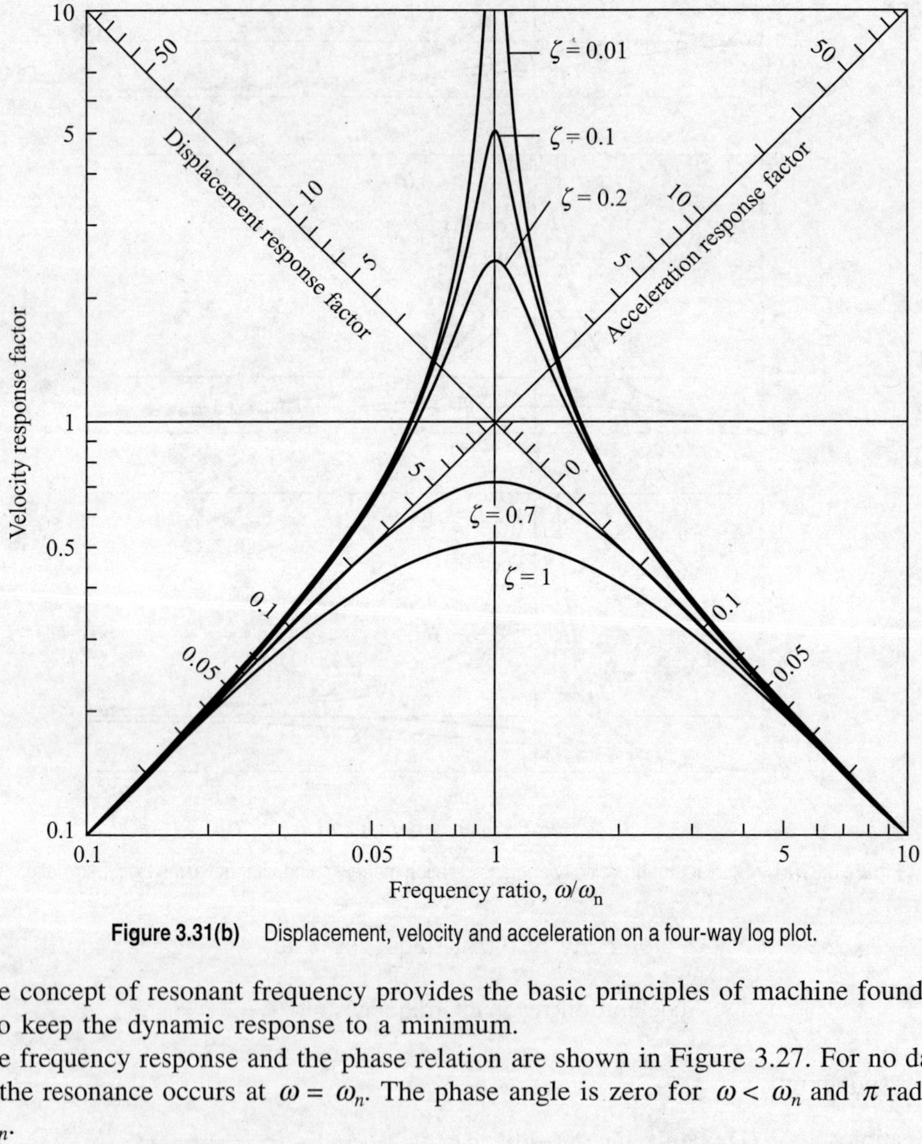

Figure 3.31(b) Displacement, velocity and acceleration on a four-way log plot.

The concept of resonant frequency provides the basic principles of machine foundation in order to keep the dynamic response to a minimum.

The frequency response and the phase relation are shown in Figure 3.27. For no damping, $\zeta = 0$, the resonance occurs at $\omega = \omega_n$. The phase angle is zero for $\omega < \omega_n$ and π radians for $\omega > \omega_n$.

For $\phi = \pi/2$, again $\omega = \omega_n$.

The dynamic factor $H(\omega)$ (often called the magnification factor M_z) is maximum when the denominator in the equation is minimum. Expressing the denominator as

$$\phi(r) = (1 - r^2)^2 + (2\zeta r)^2$$

For minima $$\frac{d\phi(r)}{dr} = 0$$

and hence, $$0 = 4r^3 - 4r + 8\zeta^2 r$$

Solving, $$1 - 2\zeta^2 = r^2, \qquad r = \sqrt{1-2\zeta^2}$$

Thus, $$H(\omega)_{\text{maximum}} = M_{z(\text{maximum})} = \frac{1}{2\zeta\sqrt{1-2\zeta^2}} = \frac{1}{\sqrt{4\zeta^4 + 4\zeta^2 - 8\zeta^4}}$$

$H(\omega)_{\text{maximum}} = M_{z(\text{maximum})} = 1/2\zeta$ for small value of ζ and at $\omega_n = \omega$.

Often this is also the expression for quality factor Q of the system.

From the above expression, it can be concluded that the inertia force is less than the spring force for $r < 1$ and more than the spring force for $r > 1$. At resonance, the inertia and the spring forces are the same. Also, the damping force is same as the exciting force. The vector relationship of forces is shown in Figure 3.32 for solution of the form

$$z = Z\sin(\omega t - \phi).$$

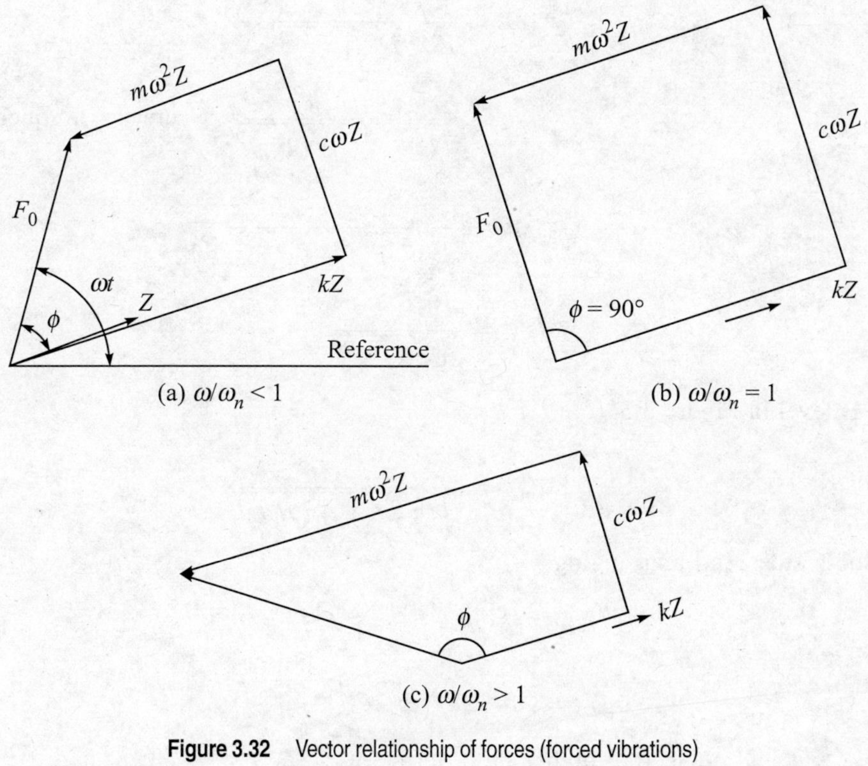

Figure 3.32 Vector relationship of forces (forced vibrations)

Half-power bandwidth (sharpness of resonance)

For forced vibration a term Q has been explained as Quality factor. For small values of ζ as per definition of the quality factor, it can be shown that

$$Q = \frac{1}{2\zeta}$$

This term Q is also a measure of the sharpness of the resonance. In Figure 3.33(a), the variation of magnification factor M_z with the frequency ratio r has been shown. At $M_z = 1/\sqrt{2}$, in the vertical axis let ω_a and ω_b be the forcing frequencies on either side of the resonant frequency, then it can be shown that

$$\frac{\omega_b - \omega_a}{\omega_n} = 2\zeta$$

so,
$$\zeta = \frac{\omega_b - \omega_a}{2\omega_n} \quad \text{or} \quad \zeta = \frac{f_b - f_a}{2f_n} \quad \text{where } f_n = \omega_n/2\pi$$

The magnification factor given by Eq. (3.78) may be equated to $1/\sqrt{2}$ times the resonant amplitude. Then,

$$\left(\frac{1}{\sqrt{2}}\right)M = \frac{1}{\sqrt{2}} \cdot \frac{1}{2\zeta\sqrt{1-\zeta^2}}$$

$$= \frac{1}{\sqrt{(1-r^2)^2 + (2\zeta r)^2}} \quad \text{where } r = \text{frequency ratio}$$

and,
$$M = \frac{1}{\sqrt{(1-r^2)^2 + (2\zeta r)^2}}$$

∴
$$M_{\max} = \frac{1}{2\zeta\sqrt{1-\zeta^2}}$$

At A and B level in Figure 3.33(b),

$$\left(\frac{1}{\sqrt{2}}\right)M_{\max} = \frac{1}{\sqrt{(1-r^2)^2 + (2\zeta r)^2}}$$

Squaring both sides and rearranging,

$$r^4 - 2(1 - 2\zeta r)r^2 + 1 - 8\zeta^2(1 - \zeta^2) = 0$$

Solving for r^2,

$$r^2 = (1 - 2\zeta r) \pm 2\zeta\sqrt{1-\zeta^2}$$

Assuming $\zeta < 1$ and neglecting higher order terms of ζ,

$$r^2 = 1 \pm 2\zeta$$

or,
$$r = (1 \pm 2\zeta)^{1/2}$$

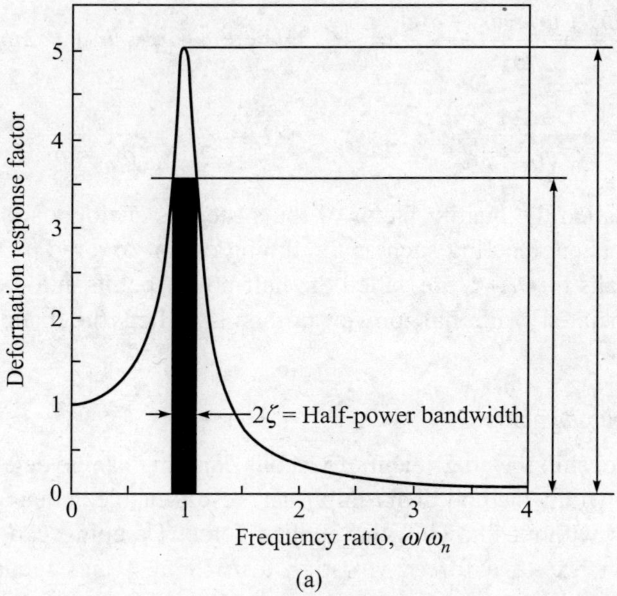

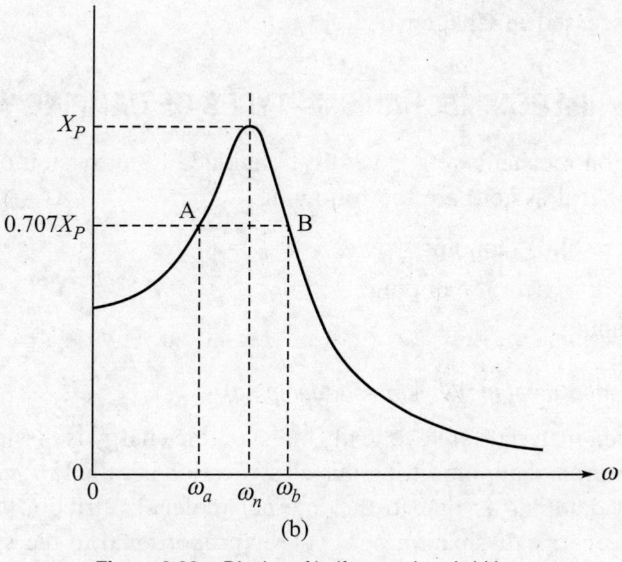

Figure 3.33 Display of half-power bandwidth.

For the first root, $r_1^2 = \left(\dfrac{\omega_a}{\omega_n}\right)^2 = 1 - 2\zeta$ and for the second root $r_2^2 = \left(\dfrac{\omega_b}{\omega_n}\right)^2 = 1 + 2\zeta$

then,
$$r_2^2 - r_1^2 = 4\zeta$$

or,
$$\left(\frac{\omega_b}{\omega_n}\right)^2 - \left(\frac{\omega_a}{\omega_n}\right)^2 = 4\zeta;\ \frac{\omega_a^2 - \omega_b^2}{\omega_n^2} = 4\zeta$$

or, $$\frac{(\omega_a + \omega_b)(\omega_a - \omega_b)}{\omega_n^2} = 4\zeta, \quad \text{where} \quad (\omega_a + \omega_b) = 2\omega_n$$

so, $$\frac{\omega_a - \omega_b}{\omega_n} = 2\zeta \quad \text{or} \quad Q \approx \frac{1}{2\zeta} = \frac{\omega_n}{\omega_a - \omega_b}$$

The factor Q, often called the quality factor of the system is analogous to similar applications in the field of electrical engineering such as the tuning circuit of a radio. The points where the amplification factor falls to $Q/\sqrt{2}$ are called the half-power points and the difference between the frequencies associated with half-power points is called the bandwidth as shown in Figure 3.33(a).

Determination of viscous damping

From the above relationship the coefficient of viscous damping can be determined. This method is known as the bandwidth method. This important result enables equation of damping from forced vibration tests without knowing the applied force. Damping can be determined from either a free-vibration test or a forced-vibration test. IS 5249 has recommended the Block Vibration test based on this bandwidth method for the determination of damping. The details of this test shall be discussed in Chapter 6.

3.8 ENERGY DISSIPATION MECHANISM—TYPES OF DAMPING

The energy dissipation mechanism is generally felicitated by provision of damping. The types of damping in a physical system are the following:

- Structural (or solid) damping
- Coulomb or dry friction damping
- Viscous damping

Structural damping (solid damping)/(hysteresis damping)

During vibration when materials are cyclically stressed, the energy is dissipated internally within the material itself. Internal damping fitting this classification is called *structural damping* or *solid damping*. Structural damping is due to the internal molecular friction of the material of the structure. When the energy dissipation per cycle is proportional to the square of the vibration amplitude, the loss coefficient is a constant. As such the shape of the hysteresis curve remains unchanged with amplitude and is independent of the strain rate. The stress-strain diagram as shown in Figure 3.34 for a vibrating system is not a straight line but forms a hysteresis loop, the area of which represents the energy dissipated due to molecular friction per cycle per unit volume. The size of the loop depends upon the material of the vibrating system, the frequency and the amount of dynamic stress as shown in Figure 3.34.

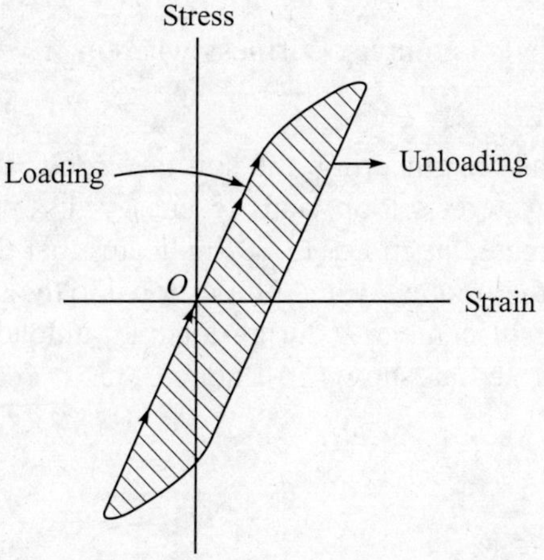

Figure 3.34 Stress-strain for structural materials.

All this means is that more work is done on the system while straining, it is then recovered during its relaxation. This type of damping is also called hysteresis damping.

Energy dissipated by structural damping may be expressed as

$$W_d = \alpha Z^2$$

where,

α = a constant with the unit of force/displacement.

Z = amplitude.

Using the concept of equivalent viscous damping, equating the energy, we have

$$\pi C_{eq}\ \omega Z^2 = \alpha Z^2 \tag{3.80}$$

or,

$$C_{eq} = \frac{\alpha}{\pi\omega} \tag{3.81}$$

Thus, the equation of motion for a SDF system with structural damping is

$$m\ddot{z} + \left(\frac{\alpha}{\pi\omega}\right)\dot{z} + kz = F(t) \tag{3.82}$$

In case of structural damping, it is customary to represent the stiffness and damping forces together, then

$$F_r = \left(\frac{\alpha}{\pi\omega}\right)\dot{z} + kz$$

Expressing harmonic motion in exponential form,

$$z = Ze^{i\omega t}$$

or,

$$F_r = \left[k + \frac{i\alpha}{\pi}\right]Ze^{i\omega t}$$

$$= k\left[1 + \frac{i\alpha}{\pi k}\right]Ze^{i\omega t} = k[1 + i\lambda]Z\cdot e^{i\omega t} \tag{3.83}$$

The quantity $k\left[1+\dfrac{i\alpha}{\pi k}\right]$ is called complex stiffness wherein $\lambda = \alpha/\pi k$ represents the damping factor or loss factor.

In geotechnical earthquake engineering, the soil undergoes a cyclic load deformation process. The existence of the hysteresis loop leads to energy dissipation from the system during each cycle. The energy dissipated as in Eq. (3.83) indicates that the energy distortion per cycle of motion is independent of the frequency and proportional to the square of the amplitude. When force F_r is plotted against displacement Z during loading, unloading and reloading in a cyclic manner, a closed loop is formed as shown in Figure 3.35.

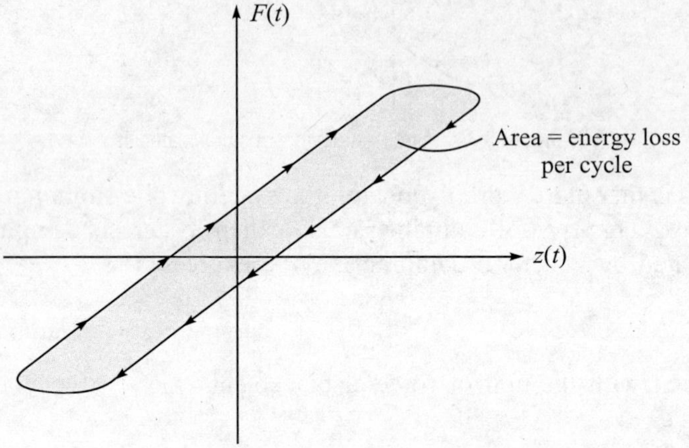

Figure 3.35 Hysteresis loop for damping.

The differential equation of motion for forced vibration of a SDF system with hysteresis damping may be written as

$$m\ddot{z} + k(1 + i\alpha/\pi k)\dot{z} = F_0 e^{i\omega t} \tag{3.84}$$

Putting $\lambda = \alpha/\pi k$ = damping factor or loss factor, the solution of Eq. (3.84) may be obtained as

$$z = Ze^{i(\omega t - \phi)}$$

where,

$$Z = \frac{F_0}{(k - m^2k^2) + (\lambda k)^2}$$

$$\phi = \tan^{-1}\frac{\lambda k}{k - m\omega^2}$$

$$H(\omega) = \frac{Z}{F_0/k} = \text{magnification factor}$$

$$= \frac{1}{\sqrt{(1-\beta^2)^2 + \lambda^2}}$$

where,

$$\beta = \frac{\omega}{\omega_n}$$

$$\phi = \tan^{-1} \frac{\lambda}{1-\beta^2}$$

The variation of $H(\omega)$ with λ has been shown in Figure 3.36.

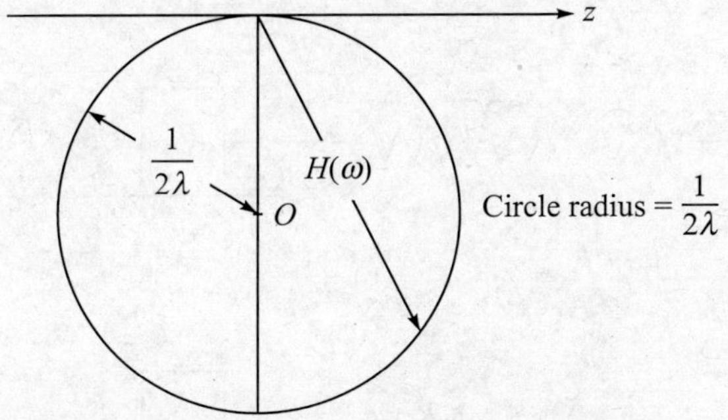

Figure 3.36 Variation of $H(\omega)$ with damping factor λ.

Coulomb damping (dry friction damping)

When one body is allowed to slide over the other, the surface of one body offers some resistance to the movement of the other body on it. This resisting force is called the *force of friction*. Thus, the force of friction arises only because of relative movement between the two surfaces. Some amount of energy is wasted in overcoming this friction, as the surfaces are dry.

So, it is sometimes known as dry friction. The general expression for coulomb damping is

$$F = \mu R_N \tag{3.85}$$

where μ is the coefficient of friction and R_N is the normal reaction. Friction force F is proportional to the normal reaction R_N on the mating surface. The dependence of coefficient μ on relative velocity is shown in Figure 3.37.

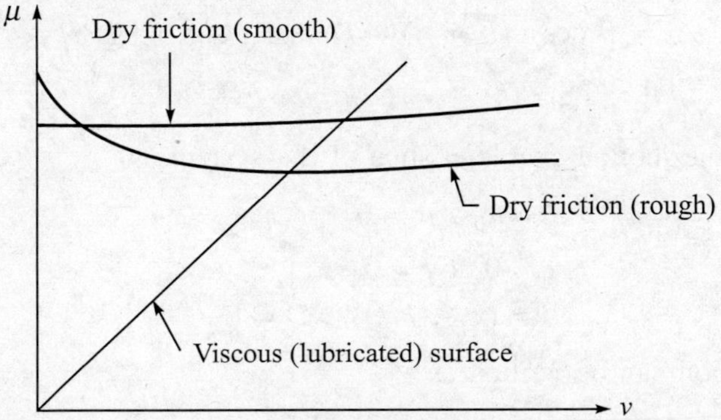

Figure 3.37 Variation of coefficient μ with velocity v [coulomb damping]

The friction force acts in a direction opposite to the direction of velocity. The damping resistance is almost constant and does not depend on the rubbing velocity. The three possible conditions of coulomb damping are shown here with mathematical expressions.

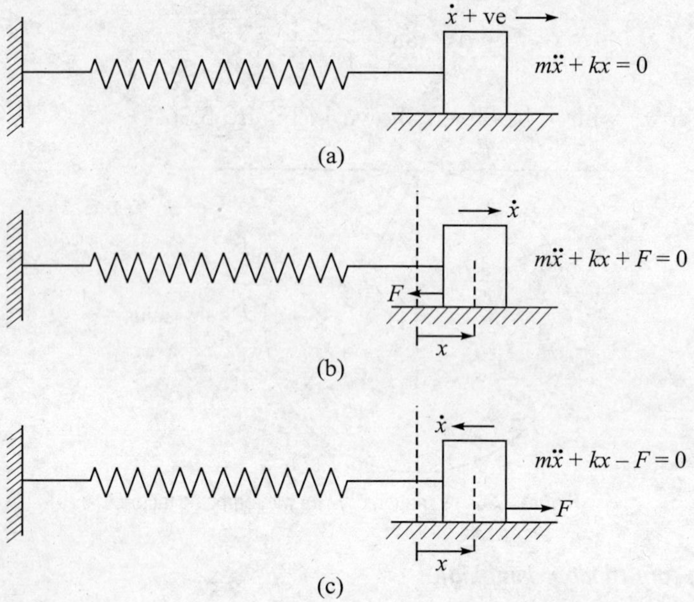

Figure 3.38 (a) Equilibrium position: $m\ddot{x} + kx = 0$, (b) Mass is displaced towards right and moving towards right: $m\ddot{x} + kx + F = 0$, and (c) mass is displaced towards right and moving towards left: $m\ddot{x} + kx - F = 0$.

Let us consider the leftward movement of the body, the equation for which can be written as

$$m\ddot{x} + kx = F \tag{3.86}$$

The friction force on the body acts towards the right in the direction, because the body is moving towards the left. This is shown in Figure 3.33(c).
The solution of the above equation can be written as

$$x = B\cos(\sqrt{k/m}\cdot t) + D\sin(\sqrt{k/m}\cdot t + (F/k)) \tag{3.87}$$

where $\omega = \sqrt{k/m}$.

Let us assume the motion characteristics of the system as

$$x = x_0 \text{ at } t = 0$$

$$\dot{x} = 0 \text{ at } t = 0$$

Thus, we get $B = (x - (F/k)), \quad D = 0$

So, the above equation can be written as

$$x = (x_0 - (F/k))\cos(\sqrt{k/m}\cdot t + (F/k)) \tag{3.88}$$

This solution holds good for half the cycle. When $t = \pi/\omega$, half the cycle is complete. So displacement for half the cycle can be obtained from the above equation, i.e.,

$$x = (x_0 - (F/k))\cos(\pi + (F/k))$$

$$= -\ (x_0 - (F/k)) + (F/k)$$

$$= -\ (x_0 - (2F/k)) \tag{3.89}$$

This is the amplitude for the left extreme position of the body. It is clear that the initial displacement x_0 is reduced by $2F/k$. In the next half cycle when the body moves to the right the initial displacement will be reduced by $2F/k$. So in one complete cycle the amplitude reduces by $4F/k$. The amplitude decay for coulomb damping is shown in Figure 3.39. The natural frequency of the system remains unchanged in coulomb damping.

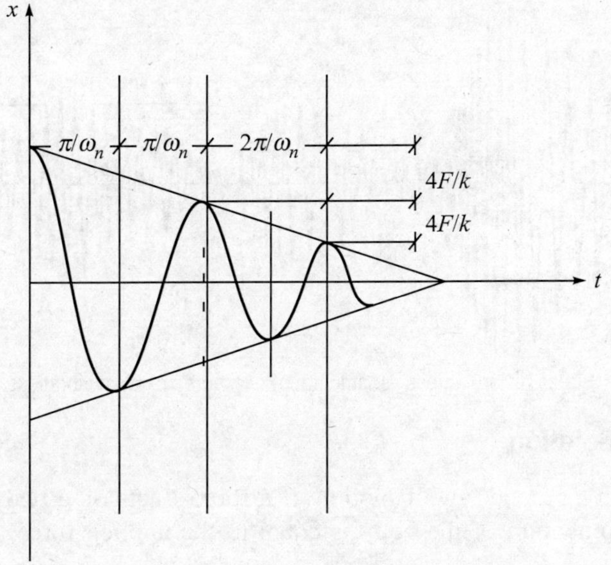

Figure 3.39 Displacement–time trace (amplitude decay) of a system with coulomb damping.

The frequency of vibration for a system having coulomb damping is the same as that of the undamped system, i.e.,

$$\omega_n = \sqrt{\frac{k}{m}}$$

and time period
$$T = \frac{2\pi}{\omega_n}$$

The amplitude loss per cycle is $4F/k$. Comparing with the case of viscous damping, the ratio of any two successive amplitudes is constant and the envelope of the maximas of the displacement–time trace is an exponential curve, whereas in the case of coulomb damping, the difference between any two successive amplitudes is constant and the envelope of the maximas of the displacement–time trace is a straight line.

3.9 SYSTEM UNDER IMPULSE AND TRANSIENT LOADING

Periodic and harmonic excitations are ideal excitations, which seldom occur in practice. However, at times the excitation is of the periodic nature like a shock pulse or an impulse or a transient excitation, the response of the system is purely transient. Many excitations are of short-duration. For such short-duration responses the maximum response may occur after the excitation has ceased. After the excitations are removed, damping causes the system to return to their equilibrium position, resulting in trivial steady states. Impulsive or shock loads frequently are of great importance in the design of certain classes of structural and substructural systems. The maximum response to an impulsive load will be reached in very short time. Damping has much less importance in controlling the maximum response of a structure to impulsive load than for periodic and harmonic loads.

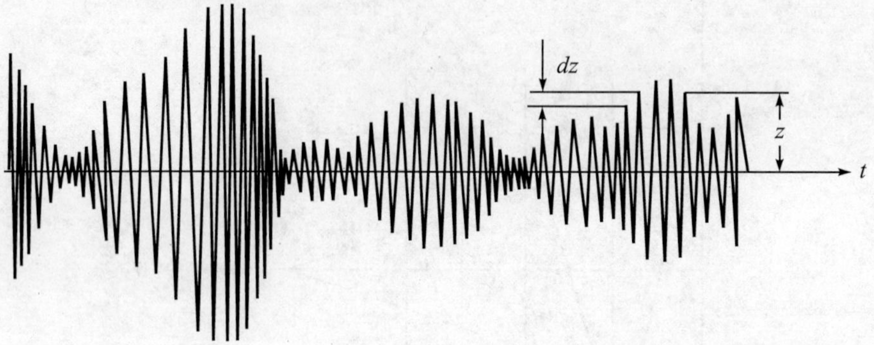

Figure 3.40 Stress history (stress vs. time trace) excited by random vibrations (SDF system).

3.9.1 Method of Solution

According to Newton's second law of motion, if a force f acts on a body of mass m, then the rate of change of momentum of the body is equal to the applied force, that is

$$\frac{d}{dt}(m\dot{z}) = f(\tau) \tag{3.90}$$

For constant mass, this equation becomes

$$f(\tau) = m \cdot \ddot{z} = \text{mass} \times \text{acceleration} \tag{3.91}$$

Integrating both sides with respect to t,

$$\int_{t_1}^{t_2} f(\tau) = m(\dot{z}_2 - \dot{z}_1) = m\Delta\dot{z} \tag{3.92}$$

or,

$$\int f(\tau)\,d\tau = \int m\left(\frac{dz}{dt}\right) \tag{3.93}$$

The integral on the left side of this equation is an impulse and the right-hand side is the momentum as product of mass and velocity. Thus, Eq. (3.93) states that the magnitude of impulse is equal to the change in momentum.

Thus, on a SDF system if the force acts for an infinitesimally short-duration $d\tau$, the spring and damper will have no time to respond. Thus, if an impulse I at $t = \tau$ is imparted to the mass, the velocity $\dot{z}(\tau)$, then from Eq. (3.93),

$$dv = d\dot{z}(t)\ d\dot{z}(\tau) = F(\tau)\ \frac{d\tau}{m} \tag{3.94}$$

For a unit impulse,

$$dv = d\,\dot{z}(\tau) = \frac{I}{m} \tag{3.95}$$

The $z(t)$ for a SDF system is given by

$$z(t) = z(0)\ \cos\ \omega t + \frac{\dot{z}(0)}{\omega}\ \sin\ \omega_n t$$

Taking $z(0) = 0$, as the displacement is zero prior to and up to the impulse and $\dot{z}(\tau) = F(\tau)\ dz$ I/m, from Eq. (3.95)

$$dz(\tau) = F(\tau)\ \frac{d\tau}{m\omega_n}\ \sin\ \omega_n(t - \tau) \qquad \text{for } t < \tau \tag{3.96}$$

Similarly, for a viscously damped SDF system, the response to unit impulse may be written as

$$\Delta z(t) = f(\tau)\ \frac{d\tau}{m\omega_n}\ e^{-\zeta\omega_n(t-\tau)} \cdot \sin(\omega_d\ (t - \tau)) \qquad \text{for } t < \tau \tag{3.97}$$

A force $f(t)$ varying arbitrarily with time can be represented as a sequence of infinitesimally short impulses as shown by the hatched portion in Figure 3.41.

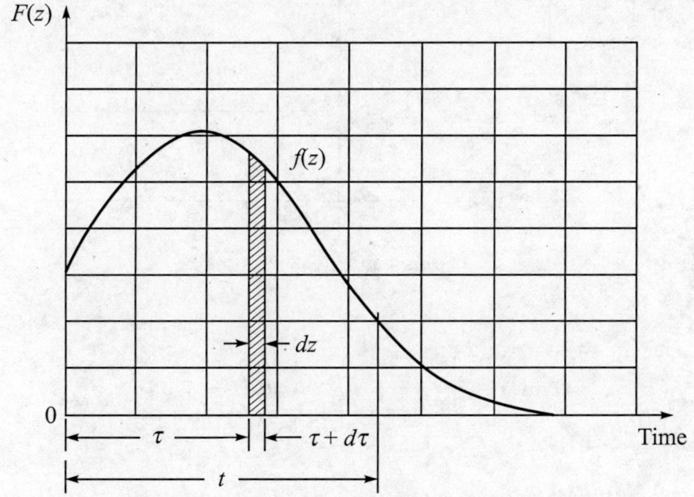

Figure 3.41 General load history (impulsive loading).

Thus, the response of the SDF system at any time t is the sum of the responses to all impulses up to that time. Therefore,

$$z(t) = \int_0^t f(\tau)\ \Delta z(t) \tag{3.98}$$

or,

$$z(t) = \int_0^t \frac{1}{m\omega_n} \cdot f(\tau)\ \sin(\omega_n\ (t - \tau))d\tau \tag{3.99}$$

Similarly, for the viscous damped SDF system,

$$z(t) = \frac{1}{m\omega_d} \int_0^t f(\tau)\, e^{-\zeta\omega_n(t-\tau)} \cdot \sin(\omega_d\,(t - \tau))\, d\tau \tag{3.100}$$

Equation (3.100) is known as Convolution Integral, which applies to any linear dynamic system. The integral in Eq. (3.99) and Eq. (3.100) is also known as Duhamel's integral.

3.9.2 Duhamel's Integral

The Duhamel's integral provides general solution for evaluating the dynamic response of a SDF system to an arbitrary force. However, the solution is restricted to linear systems as it is based on the principle of superposition.

If $f(\tau)$ is a simple function, closed form evaluation is possible, and in that case Duhamel's integral is an alternative to the classical method for solving differential equations. However, if $f(\tau)$ is a complicated function, that is described numerically, then the evaluation of the integral requires numerical methods.

Example 3.2 Find the dynamic response of the SDF system subjected to $F(\tau) = F_0$ applied suddenly, as shown in Figure 3.42.

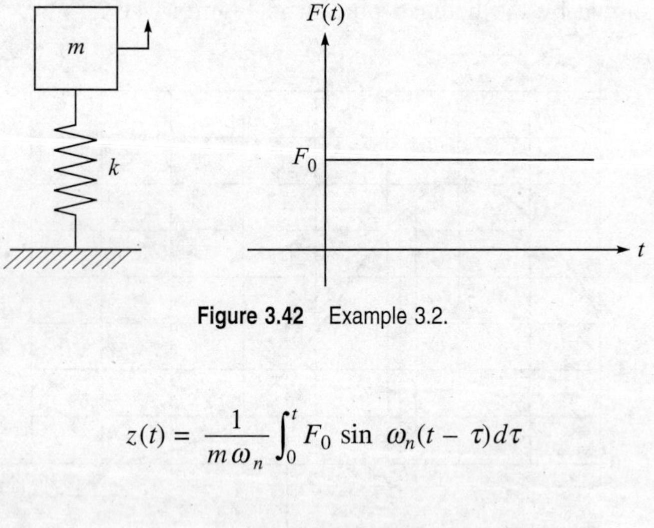

Figure 3.42 Example 3.2.

Solution:

$$z(t) = \frac{1}{m\omega_n} \int_0^t F_0 \sin\, \omega_n(t - \tau)\, d\tau$$

On integration,

$$\int_0^t \sin\, \omega_n(t - \tau) d\tau = \frac{1}{m\omega_n} \left| \cos\, \omega_n(t-\tau) \right|_0^t$$

and using $k = m\omega_n^2$

$$z(t) = \frac{F_0}{m\omega_n^2}(1 - \cos\, \omega t) = \frac{F_0}{k}(1 - \cos\, \omega t)$$

Dynamic Response or Dynamic Load Factor (DLF) $= \dfrac{z_{\text{dynamic}}}{z_{\text{static}}}$

Thus,
$$\text{DLF} = \frac{\frac{F_0}{k}(1-\cos\omega t)}{\frac{F_0}{k}} = 1 - \cos\ \omega t$$

Example 3.3 Find the response of the SDF system to instantaneous loading of the form

$$f(\tau) = f_0 \frac{t}{t_r} \text{ as shown in Figure 3.43.}$$

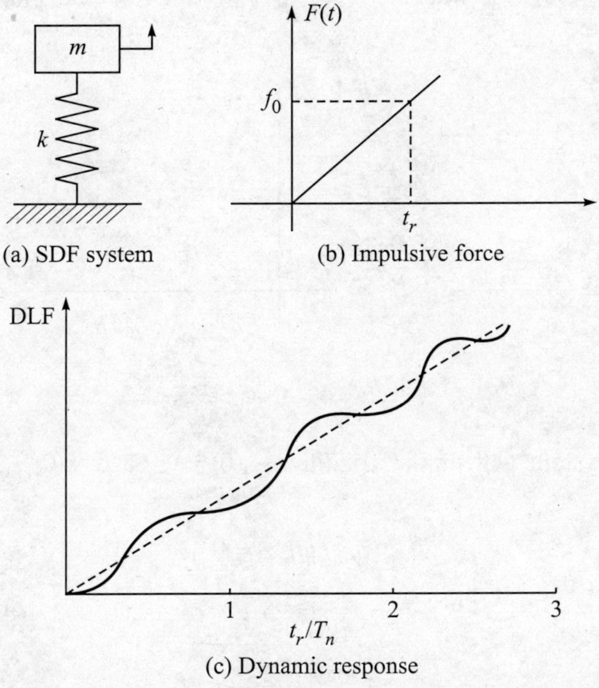

Figure 3.43 (a) SDF system, (b) impulsive force and (c) dynamic response.

Solution: The dynamic response of the SDF system is given by

$$z(t) = \frac{1}{m\omega_n}\int_0^t f(\tau)\sin\ \omega_n(t-\tau)\,d\tau$$

$$= \frac{1}{m\omega_n}\int_0^t \frac{f_0}{t_r}\,t \sin\ \omega_n(t-\tau)\,d\tau$$

$$= \frac{f_0}{k}\left[\frac{t}{t_r} - \frac{\sin\omega_n t}{\omega_n t_r}\right]$$

Thus, the dynamic factor or dynamic load factor (magnification factor) is given by

$$\text{DLF} = z(t)\ \frac{z_{dynamic}}{z_{static}}$$

$$= \frac{t}{t_r} - \frac{\sin \omega_n t}{\omega_n t_r}$$

Example 3.4 A single degree freedom system (Figure 3.44) is subjected to a rectangular pulse of the form $F(t) = f_0,\ t < \tau$, where $F(t) = 0,\ t > \tau$. Draw the phase-plane plot and the displacement-time plot.

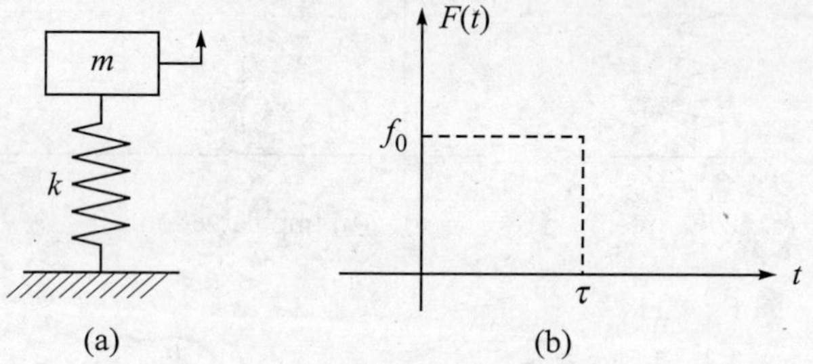

Figure 3.44 Example 3.4.

Solution: A SDF system with initial condition $z(0) = z_0$ and $\dot{z}(0) = v_0$ has its differential equation as

$$\ddot{z} + \omega_n^2 z = 0$$

where,

$$z = A \sin(\omega_n t + \phi)$$

$$A = \sqrt{z_0^2 + \frac{v_0^2}{\omega_n^2}}$$

$$\phi = \tan^{-1}\left[\frac{\omega_n}{z_0 v_0}\right]$$

This velocity is obtained as

$$\dot{z} = A\omega_n \cos(\omega_n t + \phi)$$

or,

$$\frac{\dot{z}}{\omega_n} = A \cos(\omega_n t + \phi)$$

Squaring z and $\frac{\dot{z}}{\omega_n}$ and adding, yields

$$z^2 + (\dot{z}/\omega_n)^2 = A^2$$

The above equation is a circle in a plane with coordinate axes z and $\dot{z}/\omega_n$.

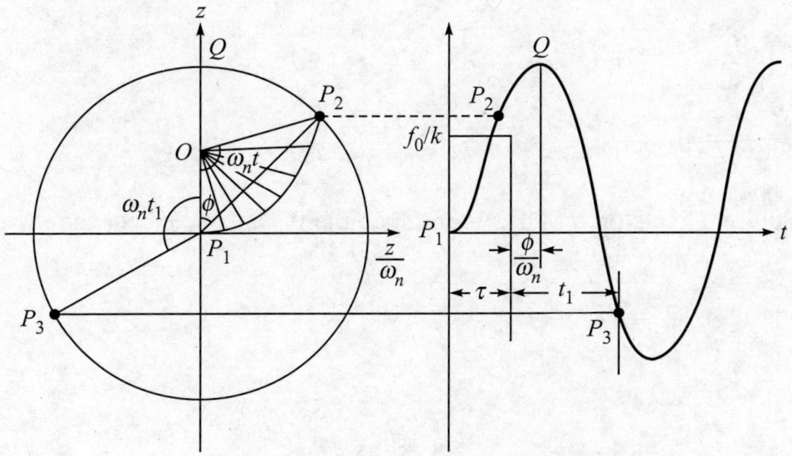

Figure 3.45 Phase-plane plot and displacement-time plot.

The construction of the phase plane is as follows:

1. As the initial conditions are zero, the phase trajectory starts from the origin.
2. For $t = 0$ to $t = \tau$, the position of map is at 0 such that $P_1O = \dfrac{f_0}{k}$.
3. With centre O and radius OP, an arc P_1P_2 is drawn.
4. Thus, $\angle P_1OP_2 = \omega_n t$ rad.
5. The phase trajectory during pulse duration is P_1P_2.
6. After $t \geq \tau$ the position of mass at the end of the pulse shifts to the origin.
7. The centre of phase trajectory also shifts to the origin.
8. P_2 becomes the initial position for next motion.
9. The phase trajectory starting from P_2 is a circle with radius equal to P_1P_2.
10. At any time $t > \tau$ the system coordinates are represented by P_3 such that $P_2P_1P_3 = \omega_n t_1$ rad
11. The displacement-time plot has been shown as $P_1P_2QP_3$.

$$\text{Maximum displacement} = \frac{\phi}{\omega_n}$$

3.9.3 Dirac Delta Function

When a force of very large magnitude acts for a very short duration $d\tau$ such as that of the shaded area under the force–time plot (Figure 3.46), then such a function is suitably represented symbolically as dirac delta function having the following properties:

(a) $\delta(t - \tau) = 0$ for all $t \neq \tau$

(b) $\int_0^{\infty} (t - \tau) d\tau = 1$

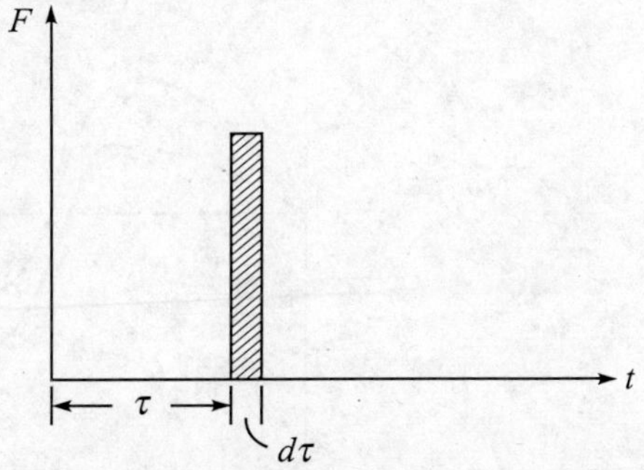

Figure 3.46 Impulse as dirac delta function.

(c) $\int_0^\infty f(\tau)\ \delta(t - \tau)d\tau = f(\tau)$

The unit impulse $\overline{F}$ is defined as $\dfrac{\overline{F}}{d\tau} \cdot d\tau = 1$ (as $d\tau \to 0$).

If this impulse $\overline{F}$ acts on a SDF system, then the instantaneous change in velocity $\dot{z}(0)$ is given by

$$d\dot{z}(0) = \frac{\overline{F}}{m}$$

Then,

$$z(t) = e^{-\zeta\omega_n t}\,\frac{\overline{F}}{m}\ \sin\ \omega_d t$$

$$\omega_d = \sqrt{1-\zeta^2}\,\omega_n$$

$$\omega_n = \sqrt{\frac{k}{m}},\quad \text{so } m\omega_n = \sqrt{m^2 \cdot \frac{k}{m}} = \sqrt{mk}$$

Substituting,

$$z(t) = \frac{\overline{F}}{\sqrt{mk}} \cdot \frac{e^{-\zeta\omega_n t}}{\sqrt{1-\zeta^2}}\ \sin\ \omega_d t$$

or,
$$z(t)\frac{\sqrt{mk}}{\overline{F}} = \frac{e^{-\zeta\omega_n t}\ \sin\ \omega_n\sqrt{1-\zeta^2}\ t}{\sqrt{1-\zeta^2}}$$

This relation may be shown as $z(t)\dfrac{\sqrt{mk}}{\overline{F}}$ vs. $\omega_n t$ (Figure 3.47).

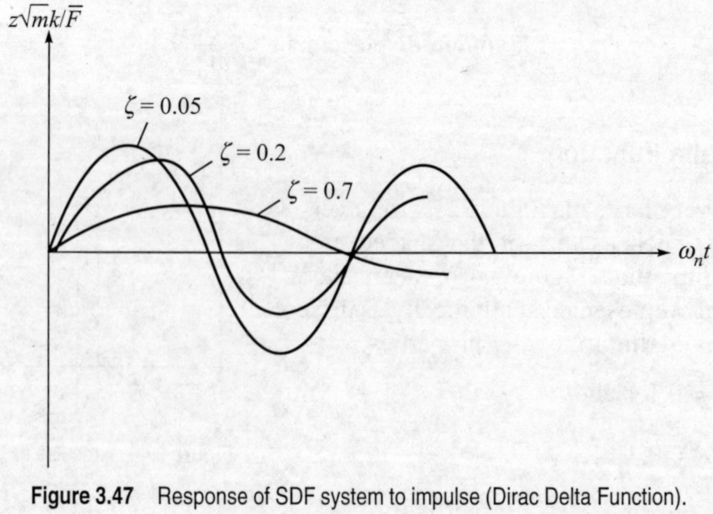

Figure 3.47 Response of SDF system to impulse (Dirac Delta Function).

3.10 TRANSMISSIBILITY

The system as shown in Figure 3.48 represents a very practical system that corresponds to a machine foundation with rotating unbalances. The governing differential equation is similar to Eq. (3.76) with the difference that $F \sin \omega t$ is replaced by $m_0 e \omega^2 \sin \omega t$ and may be expressed as

$$(m_0 + m)\ddot{z} + c\dot{z} + kz = m_0 e \omega^2 \sin \omega t$$

and the solution may be expressed $z = Z_0 \sin \omega t$ as explained in Section 3.7

such that, $$Z_0 = \frac{m_0 e \omega^2}{\sqrt{(k - M\omega^2)^2 + (c\omega)^2}}, \text{ where } M = m_0 + m$$

From the concept of transmissibility the forces imparted to the foundation through the spring and the dashpot shall be determined as follows:

The maximum force in the spring is kz_0 and the maximum force in the dashpot is $c\omega z_0$; the two forces are out-of-phase at 90° (Figure 3.48). Hence, the force transmitted F_1 to the base is

$$F_1 = \sqrt{(kz_0)^2 + (c\omega z_0)^2} \tag{3.101}$$

or, $$F_1 = kz_0\sqrt{1 + (c\omega / k)^2} \tag{3.102}$$

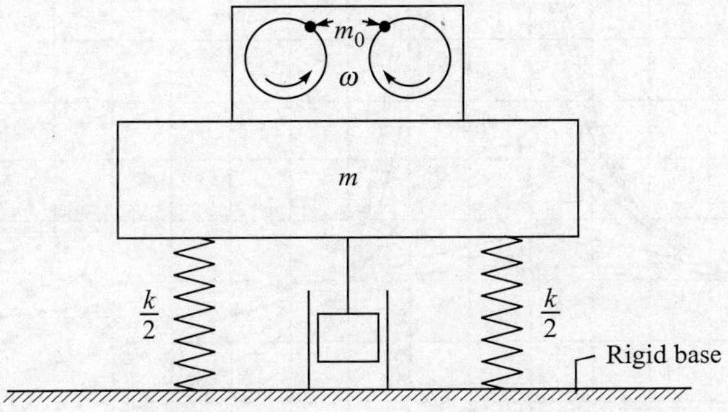

Figure 3.48 System with rotating unbalanced masses

Let $c\omega/k = 2\zeta r$ and substituting z_0 from above equation (3.101), we get transmissibility T_r defined as the ratio of the force transmitted to the force of excitation. Therefore,

$$T_r = \frac{F_1}{m_0 e \omega^2} = \frac{\sqrt{1 + (2\zeta r)^2}}{\sqrt{(1 - r^2)^2 + (2\zeta r)^2}} \tag{3.103}$$

Transmissibility T_r versus frequency ratio ω/ω_n is plotted in Figure (3.49). It will be seen that for $\zeta = 0$, T_r approaches infinite at $r = 1$. Also, all curves pass through $r = \sqrt{2}$. For $r > \sqrt{2}$, all of the curves approach the r-axis asymptotically. Figure 3.49 shows different transmissibility curves for different value of damping. Damping helps to limit these amplitudes to finite values. Three regions marked in Figure 3.49 are respectively controlled by the three parameters of the system—mass, damping and stiffness.

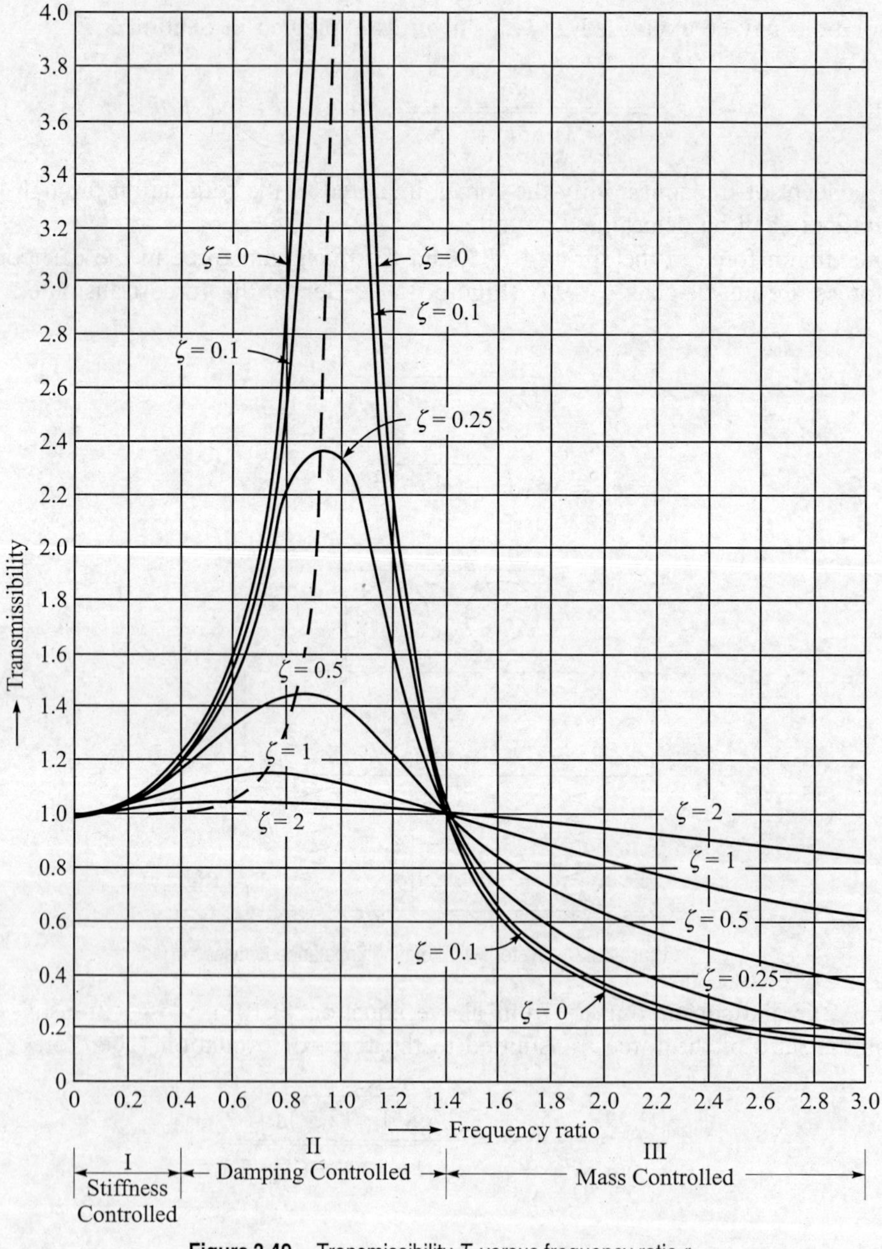

Figure 3.49 Transmissibility T_r versus frequency ratio r.

3.10.1 Transfer Function

The periodic load $Q(t)$ may be expressed as Fourier series as explained in Section 3.11, i.e.,

$$Q(t) = A_0 + \sum_{n=1}^{\infty} (A_n \cos \omega_n t + B_n \sin \omega_n t) \tag{3.105}$$

where the Fourier coefficients are

$$A_0 = \frac{1}{T_f} \int_0^{T_f} Q(t)\, dt \tag{3.106}$$

$$A_n = \frac{2}{T_f} \int_0^{T_f} Q(t) \cos \omega_n t \cdot dt \tag{3.107}$$

$$B_n = \frac{2}{T_f} \int_0^{T_f} Q(t) \sin \omega_n t \cdot dt \tag{3.108}$$

and, $$\omega_n = \frac{2\pi n}{T_f}$$

The periodic loading $Q(t)$ can further be described by Fourier series in exponential form

$$Q(t) = \sum_{n=-\infty}^{n=\infty} q_n(t) \cdot e^{i\omega_n t} \tag{3.109}$$

where, $$q_n(t) = \frac{1}{T_f} \int_0^{T_f} Q(t) \cdot e^{-i\omega_n t}\, dt$$

The dynamic response of the SDF system to periodic loading $Q(t)$ may be expressed by series solution assuming the system to be linear wherein the principle of superposition holds good.

$$z(t) = \sum_{n=-\infty}^{\infty} H(\omega_n) \cdot q_n(t) \cdot e^{i\omega_n t} \tag{3.110}$$

where $H(\omega_n)$ = transfer function.

Thus, the transfer function is defined as a function that relates one parameter to another. Here the parameters are displacement $z(t)$ and the periodic loading $Q(t)$.

From Figure 3.50, the transfer function may be regarded as an operator that operates on excitation (periodic loading) to yield the dynamic response.

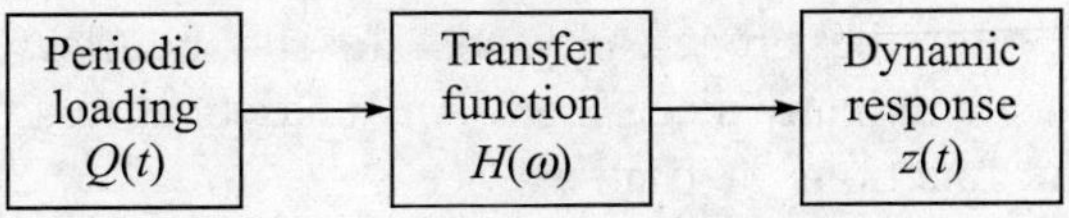

Figure 3.50 Concept of transfer function $H(\omega)$.

The dynamic response of an SDF system to periodic loading $Q(t)$ may, therefore, be obtained with the help of transfer function. The response of an SDF system to harmonic loading $Q(t)$ would be governed by equation of motion

$$m\ddot{z}_n(t) + c\dot{z}_n(t) + kz_n(t) = q_n(t)\cdot e^{i\omega_n t} \tag{3.111}$$

The response with the help of transfer function may be expressed as

$$z_n(t) = H(\omega_n)\cdot q_n(t)\cdot e^{i\omega_n t} \tag{3.112}$$

Substituting the value of $z_n(t)$ in equation of motion (3.111), we have

$$\int -m\omega_n^2 \cdot H(\omega_n)\cdot q_n(t)\cdot e^{i\omega_n t} + i(\omega_n)H(\omega_n)\cdot q_n(t)\cdot e^{i\omega_n t} + kH(\omega_n)\cdot q_n(t)\cdot e^{i\omega_n t} = q_n \cdot e^{i\omega_n t}$$

Thus,

$$H(\omega_n) = \frac{1}{-m\omega_n^2 + i(\omega_n) + k} = \frac{1}{k(-\beta_n^2 + 2i\beta_n\zeta + 1)}$$

where, $\beta_n = \dfrac{\omega_n \cdot T_f}{2\pi}$, $T_f = \dfrac{2\pi n}{\omega_n}$, ζ = damping factor

Using $B\cdot e^{i\phi} = a + ib$, $B = \sqrt{a^2 + b^2}$, $\phi = \tan^{-1}\dfrac{b}{a}$

So,

$$H(\omega_n) = \frac{\frac{1}{k}}{\sqrt{(1-\beta_n)^2 + (2\zeta\beta_n)^2}}\cdot \exp\left\{i\tan^{-1}\frac{2\zeta\beta_n}{\beta_n^2 - 1}\right\}$$

The dynamic response is given by $z(t) = \sum_{n=-\infty}^{\infty} H(\omega_n)\cdot q_n(t)\cdot e^{i\omega_n t}$

Thus, the advantage of the transfer function is that it allows computation of the response to a complicated loading pattern.

3.11 FOURIER ANALYSIS

In the previous sections the response of an SDF system to harmonic excitation has been discussed. Such ideal excitation seldom occurs in practice. Any variation from the pure sinusoid, however small, produces effects of considerable magnitude on the system response. It is therefore necessary to find some way of determining the response of a system to non-harmonic loading. Such non-harmonic motions are periodic. Any such non-harmonic periodic motion expressed by any analytical function can be represented by a series of sin or cos functions of time. The series named after the French physicist Joseph Fourier (1768–1830) is the most popular and extensively used in many disciplines. Any analytical function can be represented by a series of sine or cosine functions of time as

$$f(\omega t) = \frac{a_0}{2} + \sum_{n=1}^{\infty} a_n \cos \omega_n t + b_n \sin \omega_n t \tag{3.104}$$

where, $$a_n = \frac{1}{\pi}\int_0^{2\pi} f(\omega t) \cos\ n\omega t\ d(\omega t) \tag{3.113}$$

$$b_n = \frac{1}{\pi}\int_0^{2\pi} f(\omega t) \sin\ n\omega t\ d(\omega t) \tag{3.114}$$

The term $(1/2)a_0$ represents the average value of $f(\omega t)$ over the full point. Therefore,

$$\frac{a_0}{2} = \frac{1}{2\pi}\int_0^{2\pi} f(\omega t)dt \tag{3.115}$$

Alternatively,

$$f(\omega t) = (1/2)P_0 + P_1 \sin(\omega t + \alpha_1) + \cdots + P_n \sin(n\omega t + \alpha_n)$$

$$= (1/2)P_0 + \sum_{n=1}^{\infty} P_n \sin\ (n\omega_n t + \alpha_n)$$

where,

$$P_n = \sqrt{a_n^2 + b_n^2}$$

$$\alpha_n = \tan^{-1}\frac{a_n}{b_n}$$

An SDF undamped system subjected to general harmonic loading of the form $p \sin\ (\omega t + \alpha)$

$$mz + kz = p\ \sin(\omega t + \alpha)$$

$$\frac{z(t)}{p/k} = \frac{1}{1-r^2}\sin\ (\omega t + \alpha)$$

$$\bar{z}(t)\Big|_{\text{non-dimensional}} = \frac{1}{1-r^2}\sin\ (\omega t + \alpha)$$

Similarly, for the nth harmonic

$$\bar{z}(t) = \frac{1}{1-nr^2}\sin\ (n\omega t + \alpha_n)$$

or, $$\bar{z}(t) = \sum_{n=1}^{\infty}\frac{1}{1-nr^2}\sin\ (n\omega t + \alpha_n) \tag{3.116}$$

Thus, given any non-harmonic loading, it can be transferred into Fourier series and the dynamic response of undamped SDF system to such loading can be obtained by Eq. (3.116).

Example 3.5 A undamped SDF system is subjected to a force of the form

$$f(\omega t) = \sin^2\ \omega t \qquad 0 < \omega t < \pi$$

$$f(\omega t) = -\sin^2 \omega t \qquad \pi < \omega t < 2\pi$$

Find the Fourier transform equation.

Solution:

$$a_n = 0$$

$$a_0 = 0$$

$$b_n = \frac{1}{\pi}\int_0^{2\pi} f(\omega t)\sin n\omega t\, d(\omega t)$$

$$= \frac{1}{\pi}\int_0^{\pi} \sin^2\omega t \sin n\omega t\, d(\omega t) - \int_0^{2\pi} \sin^2\omega t \sin n\omega t\, d(\omega t)$$

$$= \frac{4(\cos n\pi - 1)}{n(n^2-4)}$$

Example 3.6 Apply Fourier's theorem to analyze the output wave from a half-wave rectifier when the input wave is of the form $E = E_0 \sin \omega t$.

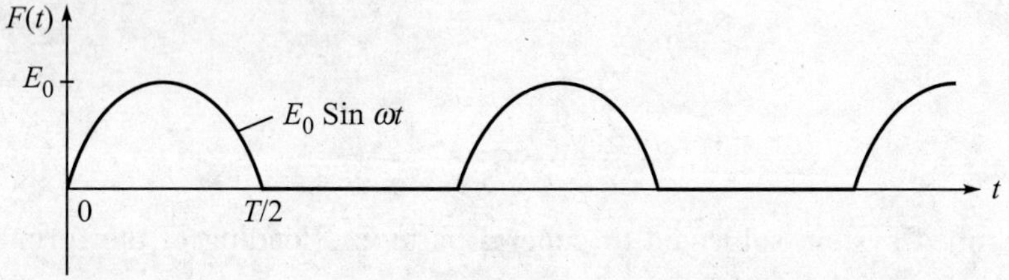

Figure 3.51 Half wave rectifier response (Example 3.6).

Solution: For Fourier's analysis of the output wave from a half-wave rectifier, a sinusoidal voltage $E = E_0 \sin \omega t$ is passed through a half-wave rectifier, which removes the negative half-cycles of the wave. The output voltage wave is of the form as shown in the Figure 3.51. This may be expressed as

$$E(t) = E_0 \sin \omega t \qquad \text{from } t = 0 \text{ to } t = \frac{T}{2} \quad \text{where} \quad T = \frac{2\pi}{\omega}.$$

$$E(t) = 0 \qquad \text{from } t = \frac{T}{2} \text{ to } t = T.$$

Let us express it as a Fourier series:

$$E(t) = A_0 + A_1 \cos \omega t + \cdots + A_r \cos r\omega t + \cdots + B_1 \sin \omega t + \cdots + B_r \sin r\omega t + \cdots$$

Let us evaluate the Fourier coefficients:

$$A_0 = \frac{1}{T}\int_0^T E(t)\, dt$$

or,

$$A_0 = \frac{1}{T}\int_0^{T/2} E_0 \sin \omega t\, dt$$

$$= \frac{1}{T}\int_0^{T/2} E_0 \sin \frac{2\pi t}{T}\, dt$$

$$= -\frac{E_0}{T}\frac{1}{2\pi T}\left[\cos\frac{2\pi t}{T}\right]_0^{T/2}$$

$$= -\frac{E_0}{2\pi}\,[\cos\ \pi - \cos\ 0]$$

$$= E_0/\pi \qquad [\because\ \cos\ \pi = -1 \text{ and } \cos\ 0 = 1]$$

Again, $$A_r = \frac{2}{T}\int_0^T E(t)\cos\ r\omega t\ dt$$

$$= \frac{2E_0}{T}\int_0^{T/2} \sin\ \omega t \cos\ r\omega t\ dt$$

$$= \frac{2E_0}{T}\int_0^{T/2} \sin\frac{2\pi t}{T}\cos\frac{2r\pi t}{T}\,dt$$

$$= \frac{2E_0}{T}\int_0^{T/2} \frac{1}{2}\left[\sin\,(1+r)\frac{2\pi t}{T} + \sin\,(1-r)\frac{2\pi t}{T}\right]dt$$

$$= \frac{E_0}{T}\left[\frac{-\cos\,(1+r)\dfrac{2\pi t}{T}}{(1+r)\dfrac{2\pi}{T}} - \frac{\cos\,(1-r)\dfrac{2\pi t}{T}}{(1-r)\dfrac{2\pi}{T}}\right]_0^{T/2}$$

$$= -\frac{E_0}{T}\left[\frac{\cos\,(1+r)\pi - \cos 0}{(1+r)\dfrac{2\pi}{T}} - \frac{\cos\,(1-r)\pi - \cos 0}{(1-r)\dfrac{2\pi}{T}}\right]$$

$$= -\frac{E_0}{2\pi}\left[\frac{\cos\,(1+r)\pi - 1}{1+r} + \frac{\cos\,(1-r)\pi - 1}{1-r}\right]$$

When r is odd, then A_r is equal to zero, because $\cos\ 0 = \cos\ 2\pi = \cos\ 4\pi = 1$.

When r is even, we have

$$A_r = \frac{E_0}{2\pi}\left[\frac{2}{1+r} + \frac{2}{1-r}\right] \qquad [\because\ \cos\ \pi = \cos\ 3\pi = \cos\ 5\pi = -1]$$

$$= \frac{E_0}{\pi}\cdot\frac{2}{(1+r)(1-r)}$$

$$= -\frac{2E_0}{\pi}\cdot\frac{1}{(1-r)(1+r)} \qquad r = 2, 4, 6, \ldots$$

$$= -\frac{2E_0}{\pi}\left[\frac{1}{(1)\,(3)}, \frac{1}{(3)\,(5)}, \frac{1}{(5)\,(7)}, \ldots\right]$$

Again, $$B_r = \frac{2}{T}\int_0^T E(t)\sin r\omega t\, dt$$

or, $$B_r = \frac{2E_0}{T}\int_0^{T/2} \sin\frac{2\pi t}{T}\sin\frac{2r\pi t}{T}dt$$

Proceeding as above, we can see that

$$B_r = E_0/2 \qquad \text{for} \qquad r = 1$$

and, $$B_r = 0 \qquad \text{for} \qquad r = 2, 3, 4,\cdots$$

Substituting these values of A_0, A_r and B_r in the above equations, $E(t)$ takes the form

$$E(t) = \frac{E_0}{\pi} + \frac{E_0}{2}\sin\omega t - 2\frac{E_0}{\pi}\left[\frac{1}{(1)\,(3)}\cos 2\omega t + \frac{1}{(3)\,(5)}\cos 4\omega t + \frac{1}{(5)\,(7)}\cos 6\omega t + \cdots\right].$$

3.12 ROTATIONAL AND TORSIONAL VIBRATION

Often the block foundations are idealized as SDF systems and they can vibrate in the following modes:

(a) Translational mode

(i) translation along the z-axis (Vertical vibrations)
(ii) translation along the x-axis (Longitudinal vibrations)
(iii) translation along the y-axis (Lateral vibrations)

(b) Rotational mode

(i) Rotation about the z-axis (Torsional vibrations)
(ii) Rotation about the x-axis (Rocking vibrations)
(iii) Rotation about the y-axis (Rocking vibrations)

It may be recalled that a mass in the Euclidean space with reference to Cartesian coordinate system has six degrees of freedom—three as translations having displacement components u, v, w in x, y and z directions and three as rotations in each of the three planes in the space as shown in Fig. 3.52(a).

The governing differential equation for vertical vibration is

$$m\ddot{z} + c\dot{z} + kz = 0$$

However, if damping is neglected, then

$$m\ddot{z} + kz = 0 \qquad \text{with} \qquad \omega_n = \sqrt{\frac{k}{m}}$$

Similarly, the equation for rocking motion (rotational vibration) with coordinate θ from the axis of rotation is

$$I_\psi\theta + k_\psi\theta = 0 \qquad \text{with} \qquad \omega_n = \sqrt{\frac{k_\psi}{I_\psi}} \tag{3.117}$$

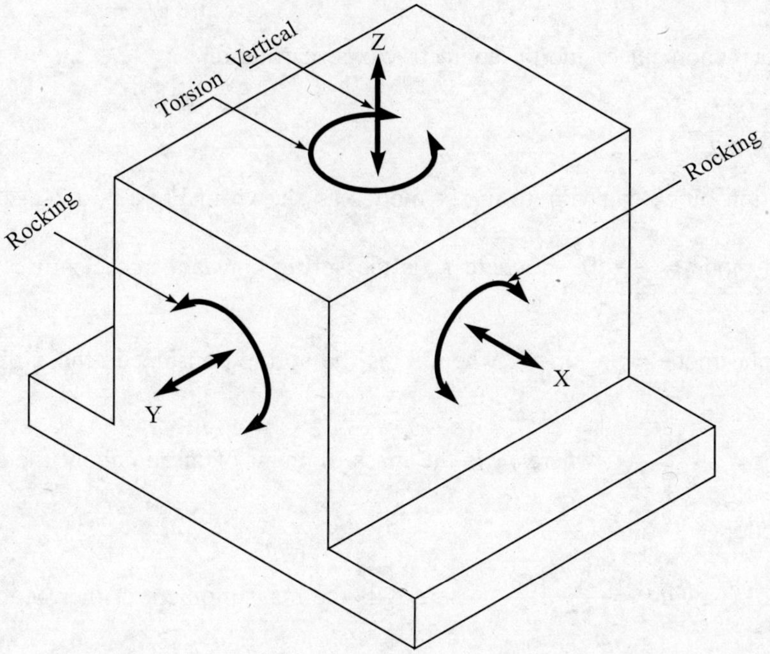

(a) Models of vibrations of a rigid block

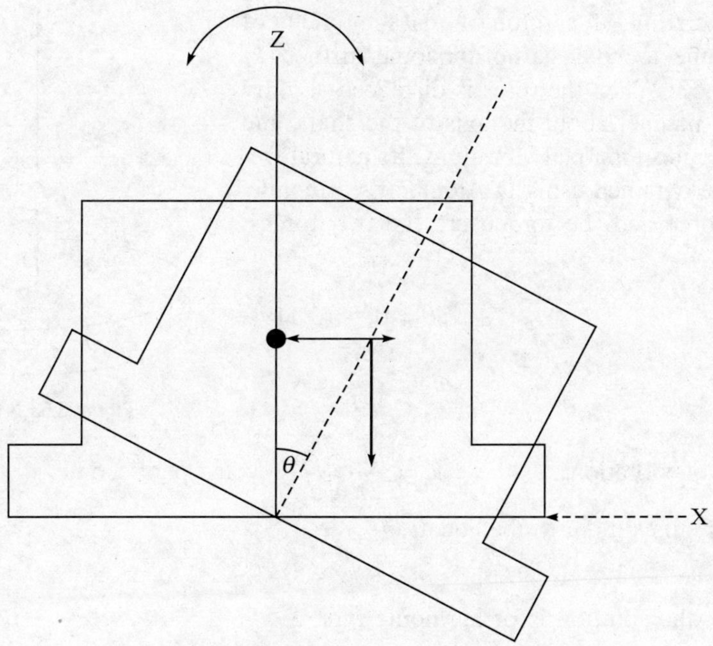

(b) Rocking vibrations of a rigid block

Figure 3.52

where,

I_ψ = mass moment of inertia about the axis of rotation

$$f_n = \frac{1}{2\pi}\sqrt{\frac{k_\psi}{I_\psi}} \tag{3.118}$$

A foundation block can vibrate in six modes as shown in Figure 3.52(a).

f_n Vertical mode $= \dfrac{1}{2\pi}\sqrt{\dfrac{k_z}{m}}$, where k_z is the spring constant in the z-direction.

f_n Horizontal mode $= \dfrac{1}{2\pi}\sqrt{\dfrac{k_x}{m}}$, where k_x is the spring constant in the x-direction.

f_n Rocking $= \dfrac{1}{2\pi}\sqrt{\dfrac{k_\psi}{I_\psi}}$, where I_ψ is the mass moment of inertia about the axis of rotation in rocking mode.

f_n Torsional (yawing) $= \dfrac{1}{2\pi}\sqrt{\dfrac{k_\theta}{I_\theta}}$, where I_θ is the mass moment of inertia about the axis of rotation in torsional mode.

Torsional vibrations

A system consisting of a rotor of mass moment of inertia I_θ is connected to a shaft of torsional stiffness k_θ as in Figure 3.53. When the rotor is displaced slightly in the angular manner about the axis of the shaft, and released it executes torsional vibrations. Its natural frequency may be obtained using D'Alembert's principle. The inertial force and the torque are the two forces, then

$$I_\theta \ddot{\theta} + k_\theta \theta = 0 \tag{3.119}$$

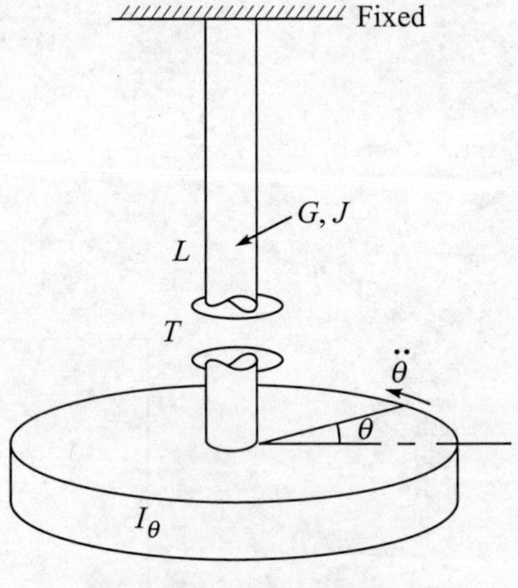

Figure 3.53 Torsional system.

So,

$$\omega_n = \sqrt{\frac{k_\theta}{I_\theta}}$$

or, frequency of vibrations, $f_n = \dfrac{1}{2\pi}\sqrt{\dfrac{k_\theta}{I_\theta}}$ cycles per second

Equation (3.119) is similar to the equation

$$m\dot{z} + kz = 0$$

and, therefore, the solution is of harmonic type

$$\theta = \theta_0 \cos \omega_n t + \frac{\ddot{\theta}_0}{\omega_n} \sin \omega_n t \tag{3.120}$$

where θ and $\ddot{\theta}_0$ are the initial angular displacement and angular acceleration, respectively. Using the analogy concept, which states that systems are analogous if they are described by the same type of differential equation, as such the theory developed for one system is applicable to its analogous system.

Combined rectilinear and rotational modes

The vibrations of a system having combined rectilinear (translation) and rotational modes have great relevance in soil dynamics, for such a coupled mode of vibrations affects block foundations.

Consider a body of mass m and moment of inertia $I = m \cdot k_g^2$ where k_g is the radius of gyration about the c.g. of the body and the body is capable of oscillating in directions z and θ.

Let at any instant, the body be displaced through a rectilinear distance z and angular distance θ as shown in Figure 3.54. Taking θ to be small, using D'Alembert's principle, the differential equation for motion may be written as below by considering the forces

$$m\ddot{z} + k_1(z - a\theta) + k_2(z + b\theta) = 0 \tag{3.121}$$

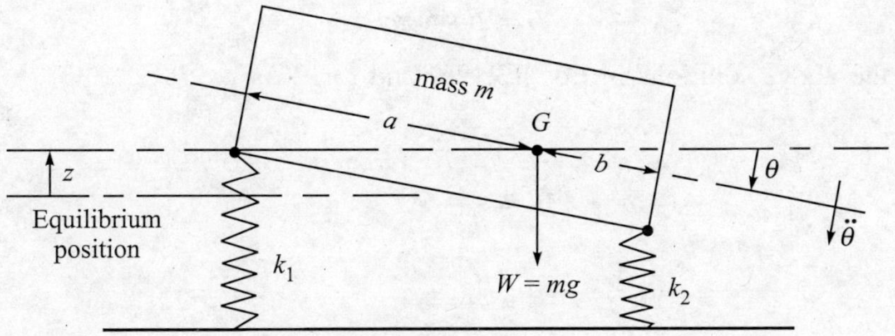

Figure 3.54 Analysis of system under combined rectilinear and angular modes.

Similarly, considering the moments in the respective directions acting on the system

$$I\ddot{\theta} - k_1(z - a\theta)a + k_2(z + b\theta)b = 0 \tag{3.122}$$

Let
$$\frac{k_1 + k_2}{m} = p$$

and,
$$\frac{k_1 a - k_2 b}{m} = q$$

and,
$$\frac{k_1 a^2 + k_2 b^2}{m} = r \tag{3.123}$$

Equations (3.121) and (3.122) reduce as follows:

$$z + pz = q\theta \tag{3.124}$$

$$\ddot{\theta} + r\theta = \frac{q}{k_g^2} \cdot z \tag{3.125}$$

where, $$k_g^2 = \frac{I}{m}$$

If $q = 0$, then $k_1 a = k_2 b$ and the two motions rectilinear (translational) and angular (rotational) can exist independently of each other with their relative natural frequencies $\sqrt{p}$ and $\sqrt{q}$. Thus, for the case of uncoupled system when $b = 0$, the natural frequencies in the translational mode and the rotation mode are

$$\omega_{n1} = \frac{k_1 + k_2}{m}$$

$$\omega_{n1} = \frac{k_1 a^2 + k_2 b^2}{I} \tag{3.126}$$

These two natural frequencies could have been obtained or written straightaway considering the system successively in rectilinear and angular mode only. But for the coupled vibrations, let the following be the solution of principal mode

$$z = Z \sin \omega t$$
$$\theta = \beta \sin \omega t \tag{3.127}$$

Substituting the above solutions in Eqs. (3.126) and (3.127),

$$(-\omega^2 + p)Z = q\beta$$

$$(-\omega^2 + r)\beta = \frac{p}{k_g^2} \cdot Z$$

Thus, $$\frac{Z}{\beta} = \frac{q}{p - \omega^2}$$

and, $$\frac{Z}{\beta} = \frac{\dfrac{r}{k_g^2} - \omega^2}{\dfrac{q}{k_g^2}}$$

Hence, $$\frac{q}{p - \omega^2} = \frac{\dfrac{r}{k_g^2} - \omega^2}{\dfrac{q}{k_g^2}}$$

which yields the frequency equation as

$$\omega^4 - \frac{p + r}{k_g^2}\omega^2 + \left(p - \frac{q}{k_g^2}\right) = 0$$

or, $$\omega_{n1,2}^2 = \left(\frac{1}{2}\right) p + \frac{r}{k_g^2} \pm \sqrt{\frac{1}{4}\left(p - \frac{r}{k_g^2}\right)^2 + \frac{q^2}{k_g^2}} \tag{3.128}$$

and,
$$\frac{Z}{\beta} = \frac{q}{p-\omega^2} \tag{3.129}$$

Example 3.7 A 2 kg mass is supported by two springs k_1 and k_2 placed at a distance 1.2 m and 1.6 m, respectively from the centre line passing through the c.g. of the mass as shown in Figure 3.55. The radius of gyration is 1.2 m^2. Determine the frequencies of vibrations both in coupled vertical mode and rotational mode. Determine their amplitude ratios as well. The values of k_1 = 45 N/m and k_2 = 55 N/m.

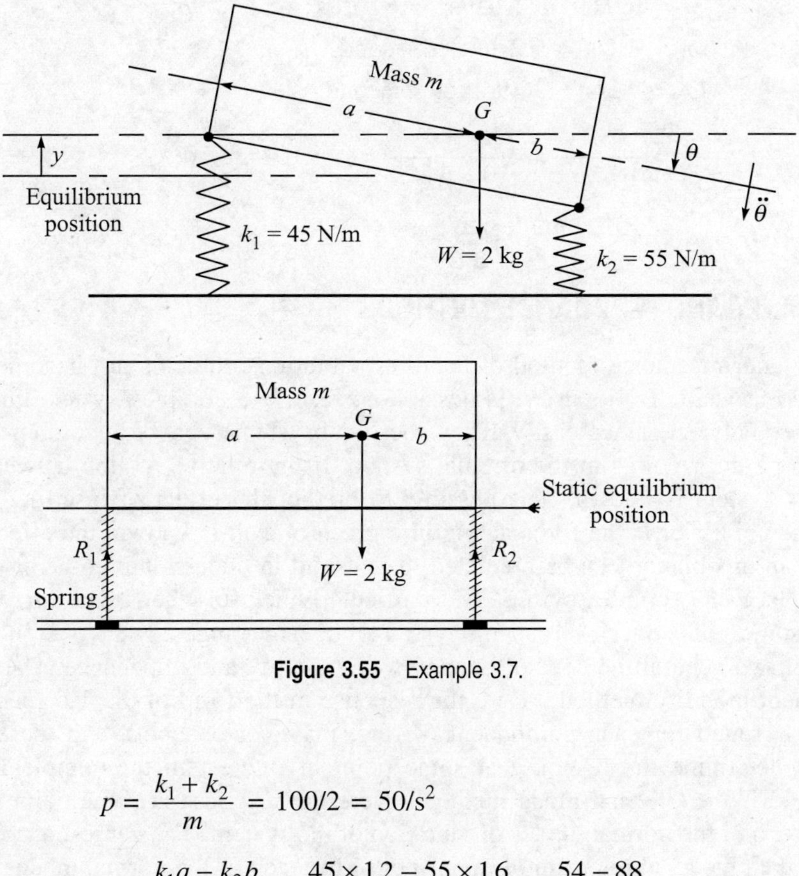

Figure 3.55 Example 3.7.

Solution:

$$p = \frac{k_1 + k_2}{m} = 100/2 = 50/\text{s}^2$$

$$q = \frac{k_1 a - k_2 b}{m} = \frac{45 \times 1.2 - 55 \times 1.6}{2} = \frac{54 - 88}{2} = -16.5 \text{ m/s}^2$$

$$r = \frac{k_1 a^2 + k_2 b_2}{2} = \frac{45(1.2 \times 1.2) + 55(1.6 \times 1.6)}{2} = 102.8 \text{ m}^2/\text{s}^2$$

$$\frac{r}{k_g^2} = \frac{102.8}{1.22} = 71.38/\text{s}^2$$

Hence,

$$[\omega_{n1,2}^2] = \left(\frac{1}{2}\right)p + \frac{r}{k_g^2} \pm \sqrt{\frac{1}{4}\left(p - \frac{r}{k_g^2}\right)^2 + \left(\frac{q}{k_g^2}\right)^2}$$

$$= (1/2)121.38/\text{s}^2 \pm \sqrt{-114.27/\text{s}^4 + 131.19/\text{s}^4}$$

$$= 60.69/\text{s}^2 \pm \sqrt{16.92/\text{s}^4}$$

$$= [60.69 \pm 4.10]/\text{s}^2 = 64.79/\text{s}^2 \text{ or } 56.59/\text{s}^2$$

$\therefore$ $\omega_1 = 8.1/\text{s} = 1.2$ cycles/second

$\omega_2 = 7.5/\text{s} = 1.19$ cycles/second

Also, $$\frac{Z}{\beta} = \frac{q}{p - \omega^2} = \frac{16.5}{50 - 64.79} = -\ 1.104 \text{ m/rad}$$

$$= 19.2 \text{ mm/degree}$$

3.13 MOBILITY AND IMPEDANCE METHODS

The mobility and impedance method of handling vibration problems has become increasingly popular in recent years. By its use it is possible to resolve a complex system into subsystems which can be analyzed rather easily. Then these subsystems can be recombined to find the response at a point or points in the original system. If the relative motion between two points of a complex system is known, it is possible to predict the effect of inserting a component between these points; or if the force acting in a given branch is known, the effect of inserting a component in this branch can be predicted. It is useful in understanding the physical interactions taking place in a complex system. It is particularly helpful when it is desired to obtain the response at some point over a broad impressed frequency range.

There are two general methods of dealing with mobility and impedance. The first is called the "component mobility method." Here the response in the form of displacement, velocity or acceleration is found for each component (spring, mass, or dashpot), and these can then be combined to determine the response at some point of interest in the complete system. The second method is the "normal mode method". Here the response at some point in the system is found for each of the normal modes of vibration of the system. The net response for any given frequency is then the algebraic sum of the response for each of the normal modes. This method is particularly suitable when it is desired to plot a response spectrum over a broad impressed-frequency range. Church, A.H. (1963) has presented a detailed use of these methods.

The discussion given here stresses the response to a steady state single harmonic force or torque. However, as outlined in Section 3.11, any periodic force or torque can be reduced to a number of harmonic excitations of integer multiple frequencies by means of harmonic analysis to express it in terms of a Fourier series. The response can be found for each of these terms acting individually; and each of these separate responses can be added algebraically to obtain the resultant periodic response. This procedure greatly extends the scope of the problems that can be handled by the mobility method.

Since rectilinear and torsional systems are completely analogous, we can use one set of symbols for both, and this will be done when discussing mobility and impedance methods. Thus, F designates either a force or a torque, m is a mass or a mass moment of inertia, x is either a rectilinear or angular displacements, and so on. This avoids needless repetition and makes the discussion more general from the point of view of explanations, diagrams and equations.

An important basic assumption of the mobility method is that the systems are linear at all times. That is, the motion response at any point is always proportional to the magnitude of the impressed harmonic force.

Two general types of response are important. One is "driving-point response", in which the response is found at the point where the excitation force acts. The other is "transfer response", which is the response at one point when excitation is applied at another point. The force–current analogies have been shown in Figure 3.56.

Definitions and principles

Mobility is defined as the ratio of the maximum value of the response of a point in a system produced by a harmonic force to the maximum value of that force, F. If the maximum value of the harmonic response is R, mobility $M = R/F$. Impedance is the inverse of mobility, or $Z = F/R$.

Mobility may be considered as the response per unit force, or as a measure of the readiness of a point to respond to a harmonic force. Impedance may be thought of as the force required to produce a unit response, or as a measure of the resistance of a point in the system to respond.

The form of the motion response R to be used depends somewhat on the personal preference of the analyst, the type of the problem and the impressed frequency. Upto this point in the text, the response has been measured in terms of the displacement z. However, in this section we denote it by x. For low frequencies, stress is proportional to displacement, which can be measured easily. Hence, displacement mobility M^D, which is the ratio of maximum displacement to maximum impressed force, or X/F may be used. This ratio is also known as "receptance" or "mechanical admittance".

Current emphasis is on velocity mobility (M^V), which is given as $M^V = V/F$. Velocity mobility is particularly applicable for higher frequencies of vibration where stress is more nearly a function of the velocity of response, or when impact loading may occur, since impact is a function of mass and velocity. When inertial loadings predominate, the use of acceleration mobility, $M^A = A/F$, may be desirable.

Forces and motions of systems having damping are conveniently expressed in terms of complex numbers, which generally simplify the calculations. The motion can then be described as

$$x = Xe^{j\omega t} \tag{3.130}$$

$$v = \dot{x} = j\omega Xe^{j\omega t} = Ve^{j\omega t} \tag{3.131}$$

$$\alpha = \dot{v} = j^2\omega^2 Xe^{j\omega t} = -\omega^2 Xe^{j\omega t} \tag{3.132}$$

where,

$$e^{j\omega t} = \cos\omega t + j\sin\omega t \tag{3.133}$$

Initially, mobility and impedance will be considered on a velocity basis. Later, these principles will be extended to include displacement and acceleration responses.

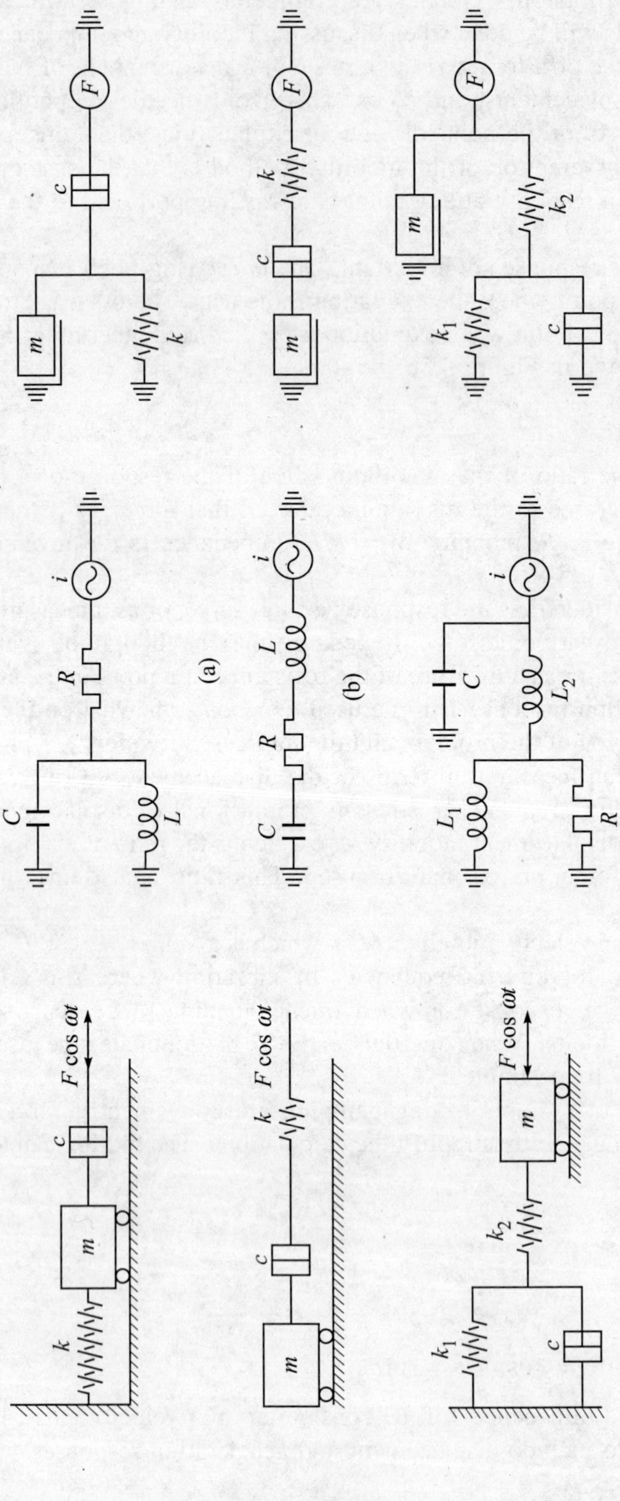

Figure 3.56 Force-current analogies.

When damping is present in a system, the force vectors are no longer collinear, and a phase angle becomes evident. Mobility is then a complex number. When a motion vector described by complex numbers is differentiated, the vector is multiplied by the product, $j\omega$, as demonstrated by Eqs. (3.130), (3.131) and (3.132). This operation multiplies the length of the vector by ω and the j term rotates the vector forward (in the direction of rotation) by 90 deg. From Eqs. (3.130), (3.131) and (3.132) then,

$$x = \frac{v}{j\omega} = \frac{\alpha}{j^2\omega^2} = -\frac{\alpha}{\omega^2} \tag{3.134}$$

By definition, the velocity mobility M^V is the ratio V/F, which generally involves a phase angle ψ. Thus, if the excitation is $f = F\cos\omega t$ and the response is $v = V\cos(\omega t + \psi)$, then $M^V = |V/F|(\psi)$, where M^V is a complex number whose modulus or absolute value is $|V/F|$ and whose argument, or phase angle, ψ, is the angle by which V leads F.

Mobility may be found for each of the various types of components. Mobility of a spring or dashpot is determined across the component. Thus, mobility is the motion of one end of the component relative to the other end divided by the maximum excitation force acting across or through the component. The mobility of a mass is the ratio of its maximum sinusoidal response relative to inertial space to the maximum sinusoidal excitation that causes the response.

Velocity response of components

Velocity mobility of springs, masses and dampers can be readily determined by straightforward procedures.

Springs: The maximum displacement of a spring is $X = F/k$. From Eq. (3.131), the maximum velocity is $V = j\omega X$, which may also be written as $V = j\omega F/k$. Hence,

$$M_S^V = \frac{V}{F} = \frac{j\omega F}{kF} = \frac{j\omega}{k} \tag{3.135}$$

A spring does not dissipate vibrational energy but merely stores it in potential form, which is recoverable. Note that the term for M_S^V is imaginary.

Since impedance of an element is the inverse of mobility, or F/V,

$$Z_S^V = \frac{k}{j\omega} = -\frac{jk}{\omega}$$

Masses: The relationship between force and motion is $F = mA$. But from Eq. (3.132), $A = j\omega V$. Hence, $V = A/j\omega = F/mj\omega$, and

$$M_m^V = \frac{F}{mj\omega F} = \frac{1}{j\omega m} = -\frac{j}{m\omega} \tag{3.136}$$

A mass in motion does not dissipate vibrational energy but stores it in kinetic form, which is recoverable. Again, note that the velocity mobility is imaginary.

The velocity impedance of the mass then becomes

$$Z_m^V = \frac{m\omega}{-j} = jm\omega$$

Dampers: If damping is viscous, $V = F/c$, and

$$M_d^V = \frac{F}{cF} = \frac{1}{c} \tag{3.137}$$

A damper dissipates vibrational energy in the form of heat, which is not recoverable. Because it does not contain j, this mobility term is real.
The impedance is

$$Z_d^V = c$$

Thus, velocity mobilities are complex numbers in which the imaginary terms represent energy storage and real terms represent energy dissipation.

Displacement response of components

As mentioned previously, response to excitation may be analyzed in terms of displacement rather than velocity. The ratio of maximum displacement to maximum value of the impressed force is common and will be called "displacement mobility," M^D. Equations for displacement mobility can be found by a process similar to that used for velocity mobility.

For a spring, $X = F/k$, so,

$$M_S^D = \frac{X}{F} = \frac{F}{kF} = \frac{1}{k} \tag{3.138}$$

For a mass, $F = mA$. But from Eq. (3.32), $A = -\omega^2 X$. Therefore, $F = -m\omega^2 X$, or $X = \dfrac{F}{-m\omega^2}$, and

$$M_m^D = \frac{X}{F} = -\frac{F}{(-m\omega^2)F} = -\frac{1}{m\omega^2} \tag{3.139}$$

For a damper, $V = F/c$. But $V = j\omega X$ or $X = \dfrac{F}{(cj\omega)F}$. Then,

$$M_d^D = \frac{X}{F} = \frac{F}{(cj\omega)F} = \frac{j}{c\omega^2} \tag{3.140}$$

In these expressions for response, the imaginary part represents dissipation of vibrational energy while the real part represents energy storage. These results are just the reverse of those obtained for velocity mobilities.

Impedance forms of Eqs. (3.138), (3.139) and (3.140) are:

$$Z_S^D = k$$
$$Z_m^D = -m\omega^2$$
$$Z_d^D = cj\omega$$

Acceleration response of components

The third possible method of calculating response is to find the ratio of maximum acceleration to maximum applied force. This method is designated "acceleration mobility," M^A.

From Eq. (3.132), $A = j^2\omega^2 X = -\omega^2 X$, or $A = j\omega V$. With these relationships, equations for acceleration mobility of the components become:

$$M_S^A = \frac{A}{F} = -\frac{\omega^2 X}{kX} = -\frac{\omega^2}{k} \tag{3.141}$$

$$M_m^A = \frac{A}{F} = -\frac{A}{mA} = \frac{1}{m} \tag{3.142}$$

$$M_d^A = \frac{A}{F} = \frac{j\omega V}{cV} = \frac{j\omega}{c} \tag{3.143}$$

In terms of acceleration impedance,

$$Z_S^A = -\frac{k}{\omega^2}$$

$$Z_m^A = m$$

$$Z_d^A = \frac{c}{j\omega} = -\frac{cj}{\omega}$$

In these expressions for acceleration response, the imaginary part represents the dissipation of vibrational energy while the real part is storage of energy. Hence, acceleration response is similar, in this respect, to displacement response.

The relationships just found are collected in Figure 3.57(a) for easy reference. Their comparison points out the relationship developed earlier, that is, for the various components, $M^V = j\omega M^D$ and $M^A = j\omega M^V = -\omega^2 M^D$.

Schematic diagrams

As already explained, a series combination has the same force acting through all the elements, and the total response is the sum of the response of all the elements. Combined all in terms of mobility as

$$M_C = \frac{R_C}{F_C} = \frac{R_1}{F_C} + \frac{R_2}{F_C} + \cdots = M_1 + M_2 + \cdots \tag{3.144}$$

A pure form of this situation is shown in Figure 3.57(b).

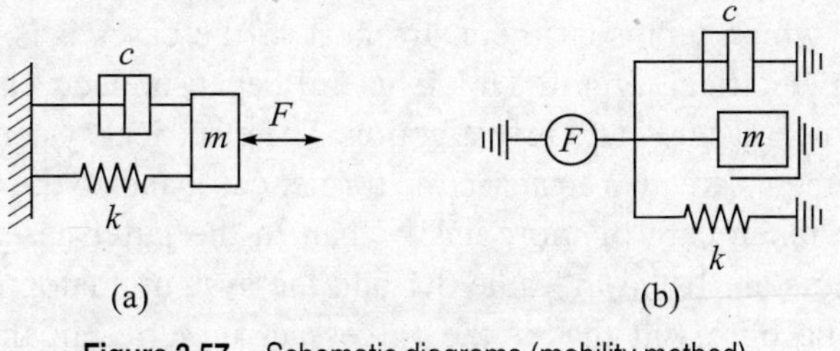

Figure 3.57 Schematic diagrams (mobility method)

A parallel combination has the same response for all the components, and the total force acting on the combination is the sum of the forces acting on or through the components.

Consequently obtaining in terms of mobility as

$$M_C = \frac{R_C}{F_C} = \frac{R_C}{F_1 + F_2 + \cdots}$$

$$= \frac{1}{\dfrac{1}{M_1} + \dfrac{1}{M_2} + \cdots} \tag{3.145}$$

Equation (3.145) can be simplified by using impedance to give

$$M_C = \frac{1}{Z_C} = \frac{1}{Z_1 + Z_2 + \cdots}$$

or,

$$Z_C = Z_1 + Z_2 + \cdots \tag{3.146}$$

Before writing equations, it is generally best to draw a schematic diagram or dynamic circuit based on the force-current analogy to show whether the elements are combined in series or parallel. The value and use of these diagrams become more evident with complex systems.

Components that are attached to a fixed support are shown "grounded". The ground connection also applies to mass elements as they have an acceleration with respect to inertial space, that is, relative to the ground or a fixed support.

Figure 3.57 shows the schematic diagrams to correspond to the mechanical systems. The force–current circuits in between are identical with the schematic diagrams except for the symbols of the components and the identifying letters.

It is important to realize that when the impedance at a point is zero, a resonant condition exists. Then $F/R = 0$ and a zero force produces a finite amplitude, or a finite force produces an infinite response. This information is quite useful in determining the natural frequency of a system by equating Z_C in Eq. (3.146) to zero and solving for the frequency. Although resonance also occurs when the mobility, or R/F is infinite, the resulting equation cannot be solved for the corresponding frequency.

3.14 ANALOGUE METHOD

The analogy between electrical and mechanical (physical) systems is through the similarity of differential equations, which represent them. In most of the cases it is possible to represent a physical system by an electrical circuit. This is useful because electrical analogies yield much more readily to experimental study and investigation. The electrical system can be assembled and installed. The measurements are also easier in the former case, and so the alterations in the values of the element can be taken care of more easily than in the latter cases.

The analogous terms are listed in Table 3.1 and the system analogies have been presented in Table 3.2. It may be observed that if the forces act in series in the physical system, the electrical elements representing these forces are in parallel.

Table 3.1 Analogous terms

Electrical system	*Physical/vibrational system*
Electromagnetic energy	Kinetic energy
Electrostatic energy	Potential energy
Charge Q, coulomb	Displacement, cm
Current, ampere	Velocity, cm/s
Voltage, volt	Force, newton
Inductance, henry	Mass, kg-s^2/cm
Resistance, ohm	Damping coefficient, kg-s/cm
Capacitance, farad	Flexibility
1/capacitance	Stiffness, kg/cm
Loop	Degree of freedom
Kirchhoff's law	D'Alembert's principle
Switch closed	Force applied
Element common to two loops	Coupling element

Table 3.2 Mechanical–electrical analogies

Mechanical			*Electrical equivalents*			
			Force–voltage analogy		*Force–current analogy*	
	Rectilinear	Torsional				
Force or torque	f	f	Voltage	e	Current	i
Damping	c	ct	Resistance	R	Conductance	$1/R$
Mass or inertia	m	J	Inductance	L	Capacitance	C
Compliance	$1/k$	$1/kt$	Capacitance	C	Inductance	L
Velocity	u	0	Current	i	Voltage	e
Displacement	x	0	Charge	q		

Damping

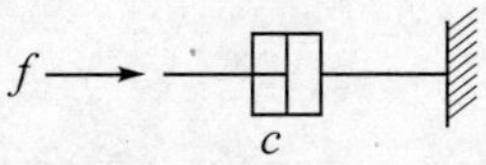

$f = cu = c\dot{x}$

Power = $fu = cu^2$

Resistance

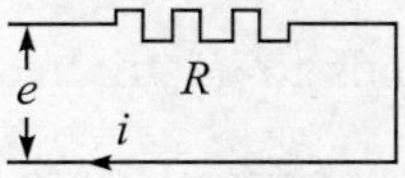

$e = Ri = R\dot{q}$

Power = $ei = Ri^2 = R\dot{q}^2$

Resistance

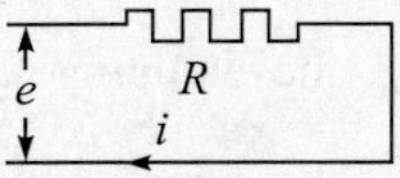

$i = e/R$

Power = $ei = e^2/R$

Mass

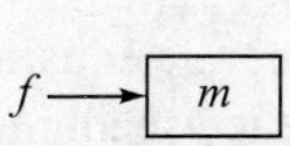

$f = m\dot{u} = m\ddot{x}$

Kinetic energy = $\frac{1}{2}mu^2$

Inductance

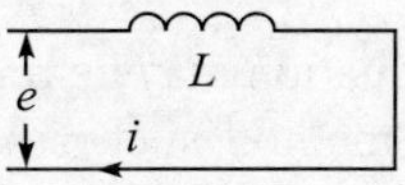

$e = L\dfrac{di}{dt} = L\ddot{q}$

Energy = $\frac{1}{2}Li^2$

Capacitance

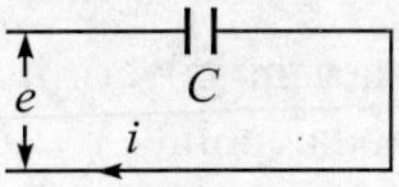

$i = C\dot{e}$

Energy = $\frac{1}{2}Ce^2$

(*Contd.*)

Table 3.2 Mechanical-electrical analogies (Contd.)

Mechanical	*Electrical Equivalents*	
	Force–voltage analogy	*Force–current analogy*
Spring	Capacitance	Inductance
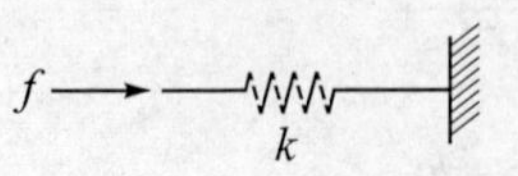	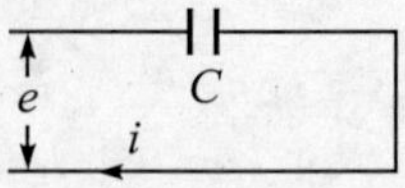	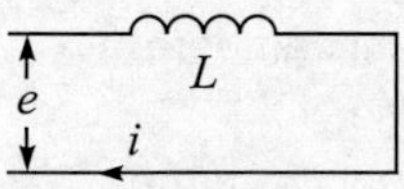
$f = kx = k\int u\,dt$	$e = (1/C)q = \frac{1}{c}\int i\,dt$	$i = (1/L)\int e\,dt$
Potential energy $= \frac{1}{2}F^2/k$	Energy $= \frac{1}{2}Ce^2$	Energy $= \frac{1}{2}Li^2$
Rectilinear System Torsional	Circuit	Circuit
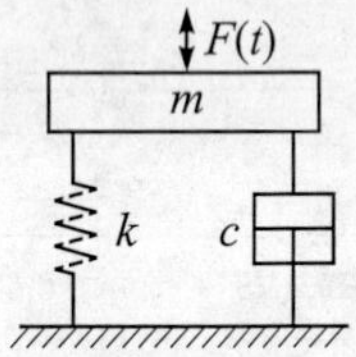 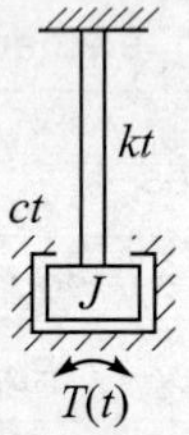	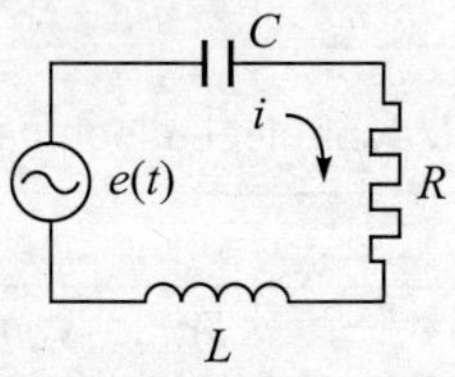	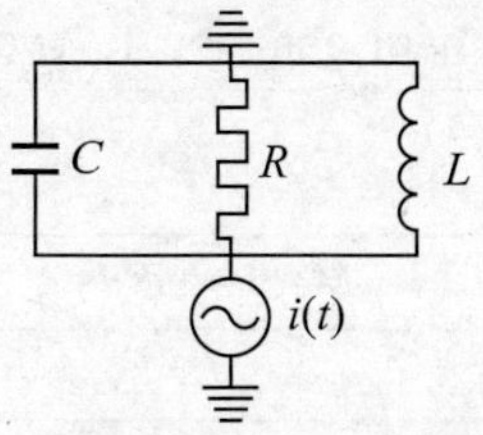
At point: $\Sigma f = 0$; $\Sigma t = 0$	For loop; $\Sigma (\Delta e) = 0$	At point : $\Sigma i = 0$
$m\dot{u} + cu + k\int u\,dt = F(t)$	$L\frac{di}{dt} + Ri + \frac{1}{c}\int i\,dt = e(t)$	$C\dot{e} + \frac{e}{R} + \frac{1}{L}\int e\,dt = i(t)$
$mx + cx + kx = F(t)$	$L\ddot{q} + R\dot{q} + \frac{q}{c} - e(t)$	

- $\frac{W}{g} = m$, the mass corresponds to L, the inductance
- e, the damping factor corresponds to R, the resistance
- k, the spring constant (stiffness) corresponds to $\frac{1}{C}$, the reciprocal of the capacitance.
- F, the force corresponds to E, the voltage
- ω, the forcing frequency of the vibration system correspond to ϕ, the electrical forced frequency

Further the governing differential equation may be replaced by the algebraic equation at each of the nodal points by any numerical method such as the finite difference technique. In the electrical analogy each of the loading terms will be represented in finite differences form as the current flowing into each of the nodal points. Thus, in the circuit analogy the independent variable time is represented continuously as such while the space variable is represented at discrete values by the nodal points.

3.14.1 Dimensional Analysis

Mathematical modelling is often required for the solution of an engineering problem. The modelling procedure is the same for different engineering disciplines, although the details of the modelling vary between disciplines. Certain assumptions are used in the modelling of must physical systems.

Force, length and time are involved in vibration problems as the fundamental units. In this context, the famous Buckingham π-theorem states that a system with n independent variables and j fundamental unit will have $(n - j)$ dimensionless parameters in its solution.

Consequently the method consists in selecting those variables which contain all three of the fundamental dimensions among them. The similarity principle can be interpreted as meaning that the non-dimensional factors in the physical vibrational system should equal to those in the analogous electric system. The problem of finding out the numerical values of electrical elements to be analogous to mechanical elements, then reduces to identifying the non-dimensional groups. This may be facilitated by the π-theorem. Several methods for determining a set of π-groups for a particular problem are now available [see Section (4.14)].

3.15 NONLINEAR VIBRATIONS

As explained, if the basic components associated with the vibration analysis, such as the spring, mass and the dashpot behave linearly, the resulting vibration is referred to as linear vibration. On the other hand, if one or more of the basic components behaves in a nonlinear manner, the resulting vibration is said to be non-linear vibration. As such the analysis deals with nonlinear differential equation of motion. The sources of nonlinearity of the system can be identified as the following:

(a) Geometric nonlinearity
(b) Material nonlinearity
(c) Nonlinear force–displacement relationship for other reasons.

Most physical systems can be represented by linear differential equations, the types of which have been dealt with previously. A general equation of motion is of the type

$$m\ddot{z} + c\dot{z} + kz = F(t) \tag{3.147}$$

In this equation for linear systems, the inertia force, the damping force and the spring force are linear function of z, $\dot{z}$ and $\ddot{z}$, respectively. This is not in the case of nonlinear systems. A general equation for a nonlinear system is

$$m\ddot{z} + \phi(\dot{z}) + f(z) = F(t) \tag{3.148}$$

in which the damping force and the spring force are nonlinear function of $\dot{z}$ and z. The analysis of motion of a nonlinear system to external excitation is an extremely difficult exercise. Daniel Bernoulli principle of superimposition no longer holds good, so that the general solution cannot be found from the synthesis of particular solution.

The general equation of motion for a SDF system may also be expressed as

$$m\ddot{z} + \phi(z, \dot{z}, t) = 0 \tag{3.149}$$

The systems with nonlinear characteristics are treated as nonlinear systems and their motion are called nonlinear vibrations. The differential equation for a pendulum (having weight, w and length, l) is given by

$$I\,\ddot{\theta} + wl \sin\theta = 0$$

or,
$$\ddot{\theta} + \frac{g}{l}\sin\theta = 0 \tag{3.150}$$

In the linear theory, which may be regarded as an approximation of zero coder, it is assumed that for small angles $\sin\theta = \theta$. The well-known solution for the period

$$T = 2\pi\sqrt{\frac{l}{g}}$$

is obtained. It is to be noted that the oscillations are isochronic under this assumption. The isoclines are defined as the lines of equal slope and such graphical representations are the principal methods used in non-linear analysis. If the amplitude is not small, the restoring moment is proportional to $\sin\theta$, which can be approximated by a power series. Taking only the first two terms, the differential equation takes the form (see Figure 3.58)

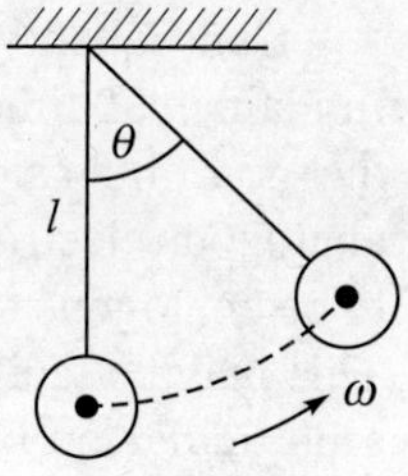

Figure 3.58

$$\ddot{\theta} + \frac{g}{l}\left(\theta - \frac{\theta^3}{6}\right) = 0 \tag{3.151}$$

Taking the following as angular frequency ω_1 and θ_0 to be angular amplitude,

$$\omega_1^2 = \frac{g}{l} - \frac{g}{gl}\left(\theta_0^2\right) \tag{3.152}$$

The period of oscillation is

$$\tau = \frac{2\pi}{\omega_1} = 2\pi\sqrt{\frac{l}{g}}\,\frac{1}{\sqrt{1-\dfrac{1}{8}\theta_0^2}}$$

$$\tau = 2\pi\sqrt{\frac{l}{g}}\left(1+\frac{1}{16}\theta_0^2\right)$$

The nonlinearity due to the spring force can also be identified with non-Hookean or nonlinear springs. In such springs when the restoring force deformation relationship is as shown in Figure 3.59(a), it is called hardening spring and if such relationship is as shown in Figure 3.59(b), it is called softening spring.

One important type of nonlinearity arises when the restoring force of a spring is not proportional to its deformation. Figure 3.59(a) shows the static load–displacement curve for a nonlinearly elastic "hardening spring", where the slope increases as the load increases. The dashed line in Figure 3.59(a) is tangent to the curve at the origin, and its slope k represents the initial stiffness of the spring. Similarly, Figure 3.59(b) depicts the load–displacement curve for a nonlinearly elastic "softening spring", where the slope decreases as the load increases. In each of these Figures the curve is symmetric with respect to the origin, and the spring is said to have a symmetric restoring force. If the load–displacement curve is not symmetric with respect to the origin, the spring is said to have an unsymmetric restoring force.

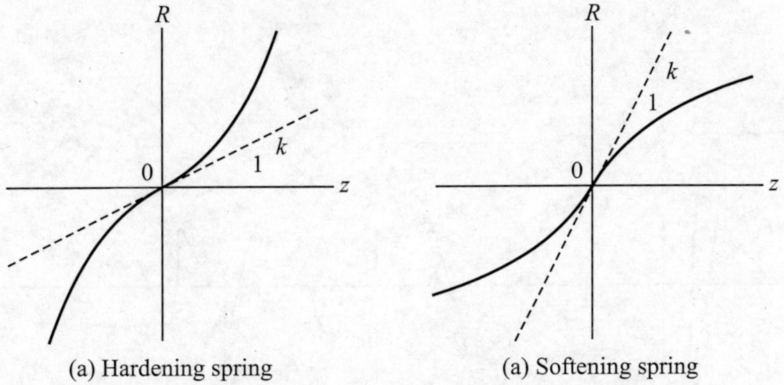

Figure 3.59 Force-displacement relationship for hardening and softening spring.

3.16 RANDOM VIBRATIONS

In the study of vibrations in the previous sections, three types of excitation functions, namely, harmonic, periodic and non-periodic, have one common characteristic. These functions are such that their mathematical expressions can be written in advance for any time that will determine their values and as such these functions are said to be deterministic. The response of systems to deterministic excitation is also deterministic. For linear systems, there is no difficulty in expressing the response to any arbitrary deterministic excitation in some closed form, such as the convolution integral as discussed in Section 3.8.1. However, there are a number of physical phenomena that result in non-deterministic manner where future instantaneous values cannot be written in a deterministic sense. Such phenomena have one thing in common: the unpredictability of their value at any future time. Meirovitch [1975] has explained the state of art for random vibrations with great excellence.

Random phenomenon

There are many physical phenomena, however, that do not exhibit time description. Examples of such phenomena are the height of waves in a rough sea, the intensity of an earthquake, etc. If the intensity of earth tremors is measured as a function of time, then the record of one tremor will be different from that of another one. In other words, if an experiment is conducted several times with all the variables remaining the same in each case, and the outputs or records continually differ from each other, then the process is random. The degree of randomness in a process depends upon

(a) The understanding of the variable parameter associated with the experiments.
(b) The ability to control them.

The reasons for the difference are many and varied, and have little or nothing to do with the measuring instrument. Phenomena whose outcome at a future instant of time cannot be predicted are classified as non-deterministic, and referred to as random. A typical random function is shown in Figure 3.60.

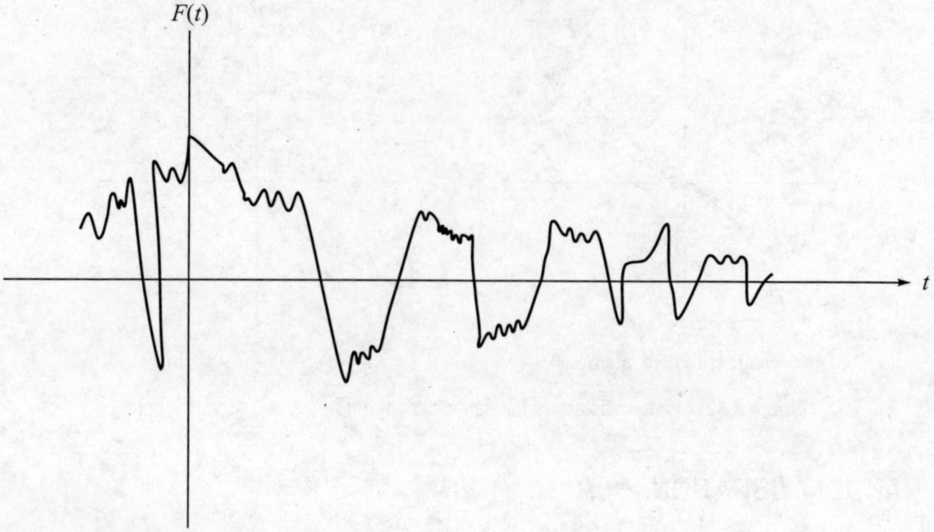

Figure 3.60 A typical random function.

The response of a system to a random excitation is also a random phenomenon. Because of the complexity involved, the description of random functions in terms of time does not appear as a particularly meaningful approach, and new methods of analysis must be adopted. Many random phenomena exhibit a certain pattern, in the sense that the data can be described in terms of certain averages. This characteristic of random phenomena is called statistical regularity. If the excitation exhibits statistical regularity, so does the response. In such cases it is more feasible to describe the excitation and response in terms of probabilities of occurrence rather than a deterministic description. In such situations certain averaging procedure can be applied to establish gross characteristics, which are hopefully useful in earthquake resistant design. The earthquake excitation is essentially a random function of time, and hence the response should be obtained in probabilistic terms using random vibration theory.

Ensemble averages

The total description of a random process may be expressed through its ensemble. Any example can yield several profiles generated by repeated trials to give a large number of samples expressed by $x_i(t)$. On different samples $x_1(t_1)$, $x_2(t_1)$, ..., $x_n(t_1)$ the instantaneous value at any arbitrary time t_1 are given by

$$E[x]_{t=t1} = \lim_{n \to \infty} \frac{1}{n} \cdot \sum_{i=1}^{n} x_i(t_1)$$

$$E[x^2]_{t=t1} = \lim_{n \to \infty} \frac{1}{n} \cdot \sum_{i=1}^{n} x_i^2(t_1)$$

$$E[x^m]_{t=t1} = \lim_{n \to \infty} \frac{1}{n} \cdot \sum_{i=1}^{n} x_i^m(t_1) \qquad (3.183)$$

Stationary random process

If the averages in equation (3.183) are independent of time, i.e.,

$$E[x^m]_{t=t_i} = E[x^m]_{t=t_j} \qquad m = 1, 2, 3, \ldots, \infty \tag{3.184}$$

Then the random process is stationary. In other words, a random process is said to be stationary, if its probability distributions are invariant under a shift of time scale, that is independent of origin.

Temporal averages

The various averages can be calculated from any one time history, say the kth one and then such time averages are called temporal averages and can be expressed as

$$x_m = \lim_{T \to \infty} \frac{1}{T} \int_{-T/2}^{T/2} x_k^m(t)\, dt \qquad m = 1, 2, \ldots, \infty \tag{3.185}$$

Ergodic random process

If the temporal averages calculated from various time histories for $k = 1, 2, \ldots$ are found to be identical, any single time history from the ensemble provides complete information about the characteristics of the random process. Such a process is called ergodic. An ergodic process must necessarily be stationary, while a stationary process need not necessarily be ergodic.

Crosscorrelation function

Correlation is a measure assessing the relationship between two functions as shown in Figure 3.61(a). Considering two functions $x(t)$ and $f(t)$, the crosscorrelation between them is given by the average of the product $x(t) \cdot f(t + \tau)$

$$R(t) = \lim_{T \to \infty} \frac{1}{T} \int_{-T/2}^{T/2} x(t) \cdot f(t + \tau)\, d\tau \tag{3.186}$$

Autocorrelation function

The autocorrelation function $R(t)$ refers to a function $f(t)$ which correlates with itself. Thus,

$$R(t) = \lim_{T \to \infty} \frac{1}{T} \int_{-T/2}^{T/2} f(t) \cdot f(t + \tau)\, d\tau \tag{3.187}$$

For statistical study a function is introduced and defined as autocorrelation function. This function describes (on the average) how a particular instantaneous amplitude value depends upon the previously occurring instantaneous amplitude values, where $f(t)$ and $f(t + \tau)$ are the random variables at two different times t and $(t + \tau)$. See Figure 3.61(b).

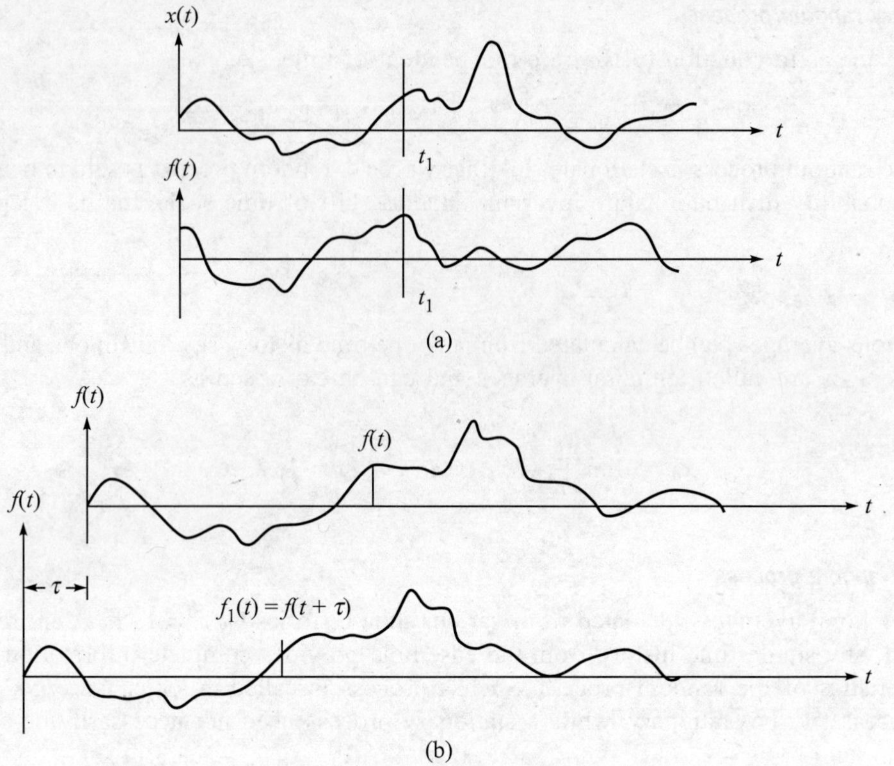

Figure 3.61 (a) Correlation between $x(t)$ and $f(t)$; and (b) definition of autocorrelation function.

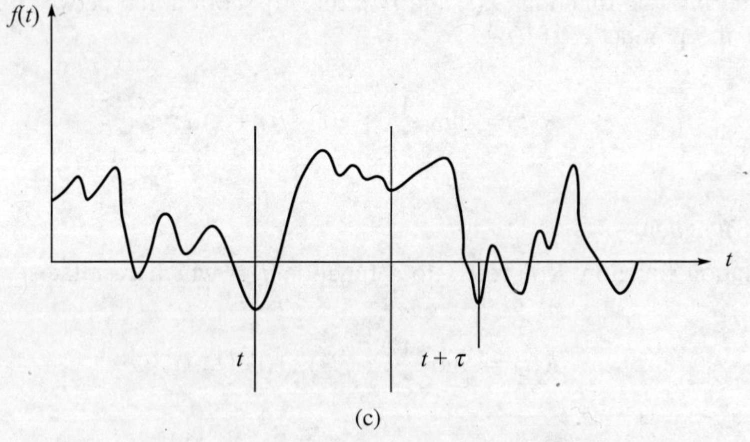

Figure 3.62 Evaluation of autocorrelation function.

If the random process is stationary, and if the temporal mean value and the temporal autocorrelation function are the same, irrespective of the time history over which these averages are calculated, the process is said to be ergodic. Hence, for ergodic processes the temporal mean value and autocorrelation function (see Fig. 3.62) calculated over a representative sample function must by necessity be equal to the ensemble mean value and autocorrelation function, respectively.

Random vibrations are met rather frequently in nature and may be characterized as vibratory processes in which the vibration particles undergo irregular motion cycles that never repeat themselves exactly, see Figure 3.63. To obtain a complete description of the vibration, an infinitely long time record is thus theoretically necessary. This is of course an impossible requirement, and finite time records would have to be used in practice. Even so, if the time record becomes too long it will also become a very inconvenient means of description and other methods have therefore been devised and are commonly used. These methods have their origin in statistical mechanics and communication theory and involve concepts such as amplitude probability distributions and probability densities, and continuous vibration frequency spectra in term of mean square spectral densities.

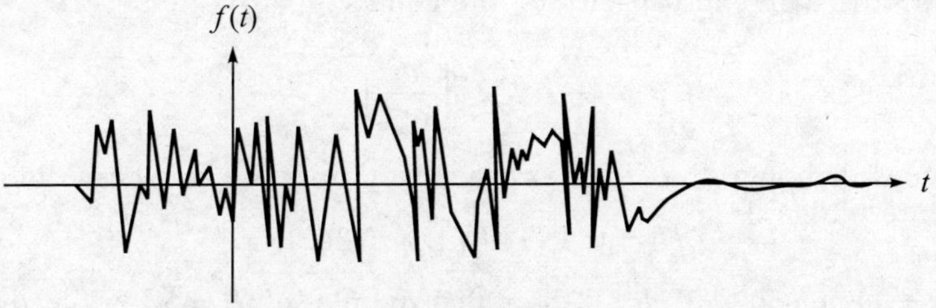

Figure 3.63 Example of a random vibration signal.

PROBLEMS

3.1 For a pure harmonic motion show that the average response is given by

$$Z_{\text{average}} = (1/\text{T})\int_0^{\tau} |z| \, d\tau$$

If Z_{rms} is the root mean square value, show that

$$Z_{\text{rms}} = \frac{\pi}{2\sqrt{2}} Z_{\text{average}}$$

3.2 What is the significance of form factor F_f and crest factor F_c in identifying a wave shape. For plane harmonic motion, show that

(a) $F_f = \dfrac{\pi}{2\sqrt{2}} = 1.11$

(b) $F_c = \sqrt{2} = 1.414$

3.3 The one-dimensional displacement of a system is given by

$$z(t) = 0.5e^{-0.2t}\sin 15t.$$

What is the maximum acceleration?

3.4 Show that the energy dissipated per cycle for viscous friction can be expressed by

$$W_d = \frac{\pi F_0^2}{k} \frac{2\zeta(\omega/\omega_n)}{\left[1-\left(\frac{\omega}{\omega_n}\right)^2\right]^2 + \left[2\zeta\left(\frac{\omega}{\omega_n}\right)\right]^2}$$

3.5 A spring-mass system has spring constant k (kg/cm) and the weight of the mass is W (kg). It has natural frequency of vibration as 12 Hz. When an extra 2 kg weight is coupled to W, the natural frequency reduces by 2 Hz. Find k and W.

3.6 Show that for viscous damping the energy loss factor is independent of the amplitude and proportional to the frequency. Find the equation for free vibration of SDF system in terms of energy loss factor η at resonance.

3.7 Logarithmic decrement δ for small damping is equal to $\delta \cong 2\pi\zeta$. Show that δ is related to the specific damping capacity by the equation

$$\frac{W_d}{U} = 2\delta\left(\frac{\omega}{\omega_n}\right)$$

3.8 For viscous damping, the complex frequency response can be written as

$$H(r) = 1/[1 - r^2) + i(2\zeta r)]$$

where $r = \omega/\omega_n$ and $\zeta = c/c_{cr}$. Show that the plot of $H = x + iy$ leads to the equation

$$x^2 + (y + 1/4\,\zeta r)^2 = (1/4\,\zeta r)^2$$

which cannot be a circle since the centre and the radius depend on the frequency ratio.

3.9 Find the Fourier series representation of the periodic force given by a sine curve as shown in Figure P3.9. Draw the frequency spectrum.

Ans. $F(t) = 2A/\pi + \sum_{n=2,4}^{\infty} - 4A/\pi\ \{1/(n^2 - 1)\}\ \cos n\omega t$

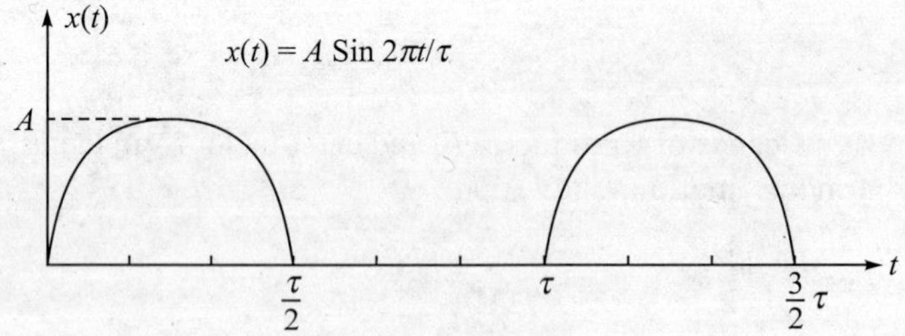

Figure P3.9

3.10 A weight W = 15 N is vertically suspended by a linear spring having stiffness k = 2 N/mm. Find the natural frequencies of free vibration. If the initial displacement is 2.5 cm and the initial velocity is 2.5 cm/s, what will be the amplitude, velocity and acceleration at t = 1.5 s.

3.11 The response of an SDF damped system (m, c and k) is given by $x = X\cos(\omega_n t + \beta)$ and the phase diagram is shown in Figure P 3.11. Discuss the relative motion and phase difference of displacement, velocity and acceleration. If OB, OC and OD are their maximum values, then how will you obtain their instantaneous values in phase diagram.

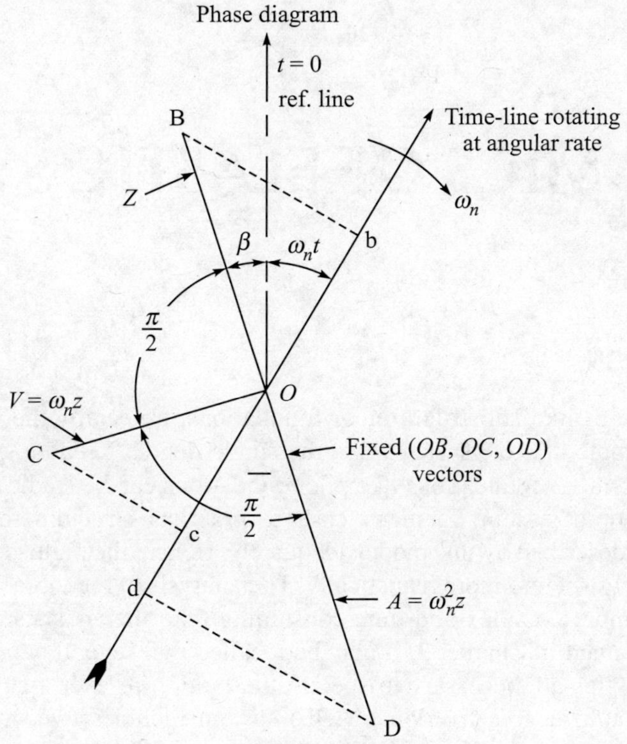

Figure P3.11

3.12 Determine the mean value and the mean square value for a sine wave with steady component of the form

$$x(t) = 50 + 100 \sin 3t$$

3.13 An SDF system is excited by the dynamic force of the form

$$F(t) = 500 \cos 5t + 1000 \cos 10t + 1500 \cos 15t \quad \text{[kg]}$$

Determine the spectral function and the mean square value of the response. The system consists of mass 50 kg-s^2/cm, stiffness 500 kg/cm and damping of 5%.

3.14 Determine the autocorrelation function of an ergodic random process $x(t)$. Each sample function is a square value of amplitude Af and period Tf.

4

DYNAMICS OF ELASTIC SYSTEM

4.1 INTRODUCTION

The practical structures like substructures or foundations, especially the machine foundations, can be treated as simple discrete systems like the single degree freedom system as discussed in Chapter 3. However, many foundations like pile foundations can be studied more realistically as a multi-degree freedom system. In many cases a practical structure represented by a SDF system does not get described by this model adequately. Hence, the multi-degree freedom system represents a practical structure more realistically. The analysis of the vibrations of a multi-degree freedom system is more complex and time-consuming than the analysis of the vibrations of a SDF system as explained in Chapter 3. It may be recalled that several numbers of kinematically independent coordinates are necessary to specify the motion of every particle contained in the system. Thus, the multi-degree freedom (MDF) system includes two, three and n-degree of freedom systems or in other words the MDF system requires two or more coordinates to describe its motion.

Modelling with a finite number of degrees of freedom provides an approximation to the behaviour of the system. All substructures or foundations or soil retaining structures are continuous systems with an infinite number of degrees of freedom. In this chapter, the vibrations of multi-degree freedom system and continuous system will be discussed with special reference to various problems of soil dynamics. The reason as to why a practical structure is reduced to a discrete system is due to the fact that the analysis of the continuous system is much more involved and complex, as for such cases the mass is inseparable from the elasticity of the system.

The study of dynamics of the elastic system may be considered as the study of a body or structure in dynamic equilibrium. In structural dynamics, the superstructures subjected to dynamic loads including seismic loads are considered. In soil dynamics the substructure or foundation supporting the superstructure is studied for dynamic equilibrium. The various foundations or their structural parts are idealized as single degree freedom system, two degree freedom system, multi-degree freedom system, or continuous system as shown in Figure 4.1. In structural dynamics, the study of beams, plates and shells under dynamic equilibrium is important, however, in soil dynamics the study of beams plates and shells resting on foundations is relevant. As such to solve problems of geotechnical earthquake engineering, the dynamic equilibrium of beams, plates and shells on elastic foundation will be presented.

The structural elements resting on suitable foundations have to be idealized as beam on elastic foundation, or plate on elastic foundation, or shell on elastic foundation. Winkler (1867) developed equations and also provided solutions for static analysis of beams on elastic

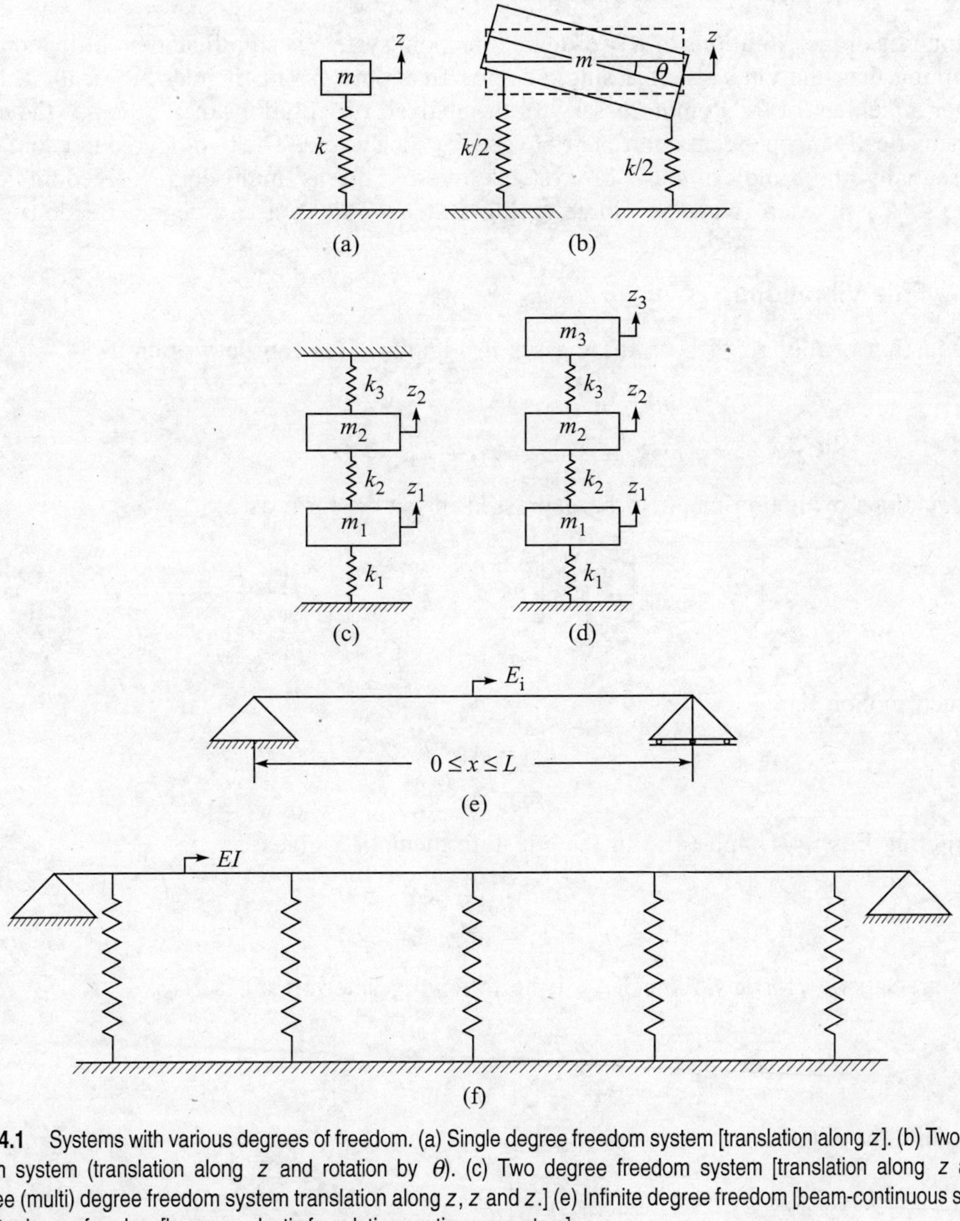

Figure 4.1 Systems with various degrees of freedom. (a) Single degree freedom system [translation along z]. (b) Two degree freedom system (translation along z and rotation by θ). (c) Two degree freedom system [translation along z and z]. (d) Three (multi) degree freedom system translation along z, z and z.] (e) Infinite degree freedom [beam-continuous system]. (f) Infinite degree freedom [beam on elastic foundation continuous system].

foundations (Winkler model). In soil dynamics and earthquake engineering, the dynamic analysis of beam and plate on elastic foundation will be presented. The formulation or representation of foundation as linear elastic model or as a complex model representing the ground conditions more realistically, depends on the importance of the foundation under study. However, for mathematical simplicity the linear elastic model (Winkler type) is more popular.

4.2 Vibrations of Two-Degree Freedom System

The analysis of the vibrations of a two-degree freedom system is significantly a little more time-consuming than the analysis of a single-degree freedom system. It may be recalled that the number of degrees of freedom necessary for the analysis of vibrations of a system is the number of kinematically independent coordinates to specify the system. The two-degree freedom system is essentially the simplest class of systems referred to as multi-degree freedom system. Figure 4.1(c) shows a two mass, three spring system. This is a two-degree freedom system.

4.2.1 Free Vibrations

The equation of motion of both the masses in Figure 4.1(c) can be written as

$$m_1\ddot{z}_1 + k_1 z_1 + k_2(z_1 - z_2) = 0 \tag{4.1}$$

$$m_2\ddot{z}_2 + k_2(z_2 - z_1) + k_3 z_2 = 0 \tag{4.2}$$

The equations of motion can also be expressed in matrix form as

$$\begin{bmatrix} m_1 & 0 \\ 0 & m_2 \end{bmatrix}\begin{bmatrix} \ddot{z}_1 \\ \ddot{z}_2 \end{bmatrix} + \begin{bmatrix} k_1+k_2 & -k_2 \\ -k_2 & k_2+k_3 \end{bmatrix}\begin{bmatrix} z_1 \\ z_2 \end{bmatrix} = 0 \tag{4.3}$$

or,

$$[m]\{\ddot{z}\} + [k]\{z\} = \{0\}$$

For such motion, let

$$z_1 = A_1 e^{i\omega t} \tag{4.4}$$

$$z_2 = A_2 e^{i\omega t} \tag{4.5}$$

Substituting Eqs. (4.4) and (4.5) in the equation of motion gives

$$(k_1 + k_2 - \omega^2 m_1)A_1 - k_2 A_2 = 0$$

$$-k_2 A_2 + (k_2 + k_3 - \omega^2 m_2)A_2 = 0$$

which are satisfied for any A_1 and A_2 if the following determinant is zero

$$\begin{vmatrix} k_1+k_2-\omega^2 m_1 & -k_2 \\ -k_2 & k_2+k_3-\omega^2 m_2 \end{vmatrix} = 0 \tag{4.6}$$

Expanding the determinant and rearranging,

$$\omega^4 - \left[\frac{k_1+k_2}{m_1} + \frac{k_2+k_3}{m_2}\right]\omega^2 + \frac{k_1k_2 + k_2k_3 + k_3k_1}{m_1 \cdot m_2} = 0$$

The roots of the above equation are

$$\omega_{1,2}^2 = \frac{k_1 + k_2}{2m_1} + \frac{k_2 + k_3}{2m_2} \pm \sqrt{\left[\frac{k_1 + k_2}{2m_1} + \frac{k_2 + k_3}{2m_2}\right]^2 - \frac{k_1 k_2 + k_2 k_3 + k_3 k_1}{m_1 m_2}}$$

Thus, the two degree freedom system will vibrate with frequency ω_1 or ω_2. While vibrating with frequency ω_1 or ω_2, the deflected slope or displacement configuration in space will be presented by mode shapes. Thus, there shall be two mode shapes. The motion is now represented by

$$z_1 = A_1 \cdot e^{i\omega_1 t} \quad \text{and} \quad z_2 = A_2 \cdot e^{i\omega_2 t}$$

The term A_1 or A_2 represent the displacement configuration in space, i.e., the mode shapes, whereas $e^{i\omega_1 t}$ or $e^{i\omega_2 t}$ represent its variation with time.

Further for mode shapes, the two values of $[A_1/A_2]$ are

$$\left[\frac{A_1}{A_2}\right]^1 = \frac{k_2}{k_1 + k_2 - m_1\omega^2}$$

$$\left[\frac{A_1}{A_2}\right]^2 = \frac{k_2}{k_2 + k_3 - m_2\omega^2}$$

4.2.2 Damped Vibrations

If damping is also included in the system, the equation of motion in the matrix form can be written as

$$[m]\ddot{z} + [c]\dot{z} + [k]\,z = \{F\} \tag{4.7}$$

where $[m]$, $[c]$ and $[k]$ are 2×2 type matrices known as mass matrix, damping matrix and stiffness matrix and $\{F\}$ represents the external dynamic force vector, which is a 2×1 type matrix. As in the SDF system, forced vibrations of a two-degree freedom system will take place at the frequency of the excitation. When the excitation frequency coincides with one of the natural frequencies of the system, a condition of resonance will be encountered with large amplitudes. In the absence of external dynamic force, the system will vibrate freely.

Semi-definite systems

The systems having one of their natural frequencies equal to zero are known as semi-definite systems. Considering the two masses m_1 and m_2 connected by a spring k as shown in Figure 4.2, their equations of motion can be written as

$$m_1\ddot{z}_1 + k(z_1 - z_2) = 0 \tag{4.8}$$

$$m_2\ddot{z}_2 + k(z_2 - z_1) = 0 \tag{4.9}$$

For free vibrations, let us assume the motion to be harmonic, i.e.,

$$z_1 = A_1 \sin(\omega t + \phi) \tag{4.10}$$

$$z_2 = A_2 \sin(\omega t + \phi) \tag{4.11}$$

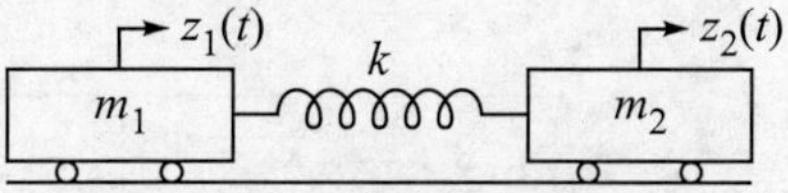

Figure 4.2 Semi-definite system.

Differentiating twice w.r.t time, $\ddot{z}_1$ and $\ddot{z}_2$ may be expressed as

$$\ddot{z}_1 = -A_1\omega^2 z_1 \sin(\omega t + \phi) \tag{4.12}$$

$$\ddot{z}_2 = -A_2\omega^2 z_2 \sin(\omega t + \phi) \tag{4.13}$$

Substituting the above terms in Eqs. (4.8) and (4.9), we get

$$\begin{aligned}(k - m_1\omega^2)A_1 - kA_2 &= 0 \\ -kA_1 + (k - m_2\omega^2)A_2 &= 0\end{aligned} \tag{4.14}$$

The determinant from the above equations can be written as

$$\begin{vmatrix} k - m_1\omega^2 & -k \\ -k & k - m_2\omega^2 \end{vmatrix} = 0 \tag{4.15}$$

The expansion of the determinant gives

$$m_1 m_2 \omega^4 - k(m_1 + m_2)\omega^2 = 0 \tag{4.16}$$

or,

$$\omega^2[m_1 m_2 \omega^2 - k(m_1 + m_2)] = 0$$

The roots of the quadratic equation are

$$\omega_1 = 0, \quad \omega_2 = \sqrt{\frac{k(m_1 + m_2)}{m_1 m_2}} \tag{4.17}$$

From Eq. (4.17) it can be seen that one of the natural frequencies of the system is equal to zero, i.e., the system is not oscillating. There is no relative motion between m_1 and m_2 and the system can move only as a rigid body.

Thus, the first mode of the system consists of a rigid body motion that counters no resistance. The natural frequency of such a rigid body motion is zero and its period is infinite. Characteristic equations having only positive roots are referred to as positive definite, while those with one or more zero roots are called positive semi-definite. Thus, if the system is physically supported in such a manner that only rigid body motion takes place, no elastic deformation takes place. As the potential energy is due to elastic effects alone, so the potential energy is zero without all coordinates being identically equal to zero. For this reason, vibrational systems with one or more rigid body modes are called semi-definite systems. They are also referred as unrestrained or degenerate systems.

Example 4.1 For a two-degree freedom system shown in Figure 4.3(a), where $k_1 = k_2 = k_3 = k$ and $m_1 = m$ and $m_2 = 2m$, determine the natural frequencies, mode shapes and the equation of motion.

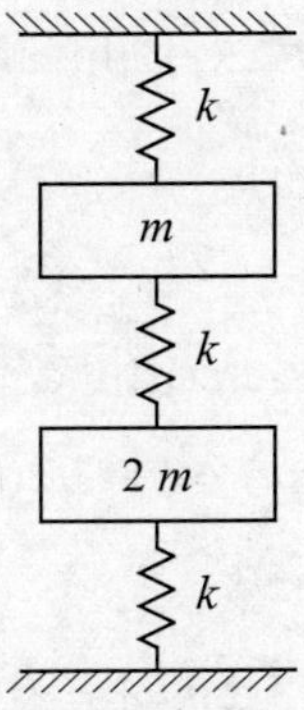

Figure 4.3(a) Two degree freedom system (Example 4.1).

Solution: The differential equations from the force system in Figure 4.3(b) are

$$m\ddot{z}_1 + k(z_1 - z_2) + kz_1 = 0$$

$$2m\ddot{z}_2 + k(z_2 - z_1) + kz_2 = 0$$

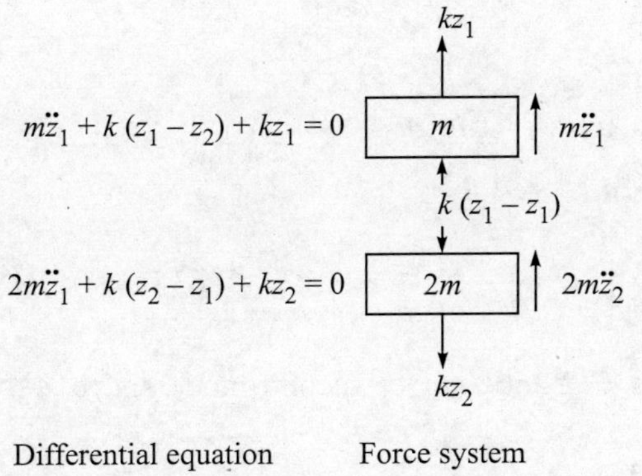

Figure 4.3(b) Two degree freedom system (Example 4.1).

In matrix form these equations may be written as

$$\begin{bmatrix} m & 0 \\ 0 & 2m \end{bmatrix}\begin{bmatrix} \ddot{z}_1 \\ \ddot{z}_2 \end{bmatrix} + \begin{bmatrix} 2k & -k \\ -k & 2k \end{bmatrix}\begin{bmatrix} z_1 \\ z_2 \end{bmatrix} = \begin{bmatrix} 0 \\ 0 \end{bmatrix}$$

For such motion let

$$z_1 = A_1 e^{i\omega t}$$

$$z_2 = A_2 e^{i\omega t}$$

Substituting these into the above differential equation,

$$(2k - \omega^2 m)A_1 - kA_2 = 0$$

$$-kA_1 + (2k - 2\omega^2 m)A_2 = 0$$

which are satisfied for any A_1 and A_2, if the following determinant is zero

$$\begin{bmatrix} 2k-\omega^2 m & -k \\ -k & 2k-2\omega^2 m \end{bmatrix} = 0$$

The above determinant leads to the characteristic equation

$$m\omega^4 - (3k/m)\omega^2 + 3/2(k/m)^2 = 0$$

The roots of the characteristic equation are

$$\omega_1^2 = (3/2 - \sqrt{3}/2)k/m = 0.634k/m$$

$$\omega_2^2 = (3/2 + \sqrt{3}/2)k/m = 2.366k/m$$

$$\therefore \quad \omega_1 = \sqrt{0.634\, k/m} = 0.796\sqrt{k/m}$$

and,

$$\omega_2 = \sqrt{2.366\, k/m} = 1.538\sqrt{k/m}$$

Accordingly, for mode shapes, ϕ with ω_1

$$\left[\frac{A_1}{A_2}\right]^1 = \frac{k}{2k-\omega_1^2 m}$$

Substituting $\omega_1^2 = 0.634k/m$

$$\left[\frac{A_1}{A_2}\right]^1 = \frac{k}{2k-\omega_1^2 m} = 1/(2-0.634) = 0.731$$

Similarly, substituting $\omega_2^2 = 2.366k/m$, for second mode shape ϕ_2 with ω_2

$$\left[\frac{A_1}{A_2}\right]^2 = \frac{k}{2k-\omega_2^2 m} = 1/(2-2.366) = -2.73$$

The amplitude ratios corresponding to the first and second natural frequencies have been mentioned but these values are not absolute values. If any of the amplitudes is taken to be unity, the ratio is normalized to that number. The normalized amplitude ratio is then called the normal mode and is designated by $\phi_i(z)$.

The two normal modes of this problem, which are now called eigenvectors are

$$\phi_1(z) = \begin{vmatrix} 0.731 \\ 1.000 \end{vmatrix} \quad \text{and} \quad \phi_2(z) = \begin{vmatrix} -2.73 \\ 1.000 \end{vmatrix}$$

According to $\phi_1(z)$, the two masses move in phase and according to $\phi_2(z)$, the negative value indicates that in the second mode the two masses move in opposition or out of phase with each other. See Figure 4.3(c).

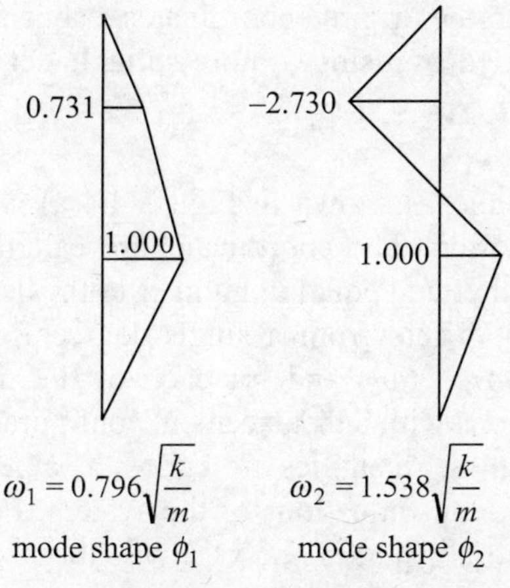

Figure 4.3(c) Mode shape for two degree freedom system.

4.3 VIBRATIONS OF MULTI-DEGREE FREEDOM SYSTEM

For obtaining the response of various civil engineering structures, like a multistoried building [Figure 4.4(a)] to dynamic loading including seismic loading such structures are often idealized as multi-degree freedom systems. The natural frequencies of vibration, the mode shapes, and their variation with time, total response or the peak displacement, velocity and acceleration are the parameters of interest and by adopting a suitable method of analysis such parameters need to be evaluated. The lateral dynamic forces on each floor of the three storied building as in [Figure 4.4(b)] are important for structural dynamic studies, however, the force acting at the base popularly known as base shear is very important from soil dynamics and geotechnical earthquake engineering standpoint. In different steps, attempts shall be made to obtain the following:

- Natural frequencies of vibration in each mode
- Different mode shapes
- Dynamic response in each mode
- Total dynamic response
- Displacement, velocity and acceleration
- Peak values of displacement, velocity and acceleration
- Evaluation of lateral dynamic forces
- Base shear

In Section 4.2, vibrations of a two-degree freedom system have been presented and having understood the equations of motion for a two-degree freedom system it will be easier to appreciate the complexity involved in a multi-degree freedom system. The procedure for analyzing a multi-degree freedom system is only an extension of the method used for analyzing a two-degree freedom system.

Additional complications arise in multi-degree systems because the number of terms increases rapidly with the number of degrees of freedom. Of course, matrix formulations prove to be very effective for the purpose of manipulating a large number of terms. More important than this consideration, however, is the fact that systems subjected to arbitrary forcing functions become extremely difficult to analyze in the original coordinates, especially in the presence of damping. These difficulties can be avoided by using a more suitable set of coordinates.

Method of analysis

The multi-degree freedom system as shown in Figure 4.4(a) is one that requires two or more coordinates to describe the motion. The coordinates are called generalized coordinates when they are independent of each other and equal in number to the degrees of freedom of the system. The n-degree freedom system differs from a single-degree freedom system in that it has n natural frequencies (ω_1, ω_2, ω_3,..., ω_n) and for each of the natural frequencies there corresponds a natural state of vibrations with a displacement configuration described as normal modes. Mathematical terms related to these quantities are known as eigenvalues and eigenvectors. They are established from n simultaneous equations of the system. For an n-degree freedom system with masses m_1, m_2,..., m_n and stiffnesses k_1, k_2,..., k_n, these equations may be written as

$$\begin{aligned} m_1 \ddot{z}_1 &= -\ k_1 z_1 - k_2(z_1 - z_2) \\ m_2 \ddot{z}_2 &= k_2(z_1 - z_2) - k_3(z_2 - z_3) \\ \vdots \quad &\quad \vdots \qquad\qquad \vdots \\ m_n \ddot{z}_n &= k_n\,(z_{n+1} - z_n) \end{aligned}$$

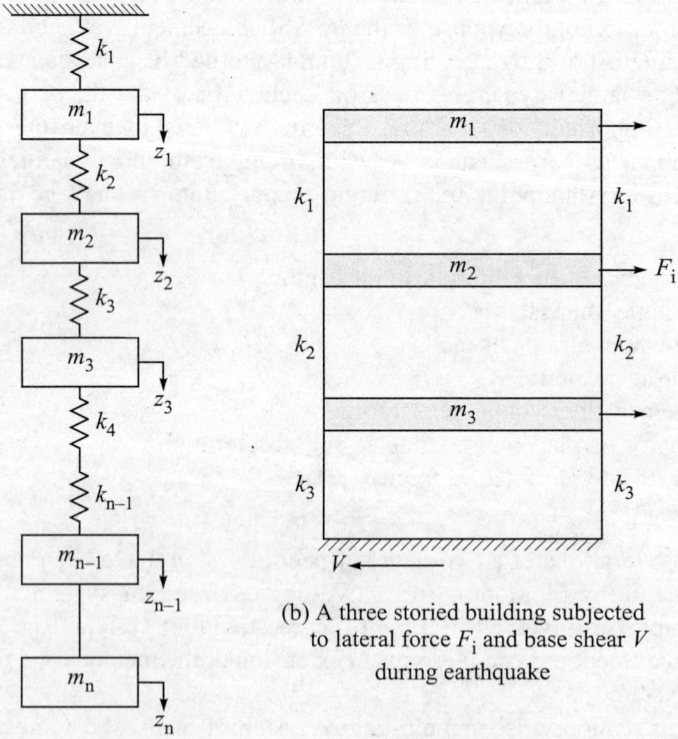

(a) MDF system

(b) A three storied building subjected to lateral force F_i and base shear V during earthquake

Figure 4.4 Free vibrations of a multi-degree freedom system.

These equations may be written in matrix form as

$$\begin{bmatrix} m_1 & 0 & 0 & \cdots & 0 \\ 0 & m_2 & 0 & \cdots & 0 \\ 0 & 0 & m_3 & \cdots & 0 \\ \vdots & \vdots & \vdots & & \vdots \\ 0 & 0 & 0 & \cdots & m_n \end{bmatrix} \begin{bmatrix} \ddot{z}_1 \\ \ddot{z}_2 \\ \ddot{z}_3 \\ \vdots \\ \ddot{z}_n \end{bmatrix} + \begin{bmatrix} k_1+k_2 & -k_2 & 0\cdots & 0 \\ -k_2 & k_2+k_3 & -k_3\cdots & 0 \\ \vdots & \vdots & \vdots & \vdots \\ 0 & 0 & 0 & k_n+k_{n+1} \end{bmatrix} \begin{bmatrix} z_1 \\ z_2 \\ z_3 \\ \vdots \\ z_n \end{bmatrix} = \begin{vmatrix} F_1(t) \\ F_2(t) \\ F_3(t) \\ \vdots \\ F_n(t) \end{vmatrix} \quad (4.18)$$

Equation (4.18) may be written in compact form as

$$[m]\{\ddot{z}\} + [k]\{z\} = \{F(t)\}$$

and if damping is included then

$$[m]\{\ddot{z}\}+[c]\{\dot{z}\}+[k]\{z\} = \{F(t)\} \quad (4.19)$$

where,

$[m]$ = mass matrix
$[c]$ = damping matrix
$[k]$ = stiffness matrix
$\{F(t)\}$ = loading matrix

If the system has an n-degrees of freedom, the size of $[m]$, $[c]$ and $[k]$ will be of the order $n \times n$. Equation (4.19) represents the general form of the equation of a system of n-degrees of freedom. The matrices in the above equation are as follows:

Mass Matrix–[m]

Many structures such as framed structures have essentially lumped masses since the mass of the column is often negligible compared to floors. As such masses are assumed to be lumped at nodal points. The mass matrix [m] of Eq. (4.19) is a lumped mass matrix.

Damping Matrix–[c]

The damping matrix as given in Eq. (4.19) may be formed by assuming the system to have viscous damping. Generally, the damping matrix is reduced to simpler forms for facilitating the analysis.

Stiffness Matrix–[k]

For a linear elastic system, the stiffness matrix $[k]$ is a symmetric matrix. The structure consists of a number of elements. The total stiffness matrix is formed by assembling the stiffness matrix of the individual elements.

Loading Matrix–[F(t)]

The dynamic loads are assumed to act at nodal points corresponding to the displacement degrees of freedom. Loads when acting in between the nodal points, or when they are distributed, are converted to equivalent values acting at respective nodal points.

Neglecting damping and considering only the free vibration, Eq. (4.19) reduces to

$$[m]\{\ddot{z}\} + [k]\{z\} = \{0\} \tag{4.20}$$

Assuming the solution in the following form

$$\{z\} = A \cdot e^{i\omega t} \cdot \{\phi\} \tag{4.21}$$

where,

A is a scalar of dimension L (length)
ω is a scalar of dimension T (time)
$[\phi]$ is a non-dimensional vector, such that

$$[\phi]^T = \{\phi_1, \quad \phi_2, \quad \phi_3,\ldots, \quad \phi_n\}$$

From Eq. (4.21), by differentiating twice with respect to time

$$\{\ddot{z}\} = -A\omega^2\, e^{i\omega t} \cdot \{\phi\} \tag{4.22}$$

Substituting the values of $\{\ddot{z}\}$ and $\{z\}$ in the differential equation (4.20)

$$-\omega^2\,[m]\phi + [k]\{\phi\} = \{0\}$$

or,

$$[m]\{\phi\} - \lambda\,[k]\{\phi\} = \{0\}$$

where,

$$\lambda = \frac{1}{\omega^2}$$

Thus,

$$|\,[m] - \lambda\,[k]\,|\,\{\phi\} = 0 \tag{4.23}$$

Equation (4.23) will have a non-trivial solution only if the determinate corresponding to $|\,[m] - \lambda\,[k]\,|$ vanishes, that is

$$|\,[m] - \lambda\,[k]\,| = 0 \tag{4.24}$$

This determinant is shown as the frequency determinant. The solution of Eq. (4.24) will give n values of λ, where n is the number of degrees of freedom. ω_1, ω_2, ω_3,..., ω_n are called natural frequencies of the system. Corresponding to each value of ω, a vector $\{\phi\}$ can be evaluated, which will provide the mode shapes.

Equation (4.23) may be rewritten in terms of eigenvalue problem, by premultiplying both sides by $[k]^{-1}$, as such

$$[k]^{-1}\,[m]\,\{\phi\} = \lambda\,[I]\,\{\phi\}$$

or,

$$[D]\,\{\phi\} = \lambda\,[I]\,\{\phi\}$$

where,

$$D = [k]^{-1}\,[m] = \text{dynamical matrix}$$

The solution may be obtained as

$$|\,[D] - \lambda\,[I]\,| = 0$$

The determinant when expanded gives a polynomial of degree n

Thus,

$$\lambda^n + A_1\lambda^{n-1} + A_2\lambda^{n-2} + \cdots + C_{n-1}\lambda + C_n = 0$$

The solution of this polynomial equation known as characteristic equation or frequency equation will yield n values of λ. Once all λs determined, next the mode shapes ϕs will be obtained. It may be noted that we obtain only the ratio among ϕs. However, there exist for each mode, a unique solution, if we assign an arbitrary value to one of them. This method will be illustrated in Example 4.2.

Response to earthquake excitation

A multistoried building, [for example, a three-storied building as shown in Figure 4.4(b)] when subjected to ground motion, the governing differential equation may be expressed as

$$[m]\{\ddot{u}_i\} + [c]\{\dot{u}_i\} + [k]\{u_i\} = -[m][\ddot{u}_g]$$

where,

$\ddot{u}_g$ = ground acceleration

u_i = displacement of ith floor

By adopting a method, the dynamic force in each floor (ith floor) can be determined, consequently the internal force specially the base shear as shown in Figure 4.4(b) can be determined.

Thus, the total seismic force that this three-storied building must resist may be written as

$$V = \alpha W \tag{4.26}$$

where,

V = total lateral force that must be resisted

α = seismic coefficient

W = weight of the building

Earthquake ground motion create inertia or lateral force by shaking the building back and forth. Since the structure moves with the ground and the earthquake ground motion will be a random motion back and forth, the eventual motion of the structure will be back and forth. The deformation of the structure at any instance will be a function of the stiffness and damping of the structure as well as that of the frequency of the ground motion. From geotechnical earthquake engineering standpoint the evaluation of base shear is important. The dynamics of elastic system as described in this chapter is to facilitate evaluation of base shear, natural frequencies of vibrations and the deflected shapes (mode shapes). As far as foundations are concerned, an emphasis is laid on a mathematical model of beam/plate/shell resting on an elastic foundation. Such analysis and studies will eventually lead to seismic resistant analysis and design of foundations, i.e., structures below the ground level.

Example 4.2 Find the natural frequencies of a three-degree freedom system as shown in Figure 4.5(a) assuming $k_1 = k_2 = k_3 = k$ and $m_1 = m_2 = m_3 = m$.

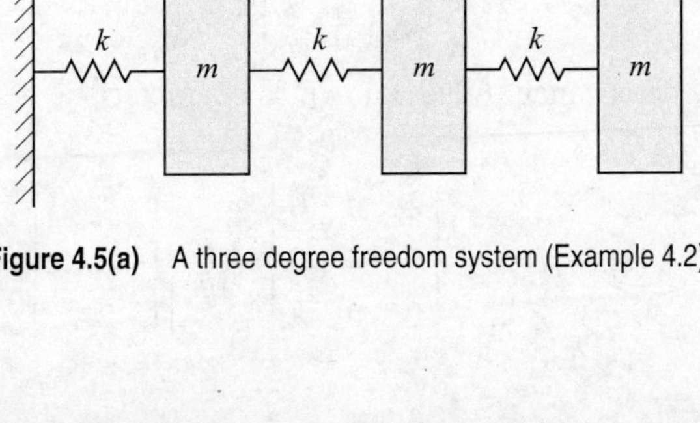

Figure 4.5(a) A three degree freedom system (Example 4.2).

Solution: The equation of motion in matrix form can be written as

$$\begin{vmatrix} m_1 & 0 & 0 \\ 0 & m_2 & 0 \\ 0 & 0 & m_3 \end{vmatrix} \begin{vmatrix} \ddot{z}_1 \\ \ddot{z}_2 \\ \ddot{z}_n \end{vmatrix} + \begin{vmatrix} k_1+k_2 & -k_2 & 0 \\ -k_2 & k_3+k_2 & -k_3 \\ 0 & -k_3 & k_3 \end{vmatrix} \begin{vmatrix} z_1 \\ z_2 \\ z_3 \end{vmatrix} = \begin{vmatrix} 0 \\ 0 \\ 0 \end{vmatrix}$$

Putting $m_1 = m_2 = m_3 = m$ and $k_1 = k_2 = k_3 = k$, we have

$$\begin{vmatrix} m & 0 & 0 \\ 0 & m & 0 \\ 0 & 0 & m \end{vmatrix} \begin{vmatrix} \ddot{z}_1 \\ \ddot{z}_2 \\ \ddot{z}_3 \end{vmatrix} + \begin{vmatrix} 2k & -k & 0 \\ -k & 2k & -k \\ 0 & -k & 2k \end{vmatrix} \begin{vmatrix} z_1 \\ z_2 \\ z_3 \end{vmatrix} = \begin{vmatrix} 0 \\ 0 \\ 0 \end{vmatrix}$$

So, mass matrix $= [m] = m\begin{bmatrix} 1 & 0 & 0 \\ 0 & 1 & 0 \\ 0 & 0 & 1 \end{bmatrix}$ and stiffness matirix $= [k] = k\begin{bmatrix} 2 & -1 & 0 \\ -1 & 2 & -1 \\ 0 & -1 & 1 \end{bmatrix}$

Obtaining inverse of the stiffness matrix,

$$[k]^{-1} = \frac{1}{k}\begin{bmatrix} 1 & 1 & 1 \\ 1 & 2 & 2 \\ 1 & 2 & 3 \end{bmatrix}$$

Thus, the frequency determinant may be written as

$$|[D] - \lambda[I]| = 0 \quad \text{or} \quad |\lambda[I] - [D]| = 0$$

where, $[D] = [k]^{-1}\,[m] =$ dynamical matrix

$$= \frac{1}{k}\begin{bmatrix} 1 & 1 & 1 \\ 1 & 2 & 2 \\ 1 & 2 & 3 \end{bmatrix} m \begin{bmatrix} 1 & 0 & 0 \\ 0 & 1 & 0 \\ 0 & 0 & 1 \end{bmatrix}$$

$$= \frac{m}{k}\begin{bmatrix} 1 & 1 & 1 \\ 1 & 2 & 2 \\ 1 & 2 & 3 \end{bmatrix}$$

and,

$$\lambda[I] = \begin{vmatrix} \lambda & 0 & 0 \\ 0 & \lambda & 0 \\ 0 & 0 & \lambda \end{vmatrix}$$

By setting the frequency determinant (also, known as characteristic determinant, Δ) to zero, we obtain the frequency equation

$$\Delta = |\lambda[I] - [D]| = \begin{bmatrix} \lambda & 0 & 0 \\ 0 & \lambda & 0 \\ 0 & 0 & \lambda \end{bmatrix} - \frac{m}{k}\begin{vmatrix} 1 & 1 & 1 \\ 1 & 2 & 2 \\ 1 & 2 & 3 \end{vmatrix} = 0$$

By dividing throughout by λ,

$$\begin{bmatrix} 1-\alpha & \alpha & \alpha \\ -\alpha & 1-2\alpha & -2\alpha \\ -\alpha & -2\alpha & 1-3\alpha \end{bmatrix} = 0, \text{ where } \alpha = \frac{m}{k\lambda}$$

The expansion of the determinant yields a cubic equation

$$\alpha^3 - 5\alpha^2 + 6\alpha - 1 = 0$$

Solving the cubic eqation

$$\alpha_1 = 0.19806, \ \alpha_2 = 1.5553, \text{ and } \alpha_3 = 3.2490$$

or,
$$\omega_1 = 0.44504\sqrt{\frac{k}{m}}, \ \omega_2 = 1.2471\sqrt{\frac{k}{m}} \text{ and } \omega_3 = 1.8025\sqrt{\frac{k}{m}}$$

Once the natural frequencies are known, the mode shapes or eigenvectors can be obtained using the equation,

$$|\lambda_i[I] - [D]|\{\vec{\phi}\}^i = \{0\}, \quad i = 1, 2, 3$$

For example, corresponding $i = 1$, $\omega = \omega_1$

$$\{\vec{\phi}\}^1 = \begin{Bmatrix} \phi_1 \\ \phi_2 \\ \phi_3 \end{Bmatrix}^1$$

Similarly, for $i = 2$, $\omega = \omega_2$

$$\{\vec{\phi}\}^2 = \begin{Bmatrix} \phi_1 \\ \phi_2 \\ \phi_3 \end{Bmatrix}^2$$

and for $i = 3$, $\omega = \omega_3$

$$\{\vec{\phi}\}^3 = \begin{Bmatrix} \phi_1 \\ \phi_2 \\ \phi_3 \end{Bmatrix}^3$$

First mode shape

$$\omega_1 = 0.44504\sqrt{\frac{k}{m}} \quad \text{and} \quad \lambda_1 = 5.0489\frac{m}{k}$$

With these values, the equation may be written as

$$\left[5.0489\frac{m}{k}\begin{bmatrix} 1 & 0 & 0 \\ 0 & 1 & 0 \\ 0 & 0 & 1 \end{bmatrix} - \frac{m}{k}\begin{bmatrix} 1 & 1 & 1 \\ 1 & 2 & 2 \\ 1 & 2 & 3 \end{bmatrix} \right] \begin{Bmatrix} \phi_1 \\ \phi_2 \\ \phi_3 \end{Bmatrix}^1 = \begin{Bmatrix} 0 \\ 0 \\ 0 \end{Bmatrix}$$

or,
$$\begin{bmatrix} 4.0489 & -1.0 & -1.0 \\ -1.0 & 3.0489 & -2.0 \\ -1.0 & -2.0 & 2.0489 \end{bmatrix} \begin{Bmatrix} \phi_1 \\ \phi_2 \\ \phi_3 \end{Bmatrix}^1 = \begin{Bmatrix} 0 \\ 0 \\ 0 \end{Bmatrix}$$

The above equation denotes a system of three homogeneous linear equations with three unknowns. Any two of these unknowns can be expressed in terms of the remaining one. Thus,

$$\phi_2 + \phi_3 = 4.0489\phi_1$$

and,
$$3.0489\phi_2 - 2.0\phi_3 = \phi_1$$

Once the above two equations are satisfied, the third row of equation containing $\{\phi_3\}^1$ will be satisfied automatically. The solution may be obtained as

$$\phi_2 = 1.8019\phi_1 \quad \text{and} \quad \phi_3 = 2.2470\phi_1$$

Taking $\phi_1 = 1.0$, the first mode shape is given by

$$\{\vec{\phi}\}^1 = \begin{Bmatrix} \phi_1 \\ \phi_2 \\ \phi_3 \end{Bmatrix}^1 = \begin{Bmatrix} 1.0 \\ 1.8019 \\ 2.2470 \end{Bmatrix}$$

Second mode shape

Taking the value of ω_2, or that of $\lambda_2 = 0.6430 \dfrac{m}{k}$, the equation now leads to

$$\left[0.6430 \frac{m}{k} \begin{bmatrix} 1 & 0 & 0 \\ 0 & 1 & 0 \\ 0 & 0 & 1 \end{bmatrix} - \frac{m}{k} \begin{bmatrix} 1 & 1 & 1 \\ 1 & 2 & 2 \\ 1 & 2 & 3 \end{bmatrix} \right] \begin{Bmatrix} \phi_1 \\ \phi_2 \\ \phi_3 \end{Bmatrix}^2 = \begin{Bmatrix} 0 \\ 0 \\ 0 \end{Bmatrix}$$

or,
$$\begin{bmatrix} -0.3570 & -1.0 & -1.0 \\ -1.0 & -1.3570 & -2.0 \\ -1.0 & -2.0 & -2.3570 \end{bmatrix} \begin{Bmatrix} \phi_1 \\ \phi_2 \\ \phi_3 \end{Bmatrix}^2 = \begin{Bmatrix} 0 \\ 0 \\ 0 \end{Bmatrix}$$

Taking the first two rows, the equations are

$$-\phi_2 - \phi_3 = 0.3570\phi_1$$

$$-1.3570\phi_2 - 2.0\phi_3 = \phi_1$$

Taking $\phi_1 = 1.0$,

$$\phi_2 = 0.4450 \text{ and } \phi_3 = -0.8020$$

Therefore,
$$\begin{Bmatrix} \phi_1 \\ \phi_2 \\ \phi_3 \end{Bmatrix}^2 = \begin{Bmatrix} 1.0 \\ 0.4450 \\ -0.8020 \end{Bmatrix}$$

Third mode shape

In order to obtain the third mode, the value of ω_3 $\left(\text{or, } \lambda_3 = 0.3078\,\dfrac{m}{k}\right)$ is substituted, and the equations are

$$\left[-0.3078\,\frac{m}{k}\begin{bmatrix}1 & 0 & 0\\ 0 & 1 & 0\\ 0 & 0 & 1\end{bmatrix} - \frac{m}{k}\begin{bmatrix}1 & 1 & 1\\ 1 & 2 & 2\\ 1 & 2 & 3\end{bmatrix}\right]\begin{Bmatrix}\phi_1\\ \phi_2\\ \phi_3\end{Bmatrix}^3 = \begin{Bmatrix}0\\ 0\\ 0\end{Bmatrix}$$

or,

$$\begin{bmatrix}-0.6922 & -1.0 & -1.0\\ -1.0 & -1.6922 & -2.0\\ -1.0 & -2.0 & -2.6922\end{bmatrix}\begin{Bmatrix}\phi_1\\ \phi_2\\ \phi_3\end{Bmatrix}^3 = \begin{Bmatrix}0\\ 0\\ 0\end{Bmatrix}$$

The first two rows of the equation can be written as

$$-\ \phi_2 - \phi_3 = 0.6922\phi_1$$

$$-\ 1.6922\phi_2 = 2.0\phi_3 = \phi_1$$

Solving $\qquad \phi_2 = -\ 1.2468\phi_1, \qquad \phi_3 = 0.5544\phi_1$

Taking $\phi_1 = 1.0$,

$$\phi_2 = -1.2468, \quad \text{and} \quad \phi_3 = 0.5544$$

Hence, the third mode shape can be written as

$$\begin{Bmatrix}\phi_1\\ \phi_2\\ \phi_3\end{Bmatrix}^3 = \begin{Bmatrix}1.0\\ -1.2468\\ 0.5544\end{Bmatrix}$$

All these mode shapes have been shown in Figure 4.5(b).

Thus, finally

frequency
$$\{\omega\} = \begin{vmatrix}\omega_1\\ \omega_2\\ \omega_3\end{vmatrix} = \sqrt{\frac{k}{m}}\begin{vmatrix}0.44504\\ 1.2471\\ 1.80250\end{vmatrix}$$

mode shapes
$$\{\phi\} = \begin{vmatrix}1.0 & 1.0 & 1.0\\ 1.8019 & 0.4450 & -1.2468\\ 2.2470 & -0.8020 & 0.5544\end{vmatrix}$$

The mode shapes have been shown in Figure 4.5 (b).

4.4 MODE PARTICIPATION FACTOR

The forced vibration of an n-degree freedom system can be represented by the equation

$$[m]\{\ddot{z}\} + [c]\{\dot{z}\} + [k]\{z\} = \{F\} \tag{4.25}$$

For example, a 40-storied building can be modelled as multi-degree freedom system with $n = 40$. Using modal analysis, the solution will lead to 40 eigenvalues and 40 eigenvectors, which

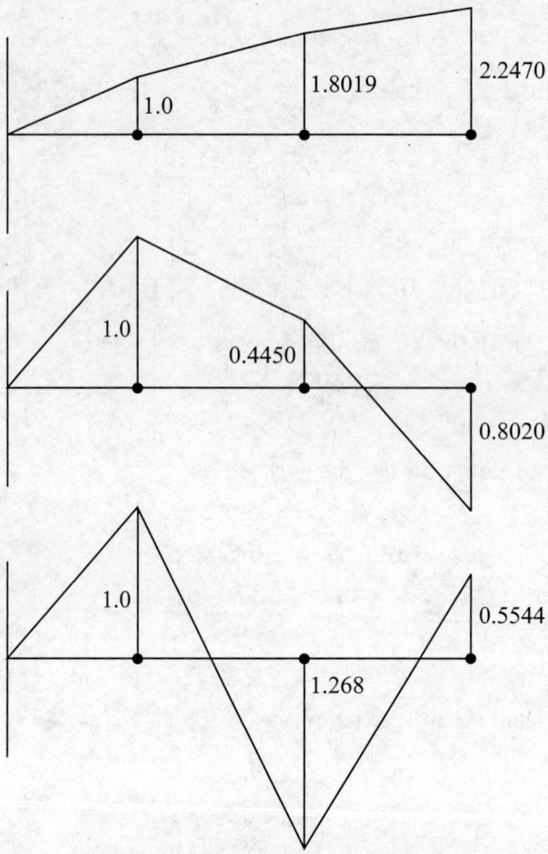

Figure 4.5(b) Mode shapes (Example 4.2).

can be routinely solved by digital computers these days. However, to cut down the size of the computations, a procedure called mode summation method is used. For example, for the 40-storied building, all the matrices like $[k]$ will be 40×40 matrix, involving 40 mode shapes and frequencies. But on practical considerations it is well-known that the excitation of the building centres around the lower frequencies. So it may be sufficient to consider only the first three modes, then the deflection under the forced excitation may be expressed as

$$z_i = \phi_1(z_i)q_1(t) + \phi_2(z_i)q_2(t) + \phi_3(z_i)q_3(t) \tag{4.26}$$

The displacement field of all n floors can be expressed in matrix form as

$$\begin{Bmatrix} z_1 \\ z_2 \\ z_3 \\ \vdots \\ z_n \end{Bmatrix} = \begin{bmatrix} \phi_1(z_1) & \phi_2(z_1) & \phi_3(z_1) \\ \phi_1(z_2) & \phi_2(z_2) & \phi_3(z_2) \\ \phi_1(z_3) & \phi_2(z_3) & \phi_3(z_3) \\ \vdots & \vdots & \vdots \\ \phi_1(z_n) & \phi_2(z_n) & \phi_3(z_n) \end{bmatrix} \begin{Bmatrix} q_1 \\ q_2 \\ q_3 \\ \vdots \\ q_n \end{Bmatrix}$$

By using the normal modes and introducing ψ as the modal matrix, if the n normal modes are assembled in a square matrix with each normal mode represented by a column, then such a matrix is called a modal matrix. For example for a three degree freedom system, let these modes be represented by

$$\{\phi_1\} = \begin{Bmatrix} z_1 \\ z_2 \\ z_3 \end{Bmatrix}^1, \ \{\phi_2\} = \begin{Bmatrix} z_1 \\ z_2 \\ z_3 \end{Bmatrix}^2, \ \{\phi_3\} = \begin{Bmatrix} z_1 \\ z_2 \\ z_3 \end{Bmatrix}^3$$

then the modal matrix

$$\psi = \begin{bmatrix} \begin{Bmatrix} z_1 \\ z_2 \\ z_3 \end{Bmatrix}^1 \begin{Bmatrix} z_1 \\ z_2 \\ z_3 \end{Bmatrix}^2 \begin{Bmatrix} z_1 \\ z_2 \\ z_3 \end{Bmatrix}^3 \end{bmatrix}$$

As such the transpose of $\{\psi\}$ may be written as

$$\{\psi\}^T = \begin{bmatrix} (z_1\ z_2\ z_3)^1 \\ (z_1\ z_2\ z_3)^2 \\ (z_1\ z_2\ z_3)^3 \end{bmatrix}, \text{ with each row corresponding to a mode.}$$

Now let us obtain $\{\psi\}^{\mathrm{T}}\{m\}\{\psi\}$, in which we find a diagonal matrix

$$\{\psi\}^T[m]\{\psi\} = \begin{bmatrix} (z_1\ z_2\ z_3)^1 \\ (z_1\ z_2\ z_3)^2 \\ (z_1\ z_2\ z_3)^3 \end{bmatrix} [m] \begin{bmatrix} \begin{Bmatrix} z_1 \\ z_2 \\ z_3 \end{Bmatrix}^1 \begin{Bmatrix} z_1 \\ z_2 \\ z_3 \end{Bmatrix}^2 \begin{Bmatrix} z_1 \\ z_2 \\ z_3 \end{Bmatrix}^3 \end{bmatrix}$$

$$= \begin{bmatrix} m_{11} & 0 & 0 \\ 0 & m_{22} & 0 \\ 0 & 0 & m_{33} \end{bmatrix}$$

Thus, the product $\{\psi\}^T[m]\{\psi\} = (3 \times n)\ (n \times n)\ (n \times 3) = (3 \times 3)$ matrix. For example, for the 40 storied building, the stiffness matrix is a (40×40) matrix and by using only three normal modes, $\{\psi\}$ is a (40×3) matrix and the product $\{\psi\}^T[m]\{\psi\}$ becomes (3×40) (40×40) $(40 \times 3) = (3 \times 3)$ matrix.

Thus, instead of solving the 40 coupled equations represented by Eq. (4.25) we need only to solve three equations, represented by

$$\{\psi\}^T[m]\{\psi\}\ddot{q} + \{\psi\}^T[c]\{\psi\}\dot{q} + \{\psi\}^T[k]\{\psi\}\ q = \{\psi\}^T[F\ (x, t)]$$

Further taking $F\ (x, t) = p(z) \cdot f(t)$ and the damping assumed to be proportional, the three equations take the form

$$\ddot{q}_i + 2\xi\ \omega_i\,\dot{q}_i + \omega_i^2\,q_i = \beta_i\,f(t),\ i = 1, 2, 3$$

where, $\beta_i = \dfrac{\sum_{j=1}^{n} \phi_j(z_j) \cdot p(z_j)}{\sum_{j=1}^{n} m_i \phi_i^2(z_j)}$ and β_i is called mode participation factor.

For a 40 storied buidlng subjected to a ground displacement $u_g(t)$, if three mode shapes, ϕ_1, ϕ_2 and ϕ_3 are considered, the governing differential equation may be written as

$$\{m\}\{Z\} + [c]\{Z\} + [x]\{Z\} = -[m][I]\ddot{u}_g(t)$$

where $[I]$ is a unit vector.

$$[I] = \begin{vmatrix} 1 \\ 1 \\ \vdots \\ 1 \end{vmatrix}$$

For this 40 storied buidling, the above equation contains 40 ordinary differential equations in n unknowns (floor displacement). The n equations are coupled and cannot be solved independently and $\{Z\}$ is a 40×1 vector.

Using three mode shapes we can make the transformation

$$\{Z\} = \{\psi\}\{q\}$$

where $\{\psi\}$ is a 40×3 vector and $\{q\}$ is a 3×1 vector.

$$\{\psi\} = \begin{bmatrix} \phi_1(z_1) & \phi_2(z_1) & \phi_3(z_1) \\ \phi_1(z_2) & \phi_2(z_2) & \phi_3(z_2) \\ \phi_1(z_3) & \phi_2(z_3) & \phi_3(z_3) \\ \vdots & \vdots & \vdots \\ \phi_1(z_n) & \phi_2(z_n) & \phi_3(z_n) \end{bmatrix} \quad \text{and} \quad \{q\} = \begin{Bmatrix} q_1 \\ q_2 \\ q_3 \end{Bmatrix}$$

The displacement of ith floor may be expressed as

$$z_i = \phi_1(z_i)\, q_1(t) + \phi_2(z_i)\, q_2(t) + \phi_3(z_i)\, q_3(t).$$

Premultiplying the governing differential equation by $\{\psi\}^T$,

$$\{\psi\}^T[m]\{\psi\}\ddot{q} + \{\psi\}^T[c]\{\psi\}\dot{q} + \{\psi\}^T[k]\{\psi\} = -\{\psi\}^T[m]\{I\}\ddot{u}_g(t)$$

Assuming $[c]$ to be proportional damping, the above equation result in three uncoupled equations

$$m_{11}\ddot{q} + C_{11}\dot{q} + k_{11}q = -\ddot{u}_g(t) \sum_{n=1}^{40} m_i\, \phi_i(z_i)$$

$$m_{22}\ddot{q} + C_{22}\dot{q} + k_{22}q = -\ddot{u}_g(t) \sum_{n=1}^{40} m_i\, \phi_2(z_i)$$

$$m_{33}\ddot{q} + C_{33}\dot{q} + k_{33}q = -\ \ddot{u}_g(t) \sum_{n=1}^{n=40} m_i\ \phi_3(z_i)$$

The first equation corresponding to ω_1 takes the form

$$\ddot{q}_1 + 2\xi\omega_1\dot{q} + \omega_1^2 q_1 = -\frac{\sum_{i=1}^{40} m_i \cdot \phi_1(z_i)}{\sum_{i=1}^{40} m_i \cdot \phi_1^2(z_i)}\ \ddot{u}_g(t)$$

where,
$$m_{11} = \sum m_i \phi_1^2$$

$$\frac{c_{11}}{m_{11}} = 2\xi_1\omega_1$$

$$\frac{k_{11}}{m_{11}} = \omega_1^2$$

Thus, for a given value of mass, stiffness and damping, these three equations can be solved for any value of ground displacement $u_g(t)$.

Example 4.3 A three-storied building as shown in Figure 4.6(a) has floor masses as $m_1 = 1.0$, $m_2 = 1.5$ and $m_3 = 2.0$ with stiffness of the columns, $k_1 = 60$, $k_2 = 120$ and $k_3 = 180$. The damping may be taken as 5%. Determine the natural frequencies, the time periods, the mode shapes and the mode participation factor.

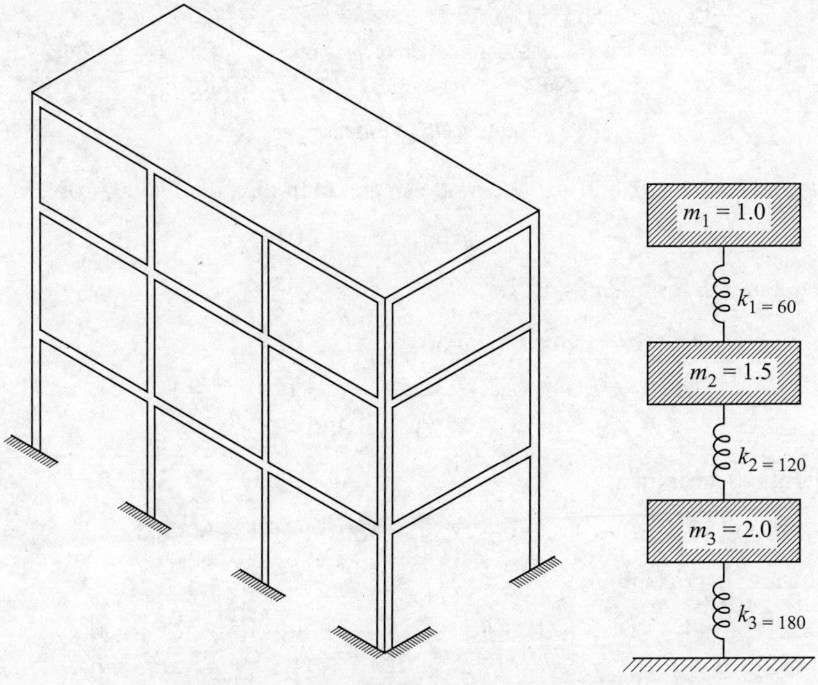

Figure 4.6(a) A three-storied building (Example 4.3).

Solution: The mass matrix for the building shown is

$$[m] = \begin{vmatrix} 1 & 0 & 0 \\ 0 & 1.5 & 0 \\ 0 & 0 & 2 \end{vmatrix}$$

The stiffness matrix may be written as shown in Figure 4.6(b),

$$k = \begin{bmatrix} k_{11} & k_{12} & k_{13} \\ k_{21} & k_{22} & k_{23} \\ k_{31} & k_{32} & k_{33} \end{bmatrix}, \text{ so } [k] = \begin{vmatrix} 60 & -60 & 0 \\ -60 & 180 & -120 \\ 0 & -120 & 300 \end{vmatrix}$$

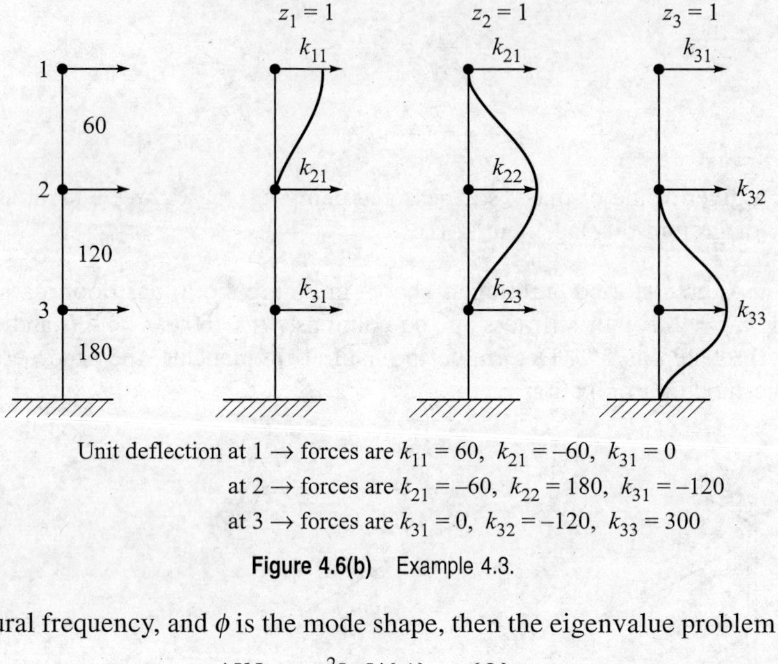

Unit deflection at 1 → forces are $k_{11} = 60$, $k_{21} = -60$, $k_{31} = 0$
at 2 → forces are $k_{21} = -60$, $k_{22} = 180$, $k_{31} = -120$
at 3 → forces are $k_{31} = 0$, $k_{32} = -120$, $k_{33} = 300$

Figure 4.6(b) Example 4.3.

If ω is the natural frequency, and ϕ is the mode shape, then the eigenvalue problem is as follows:

$$|[k] - \omega^2[m]|\{\phi\} = \{0\}$$

Then the frequency determinant will be

$$\begin{vmatrix} 60 - 1.0\omega^2 & -60 & 0 \\ -60 & 180 - 1.5\omega^2 & -120 \\ 0 & -120 & 300 - 2\omega^2 \end{vmatrix} = 0$$

By expanding the determinant

$$(60 - \omega^2)[(180 - 1.5\omega^2)(300 - 2\omega^2) - 120^2] + 60\,[-60(300 - 2\omega^2)] = 0$$

Simplify by using $\lambda = \omega^2/60$

$$(1 - \lambda)[(3 - 1.5\lambda)(5 - 2\lambda) - 4] + [-5(5 - 2\lambda)] = 0$$

or,

$$-3\lambda^3 + 16.5\lambda^2 - 22.5\lambda + 6 = 0$$

Solving the cubic equation,

$$\lambda_1 = 0.35 \Rightarrow \omega_1^2 = 21$$
$$\lambda_2 = 1.61 \Rightarrow \omega_2^2 = 96.6$$
$$\lambda_3 = 3.54 \Rightarrow \omega_3^2 = 212.4$$

Natural frequencies

$$\begin{Bmatrix} \omega_1^2 \\ \omega_2^2 \\ \omega_3^2 \end{Bmatrix} = \begin{Bmatrix} 21 \\ 96.6 \\ 212 \end{Bmatrix} \text{ and } \begin{Bmatrix} \omega_1 \\ \omega_2 \\ \omega_3 \end{Bmatrix} = \begin{Bmatrix} 4.58 \\ 9.82 \\ 14.59 \end{Bmatrix} \text{ rad/s}$$

Since time period, $T = 2\pi/\omega$, in seconds

$$\begin{Bmatrix} T_1 \\ T_2 \\ T_3 \end{Bmatrix} = \begin{Bmatrix} 1.370 \\ 0.646 \\ 0.431 \end{Bmatrix}$$

Mode Shapes

Taking first the first mode shape corresponding to $\omega_1^2 = 21$

$$\begin{vmatrix} 60 - 1.0\,\omega_1^2 & -60 & 0 \\ -60 & 180 - 1.5\,\omega_1^2 & -120 \\ 0 & 120 & 300 - 2.0\,\omega_1^2 \end{vmatrix} \{\phi\}^1 = \{0\}$$

Putting the value of $\omega_1^2 = 21$,

$$\begin{vmatrix} 39 & -60 & 0 \\ -60 & 148.5 & -120 \\ 0 & 120 & 258 \end{vmatrix} \begin{Bmatrix} \phi_1 \\ \phi_2 \\ \phi_3 \end{Bmatrix}^1 = \{0\}$$

Dividing the first row by 39, the second row by 60 and the third row by 120,

$$\begin{vmatrix} 1 & -1.5385 & 0 \\ -1 & 2.475 & -2 \\ 0 & 1 & -2.15 \end{vmatrix} \begin{Bmatrix} \phi_1 \\ \phi_2 \\ \phi_3 \end{Bmatrix}^1 = \begin{Bmatrix} 0 \\ 0 \\ 0 \end{Bmatrix}$$

Adding rows 1 and 2,

$$\begin{vmatrix} 1 & -1.5385 & 0 \\ 0 & 0.9365 & -2 \\ 0 & 1 & -2.15 \end{vmatrix} \begin{Bmatrix} \phi_1 \\ \phi_2 \\ \phi_3 \end{Bmatrix}^1 = \begin{Bmatrix} 0 \\ 0 \\ 0 \end{Bmatrix}$$

Dividing the second row by 0.9365 and subtracting from the third row,

$$\begin{vmatrix} 1 & -1.5385 & 0 \\ 0 & 0.9365 & -2 \\ 0 & 0 & 0 \end{vmatrix} \begin{Bmatrix} \phi_1 \\ \phi_2 \\ \phi_3 \end{Bmatrix}^1 = \begin{Bmatrix} 0 \\ 0 \\ 0 \end{Bmatrix}$$

Thus, there are two equations and three unknowns. Assume a value for one of the unknowns as

$$\phi_3^1 = 1.0$$

then, $\phi_2^1\,(0.9365) - 2(1.000) = 0 \Rightarrow \phi_2^1 = 2.150$

Again, $\phi_1(1.000) + (-1.5385)(2.150) = 0 \Rightarrow \phi_1^1 = 3.308$

$$\begin{Bmatrix} \phi_1 \\ \phi_2 \\ \phi_3 \end{Bmatrix}^1 = \begin{Bmatrix} 3.308 \\ 2.150 \\ 1.000 \end{Bmatrix} \quad \text{or} \quad \begin{Bmatrix} \phi_1 \\ \phi_2 \\ \phi_3 \end{Bmatrix}^1 = \begin{Bmatrix} 1.000 \\ 0.644 \\ 0.300 \end{Bmatrix}$$

Similarly,

$$\begin{Bmatrix} \phi_1 \\ \phi_2 \\ \phi_3 \end{Bmatrix}^2 = \begin{Bmatrix} 1.000 \\ -0.601 \\ -0.676 \end{Bmatrix} \quad \text{and} \quad \begin{Bmatrix} \phi_1 \\ \phi_2 \\ \phi_3 \end{Bmatrix}^3 = \begin{Bmatrix} 1.000 \\ -2.570 \\ 2.470 \end{Bmatrix}$$

Thus,

$$\{\phi\} = \begin{bmatrix} 1.000 & 1.000 & 1.000 \\ 0.644 & -0.601 & -2.570 \\ 0.300 & -0.676 & 2.470 \end{bmatrix}$$

The mode shapes are shown in Figue 4.6(c).

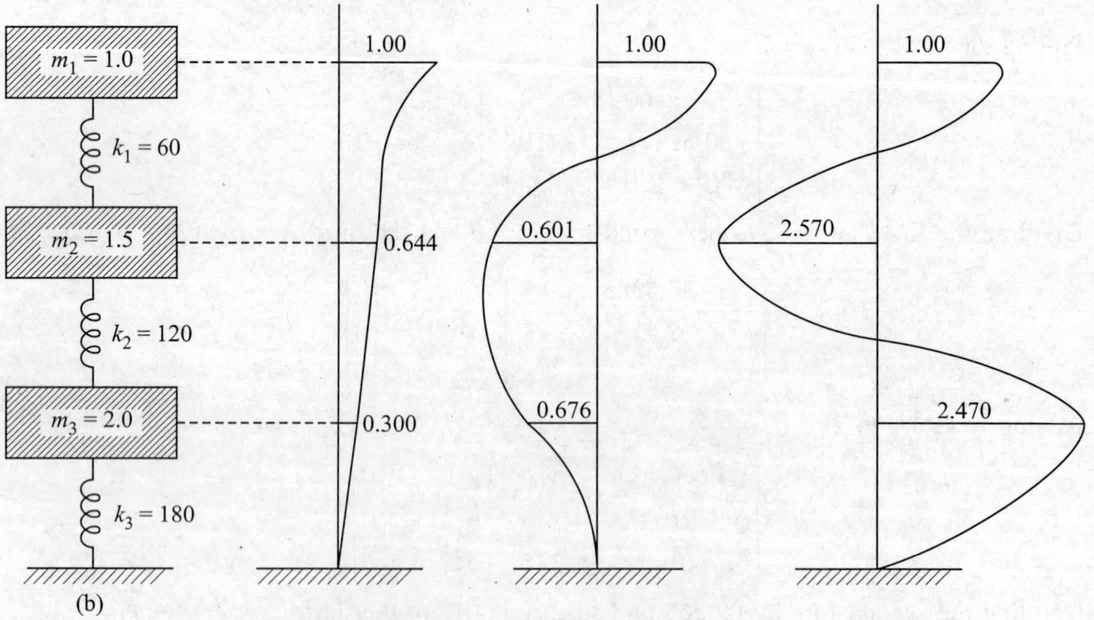

Figure 4.6(c) Mode shapes (Example 4.3).

Example 4.4 The mass and stiffness properties of a three-storied building are shown in Figure 4.7. The building is subjected to an earthquake wherein the ground acceleration during an earthquake may be taken as a stationary random process. Assume a power spectral density and damping of the structure. Explain the procedure for obtaining the mean square value of the relative displacement of the various floors of the building frame.

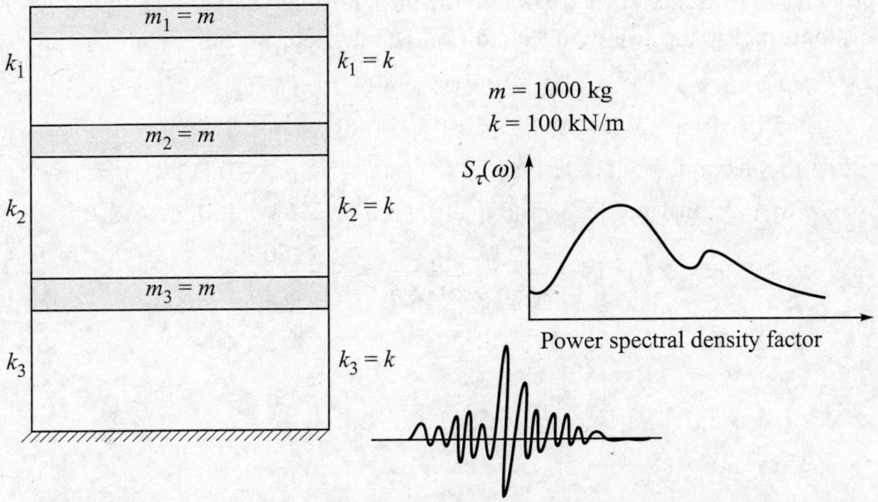

Figure 4.7 A three-storied building subjected to ground motion (Example 4.4).

Solution: The mass matrix of the building frame may be expressed as

$$[m] = \begin{bmatrix} 1000 & 0 & 0 \\ 0 & 1000 & 0 \\ 0 & 0 & 1000 \end{bmatrix} = m \begin{vmatrix} 1 & 0 & 0 \\ 0 & 1 & 0 \\ 0 & 0 & 1 \end{vmatrix} \text{ where } m = 1000 \text{ kg}$$

and stiffness matrix

$$[k] = \begin{bmatrix} 200 & -100 & 0 \\ -100 & 200 & -100 \\ 0 & -100 & 100 \end{bmatrix} = k \begin{vmatrix} 2 & -1 & 0 \\ -1 & 2 & -1 \\ 0 & -1 & 1 \end{vmatrix} \text{ where } k = 100 \text{ kN/m}$$

From Example 4.2, the natural frequencies are

$$\omega_1 = 0.44504\sqrt{\frac{k}{m}} = 0.44504\sqrt{\frac{100 \times 1000}{1000}} = 4.45 \text{ rad/s}$$

$$\omega_2 = 1.2471\sqrt{\frac{k}{m}} = 1.2471\sqrt{\frac{100 \times 1000}{1000}} = 12.471 \text{ rad/s}$$

$$\omega_3 = 1.8025\sqrt{\frac{k}{m}} = 1.8025\sqrt{\frac{100 \times 1000}{1000}} = 18.02 \text{ rad/s}$$

Similarly, the mode shape

$$\{\vec{\phi}\} = \begin{bmatrix} 1.000 & 1.000 & 1.000 \\ 1.8019 & 0.4450 & -1.2468 \\ 2.2470 & -0.8020 & 0.5544 \end{bmatrix}$$

Using the property of orthogonality, we obtain the orthonormal values. The eigenvector $\vec{\phi}$ is said to be $[m]$ orthonormal if the following condition is satisfied, i.e.,

$$\{\vec{\phi}\}^T[m][\phi] = \{1\}$$

$$\text{For the first mode:} \quad m\{\phi^1\}^2(1.0^2 + 1.8019^2 + 2.2470^2) = 1$$

$$\text{For the second mode:} \quad m\{\phi^2\}^2(1.0^2 + 0.4450^2 + (-0.8020)^2) = 1$$

$$\text{For the third mode:} \quad m\{\phi^3\}^2(1.0^2 + (-1.2468)^2 + 0.5544^2)$$

So,

$$\{\phi^1\} = \frac{1}{\sqrt{9.2959m}} = \frac{0.3280}{\sqrt{m}}$$

Similarly

$$\{\phi^2\} = \frac{0.7350}{\sqrt{m}}$$

$$\{\phi^3\} = \frac{0.5911}{\sqrt{m}}$$

Thus,

$$\vec{z}_1 = \frac{0.3280}{\sqrt{m}}\{\phi^1\} = \frac{0.3280}{\sqrt{m}}\begin{Bmatrix} 1.000 \\ 1.8019 \\ 2.2470 \end{Bmatrix} = \begin{Bmatrix} 0.01037 \\ 0.01869 \\ 0.02330 \end{Bmatrix}$$

where $\vec{z}_1$ is used to denote the mode shape instead of $\vec{\phi}_1$ since $\vec{z}_1$ represents the relative displacement instead of the absolute displacement. It $\phi_i(t)$ is the absolute displacement and $u_g(t)$ is the ground motion, then

$$z_i(t) = \phi_i(t) - u_g(t)$$

The equation of motion for the three-storied building

$$[m]\ddot{\phi}(t) + c\dot{z}(t) + kz = \vec{0}$$

$$[m]\ddot{z}(t) + c\dot{z}(t) + kz = -m\ddot{u}_g(t)$$

where,

$$\ddot{u}_g(t) = \text{ground acceleration}$$

The above equation is a coupled equation of motion and as such the uncoupled equation of motion shall be obtained by expressing the displacement vector $\vec{z}$ in terms of normal modes

$$\vec{z} = [\psi]\vec{q}$$

where $[\psi]$ = modal matrix.

By substituting $\vec{z}$ in the equation of motion and premultiplying the resulting equation with $[\psi]^T$, we usually derive the uncoupled equation of motion.

Assuming a damping of ζ in mode i, the uncoupled equation of motion may be expressed as

$$\ddot{q}_i + 2\zeta_i\omega_i q_i + \omega_i^2 q_i = Q_i, \quad i = 1, 2, 3$$

where,

$$q_i = \text{modal damping ratio}$$

$$Q_i = \sum_{j=1}^{n} z_j^i \, F_j(t)$$

and,
$$F_j(t) = -m_j\ddot{u}_g(t) = m\ddot{u}_g(t)$$

with $m_j = m$ denoting the mass of the jth floor. Again representing $F_j(t)$ as

$$F_j(t) = f_j \, \tau(t)$$

so,
$$f_j = -m_j = -m$$

and,
$$\tau(t) = \ddot{u}_g(t)$$

The uncoupled equation of motion can also be obtained using the method of mode participation (see section 4.3).

The equation of motion can be decoupled as

$$\ddot{q}_i + 2\zeta_i \, \omega_i \, \dot{q}_i + \omega_i^2 \, q_i = -\beta_r \, \ddot{u}_g, \quad \text{where} \quad r = 1, 2, 3, \ldots, x$$

where,
$$\beta_r = \text{mode participation factor} = \frac{\sum_{i=1}^{n} m_i \, \phi_{ir}}{\sum_{i=1}^{n} m_i (\phi_{ir})^2}$$

Again by assuming that the spectral displacement ordinate for frequency ω_r and damping ζ_r in a given spectra may be taken as $S_d(\omega_r \cdot \zeta_r)$, the maximum response to the rth modal coordinate $q_i(\max)$ may be obtained as

$$q_i(\max) = \beta_r S_d(\omega_r \cdot \zeta_r) \quad \text{where} \quad r = 1, 2, 3, \ldots, x$$

Assuming that the maximum of each modal coordinate occurs at the same instant of time, then

$$\vec{Z}_i(\max) = \sum_{r=1}^{n} q_i(\max) \cdot \phi_{ir}$$

The maximum floor displacement for this three storied building may be expressed as

$$\vec{Z}_i(\max) = \left[\sum_{r=1}^{3} |\, \phi_{ir} \cdot \beta_r \cdot \{S_d(\omega_r \cdot \zeta_r)\}^2 \right]^{0.5}$$

The maximum inter-storage drifts may be obtained as

$$\Delta_{ij}(\max) = \left[\sum_{r=1}^{3}\left\{(\phi_i^r - \phi_j^r)\beta_r \cdot \left(\frac{S_{ar}}{\omega_r^2}\right)\right\}\right]^{0.5}$$

Maximum storey shear may be obtained as

$$V_j(\max) = \left[\sum_{r=1}^{3}\left\{\beta_r \cdot S_{ar} \sum m_{ii}\,\phi_i^r\right\}^2\right]^{0.5}$$

The mean square value $\vec{Z}_i^2(t)$ may be expressed as

$$\text{Mean square value} = \sum_{r=1}^{3}[Z_{ir}]^2\left(\frac{N_r^2}{\omega_r^3}\right)\left(\frac{\Pi}{2\zeta_r}\right)S_d\,(\omega_r \cdot \zeta_r)$$

And square root of sum of squares (SRSS) value may be obtained as

$$\text{SRSS value} = \left[\sum_{i=1}^{3}[(q_i\,(\max) \cdot \phi_i^r)^2]\right]^{0.5}$$

Summary of the procedure (Example 4.4)

The response of an idealized multistoried building under consideration to earthquake ground motion can be obtained by the following procedure:

- Define the ground acceleration $\ddot{u}_g(t)$ by the numerical ordinates of the accelerogram.
- Compute the mass and stiffness matrices, $[m]$ and $[k]$.
- Solve the eigen-problem to determine the natural frequencies ω_n and mode shape ϕ_n of vibration.
- Obtain the uncoupled equation of motion with mode participation factor.
- Compute the modal response.
- Compute the floor displacement.
- Compute the storey drifts.
- Compute the internal force-storey shears.
- Compute the base shear.
- Compute the base moments.

For further details, Dynamics of Structure by Chopra (2001) may be referred to.

4.5 VIBRATIONS OF CONTINUOUS SYSTEMS

In the previous sections, free and forced vibrations of discrete systems have been presented whereas in this section free and forced vibrations of continuous systems will be discussed. However, this should not lead to an interpretation that the discrete and continuous systems exhibit or represent dissimilar dynamical characteristics. On the contrary, the discrete and continuous systems represent merely two mathematical models of the same physical system.

The mathematical formulation for a given continuous system is derived as a limiting case of that of a discrete system.

Furthermore, in the discrete systems as explained in the previous sections, mass, damping and elasticity were assumed to be present only at certain discrete points (nodal points) in the system. In the continuous system, it is not possible to identify discrete masses, dampers or springs. Owing to continuous distribution of mass, damping and elasticity, each of the infinite number of points of the system can vibrate. That is why, a continuous system is called a system of infinite degrees of freedom.

For analysis and design of a foundation, its components may be treated as discrete systems for convenience and simplicity of computation, therefore the results obtained as such can only be approximate. But the results are, however, sufficiently accurate for most practical cases. The reason for discretization is due to the fact that the analysis of a continuous system is much more involved. For all systems, the masses of the members are continuously distributed. As such specifying the displacement at every point in the system will require an infinite number of coordinates. The continuous system will thus have an infinite number of degrees of freedom. For such systems the mass is inseparable from the elasticity of the system. The continuous models of vibrating systems are indeed more realistic since the structural properties are distributed rather than concentrated at discrete points. The equation of motion is a partial differential equation for continuous system, whereas for discrete systems there have been only ordinary differential equations.

The continuous systems under consideration enjoy infinite degrees of freedom and they lead to a frequency equation, which is transcendental in nature with an infinite number of roots corresponding to the infinite number of degrees of freedom possessed by the system.

The essential difference between these two types of motions is, in fact, the same difference that exists between an oscillation and a wave motion. An oscillation takes place in the time domain and is completely specified by the initial values at one point, i.e., $t = 0$. A wave motion, on the other hand, takes place both in time and space domains. We require a "time-table" indicating both time and space to specify the configuration of the system at any instant. In addition to the initial values at $t = 0$, we require boundary conditions at the ends to predict the motion of the system. It is like the case of a discrete system, where we restrict our discussion to small motion in order to keep the equations of motion linear and to render the analysis easier.

The equations of motion of continuous systems are more easily derived from Hamilton's principle rather than from the Lagrange's equation. In the case of Lagrange's equation, we use the principle of virtual work, which refers to an instantaneous state of the system and is differential in nature. The Hamilton principle, on the other hand, is an integral principle. It states that of all the possible paths open to a mechanical system, between two instants of the t_1 and t_2, the actual path taken by the system is one which renders a stationary value to an integral I, called the action integral, i.e.,

$$I = \int_{t_1}^{t_2} L dt \tag{4.27}$$

where,

$$L = T - V = L(q_1, q_2, ..., q_n, ..., \dot{q}_1, \dot{q}_2, \dot{q}_n, t) \tag{4.28}$$

is the Lagrangian of the system.

Here,

T = kinetic energy,
V = potential energy, and
q_1 = generalized coordinates representing translation of mass m_i [i = 1,2,3,...,n].

The condition for the stationarity of the integral is that its variation between the two fixed instants should vanish, i.e.,

$$dI = d\int_{t_1}^{t_2} L dt = d\int_{t_1}^{t_2} L(q_1, q_2,..., q_n, \dot{q}_1, \dot{q}_2,..., \dot{q}_n, t)dt = 0 \tag{4.29}$$

where q represents the generalized coordinates. Srinivasan (1982) has given a detailed procedure for vibrations of continuous systems.

Generalized coordinates

The generalized coordinates represent the degrees of freedom of the system. The generalized coordinates should be finite, single-valued, continuous and differentiable entities. They are denoted by $\{q\}$, i = 1,2,3,...,n. They are by no means unique. For a single particle (i = 1 to 3), q_1, q_2, q_3 can represent the Cartesian coordinates (x, y, z), spherical polar coordinates (r, θ, ϕ), cylindrieal polar coordinates (r, θ, z), and so on.
For Cartesian coordinates,

$$q_1 = x, \quad q_2 = y, \quad q_3 = z$$

For Spherical polar coordinates,

$$q_1 = r, \quad q_2 = \theta, \quad q_3 = \phi$$

For Cylindrical polar coordinates

$$q_1 = r, \quad q_2 = \theta, \quad q_3 = z$$

For continuous systems in this section, the vibration analysis is limited to the following:

- vibrations of beams
- vibrations of beams on elastic foundations
- vibrations of plates
- vibrations of plates on elastic foundations

For vibrations of other continuous systems like strings, membranes, rings and shells, textbooks on Vibrations/Structural dynamics may be referred to.

4.6 VIBRATIONS OF BEAMS

Consider an Euler–Bernouli beam ($0 \le x \le L$) as shown in Figure 4.8 with lateral deflection (x, t). The strain energy is

$$U = (1/2)\int_0^L EI(x)\left(\frac{\delta^2 w}{\delta x^2}\right)^2 dx$$

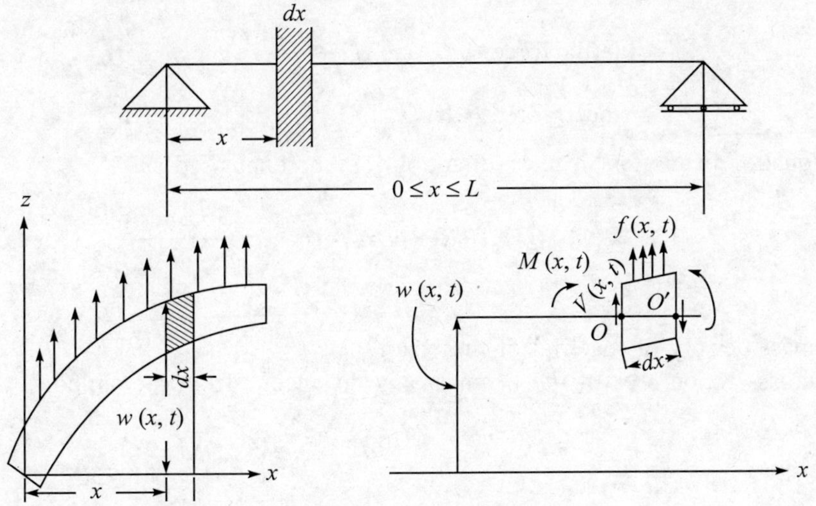

Figure 4.8 Flexure of an Euler–Bernoulli beam.

and the kinetic energy of the beam

$$T = (1/2)\int_0^L \rho A(x) \cdot \left(\frac{dw}{dt}\right)^2 dx$$

$$Q = \int_0^L F \cdot W \cdot dx$$

Using Hamilton principle,

$$\delta \int_{t_1}^{t_2} (U - T - W)dt = 0 \tag{4.30}$$

Substituting these values in Eq. (4.30), the governing differential equation is obtained as

$$\frac{\delta^2}{\delta^2 x}\left[EI(x)\frac{\delta^2 w}{\delta^2 x}(x,t)\right] + \rho A(x)\ \frac{\delta^2 w}{\delta^2 t}(x,t) = Q(x,t) \tag{4.31}$$

Thus, using Hamilton's principle, it is very convenient to obtain the equation of motion for various continuous systems like plates resting on an elastic foundation. However, there are other methods to obtain the equation. One of them is discussed now, and this may appear more familiar as moment and force equilibrium equations are considered on an elemental part of the beam ($0 \le x \le L$).

Considering the free-body diagram of the beam shown in Figure 4.8, where $M(x,t)$ is the bending moment, $V(x,t)$ is the shear force and $Q(x,t)$ is the external dynamic force per unit length of the beam. The forces acting on the element of beam are shown in Figure 4.8 and they are

$$\text{inertia force} = \text{mass} \times \text{acceleration}$$

$$\text{Inertia force} = \rho A(x)\, dx \frac{\partial^2 w(x,t)}{\partial t^2}$$

$$\text{Shear force} = dV(x,t)$$

The force equation of motion in the z-direction may be expressed as

$$-(V + dV) + Q(x,t)dx + V = \rho A(x)\, dx \frac{\partial^2 w(x,t)}{\partial t^2}$$

where,

ρ = mass density of material of the beam
$A(x)$ = cross-section area of the beam at any distance x from the support.

By writing
$$dV = \frac{\partial}{\partial x} V(x, t)dx$$

$$-\left(V + \frac{\partial V}{\partial x} dx\right) + Q(x,t)dx + V = \rho A(x)dx \frac{\partial^2 w(x,t)}{\partial t^2}$$

$$-\frac{\partial V}{\partial x}(x,t) + Q(x,t) = \rho A(x) \frac{\partial^2 w(x,t)}{\partial t^2}$$

Similarly, the moment equation of motion about the y-axis passing through the point O in Figure 4.8 leads to

$$(M + dM) - (V + dV)dx + Q(x,t) \cdot dx \frac{dx}{2} - M = 0$$

Putting $dM = \dfrac{\partial M}{\partial x} dx$ and disregarding the second power index

$$\frac{\partial M(x,t)}{\partial x} - V(x,t) = 0$$

From bending theory $V = \dfrac{\partial M}{\partial x}$ and $M(x,t) = EI(x) \cdot \dfrac{\partial^2 w(x,t)}{\partial x^2}$

where,

E = Young's modulus of elasticity
$I(x)$ = moment of inertia of the cross-section of the beam about y-axis.

The equation of motion for the flexural vibration (also called the lateral vibration) of a non-uniform beam is

$$\frac{\partial^2}{\partial x^2}\left[EI(x) \frac{\partial^2 w(x,t)}{\partial t^2}\right] + \rho A(x) \frac{\partial^2 w(x,t)}{\partial t^2} = Q(x,t)$$

For a uniform beam, $A(x) = A$, $I(x) = I$, then

$$EI \frac{\partial^4 w(x,t)}{\partial x^4} + \rho A \frac{\partial^2 w(x,t)}{\partial x^2} = Q(x, t)$$

For free flexural vibrations

$$EI\frac{\partial^4 w(x,t)}{\partial x^4} + \rho A\frac{\partial^2 w(x,t)}{\partial x^2} = 0$$

Boundary conditions and initial conditions

Since the governing differential equation of motion involves a fourth-order derivative with respect to x and a second-order derivative with respect to time, four boundary conditions and two initial conditions are needed for finding a unique solution for $w(x, t)$.

Natural frequencies of vibrations

The free vibrations solution can be found using the method of separation of variables as

$$w(x, t) = \phi(x)\cdot q(t)$$

Substituting this into differential equation and rearranging terms,

$$\frac{EI}{\rho A}\cdot\frac{1}{\phi(x)}\frac{d^4\phi(x)}{dx^4} = -\frac{1}{q(t)}\frac{d^2q(t)}{dt^2} = a = \omega^2$$

where $a = \omega^2$ is a positive constant. The above equation can be written as two equations

$$\frac{d^4\phi(x)}{dx^4} - \lambda^4\phi(x) = 0$$

$$\frac{d^2q(t)}{dt^2} + \omega^2 q(t) = 0$$

where,

$$\lambda^4 = \frac{\rho A}{EI}\omega^2$$

The solution may be expressed as

$$q(t) = A\cos\omega t + B\sin\omega t$$

The constants A and B can be found from the initial conditions. The solution of the other equation may be taken as

$$\phi(x) = A\cdot e^{sx}$$

where A and S are constants and we derive the auxiliary equation as

$$S^4 - \lambda^4 = 0$$

The roots of this equation are

$$S_{1,2} = \pm\lambda \quad \text{and} \quad S_{3,4} = \pm i\lambda$$

Hence, the solution becomes

$$\begin{aligned}\phi(x) &= A_1e^{\lambda x} + A_2e^{-\lambda x} + A_3e^{+i\lambda x} + A_4e^{-i\lambda x}\\ &= A_1\cos\lambda x + A_2\sin\lambda x + A_3\cosh\lambda x + A_4\sinh\lambda x\end{aligned}$$

These constants can be obtained by using four boundary conditions of the beam ($0 \le x \le L$). The natural frequency is expressed as

$$\omega = \lambda^2 \sqrt{\frac{EI}{\rho A}} = (\lambda L)^2 \sqrt{\frac{EI}{\rho A L^4}}$$

The function $\phi(x)$ is known as the normal mode or characteristic function of the beam and ω is called the natural frequency of vibration. For any beam there will be an infinite number of normal modes with one natural frequency associated with each normal mode.

Let the solution be assumed as summation of modal components

$$w(x,t) = \sum_{n=1}^{n} \phi_n(x)\cdot q_n(t) \tag{4.32}$$

Example 4.5 A pile having axial force is subjected to flexural vibrations. Discuss the effect of an axial force 'N' on the free flexural vibrations of a pile treating the pile as a uniform simply supported beam ($0 \le x \le L$).

Solution: Treating the pile as a simply supported beam with axial force as shown in Figure 4.9, the governing differential equation for free flexural vibrations of a uniform simply supported beam may be expressed as

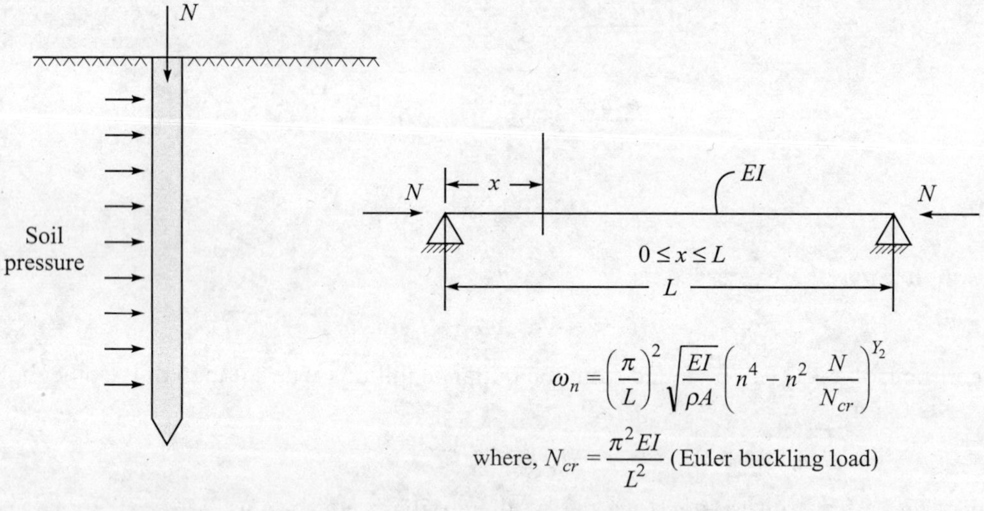

Figure 4.9 Effect of axial force N on flexural vibration of piles [Example 4.5].

$$EI\frac{\delta^4 w}{\delta x^4} + \rho A\frac{\delta^2 w}{\delta t^2} = 0$$

In addition to the elastic force and inertia force, the tensile force N will provide another force equal to

$$-N\frac{\delta^2 w}{\delta x^2}$$

However, the sign will have to be reversed, if axial compression is applied. Finally, the governing differential equation takes the form

$$EI\frac{\delta^4 w}{\delta x^4} - N\frac{\delta^2 w}{\delta x^2} + \rho A\frac{\delta^2 w}{\delta t^2} = 0$$

The solution may be assumed as

$$w(x, t) = \sum_{n=1}^{\infty} C \sin\frac{n\pi x}{L} \sin(\omega_n - \alpha)$$

Substituting the solution in the governing differential equation,

$$EI = \left(\frac{n\pi}{L}\right)^4 - N\left(\frac{n\pi}{L}\right)^2 - \rho A\omega_n^2 = 0$$

or, $$\omega_n = (n\pi)^2\sqrt{\left[\left\{1 - \frac{NL^2}{n^2\pi^2 EI}\right\} \cdot \frac{EI}{\rho AL^2}\right]} = \left(\frac{\pi}{L}\right)^2\sqrt{\frac{EI}{\rho A}}\left(n^4 - n^2\frac{N}{N_{cr}}\right)^{1/2}$$

where, $n = 1, 2, 3, \ldots,$ and $N_{cr} = \pi^2\frac{EI}{L^2}$

when $N = 0$, the above equation will give solution for a simply supported beam undergoing flexural free vibrations.

The first natural fundamental frequency may be expressed as

$$\omega = \pi^2\sqrt{\left[\left\{1 - \frac{N}{N_{cr}}\right\}\frac{EI}{\rho AL^4}\right]}$$

where, $N_{cr} = \dfrac{\pi^2 EI}{L^2}$ is the Euler critical (buckling) load for the beam. Thus, the effect of axial compression is to reduce the natural frequencies and the axial tension will increase the natural frequencies.

Example 4.6 Discuss the free flexural vibrations of a simply supported beam ($0 \le x \le L$) and obtain a general solution of the dynamic displacement.

Solution: The governing differential equation for a beam under free flexural vibrations may be expressed as

$$\frac{\delta^2}{\delta x^2}\left[EI(x)\frac{\delta^2 w}{\delta x^2}\right] + \rho A(x)\frac{\delta^2 w}{\delta t^2} = 0$$

For a uniform beam, $EI(x) = EI$, $\rho A(x) = \rho A$, therefore,

$$EI\frac{\delta^4 w}{\delta x^4} + \rho A\frac{\delta^2 w}{\delta t^2} = 0$$

The solution may further be written as

$$w(x,t) = \phi(x).\ (A_1 \cos\ \omega_n t + A_2 \sin\ \omega_n t)$$

where,
$$q_n(t) = A_1 \cos\ \omega_n t + A_2 \sin\ \omega_n t$$

and $\phi(x)$ is a function of x only and represents the mode shape.

Substituting for $w(x, t)$ in the governing differential equation,

$$EI\frac{d^4\phi(x)}{dx^4} - \rho A \cdot \omega_n^2 \cdot \phi(x) = 0$$

or,
$$\frac{d^4\phi(x)}{dx^4} - \lambda^4 \phi(x) = 0$$

where,
$$\lambda^4 = \frac{\rho A \omega_n^2}{EI}$$

Thus,
$$\phi(x) = A_1 \sin\ \lambda x + A_2 \cos\ \lambda x + A_3 \sinh\ \lambda x + A_4 \cosh\ \lambda x$$

where A_1, A_2, A_3 and A_4 are constants and depend upon the boundary conditions. The boundary conditions are

At
$$x = 0,\ \phi(0) = 0,\ \frac{d^2\phi}{dx^2} = 0, \quad \text{(deflection and moment are zero at } x = 0)$$

$$x = L,\ \phi(L) = 0,\ \frac{d^2\phi}{dx^2} = 0, \quad \text{(deflection and moment are zero at } x = L)$$

Applying the boundary conditions,

$$A_2 + A_4 = 0$$
$$-\lambda^2 A_2 + \lambda^2 A_4 = 0$$

Thus,

$$A_2 = A_4 = 0$$
$$A_1 \sin\ \lambda L + A_3 \sinh\ \lambda L = 0$$
$$\lambda\ (-A_1 \sin\ \lambda L + A_3 \sinh\ \lambda L) = 0$$

which yields,

$$A_3 \sinh\ \lambda L = 0$$
$$A_1 \sin\ \lambda L = 0$$

As λ is not zero, $\sin\ \lambda L = 0$ for all values of λ. Therefore, $A_3 = 0$

If we assume $A_1 = 0$, then there cannot be any vibrations of the system, therefore

$$\sin\ \lambda L = 0$$

which means,

$$\lambda L = n\pi,\ n = 1, 2, 3, \ldots$$

or,
$$\lambda = \frac{n\pi}{L}$$

Thus, the natural frequencies are

$$\omega_n = n^2\pi^2\sqrt{\left[\frac{EI}{\rho AL^2}\right]}, \text{ where } n = 1, 2, 3,...$$

The mode shape of the beam is given by

$$\phi(x) = A_n \sin \frac{n\pi x}{L}$$

Thus, the general solution may be expressed as

$$w(x, t) = \sum_{n=1}^{\infty} \sin \frac{n\pi x}{L} (A_n \cos \omega_n t + B_n \sin \omega_n t)$$

where the constants A_n, B_n depend upon the initial conditions. The three mode shapes and the corresponding frequencies are shown in Figure 4.10.

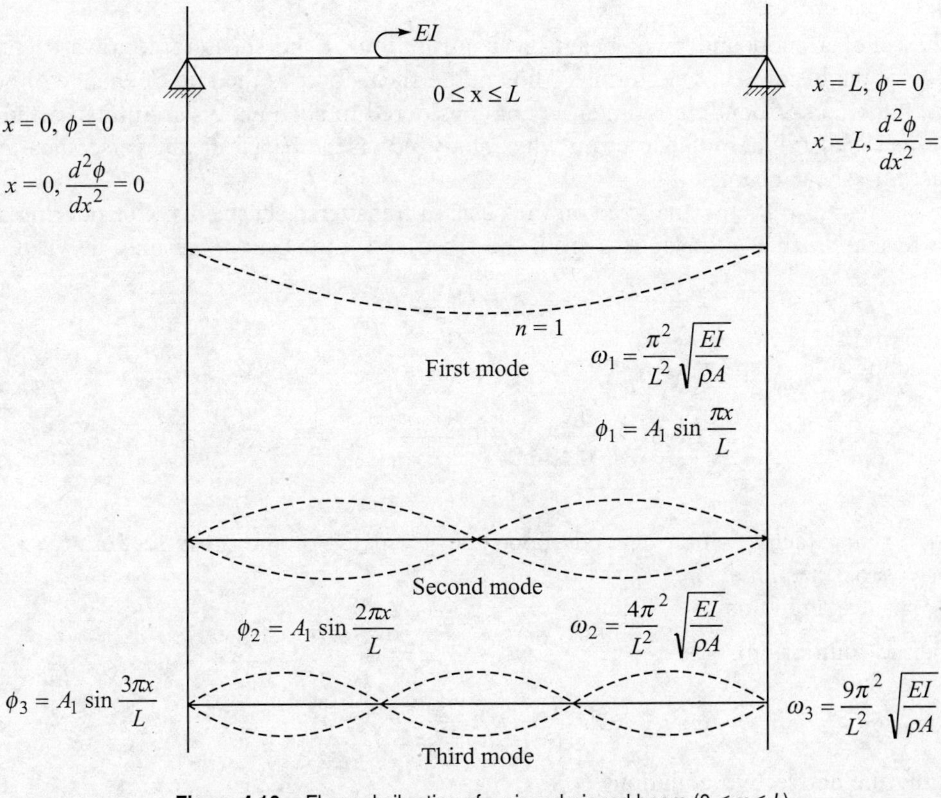

Figure 4.10 Flexural vibration of a pinned-pinned beam ($0 \le x \le L$)

Example 4.7 Explain the difference between shear vibrations of beam and its flexural vibrations. Discuss the free vibration of a shear beam ($0 \le x \le L$) with one end fixed and the other end free.

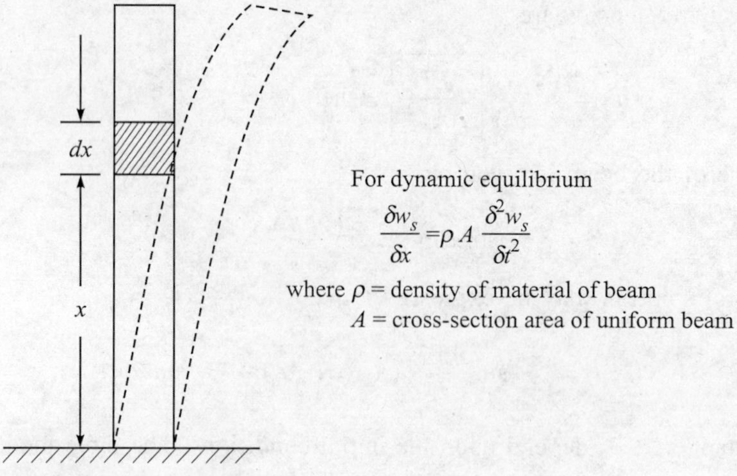

Figure 4.11 A shear beam ($0 \le x \le L$).

Solution: For a short and thick beam the contribution of the shear force towards the total deflection of the beam is not negligible. Though the shear force is the rate of change of bending moment, the effect of bending moment is not considered in such analysis. Similarly, the rotary inertia effect is also ignored. The beam when analyzed on the basis of transverse shear only, is referred to as shear beam.

The beam ($0 \le x \le L$) is analyzed on the basis of transverse shear only. Considering a small element of length dx at a distance a from the fixed end, the shear force may be written as

$$S = \mu AG \frac{\delta w_S}{\delta x},$$

Differentiating with respect to x.

$$\frac{\delta S}{dx} = \mu AG \frac{\delta^2 w_S}{\delta x^2}$$

where,

μ = shape factor which depends upon the geometry of the cross-section
A = cross-sectional area
G = shear modulus of elasticity

For dynamic equilibrium,

$$\frac{\delta S}{\delta x} = \rho A \frac{\delta^2 w_S}{\delta t^2}$$

Combining the above two equations,

$$\frac{\delta S}{\partial x} = \mu A G \frac{\delta^2 w_S}{\partial x^2} = \rho A \frac{\delta^2 w_S}{\partial t^2}$$

or,

$$\frac{\delta^2 w_S}{\delta x^2} = \frac{1}{\lambda^2} \frac{\delta^2 w_S}{\delta t^2}$$

where, $$\lambda^2 = \frac{\mu G}{\rho}$$

Solutions of the above differential equation may be written as

$$w_S(x,\ t) = (A_1 \sin\ \alpha x + A_2 \cos\ \alpha x)(A_3 \sin\ \omega t + A_4 \cos\ \omega t)$$

where, $$\alpha^2 = \frac{\omega^2}{\lambda^2}$$

Using the boundary conditions

At $$x = 0,\ w_S = 0 \quad \text{(deflection is zero)}$$

At $$x = L,\ \frac{\delta w_S}{\delta x} = 0 \quad \text{(slope is zero)}$$

which yields $C_2 = 0$ and $\cos\ \alpha L = 0$.
And the frequency equation becomes

$$\cos\ \alpha L = 0$$

So, $$\alpha = \frac{2n-1}{2L}\pi,\quad n = 1,\ 2,\ 3,...$$

Thus, the frequency of vibrations of beam in shear mode is

$$\omega_n = \frac{2n-1}{2L}\cdot\pi\sqrt{\frac{\mu G}{\rho}},\quad n = 1,\ 2,\ 3,...$$

The general solution of the dynamic displacement $w_S(x,\ t)$ becomes

$$w_S(x,\ t) = \sum_{n=1}^{\infty} \sin\ \alpha x[A_3 \sin\ \omega_n t + A_4 \cos\ \omega_n t]$$

4.7 VIBRATIONS OF BEAMS ON ELASTIC FOUNDATION

When stiffness of the foundation is taken into account, a solution is used that is based on some form of beam on elastic foundation. This may be of the classical Winkler (1867) solution in which foundation is considered as a bed of springs having stiffness *k*. Bridge piers and laterally loaded piles are usually designed as beam on elastic foundation. Ring foundations are generally used for water tower structures, transmission towers, TV towers and various other possible superstructures. The ring foundation may be considered as a relatively narrow circular beam resting on elastic foundation. Pile foundations specially piles under lateral load when subjected to ground motion due to earthquakes, may be idealized as beam on elastic foundation representing the soil-pile system. The questions of dynamic deformations under earthquake excitation for a class of foundations can be obtained by using the theory of vibration of beams on elastic foundation.

However, the two-parameter foundation models represent more accurately the foundation characteristics compared to the simple, single parameter model (Winkler model). The widely used two-parameter foundation model is the Pasternak foundation model. Further, beams on elastic foundation exhibit an interesting phenomenon of changing mode shapes (from the first mode to the second mode, and so on) for both buckling and free vibration problems at specific foundation stiffnesses parameter(s). While evaluating the foundation stiffness parameter for beams on Winkler foundation for both the buckling and vibration problems is easy, the procedure is more involved in the case of the one or two parameter, uniform or variable foundation. Further, most of the practising engineers are very familiar with the Winkler foundation than with the two parameter foundation. Hence, it will be very useful and elegant if one obtains an equivalent uniform Winkler foundation to represent the uniform or variable two-parameter elastic foundation. In this context the fundamentals of vibrations of beam on elastic foundation (Winkler model) are presented to have a mathematical tool for seismic analysis and earthquake resistant design of foundations.

We assume a beam with hinged ends and supported along its length by a continuous elastic foundation as shown in Figure 4.12, the rigidity of which is given by k, the modulus of foundation (k is the load per unit length of the beam to produce a compression in the foundation equal to unity). If the mass of the foundation can be neglected, the equation of motion of such a beam can be set up by the Hamilton's principle. It is only necessary in calculating the potential energy of the system to add to the energy of bending, the energy of deformation of the elastic foundation, i.e.,

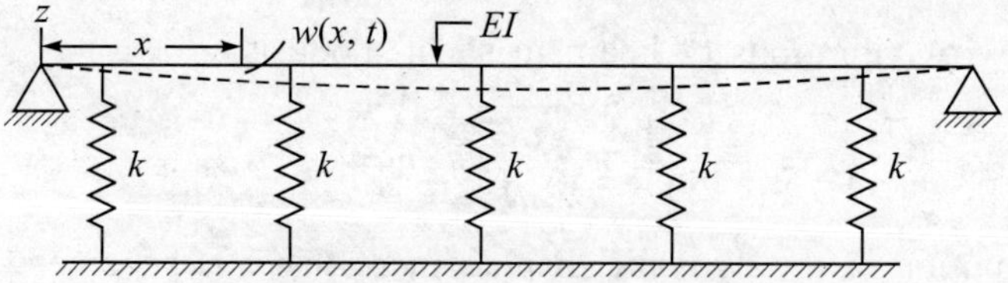

Figure 4.12 A beam on elastic foundations.

$$V = EI/2\int_0^L (d^2w/dx^2)^2\,dx + (k/2)\int_0^L w^2\,dx \tag{4.33}$$

and the kinetic energy T is given by

$$T = (\rho A/2)\int_0^L \left(\frac{\partial w}{2t}\right)^2 dx \tag{4.34}$$

It may be noted that massless foundations have no kinetic energy.

Forming the Lagrangian L and applying the Hamilton's principle, we have

$$\delta\int_{t_1}^{t_2}\left[\int_0^L \rho A/2\left(\frac{\partial w}{\partial t}\right)^2 dx - EI/2\int_0^L (d^2w/dx^2)^2\,dx - k/2\int_0^L w^2 dx\right] = 0 \tag{4.35}$$

Performing the variation on Eq. (4.35) as before, the only additional term being

$k/2\int_0^L w^2\,dx$, we obtain the equation of motion as

$$EI\,\frac{\delta^4 w}{\delta x^4} + \rho A\,\frac{\delta^2 w}{dt^2} + kw = 0 \tag{4.36}$$

The equation of motion contains

- elastic force in beam as EI $\dfrac{\delta^4 w}{\delta x^4}$
- inertia force in beam as $\rho_A\,\dfrac{\delta^2 w}{\delta t^2}$
- force in springs (representing foundation) as $k.w$

The governing differential Eq. (4.36) is for a uniform beam (constant *EI*) resting on Winkler model of elastic foundation, wherein the effects of shear deformation and rotatory inertia are neglected. The solution of Eq. (4.36) may be written as

$$w(x,\,t) = \sum_{n=1}^{\infty} X_n(n)\cdot q_n(t)$$

Separating the variables in Eq. (4.36),

$$\frac{d^4 X_n(x)}{dx^4} - \alpha_n^4 X_n(n) = 0$$

and,
$$\frac{d^2 q_n(t)}{dt^2} + \omega_n^2\cdot q_n(t) = 0$$

where, $\alpha_n^4 = \dfrac{\omega_n^2}{a^2} - \dfrac{k}{EI}$

$a^2 = \dfrac{EI}{\rho A}$

x_n = eigenfunction
ω_n = eigenfrequency
α_n = eigenvalue

The above two equations yield harmonic rather than an exponential solution which is consistent with the fact that a conservative system has constant total energy. The equation in space is a fourth-order homogeneous ordinary differential equation and as such must be supplemented by four boundary conditions, i.e., two boundary conditions for each end. The boundary conditions resulting from pure geometric (slope or deflection) compatibility are called geometric boundary conditions. The boundary conditions resulting from moment or shearing force balance are called natural boundary conditions.

Natural frequencies of vibrations

The frequency equation may be obtained by using the boundary conditions. In general, the end conditions of the beam–foundation system are

- Clamped free ends
- Free-free ends
- Pinned-pinned ends
- Clamped-clamped ends
- Clamped-pinned ends

For clamped-free ends

For clamped-free ends the boundary conditions are deflection and moments being zero at ends, and they may be expressed as

$$w(0,t) = 0, \qquad \frac{\partial w(0,t)}{\partial x} = 0$$

and,

$$\frac{\partial^2 w(L,t)}{\partial x^2} = 0, \qquad \frac{\partial^3 w(L,t)}{\partial x^3} = 0$$

The solution of the differential equation yields

$$X_n(x) = C_1 \sin \alpha_n x + C_2 \cos \alpha_n x + C_3 \sinh \alpha_n x + C_4 \cosh \alpha_n x$$

With the help of four boundary conditions, a set of four equations will have a non-trivial solution if the determinant of the coefficients of C_1, C_2 ... C_4 is zero. Thus, by expanding the 4 × 4 determinant, the frequency equation may be obtained. For the above mentioned boundary conditions, the determinant becomes

$$\begin{vmatrix} 0 & 1 & 0 & 1 \\ 1 & 0 & 1 & 0 \\ -\cos \alpha_n L & \sin \alpha_n L & \cosh \alpha_n L & \sinh \alpha_n L \\ -\sin \alpha_n L & -\cos \alpha_n L & \sinh \alpha_n L & \cosh \alpha_n L \end{vmatrix} = 0$$

The frequency equation takes the form

$$\cos \alpha_n L \cdot \cosh \alpha_n L + 1 = 0$$

The above transcendental equation may be solved numerically and yields an infinite solution $\alpha_n L$. This may be expressed as

n	$\alpha_n L$
1	1.875104069
2	4.694091133
3	7.85475743
4	10.99554073
5	14.13716839

and for higher order, i.e., $n > 5$

$$\alpha_n \cdot L = (2n - 1)\frac{\pi}{2}$$

The fundamental frequency ω_n is obtained as

$$\omega_n = \sqrt{\frac{EI}{\rho A}} \cdot \left(\frac{\pi}{L}\right)^2 \left[\alpha_n^4 + \frac{KL^4}{\pi^4 EI}\right]$$

The frequency of vibration depends not only on flexural rigidity of the beam but also on stiffness of the foundation.

Further the in-depth study of dynamics of beams on elastic foundation is being presented in Chapter 15.

Free-free beam on elastic foundation

For free-free beam on elastic foundation ($0 \le x \le L$) as shown in Figure 4.13, the boundary conditions are due to moment and shear being zero at both ends. This may be expressed as

$$EI(0)\frac{\partial^3 w(0,t)}{\partial x^3} = 0, \qquad EI(0)\frac{\partial^2 w(0,t)}{\partial x^2} = 0$$

$$EI(L)\frac{\partial^3 w(L,t)}{\partial x^3} = 0, \qquad EI(L)\frac{\partial^2 w(L,t)}{\partial x^2} = 0$$

Substituting these conditions into the governing differential equation, the frequency equation takes the form

$$\cos \alpha_n L \cdot \cosh \alpha_n L = 1$$

The above transcendental equation may be solved numerically for the given values

n	$\alpha_n L$
0	00.00000
1	0.00000
2	4.730040
3	7.85320462
4	10.99560783
5	14.13716549
6	17.27875965

and for higher values of x ($n > 6$)

$$\alpha_n L = \frac{2n-1}{2}\pi$$

So,

$$\omega_n = \sqrt{\frac{EI}{\rho A}}\left(\frac{\pi}{L}\right)^2 [\alpha_n^4 + \lambda^4]^{1/2}, \qquad \text{where } \lambda^4 = \frac{kL^4}{EI\pi^4}$$

The mode shapes are shown in Figure 4.13.

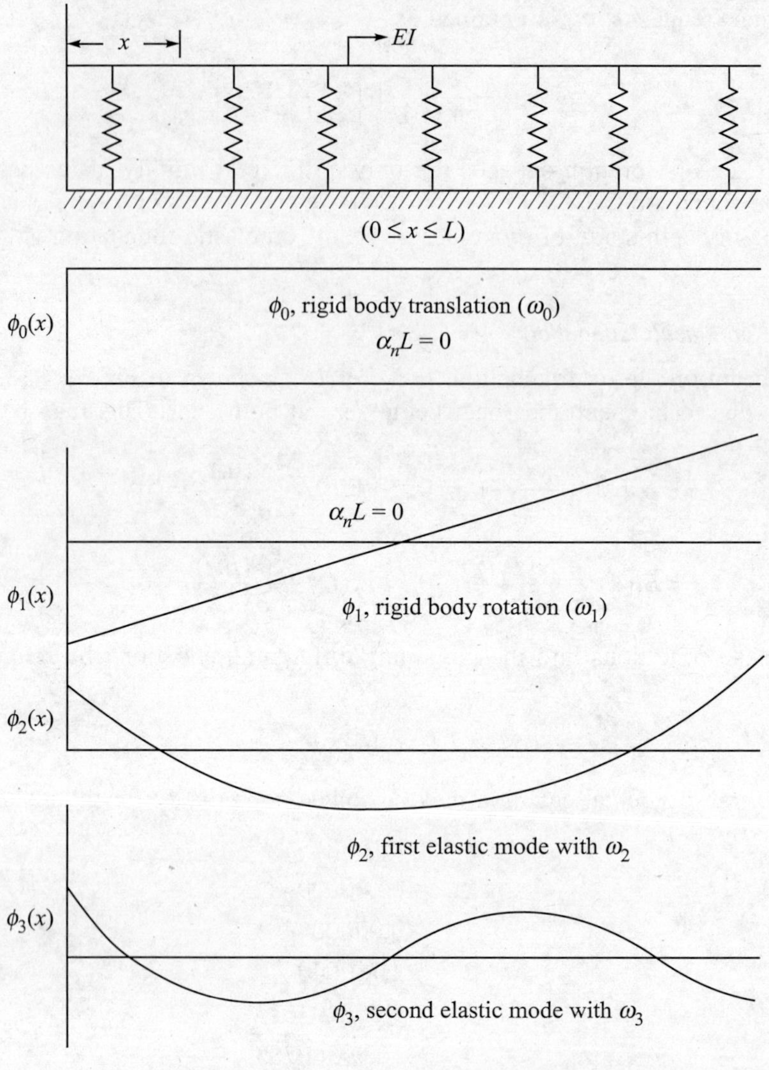

Figure 4.13 Free-free beam on elastic foundation.

4.8 VIBRATION OF PLATES

The plate forms a very important representative modal element in structural engineering design and construction. Mat or raft foundation and rigid pavements of airfield/highways are well represented by plates on elastic foundation. But let us first discuss the vibrations of plates. The differential equation for the vibration of a plate can be derived by Hamilton's principle. It is assumed that the plate is thin, i.e., the thickness of the plate is small compared to its other dimensions. The stretching of the middle surface of the plate is neglected in order to keep the equations of motion linear. The deflections considered are also small (in comparison with the thickness) for the same reason. It is assumed that the plane cross-sections before and after deformation remain plane, Srinivasan (1982).

The coordinate axes x and y are taken in the middle plane of the plate and the z-axis is taken perpendicular to that plane as shown in Figure 4.14. We consider a small element cut out by two pairs of planes parallel to the x-z and y-z planes. If w is the transverse deflection of the plate, the elementary kinetic energy dT of the shaded element of the plate is given by

$$dT = (1/2)\rho\left(\frac{\delta w}{\delta t}\right)^2 dx\ dy\ dz \tag{4.37}$$

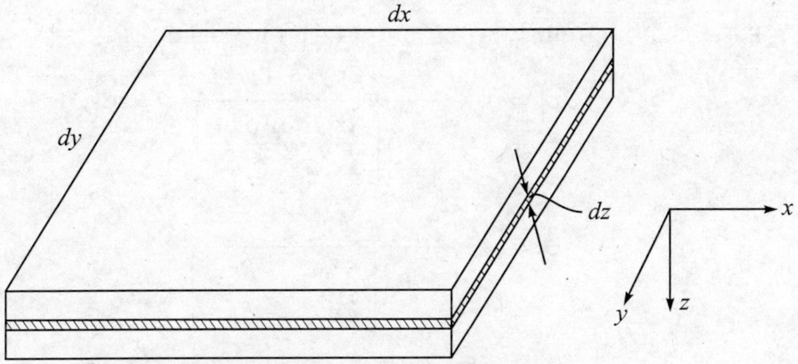

Figure 4.14 Coordinate system for a plate.

The total kinetic energy T of the plate is given by integrating Eq. (4.37), i.e.

$$T = (1/2)\int_0^a\int_0^b\int_{-h/2}^{+h/2} \rho\left(\frac{\delta w}{\delta t}\right)^2 dx\ dy\ dz$$

Integrating over z, we have

$$T = (1/2)\int_0^a\int_0^b\int_{-h/2}^{+h/2} \rho\left(\frac{\delta w}{\delta t}\right)^2 zdx\ dy \tag{4.38}$$

The elementary potential energy associated with the small shaded element is given by

$$dV = (1/2)(\sigma_x\ dy\ dz)[\varepsilon_x\ dx] + (1/2)(\sigma_y\ dx\ dz)[\varepsilon_y\ dy] + (1/2)(\tau_{xy}\ dy\ dz)[\gamma_{xy}\ dx]$$

From the strength of materials, we have the following stress–strain relations for the plate, viz.

$$\varepsilon_x = -z\frac{\delta^2 w}{\delta x^2} \tag{4.39}$$

$$\varepsilon_y = -z\frac{\delta^2 w}{\delta y^2} \tag{4.40}$$

$$\gamma_{xy} = -2z\frac{\delta^2 w}{\delta x\,\delta y} \tag{4.41}$$

Using Hooke's law for plane stresses, we have

$$\sigma_x = \frac{E}{(1-\nu^2)(\varepsilon_x + \nu\varepsilon_y)} \tag{4.42}$$

$$\sigma_y = \frac{E}{(1-\nu^2)(\varepsilon_y + \nu\varepsilon_x)} \tag{4.43}$$

$$\tau_{xy} = G \cdot \gamma_{xy} \tag{4.44}$$

Substituting for ε_x, ε_y and γ_{xy} from Eq. (4.39) to (4.41), we obtain

$$\sigma_x = -\frac{Ez}{(1-\nu^2)\left[\dfrac{\delta^2 w}{\delta x^2} + \dfrac{\nu\delta^2 w}{\delta y^2}\right]} \tag{4.45}$$

$$\sigma_y = -\frac{Ez}{(1-\nu^2)\left[\dfrac{\delta^2 w}{\delta y^2} + \nu\dfrac{\delta^2 w}{\delta x^2}\right]} \tag{4.46}$$

$$\tau_{xy} = -2Gz\frac{\delta^2 w}{\delta x\,\delta y} = -\frac{Ez}{1+\nu}\frac{\delta^2 w}{\delta y\,\delta y} \tag{4.47}$$

Substituting the expressions for σ_x, σ_y, τ_{xy}, ε_x, ε_y and γ_{xy} into the above equations, the elementary potential energy dV is given by

$$dV = \frac{Ez^2}{2(1-\nu^2)}\left[\left(\frac{\delta^2 w}{\delta x^2}\right)^2 + \left(\frac{\delta^2 w}{\delta y^2}\right)^2 + 2\frac{\delta^2 w}{\delta x^2}\frac{\delta^2 w}{\delta y^2} + 2(1-\nu)\left(\frac{\delta^2 w}{\delta x\,\delta y}\right)^2\right]dx\,dy\,dz \tag{4.48}$$

where,

$$D = \frac{Eh^3}{12(1-\nu^2)} \tag{4.49}$$

is called the plate constant. Forming the Lagrangian $L = T - V$ and applying the Hamilton's principle,

$$\delta\int_{t_1}^{t_2} L\,dt = 0, \text{ we obtain}$$

$$(1/2)\delta\int_{t_1}^{t_2}\int_0^a\int_0^b [\rho h w_t^2 - D\{w_{xx}^2 + w_{yy}^2 + 2\nu w_{xx}\,w_{yy} + 2(1-\nu)\,w_{xy}^2\}]\,dx\,dy = 0$$

Where, $$w_t = \frac{\partial w}{\partial t}$$

Performing the operations as per the rules of the calculus of variation term by term, we obtain the equation of motion of the vibrating plate as

$$D\nabla^4 w + \rho h w = 0 \tag{4.50}$$

where,

$$\nabla^4 w = \left(\frac{\delta^2}{\delta x^2} + \frac{\delta^2}{\delta y^2}\right)\left(\frac{\delta^2 w}{\delta x^2} + \frac{\delta^2 w}{\delta y^2}\right)$$

$$= \frac{\delta^4 w}{\delta x^4} + 2\frac{\delta^4 w}{\delta x^2\,\delta y^2} + \frac{\delta^4 w}{\delta y^4}$$

$w = \dfrac{\delta^2 w}{\delta t^2}$, D = plate constant, h = thickness of the plate and ρ = mass density of the plate.

The problem of the vibrating plate has important applications in dynamic and seismic resistant analysis and design of raft and other foundations, as well as in rigid pavement analysis for airfield and highways.

4.9 VLASOV AND LEONTEV METHOD FOR VIBRATION ANALYSIS

Vlasov and Leontev (1966) presented a variational approach for a two-parameter model of beam on elastic foundation.

The equation of motion for the transverse vibrations of an Euler–Bernoulli beam on a generalized two parameter elastic foundation (Figure 4.15) can be written as

$$E_b J\,\frac{\delta^4 v}{\delta x^4} - 2t\frac{\delta^2 v}{\delta x^2} + kv + (m_1 + m_0)\frac{\delta^2 v}{\delta t^2} = p(x, t) \qquad (4.51)$$

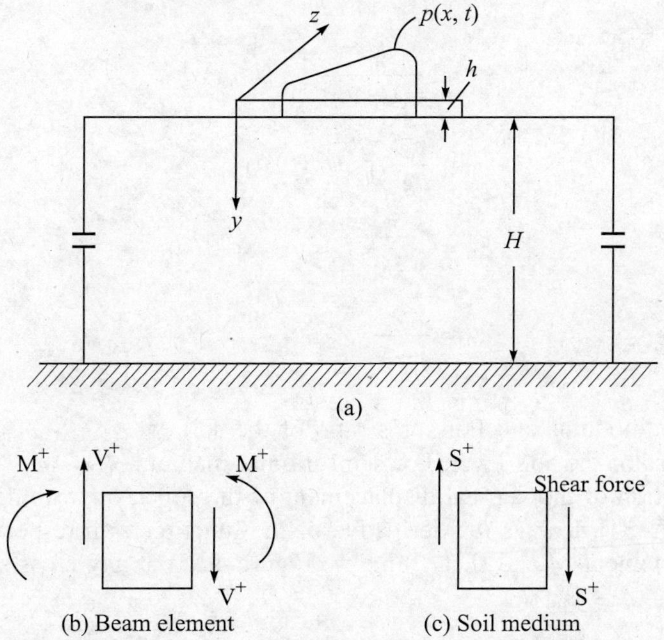

Figure 4.15 Beam on an elastic foundation [Vlasov and Leontev model].

where,

$E_b J$ = flexural rigidity of the beam

$$= \frac{E_b \delta h^3}{12(1-\nu_b^2)} \approx \frac{E_b \delta h^3}{12} = E_b I \qquad (4.52)$$

δ, h = width and depth of the beam, respectively

E_b, ν_b = Young's modulus and Poisson's ratio of the beam material

v = displacement of the beam in y-direction

I = moment of inertia of beam = $\dfrac{\delta h^3}{12} \approx J$

t, k = the two parameters of the foundation model

$$m_1 = \text{mass of the beam per unit length} = \frac{\gamma \delta h}{g} \qquad (4.53)$$

m_0 = equivalent mass of the soil participating in vibrations

$p(x,t)$ = external load on the beam

γ = unit weight of the beam material

g = acceleration due to gravity.

Using the Vlasov and Leontev's variational approach and reducing the foundation to a two-parameter mode, by neglecting the horizontal displacements of the soil medium, the expressions for the parameters can be obtained as

$$k = \frac{E_0 \delta}{(1-\nu_0^2)} \int_0^H \psi'^2(y)\, dy \qquad (4.54)$$

$$t = \frac{E_0 \delta}{4(1+\nu_0)} \int_0^H \psi^2(y)\, dy \qquad (4.55)$$

$$m_0 = \frac{\nu_0 \delta}{g} \int_0^H \psi^2(y)\, dy \qquad (4.56)$$

$$E_0 = \frac{E}{1-\nu^2} \qquad (4.57)$$

$$\nu_0 = \frac{\nu}{1-\nu} \qquad (4.58)$$

where,

E, ν = Young's modulus and Poisson's ratio of the soil

H = thickness of the soil layer. For semi-infinite medium, $H = \infty$

$\psi(y)$ = distribution of the vertical displacement of the soil layer with depth, and the prime in Eq. (4.54) denotes the derivative of the function with respect to y.

The bending moment M and shear force V (Figure 4.15) at any cross-section of the beam are given by

$$M = -\frac{E_b J \delta^2 v}{\delta x^2} \qquad (4.59)$$

$$V = -\frac{E_b J \delta^3 v}{\delta x^3} \tag{4.60}$$

The resultant shear force in the soil (Figure 4.15) is given by

$$S(x) = \int_0^H \tau_{xy} \psi \cdot \delta y = 2t \frac{\delta v}{\delta x} \tag{4.61}$$

where τ_{xy} is the shear stress in the soil at any section of the soil medium.

It may be noted that at any cross-section of the beam-foundation system the total shear force will be the sum of the shear force in the beam and the resultant shear force in the foundation soil layer, i.e.

$$Q = V + S \tag{4.62}$$

where,

Q = total shear at any cross-section of the system
V = shear force in the beam, given by Eq. (4.60)
S = resultant shear force in soil layer, given by Eq. (4.61).

While Eq. (4.51) can be used for any two-parameter model, the Vlasov and Leontev's model facilitates determination of these foundation parameters rationally, using Eqs. (4.54) to (4.58) and field and laboratory tests on soil. By putting $t = 0$ the foundation model reduces to the well-known Winkler model where k is the spring constant of the soil. Eq. (4.51) can be written as

$$\frac{\delta^4 v}{\delta x^4} - 2r^2 \frac{\delta^2 v}{\delta x^2} + s^4 v + m^* \frac{\delta^2 v}{\delta t^2} = \frac{p(x,t)}{EJ} \tag{4.63}$$

where,

$$r^2 = \frac{t}{E_b J} \tag{4.64}$$

$$s^4 = \frac{k}{E_b J} \tag{4.65}$$

$$m^* = \frac{m_1 + m_0}{E_b J} \tag{4.66}$$

4.9.1 Free Vibrations of Beams on Elastic Foundation

By putting $p(x, t) = 0$ in Eq. (4.63), the resulting homogeneous equation represents the free vibrations of the beam on elastic foundation. Using the separation of variables technique, $v(x, t)$ can be expressed as

$$v(x,t) = X(x)T(t) \tag{4.67}$$

Substituting Eq. (4.67) in Eq. (4.63), we get two uncoupled ordinary differential equations in X and T which can be easily solved as

$$T = A \sin \omega t + B \cos \omega t \tag{4.68}$$

$$X = C_1 \sinh \alpha x + C_2 \cosh \alpha x + C_3 \sin \beta x + C_4 \cos \beta x \tag{4.69}$$

where A, B, C_1, C_2, C_3, C_4 are arbitrary constants to be determined from the initial and boundary conditions of the beam foundation system, and

$$\alpha^2 = \lambda^2 + r^2; \quad \beta^2 = \lambda^2 - r^2 \tag{4.70}$$

$$\lambda^4 = r^4 - s^4 + m^* \omega^2 \tag{4.71}$$

The general solution expressed in Eq. (4.69) involves the four arbitrary constants C_1, C_2, C_3, C_4 and the frequency parameter λ (which depends on ω), which have to be determined from the boundary conditions at the two ends of the beam, two at each end. These boundary conditions being homogeneous and with no external load acting as it is a free vibration problem, the four arbitrary constants C_1 to C_4 can be solved from the four equations coming from the end conditions at $x = 0$, and l. Since λ is still unknown, the above homogeneous set of equations will lead to an eigenvalue problem. The solution to the resulting eigenvalue problem gives the infinite eigenvalues λ_1^2, λ_2^2, ..., λ_n^2, each one of which is associated with a corresponding frequency ω.

Thus, these infinite natural frequencies of the beam–foundation system can be obtained from Eq. (4.71) as

$$\omega_n^2 = \frac{\lambda_n^4 + s^4 - r^4}{m^*}, \quad n = 1, 2, ..., \infty \tag{4.72}$$

The corresponding values $X_n(x)$, (n = 1, 2, ..., ∞) are the eigenvectors or eigenfunctions.

It may be noted that ω_n and X_n, 1, 2, ..., ∞ depend on the boundary conditions of the beam–foundation system. Thus, the free vibration solution for a beam on an elastic foundation can be written from Eq. (4.67) as

$$v(x,t) = \sum_{n=1}^{\infty} X_n (A_n \sin \omega_n t + B_n \cos \omega_n t)$$

$$= \sum_{n=1}^{\infty} X_n C_n \sin \omega_n (t - \phi_n) \tag{4.73}$$

where, A, B or C, f are arbitrary constants. The above expressions represent the principal modes of the transverse vibrations of the beam on an elastic foundation. The first and second time derivatives of $v(x,t)$ give the velocity and acceleration at any cross-section of the beam. The bending moment, the shear force in the beam and the resultant shear force in the foundation can be expressed from Eqs. (4.59) to (4.61) as

$$M(x,J) = -EJ \sum_{n=1}^{\infty} X_n'' (A_n \sin \omega_n t + B_n \cos \omega_n t) \tag{4.74}$$

$$V(x,t) = -EJ \sum_{n=1}^{\infty} X_n'' (A_n \sin \omega_n t + B_n \cos \omega_n t) \tag{4.75}$$

$$S(x,t) = 2t \sum_{n=1}^{\infty} X_n' (A_n \sin \omega_n t + B_n \cos \omega_n t)$$

$$= 2EJ\, r^2 \sum_{n=1}^{\infty} X_n (A_n \sin \omega_n t + B_n \cos \omega_n t) \tag{4.76}$$

where the soil parameter t is given by Eq. (4.90) and primes denote derivatives with respect to x.

By putting $s = r = 0$, in the above Eqs. (4.86) to (4.96), the resulting expression can be noted to correspond to beam without foundation. By putting $r = 0$, we obtain the expressions applicable to beams on Winkler foundation of modulus k (single parameter).

Further, the response of the beam foundation system due to sudden impulse of intensity $p(x)$ per unit length acting for a very short duration may be expressed as

$$v(x,t) = \frac{1}{m} \sum_{n=1}^{\infty} \frac{\int_0^L p(x) X_n dx}{w_n \int_0^L (x_n^2) dx} . X_n \sin w_n t \tag{4.77}$$

4.10 VIBRATION OF PLATES ON ELASTIC FOUNDATION

A plate on an elastic foundation is shown in Figure 4.16. The convention of moments and shears for a thin plate is also given in Figure. Kameswara Rao (1998) has presented such analysis in great depth.

The governing equation of motion of a thin plate on a two-parameter elastic soil layer of thickness H (for a semi-infinite layer $H = \infty$) can be obtained similar to Eq. (4.51) as

$$\mathrm{D}\nabla^4 w - 2t\nabla^2 w + kw + (m_1 + m_0)\frac{\delta^2 w}{\delta t^2} = p(x, y, t) \tag{4.78}$$

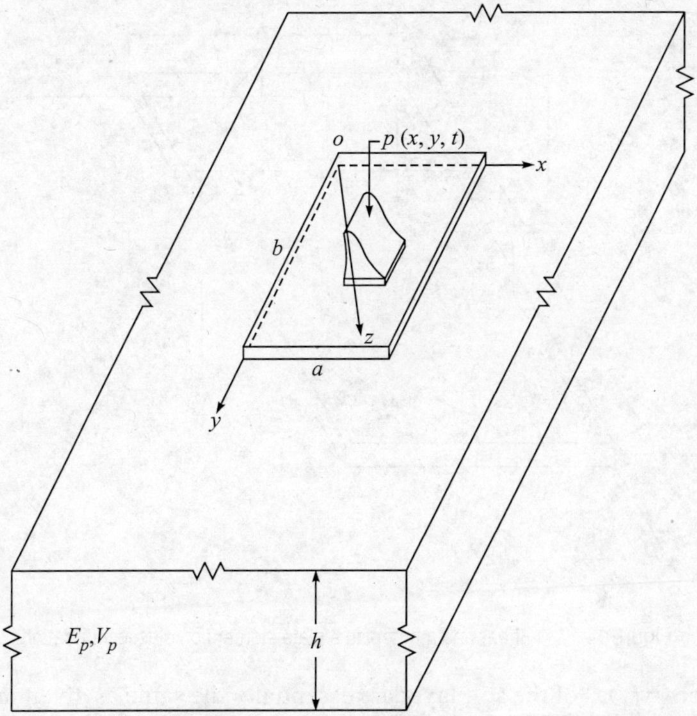

Figure 4.16 Plate on elastic foundation.

where,

$$D = \text{flexural rigidity of the plate} = \frac{E_p h^3}{12(1-\nu_p^2)} \tag{4.79}$$

E_p, ν_p = Young's modulus of elasticity and Poisson's ratio of plate
h = thickness of the plate

$$\nabla^2 = \text{Laplace operator} = \frac{\delta^2}{\delta x^2} + \frac{\delta^2}{\delta y^2}$$

$$\nabla^4 = \text{biharmonic operator} = \nabla^2\,\nabla^2 = \frac{\delta^4}{\delta x^4} + \frac{2\delta^4}{\delta x^2\,\delta y^2} + \frac{\delta^4}{\delta y^4}$$

w = vertical deflection of plate = $w(x, y, t)$

$$m_1 = \text{mass per unit plate area} = \frac{\gamma h}{g} \tag{4.80}$$

γ = unit weight of plate material
$p(x, y, t)$ = external loads on the plate
t, k = foundation parameters
m_0 = equivalent mass of soil participating in vibrations
g = acceleration due to gravity.

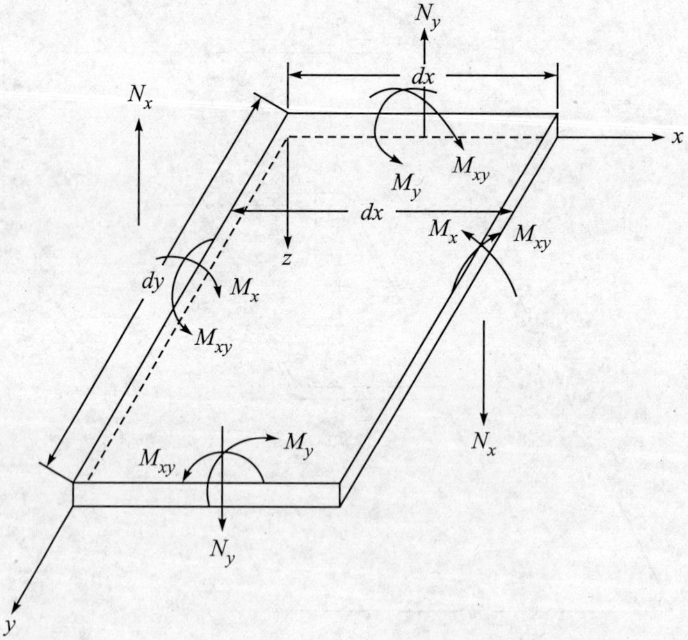

Figure 14.17 Shear and moment in a plate element on elastic foundation.

The parameters t, k, m_0 of the soil layer are essentially the same as given in Eq. (4.78) except that the depth coordinate now is z and $\delta = 1$ and $\psi(z)$ is the distribution function of vertical deflection along depth z. It may be reiterated that the horizontal deflections of the soil u and v are taken as zero in developing the two-parameter foundation model by Vlasov and Leontev (1966). These parameters can be expressed as

$$k = \frac{E_0}{1 - v_0^2} \int_0^H \psi'^2(z)\, dz \tag{4.81}$$

$$t = \frac{E_0}{4(1 + v_2)} \int_0^H \psi^2(z)\, dz \tag{4.82}$$

$$m_0 = \frac{v_0}{g} \int_0^H \psi^2(z)\, dz \tag{4.83}$$

$$E_0 = \frac{E}{1 - v^2} \tag{4.84}$$

$$v_0 = \frac{v}{1 - v} \tag{4.85}$$

$$\psi' = \frac{d\psi}{dz}$$

where E, v = Young's modulus of elasticity and Poisson's ratio of soil, respectively.

Equation (4.78) can be written as

$$\nabla^4 w - 2r^2\, \nabla^2 w + s^4\, w + m^* \frac{\delta^2 w}{\delta t^2} = \frac{p(x, y, t)}{D} \tag{4.86}$$

where r^2, s^4, m^* are similar to the expressions given in Eqs. (4.64) to (4.66) and can be expressed as

$$r^2 = \frac{t}{D} \tag{4.87}$$

$$s^4 = \frac{k}{D} \tag{4.88}$$

$$m^* = \frac{m_1 + m_0}{D} \tag{4.89}$$

The bending moments and shear forces shown in Figure 4.9 can be expressed using thin plate theory as obtained by Kameswara Rao (1998)

$$M_x = -\, D\left[\frac{\delta^2 w}{dx^2} + v_p \frac{\delta^2 w}{dy^2}\right] \tag{4.90}$$

$$M_y = -\, D\left[\frac{\delta^2 w}{dy^2} + v_p \frac{\delta^2 w}{dx^2}\right] \tag{4.91}$$

$$H = H_x = -\, H_y = -\, D(1 - v_p)\, \frac{\delta^2 w}{\delta x\, \delta y} \tag{4.92}$$

$$N_x = -D\frac{\delta}{\delta x}\left[\frac{\delta^2 w}{\delta x^2}+\frac{\delta^2 w}{\delta y^2}\right] \tag{4.93}$$

$$N_y = -D\frac{\delta}{\delta y}\left[\frac{\delta^2 w}{\delta x^2}+\frac{\delta^2 w}{\delta y^2}\right] \tag{4.94}$$

The quantities N_x, N_y and H can be combined to get the Kirchhoff shear as

$$Q_x = -D\left[\frac{\delta^3 w}{\delta x^3}+(2-v_p)\frac{\delta^3 w}{\delta x\,\delta y^2}\right] \tag{4.95}$$

$$Q_y = -D\left[\frac{\delta^3 w}{\delta y^3}+(2-v_p)\frac{\delta^3 w}{\delta x^2\,\delta y}\right] \tag{4.96}$$

In addition to the above, the resultant shear forces S_{XZ} and S_{YZ} from the foundation soil layer below the plate need to be considered in the analysis at any cross-section y = constant, or x = constant, respectively.

Along any cross-section, y = constant

$$S_{XZ} = \int_0^a dx \int_0^H \tau_{XZ}\,\psi\, dz \tag{4.97}$$

Along any cross-section, x = constant

$$S_{YZ} = \int_0^b dy \int_0^H \tau_{YZ}\,\psi\, dz \tag{4.98}$$

where a and b are dimensions of the plate along x- and y-directions as shown in Figure 4.17.

Navier type solutions can be obtained for Eq. (4.78), which are exact but applicable to simply supported boundary conditions along either all the plate edges or at least a pair of opposite edges. For plates with arbitrary boundary conditions, many approximate methods such as Rayleigh, Rayleigh-Ritz, Galerkin and Kantorovich have to be used besides numerical methods and finite element method. Vlasov and Leontev (1966), Kameswara Rao (1998) have used the variational method which is similar to Kantorovich method, to solve free and forced vibration problems in a few cases.

The governing equation for vibrations of plates on elastic foundations as given by Eq. (4.78) is in the invariant form since Laplace operator is an invariant which does not depend on the coordinate system and expressions of Laplace operators in other coordinate systems are available in several books. Of particular interest could be circular plates for which cylindrical polar coordinates can be used for the analysis.

By putting $s = 0$, $r = 0$, the governing Eqs. (4.78) and (4.86) become applicable to vibrations of plates without foundation. By putting $r = 0$, the above equations correspond to plates on Winkler type foundation with foundation modulus k (single parameter model).

4.11 NUMERICAL METHODS

Quite often continuous systems lead to eigenvalue problems which for all practical purposes are impossible to solve. This is frequently the case when the stiffness or mass distribution of the system is non-uniform or the shape of the boundary curves cannot be described in terms of known functions. Yet it may be imperative to obtain information about the physical system and in particular about the natural frequencies. Quite often it is sufficient to know the values of only a limited number of lowest frequencies rather than all the frequencies. The higher frequencies cannot be taken too seriously anyway if an exact solution of the eigenvalue problem is obtained, because the nature of the assumptions employed in defining the models in most elementary theories restricts the validity of the solutions to the lower modes only.

Thus, there are several methods to obtain the estimate of the fundamental frequency without solving the eigenvalue problem. They may be listed as :

1. Rayleigh's method
2. Rayleigh–Ritz method
3. Galerkin's method
4. Collocation method
5. Integral Formulation method
6. Holzer's method
7. Myklestad's method
8. Lumped–parameter method
9. Dunkerley's method
10. Southwell's method
11. Kantorovich method

Of all these, the Rayleigh's method is very popular owing to its simplicity. In this method a form of the deflection is assumed. The maximum potential energy and the maximum kinetic energy as found, are equated. The resulting equation gives the fundamental frequency of vibration with a fair degree of accuracy. It is of course always higher than the true value, though the difference is small.

Considering a non-uniform beam on elastic foundation undergoing flexural vibrations, the governing differential equation may be expressed as

$$\frac{\delta^2}{\delta x^2}\left[EI(x)\frac{\delta^2 w}{\delta x^2}\right] + m(x)\frac{\delta^2 w}{\delta t^2} + K(w)\,w = 0 \tag{4.99}$$

in which I, m, K are certain functions of x.

Applying the Rayleigh's method to such a non-uniform beam-foundation system

$$\text{kinetic energy} = T_{\max} = \int_0^L \left\{\frac{1}{2}\rho A(x)\left(\frac{\delta w}{\delta t}\right)^2\right\} dx \tag{4.100}$$

$$\text{potential energy} = V_{\max} = \frac{E}{2}\int_0^L I(x)\left(\frac{\delta^2 w}{\delta x^2}\right)^2 dx + \frac{1}{2}\int_0^L K(x)w^2\,dx \tag{4.101}$$

Equating $T_{max} = V_{max}$, we obtain

$$\omega^2 = \frac{\left[E\int_0^L I(x)\left(\frac{d^2w}{dx^2}\right)^2 dx + \int_0^L K(x)w^2 dx\right]}{\left[\int_0^L \rho A(x)\, w^2\, dx\right]} \tag{4.102}$$

Further, assuming the deflection curve to be

$$w = \sum_{n=1}^{\infty} q_n(t) \sin\frac{n\pi x}{L} \tag{4.103}$$

$$\sqrt{\max} = \frac{EI\pi^4}{4L^3}\sum_{n=1}^{\infty} n^4 q_n^2(t) + K\frac{1}{4}\sum_{n=1}^{\infty} q_n(t)^2 \tag{4.104}$$

and,

$$T_{max} = \frac{\rho AL}{4}\sum_{n=1}^{\infty} q_n(t)^2 \tag{4.105}$$

Further assuming $q_n(t) = e^{ixnt}$, equating $T_{max} = V_{max}$ yields

$$w_n^2 = \frac{EI\pi^4}{4L^3}\frac{(n^4 + K)\cdot 4}{\rho AL} = \left(\frac{EI}{\rho A}\right)\left(\frac{\pi}{L}\right)^4\left(n^4 + \frac{kL^4}{EI\pi^4}\right) \tag{4.106}$$

putting $a^2 = EI/\rho A$, $w_n^2 = \dfrac{a^2\pi^4}{L^4}(n^4 + \lambda)$ where $\lambda = \dfrac{KL^4}{EI\pi^4}$

4.12 DIMENSIONAL ANALYSIS

Force, length and time are involved in vibration problems as fundamental units. The method consists in selecting variables, which contain all three of the fundamental dimensions among them. The similarity principle can be interpreted as meaning that the non-dimensional factors in the mechanical system should equal the non-dimensional factors in the analogous electric system. The problem of finding out the numerical values of electrical elements to be analogous to mechanical elements then reduces to identifying the non-dimensional groups. This may be facilitated by the π-theorem.

Buckingham's π-theorem states that a system with n independent variables and j fundamental units will have $(n - j)$ dimensionless parameters in its solution.

Forces in parallel in mechanical systems are represented by electric elements in series.

The beam resting on Winkler foundation and the Newtonian dashpot may be represented by the differential equation.

$$EI\frac{\delta^4 w}{\delta x^4} + m\frac{\delta^2 w}{\delta t^2} + kw + C\frac{\delta w}{\delta t} = F(x,\ t) \tag{4.107}$$

or,

$$Q_c + Q_i + Q_f + Q_d = F(x,\ t) \tag{4.108}$$

where,

Q_c = elastic force in the beam
Q_i = inertia force
Q_f = foundation reaction force
Q_d = damping force
$F(x, t)$ = external load

Substituting,

$$w = \frac{F_0}{k}w, \qquad t = \sqrt{\frac{m}{k}}\,\tau, \qquad \frac{\delta}{\delta t} = \sqrt{\frac{k}{m}}\frac{\delta}{\delta\tau}, \qquad x = 4\sqrt{\frac{EI}{k\cdot j}}$$

$$\frac{\delta^4 w}{\delta j^4} + \frac{\delta^2 w}{\delta\tau^2} + \frac{c}{\sqrt{(mk)}}\frac{\delta w}{\delta j} + w = \frac{F}{F_0}\left\{4\sqrt{\frac{EI}{kj}}, \left(\sqrt{\frac{m}{k}}\right)\tau\right\} \tag{4.109}$$

Considering the case of forced vibrations of a beam on elastic foundation, the independent quantities are displacement w, flexural rigidity EI, foundation stiffness k, mass of beam m, the damping coefficient c, the exciting force F_0 and its frequency ω. Thus,

$$w = \phi\,[EI, k, m, c, F_0, \omega] \tag{4.110}$$

Thus, the total number of independent quantities is 7. The fundamental units are 3. Thus, the number of non-dimensional units is (7 – 3) = 4. These may be identified as

$$\pi_1 = \sqrt[4]{\frac{k}{EI}}\cdot L, \qquad \pi_2 = \frac{m\omega^2}{EIL^4}$$

$$\pi_3 = \frac{C}{\sqrt{mk}}, \qquad \pi_4 = \frac{k}{F_D}w$$

So,

$$\pi_4 = \phi\{\pi_1, \pi_2, \pi_3\}$$

i.e.,

$$\frac{k}{F_D}w = \phi\left\{\sqrt[4]{\frac{k}{EI}}\cdot L, \frac{m\omega^2}{EIL^4}, \frac{C}{\sqrt{mk}}\right\} \tag{4.111}$$

4.13 ANALOGUE METHOD

The analogy between electrical and mechanical systems is through the similarity of differential equations, which represent them. In most of the cases it is possible to represent a physical system by an electrical circuit. This is useful because electrical analogies yield much more readily to experimental study and investigation. The electrical system can be assembled and installed. The measurements are also easier in the former case, and so also are the alterations in the values of the elements.

The table of analogous terms are given in Table 4.1. If the forces act in series in the physical system, the electric elements representing these forces are connected in parallel.

Table 4.1 Table of analogous terms

Physical/Vibrational system	*Electrical system*
Kinetic energy	Electromagnetic energy
Potential energy	Electrostatic energy
Displacement, cm	Charge Q, coulomb
Velocity, cm/s	Current, ampere
Force, newton	Voltage, volt
Mass, kg-s^2/cm	Inductance, henry
Damping coefficient, kg-s/cm	Resistance, ohm
Flexibility	Capacitance, farad
Stiffness, kg/cm	1/capacitance
Degree of freedom	Loop
D'Alembert's principle	Kirchhoff's law
Force applied	Switch closed
Coupling element	Element common to two loops

In many cases it has been observed that the physical vibrating system and the electric system have the same form in the differential equation. For example,

$$\frac{md^2z}{dt^2} + \frac{c\,dz}{dt} + kz = F \sin \omega t \tag{4.112}$$

$$\frac{Ld^2Q}{dt^2} + \frac{RdQ}{dt} + \frac{1}{C}Q = E \sin \phi t \tag{4.113}$$

Mathematically, both equations have the same solution, the dimensions however are entirely different. From the above two equations, the following comparisons are obvious:

(a) z the displacement corresponds to Q, the charge.
(b) dz/dt the velocity corresponds to dQ/dt, the current.
(c) d^2z/dt^2 the acceleration corresponds to d^2Q/dt^2.
(d) $W/g = m$, the mass corresponds to L, the inductance.
(e) c the damping factor corresponds to R, the resistance.
(f) k the spring constant (stiffness) corresponds to $1/C$, the reciprocal of the capacitance.
(g) F the force corresponds to E, the voltage.
(h) ω the forcing frequency of the vibration system corresponds to ϕ, the electric forced frequency.

The problem of finding out the numerical values of electrical elements to be analogous to mechanical elements, then reduces to identifying the non-dimensional groups. This may be facilitated by the π-theorem.

The governing differential equation may be replaced by an algebraic equation at each of the modal points by the finite difference technique. In the electrical analogy, each of the loading terms will be represented in finite differences form as current flowing into each of the nodal points. Thus, in the circuit analogy the independent variable, time, is represented continuously as such while the space variables are represented as discrete values by the nodal points.

Criner (1953) used the electric analogue computer technique for the analysis of beams on elastic foundations. In the USA in the project SNORT, the Naval Ordnance Test Station, Inyokern, designed a high speed test track capable of carrying relatively high carriage loads at very high speeds. This experiment was conducted in the early 1950s. A circuit was suggested to represent the non-linearities of foundation reaction at each nodal point.

To include the various parameters controlling the general dynamics of beams on elastic foundations, different circuit analogies can be thought and tested with high-speed computational computers. For a general analysis, the following are made use of.

General similarity principles, state that the solution of a physical system must be valid in all systems of units. In equations written in terms of non-dimensional units, two systems having non-dimensional parameters with the same numerical values are similar.

To establish correspondence between mechanical and electrical non-dimensional factors, these may be listed as

Mechanical	**Electrical**
$W(K/F_0)$	Q/CE
$4\sqrt{(k/EI)}\cdot L$	L/E
$\omega^2 m/EIL^4$	$\omega_e^2\, LC$
$\zeta/\sqrt{mk}$	$R/\omega_{e,n}\ L$

where, Q = change in coulomb
C = capacitance in farad
L_c = inductance in henry
R = resistance in ohm
ω_e = frequency of electrical circuit
E = voltage in volts

PROBLEMS

4.1 Figure P4.1 shows a two-storied structure. The two degree freedom system representing the two-storied structure undergoes true vibrations with an initial displacement of 12 cm to the top storey. Determine the frequencies of vibrations and show the mode shapes. Express the equation of motion candidly.

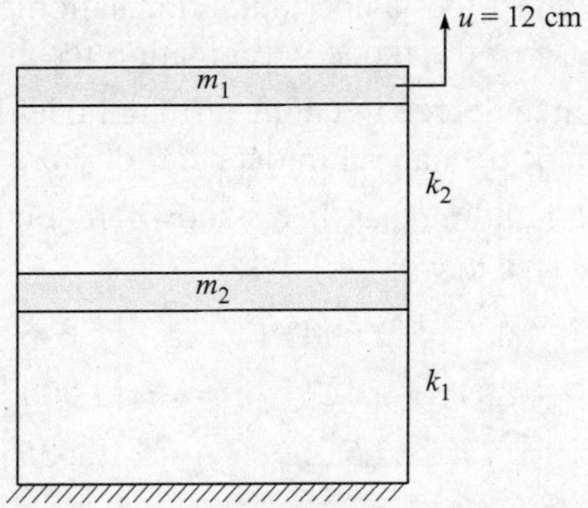

Figure P4.1

4.2 A two degree freedom system as shown in Figure P4.1 is under forced vibration. Show that the equation of motion for the system is given by

$$\begin{bmatrix} m_1 & 0 \\ 0 & m_2 \end{bmatrix} \begin{Bmatrix} \ddot{z}_1 \\ \ddot{z}_2 \end{Bmatrix} + \begin{bmatrix} 2k & -k \\ -k & 2k \end{bmatrix} \begin{Bmatrix} z_1 \\ z_2 \end{Bmatrix} = \begin{Bmatrix} F \\ 0 \end{Bmatrix} \sin \omega t$$

Also find the frequencies of vibration and draw the forced response as zk/F versus ω/ω_n.

4.3 Write the equation of motion for the system shown in Figure P4.3. Determine its natural frequencies and mode shapes.

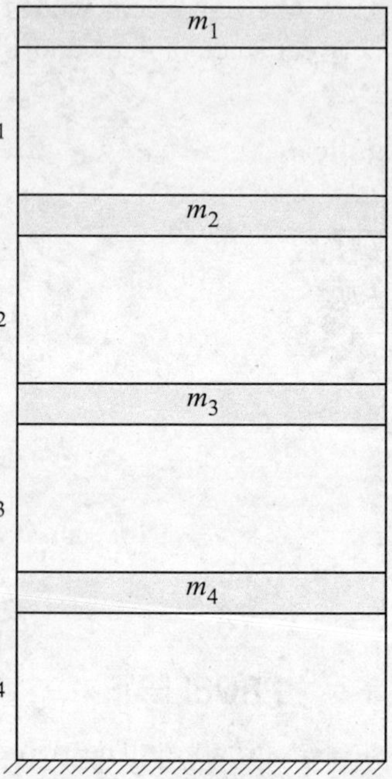

Figure P4.3 A four-storied building.

4.4 In a ten-storied building of equal rigid floors and equal interstorey stiffness, if the foundation of the building undergoes a horizontal translation $u_0(t)$, determine the response of the building making use of the mode participation technique.

4.5 Draw the equivalent three degree freedom system model for the framed structure shown in Figure P4.5. Determine the natural frequencies of vibration and draw the mode shapes.

4.6 A three-storied shear building has been shown in Figure P4.5. The frequencies of vibration and mode shapes are as follows:

$$\phi = \begin{bmatrix} 1.000 & 1.000 & 1.000 \\ 0.759 & -0.804 & -2.427 \\ 0.336 & -1.175 & 2.512 \end{bmatrix}, i \quad \omega = \begin{bmatrix} 0.695 \\ 1.900 \\ 2.635 \end{bmatrix} \sqrt{\frac{k}{m}} \text{ rad/s}$$

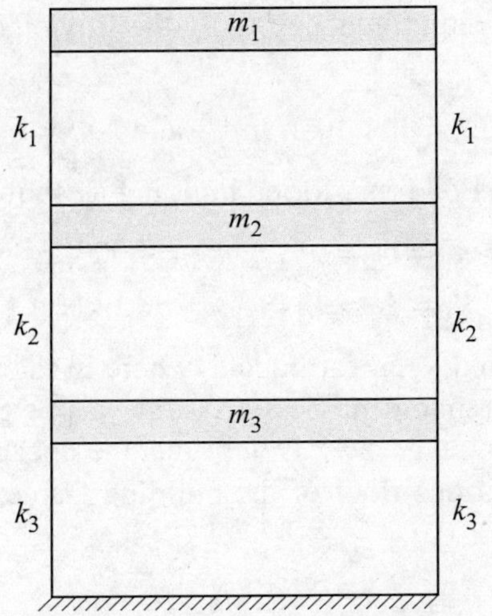

Figure P4.5 A three-storied building.

The structure is set into true vibration by displacing the floor $u_1 = 7.5$ mm, $u_2 = -10$ mm and $u_3 = 7.5$ mm and then, releasing it suddenly at time $t = 0$. Determine the displacement shape at time $= 2\pi/\omega_1$ assuming no damping of the system.

4.7 A simply supported beam on elastic foundation has a distributed load whose variation with time is shown in Figure P4.7. Derive the expression for the dynamic deflection. The beam has mass density ρ, flexural rigidity EI and a uniform cross-sectional area A. The stiffness of the foundation is K.

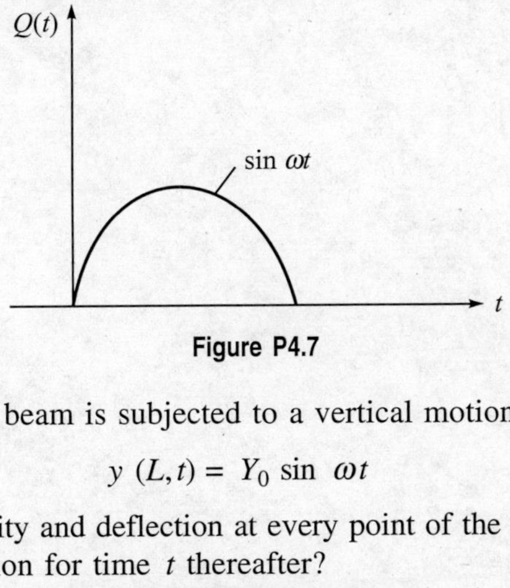

Figure P4.7

4.8 A simply supported beam is subjected to a vertical motion of right-support of amount

$$y\ (L, t) = Y_0 \sin\ \omega t$$

If at $t = 0$, the velocity and deflection at every point of the beam are zero, what will be the deflection equation for time t thereafter?

4.9 The elastic curve of a cantilever beam $(0 \le x \le L)$ is given by

$$y = y_0^2\ (1 - \cos\ \pi x/L)$$

Determine the fundamental frequency using the Rayleigh method. The specifications of the beam are

$$L = 30 \text{ m}, \; EI = 4 \times 10 \text{ N}\cdot\text{m}^2 \quad \text{and} \quad m = 6 \times 10^4 \text{ kg}$$

4.10 The mass and stiffness of a four-storied building as shown in Figure P4.3 are as follows

$$m_1 = m_2 = m_3 = m_4 = 3000 \text{ kg}$$
$$k_1 = k_2 = k_3 = k_4 = 300 \text{ kN/m}$$

The building is subjected to an earthquake where in the acceleration–time trace may be taken as a stationary random process. Assume a power spectral density of 0.05 and damping of the structure as 5%. Determine the mean square value of the relative displacement of the various floor of the building frame.

5

WAVE PROPAGATION

5.1 INTRODUCTION

We have seen in the previous chapters that the vibration problem of discrete systems, with finite degrees of freedom, leads to a frequency equation which is a polynomial in ω_n^2 with as many roots as the number of degrees of freedom possessed by the system. The continuous system enjoys infinite degrees of freedom and leads to a frequency equation which is transcendental in nature with an infinite number of roots corresponding to the infinite number of degrees of freedom possessed by the system. The essential difference between these two types of motion is in fact the same difference that exists between an oscillation and a wave motion. An oscillation takes place in the time domain and is completely specified by the initial values at one point, i.e., $t = 0$. A wave equation, on the other hand, takes place in time and space domain. Thus, a time-table indicating both time and space is required to specify the configuration of the system at any instant of time.

The elastic body when subjected to sudden impulse or loading does not experience disturbance in the entire body instantly, rather only that part of the elastic body which in close contact with the external force agency is affected first, and then the deformations produced subsequently spread throughout closely in the form of waves. In this process of propagation, elasticity and inertia of the body play important roles. Thus, wave is essentially a form of disturbance, which travels from one part of the elastic body to the other through the oscillatory motion of the particles of the elastic medium. In other words, waves are generated in a medium due to a disturbance in the medium.

Seismic waves are classified from the earthquake engineering angle as P-wave and S-wave. A P-wave is the first to reach the earth's surface. A P-wave arrives longitudinally at short distances and, therefore, is also called a longitudinal wave. Physically, a P-wave is compressive in nature and propagates generating vibrations parallel to the direction of propagation. An S-wave reaches after a P-wave and is physically a shear or torsion wave. An S-wave oscillates (vibrates) in a direction normal to the direction of propagation. At short distances its motion is primarily transverse. Hence, an S-wave is called a transverse wave. An S-wave is further classified as one having apparently only a horizontal component, which is called the SH wave, and one having only a vertical component, called the SV wave. The propagation velocity of P- and S-waves, V_P, V_S, is determined by the modulus of elasticity of the propagation medium.

If the medium is purely elastic, both P- and S-waves can propagate at whatever depth. From this point of view, these waves are called body waves. However, since there are a number of discontinuities in the earth's crust, these body waves are subject to complex phenomena such as reflection, refraction, diffraction, scattering, amplification, damping, etc. Reflection or refraction of P- and S-waves at these discontinuities follow Snell's law just as light rays do. The mechanics of wave propagation is presented in greater detail in this chapter.

Let the body under consideration consist of particles forming a linear isotropic homogeneous elastic medium. If particle in contact with an external agency is disturbed and set into oscillation, the disturbance is handed over from particle to particle due to elasticity. That is, essentially such a particle is set in oscillation, but owing to inertia, a little later than the preceding particle. The phase of oscillation changes from particle to particle. Thus, it is the phase relationship of the medium particles, which is observed as waves. In this way all materials like solids, liquids and gases can carry energy. For example, the ocean tide in sea beaches is a good demonstration of the energy transported by water waves. Let us consider another example of a stone being dropped suddenly into a quiet lake in which there is a floating piece of wood. From the point of contact of the stone thrown in water, circular waves spread and travel over the surface of the lake. When the waves reach close to the piece of wood, they set in up and down motion of the wooden piece. This demonstrates that the waves have transferred energy to the piece of wood. But after elapse of certain time, the lake returns to its previous quiet and motionless look, which demonstrates the dissipation of energy.

Thus, there are essentially two ways of transporting energy from one place to another. One way of energy propagation involves the actual transport of matter. For example, when a bullet is fired from the pistol, the bullet carries its kinetic energy with it which can be used at another location. In this case there is transport of matter along with energy propagation. The second method by which energy can be transported is by virtue of wave propagation. Some of the different situations of wave propagation may be listed as follows:

- When a drummer beats the drum, its sound is heard at a distant place since energy in the form of sound can move the diaphragm of the ear.
- When a stone is dropped in the still water in a pond, water waves moves steadily along the water.
- When an electrical signal is transmitted from one place to another.
- When an earthquake takes place in deep ocean, the energy is propagated through water waves (tsunami).
- Due to earthquakes, the acceleration of the ground surface takes place owing to generation and propagation of seismic waves.

The elastic rebound theory states that the earth crest behaves like an elastic medium, that is, soil or rock can be deformed and will return to its original shape after the stresses are released. So, soil/rocks under pressure will accumulate strain energy until the pressure becomes greater than friction and the movement occurs. After the sudden movement, the earthquake produces motion of the ground by generating stress waves that originate from the rupture of the stressed earth mass. Even by artificial means like bomb-blast or aerial bombardment, waves may be generated on the surface as well as within the earth mass. The parts close to the source of disturbance are affected first and subsequently the disturbance spreads through the body in the form of stress waves. Thus, the phenomenon of wave propagation in an elastic medium gains importance in geotechnical earthquake engineering (see Figures 5.1 and 5.2).

The wave propagation basic studies include axial or longitudinal vibration of strings, axial or longitudinal vibration of prismatic bars or rods. The governing equations of motion in one-, two- or three-dimensional cases are derived from Newton's laws of motion.

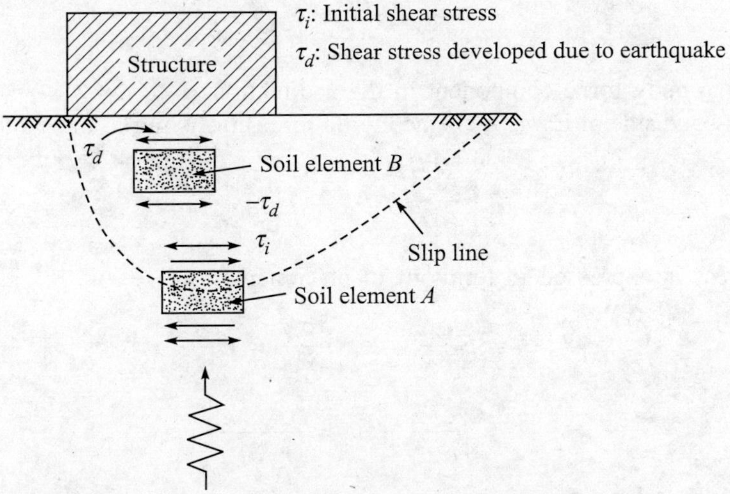

Figure 5.1 Shear wave due to earthquake (soil elements A and B).

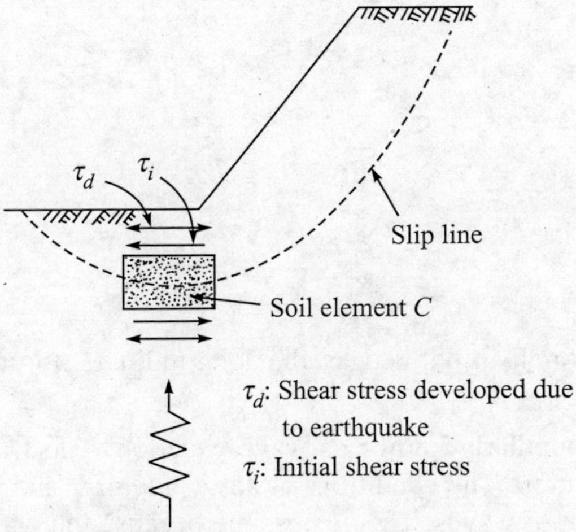

Figure 5.2 Shear wave due to earthquake (soil element C).

5.2 ONE-DIMENSIONAL WAVE MOTION

If the forces, displacement and stress are represented by the single coordinate x, we have from

Newton's law

$$\Sigma F_x = \rho \frac{\delta^2 u}{\delta t^2} \tag{5.1}$$

where u is displacement field in the x-direction.

In term of stress, the left-hand side may be written as

$$F_x = \frac{\delta \sigma_x}{\delta x} + B_x \tag{5.2}$$

where B_x is the body force component in the x-direction.

The right-hand side of Eq. (5.1) is the inertia force consisting of mass and acceleration, then

$$\frac{\delta \sigma_x}{\delta x} + B_x = \rho \frac{\delta^2 u}{\delta t^2} \tag{5.3}$$

If the stress is expressed in terms of displacements

$$\sigma_x = E \frac{\delta u}{\delta x} \quad \text{so,} \quad \frac{\delta \sigma_x}{\delta x} = E \frac{\delta^2 u}{\delta x^2}$$

$$\therefore \quad E \frac{\delta^2 u}{\delta x^2} + B_x = \frac{\sigma}{g} \frac{\delta^2 u}{\delta t^2}$$

As a special case, taking $B_x = 0$,

$$E \frac{\delta^2 u}{\delta x^2} = \rho \frac{\delta^2 u}{\delta t^2}$$

where ρ = mass density

$$\therefore \quad \frac{\delta^2 u}{\delta t^2} = \frac{E}{\rho} \frac{\delta^2 u}{\delta x^2} \tag{5.4}$$

or,

$$\frac{\delta^2 u}{\delta t^2} = V_P^2 \frac{\delta^2 u}{\delta x^2} \tag{5.5}$$

where $V_P = \sqrt{E/\rho}$ which is defined as the longitudinal (primary or compressional) wave propagation velocity.

Equation (5.5) is popularly known as wave equation. The wave equation is a hyperbolic partial differential equation. The conditions at any value of x and t may be studied by making use of the superposition of waves. This is possible because the wave equation is linear. Hence, the sum of two solutions or the sum of series solution is also a solution. Obviously, displacement u is a function of x and t and its solution in space–time domain is obtained as

$$u = f_1(x - V_P t) + f_2(x + V_P t) \tag{5.6}$$

where f_1 and f_2 are the arbitrary functions.

In Eq. (5.6), the term $f_1(x)$ represents the wave travelling in the positive x-direction and the second term $f_2(x)$ represents the wave travelling in the negative x-direction.

5.3 AXIAL WAVE PROPAGATION

The wave propagation mechanism is easily understood by considering the forward propagation wave in positive direction as in Eq. (5.6) at time intervals $t = 0$ and $t = \Delta t$.

Then the new position variable is $x' = x - V_P\Delta t$, expressed as

$$F_1(x - V_P t) = f_1(x') \quad (5.7)$$

The shape of the wave relative to variable x' in Figure 5.3(b) is the same as the shape relative to x in Figure 5.3(a). Thus, the wave has merely advanced a distance of $V_P\Delta t$ during the time Δt, with no change in shape, the velocity of wave propagation being $V_P = \sqrt{E/\rho}$. By similar reasoning it can be shown that the second term in Eq. (5.6) represents a waveform moving in the negative x-direction.

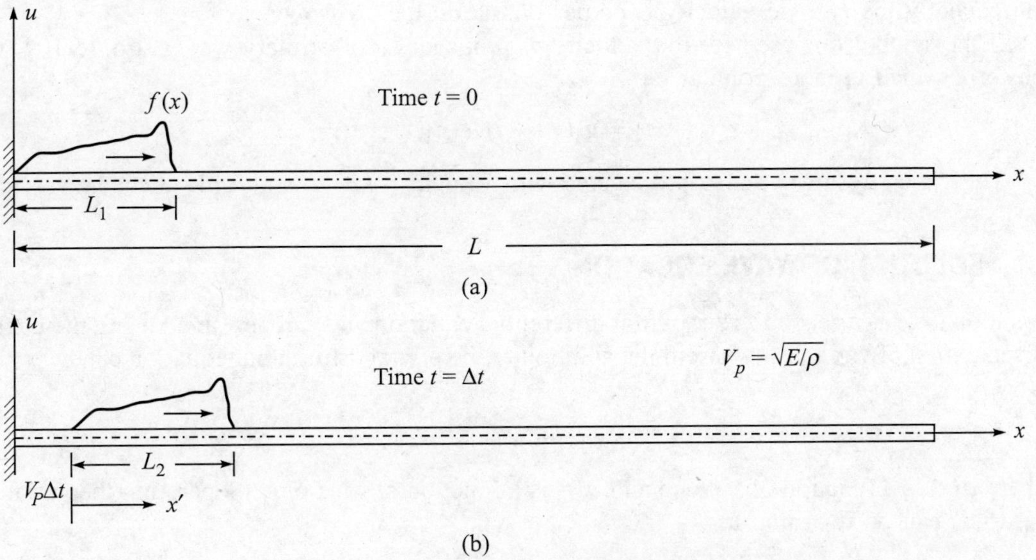

Figure 5.3 Propagation of wave during time interval Δt.

The phase velocity $V_P = \sqrt{E/\rho}$ is the property of the conducting material or medium but not of its shape. This velocity is commonly referred to as the velocity of sound in that material.

Equation (5.7) represents wave propagation in many physical systems like

(a) Axial displacements during longitudinal motion of beams or rods or a bar.
(b) Free vibrations of a taut string.
(c) Torsional vibrations of circular rods.
(d) Propagation of surface waves of two fields velocity potential for supersonic flow in an ideal field.

The dynamic behaviour of the bar can also be expressed in terms of its stress distribution rather than with respect to its displacements. Using Hooke's law,

$$\sigma = \mathrm{E} \cdot \varepsilon \quad \text{where} \quad \varepsilon = \frac{\delta u}{\delta x} \tag{5.8}$$

$$\therefore \qquad \sigma(x, t) = E\frac{\delta u}{\delta x}$$

$$= E\frac{\delta f_1}{\delta x}(x - V_P t) + E\frac{\delta f_2}{\delta x}(x + V_P t) \tag{5.9}$$

Alternatively, the stress wave can be written as

$$\sigma(x, t) = g_1(x - V_P t) + g_2(x + V_P t) \tag{5.10}$$

From Eq. (5.10) it is obvious that during wave propagation, the compressional wave travelling in the positive x-direction, an identical tension wave will be travelling in the negative x-direction. As a result, at the crossover zone where the two waves pass each other, there is zero stress and consequently the particle velocity becomes double in the crossover zone.

If the symbol c is used for the velocity of propagation of primary waves, Eq. (5.10) takes the form which is more popular, i.e.,

$$\sigma(x, t) = g_1(x - ct) + g_2(x + ct)$$

Thus, the stress wave also propagates with velocity $\sqrt{E/\rho}$ and with unchanging shape.

5.4 SOLUTION OF WAVE EQUATION

Equation (5.5) is a second-order partial differential equation and can be solved using the method of separation of variables, wherein the solution to the partial differential equation can be written as

$$u(x, t) = \phi(x).q(t) \tag{5.11}$$

where $\phi(x)$ is a function of x-alone and $q(t)$ is a function of time only. Separating the variables, Eq. (5.5) can be rewritten as

$$\frac{d^2\phi(x)}{dx^2} + \frac{\omega_n^2}{V_r^2}\phi = 0 \tag{5.12}$$

$$\frac{d^2 q(t)}{dt^2} + \omega_n^2 q(t) = 0 \tag{5.13}$$

where ω_n is the natural frequency of vibration and $V_r = \sqrt{E/\rho}$ is the velocity of propagation.

The solution of Eqs. (5.12) and (5.13) may be written as

$$\phi(x) = A\cos\left(\frac{\omega_n}{V_r}\right)x + B\sin\left(\frac{\omega_n}{V_r}\right)x \tag{5.14}$$

$$q(t) = C\cos\omega_n t + D\sin\omega_n t \tag{5.15}$$

Thus, the complete solution to the wave equation can be written as

$$u(x, t) = \sum_{n=1}^{\infty} \left\{ A \cos \omega_n \frac{x}{V_r} + B \sin \omega_n \frac{x}{V_r} \right\} (C \cos \omega_n t + D \sin \omega_n t) \tag{5.16}$$

wherein the constants A, B, C, D can be evaluated from the boundary conditions and the initial conditions. This can be applied to various physical problems as illustrated below.

Example 5.1 Discuss the transverse vibration of a taut string of length L fixed at $x = 0$ and $x = L$.

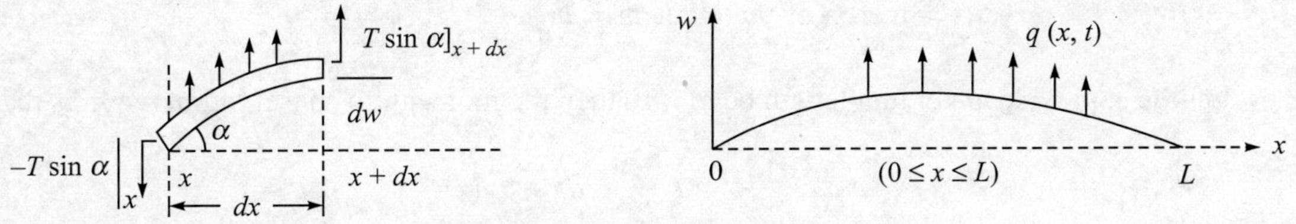

Figure 5.4 Taut string (Example 5.1).

Solution: The taut string of length L $(0 \le x \le L)$ as shown in Figure 5.4 is stretched between two fixed points of $x = 0$ and $x = L$. Let the cross-section area be A. If there is an initial stretching such that L increases to $L + \Delta L$, the resulting tension T in the string may be expressed using Hooke's law as

$$T = EA\frac{\Delta L}{L}$$

By the law of conservation of transverse momentum, the total lateral force on the string element must be balanced by the inertia force. If $w(x, t)$ is the lateral displacement at any distance x, then considering a differential small element between x and $x + dx$,

$$\text{the net force on the element} = [T \sin \alpha]_{x+dx} - [T \sin \propto]_x$$

where,

$$\sin \alpha = \frac{dw}{\sqrt{dx^2 + dw^2}} = \frac{\dfrac{\partial w}{\partial x}}{\sqrt{1 + \left(\dfrac{\partial w}{\partial x}\right)^2}}$$

By assuming the displacement to be small such that $\dfrac{\partial w}{\partial x} << 1$, the higher-order (2nd order and above) terms may be neglected and sin α may be expressed as

$$\sin \alpha = \frac{\partial w}{\partial x} \quad \text{or} \quad \sin \alpha = \frac{\partial w}{\partial x} + 0\left(\frac{\partial w}{\partial x}\right)^2$$

The instantaneous length $l(x, t)$ of the string from 0 to x is

$$l(x, t) = \int_0^x dx \left[1 + \left(\frac{\partial w}{\partial x}\right)^2\right]^{1/2} = x + \text{higher order} \geq \left(\frac{\partial w}{\partial x}\right)^2$$

Net tension in the string element $= T\dfrac{\partial^2 w}{\partial x^2}\, dx$

If m is the mass per unit length, then

$$\text{inertia of the element} = m\left(\frac{\partial^2 w}{\partial t^2}\right) dx$$

Let the applied load per unit length be $q(x, t)$, then the momentum conservation requires that

$$m\,dx\frac{\partial^2 w}{\partial t^2} = T\frac{\partial^2 w}{\partial x^2}\, dx + q\,dx$$

Eliminating dx and taking the limit of $dx \to 0$,

$$\frac{m}{T}\frac{\partial^2 w}{\partial t^2} - \frac{\partial^2 w}{\partial x^2} = \frac{q}{T}$$

or,

$$\frac{1}{c^2}\frac{\partial^2 w}{\partial t^2} - \frac{\partial^2 w}{\partial x^2} = \frac{q}{T}$$

where $c = \sqrt{\dfrac{T}{m}}$.

The dimension of the coefficient $\dfrac{T}{m}$ is

$$\left[\frac{T}{m}\right] = \frac{\text{MLT}^{-2}}{\text{M/L}} = \left(\frac{L}{T}\right)^2 = [\text{velocity}]^2$$

Thus, $c = \sqrt{\dfrac{T}{m}}$ represents the characteristic velocity of the physical problem. For $q = 0$, the equation reduces to

$$\frac{1}{c^2}\frac{\partial^2 w}{\partial t^2} = \frac{\partial^2 w}{\partial x^2}$$

which is called the wave equation.

Example 5.2 Describe the wave propagation during the longitudinal vibrations of a uniform cantilever rod ($0 \leq x \leq L$). Also explain the significance of the wave equation and its solution.

Solution: To describe the longitudinal vibrations of a rod ($0 \le x \le L$) as shown in Figure 5.5 in mathematical terms, the basic laws that govern the vibrations must be used. In continuum mechanics, there are conservation laws for mass and momentum. In addition, constitutive laws are often needed to relate certain unknown variables, examples are equation of state, Hooke's law between stress and strain, etc. To derive the vibration problem in mathematical terms, let us consider an infinitesimal element of the rod shown as dx in Figure 5.5.

Figure 5.5 Cantilever rod (Example 5.2).

For the cantilever rod ($0 \le x \le L$), let $u(x, t)$ represent the longitudinal displacement from the equilibrium position, then the strain at any distance x is given by

$$\lim_{\Delta x \to 0} \frac{\Delta u}{\Delta x} = \frac{\partial u}{\partial x}$$

By Hooke's law, the force (tension) at x is

$$\text{force} = \text{stress} \times \text{area} = (E)\ (\text{strain})\ (\text{Area}) = EA\frac{\partial u}{\partial x}$$

where A is the cross-sectional area of the uniform rod.

The net tension on the rod element from x to $x + dx$ is

$$\left[EA\frac{\partial u}{\partial x}\right]_{x+dx} - \left[EA\frac{\partial u}{\partial x}\right]_x = dx\ \frac{\partial}{\partial x}\left[EA\frac{\partial u}{\partial x}\right] + \text{terms containing } (dx)^2 \text{ and higher-order terms}$$

The term $(dx)^2$ is very small and hence may be neglected along with higher-order terms, and such approximations may be expressed symbolically by $0.(dx)^2$.

Taking the externally applied longitudinal force as $q(x, t)$ per unit length, the conservation of momentum requires that

$$\rho A\frac{\partial^2 u}{\partial t^2}\,dx = \frac{\partial}{\partial x}\left(EA\frac{\partial u}{\partial x}\right)dx + q\,dx + 0(dx)^2$$

where ρ = density of the rod.

In the limit of vanishing dx, we get the differential equation

$$\rho A \frac{\partial^2 u}{\partial t^2} = \frac{\partial}{\partial x}\left(EA \frac{\partial u}{\partial x} \right) + q$$

For uniform cross-section, A = constant and for no external load, $q = 0$, then

$$\frac{1}{c^2}\frac{\partial^2 u}{\partial t^2} = \frac{\partial^2 u}{\partial x^2}, \quad \text{where} \quad c = \sqrt{\frac{E}{\rho}}$$

The solution of the above equation which is essentially a wave equation, may be written as

$$u(x, t) = f(x - ct) + g(x + ct)$$

where f, g in above equation represent arbitrary functions. In x, t (space, time), the plane $f(x - ct)$ is constant along the straight line $x - ct$ = constant. Thus, to the observer (x, t) who moves at the steady speed c along the positive x-axis, the function f is stationary.

Thus, to an observer moving from left to right at the speed c, the signal described initially by $f(x)$ at $t = 0$ remains unchanged in form as t increase, i.e., f propagates to the right at the speed c. Similarly, g propagates to the left at the speed c. The lines $x - ct$ = constant and $x + ct$ = constant are often called characteristic curves along which signals propagate.

Example 5.3 Discuss the flexural wave generation and propagation in a beam on elastic foundation.

Solution:

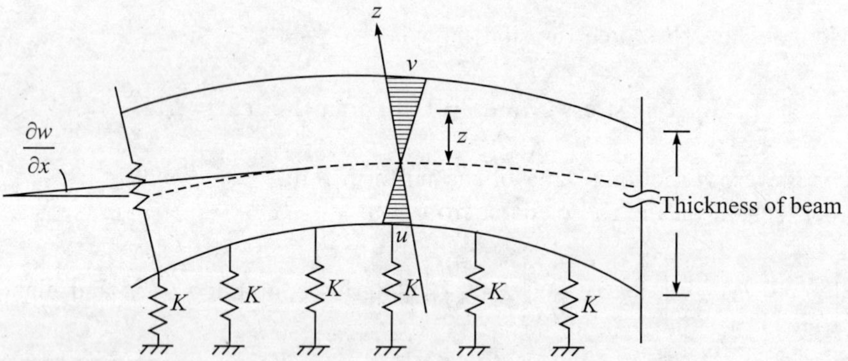

Figure 5.6 Beam on elastic foundation (Example 5.3).

Assume a uniform beam, having a constant cross-section area A as well as a constant EI, resting on Winkler model of elastic foundation having stiffness K. As the thickness of the beam is small compared to the length and consequently the deflection is small compared to the thickness. As such, a plane cross-section remains approximately plane after deformation. If u denotes the longitudinal displacement at section x and height z about the axis, then

$$u \cong -z\frac{\partial w}{\partial x}$$

Consequently the strain

$$\varepsilon_x = \frac{\partial u}{\partial x} = -z\frac{\partial^2 w}{\partial x^2}$$

and the longitudinal stress is

$$\sigma_x = E\,\varepsilon_x = E\frac{\partial u}{\partial x}$$

$$= -\,Ez\frac{\partial^2 w}{\partial x^2}$$

Then the total moment about the mid-section, $z = 0$

$$M = \int_{-h/2}^{h/2} \sigma_x^2\, dz = E\frac{\partial^2 w}{\partial x^2}\int_{-h/2}^{h/2} z^2\, dz = EI\frac{\partial^2 w}{\partial x^2}$$

where $I = \int_{-h/2}^{h/2} z^{2\mathrm{m}}\, dz$ = moment of inertia of the cross-section with respect to its mid-section ($z = 0$).

Considering a length element of the beam from x to $x + dx$, as shown in the Figure 5.7, the balance of angular momentum about the centre of the element requires that

$$M + \frac{\partial M}{\partial x}dx - M + \left(S + \frac{\partial S}{\partial x}dx\right)\frac{dx}{2} + S\frac{dx}{2} = \rho\, J\, dx\frac{\partial^3 w}{\partial x \partial t^2}$$

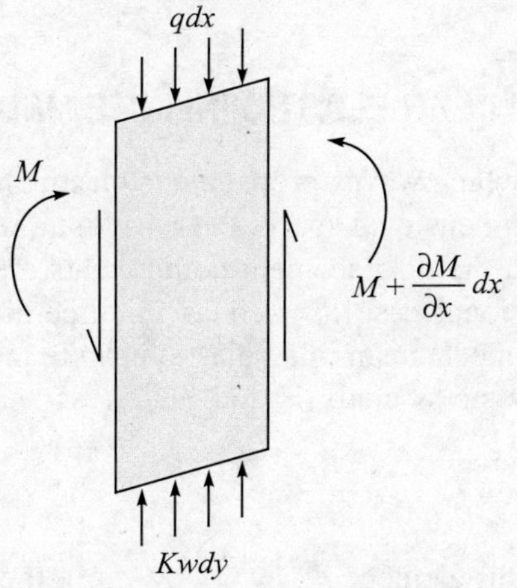

Figure 5.7 Example 5.3

where,

ρ = mass of the beam per unit length

ρJ = rotatory moment of inertia.

Thus,

$$S = -\frac{\partial M}{\partial x} + \rho J \frac{\partial^3 w}{\partial x dt^2} = -EI\frac{\partial^3 w}{\partial x^3} + \rho J \frac{\partial^3 w}{\partial x dt^2}$$

On the other hand, the balance of vertical forces requires that

$$S + \frac{\partial S}{\partial x} dx - S = qdx + Kwdx + \rho \frac{\partial^2 w}{\partial t^2}$$

where,

K = stiffness of foundation

$q(x, t)$ = uniformly distributed load on beam

Combining the above two equations for S,

$$\rho\left(1 - J\frac{\partial^2 w}{\partial x^2}\right)\frac{\partial^2 w}{\partial t^2} + EI\frac{\partial^4 w}{\partial x^4} + Kw = -q(x, t)$$

which is a fourth-order partial differential equation, derived first by Lord Rayleigh. If the wavelength L is much greater than the beam height ($h/L << 1$), then the term representing the rotatory inertia is negligible and the above equation reduces to

$$\rho \frac{\partial^2 w}{\partial t^2} + EI\frac{\partial^4 w}{\partial x^4} + Kw = -q(x, t)$$

The above differential equation represents the wave generation and its propagation in beams on elastic foundations.

5.5 WAVE PROPAGATION IN AN ELASTIC INFINITE MEDIUM

The infinite medium through which waves propagate is elastic, homogeneous and isotropic and this can be analyzed in three-dimensional space. Let us adopt a rectangular Cartesian coordinate system x, y and z along which the displacement components are u, v and w, respectively. The stress analysis in two-dimensional case as well as in three-dimensional case is essential to express the wave propagation mechanism in an elastic infinite medium. Thus, before discussing the wave propagation, the 2D stress analysis and 3D stress analysis are presented first.

5.5.1 2D Stress Analysis

The static stress field in the soil element dx by dy by dz in the 2D case is given by

$$\sigma_{ij} = \begin{vmatrix} \sigma_{xx} & \tau_{xy} \\ \tau_{yx} & \sigma_{yy} \end{vmatrix} \tag{5.17}$$

where $i = (x, y)$ and $j = (y, x)$.

$$\text{Stress invariant, } I_1 = \sigma_1 + \sigma_2 = \sigma_{xx} + \sigma_{yy}$$

where σ_1 and σ_2 being the principal stresses on planes having shear stress zero are given by the determinant

$$\begin{vmatrix} \sigma_{xx} - \sigma_n & \tau_{xy} \\ \tau_{xy} & \sigma_{yy} - \sigma_n \end{vmatrix} = 0 \tag{5.18}$$

Expanding the determinant and rearranging the terms, and obtaining the two roots of σ_n as

$$\sigma_1 = \frac{\sigma_{xx} + \sigma_{yy}}{2} + \sqrt{[(\sigma_{xx} - \sigma_{yy})^2 + \tau_{xy}^2]} \tag{5.19}$$

$$\sigma_2 = \frac{\sigma_{xx} + \sigma_{yy}}{2} - \sqrt{[(\sigma_{xx} - \sigma_{yy})^2 + \tau_{xy}^2]} \tag{5.20}$$

These values are shown on Mohr circle in Figure 5.8, where

major principal stress = $\sigma_1 = OC + CD$

minor principal stress = $\sigma_2 = OC - CD$

The displacements u, v in the x and y directions and rotation θ in the xy-plane are given by

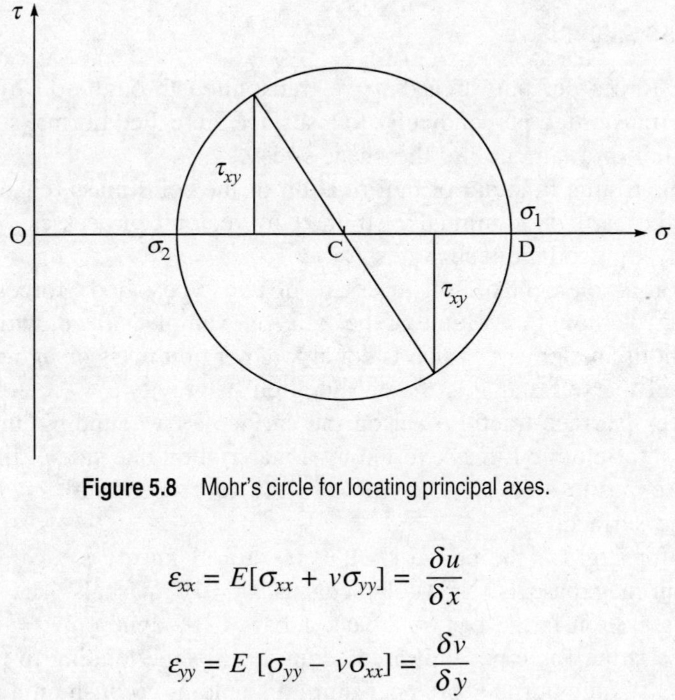

Figure 5.8 Mohr's circle for locating principal axes.

$$\varepsilon_{xx} = E[\sigma_{xx} + v\sigma_{yy}] = \frac{\delta u}{\delta x} \tag{5.21}$$

$$\varepsilon_{yy} = E\ [\sigma_{yy} - v\sigma_{xx}] = \frac{\delta v}{\delta y} \tag{5.22}$$

$$\gamma_{xy} = \frac{\delta u}{\delta y} + \frac{\delta v}{\delta x} \tag{5.23}$$

$$\theta = (1/2)\left(\frac{\delta v}{\delta x} - \frac{\delta u}{\delta y}\right) \tag{5.24}$$

By introducing a new function which is known as stress function ϕ such that

$$\sigma_{xx} = \frac{\delta^2\phi}{\delta y^2} - B_x \tag{5.25}$$

$$\sigma_{yy} = \frac{\delta^2\phi}{\delta x^2} - B_y \tag{5.26}$$

$$\gamma_{xy} = \frac{-\delta^2\phi}{\delta x \delta y} \tag{5.27}$$

where B_x and B_y are components of body forces.

The equilibrium sequations of compatibility can be taken care of by a single differential equation. This is known as bi-harmonic equation and is very useful for the solution of a two-dimensional problem. Then the stress function ϕ must satisfy

$$\frac{\delta^4\phi}{\delta x^4} + 2\frac{\delta^4\phi}{\delta x^2 \delta y^2} + \frac{\delta^4\phi}{\delta y^4} = 0 \tag{5.28}$$

This function was introduced in the solution of 2D problems by G.B. Airy (1862) and is also called Airy's stress function.

5.5.2 3D Stress Alalysis

The stresses are forces per unit area that are transmitted through soil by inter-atomic fields. Stresses that are transmitted perpendicular to a surface are called normal stresses; those that are transmitted parallel to a surface are the shear stresses.

Stress in soils results in strain or deformation of the solid mass (change in shape or size). The surface of the earth is continually strained by tectonic processes. These strains can be measured directly by geodetic techniques.

The forces on an element of soil mass are of two types: body forces and surface forces. Body forces act throughout the volume of the soil. An example is the downward force of gravity, that is, the weight of an element, which is equal to g per unit mass or ρg per unit volume where ρ is the density of the soil and g is the acceleration of gravity.

Surface forces (surface traction) act on the surface area bounding an element of volume. They arise from inter-atomic forces exerted by a material on one side of the surface on another material on the opposite side. Surface force is the force per unit area and depends on the orientation of the surface.

The SI unit for stress is the pascal (Pa); 1 pascal = 1 kg m^{-1} s^{-2}. Stresses in the earth are normally given in megapascals (MPa); 1 megapascal = 10^6 pascals. The unit for stress in the earth in the c.g.s system is the bar (or kbar); 1 bar = 10^6 dyne cm^{-2} = 10^5 Pa.

Tensile stress is the force per unit area acting on a plane tending to pull on such a plane. Compressive stress is the normal force per unit area tending to push on a plane. Generally, the compressive stresses are considered the negative and tensile stresses positive—a convention generally adopted in elasticity. Mal and Singh (1991).

Let us consider a part of soil mass occupying a region R enclosed by the surface S in the deformed state as shown in Figure 5.9. The boundary S is acted upon by surface forces caused by the action of the material exterior to R on that within R. The forces acting across an element

δS of S about a point P due to the action of the material outside R can be represented by a single force δp and that the limit exists.

$$\lim_{\delta S \to 0} \frac{\delta p}{\delta S}$$

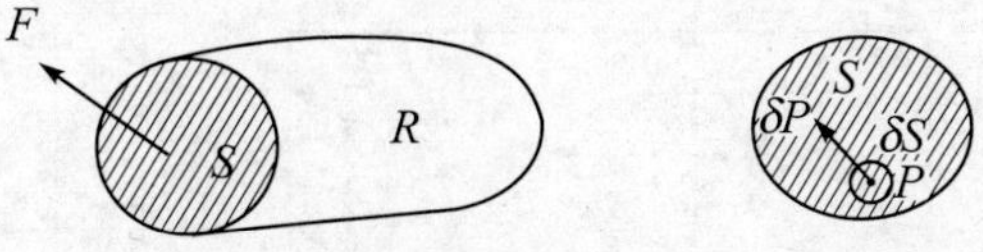

$$\sigma = \lim_{\delta s \to 0} \frac{\delta p}{\delta S}$$

Figure 5.9 Concept of stress.

This limit is known as the traction, or the stress vector on ΔS at the point P. In general, the traction is inclined to the area ΔS on which it acts, and can be resolved into two components: a normal stress perpendicular to the area and a tangential or shear stress acting in the plane of the area δS.

Let the outward drawn unit normal to δS be n. In this notation, the first suffix indicates the direction of the normal to the face and the second suffix indicates the direction of the traction component. In all, there are nine components, known as the components of stress, which can be displayed in the form of a matrix as

$$\sigma_{ij} = \begin{vmatrix} \sigma_{xx} & \tau_{xy} & \tau_{xz} \\ \tau_{yx} & \sigma_{yy} & \tau_{yz} \\ \tau_{zx} & \tau_{zy} & \sigma_{zz} \end{vmatrix} \tag{5.29}$$

where $i = (x, y, z)$ and $j = (x, y, z)$.

Figure 5.10 shows a small parallelepiped having the dimension dx by dy by dz in which on its six faces the stress components have been shown. From the knowledge of mechanics of the material, the forces on each of the faces can be assigned and considered for the equilibrium of forces.

In the x-direction, the equilibrium equation is

$$\left(\sigma_{xx} + \frac{\delta\sigma_{xx}}{\delta x} dx\right) dydz - \sigma_{xx} dydz + \left(\tau_{xy} + \frac{\delta\tau_{xy}}{\delta y} dy\right) dxdz - \tau_{xy} dxdz + \left(\tau_{xz} + \frac{\delta\tau_{xz}}{\delta z} dz\right) dxdy - \tau_{xz} dxdy = 0 \tag{5.30}$$

Similar equations can be written in the y and z directions. According to Newton's second law the equilibrium equation in the x-direction by neglecting the body forces may be expressed as

$$\left(\frac{\delta\sigma_{xx}}{\delta x} + \frac{\delta\tau_{xy}}{\delta y} + \frac{\delta\tau_{xz}}{\delta z}\right) dx\ dy\ dz = \rho(dx\ dy\ dz)\ \frac{\delta^2 u}{\delta t^2}$$

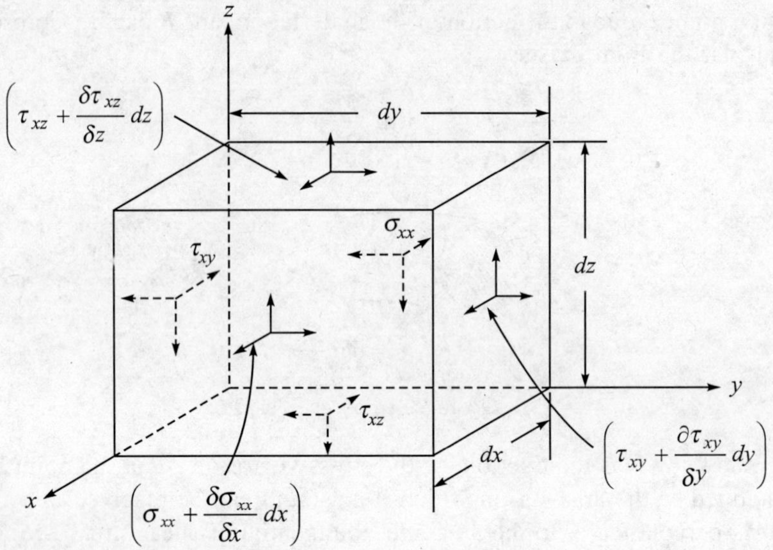

Figure 5.10 Stress components in 3D space.

$$\therefore \quad \left(\frac{\delta \sigma_{xx}}{\delta x} + \frac{\delta \tau_{xy}}{\delta y} + \frac{\delta \tau_{xz}}{\delta z}\right) = \rho \frac{\delta^2 u}{\delta t^2} \tag{5.31a}$$

Similarly,

$$\left(\frac{\delta \tau_{yx}}{\delta x} + \frac{\delta \sigma_{yy}}{\delta y} + \frac{\delta \tau_{yz}}{\delta z}\right) = \rho \frac{\delta^2 v}{\delta t^2} \tag{5.31b}$$

$$\left(\frac{\delta \tau_{zx}}{\delta x} + \frac{\delta \tau_{xy}}{\delta y} + \frac{\delta \sigma_{zz}}{\delta z}\right) = \rho \frac{\delta^2 w}{\delta t^2} \tag{5.31c}$$

The stress can also be expressed in terms of displacements. Thus,

$$\begin{aligned}
\sigma_{xx} &= \lambda e + 2G\varepsilon_{xx} \\
\sigma_{yy} &= \lambda e + 2G\varepsilon_{yy} \\
\sigma_{zz} &= e\lambda + 2G\varepsilon_{zz} \\
\tau_{xy} &= \tau_{yx} = G.\gamma_{xy} \\
\tau_{yz} &= \tau_{xy} = G.\gamma_{yz} \\
\varepsilon_{zx} &= \varepsilon_{xz} = G.\gamma_{xz}
\end{aligned} \tag{5.32}$$

where, $G = \dfrac{E}{2(1+\nu)}, \ \lambda = \dfrac{\nu E}{(1+\nu)(1-2\nu)}$

with ν = Poisson's ratio

E = Young's modulus of elasticity

λ, G = Lame's constants

$e = \varepsilon_{xx} + \varepsilon_{yy} + \varepsilon_{zz}$

For $i = 1, 2, 3$ and $j = 1, 2, 3$

$$\sigma_{ij} = \begin{vmatrix} \sigma_{11} & \tau_{12} & \tau_{13} \\ \tau_{21} & \sigma_{22} & \tau_{23} \\ \tau_{31} & \tau_{32} & \sigma_{33} \end{vmatrix} \tag{5.33}$$

The components σ_{xx}, σ_{yy} and σ_{zz} represent normal stresses and the components τ_{xy}, τ_{yx}, etc. represent shear stresses. The sign convention used is depicted in Figure 5.11.

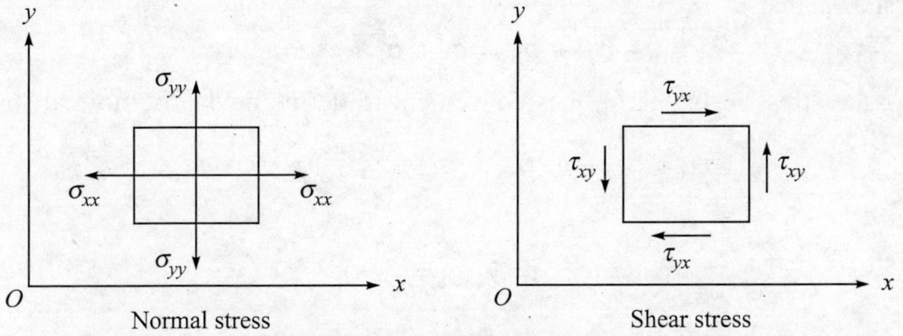

Figure 5.11 Sign convention for normal and shear stresses in two dimensions.

Let a second system of coordinates (x', y', z') with the same origin be obtained from the system of coordinates (x, y, z) by a rotation of axes so that

$$\begin{vmatrix} x' \\ y' \\ z' \end{vmatrix} = [a] \begin{vmatrix} x \\ y \\ z \end{vmatrix} \tag{5.34}$$

where $[a]$ is the matrix of direction cosine, i.e.,

$$[a] = \begin{vmatrix} a_{11} & a_{12} & a_{13} \\ a_{21} & a_{22} & a_{23} \\ a_{31} & a_{32} & a_{33} \end{vmatrix} \tag{5.35}$$

Here, for example, (a_{11}, a_{12}, a_{13}) are the direction cosines of $O_{x'y'z'}$ relative to the system (O_{xyz}). The components of stress in the two systems are connected through the matrix relation.

$$[\sigma_{ij}]' = [a]^T[\sigma_{ij}][a] \tag{5.36}$$

where $[a]^T$ is the transpose of the matrix $[a]$.

It is always possible to find three mutually orthogonal axes such that the corresponding shear stresses vanish. These axes are known as principal axes of stress and the planes normal to the principal axes of stress are called principal planes of stress. Thus, there are no shear stresses on the principal planes of stress and the corresponding normal stresses are called principal stresses, and are usually denoted by $\sigma_1 > \sigma_2 > \sigma_3$. By convention, these are chosen such that $\sigma_1 > \sigma_2 > \sigma_3$. σ_1 is the maximum principal stress, σ_3 is the minimum principal stress and σ_2 is the intermediate principal stress.

When the three principal stresses are equal, the state of stress is isotropic and

$$\sigma_1 = \sigma_2 = \sigma_3 = -p \tag{5.37}$$

where p is the pressure. This is referred to as the hydrostatic state of stress. When the principal stresses are not equal, the pressure is defined as

$$p = -(1/3)(\sigma_1 + \sigma_2 + \sigma_3) \tag{5.38}$$

It can be shown that

$$\sigma_{xx} + \sigma_{yy} + \sigma_{zz} = \sigma_1 + \sigma_2 + \sigma_3 = -3p \tag{5.39}$$

In studying the stress in the earth, it is convenient to define deviatoric normal stresses as

$$\bar{\sigma}_{xx} = \sigma_{xx} + p,\ \bar{\sigma}_{yy} = \sigma_{yy} + p,\ \bar{\tau}_{zz} = \sigma_{zz} + p \tag{5.40}$$

so that,

$$\bar{\sigma}_{xx} + \bar{\sigma}_{yy} + \bar{\sigma}_{zz} = 0. \tag{5.41}$$

It can be shown that the largest possible value of shear stress is $(\sigma_1 - \sigma_3)/2$ and it acts on a plane bisecting the angle between the two principal planes of stress having the principal stresses σ_1 and σ_3.

We consider the equilibrium of a volume element in the form of a small rectangular parallelepiped with its centre at (x, y, z) and edges δx, δy, δz parallel to the coordinate axes. Assuming that the external body force per unit volume has components (f_x, f_y, f_z) and space diagonal—a direction equally inclined to all axes, i.e., $l = m = n$ and $l^2 + m^2 + n^2 = l$, $3l^2 = 1$, $l = \pm 1/\sqrt{3}$—equating to zero the sum of all the forces acting on the volume element along the coordinate axes, we obtain the equations of equilibrium as

$$\frac{\delta}{\delta x}\sigma_{xx} + \frac{\delta}{\delta y}\tau_{yx} + \frac{\delta}{\delta z}\tau_{zx} + f_x = 0$$

$$\frac{\delta}{\delta x}\tau_{xy} + \frac{\delta}{\delta y}\sigma_{yy} + \frac{\delta}{\delta z}\tau_{zy} + f_y = 0$$

$$\frac{\delta}{\delta x}\tau_{xz} + \frac{\delta}{\delta y}\tau_{yz} + \frac{\delta}{\delta z}\sigma_{zz} + f_z = 0 \tag{5.42}$$

Similarly, equating to zero the moments of all the forces about the coordinate axes, and the symmetry relations, yields

$$\tau_{yz} = \tau_{zy},\quad \tau_{zx} = \tau_{xz},\quad \tau_{xy} = \tau_{yx}. \tag{5.43}$$

By considering the equilibrium of a volume element in the form of a tetrahedron with one of its vertices at a given point $P(x, y, z)$ and three of its edges parallel to the coordinate axes, it can be shown that the components of the stress vector acting on any plane through P with unit normal n (l, m, n) can be expressed as a linear combination of the stress components at P [Figure 5.9] in the form:

$$\begin{vmatrix} T_x \\ T_y \\ T_z \end{vmatrix} = \begin{vmatrix} \sigma_{xx} & \tau_{xy} & \tau_{xz} \\ \tau_{yx} & \sigma_{yy} & \tau_{yz} \\ \tau_{zx} & \tau_{zy} & \sigma_{zz} \end{vmatrix} \begin{vmatrix} l \\ m \\ n \end{vmatrix} \tag{5.44}$$

Strain

To begin with, let us consider deformation that is independent of the Cartesian coordinate, say z. Let a point $P(x, y)$ in the undeformed state be displaced to the point $P'(x', y')$ due to deformation. Then

$$u = x' - x, \; v = y' - y$$

are known as displacement components. Therefore,

$$x' = x + u, \; y' = y + v$$

Consider a small rectangular element $PQRS$ in the undeformed state, with sides (dx, dy) parallel to coordinate axes (see Figure 5.12). Let the points P, Q, R and S move, respectively, to the points P', Q', R' and S' after deformation. The point $Q(x + dx, y)$ is displaced to the point Q' $(x' + dx', y' + dy')$, where

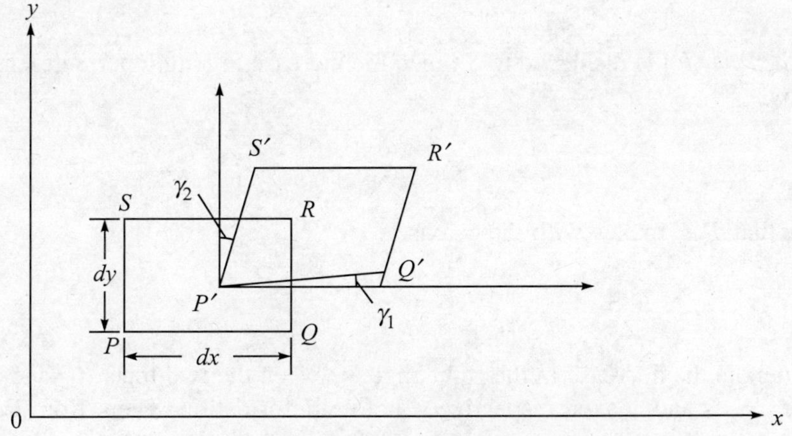

Figure 5.12 Deformation of a rectangular element

$$dx' = dx + du = dx + \frac{\delta u}{\delta x} dx = \left(1 + \frac{\delta u}{\delta x}\right) dx$$

$$dy' = dv = \frac{\delta v}{\delta x} dx \tag{5.45}$$

We thus have

$$P'Q' = [(dx')^2 + (dy')^2]^{1/2}$$

$$= \left[\left(1 + \frac{\delta u}{\delta x}\right)^2 + \left(\frac{\delta v}{\delta x}\right)^2\right]^{1/2} dx \tag{5.46}$$

$$= \left(1 + \frac{\delta u}{\delta x}\right) dx$$

If $\delta u/\delta x$ and $\delta v/\delta x$ are small, the increase in length per unit length of the line PQ, denoted by ε_{xx}, is given by

$$\varepsilon_{xx} = \frac{P'Q' - PQ}{PQ} = \frac{\delta u}{\delta x} \tag{5.47}$$

If γ_1 is the angle that $P'Q'$ makes with the x-axis, we have (Figure 5.8)

$$\tan \gamma_1 = \frac{dy'}{dx'} = \frac{\dfrac{\delta v}{\delta x}}{1 + \dfrac{\delta u}{\delta x}}$$

using Eq. (5.47). Assuming that the angle γ_1 (measured in radians) is small and neglecting small quantities of the second and higher orders in $\delta u/\delta x$ and $\delta v/\delta x$, we find

$$\gamma_1 = \frac{\delta v}{\delta y} \tag{5.48}$$

Similarly, if $S(x, y + dy)$ is displaced to S', then the increase in length per unit length of the line PS is given by

$$\varepsilon_{yy} = \frac{\delta v}{\delta y} \tag{5.49}$$

and the angle that $P'S'$ makes with the y-axis is

$$\gamma_2 = \frac{\delta u}{\delta y} \tag{5.50}$$

Let $2\varepsilon_{xy}$ denote the decrease in the right angle between the two lines PQ and PS, which are parallel to the x-axis and y-axis, respectively, before deformation. Then, from Eqs. (5.49) and (5.50),

$$2\varepsilon_{xy} = \gamma_1 + \gamma_2 = \frac{\delta v}{\delta x} + \frac{\delta u}{\delta y}$$

By definition, $\varepsilon_{xy} = \varepsilon_{yx}$.

The quantities ε_{xy} and ε_{yy} are known as normal strains and γ_{xy} as shear strain.

In the sign convention adopted here, the normal component of strain ε_{xx} is positive if the deformation results in an increase in the length of a line parallel to the x-axis before deformation. Similarly, the shear strain γ_{xx} is positive if the deformation results in a decrease in the right angle between two lines parallel to the x-axis and y-axis before deformation. However, in the sign convention adopted by Turcotte and Schubert (1982), a positive normal strain implies a decrease in length and a positive shear strain implies an increase in the right angle. S.N. Rai (2002).

If $\gamma_1 = \gamma_2$, the deformation is known as pure shear. However, if $\gamma_1 = 0$, the deformation is known as simple shear parallel to the x-axis. Simple shear is often associated with strike-slip faulting.

We have seen that in the two-dimensional problem, there are two normal strains and one shear strain. In the general case of three-dimensional deformation, there are three normal strains, ε_{xx}, ε_{yy} and ε_{zz} and three shear strains, γ_{yz}, γ_{zx} and ε_{xy}. The strain displacement relations (Timoshenko and Godier, 1951) are

$$\varepsilon_{xx} = \frac{\delta u}{\delta x},\ \varepsilon_{yy} = \frac{\delta v}{\delta y},\ \varepsilon_{zz} = \frac{\delta w}{\delta z},$$

$$\gamma_{yz} = \gamma_{zy} = (1/2)\left(\frac{\delta v}{\delta z} + \frac{\delta w}{\delta y}\right),$$

$$\gamma_{xz} = \gamma_{zx} = (1/2)\left(\frac{\delta w}{\delta x} + \frac{\delta u}{\delta z}\right),$$

$$\gamma_{xy} = \gamma_{yx} = (1/2)\left(\frac{\delta u}{\delta y} + \frac{\delta v}{\delta x}\right) \tag{5.51}$$

The quantity

$$e = \frac{\delta u}{\delta x} + \frac{\delta v}{\delta y} + \frac{\delta w}{\delta z} \tag{5.52}$$

is known as *dilatation* and represents the increase or decrease (i.e., change) in volume per unit volume.

We denote the strain matrix by ε so that

$$\varepsilon_{ij} = \begin{vmatrix} \varepsilon_{xx} & \varepsilon_{xy} & \varepsilon_{xz} \\ \varepsilon_{yx} & \varepsilon_{yy} & \varepsilon_{yz} \\ \varepsilon_{zx} & \varepsilon_{zy} & \varepsilon_{zz} \end{vmatrix} \tag{5.53}$$

$$= \begin{vmatrix} \varepsilon_{xx} & \gamma_{xy} & \gamma_{xz} \\ \gamma_{yx} & \varepsilon_{yy} & \gamma_{yz} \\ \gamma_{zx} & \gamma_{zy} & \varepsilon_{zz} \end{vmatrix} \tag{5.54}$$

As in the case of components of stress, the components of strain relative to a rotated frame of axes ($O_{x'y'z'}$) can be obtained through the matrix relation

$$\varepsilon = \begin{vmatrix} \varepsilon'_{xx} & \varepsilon'_{xy} & \varepsilon'_{xz} \\ \varepsilon'_{yx} & \varepsilon'_{yy} & \varepsilon'_{yz} \\ \varepsilon'_{zx} & \varepsilon'_{zy} & \varepsilon'_{zz} \end{vmatrix} \tag{5.55}$$

Increase in length per unit length of a line with direction cosines (l, m, n) in the undeformed state can be expressed in terms of the components of strain in the form

$$
\begin{aligned}
\bar{e} &= \text{extension} \\
&= \text{increase in length per unit length} \\
&= l^2\varepsilon_{xx} + m^2\varepsilon_{yy} + n^2\varepsilon_{zz} + 2lm\gamma_{xy} + 2ln\gamma_{xz} + 2mn\gamma_{yz}
\end{aligned} \tag{5.56}
$$

It is always possible to find three mutually orthogonal axes such that the corresponding shear strains vanish. These axes are known as the principal axes of strain and the corresponding normal strains as the principal strains, which are denoted by ε_1, ε_2 and ε_3. For an isotropic solid, the principal axes of strain coincide with the principal axes of stress.

As the stress–strain relationship is linear, the principle of superposition can be applied. Thus, if an element of volume is subjected simultaneously to the action of normal stresses σ_{xx}, σ_{yy} and σ_{zz} uniformly distributed over the faces of the cube, the resultant components of strain can be obtained from the above as

$$
\begin{aligned}
\varepsilon_{xx} &= (1/\mathrm{E})[\sigma_{xx} - \nu(\sigma_{yy} + \sigma_{zz})] \\
\varepsilon_{yy} &= (1/\mathrm{E})[\sigma_{yy} - \nu(\sigma_{zz} + \sigma_{xx})] \\
\varepsilon_{zz} &= (1/\mathrm{E})[\sigma_{zz} - \nu(\sigma_{xx} + \sigma_{yy})]
\end{aligned} \tag{5.57}
$$

Sometimes it is desirable to express the stress components in terms of the strain components. This is easily achieved through inversion of Eq. (5.57) and the results can be expressed in the form

$$
\begin{aligned}
\sigma_{xx} &= \lambda\Delta + \frac{E}{(1+\nu)\varepsilon_{xx}} \\
\sigma_{yy} &= \lambda\Delta + \frac{E}{(1+\nu)\varepsilon_{yy}} \\
\sigma_{zz} &= \lambda\Delta + \frac{E}{(1+\nu)\varepsilon_{zz}} \\
\tau_{xy} = 2\mu\varepsilon_{xy},\ \ \tau_{yz} &= 2\mu\varepsilon_{yz},\ \ \tau_{xz} = 2\mu\varepsilon_{xz}
\end{aligned} \tag{5.58}
$$

where,

$$
\Delta = \varepsilon_{xx} + \varepsilon_{yy} + \varepsilon_{zz} \tag{5.59}
$$

and,

$$
\lambda = \frac{\nu E}{(1+\nu)(1-2\nu)} \tag{5.60}
$$

The three material constants E, G and ν are related by the equation

$$
G = \frac{E}{2(1+\nu)} \tag{5.61}
$$

where $G = \mu$. Hence, of the three constants introduced above, only two are independent, that is, an isotropic material is characterized by two independent material constants. The interrelationships between the elastic constants, namely, modulus of elasticity E, Poissons ratio ν, bulk modulus K, Lame's constants λ and μ may be expressed as

1. $E = \dfrac{\mu(3\lambda + 2\mu)}{\lambda + \mu} = \dfrac{\lambda(1+\nu)(1-2\nu)}{\nu} = \dfrac{9k\mu}{3K + \mu} = 3K(1 - 2\nu)$

2. $\nu = \dfrac{\lambda}{2(\lambda + \mu)} = \dfrac{3K - 2\mu}{2(3K + \mu)} = \dfrac{E}{2\mu} - 1$

3. $K = \lambda + \dfrac{2\mu}{3} = \dfrac{\lambda(1+\nu)}{3\nu} = \dfrac{\mu E}{3(3\mu - E)} = \dfrac{E}{3(1-2\nu)}$

4. $\mu = \dfrac{\lambda(1-2\nu)}{2\nu} = \dfrac{3K(1-2\nu)}{2(1+\nu)} = \dfrac{E}{2(1+\nu)}$

5. $\lambda = K - \dfrac{2\mu}{3} = \dfrac{\mu(E - 2\mu)}{3\mu - E} = \dfrac{3K\nu}{1+\nu} = \dfrac{E\nu}{(1+\nu)(1-2\nu)}$

Using Eq. (5.58), we may write the stress–strain relationship in the form

$$\sigma_{xx} = \lambda e + 2\mu\varepsilon_{xx}, \quad \sigma_{yy} = \lambda e + 2\mu\varepsilon_{yy}, \quad \sigma_{zz} = \lambda e + 2\mu\varepsilon_{zz},$$
$$\tau_{yz} = 2\mu\varepsilon_{yz}, \quad \tau_{zx} = 2\mu\varepsilon_{zx}, \quad \tau_{xy} = 2\mu\varepsilon_{xy} \tag{5.62}$$

The constants λ and $\mu(G = \mu)$ are known as Lame constants. From Eqs. (5.60) and (5.58), we can express E, ν in terms of λ, μ as

$$E = \frac{\mu(3\lambda + 2\mu)}{\lambda + \mu}, \quad \nu = \frac{\lambda}{2}(\lambda + \mu) \tag{5.63}$$

In the case of a uniform hydrostatic pressure of magnitude p, we have

$$\tau_{xy} = \tau_{yy} = \tau_{zz} = -p$$
$$\tau_{yz} = \tau_{zx} = \tau_{xy} = 0$$
$$\sigma_{xx} = \sigma_{yy} = \sigma_{zz} = -3p$$

From the above equations,

$$e = -\frac{3(1-2\nu)}{E}p = \left(-\frac{1}{K}\right)p \tag{5.64}$$

where,

$$K = \frac{E}{3(1-2\nu)} = \lambda + (2/3)\mu \tag{5.65}$$

The material constant K is known as the *bulk modulus* or incompressibility. As a hydrostatic pressure leads to a decrease in volume, $K > 0$. Equations (5.63), (5.64) and (5.65) in conjunction with the conditions $E > 0$, $\mu > 0$, imply that $-1 < \nu < 1/2$. However, the negative values of ν are unknown in reality. [Rao, S.N. (2002)]

For a perfect fluid, $\mu = 0$ and $[K = \lambda]$ is finite. For an incompressible solid, both K and λ are infinite but μ is finite. It can be easily seen that for both of these special cases, $\nu = 1/2$. In soil mechanics, $\nu = 1/2$ correspond to undrained conditions where in there is no change in volume during shear testing or in situ conditions.

Any two of the five constants λ, G, E, ν and K may be used to characterize a given isotropic material and the remaining three constants can be expressed in terms of these two. Further,

$$\sigma_{ij} = \begin{vmatrix} \sigma_{xx} & \tau_{xy} & \tau_{xz} \\ \tau_{yz} & \sigma_{yy} & \tau_{yz} \\ \tau_{zx} & \tau_{zy} & \sigma_{zz} \end{vmatrix} \quad (5.66)$$

where $i = x$, y and z and $j = x$, y and z.

Further the strains are

$$\varepsilon_{xx} = \frac{\delta u}{\delta x} \text{ and } \gamma_{xy} = \frac{\delta v}{\delta x} + \frac{\delta u}{\delta y}$$

$$\varepsilon_{yy} = \frac{\delta v}{\delta y} \text{ and } \gamma_{yz} = \frac{\delta w}{\delta y} + \frac{\delta v}{\delta z}$$

$$\varepsilon_{zz} = \frac{\delta w}{\delta z} \text{ and } \gamma_{zx} = \frac{\delta u}{\delta z} + \frac{\delta w}{\delta x} \quad (5.67)$$

The rotations are

$$2\theta_x = \frac{\delta w}{\delta y} - \frac{\delta v}{\delta z}$$

or,

$$\theta_x = \frac{1}{2}\left[\frac{\delta w}{\delta y} - \frac{\delta v}{\delta z}\right]$$

$$2\theta_y = \frac{\delta u}{\delta z} - \frac{\delta w}{\delta x}$$

$$2\theta_z = \frac{\delta v}{\delta x} - \frac{\delta u}{\delta y} \quad (5.68)$$

where θ_x, θ_y and θ_z are the rotations about the x-, y- and z-axes, respectively.

By substituting Eqs. (5.67) and (5.68), the equilibrium equation takes the form

$$\rho \frac{\partial^2 u}{\partial t^2} = (\lambda + G) \frac{\delta e}{\delta x} + G\nabla^2 u \quad (5.69)$$

$$\rho \frac{\partial^2 v}{\partial t^2} = (\lambda + G) \frac{\delta e}{\delta y} + G\nabla^2 v \quad (5.70)$$

$$\rho \frac{\partial^2 w}{\partial t^2} = (\lambda + G) \frac{\delta e}{\delta z} + G\nabla^2 w \quad (5.71)$$

The above governing equation in terms of displacements are known as Navier's Equation.

where,

$$\nabla^2 = \left(\frac{\delta^2}{\delta x^2} + \frac{\delta^2}{\delta y^2} + \frac{\delta^2}{\delta z^2}\right) = \text{Laplace operator}$$

$$e = \varepsilon_{xx} + \varepsilon_{yy} + \varepsilon_{zz}$$

These three equations [Eqs. (5.69), (5.70) and (5.71)] are the equations of motion of an infinite homogeneous isotropic elastic medium. It can be seen from these equations that the process of propagation of deformation in elastic medium is a process, which develops at finite speed.

Let the following three solutions satisfy Eqs. (5.69), (5.70), (5.71) in their homogeneous form,

$$u = \phi_1(x - at)$$
$$v = \phi_2(x - bt)$$
$$w = \phi_3(x - ct) \tag{5.72}$$

Equation (5.72) corresponds to three plane waves of deformations of arbitrary profile propagated in the medium in the direction of x with velocities a, b and c, respectively. Substituting Eq. (5.72) in the above three equations of motion (5.69), (5.70) and (5.71), it yields

$$\rho a^2 = \lambda + 2\mu$$
$$\rho b^2 = \mu$$
$$\rho c^2 = \mu \tag{5.73}$$

from which it follows that the set of equations (5.24), (5.25), (5.26) do in fact admit a solution of the form (5.72), which proves that in a linearly elastic isotropic medium there can be transmitted plane waves of deformation of two distinct types, namely, those in which displacements take place in the direction of propagation of the wave and those in which displacements take place perpendicular to this direction. Then, from Eq. (5.73) the propagation velocities of longitudinal and transverse waves are given by the formulae

$$a = \sqrt{\frac{\lambda + 2\mu}{\rho}}, \quad b = c = \sqrt{\frac{\mu}{\rho}}, \tag{5.74}$$

from which it is obvious that $a > b$. The constants λ and $\mu (G = \mu)$ are Lame constants.

This in itself is, of course, only a particular solution, since Eq. (5.72) cannot be subjected to arbitrarily chosen boundary and initial conditions for any specific body. It is the simplest of all to interpret these formulae as referring to the case of an infinitely large body, so that the question of subjecting the solutions to boundary conditions does not, in general, arise.

5.5.3 Solution for Equation of Motion—Primary Wave

There are two solutions for these equations. One solution describes the propagation of an irrotational wave while the other solution describes the propagation of a wave of pure rotation. Referring to differential equations (5.69), (5.70) and (5.71), and differentiating with respect to x, y and z, respectively and adding, we have

$$\rho \frac{\delta^2}{\delta t^2}\left(\frac{\delta u}{\delta x} + \frac{\delta v}{\delta y} + \frac{\delta w}{\delta z}\right) = (\lambda + G)\left(\frac{\delta^2 e}{\delta x^2} + \frac{\delta^2 e}{\delta y^2} + \frac{\delta^2 e}{\delta z^2}\right) + G\nabla^2\left(\frac{\delta u}{\delta x} + \frac{\delta v}{\delta y} + \frac{\delta w}{\delta z}\right)$$

or,

$$\rho \frac{\delta^2 e}{\delta t^2} = (\lambda + G)\, \nabla^2 e + G\nabla^2 e = (\lambda + 2G)\nabla^2 e$$

Hence, $$\frac{\delta^2 e}{\delta t^2} = V_P^2 \nabla^2 e \tag{5.75}$$

This is simplified form of Navier's Equation (attributed to Prof. Lamb)

where, $$V_P^2 = \frac{(\lambda + 2G)}{\rho}$$

Here V_P is the velocity of compression waves which are also referred to as Primary wave (P-wave). This may also be expressed as

$$V_P^2 = \frac{E(1-\nu)}{\rho(1+\nu)(1-2\nu)} \tag{5.76}$$

Therefore, $$\text{for } \nu = 0, \quad V_P = \sqrt{\frac{E}{\rho}} \tag{5.77}$$

and equals the velocity of compressive wave propagation in rods. For $\nu > 0$, V_P is always greater than $\sqrt{\frac{E}{\rho}}$.

5.5.4 Solution for Equation of Motions—Shear Waves

The solution for shear waves can be obtained by differentiating Eq. (5.70) with respect to z and Eq. (5.71) with respect to y, and then may be expressed as

$$\rho \frac{\delta^2}{\delta t^2}\left(\frac{\delta v}{\delta z}\right) = (\lambda + G)\frac{\delta^2 e}{\delta y \delta z} + G\nabla^2 \frac{\delta v}{\delta z} \tag{5.78}$$

$$\rho \frac{\delta^2}{\delta t^2}\left(\frac{\delta w}{\delta y}\right) = (\lambda + G)\frac{\delta^2 e}{\delta y \delta z} + G\nabla^2 \frac{\delta w}{\delta y} \tag{5.79}$$

By subtracting Eq. (5.78) from equation (5.79), we get

$$\rho \frac{\delta^2}{\delta^2 t}\left(\frac{\delta w}{\delta y} - \frac{\delta v}{\delta z}\right) = G\nabla^2\left(\frac{\delta w}{\delta y} - \frac{\delta v}{\delta z}\right)$$

But from Eq. (5.68), $2\theta_x = \frac{\delta w}{\delta y} - \frac{\delta v}{\delta z}$, therefore

$$\rho \frac{\delta^2 \theta_x}{\delta t^2} = G\nabla^2 \theta_x$$

or, $$\frac{\delta^2 \theta_x}{\delta t^2} = \frac{G}{\rho} \nabla^2 \theta_x$$

Thus,
$$\frac{\delta^2\theta_x}{\delta t^2} = V_S^2\nabla^2\theta_x \tag{5.80}$$

Similarly,
$$\frac{\delta^2\theta_y}{\delta t^2} = V_S^2\nabla^2\theta_y \tag{5.81}$$

and,
$$\frac{\delta^2\theta_z}{\delta t^2} = V_S^2\nabla^2\theta_z \tag{5.82}$$

These indicate that the rotation or distortion S-wave propagates with velocity) V_S which is equal to $\sqrt{\frac{G}{\rho}}$. Thus,

$$V_S = \sqrt{\frac{G}{\rho}} \tag{5.83}$$

or,
$$V_S = \sqrt{\frac{2E}{\rho(1+\nu)}}$$

Thus,
$$\frac{V_P}{V_S} = \sqrt{\frac{2(1-\nu)}{1-2\nu}} \tag{5.84}$$

Thus, from the above analysis and expressions as in Eq. (5.84) the variation of velocity of propagation of primary as well as secondary waves depends upon the Poisson's ratio of the soil mass. Figure 5.13 shows the variation of the velocity ratio V_P/V_S with Poisson's ratio ν. It may be observed that for $\nu = 0.5$, this ratio approaches infinity.

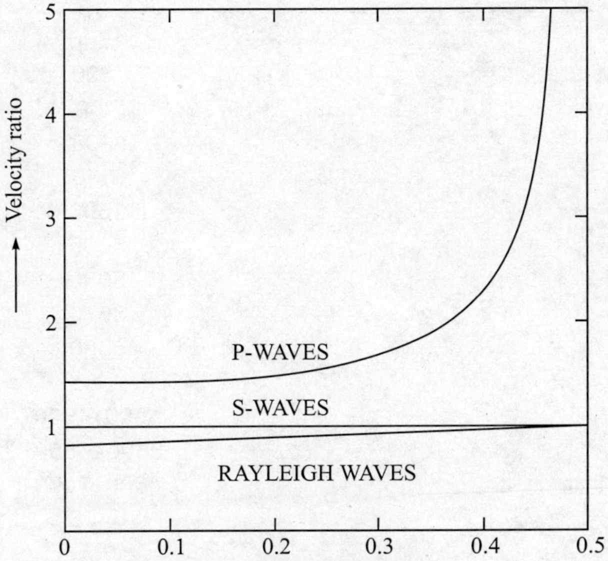

Figure 5.13 Variation of velocity of propagation of waves with the Poisson ratio (After Richart 1962).

Further from the studies of propagation of waves in elastic medium, it is evident that:

(a) Compression wave, which is also called the P-wave is basically the dilatational wave of irrotational nature.
(b) Shear wave, which is also called the S-wave is basically the distortional wave of rotational nature.
(c) In the infinite medium, the velocity of propagation of P-wave is $V_P = \sqrt{\lambda + (2G/\rho)}$ whereas in the rods $V_r = \sqrt{E/\rho}$ which means that the P-wave moves faster in the infinite medium because of no lateral displacements.
(d) The shear wave (distortional) propagates with the same velocity $V_S = \sqrt{G/\rho}$ in the rod as well as in the infinite medium.
(e) For Poisson ratio, $\nu = 0.5$ for no volume change, $V_P \to \infty$ and so $K \to \infty$.
(f) For water-saturated soil, the velocity of propagation of P–waves will not be the representative velocity for soil but will be for water. This is due to the fact that water is relatively incompressible ($K \to \infty$) compared to the soil skeleton.
(g) Since water has no shear strength, the velocity of shear waves in water-saturated soils represents the basic soil property only.

The values of V_P, V_S for different soils are listed in Table 5.1.

Table 5.1 Values of V_P, V_S for different soils

Material/soil	*Poisson ratio ν*	*Velocity P-waves, V_P, m/s*	*Velocity S-wave, V_S m/s*
Moist clay	0.50	1500	150
Loess at natural moisture	0.44	800	260
Dense sand and gravel	0.35	480	250
Fine-grained sand	0.30	300	110
Medium-grained	0.35	550	160
Loose sand	–	1800	500
Steels	0.29	5000	3220
Aluminium alloys	0.34	5030	3100
Copper	0.34	3670	2250
Glass	0.25	5300	3350
Concrete	0.20	–	–
Sandstone	–	1500–4500	750–2200
Limestone	–	3500–6500	1800–3800
Granite	–	4600–7000	2500–4000
Basalt	–	6400	3200
Medium-sized gravel	–	750	180

5.6 LAMB THEORY FOR WAVE PROPAGATION

In 1885, Lord Rayleigh considered wave propagation in a semi-infinite half space, which is popularly known as elastic half space. Since all foundations are supported on soil—the infinite mother earth—so the boundary conditions are more realistically close to an approximation of elastic half space. Lord Rayleigh candidly gave a picture of another wave in addition to P- and S-waves which is called Rayleigh wave. The motion of a Rayleigh wave is confined to a zone near the boundary of the half space as shown in Figure 5.14.

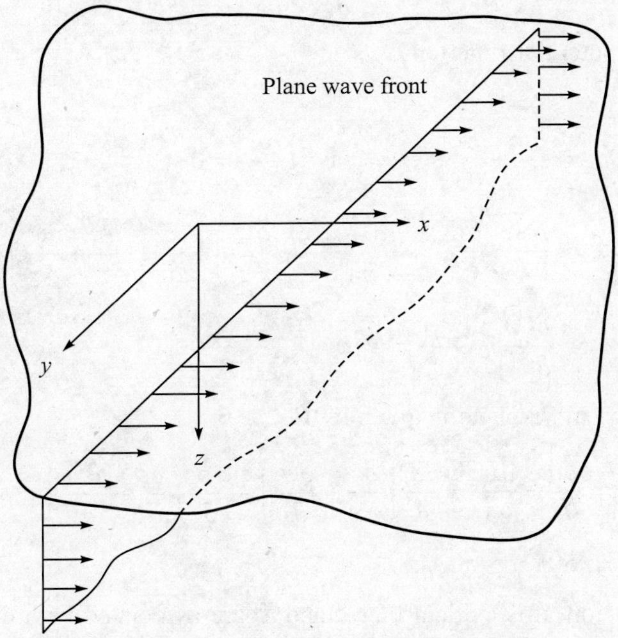

Figure 5.14 Wave propagation in elastic half space.

The detailed solution for propagation of Rayleigh wave was given by Sir Horace Lamb in 1904. In dynamics of waves and foundations, the waves propagating in a zone close to the surface are of practical importance and consequently the effect of the free surface of soil on wave propagation needs investigation.

Select a Cartesian coordinate system *x-y-z* with half space being represented in *xy*-plane with *z* assuming positive in the domain direction as shown in Figure 5.15. If *u*, *v*, *w* are the displacements in *x*-, *y*- and *z*- directions, *V* is taken as zero, such that

$$\frac{\delta u}{\delta y} = \frac{\delta w}{\delta y} = 0$$

and then the equation of motion as in Eq. (5.69) and Eq. (5.71) becomes

$$\rho \frac{\delta^2 u}{\delta t^2} = (\lambda + G)\frac{\delta e}{\delta x} + G\nabla^2 u \tag{5.85}$$

$$\rho \frac{\delta^2 w}{\delta t^2} = (\lambda + G)\frac{\delta e}{\delta z} + G\nabla^2 w \tag{5.86}$$

The solution for u and w may be selected as

$$u = \frac{\delta \phi}{\delta x} + \frac{\delta \psi}{\delta z} \tag{5.87}$$

$$w = \frac{\delta \phi}{\delta z} - \frac{\delta \psi}{\delta x} \tag{5.88}$$

where ϕ and ψ are potential functions.
Again,

$$e = \frac{\delta u}{\delta x} + \frac{\delta v}{\delta y} + \frac{\delta w}{\delta z} \quad \text{as} \quad \frac{\delta v}{\delta y} = 0$$

$$e = \frac{\delta u}{\delta x} + \frac{\delta w}{\delta z} = \frac{\delta}{\delta x}\left(\frac{\delta \phi}{\delta x} + \frac{\delta \psi}{\delta z}\right) + \frac{\delta}{\delta z}\left(\frac{\delta \phi}{\delta z} - \frac{\delta \psi}{\delta x}\right)$$

$$= \frac{\delta^2 \phi}{\delta x^2} + \frac{\delta^2 \phi}{\delta z^2} = \nabla^2 \phi \tag{5.89}$$

Similarly, the rotation in xz-plane is given by Eq. (5.68)

$$2\theta_y = \frac{\delta u}{\delta z} - \frac{\delta w}{\delta x} = \frac{\partial}{\partial z}\left(\frac{\delta \phi}{\delta x} + \frac{\delta \psi}{\delta z}\right) - \frac{\delta}{\delta z}\left(\frac{\delta \phi}{\delta z} - \frac{\partial \psi}{\partial x}\right) = \frac{\delta^2 \psi}{\partial z^2} + \frac{\delta^2 \psi}{\partial x^2} = \nabla^2 \psi$$

$$2\theta_y = \nabla^2 \psi$$

from the above equation, it is evident that ϕ and ψ are associated with dilatation and rotation. As such Rayleigh waves are generated by combination of P-waves and S-waves as shown in Figure 5.14.

Equations (5.85) and (5.86) as equations of motion take the form as follows after substitution of u, w, e from Eqs. (5.87), (5.88) and (5.89)

$$\rho \frac{\delta}{\delta x}\left(\frac{\delta^2 \phi}{\delta t^2}\right) + \rho \frac{\delta}{\delta z}\left(\frac{\delta^2 \psi}{\delta t^2}\right) = (\lambda + 2G)\frac{\delta}{\delta x}(\nabla^2 \phi) + G\frac{\delta}{\delta z}(\nabla^2 \psi) \tag{5.90}$$

$$\rho \frac{\delta}{\delta z}\left(\frac{\delta^2 \phi}{\delta t^2}\right) - \rho \frac{\delta}{\delta z}\left(\frac{\delta^2 \psi}{\delta t^2}\right) = (\lambda + 2G)\frac{\delta}{\delta z}(\nabla^2 \phi) - G\frac{\delta}{\delta x}(\nabla^2 \psi) \tag{5.91}$$

Equations (5.90) and (5.91) satisfy

$$\frac{\delta^2 \phi}{\delta t^2} = \frac{(\lambda + 2G)}{\rho}\nabla^2 \phi = V_P^2 \nabla^2 \phi \tag{5.92}$$

$$\frac{\delta^2 \psi}{\delta t^2} = \frac{G}{\rho}\nabla^2 \psi = V_S^2 \nabla^2 \psi \tag{5.93}$$

Considering a sinusoidal wave travelling with frequency ω in the soil in space $z > 0$, the solutions are assumed as

$$\phi = e^{i\omega t}/e^{inx}\ F(z) \tag{5.94}$$

$$\psi = e^{i\omega x}/e^{inx}\ G(z) \tag{5.95}$$

where $F(z)$ and $G(z)$ are the functions which describe the variation in amplitude of the wave with depth and n is the wave number given by

$$n = \frac{2\pi}{\lambda}$$

where λ is the wavelength. Alternatively,

$$\begin{aligned} \text{wavelength} &= \lambda \\ &= \frac{2\pi}{n} \\ &= \frac{\text{velocity of wave}}{\omega/2\pi} \\ &= \frac{V_R \cdot 2\pi}{\omega} \end{aligned}$$

$$\therefore \qquad n = \frac{\omega}{V_R} \tag{5.96}$$

where V_R = Rayleigh wave velocity.

Again at the boundary surface of the half space, stress components at $z = 0$ are

$$\sigma_{zz} = 0,\ \tau_{zx} = 0,\ \tau_{zy} = 0$$

Therefore $$\sigma_{zz} = \lambda e + 2G\ \varepsilon_{zz} = \lambda e + 2G\frac{\partial w}{\partial z} = 0$$

and $$\tau_{xz} = G \cdot \gamma_{xz} = G\left(\frac{dw}{dx}\ \frac{du}{dz}\right) = 0$$

Using suitable potention functions of u and w, the velocities and displacement pattern of Rayleigh waves can be determined.

5.7 RAYLEIGH WAVES—WAVE PROPAGATION IN ELASTIC HALF SPACE

The primary P waves and shear S waves as discussed propagate in the interior of the elastic continuum, however in the neighbourhood of its surfaces, waves of different types are possible. Thus, near the boundary of the half space the Rayleigh wave is propagated. This was first obtained by Rayleigh (1885) and the fact that it can play a role in the succession of waves received from an earthquake was established by Lamb (1904) in a paper, which provides the basis of modern seismology. Since then, the surface waves have played an important role in earthquakes and in collision of elastic solids. The study of records of seismic waves have supported Rayleigh's expectations.

At a great distance from the source, the deformation produced by these waves is essentially a two-dimensional one.

The plane x–y is taken to be free surface of a uniform half space, which extends in the z-direction as shown in Figure 5.15. The z-direction has been taken pointed in the downward direction in the half space (xy-plane). For a plane wave travelling in the x-direction, particle displacement is independent of the y-direction. If u, v, w are the displacement components in x, y and z-directions, then

$$v = 0 \quad \text{and} \quad \frac{\delta v}{\delta y} = \frac{\delta w}{\delta y} = 0$$

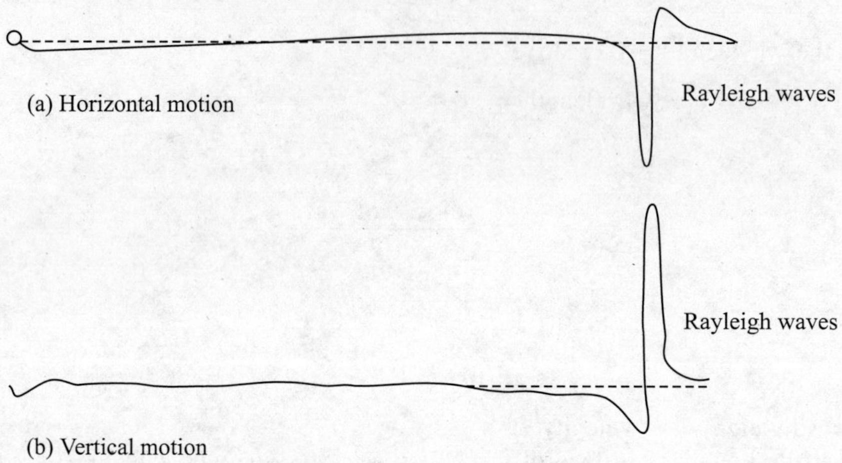

Figure 5.15 Propagation of Rayleigh waves.

Neglecting the body forces, the equations of wave propagation are

$$(\lambda + \mu)\frac{\delta e}{\delta x} + \mu \nabla^2 u = \rho \frac{\delta^2 u}{\delta t^2} \tag{5.97}$$

$$(\lambda + \mu)\frac{\delta e}{\delta z} + \mu \nabla^2 w = \rho \frac{\delta^2 w}{\delta t^2} \tag{5.98}$$

As u and w represent the displacements in the directions of x and z, respectively and are independent of y, by introducing ϕ and ψ as two potential functions, u and w may be taken as

$$u = \frac{\delta \phi}{\delta x} + \frac{\delta \psi}{\delta z} \tag{5.99}$$

$$w = \frac{\delta \phi}{\delta z} - \frac{\delta \psi}{\delta x} \tag{5.100}$$

$$e = \frac{\delta u}{\delta x} + \frac{\delta v}{\delta y} + \frac{\delta w}{\delta z} \quad \text{as} \quad \frac{\delta v}{\delta y} = 0, \text{ then}$$

$$e = \frac{\delta u}{\delta x} + \frac{\delta w}{\delta z} = \left(\frac{\delta^2 \phi}{\delta x^2} + \frac{\delta^2 \psi}{\delta x \delta z}\right) + \left(\frac{\delta^2 \phi}{\delta z^2} - \frac{\delta^2 \psi}{\delta x \delta z}\right)$$

$$e = \left(\frac{\delta^2 \phi}{\delta x^2} + \frac{\delta^2 \phi}{\delta z^2}\right) = \nabla^2 \phi \tag{5.101}$$

The rotation in the x–z plane is given by

$$2\theta_y = \frac{\delta u}{\delta z} - \frac{\delta w}{\delta x} = \frac{\delta^2 \psi}{\delta x^2} + \frac{\delta^2 \psi}{\delta z^2} = \nabla^2 \psi \tag{5.102}$$

Substituting u and w from Eqs. (5.99) and (5.100) in the equation of wave propagation Eqs. (5.97) and (5.98)

$$(\lambda + \mu)\frac{\delta}{x}[\nabla^2\phi] + \mu\frac{\delta}{\delta x}\nabla^2\phi = \rho\frac{\delta}{\delta x}[\delta^2\phi/\delta t^2] + \rho\frac{\delta}{\delta z}[\delta^2\psi/\delta t^2] + \mu\frac{\delta}{\delta z}\nabla^2\psi\,(\lambda + 2\mu)$$

$$\frac{\delta}{\delta z}[\nabla^2\phi] + \mu\frac{\delta}{\delta z}\nabla^2\psi = \rho\frac{\delta}{\delta x}[\delta^2\phi/\delta t^2] + \rho\frac{\delta}{\delta z}[\delta^2\psi/\delta t^2] \tag{5.103}$$

and, $$(\lambda + 2\mu)\frac{\delta}{\delta z}[\nabla^2\phi] - \mu\frac{\delta}{x}\nabla^2\psi = \rho\frac{\delta}{\delta z}[\delta^2\phi/\delta t^2] + \rho\frac{\delta}{\delta z}[\delta^2\psi/\delta t^2] \tag{5.104}$$

The above Eqs. (5.103) and (5.104) are satisfied if

$$\frac{\delta^2\phi}{\delta t^2} = \frac{(\lambda + 2\mu)}{\rho}\nabla^2\phi$$

$$\frac{\delta^2\phi}{\delta t^2} = V_\rho^2\,\nabla^2\phi \tag{5.105}$$

$$\frac{\delta^2\psi}{\delta t^2} = \frac{\mu}{\rho}\nabla^2\psi = V_S^2\,\nabla^2\psi \tag{5.106}$$

or, $$\nabla^2\phi = \frac{1}{V_P^2}\frac{\delta^2\phi}{\delta t^2} \tag{5.107}$$

$$\nabla^2\psi = \frac{1}{V_S^2}\frac{\delta^2\psi}{\delta t^2}$$

The preceding two equations represent harmonic waves in the x-direction and which waves decreases with z. Such solutions will be of the form

$$\phi = A.\ \exp[ik(V_R t \pm az - x] \tag{5.108}$$

where A and a are real and imaginary constants, and k and V_R the surface wave number and velocity, respectively.

From Eq. (5.108), ϕ and ψ may be written as

$$\phi = A\exp\left[ik\,(V_R t \pm \sqrt{\left\{\left(\frac{V_R^2}{V_P^2}\right) - 1\right\}}\,z - x\right] \tag{5.108a}$$

$$\psi = B \exp\left[ik\,(V_R t \pm \sqrt{\left\{\left(\frac{V_R^2}{V_S^2}\right) - 1\right\}}\; z - x \right]$$

provided, $\sqrt{\{(V_R^2/V_P^2) - 1\}}$ and $\sqrt{\{(V_R^2/V_P^2) - 1\}}$ are imaginary quantities and the proper sign is chosen. Obviously, therefore

$$V_R < V_S < V_P$$

The boundary conditions are the vanishing of normal and shear stresses at $z = 0$, then

$$\sigma_{zz} = \lambda\,\nabla^2\phi + 2\mu\left(\frac{\delta^2\phi}{\delta z} + \frac{\delta^2\psi}{\delta x\,\delta z}\right) = 0$$

$$\tau_{xz} = \mu\left(2\frac{\delta^2\phi}{\delta x\,\delta z} + \frac{\delta^2\psi}{\delta x^2} - \frac{\delta^2\psi}{\delta z^2}\right) = 0 \qquad (5.109)$$

Substituting for ϕ and ψ from Eq. (5.108a) and putting $z = 0$ yields

$$\left(2 - \frac{V_R^2}{V_S^2}\right) A \pm 2\sqrt{\left\{\left(\frac{V_R^2}{V_S^2}\right) - 1\right\}}\; B = 0$$

$$\pm 2\sqrt{\left\{\left(\frac{V_R^2}{V_P^2}\right) - 1\right\}}\; A \pm \left(2 - \frac{V_R^2}{V_S^2}\right) B = 0$$

If A and B are not to vanish,

$$\left(2 - \frac{V_R^2}{V_S^2}\right)^2 = 4\left(1 - \frac{V_R^2}{V_P^2}\right)^{\frac{1}{2}}\left(1 - \frac{V_R^2}{V_S^2}\right)^{\frac{1}{2}}$$

or,
$$\frac{V_R^2}{V_S^2}\left[\frac{V_R^6}{V_S^6} - 8\frac{V_R^4}{V_S^4}\right] + V_R^2\left(\frac{24}{V_S^2} - \frac{16}{V_P^2}\right) - 16\left(1 - \frac{V_S^2}{V_P^2}\right) = 0$$

One value of V_R such that $0 < V_R < V_S$ can always be found.

Selecting, $\lambda = \mu$ and $V_P = \sqrt{3}\cdot V_S$, i.e., $\nu = 1/4$, then

$$3\frac{V_R^6}{V_S^6} - 24\frac{V_R^4}{V_S^4} + 56\frac{V_R^6}{V_S^2} - 32 = 0$$

which has roots $\frac{V_R^2}{V_S^2} = 4,\ 2 + 2\sqrt{3},\ 2 - 2/\sqrt{3}$ of which only the last is acceptable, hence

$$V_R = \left(2 - \frac{2}{\sqrt{3}}\right)^{1/2} V_S,\ V_R = 0.9194 V_S = 0.9194\sqrt{(G/\rho)}\ 4$$

A wave with the above characteristics is known as Rayleigh wave which moves with velocity $V_R = \left(2 - \frac{2}{\sqrt{3}}\right)^{1/2} \sqrt{G/\rho}$ for $\nu = 0.25$. The horizontal and vertical motion has been shown in Figure 5.15.

5.7.1 Mechanism of Wave Propagation at the Surface

Lamb (1904) presented in great detail the propagation of waves in a homogeneous isotropic elastic medium. He has candidly shown that how a disturbance spreads out from the point source in the form of a wave system. If a point source acts at the surface of an elastic half space, the disturbance spreads out in the form of symmetrical annular waves. The initial form of these waves will depend on the input impulse; if the input is of short duration, the characteristic waves shown in Figure 5.15 will develop. These waves have three salient features that correspond to the arrivals of the P-wave, S-wave and Rayleigh (R) wave. The horizontal and vertical components of particle motion are shown separately in Figure 5.15(a) and Figure 5.15(b), respectively.

A particle at the surface first undergoes an oscillatory lateral displacement on the arrival of the P-wave, followed by a relatively quiet period leading up to another oscillation at the arrival of the S-wave. This is followed by an oscillation of much larger magnitude when the R-wave arrives.

The time interval between wave arrivals becomes greater and the amplitude of the oscillations becomes smaller with increasing distance from the source. In addition, the amplitude of P-wave and S-wave decays more rapidly than that of an R-wave. Therefore, the R-wave is the most significant disturbance along the surface of an elastic half space and, at large distances from the source may be the only clearly distinguishable wave. The relative percentages of the total energy carried by these three waves are as follows:

Type	*Per cent of total energy*
R-wave	67
S-wave	26
P-wave	7

Further, the distribution of displacement waves from a circular footing on a homogeneous, isotropic elastic half space has been shown in Figure 5.16. In such a situation, the energy dissipation is called *geometric damping*. The concept of geometric damping has great applications in the design of vibrating footings.

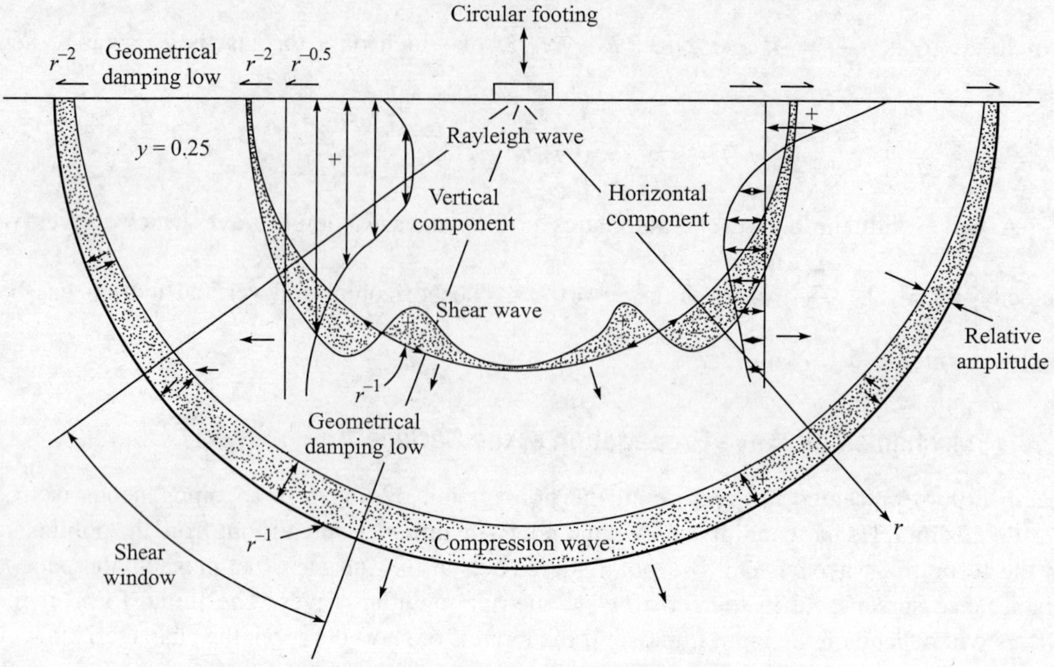

Figure 5.16 Concepts of geometrical damping (After Wood 1968).

5.7.2 Love Waves

The propagation of Rayleigh waves have been presented in the previous section. It has been shown that particle move in the direction of plane of propagation of Rayleigh waves. These waves produce both vertical and horizontal displacement of the ground as the surface wave propagate outward. The dilatational and equivoluminal waves as discussed in Section 5.5.3 take place in the interior of an elastic body. However in the neighourhood of its surface waves of different type are possible (Rayleigh waves), while near the surface of two elastic bodies or elastic layers special waves occur which is known as Love Waves. It is named after A.E.H Love, the English mathematician who discovered it. Waves associated with a elastic layer as Love Waves may be viewed as the result of constructive interface of plane waves successively from the top and bottom of the elastic layer. Love () showed that SH surface wave analogus to Transverse S waves propagate in homogenous upper layer overlying another homogenous half space of another medium.

5.8 CONCEPTS OF PHASE VELOCITY AND GROUP VELOCITY

5.8.1 Phase Velocity

When a wave travels through a medium, its velocity of advancement in the medium is called the *wave velocity*. For example, a plane harmonic wave travelling along the + x-direction may be represented by

$$y = B \sin(\omega t - kx) \tag{5.110}$$

where B is the maximum amplitude, ω is the angular frequency and k is the propagation constant of the wave. Then by definition, the ratio of the angular frequency ω to the propagation constant k is the wave velocity V. That is,

$$V = \frac{\omega}{k} \tag{5.111}$$

The term $(\omega t - kx)$ is the phase of the wave motion. Therefore, the planes of constant phase (wave fronts) are defined by

$$\omega t - kx = \text{constant}$$

Differentiating with respect to time,

$$\omega - k\frac{dx}{dt} = 0$$

or,

$$\frac{\omega}{k} = \frac{dx}{dt} = V \tag{5.112}$$

Thus, ω/k is the wave velocity. It is the velocity with which the planes of constant phase advance through the medium. The wave velocity is also called *phase velocity*, and it is also called the *wave number*. For example, the phase velocity is that which is obtained from the equation of motion like the value of $V_R = 0.9194\sqrt{(G/\rho)}$ for the velocity of Rayleigh wave when the Poisson ratio is 0.25.

5.8.2 Group Velocity

In situations where a pulse consists of waves differing slightly from one another in frequency, a superposition of these waves is called a wave packet or a wave group. When such a group travels in a medium, the velocities of its different components are different. The observed velocity is however the velocity which is the maximum. This is called the group velocity. In other words, the group velocity is the velocity with which the energy is propagated or transmitted. The individual waves travel inside the group with their phase velocities. Thus, the group velocity is essentially that which with the radiated energy level, controls the response of the recording instruments and is more likely to the found correctly from the seismograms rather than from the phase velocity, specially when the two differ significantly.

Consider a wave group consisting of two components of equal amplitudes B but of slightly different angular frequencies ω_1 and ω_2, with propagation constants k_1 and k_2 as shown in Figure 5.17. Their displacements are expressed as

$$y_1 = B\sin(\omega_1 t - k_1 x)$$
$$y_2 = B\sin(\omega_2 t - k_2 x) \tag{5.113}$$

Their algebraic summation yields

$$y = y_1 + y_2 = B[\sin(\omega_1 t - k_1 x) + \sin(\omega_2 t - k_2 x)]$$

Using the trigonometric relation, $\sin\alpha + \sin\beta = 2\sin(\alpha + \beta)/2.\ \cos(\alpha - \beta)/2$,

$$y = 2B\cos[(\omega_1 - \omega_2)t/2 - (k_1 - k_2)x/2].\ \sin[(\omega_1 + \omega_2)t/2 - (k_1 + k_2)x/2] \tag{5.114}$$

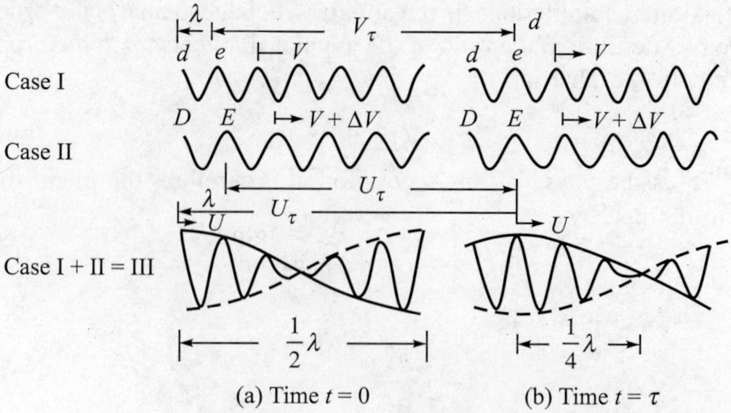

Figure 5.17 Concept of phase velocity and group velocity.

Comparing Eq. (5.114) with Eq. (5.113) it is obvious that Eq. (5.114) represents a wave system with a frequency = $(\omega_1 + \omega_2)/2$ which is very close to the frequency of either component but with an amplitude

$$2B\cos[(\omega_1 - \omega_2)t/2 - (k_1 - k_2)x/2]$$

Thus, the amplitude of the wave group is modulated both in space and time by a very slowly varying envelope of frequency $(\omega_1 - \omega_2)/2$ and propagation constant $(k_1 - k_2)/2$ and as a maximum value of $2B$. The velocity with which the envelope moves is the velocity of the maximum amplitude of the group and is given by

$$U = (\omega_1 - \omega_2)/(k_1 - k_2) = \Delta w/\Delta k \tag{5.115}$$

But if a group contains a number of frequency components in an infinitely small frequency level, then the expression for the group velocity may be written as

$$U = \frac{dw}{dk}$$

5.8.3 Relationship of Group Velocity with Phase Velocity

The phase velocity, V is expressed as in Eq. (5.112), so

$$\omega = k.V,$$

then,
$$U = \frac{dw}{dk} = \frac{d}{dk}(k.V) = V + k\frac{dV}{dk}$$

Now $k = 2\pi/\lambda$, where λ is the wavelength, so

$$U = V + \frac{2\pi}{\lambda}\frac{dV}{d(2\pi/\lambda)}$$

$$= V + \frac{1}{\lambda}\frac{dV}{d(1/\lambda)}$$

But $d(1/\lambda) = -(1/\lambda^2)\, d\lambda$, therefore

$$\therefore \qquad U = V - \lambda \frac{dV}{d\lambda} \tag{5.116}$$

Stephen and Bate (1950) obtained similar expression for group velocity as given in Eq. (5.116).

Example 5.4 A simply supported beam ($0 \le x \le L$) is under flexural vibration. The uniform beam has flexural rigidity *EI* and a mass *m* per unit length. Obtain the values of phase velocity and group velocity for the flexural pulse.

Solution: The governing differential equation is

$$\frac{\delta^2}{\delta x^2}\left[EI\,\frac{\delta^2 w(x,t)}{\delta x^2}\right] + m\frac{\delta^2 w(x,t)}{\delta t^2} = 0$$

Substituting $a^2 = EI/m$, then

$$a^2\frac{\delta^4 w(x,t)}{\delta x^4} + \frac{\delta^2 w(x,t)}{\delta t^2} = 0 \qquad \text{or} \qquad \frac{\delta^4 w(x,t)}{\delta x^4} = -\frac{1}{a^2}\frac{\delta^2 w(x,t)}{\delta x^2}$$

and by comparing with the wave equation of the form

$$\frac{\delta^2 w(x,t)}{\delta x^2} = (1/V_P^2)\frac{\delta^2 w(x,t)}{\delta t^2}$$

it is observed that is has a fourth-order derivative with respect to x instead of the second-order derivative and the constant $a = \sqrt{(EI/m)}$ does not possess the dimensions of velocity. Thus, it can be easily verified that the general solution of the wave equation is of the form

$$w(x,\, t) = F_1(x - V_P t) + F_2(x + V_P t)$$

Let the solution of the governing differential equation for flexural vibration of the beam ($0 \le x \le L$) be in the form of a simple harmonic travelling with velocity V in the positive x-direction and let it be represented by

$$w(x,\, t) = A\cos\,(2\pi/\lambda)\,(x - Vt) \tag{5.117}$$

where λ is the wavelength.

Substituting the above solution in the differential equation, the velocity of propagation of the sinusoidal wave is

$$V = 2\pi a/\lambda = \frac{2\pi}{\lambda}\sqrt{\frac{EI}{m}}$$

Differentiating w.r.t. λ, $\qquad \dfrac{dV}{d\lambda} = -2\pi a\cdot\dfrac{1}{\lambda^2} \quad \text{as} \quad \dfrac{d}{d\lambda}\left(\dfrac{1}{\lambda}\right) = -\dfrac{1}{\lambda^2}$

The group velocity is
$$U = V - \lambda \frac{dV}{d\lambda}$$
$$= \frac{2\pi}{\lambda}\sqrt{\frac{EI}{m}} - \lambda \cdot \frac{dV}{d\lambda}$$
$$= \frac{2\pi}{\lambda}\sqrt{\frac{EI}{m}} - \lambda\left[\frac{-2\pi}{\lambda^2}\sqrt{\frac{EI}{m}}\right] = \frac{2\pi}{\lambda}\sqrt{\frac{EI}{m}} + \frac{2\pi}{\lambda}\sqrt{\frac{EI}{m}}$$

Therefore,
$$U = \frac{4\pi}{\lambda} \cdot \sqrt{\frac{EI}{m}}$$

5.9 PROPAGATION OF FLEXURAL WAVES IN BEAMS ON ELASTIC FOUNDATIONS

The study of wave motion of a beam on elastic foundation in the flexural vibration as a dynamic soil–structure interaction problem, has numerous engineering applications, namely, landing of aircraft on runways and impact loadings on nuclear reactors, naval structures and reinforcing wires in composite materials. Although considerable work has been done towards the study of wave motion of a beam in flexural vibration by J. Micklowitz (1960), very little work has been reported on propagation of flexural waves in beams on elastic foundations. Generally, in such studies, the velocity of propagation of a simple harmonic wave is called *wave velocity*, and the phase velocity with which such a group of waves is propagated is called *group velocity*. Hence, group velocity is the velocity with which energy is propagated. A medium like a beam on elastic foundation exhibiting a wave velocity depending upon wavelength is called a dispersive medium. It has been generally accepted that the wave propagation in a dispersive medium is a much more complex phenomenon than that in a non-dispersive medium.

5.9.1 Equation of Wave Motion

If the shear deformation and rotatory inertia effects are neglected, the differential equation for the free transverse vibration of a uniform Euler-Bernoulli beam ($0 \le x \le L$) resting on Winkler foundation is given by

$$\frac{\delta^4 w(x,t)}{\delta x^4} + \rho A \frac{\delta^2 w(x,t)}{\delta t^2} + Kw(x,t) = 0 \qquad (5.118)$$

in which

$w(x, t)$ = transverse displacement

$\rho A = m$ = mass per unit length of the beam

EI = flexural rigidity of the beam

K = foundation modulus

Assuming that the solution of Eq. (5.118) is a simple harmonic wave travelling with velocity, V, in the positive x-direction, so that its form is

$$W(x, t) = B' \cos \frac{2\pi}{\lambda}(x - Vt) \tag{5.119}$$

in which

B' = a constant
λ = wavelength

Substituting Eq. (5.119) in Eq. (5.118), the velocity of propagation of the wave, i.e., phase velocity, V, is obtained as

$$V = \frac{2\pi}{\lambda}\sqrt{\frac{EI}{\rho A}}\sqrt{\left(1 + \frac{K}{EI}\cdot\frac{\lambda^4}{16\pi^4}\right)} \tag{5.120}$$

For $K = 0$, i.e., for the beam without foundation

$$V = \frac{2\pi}{\lambda}\cdot\sqrt{\frac{EI}{\rho A}} \tag{5.121}$$

From Eqs. (5.120) and (5.121), it is evident that waves move faster in beam on elastic foundation. Further, the phase velocity is not constant but varies with the wavelength. Thus, this beam on elastic foundation is a dispersive medium for wave propagation. A pulse consisting of a group of harmonic waves is called a *wave packet* and the velocity with which such a group of waves is propagated is called the *group velocity* denoted by U. It has been shown in Eq. (5.116) that U is given by

$$U = V - \lambda\frac{dV}{d\lambda} \tag{5.122}$$

Thus, the velocity with which the energy will be propagated in the beam on elastic foundation under consideration is obtained as

$$U = V\left[1 + \left\{\frac{\left(\frac{2\pi}{\lambda}\right)^2\frac{EI}{\rho A} - \frac{K}{A}\left(\frac{\lambda}{2\pi}\right)^2}{\left(\frac{2\pi}{\lambda}\right)^2\frac{EI}{\rho A} + \frac{K}{A}\left(\frac{\lambda}{2\pi}\right)^2}\right\}\right] \tag{5.123}$$

For $K = 0$, i.e., for beams without foundation

$$U = 2V \tag{5.124}$$

However, for beams on elastic foundation as in Eq. (5.123),

$$U < 2V \tag{5.125}$$

Effects of Rotary Inertia on Wave Propagation.

From Eqs. (5.124) and (5.125), it may be observed that as the wavelength approaches zero, both the phase and group velocities approach infinity. The group velocity is the velocity with which the energy propagates. So, it can be concluded that for a pulse consisting primarily of short waves, the energy is propagated with infinite velocity, which does not seem physically possible. The reason for this puzzling result can possibly be explained by the fact that for very short waves the rotatory inertia cannot be neglected. When the wavelength is of the same order of magnitude as the depth of the beam, the rotary inertia effects should be considered. Thus, neglecting shear deformations but considering the effects of rotatory inertia, the differential equation of motion for a uniform beam ($0 \le x \le L$) resting on an elastic foundation becomes

$$EI\left[\frac{\partial^4 W(x,t)}{\partial x^4} + \frac{\rho A}{EI}\frac{\partial^2 W(x,t)}{\partial t^2} + \frac{KW(x,t)}{EI} - r_g^2\frac{(\rho A)}{EI}\frac{\partial^4 W(x,t)}{\partial x^2\,\partial t^2}\right] = 0 \tag{5.126}$$

where r_g = radius of gyration. Assuming a solution in the form given by Eq. (5.119), the phase velocity can be obtained as

$$V = \sqrt{\frac{E}{\rho}}\cdot\left[1+\frac{\lambda^2}{4\pi^2 r_g^2}\right]^{-\frac{1}{2}}\left[1+\frac{K}{EI}\cdot\frac{\lambda^4}{16\pi^4}\right] \tag{5.127}$$

where $\sqrt{E/\rho}$ is the velocity of propagation of longitudinal waves. For $K = 0$, i.e., for the beam without foundation,

$$V = \sqrt{\frac{E}{\rho}}\left[1+\frac{\lambda^2}{4\pi^2 r_g^2}\right]^{-\frac{1}{2}} \tag{5.128}$$

Similarly, the group velocity for the beam-foundation system, U may be obtained from

$$U = V - \lambda\frac{dV}{d\lambda}$$

For $K = 0$, i.e., for beams without foundation, the group velocity is given by

$$U = \sqrt{\frac{E}{\rho}}\left[1+\frac{\lambda^2}{4\pi^2 r_g^2}\right]^{-\frac{1}{2}}\left[1+\frac{1}{1+4\pi^2 r_g^2/\lambda^2}\right] \tag{5.129}$$

Here the above equations candidly show that when $\lambda \to 0$, the phase velocity and group velocities tend to be of constant value $\sqrt{E/\rho}$.

Example 5.5 On a beam on elastic foundation ($0 \le x \le L$), a striker falls in a transverse direction, giving rise to local bending which propagates from the point of strike in the form of a bending wave. If the transverse displacement is assumed to be sufficiently small, so that the neutral line of the beam may be considered one of constant length, find the group velocity and the phase velocity of propagation of flexural waves.

Solution: Consider a small segment Δx along length of the beam. The shear force $F(x)$ acts from one side of the section to the other side. Newton's second law for the element Δx may be expressed in the form

$$\rho A \Delta x \frac{\delta^2 w}{\delta t^2} = F\,(X + \Delta x)$$

$$= \Delta x \frac{\delta F}{\delta x} + Kw$$

$$= - F(x) + Kw$$

where,

ρ = density of the beam

A = cross-sectional area of the beam.

Considering the balance of moments in the rod

$$M(x) = EI \cdot \frac{\delta^2 w}{\delta x^2}$$

The total moment acting on the element x is the difference

$$M(x + \Delta x) - M(x) = \Delta x \cdot \frac{\delta M}{\delta x}$$

Thus, the moment balance equation takes the form,

$$EI \, . \, \frac{\delta^3 w}{\delta t^2 \delta x} = \frac{\delta M}{\delta x} + F + Kw$$

Differentiating with respect to x and substituting for M

$$EI \cdot \frac{\delta^4 w}{\delta t^2 \delta x^2} = EI \cdot \frac{\delta^4 w}{\delta x^4} + \rho A \cdot \frac{\delta^2 w}{\delta t^2} + Kw$$

Adopting $a^2 = \dfrac{EI}{\rho A}$, $\beta = \dfrac{K}{EI}$ and $r_g^2 = \dfrac{I}{A}$

$$\frac{r_g^2}{a^2} \cdot \frac{\delta^4 w}{\delta t^2 \delta x^2} = \frac{\delta^4 w}{\delta x^4} + \frac{1}{a^2} \cdot \frac{\delta^2 w}{\delta t^2} + \beta w$$

The solution may be assumed for a simple harmonic wave

$$w = A \cos \frac{2\pi}{\lambda} (x - V_{ph} t)$$

where,

V_{ph} = wave velocity

λ = wavelength. Thus, the wave velocity V_{ph} may be obtained by using Eq. (5.127) as

$$V_{ph} = \sqrt{\frac{E}{\rho}} \cdot \left\{ \left[1 + \frac{\lambda^2}{4\pi^2 r_g^2} \right]^{-\frac{1}{2}} \right\} \left\{ 1 + \beta \cdot \frac{\lambda^4}{16\pi^4} \right\}$$

For $K = 0$ (beam only)

$$V_{ph} = \sqrt{\frac{E}{\rho}} \cdot \left[1 + \frac{\lambda^2}{4\pi^2 r_g^2}\right]^{-\frac{1}{2}}$$

and,
$$\frac{V_{ph}\text{ (Beam on elastic foundation)}}{V_{ph}\text{ for beam}} = \left\{1 + \beta \frac{\lambda^4}{16\pi^2}\right\}$$

In the above equations, if $\lambda \to 0$, then the wave velocity tends to the constant value $\sqrt{E/\rho}$. The group velocity with which the energy propagates may be expressed by

$$V_g = V_{ph} - \lambda \frac{dV}{d\lambda}$$

For $K = 0$ (for beam only), group velocity

$$V_g = \sqrt{\frac{E}{\rho}} \left\{1 + \frac{\lambda^2}{4\pi^2 r_g^2}\right\}^{-\frac{1}{2}} \left[1 + \frac{1}{1 + 4\pi^2 r_g^2/\lambda^2}\right]$$

Thus, for $\lambda \to 0$, both the phase and group velocity tend to the constant value $\sqrt{E/\rho}$.

PROBLEMS

5.1 Show that the transverse free vibration displacement $w(x, t)$ of a string $(0 \le x \le L)$ fixed at both ends may be expressed as

$$w(x, t) = \sum_{n=1}^{\infty} C_n \sin \frac{n\pi x}{L} \frac{\cos n c\pi t}{L}$$

where $c = \sqrt{E/\rho}$ and C_n = constant.

If the string is plucked at the mid-point by an amount h and then released, show that the motion may be represented by

$$w(x, t) = \frac{8h}{\pi^2}\left[\left\{\frac{\sin \pi x}{L}\right\}\left[\frac{\cos \pi ct}{L}\right] - \frac{1}{9}\left\{\sin \frac{3\pi x}{L}\left[\frac{\cos 3\pi ct}{L}\right]\right\} + \ldots\right]$$

5.2 When a plane wave traverses a medium having density ρ, the maximum displacement of particles is given by

$$\phi(x, t) = 0.01 \sin\left\{\frac{2\pi}{100}(200t - x)\right\}$$

Find the amplitude, wavelength and the frequency of wave.

5.3 During an earthquake, the water in a reservoir exerts hydrodynamic pressure on the dam. Formulate the dam-reservoir interaction problem when the reservoir is infinitely long and has a uniform rectangular cross-section.

5.4 Show that there are three principal waves in an elastic medium. What are the velocities of propagation of these waves? Describe with a suitable sketch the propagation of Rayleigh waves.

5.5 Discuss the relationship of Poisson ratio with propagation of various principal waves in an elastic medium. What is the special significance of velocity ratios when the Poisson ratio of soil is 0.5?

5.6 Discuss the propagation of energy from one place to another in an elastic medium. If P-waves, S-waves and Rayleigh waves are generated and propagated during earthquake, what will be the percentage of total energy shared by P-waves, S-waves and Rayleigh waves? Derive expressions for group velocity with which energy propagates in an elastic medium.

6

DYNAMIC SOIL PROPERTIES

6.1 INTRODUCTION

The dynamic soil properties are of great importance in the study of dynamic behaviour of foundations, substructures, soil retaining structures, and other soil structures, such as earth and rock fill dam, during earthquakes and also for their earthquake resistant design. The soil is a polyphase material consisting of solid soil particles, water and air. The dynamic soil properties should be determined by the properties of the solid particles, the mixture of soil, water and air.

The dynamic soil properties are strain level dependent. The dynamic properties change with the magnitude of deformation or the loading vibration frequency. Thus, the effect of strain level is quite significant. When the strain level is less than $10^{-4}\%$ it indicates elastic properties represented by wave propagation, while the larger strain levels reflect changes in deformation modulus, damping ratio or in pore water pressure or volume. For the sake of simplicity in applying mathematical theory of elasticity, soil is generally considered as a linear mass and with such assumptions, the dynamic soil properties can be theoretically handled easily. Up to a strain level of $10^{-4}\%$, beyond which the non-linear behaviour becomes prominent, an approximate analysis is carried out using the so called "equivalent linearization method", which takes into consideration the changes in deformation coefficient and damping ratio. Recently, soil has been represented by an elastoplastic model to enable the realistic analysis of failure phenomenon such as liquefaction with desired accuracy.

In soil dynamics and for geotechnical earthquake engineering, solutions to various problems depend greatly upon dynamic properties. These properties are evaluated by various tests in the laboratory as well as under *in situ* conditions.

In soil dynamics not only the time dependent motions but also the constitutive laws for soil as engineering material need to be taken into consideration. Thus, for soil under cyclic state of stress at low strains or even high strains, there is involvement of energy dissipation. For studying such a dynamic soil behaviour, it is adequate to simplify this type of behaviour in terms of two equivalent parameters that characterize the stiffness and damping of the soil.

The case studies of earthquake damage reveal that the nature, distribution and quantum of damage also depend upon the response of soil to dynamic loading, especially the cyclic loading.

The classical shear parameter of soil given by Coulomb's equation is

$$\tau = c + \sigma_n \tan \phi \tag{6.1}$$

The above relationship holds good even during earthquake excitation, but then it is essential to determine the values of c, ϕ and shear modulus in dynamic state. During such state of dynamic loading, the soil properties and their determination are strain level dependent as shown in Table 6.4. This has been recommended by (Ishihara, 1971). Tests are done in the field as well as in the laboratory to determine the dynamic soil properties. This chapter presents a variety of methods by which low strain and high strain soil behaviour can be measured in the field and in the laboratory.

The dynamic soil properties in state of cyclic loading depend upon state of stress in the soil prior to loading. In order to study the dynamic stress-strain relationship, it will be first convenient to describe stresses and their concepts. The stresses at any point in a soil mass can be described by normal and shear stresses acting on a particular plane as shown in Figure 6.1. The stresses are forces per unit area that are transmitted through soil by inter-atomic fields. Stresses that are transmitted perpendicular to a surface are normal stresses; those that are transmitted parallel to a surface are shear stresses. A Mohr' circle describes the 2D stresses at a point in soil mass.

6.2 REPRESENTATION OF STRESS CONDITION BY MOHR' CIRCLE AND STRESS PATH

A soil element, as in Figure 6.1 referred in Cartesian coordinate system x, y and z, may be considered to be subjected to a series of cyclic shear strains or stresses that may reverse many times during an earthquake. In case of horizontal ground surface there are no shear stresses on the horizontal plane before the earthquake. During earthquake the normal stresses on this plane remain constant while cyclic shear stresses are induced during the period of shaking. Before we analyze the effects of shaking on soil element under consideration, let us revise the fundamentals of stress analysis.

Let σ_{xx} and σ_{yy} represent the normal stresses and τ_{xy} represent the shear stresses on two perpendicular faces of the soil element as shown in Figure 6.1(b). In Mohr's circle normal stresses and shear stresses are plotted as shown in the Figure 6.1(c) wherein points C and D represent (σ_{xx}, τ_{xy}) and (σ_{yy}, τ_{xy}). The orientation of principal planes and principal stresses can be obtained graphically by Mohr's circle as shown in Figure 6.2 or they are determined analytically by applying the methods of theory of elasticity, as discussed in Section 5.

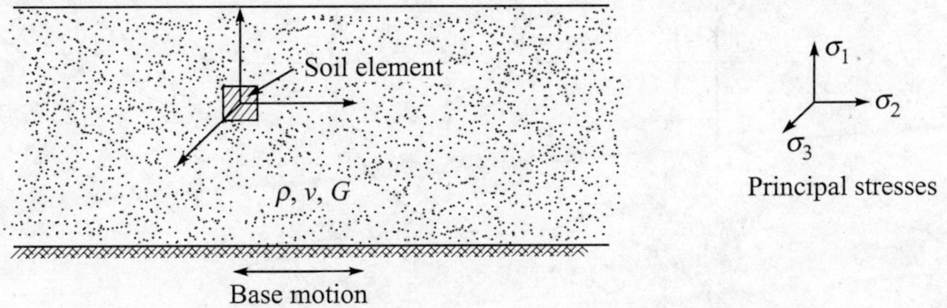

(a) Soil mass

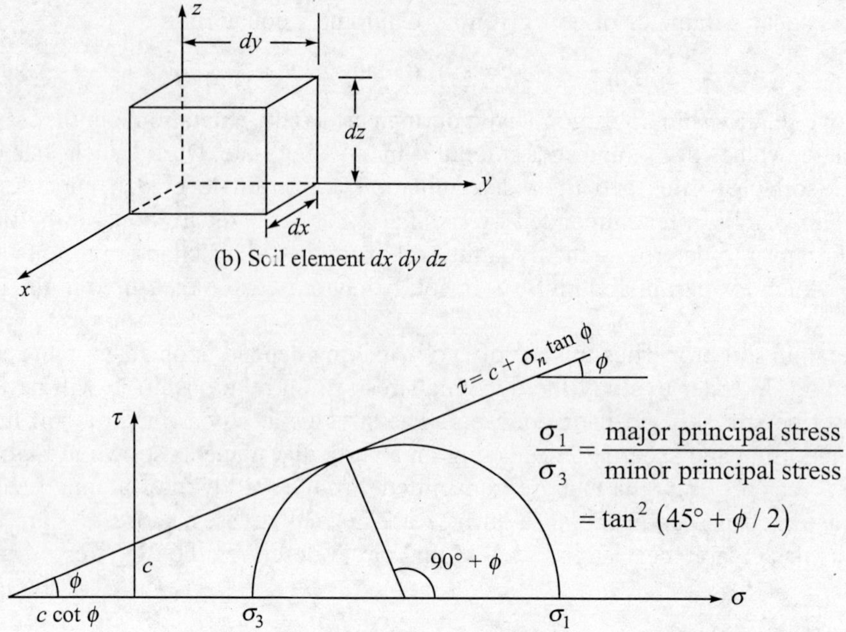

(b) Soil element $dx\ dy\ dz$

(c) Stresses in soil at limiting equilibrium

Figure 6.1 Relationship of stress in soil at limiting equilibrium.

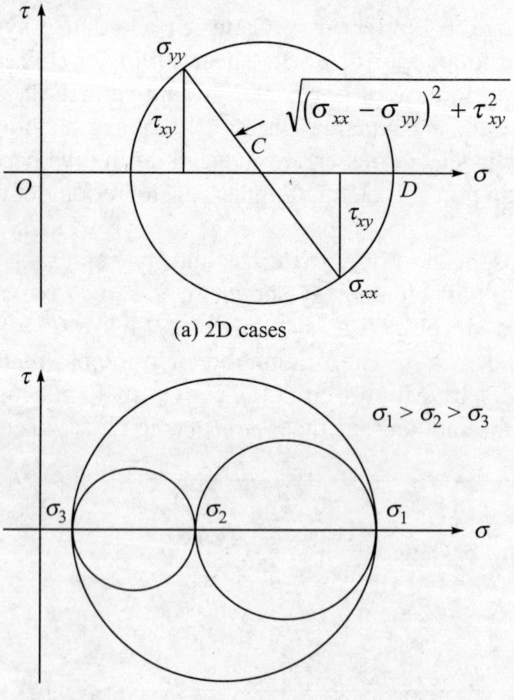

(a) 2D cases

(b) 3D cases

Figure 6.2 Mohr's circle for locating principal stresses.

According to Coulomb, the shearing strength of a soil can be expressed as a linear function of normal stress σ,

$$\tau = c + \sigma \tan \phi$$

where,

c = cohesion

ϕ = angle of internal shearing resistance or angle of internal friction.

The above relationship is a straight line in σ–τ plane similar to $y = mx + c$ representation of a straight line in x–y plane. For soil conditions having $\sigma = 0$,

$$c = \tau_0 \tag{6.2}$$

Therefore, the shearing strength may be expressed as

$$\tau = \tau_0 + \sigma \tan \phi \tag{6.3}$$

Equations (6.1) and (6.3) have the same form. According to Coulomb the initial shearing strength τ_0 is called cohesion. The shearing strength τ consists of resistances, which act during a macro-dilatancy τ_m, during which the grains in the shearing zone rise in such a way that the grains can roll over each other and a rolling friction is induced in the soil. Then there are the resistances during micro-dilatancy τ_{mm} where the grains, which slide over each other and are wedged together between protrusions, partially rise and then the protrusions are broken off. The resistance encountered in the breaking away of the grains can be designated τ_1. Further, there are the resistances caused by the speed of the deformation of the soil during shearing τ_p. Lastly, there are the resistances due to the mutual attraction of the grains τ_c. This depends on the type of the minerals, the degree of electrochemical saturation, the polarity of the adsorbed ions, the digenetic strengthening, etc. Therefore, the shearing strength may be expressed as

$$\tau = \tau_m + \tau_{mm} + \tau_1 + \tau_p + \tau_c \tag{6.4}$$

The principal stresses σ_1, σ_2 and σ_3 on principal planes of a soil element in 3D cases and σ_1 and σ_3 are shown in Figure 6.3 for 2D cases. It is interesting to recall the various practical problems of geotechnical engineering such as those of bearing capacity, active earth pressure and passive earth pressure. In these cases there is an interesting relationship between three principal stresses σ_1, σ_2 and σ_3.

Class of geotechnical problems

- ***Bearing capacity problem***— σ_1 and σ_3 remaining constant, σ_1 increases due to load of the newly constructed structure on the soil.
- ***Active earth pressure problem***— σ_1, σ_2 remaining constant, σ_3 decreases as the retaining structures moves away from the backfill to attain limiting equilibrium.
- ***Passive earth pressure problem***— σ_1, σ_2 remaining constant, σ_3 increases as the retaining structure moves towards the backfill to attain limiting equilibrium.

Stress path

Mohr circles are used to represent the state of stress such as that of bearing capacity, active earth pressure and passive earth pressure soil element. It is at times more convenient to use

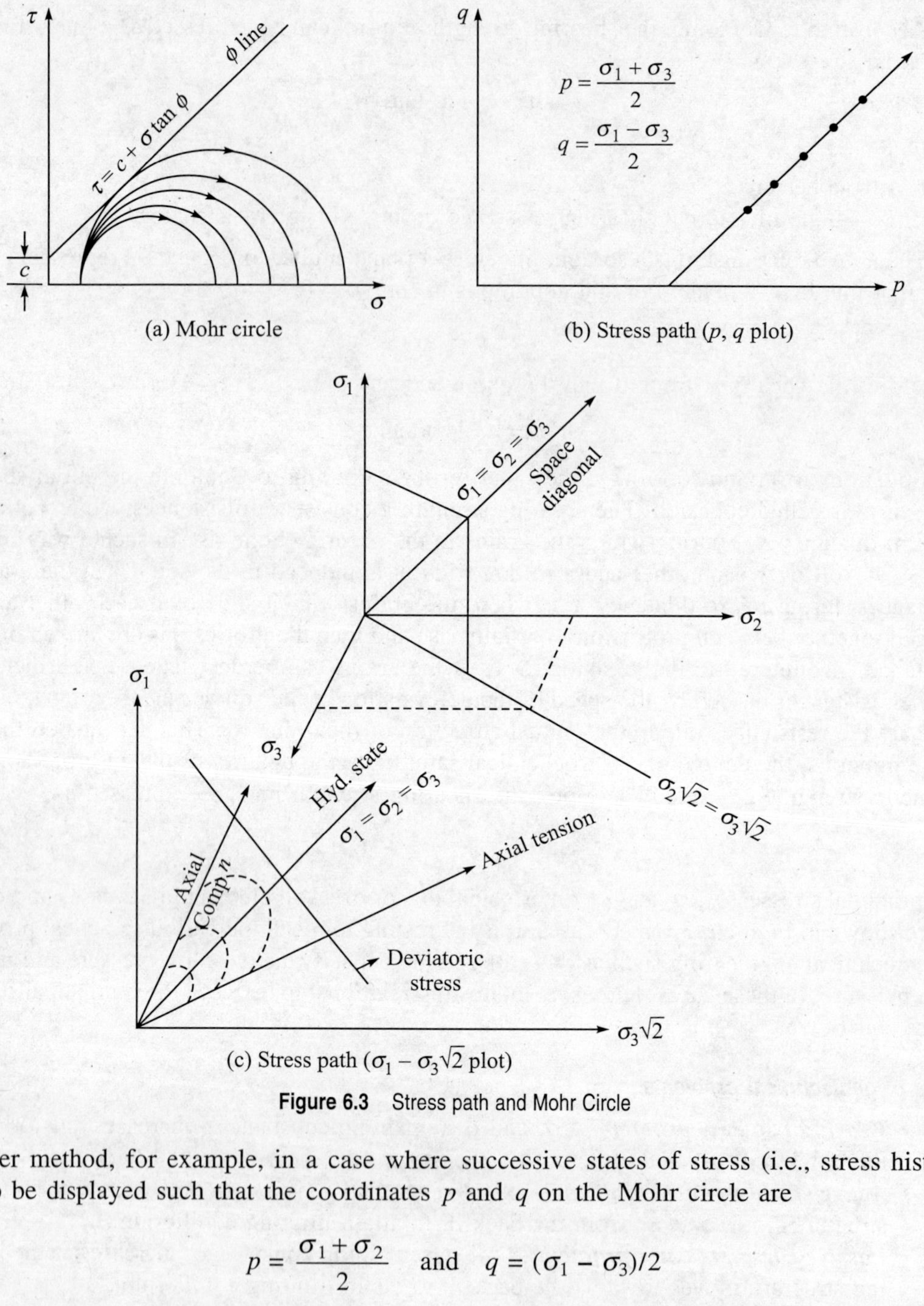

Figure 6.3 Stress path and Mohr Circle

another method, for example, in a case where successive states of stress (i.e., stress history) are to be displayed such that the coordinates p and q on the Mohr circle are

$$p = \frac{\sigma_1 + \sigma_2}{2} \quad \text{and} \quad q = (\sigma_1 - \sigma_3)/2$$

In recent past, major developments have taken place in geotechnical earthquake engineering wherein advanced mathematical methods have been used for dynamic behaviour of foundations, earth retaining structures and earth structures. Such complex analysis hopefully leads to more realistic earthquake resistant design. The design process is essentially a synthesis of the methods of analysis and experimental soil data and soil properties. Any attempt for the improvement of methods of analysis with disregard to experimentally obtained soil properties may ultimately reduce the accuracy of the results. Thus, the measurement of dynamic soil properties is very important and deserves high degree of accuracy for seeking various geotechnical earthquake engineering problems. Field as well as laboratory tests are required for the measurement of dynamic soil properties with a motive to replicate the ground conditions of a real problem. Soil tests are strain level dependent. The choice of test is dependent on low strain phenomenon and high strain phenomenon of soil behaviour. Table 6.4 as suggested by Ishihara (1971) is a good guide. The soils may be tested under monotonically loading conditions or cyclically loading conditions.

6.3 DYNAMIC STRESS-STRAIN RELATIONSHIP

Figure 6.4(a) shows a typical shear stress-strain curve of soil subjected to cyclic loading. The linear shear modulus G is a function of the cyclic strain γ_c and is described in terms of a back-bone curve. The maximum value of shear modulus $G_{\max}$ represents the tangent stiffness at low strain (less than $10^{-4}\%$). As far as energy dissipation is concerned, it is expressed in terms of equivalent damping ratio η, which can be expressed as

$$\eta = \frac{1}{4\pi} \cdot \frac{E_D}{E_{so}} = \frac{1}{2\pi} \cdot \frac{E_D}{G\gamma_c^2} \tag{6.5}$$

where,

E_D = energy dissipated within the hysteresis loop

E_{so} = strain energy stored in soil at a cyclic strain of γ_c.

The determination of G from the stress-strain response of the laboratory specimen is at times difficult, and then in that case G is obtained from the shear wave velocity measurements either in the field or in the laboratory specimen using the relationship

$$G_{\max} = V_S^2 \rho \tag{6.6}$$

where,

ρ = mass density of soil in kg s^2/m^2

V_s = velocity of shear waves, in m/s.

Figure 6.4(b) shows the variation of modulus ratio $G/G_{\max}$ with shear strain and the resulting curve is known as modulus reduction curve. This curve presents the same information as the backbone curve as shown in Figure 6.4(a), whereas Figure 6.4(c) shows the variation of the damping ratio η with shear strain.

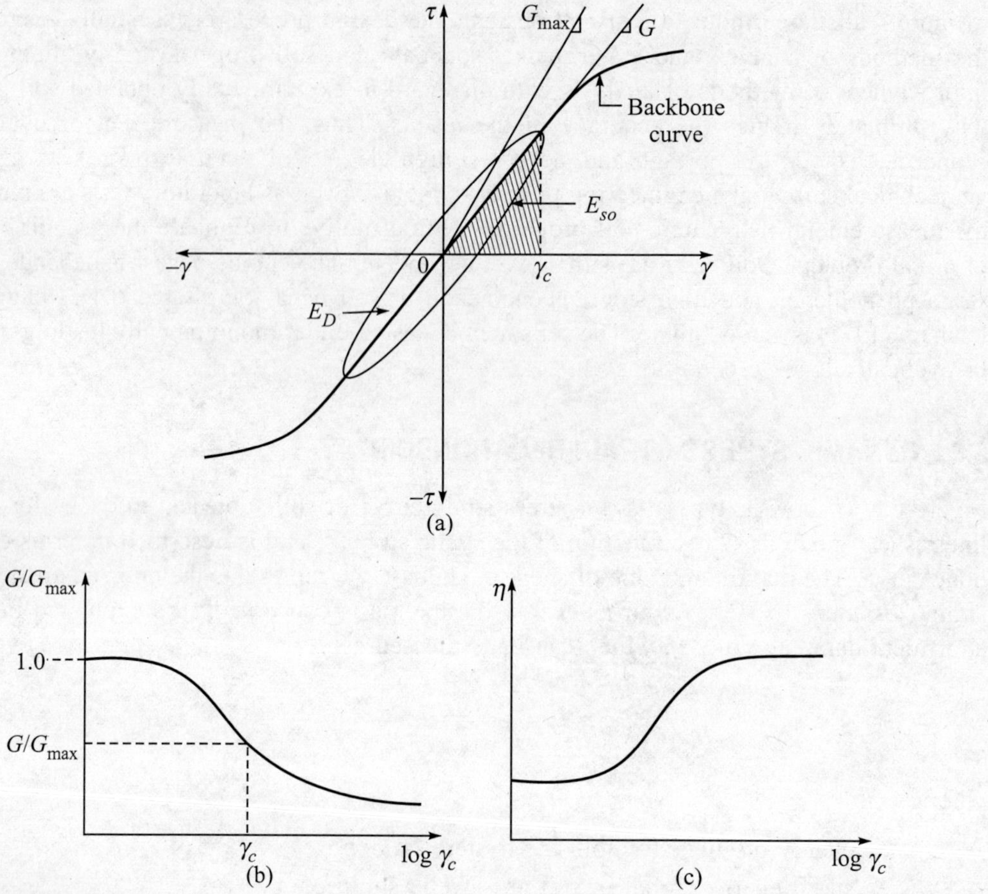

Figure 6.4 (a) Dynamic stress–strain relationship of soil, (b) Modulus reduction curve, (c) Damping curve.

6.4 DETERMINATION OF DYNAMIC SOIL PROPERTIES

The various dynamic soil properties may be determined in terms of the following parameters:

Dynamic elastic coefficient

- Coefficient of elastic uniform shear
- Coefficient of elastic uniform compression
- Coefficient of non-uniform shear
- Coefficient of non-uniform compression

Dynamic modulus

- Young's modulus
- Shear modulus
- Bulk modulus

Liquefaction parameters

- Cyclic stress ratio
- Dynamic pore pressure response

Velocity of wave propagation

- Velocity of P waves
- Velocity of S waves
- Velocity of R waves

Poisson's ratio

Damping ratio

Since these dynamic properties are very sensitive to values of strain, they are classified into large strain field and small strain field as listed in Table 6.4.

6.4.1 Field Tests

Seismic reflection test

This test is based on reflection of seismic waves in elastic medium.

The test in the field is performed by providing an impulse at A the source, and the arrival time is recorded at B which is the receiver end. In general, the stress wave produced by the impact at the source A radiate with hemispherical wave front as shown in the Figure 6.5(a). However, some of the wave energy follows the linear path $\overrightarrow{AB}$ as shown in Figure 6.5(a). Let us assume a horizontal reflecting interface with a depth H below the earth's surface. The seismic velocity above the interface is V_1 and that below it is V_2.

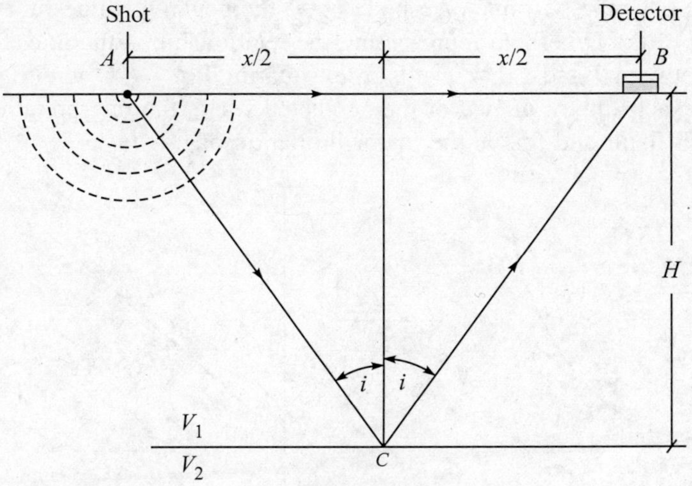

Figure 6.5(a) Reflection of seismic waves.

The time of travel from A to B is expressed

$$T_{AB} = \frac{x}{V_1}$$

By measuring the time T_{AB} and the distance x the primary wave velocity of the upper layer V_1 can be determined.

The total path length L is related to x and H as

$$L = 2\sqrt{H^2 + \left(\frac{x}{2}\right)^2} = V_1 T \quad (6.7)$$

For a horizontal reflector having H constant, it is evident that the relation of T to x is hyperbolic. Similarly,

$$H = \frac{1}{2}\sqrt{(V_1 T)^2 - x^2}$$

Some of the waves travel along AC and get reflected towards the ground and arrive at the receiver B. In such a case the angle of incidence i is obviously

$$i = \tan^{-1}\frac{x}{2H}$$

The distance of travel = $AC + CB = 2\sqrt{\left(\frac{x}{2}\right)^2 + H^2}$

The time of travel $\quad T = \sqrt{\dfrac{4H^2 + x^2}{V_1}} \quad (6.8)$

Figure 6.5(b) and Figure 6.5(c) show the relation between travel time and horizontal distances. Figure 6.5(b), T versus x curve, is a symmetrical one, as it holds for negative as well as positive values of x, the portion corresponding to the negative values not being shown. The axis of symmetry of the hyperbola representing the relationship is the line $x = 0$. Figure 6.5(c) shows linearity between T and x that results from squaring Eq. 6.7. Often it is difficult to record and identify the exact time of arrival of the neglected wave, the interpretation of result in this method is often difficult and this is the major limitation of this test.

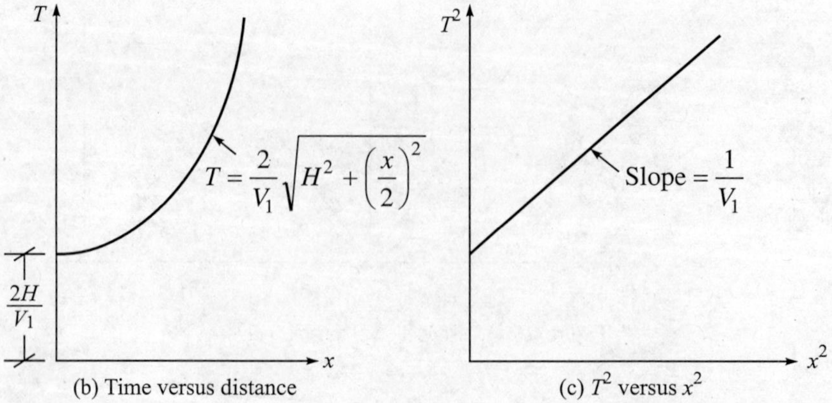

Figure 6.5(b) and (c) Reflected seismic wave.

Seismic refraction test

In this method shock waves are created into the soil, at ground level or at a certain depth below it, by striking a plate on the soil with a hammer or by exploding small charges in the soil. The radiating shock waves are picked up by the vibration detector (geophone), where the time of travel gets recorded. Either a number of geophones are arranged in a line, or the shock-producing device is moved away from the geophone to produce shock waves at given intervals. Some of the waves, known as direct or primary waves, travel directly from the shock point along the ground surface and are picked up first by the geophone. If the subsoil consists of two or more distinct layers, some of the primary waves travel downwards to the lower layer and get refracted at its surface. If the underlying layer is denser, the refracted waves travel much faster. As the distance between the shock point and the geophone increases, the refracted waves reach the geophone earlier than the direct waves.

Consider a subsurface consisting of two media, each with uniform elastic properties, the upper layer separated from the lower one by horizontal interface at a depth H as shown in Figure 6.6. The velocity of seismic waves in the upper layer is V_1 and in the lower $V_2 > V_1$. A seismic wave is generated at point S on the surface, and the energy travels out from it in the hemispherical wave fronts. A receiving instrument is located at a point D at a distance x from S.

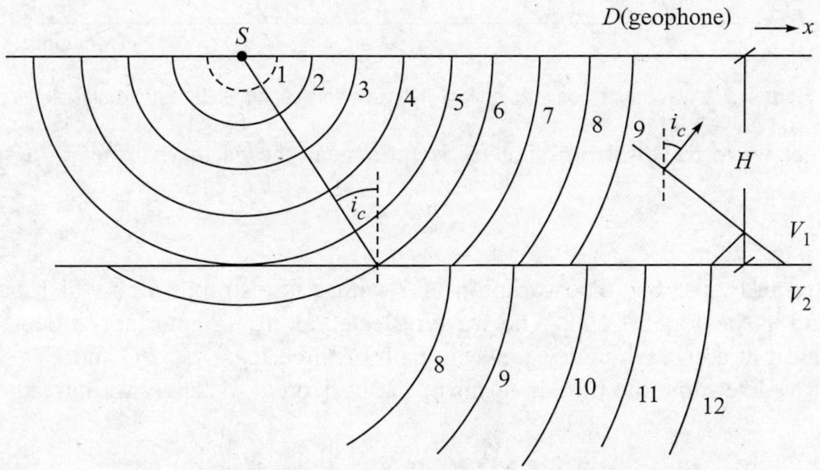

Figure 6.6 Mechanism for transmission of refracted waves in two-layered earth [After Dix, 1939].

The waves handled by the refraction method are mostly P waves. If we consider only the P waves for simplicity, then assuming that waves arrive from medium I at the boundary with an angle of incidence i_1, we can use Snell's law to ascertain the refraction angle i_2 as follows:

$$\frac{\sin i_1}{\sin i_2} = \frac{V_1}{V_2} \tag{6.9}$$

Here V_1, V_2 are the propagation velocities of P waves in media I, II, respectively. If we denote the angle of incidence i_1 as i_c, then for refraction angle $i_2 = 90°$, we get the following expression using Eq. (6.9). This i_c is called the critical angle.

$$\sin i_c = \frac{V_1}{V_2} \tag{6.10}$$

Accordingly, the wave arriving at the critical angle propagates through the boundary layer at velocity V_2, part of it being refracted again at critical angle i_c and thus returning to medium I (velocity V_1).

In the refraction method, the time at which the wave originating from the source passes through each point of observation is measured. The travel-time curve is drawn and the ground structure is analyzed from this curve.

Figure 6.7 presents a two media case wherein the time-distance relationship is to be determined with respective speeds of V_1 and V_2 in these media, separated by a horizontal discontinuity at depth H.

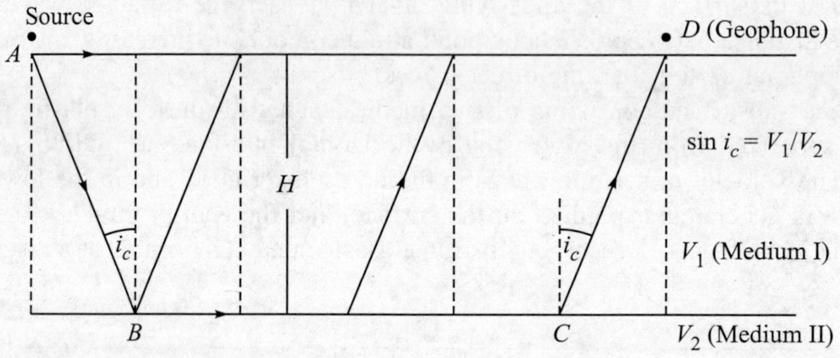

Figure 6.7 Wave refraction ray path for two-layer media separated by a horizontal interface.

The direct wave travels from shot to detector near the earth's surface at a speed V_1. So,

$$T = \frac{x}{V_1} \tag{6.11}$$

as represented in Figure 6.7. The variation of T with x is a straight line which passes through the origin and has a slope of $1/V_1$. The wave reflected along the interface at depth H, reaching it and leaving it at a critical angle i_c takes a path of three legs AB, BC and CD.

In order to determine the time in terms of the horizontal distance, we have the relationship

$$\sin i_c = \frac{V_1}{V_2}, \quad \cos i_c = \left(1 - \frac{V_1^2}{V_2^2}\right)^{1/2}$$

and,

$$\tan i_c = \frac{\sin i_c}{\cos i_c} = \frac{V_1}{\sqrt{V_2^2 - V_1^2}}$$

The total time along the refraction path $ABCD$ is

$$\begin{aligned} T &= T_{AB} + T_{BC} + T_{CD} \\ &= \frac{H}{V_1 \cos i_c} + \frac{(x - 2H)\tan i_c}{V_2} + \frac{H}{V_1 \cos i_c} \\ &= \frac{2H}{V_1 \cos i_c} - \frac{2H \sin i_c}{V_2 \cos i_c} + \frac{x}{V_2} \end{aligned}$$

Further simplifying

$$T = \frac{2H}{V_1 \cos i_c}(1 - \sin^2 i_c) + \frac{x}{V_2} = \frac{x}{V_2} + \frac{2H \cos i_c}{V_1}$$

$$= \frac{x}{V_2} + 2H\frac{\sqrt{1-(V_1/V_2)^2}}{V_1}$$

and finally,

$$T = \frac{x}{V_2} + \frac{2H\sqrt{V_2^2 - V_1^2}}{V_1 V_2} \tag{6.12}$$

In Figure 6.8, the plot of T versus x, represents a straight line with a slope of $1/V_2$, and which intercepts the time axis at $x = 0$ at a time

$$T_i = 2H\frac{\sqrt{V_2^2 - V_1^2}}{V_1 V_2} \tag{6.13}$$

In Figure 6.8, at a distance X_{cros} the two linear segments cross each other. At a distance less than this, the direct wave travelling at V_1 along top of the layer reaches the detector first. At distances greater than X_{cros}, the wave refracted by the interface reached the detector before the direct wave. Therefore, X_{cros} is called the crossover distance. It can be shown that

$$H = \frac{1}{2}\sqrt{\frac{V_2 - V_1}{V_1 + V_2}} \cdot X_{cros} \tag{6.14}$$

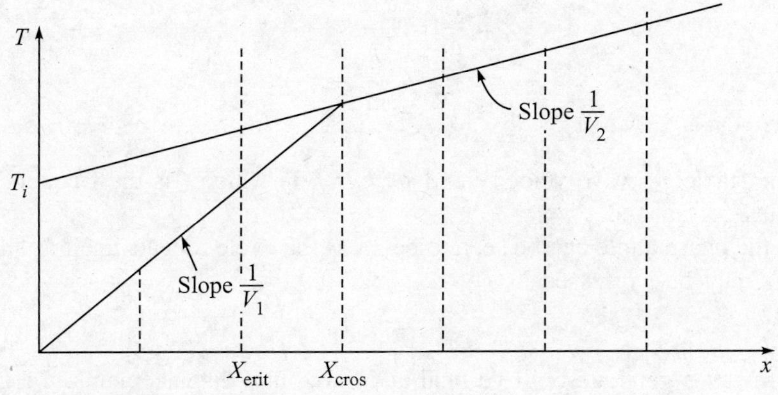

Figure 6.8 Time-distance curve for two-layered surface [refraction test]

Seismic surface wave (Rayleigh wave test)

Vertical oscillating excitations at the surface of an elastic half-space generate Rayleigh waves besides other body waves. As the majority of the energy is with Rayleigh waves, the determination of velocity of propagation of Rayleigh waves becomes very important. The vertical displacement on the surface of half-space at any instance due to harmonic (say sinusoidal) excitations is also harmonic as shown in Figure 6.9. As such the input source of the form $R \sin \omega t$ can generate vertical displacement on the surface such that the distance covered in one cycle is the Rayleigh wavelength λ_R.

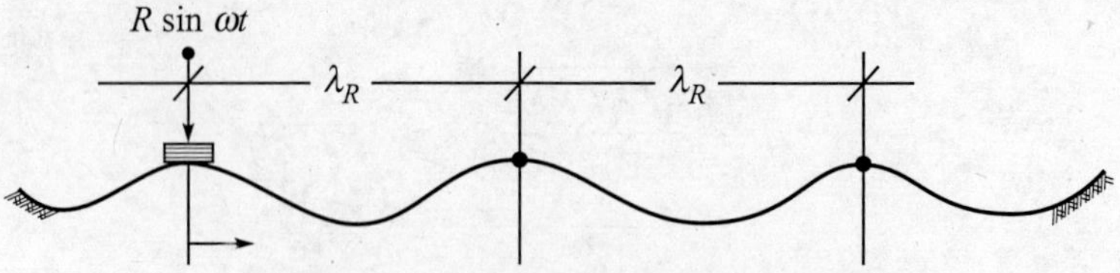

Figure 6.9 Rayleigh waves on half-space surface [After Richart et al. 1970].

The vertical displacement at the source may be expressed as

$$w_S = R_1 \sin \omega t \tag{6.15}$$

where ω is the frequency of the input source.

At a point on the surface at a distance r from the source, the displacement can again be written as

$$\begin{aligned} w_R &= R_2 \sin(\omega t - \phi) \\ &= R_2 \sin \omega\left(\frac{t-\phi}{\omega}\right) \end{aligned} \tag{6.16}$$

where ϕ = phase angle.

Thus, the time tag t_r of arrival of the Rayleigh wave at the point R,

$$t_r = \frac{\phi}{\omega} = \frac{r}{V_R} \tag{6.17}$$

$$\phi = \frac{\omega r}{\lambda_R} = \frac{2\pi f r}{V_R} \tag{6.18}$$

where V_R is the Rayleigh wave velocity and $\omega = 2\pi f$, in which f is the frequency of the input source in cycles per second.

At $r = \lambda_R$, the phase angle can be seen to be 2π as the cycle repeats itself as shown in Figure 6.9. Hence, V_R may be expressed as

$$V_R = f.\lambda_R$$

Since Rayleigh waves generate both vertical and horizontal displacements, the ground surface will, for sinusoidal loading frequency, be distorted as shown in exaggerated form in Figure 6.9. By placing a receiver at the centre of loading and moving another receiver to points at different distances from the receiver the locations of points vibrating in phase can be determined. The horizontal distances between such points are equal to the wavelength of the Rayleigh wave. By measuring the Rayleigh wavelength, the Rayleigh wave velocity V_R can be calculated. Richart et al. (1970) have observed that the measured phase velocity corresponds to the soil properties at a depth from about $\lambda_R/2$ to $\lambda_R/3$.

By varying the loading frequency ω in the field, the variation of wave velocity with depth can be estimated.

Wave propagation tests for determination of shear modulus

The wave propagation tests for the determination of shear modulus may be conducted by making seismic waves to pass through the ground by the impact of a hammer and determining the time of travel of these waves between two points at a known distance apart or by measuring the phase difference between vibrations at two points under steady vibrations.

In case of uniform soil extending up to infinite depth, the wavelength of the propagating vibrations is given by

$$\frac{\lambda}{4} = \frac{\pi S}{\pi + 2(\lambda_1 - \lambda_2)} \tag{6.19}$$

where the geophones have the same characteristics, that is, $\lambda_1 = \lambda_2$. Therefore,

$$\frac{\lambda}{4} = S$$

where,

λ = wavelength in cm
S = measured distance between geophones in cm
λ_1 = phase shift of geophones with respect to wave nearer to concrete block at the frequency of the propagating vibrations in radians
λ_2 = phase shift of the other geophone at the frequency of the propagating vibrations in radians.

Velocity of shear waves V_s is given by

$$V_s = \lambda f$$

where f is the frequency of vibration at which the wavelength has been measured.

When the test is conducted using a phase meter, the phase angle corresponding to different distances between the geophones should be recorded and a curve plotted between the phase angle and the distance. From the curve, the distance S between the geophones for a phase difference of 90 degree should be determined.

Testing procedure

The equipment required comprises a hammer to impart impact to the ground, a geophone or velocity pick-up or a time marking device to record the time of impact, an acceleration pick-up (or a geophone) to monitor the time of arrival of waves, a universal amplifier, an ink-writing oscilloscope or a timer capable of measuring time interval up to a precision of 10 seconds and a steel measuring tape. IS 5249:1992 recommendations should be followed.

A suitable location in the area where this test is to be conducted is selected and radial lines are ranged out from this point for a distance of 30 m to 40 m. Points are marked on these lines at 2 m intervals. A velocity pick-up or a geophone is fixed at the origin of the radial lines and waves are generated near this point by the impact of a 10 kg hammer falling through a height of 2 m on a steel plate of 150 mm × 150 mm resting on the surface of ground. An acceleration pick-up is placed at a known distance along one of the radial lines, the pick-ups are amplified through the universal amplifier and fed to two channels of the same pen recorder. The time taken by the waves to travel the distance between the two pick-ups can be obtained from these records. The test is repeated for different known distances between the pick-ups along all the marked lines one by one.

The test may be repeated at different locations to obtain a representative value of wave velocities in the area under investigation. Alternatively, the time taken by the waves to travel a known distance may be obtained directly by feeding the output of the pick-ups to a timer.

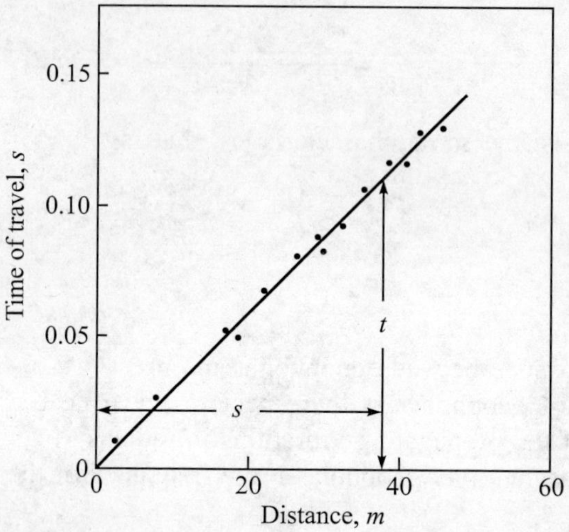

Figure 6.10 Determination of average wave velocity of stress wave propagation in soil medium (hammer test)

The values of travel time of compression waves and the corresponding distance along each selected line at a location are plotted. A straight line is fitted through these points as shown in Figure 7.10. The value of average velocity is obtained as:

$$V_s = \frac{s}{t}$$

where,

V_s = velocity of compression waves, in m/s
s = distance, in m
t = corresponding time of travel of waves in s.

Determination of elastic modulus and shear modulus of soil

The elastic modulus E is determined by the equation

$$E = V_{\text{P}}^2 \rho \; \frac{[(1+\nu)(1-2\nu)]}{1-\nu} \tag{6.20}$$

where,

V_{P} = velocity of P waves
ρ = mass density of soil
ν = Poisson's ratio of soil.

The following values for Poisson's ratio may be used:

Table 6.1

Type of soil	ν
Clay	0.5
Sand	0.30 to 0.35
Rock	0.15 to 0.25

Depending upon the nature of the medium involved, and if the distance between pick-ups is sufficiently large, both the arrival of compression and shear waves may be distinguishable from the records. In such a case both E and G can be determined independently.

$$E = 2G(1 + \nu) \tag{6.21}$$

$$G = V_s^2 \rho \tag{6.22}$$

where,

ρ = mass density of soil, in kg s^2/m^2
V_s = velocity of shear waves, in m/s
ν = Poisson's ratio of soil.

As the strain level of the propagating shear wave is low ($< 10^{-4}$%) and the elastic wave velocity is measured, it should be noted that the dynamic (small strain) shear modulus decreases with increasing strain level and can thus not be directly converted into a static (large strain) modulus value.

The values of E and G can also be obtained from the values of C_u obtained from other field tests. Alternatively, the values of C_u can be determined from E and G values obtained in wave propagation tests.

Block vibration test

A test pit of suitable size depending upon the size of the block should be made. For block size, the size of the pit may be 3 m × 6 m at the bottom and a depth preferably equal to the proposed depth of foundations. The test should be conducted above the groundwater table. In case of rock, the test may be performed on the surface of rock bed itself. The bottom of the pit should be at stable slope and may be kept vertical where possible.

Test block

A plain cement concrete block of M-15 concrete should be constructed in the test pit as shown in Figure 6.11. The size of the block should be selected depending upon the subsoil conditions. In ordinary soils it may be 1 m × 1 m × 1.5 m and in dense soils it may be 0.75 m × 0.75 m × 1 m. In boulder deposits the height may be increased suitably. The block size should be so adjusted that the mass ratio

$$B_z = \left[\left\{\frac{(1-\nu)}{4} \times \frac{m}{\rho r_0^3}\right\}\right]$$

is always more than unity. The concrete block should be cured for at least 15 days before testing. Foundation bolts should be embedded into the concrete block at the time of testing for fixing the oscillator assembly. Details of the test block are shown in Figure 6.11.

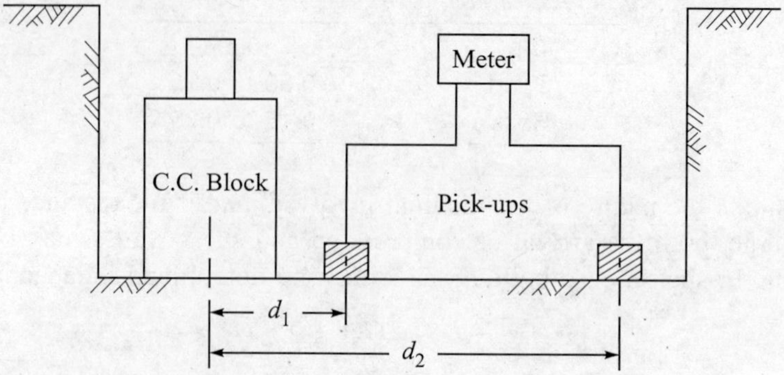

Figure 6.11 Set-up for block vibration test.

Test set-up

The vibration exciter should be fixed on the concrete block and suitable connections between power supplies, speed control unit should be made as shown in Figure 6.12. Any suitable electronic instrumentation may be used to measure the frequency and amplitude of vibrations.

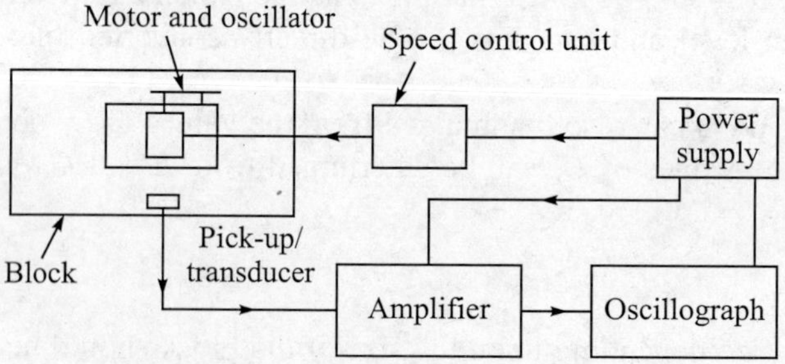

Figure 6.12 Block diagram of testing equipment for block vibration test.

Forced vibration test in vertical vibration mode

The vibration pick-up should be fixed at the top of the block as shown in Figure 6.11, such that it senses vertical motion of the block. The vibration exciter should be mounted on the block such that it generates purely vertical sinusoidal vibrations and the line of action of the vibrating force passes through the centre of gravity of the block. The exciter is operated at a constant frequency. The signals of the vibration pick-ups are fed into suitable electronic circuitry to measure the frequency and the amplitude of vibrations. The frequency of the exciter is increased in steps of small values, (1 – 4 cycles/s) up to the maximum frequency of the exciter and the signals measured. The same procedure should be repeated if necessary for different excitation levels. The dynamic force should never exceed 20 per cent of the total mass of the block and exciter assembly.

The amplitude versus frequency curve should be plotted for each excitation level to obtain the natural frequency of the soil, and the foundation block tested. A typical plot is shown in Figure 6.13.

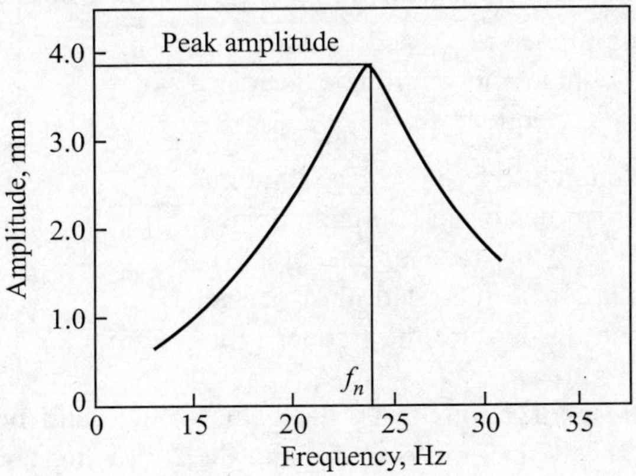

Figure 6.13 Typical amplitude versus frequency curve from vertical vibration test.

Determination of coefficient of elastic uniform compression of soil

The coefficient of the elastic uniform compression C_u of soil is given by the following equation:

$$C_u = \left(4\pi^2 f_{nz}^2 \frac{M}{A}\right) \tag{6.23}$$

where,

f_{nz} = natural frequency
M = mass of the block, exciter and motor
A = contact area of the block with the soil.

From the value of C_u obtained for the test block of contact area A, the value of C_{u_1} for the foundation having contact area A_1 may be obtained from the equation:

$$C_{u_1} = C_u \sqrt{\frac{A}{A_1}} \tag{6.24}$$

This relation is valid for small variations in base area of the foundations and may be used for areas up to 10 m^2. For actual foundation areas larger than 10 m^2, the value of C_u obtained for 10 m^2 may be used.

Determination of damping coefficient of soil

The damping coefficient of soil can be measured from the forced vibration test as well as from the free vibration test. In case of the forced vertical vibration test, the value of the damping coefficient ξ of soil is given by the following equation:

$$\xi = \frac{f_2 - f_1}{2 f_{nz}} \tag{6.25}$$

where,

ξ = damping coefficient of soil

f_2, f_1 = two frequencies at which the amplitude is equal to $X_m/\sqrt{2}$ as shown in Figure 6.14

X_m = maximum amplitude

f_{nz} = frequency at which the amplitude is maximum (resonant frequency).

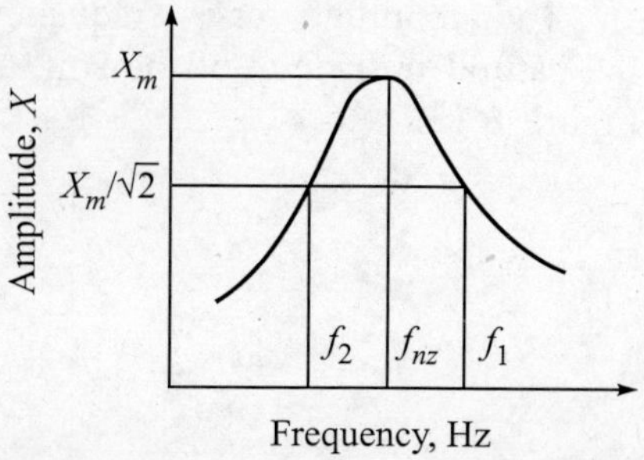

Figure 6.14 Determination of damping of soil (forced vibration test).

The block shall be excited into free vertical vibrations by the impact of sledge hammer or any suitable device, as near to the centre of the top face of the block as possible. The vibrations shall be recorded on a pen recorder or a suitable device to measure the frequency and amplitude of vibration. The test may be repeated three or four times.

In case of free vertical vibrations tests, the value of ξ_u shall be obtained from the free vibration test curves as shown in Figure 6.15 using the following equation:

$$\xi = \frac{1}{2\pi} \log_e \frac{X_m}{X_{m+1}} \tag{6.26}$$

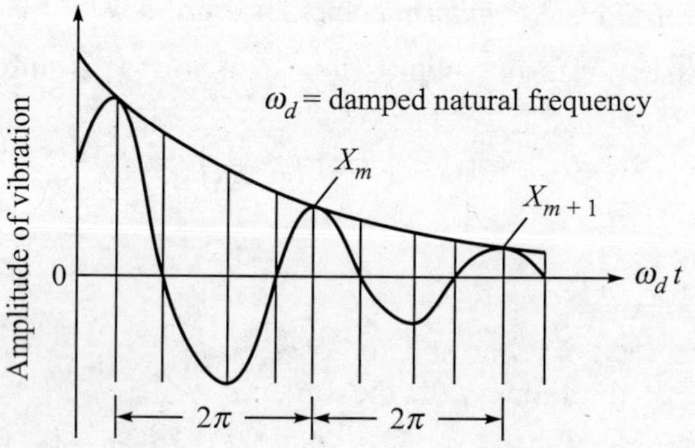

Figure 6.15 Determination of damping of soil (free vibration test).

Evaluation of coefficient of attenuation

The test set-up is same as that for the block resonance test. The pick-up fitted on the block is removed and installed at a certain distance d_1 (approximately 30 cm) from the block. The second pick-up from the centre of the block is at a distance of d_2. The amplitudes of vibration at these two locations are measured for different frequencies.

The coefficient of attenuation is calculated from the following expression:

$$A_2 = A_1\sqrt{\frac{d_1}{d_2}}\; e^{-\alpha\,(d_2 - d_1)} \tag{6.27}$$

where,

A_2 = amplitude at distance d_2

A_1 = amplitude at distance d_1

α = coefficient of attenuation (see Table 6.2).

Table 6.2 Typical values of α

Soil type	α, m^{-1}
Saturated sand or sandy silt	0.1
Saturated silty sand	0.04
Saturated sandy silty clay	0.04–0.12

Cyclic plate load test

A suitable arrangement for providing reaction of adequate magnitude depending upon the size of plate employed should be used. The load mechanism should have the facility to apply and remove the loads quickly. A hydraulic jack or any other suitable equipment may be used.

Test procedure

The equipment for the test shall be assembled according to the details given in IS 1888: 1982. The plate shall be located in a pit at a depth equal to the depth of the proposed foundation excavated as given in IS 1888: 1982.

After the set-up has been arranged the initial readings of the dial gauges should be noted and the first increment of the static load should be applied to the plate. This load shall be maintained constant throughout for a period till no further settlement occurs or the rate of settlement becomes negligible. The final readings of the dial gauges should then be recorded. The entire load is then removed quickly but gradually and the plate allowed rebound. When no further rebound occurs or the rate of rebound becomes negligible, the readings of the dial gauges should be again noted. The load shall then be increased gradually till its magnitude acquires a value equal to the proposed next higher stage of loading, which shall be maintained constant and the final dial gauge readings should be noted as mentioned earlier. The entire load should then be reduced to zero and final dial gauge reading recorded when the rate of rebound becomes negligible.

The cycles of loading, unloading and reloading are continued till the estimated ultimate load has been reached, the final values of dial gauge readings being noted each time.

The magnitude of the load increment should be such that the ultimate load is reached in five to six increments. The initial loading and unloading cycles up to the safe bearing capacity of the soil should be with smaller increments in load. The duration of each loading and unloading cycle depends upon the type of soil under investigation.

Coefficient of elastic uniform compression from cyclic plate load test

From the data obtained during the cyclic plate load test, the elastic rebound of the plate corresponding to each intensity of loading shall be obtained as shown in Figure 6.16. The load intensity versus elastic rebound shall be plotted as shown in Figure 6.17.

The value of C_u shall be calculated from the equation given below:

$$C_u = \frac{P}{S_e} \frac{\text{kgf}}{\text{cm}^3} \tag{6.28}$$

where,

P = corresponding load intensity, in kg/cm^2

S_e = elastic rebound corresponding to P in cm.

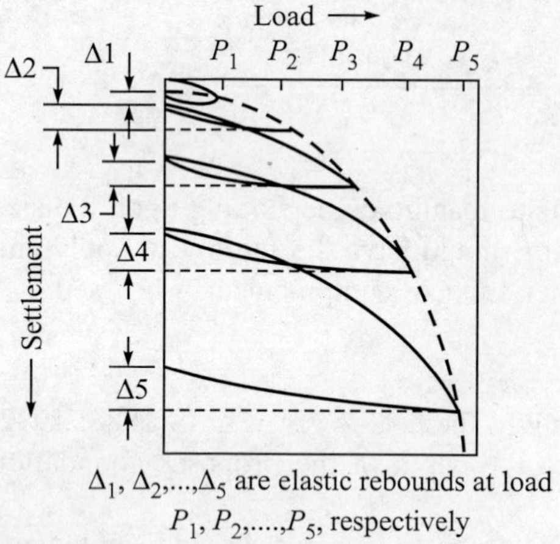

Figure 6.16 Load settlement curve for cyclic plate load test.

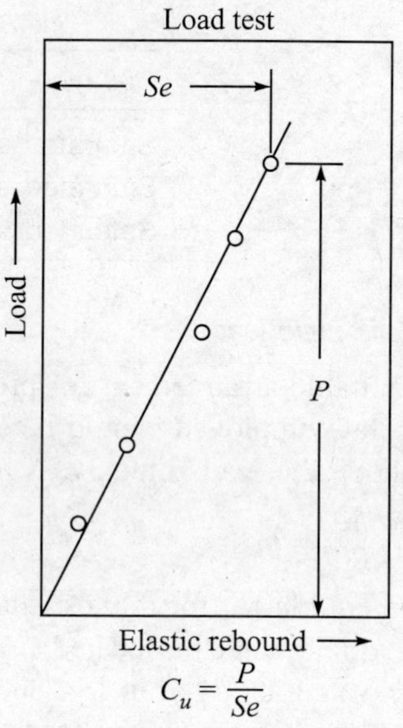

Figure 6.17 Method for obtaining value of C_u (cyclic plate load test).

Seismic Cross-Hole Test

Seismic cross-hole test can be performed in the true field using two or more boreholes to measure wave propagation velocities along horizontal paths. The two bore test as shown in Figure (6.18) is more suitable in constrained areas, where free space is restricted.

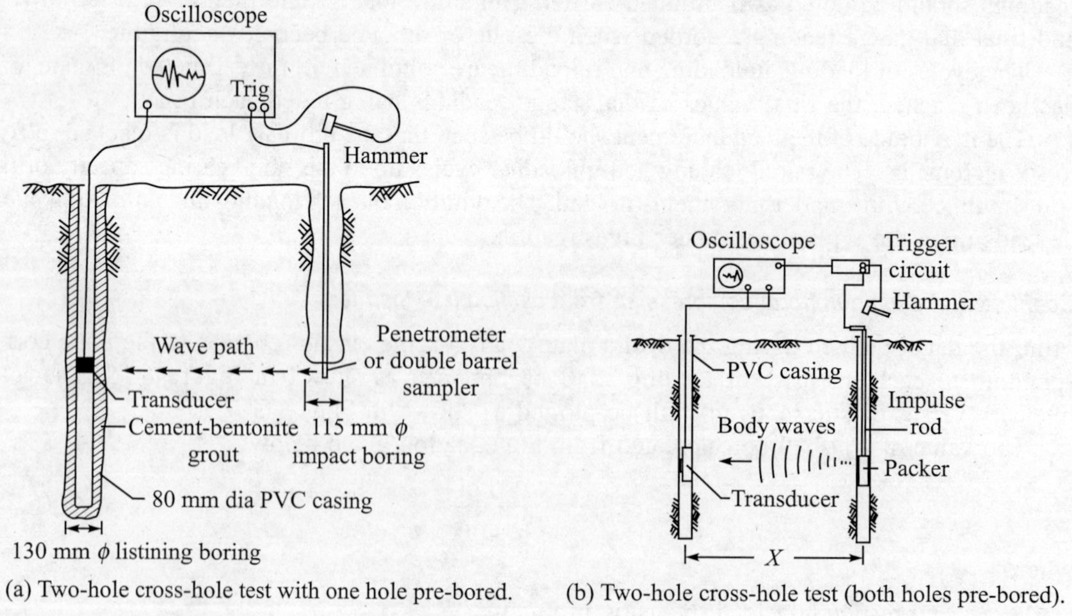

Figure 6.18 Cross-hole test.

In this, two boreholes are drilled simultaneously to the desired depth where test is being carried out. By fixing both the source and the receiver at the same depth in each borehole, the wave propagation velocity of the material between the boreholes at that depth is measured. The procedure requires accurately calibrated and oriented receivers that are well-coupled to the borehole wall. During this test the shear wave is assumed to travel horizontally through the soil to the vertical motion sensor in the second hole. The time required for a shear wave to traverse this known distance is measured. Velocity transducers (geophones) that have natural frequencies of 4 to 15 Hz are adequate for detecting (receiving) the shear waves as they arrive from the source.

Three cross-hole test

In some situations more than two boreholes are desirable to minimize possible inaccuracies. The need for precise triggering can be eliminated by using three boreholes. Figure 6.19 shows the arrangement for the three-borehole techniques using packing hammer and in-hole hammer.

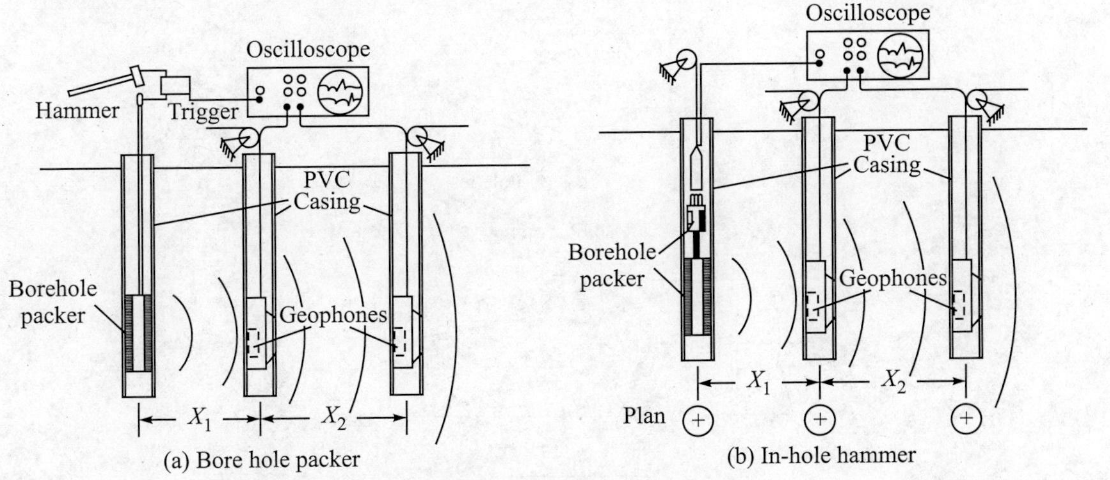

Figure 6.19 Three cross-hole test

The interval for the waves to travel between the two boreholes can be used as the travel time, and as long as both the receivers trace are displayed on the oscilloscope screen, the travel time can be determined. If triggering is also available, it is possible to make use of three different travel paths for the determination of travel time between source and near receiver, source and far receiver, and near and far receivers. It is possible to conduct the test, by preboring and casing as in the two-borehole techniques.

Up-hole or down-hole wave-propagation test

Up-hole and down-hole tests can be conducted by using one-bore hole as shown in Figure 6.20. In the up-hole method, the sensor is placed at the surface and shear waves are generated at various depths within the borehole. Similarity in the down hole method, the excitation is applied at the surface and one or more sensors are placed at different depths within the hole. In both the tests the average value of wave velocities are determined. The economy of reducing the

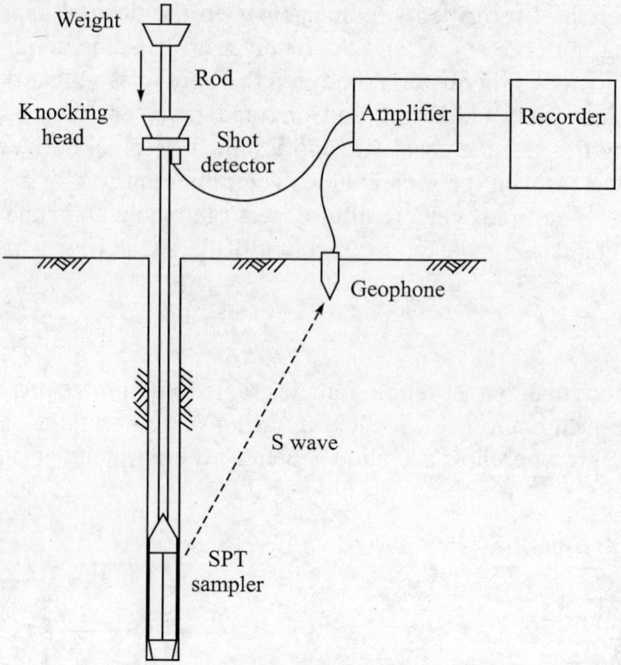

(a) Cross up-hole test.

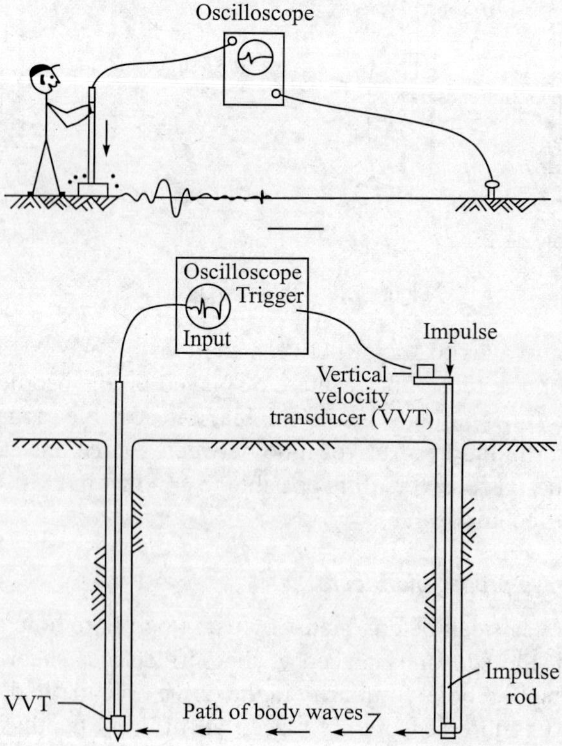

(b)

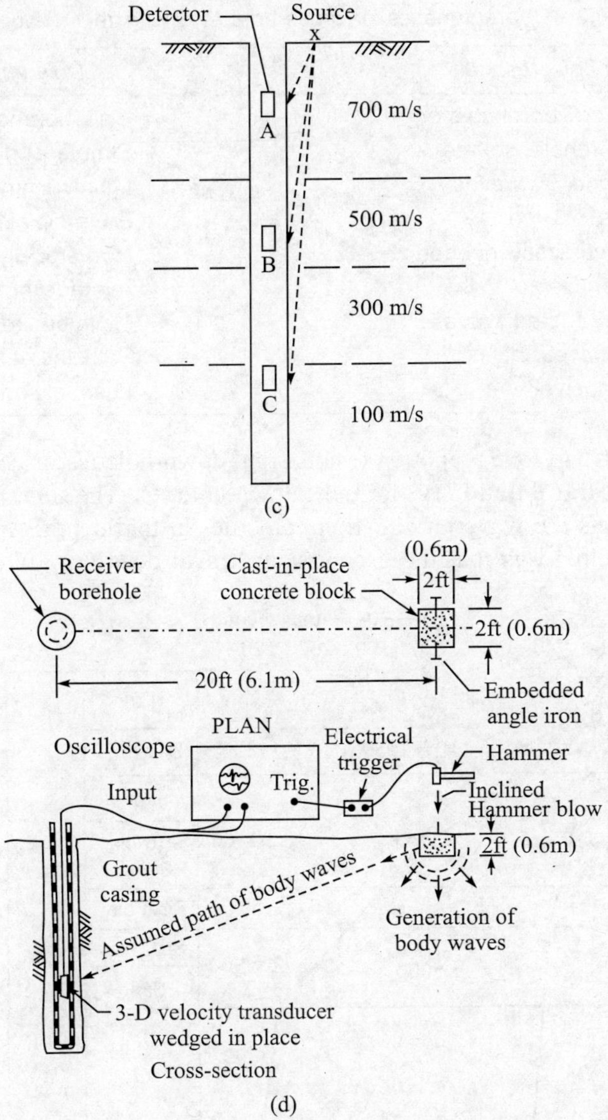

Figure 6.20 Down-hole up-hole test.

number of boreholes can be achieved by performing the down-hole technique. The arrangement for the down-hole test is shown in Figures 6.20(b) and 6.20(c). S-wave can be generated much more easily in the down-hole test than in the up-hole test; consequently, the down-hole test is more commonly used. With an SH-wave, the down-hole test measures the velocities of waves similar to those that carry most seismic energy to the ground surface.

More satisfactory down-hole data can be obtained by joining two or more receivers together at a known spacing in the borehole and obtaining the travel times between them. Table 6.3 shows the comparison of down-hole test with the cross-hole test.

Table 6.3 Characteristics of cross-hole test with the down-hole test

Cross-hole test	*Down-hole test*
Two or more boreholes	One borehole
Simple borehole source	Simple surface source
Mainly P and SV waves	Mainly P and SH waves
Reversible source	Reversible source
Borehole vertically needed	No vertically needed
Average profile	Average profile
Possible refracted waves	Minimum refraction waves
Constant travel path	Path increase with depth
More expansive	Less expansive

Schwarz and Musser (1972) reported results from down-hole test in San Francisco Bay area. Figure 6.21 shows the travel time curve from the down-hole test. The slope of the travel time curve at any depth represents the wave propagation velocities at that depth. An important advantage of the shear wave velocity is that the groundwater travel does not affect the measurements.

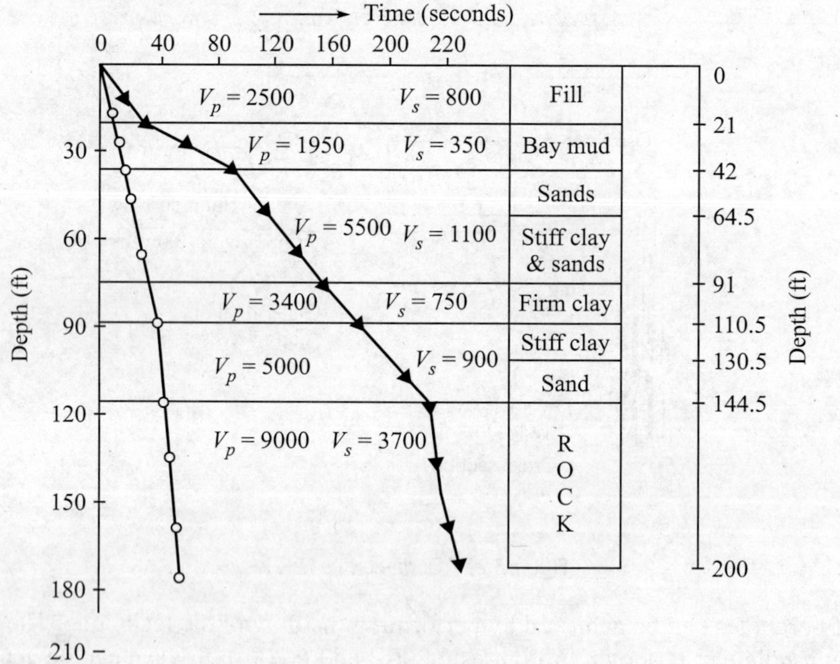

Figure 6.21 Travel time curve from down-hole test in San Francisco Bay area (After Schwarrz and Musser, 1972)

Seismic cone penetration test

This method has been introduced by Robertson et al. (1983), which is very similar to the down-hole test, except that no borehole is required. Figure 6.22 shows the schematic diagram of this test.

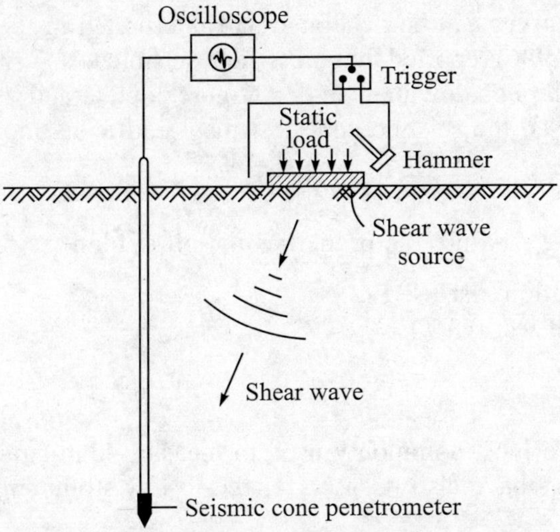

Figure 6.22 Seismic cone penetration test (SCPT)

Modulus can be obtained from the downhole position of the test while the cone penetrometer may be used to determine the other properties at the same depth, both without making a borehole. A seismic cone penetrometer consists of a piezo seismic probe with a geophone attachment as shown in Figure 6.23.

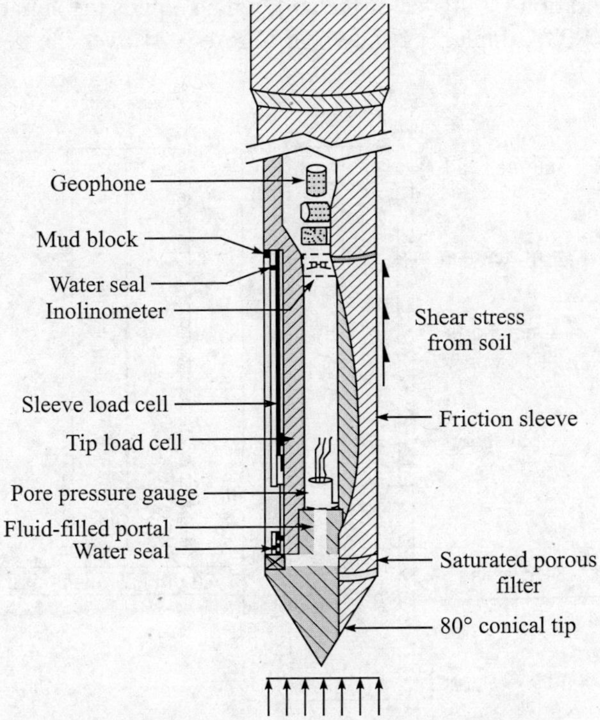

Figure 6.23 Schematic diagram of piezo seismic probe.

Travel time-depth curves can be generated and interpreted in the same way as for downhole tests. Baldi et al., (1986) reported that cross-hole seismic tests using two seismic cone have also been performed. Although downhole tests are more popular and more frequently performed to complement other tests, the seismic cone test may lead to its more common use.

High Strain Tests

The following field tests, measure the properties of soil at higher strain levels, namely:

- Standard penetration test (SPT)
- Cone penetration test (CPT)
- Dilatometer
- Pressuremeter test

While these tests are most commonly used to measure high-strain characteristics such as soil strength, their results have also been correlated to low-strain soil properties.

Standard penetration test

The standard penetration test (SPT) is probably the most widely used in the field of geotechnical engineering all along the globe. Thus, SPT is by far the oldest and most commonly used test. The standards penetration test consists of driving a thick-walked sampler into the granular soil deposit. The SPT can be especially valuable for sand deposits where the sand falls or flows out from the sampler when retrieved from the ground. The measured SPT N value (blows per foot) is defined as the penetration resistance of the soil, which equals the sum of the number of blows required to drive the SPT sampler, as shown in Figure 6.24 over the depth interval of 6 to 18

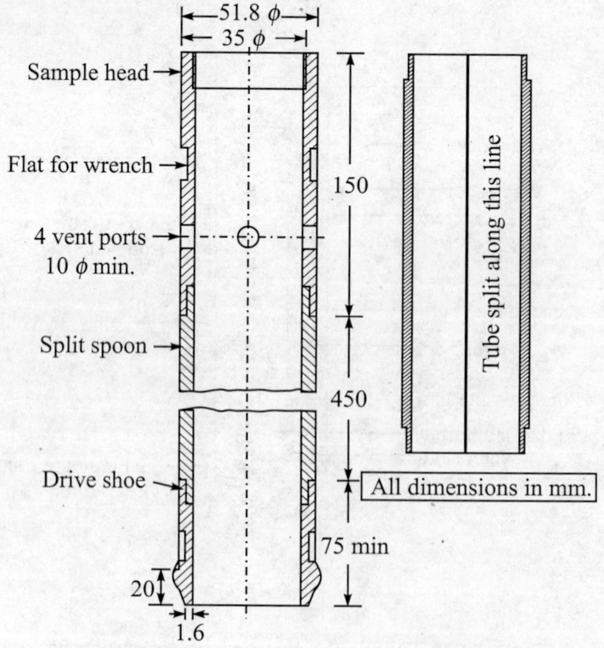

Figure 6.24 Split spoon sampler (SPT test).

inch (15 to 45 cm). The reason the number of blows required to drive the SPT sampler for the first 6 inch (15 cm) is not included in the N value is that the drilling process after disturbing the soil at the bottom of the borehole, and readings of blows pertaining to the next 6 to 18 inch (15 to 45 cm) are believed to be more representative of the in situ penetration resistance of the granular soil. However, several corrections are applied to the observed N values.

Thus, SPT is an essentially undrained test for the duration of each blow and the energy generated by the SPT hammer is principally the shearing energy. Therefore, the test results may be useful to predict the dynamic behaviour of soils. Seed et al., (1983) presented correlations between SPT and observed liquefaction. Imai (1977) reported the following relationship for measuring shear wave velocity V_s from SPT N values,

$$V_s = 91\text{N}^{0.337} \tag{6.29}$$

Even with the limitation and all corrections that must be applied to the measured N values, the standard penetration test is probably the most widely used field test worldwide. This is due to the fact that it is relatively easy to use, the test is economical compared to other types of field testing and the SPT equipment can be quickly adapted and included as part of almost any type of drilling rig.

Cone penetration test (CPT)

The idea for the cone penetration test is similar to the standard penetration test except that instead of driving a thick-walled sampler into the soil, a steel cone is pushed into the soil. There are many different types of cone penetration devices such as the mechanical cone, mechanical friction cone, electric cone and piezocone. The simplest type of cone is shown in Figure 6.25.

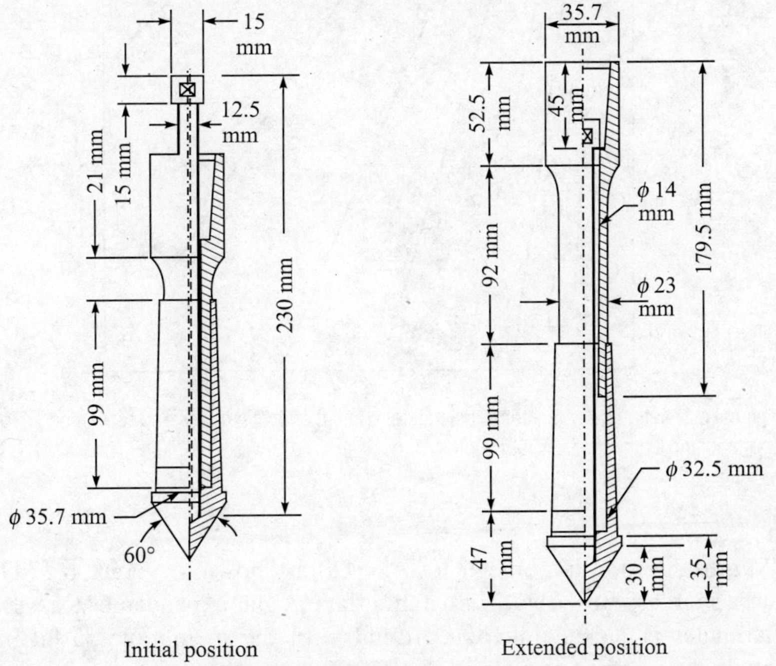

Figure 6.25 Cone penetrometer tip (Dutch cone).

First the cone is pushed into the soil to the desired depth (initial position) and then a force is applied to the inner rods which moves the cone downwards into the extended position as shown in Figure 6.25. The force required to move the cone into the extended position divided by the horizontally projected area of the cone is defined as cone resistance. By continually repeating the two-step process, the cone resistance data are obtained at increments of depth. Figure 6.26 shows an empirical correlation between the cone resistance, the vertical effective stress, and the friction angle ϕ of clean quartz sand as reported by Robertson and Campanella (1983)

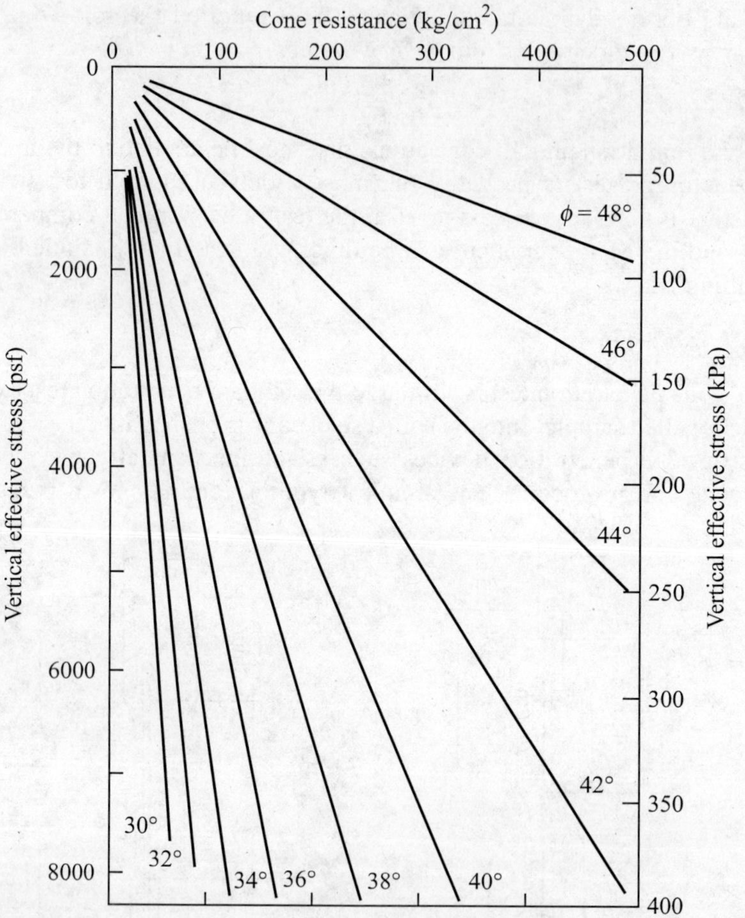

Figure 6.26 Empirical correlation between cone resistance, vertical effective stress and friction angle for clean quartz sand deposits [After Robertson and Campanella, 1983]

Dilatometer test (DMT)

Marchetti 1980 introduced a dilatometer test (DMT) as shown in Figure 6.27. This test uses a flat dilatometer, a stainless steel blade with a thin flat circular expandable steel membrane on one side. The dilatometer is jacked into the ground with the membrane surface flush with the surrounding blade surface. At intervals of 10 to 20 cm, the penetration is stopped and the membrane inflated by pressurized gas. The pressure at which the membranae moves by 0.05 mm, the lift-off pressure and the pressure at which its centre moves 1.1 mm are recorded, corrected and used with the hydrostatic pressure and the effective overburden pressure. Further, they are used to compute various indices to which the soil properties can be correlated. Baldi et al., (1986) have given details of these correlations.

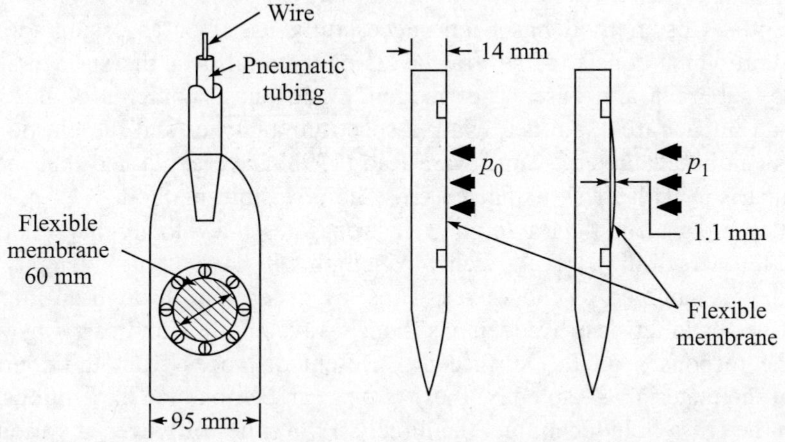

Figure 6.27 Marchetti flat dilatometer equipment [After Baldi et al., 1986]

Pressuremeter test

According to Mair and Wood (1987), the pressuremeter test is the only *in situ* test capable of measuring stress-strain, as well as strength characteristics of soil. Figure 6.28 shows the set-up for this test. The pressuremeter is essentially a cylindrical device that uses a flexible membrane to apply a uniform pressure to the wall of a borehole.

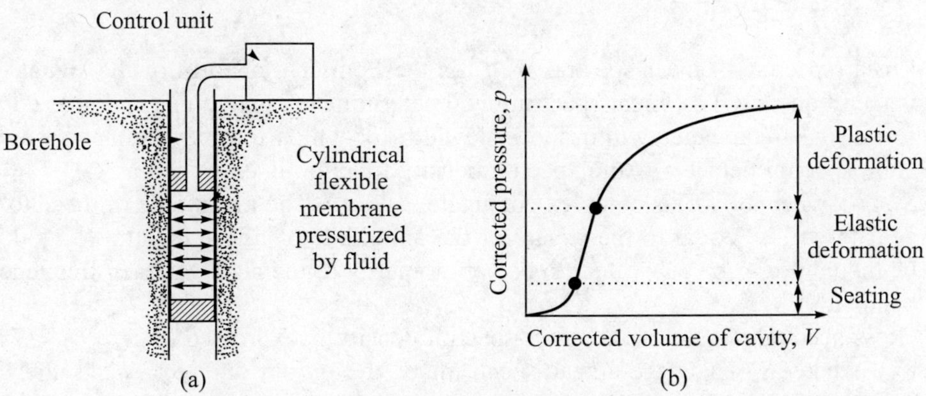

Figure 6.28 Pressuremeter test [After Mair and Wood, 1987]

Deformation of the soil can be measured by the volume of field injected into the flexible membrane or the feeler arms for pressuremeters that use compressed gas. After correcting the measured pressures and the volume changes for the system compliance, elevation differences, and membrane effects, a pressure–volume curve as shown in Figure 6.28 can be developed. From these curves the stress–strain behaviour can be computed.

Spectral analysis of surface waves (SASW)

The method using the spectra analysis of surface waves is similar to the refraction method as discussed earlier, and uses the velocities of surface waves propagating through different layers at shallow depth. This method has been successfully used for assessing the status of the pavement system which consist of several layers of material near the surface such as surface layer, subbase, subgrade and base. The frequency and phase contents of the surface waves generated by an impact are examined using a spectrum analyzer and the elastic moduli of the various layers can be evaluated. Kameswara Rao (1998) have explained that the results show excellent comparison with these using the cross-hole seismic method.

In a multilayered system, the Rayleigh wave propagates at a velocity that reflects the material properties through which they travel. Each wavelength will have a phase velocity depending on the extent of layers sampled by the waves. Thus, by generating through an impulse source, a wide range of waves in terms of frequencies (hence wavelengths), and carrying out the spectral analysis of the responses on the surface at different distances from the source, a detailed information of the materials sampled by these waves can be obtained expeditiously in one step. In contrast, in the cross-hole technique, the direct arrival times of waves at various depths have to be measured by conducting the test repeatedly at various depths which is time-consuming.

Spectral analysis

A typical response at a receiving point as a function of time is shown in Figure 6.29. A Fast Fourier Transform (FFT) of the above signal using a spectrum analyzer converts the signal in time domain to that in frequency domain as shown in Figure 6.29. This is referred to as frequency spectrum analysis or simply spectral analysis which converts the signal response in time domain into its frequency domain.

We may define the following:

- Linear spectral—Linear spectral $S_x(f)$ is the Fourier transform of the signal, where f is the frequency. The linear spectrum will determine the magnitude and phase difference for all the frequencies within the bandwidth for which the measurements are taken.
- Auto spectral density—Auto spectral density function is expanded as $G_{xx}(f) = S_x(f) \cdot S_x^*(f)$, where $S_x^*(f)$ is complex conjugate of $S_x(f)$. The magnitude of the auto spectral density will be equal to the square of the amplitude of linear spectral $[S_x(f)]$ and can be thought of to be equivalent to power or energy of the signal at each frequency in the bandwidth.
- Cross-spectral density—The cross-spectral density is expressed as $G_{yx}(f) = S_y(f) \cdot S_x^*(f)$, where $S_y(f)$ is the linear spectrum of the output and $S_x^*(f)$ is the complex conjugate of the linear spectrum of the input. The magnitude of the cross spectrum $(G_{yx}(f))$ is a measure of the powers of the two signals making it very useful to identify predominant frequencies that are present in both the input and output signals. The phase of $G_{yx}(f)$ is the relative phase between the signals at each frequency in the measured bandwidth and is used to determine the phase relationship between two signals.

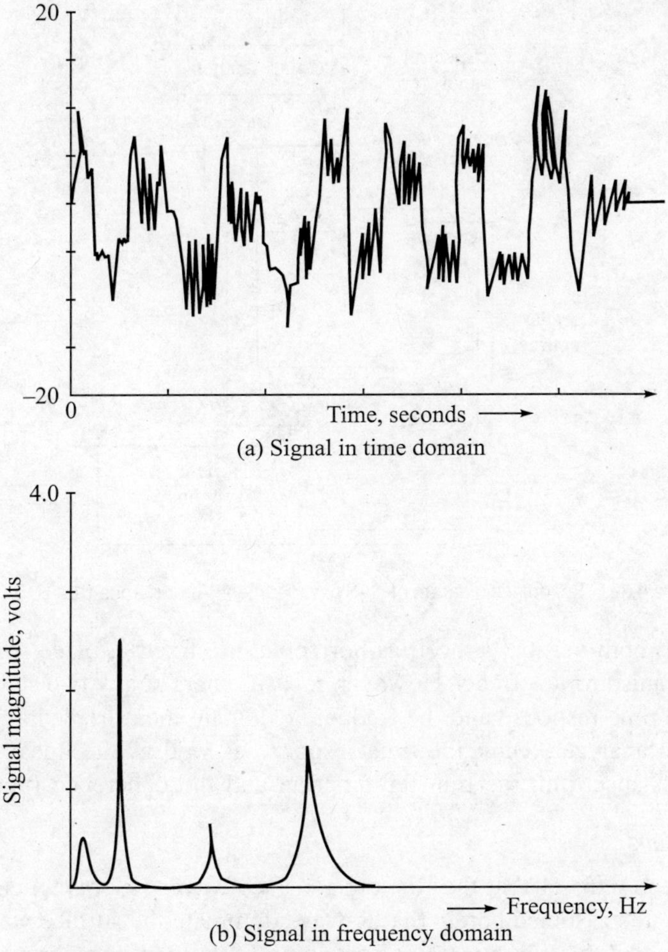

Figure 6.29 Representation of signal by spectral analysis.

- Transfer function—The transfer function H_f may be expressed as

$$H_f = \frac{\text{spectrum of the response signal}}{\text{spectrum of the input signal}}$$

It characterizes the input-output relationship of a dynamic system and is generally used to identify natural frequencies and damping coefficients of the dynamic system.

- Coherence function—The coherence function $r^2(f)$ may be expressed as

$$r^2(f) = \frac{\text{response power caused by the measured input}}{\text{total measured response power}}$$

Experimental procedure

A typical experimental set-up is shown in Figure 6.30.

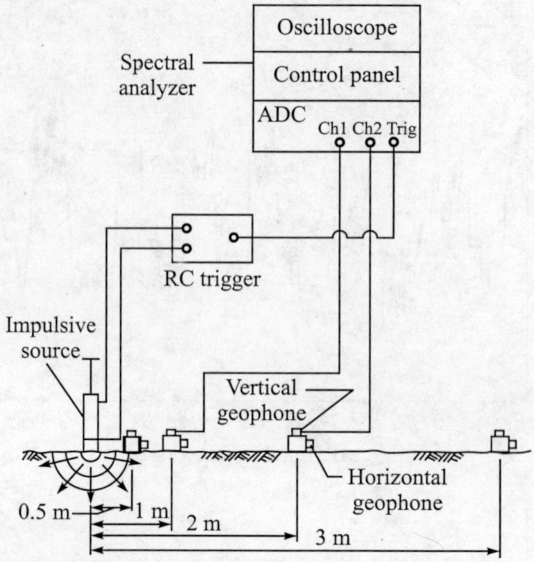

Figure 6.30 Schematic diagram of SASW experiment [After Kameswara Rao, 1998].

The geophones both vertical as well as horizontal are fixed at places on the surface along a line to minimize anisotropic effects. However, measurement at any two stations are needed for each test. Both the time response and the frequency domain measurements need to be recorded using the spectrum analyzer. Thus, the time response as well as the linear spectrum, the auto spectrum, the cross-spectrum, the transfer function and the coherence function are recorded.

Analysis of test results

As explained in Section 5.7.1 that the Rayleigh surface waves (R-waves) carry majority of the 5.7.1 input energy, it is assumed herein that R-wave dominate the surface wave propagation. As such it is sufficient that the measurements pertaining to the R-way motion are recorded. Using the following relationship for R-wave,

$$V_R = fL_R \tag{6.30}$$

where,

V_R = Rayleigh wave velocity
f = frequency in Hz
L_R = Rayleigh wavelength.

By using the phase lag given by cross-spectrum, the travel time can be calculated for each frequency generated by the impulse. The phase difference (say) θ between the input and the output signals recorded in the geophones represents the time lag or travel time Δt for an R-wave of frequency f and velocity V_R to travel over a distance Δs between the two geophones. It may be noted that the phase difference $\theta = 360°$ for a travel time equal to the period τ of the wave, hence the travel time of the R-wave between the geophones way be expressed as

$$\Delta t = \frac{\theta}{360} \cdot \tau = \frac{\theta}{360} \cdot \frac{1}{f} \tag{6.31}$$

where, $$\text{period } \tau = \frac{1}{f}$$

hence, $$V_R = \frac{\Delta s}{\Delta t} = \Delta s \, \frac{360}{\theta} f \tag{6.32}$$

where Δs is the distance between two geophones.

The wavelength L_R may be expressed as

$$L_R = \frac{V_R}{f} = \Delta s \, \frac{360}{\theta} \tag{6.33}$$

Figure 6.31 shows the variation of shear wave velocity with depth as reported by Gucunski and Woods (1991). SASW test have a number of advantages over other fields tests. They can be performed quickly, they require no borehole, they can detect low-velocity layers and they can be used to considerable depth (say) greater than 100 m. SASW testing is particularly useful at sites where drilling and sampling are difficult. On the other hand, this test does require specialized sophisticated equipment and experienced operators.

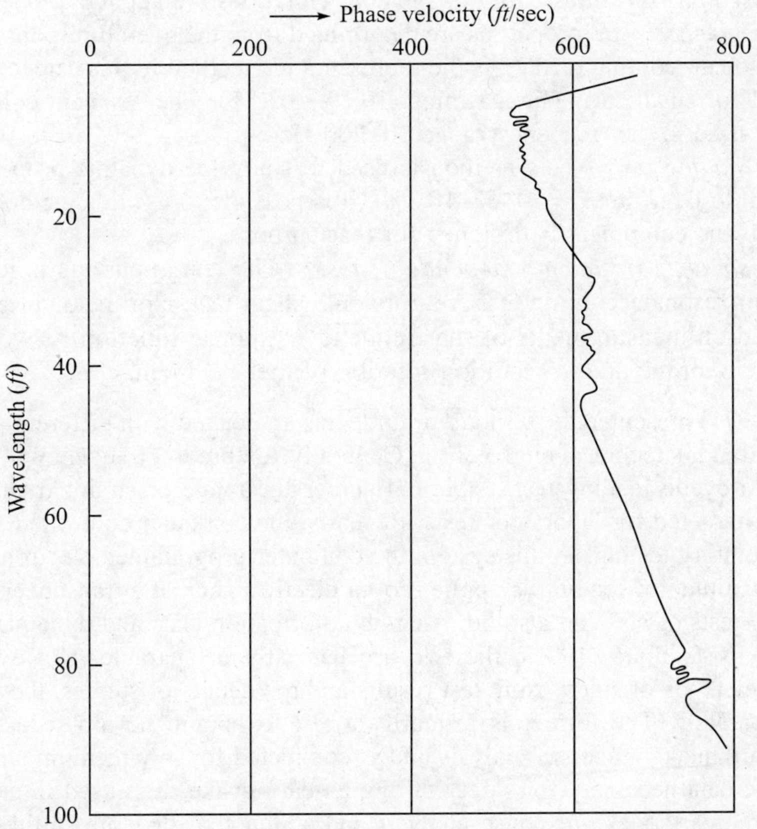

Figure 6.31 Variation of R-wave velocity with depth from SASW test [After Gucunski and Woods, 1991]

6.4.2 Laboratory Tests

Dynamic properties of soil can be established by laboratory tests and by *in situ* tests. The merits of laboratory tests include freedom of selection of confining pressure and dynamic stress level, ease of measurement and accuracy. However, the soil sampled may get disturbed while testing in the laboratory and the dynamic properties can easily change. As such, care is necessary while applying the results of laboratory tests to the soil at site. On the other hand, testing at site is generally more inconvenient and expensive than the laboratory tests. Tests at higher strain levels are difficult to carry out and precision more difficult but the merit of *in situ* testing is that it is free from disturbance during soil sampling. Thus, both methods have distinct advantages and disadvantages and the optimum method should be decided according to the purpose of testing.

According to Ishihara (1971) the dynamic laboratory test methods used these days can be classified as follows according to the type of the dynamic problem to be investigated.

- *Test based on wave propagation*: This method is used to study the elastic properties of soil at very low strains (10^{-6}–10^{-5}). The ultrasonic pulse test is widely used in this category but other methods also exist.
- *Test based on vibrations*: In this method, vibrations are applied at one end of the test specimen and the soil properties are determined from the overall response of the system. The resonant-column method is the representative method in this category. This method is used for small to medium strains (10^{-6}%–10^{-3}%). The resonant-column method is mainly used in the frequency range 10–200 Hz.
- *Cyclic loading test*: This method is used to study the dynamic properties of mainly medium to large strains (10^{-4}%–10^{-2}%). It is possible to use this method for low strain levels if the equipment is designed for that purpose.
- *Non-resonance (frequency dependant) test*: The test apparatus is identical to that used for resonant column test and torsional shear test. Non-resonance (NR) method is based on measurements of the frequency response function between the applied harmonic torque and to resulting rotation of the specimen.

Ishihara (1971) presented the various strain levels associated with different phenomena and tests in the field of geotechnical engineering. Generally, a vibration frequency of less than a few Hz is used in a cyclic loading test. Table 6.4 shows the range of strain, frequency, dynamic properties, etc. targeted for laboratory tests. Recently, the computer controlled system is being adopted in laboratory testing. In this system the computer programmed electronic signal representing the magnitude of loading is applied to an electro-pneumatic transducer.

Laboratory tests of soil can also be dynamic deformation tests and dynamic strength tests. As mentioned by Ishihara (1971), the two are basically inseparable. However, because of differences in methods of study using test results and the details of studies, they are frequently considered separately. The former is primarily a test to obtain the input data necessary for carrying out earthquake response analysis and is conducted for low-medium strain levels. The latter yields the data necessary for safety during an earthquake. It is used in conjunction with the results of the earthquake-response analysis and conducted for large strain levels.

Table 6.4 Correspondence between strain level and physical phenomena as observed in laboratory tests (After Ishihara, 1971).

Strain level	10^{-6} 10^{-5} 10^{-4}	10^{-3} 10^{-2}	10^{-1}
Phenomena	Wave propagation, vibration	Cracks, differential settlement	Slide, compaction, liquefaction
Mechanical properties	Elasticity	Elastoplasticity	Rupture/Fracture
			Effect of cyclic loading, loading rate

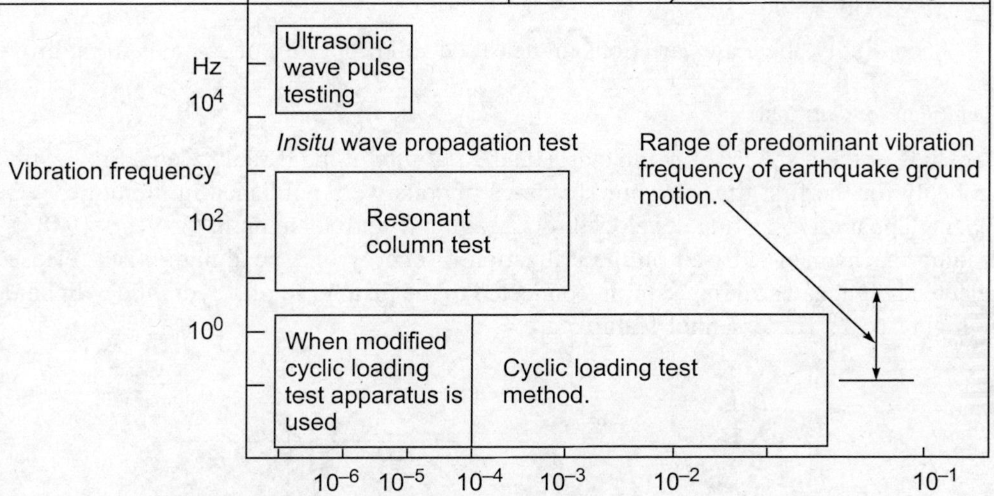

If a comparison is made with reference to static triaxial testing of soil it may be recalled that the loads and deformations are read on the load cell and the dial gauge, respectively, while the pore pressures are read on the pore pressure device. In dynamic tests, the records of all these quantities must necessarily be automatic. In the classical triaxial test, the rate of loading is approximately 1.25/minute and the soil sample may fail in 10 to 15 minutes or more. Because the dynamic loads may be applied in a fraction of a minute/second, the loading device has to be of a special design. Several different types of apparatus for testing soils under dynamic loading have been designed and fabricated from time to time. The different testing apparatus may be listed as follows:

- Pendulum loading apparatus
- Cyclic triaxial apparatus
- Cyclic direct simple shear apparatus
- Cyclic torsion shear testing apparatus
- Cyclic resonant column apparatus
- Ultrasonic pulse testing apparatus
- Shake table testing apparatus
- Non-resonance (frequency dependant) testing apparatus

Some of the above testing procedures shall be addressed in the next section. The main effort in evaluating the dynamic soil properties is to adopt to such test procedures that can as far as possible simulate the initial stress conditions and the application of dynamic loading and the cyclic loading as realistically as possible. These determination of dynamic soil properties are basically location specific as such there is inbuilt uncertainty and variation in measured dynamic properties. These are due to

- Different geologic conditions of site
- Non-homogeneity and anisotropy of soil deports
- Sampling procedures
- Testing procedures
- Interpretation errors

Accordingly, there are variations in measured soil properties of samples from different sites.

Pendulum loading test

For study of the response of bomb blasts on the stability of the Panama Canal Project in the USA, probably for the first time dynamic loadings of soils were published in literature. A pendulum loading apparatus was first developed by Casagrande and Shannon in the years 1948–1949. The loading mechanism is based on the utilization of energy of a pendulum when released from a selected height and striking a spring connected to the piston rod of a hydraulic cylinder as shown in Figure 6.32. The essential features are:

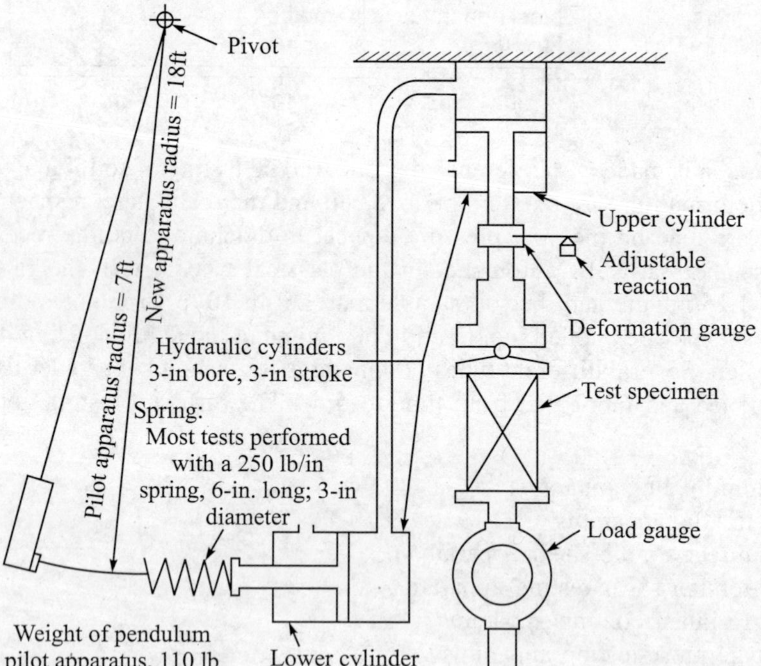

Figure 6.32 Pendulum loading apparatus (After Casagrande and Shannon, 1948b)

- The time of loading in these tests was defined as the time between the beginning of the test and the point at which the maximum compressive stress is reached.
- The time of loading for a pendulum is directly proportional to the square root of the weight of pendulum and is inversely proportional to the square root of the spring constant.
- The maximum force is directly proportional to the distance by which the pendulum is pulled back and the square root of the product of spring constant and weight of pendulum.
- The load gauge used with equipment consisted of electric-resistance strain gauges mounted on a metal ring. The strain introduced in the gauges was then in direct proportion to the load.

These load gauges can be calibrated under static loads and can be used in a dynamic test. Similarly, a deformation gauge was constructed on a cantilever metal strip with electric-resistance gauges mounted on one end while the other end rested on an unmovable support. The strain introduced in the cantilever was a measure of the deformation of the soil sample. The experimental set-up thus essentially consisted of electric-resistance strain gauges mounted on a metal ring.

Various types of soils have been tested and Casagrande and Shannon tested Cambridge clay having the following properties.

Test sample (Cambridge clay)

Natural moisture content—30–50%
Liquid limit—37–59%
Plastic Limit—20–29%

The tests were performed in the unconfined as well as in the confined state. The stress-time and strain-time trace has been reported by Casagrande and Shannon. In this test, the time of loading was defined as the time between the beginning of the test and the point at which the maximum compressive stress is reached. The time of loading in transient test has been shown in Figure 6.33.

This apparatus with a time of loading between 0.05 and 0.01 second was found to the best suited for the transient test.

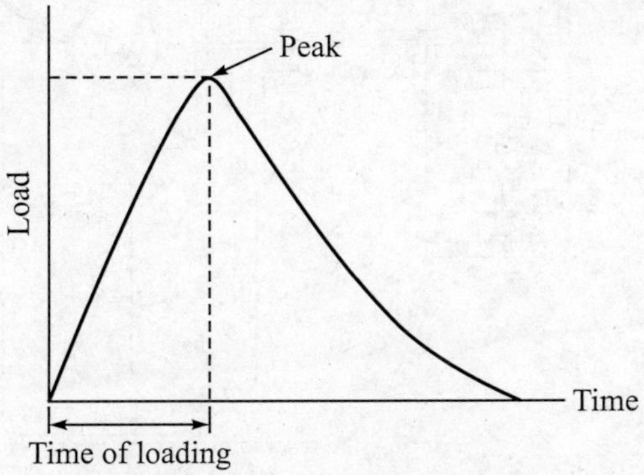

Figure 6.33 Time of loading in transient test.

Cyclic triaxial test

At high strain levels, the cyclic triaxial test has been the most commonly used test for measurement of dynamic soil properties. Figure 6.34 shows a typical cyclic test apparatus, wherein a

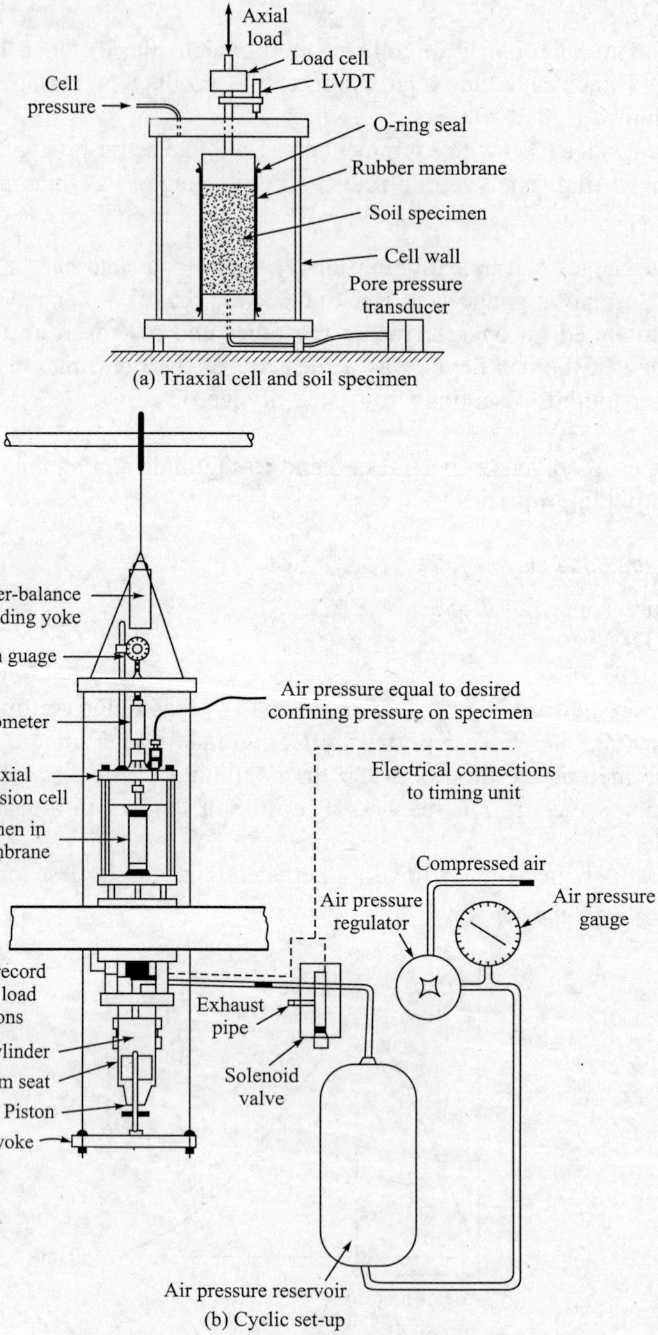

Figure 6.34 Cyclic triaxial test apparatus

cylindrical specimen surrounded by a thin rubber membrane, is placed between the top and bottom loading platens. The rubber membrane serves to isolate the water in the pores of the sample from the chamber fluid. The porous stones provide access to the sample for either pore-water pressure or pore-drainage pressure. Fluid pressure is applied within the chamber containing the sample. The chamber pressure acts uniformly on the surface of the sample, including the top and bottom loading caps.

In the cyclic triaxial test, the deviator stress is applied cyclically, either under stress-controlled conditions or under strain-controlled conditions. The stress-controlled conditions are felicitated by pneumatic or hydraulic loaders, whereas strain control is governed by servohydraulic or mechanical loaders. Cyclic triaxial tests are most commonly preferred with the radial stress held constant and the axial stress cycled at a frequency of about 1 Hz. The axial deformation is usually measured by an LVDT attached to the bottom of the load cell and abutting against the top of chamber.

Cyclic direct-shear testing apparatus

The seismic stress condition can be better reproduced in cyclic direct-shear test than in cyclic triaxial test. It can be also specifically used for liquefaction testing. By applying cyclic horizontal shear stress to the top or bottom of the specimen, the test specimen is deformed in much the same way as an element of soil subjected to vertically propagating S-waves. Figure 6.35 shows the cyclic direct shear apparatus.

A soil element in the field may be subjected to a series of cyclic shear strains or stresses that may reverse many times during an earthquake. In case of a horizontal ground surface, there are no shear stresses on the horizontal plane before the earthquake. During the earthquake, the normal stresses on this plane remain constant while cyclic shear stresses are induced during the period of shaking. The simple shear test, however, applies shear stresses only on the top and bottom surfaces of the specimen. Simple shear apparatuses are limited by their inability to impose initial stresses other than those corresponding to K_o condition.

Test data from cyclic direct shear test are generally analyzed to determine shear parameters, soil moduli, damping and liquefaction potential of loose sands.

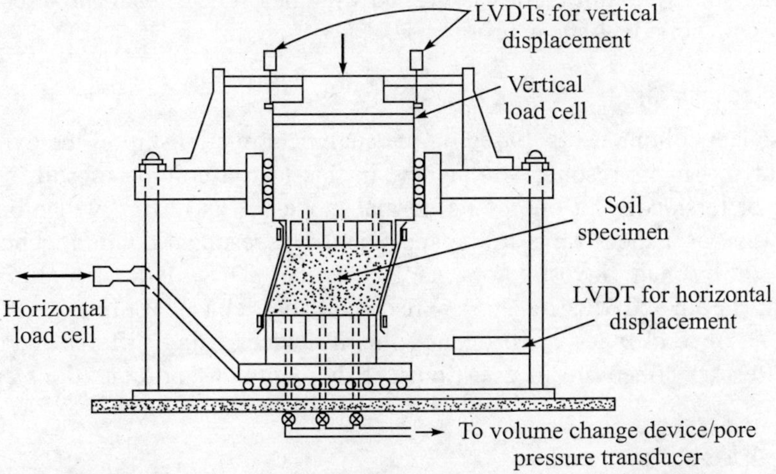

Figure 6.35 Cyclic direct shear test apparatus [After Airey and Woods, 1987]

Cyclic torsion shear test

Ishihara and Li (1972) developed a torsional triaxial test. Dobry et al., (1975) used strain-controlled cyclic torsional loading along with stress-controlled axial loading of solid specimen that has proven effective for measurement of liquefaction behaviour. However, Drnevich (1967, 1972) developed hollow cylinder cyclic torsional shear apparatus as shown in Figure 6.36.

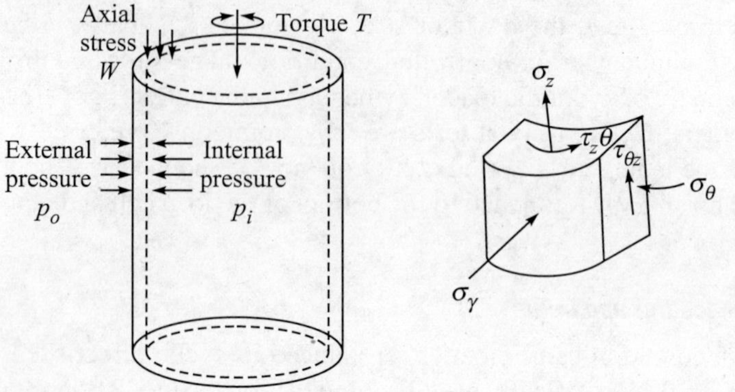

Figure 6.36 Hollow cylindrical cyclic torsion shear apparatus (After Drnevich, 1972).

A hollow cylinder sample subjected to loads/pressures exerted by a torsional shear apparatus is shown in Figure 6.36. It can be seen that there are four independent loads/pressures acting on the sample, namely:

- Outer chamber pressure p_o
- Inner chamber pressure p_i
- An axial load W
- A torque T

Combining the four loading components, various stress paths can be induced in the sample to simulate a variety of *in situ* loading conditions. To compare with the triaxial shearing of only the degrees of freedom (p and q), the torsional device is much more versatile and powerful. Many of the limitations and difficulties associated with the cyclic triaxial and direct shear test can be overcomed with torsional shear test.

Cyclic resonant column test

The cyclic resonant column test is based on the analytical relationship of the dynamic modulus of a column of soil to its resonant frequency. In this test, a column of soil is excited either longitudinally or torsionally in one of its normal modes. This in turn will propagate either a compression wave or a shear wave in the specimen. The resonant column technique was used for testing of soil by many investigators, namely, Wood (1978), Drnevich (1967), Hardin and Richart (1963), Hardin and Music (1965), Wilson and Dietrich (1960), Iida (1940) and Ishimoto and Iida (1937). Several types of resonant column devices using different end conditions to constraint the test specimen are in use. Some of the common end conditions are:

- Fixed-free end
- Spring-base and free end
- Partially fixed base and free end

For calibration, a test is performed on a calibration bar to compute its resonant frequency in torsion and longitudinal compression. This is achieved by calibrating the apparatus by substituting metal calibration bars in place of the specimen whose mechanical properties are known.

The column specimen as shown in Figure 6.37 is suitably prepared and consolidated.

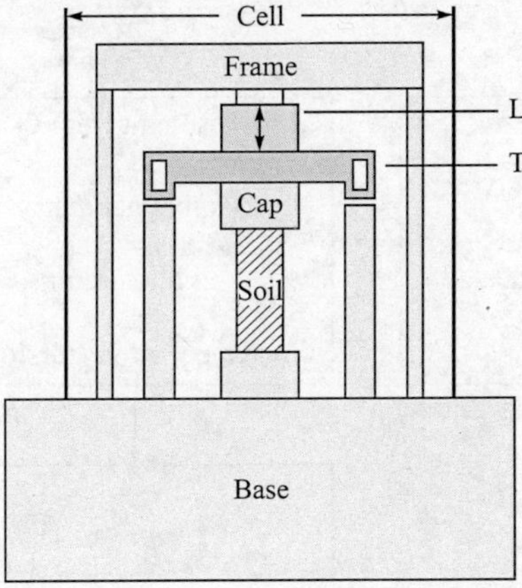

(a)

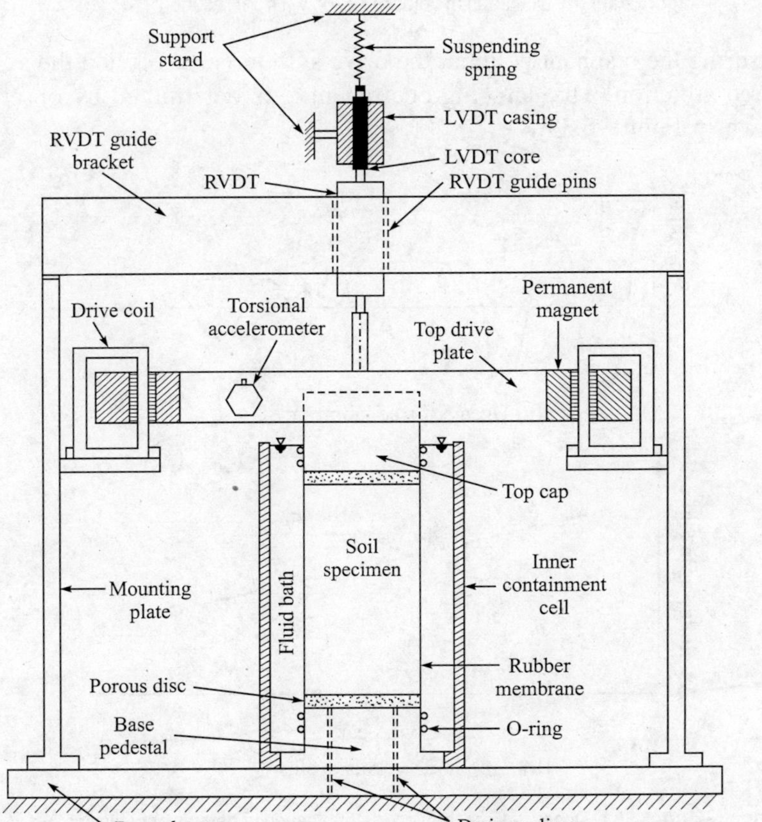

(b)

Figure 6.37 Resonant-column test apparatus

The frequency of the electromagnetic drive is gradually increased until all the first mode resonant conditions are encountered. With the known value of the resonant frequency it is possible to back calculate the velocity (V_P or V_S) of the wave propagation, modulus of elasticity E and modulus of rigidity G of the soil as shown in Figure 6.38.

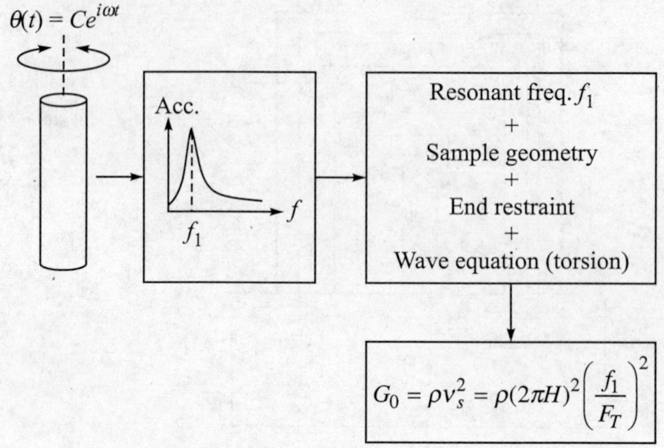

Figure 6.38 Calculation of velocity (V_P or V_S) from resonant-frequency.

After measuring the resonant column, the drive system is cut off and the system is brought to a state of free vibration. The damping coefficient ξ is determined by observing the decay pattern as shown in Figure 6.39.

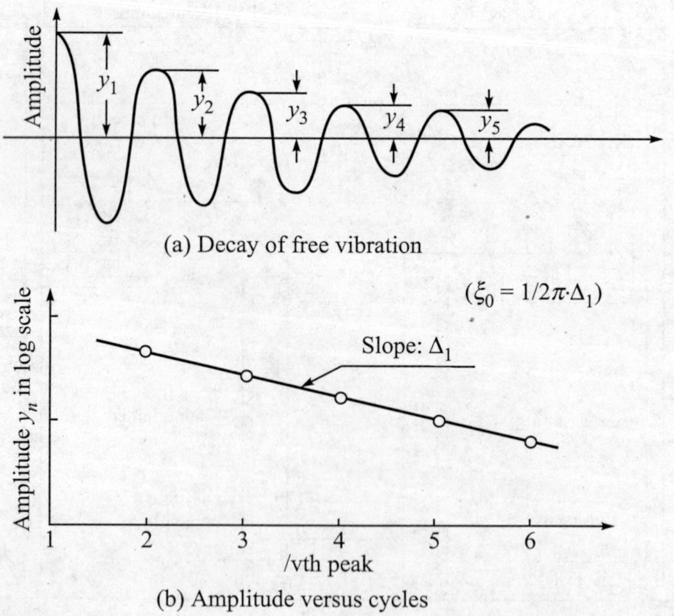

Figure 6.39 Decay of free vibration (determination of damping coefficient-ξ).

The variation of shear modulus ratio G/G_0 with shear strain and variation of damping ratio with shear strain are both shown in Figure 6.40.

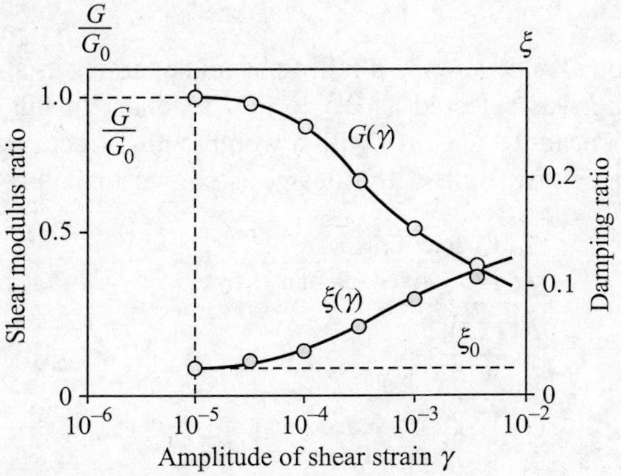

Figure 6.40 Variation of shear modulus ratio with shear strain.

Thus, the resonant column test is a useful tool for evaluating the strain-dependant modulus and the damping properties of soils. However, there are some limitations on its use. The test is basically based on a back analysis procedure. The output is not the response of the specimen itself, but contains combined effect of the soil and its attached apparatus. Great caution must be exercised in order to obtain reliable data. The test is eventually useful for obtaining data on dynamic properties of soil within the range of shear strain less than about $5 \times 10^{-2}\%$.

Ultrasonic pulse testing apparatus

Lawrence (1963) described the basic apparatus required to measure the wave propagation velocity V_P and V_S through sand. Stephenson (1978) further developed an equipment for conducting the ultrasonic tests. The testing equipment includes a pulse generator, an oscilloscope, and the ultrasonic probes each for transmitter and receiver. Stephenson (1978) carried out the ultrasonic pulse tests on silty clay samples. One of the main strengths of this type of testing is that it can be performed on very soft sea floor sediments while still being retained in the core liner. However, the main limitation lies in the difficulties to identify the exact wave arrival times. Secondly, the strain amplitudes which can be achieved by this method are also in a low range.

In this test the transmitters and receivers essentially consist of piezoelectric crystals which exhibit changes in dimensions when subjected to a voltage across their faces and which produce a voltage across their faces when distorted. Wood (1978) presented the general procedure for ultrasonic testing of soft clays. When a high frequency electrical pulse is applied to the transmitter, it produces a stress wave that travels through the specimen towards the receiver. When the stress wave reaches the receiver, it generates a voltage pulse that is measured. The wave propagation is obtained by

$$\text{Wave propagation velocity} = \frac{\text{distance between the transmitter and receiver}}{\text{time difference between voltage pulses}}$$

According to Wood (1978), this test is particularly useful for very soft soil such as seafloor clays, since it can be performed while soil is still in sampling tubes.

Falling beam apparatus

Casagrande and Shannon also developed a falling beam apparatus as shown in Figure 6.41. The range of time of loading was selected as 0.5 to 300 seconds. In this set-up the falling beam apparatus consisted of a beam ($0 \leq x \leq L$) with a weight and a rider. The dashpot as shown was added to control the velocity of fall of the beam, as shown in Figure 6.41.

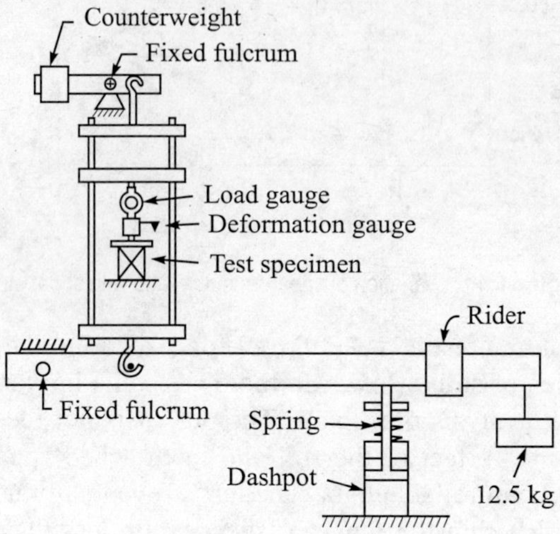

Figure 6.41 Falling beam apparatus

Hydraulic loading apparatus

The hydraulic loading apparatus was also developed by Casagrande and Shannon for studying the response of bomb blast on the stability of the Panama Canal Project in the USA. This hydraulic apparatus consisted essentially of a constant volume vane type hydraulic pump connected to a hydraulic cylinder through valves by which either the pressure in the cylinder or the volume of the liquid delivered to the cylinder can be controlled. The essential feature of this device is the increase in peak load produced compared to the other two types, namely, Pendulum Loading Apparatus or Falling Beam Apparatus.

6.4.3 Interpretation of Test Results

The value of the dynamic shear modulus G is affected by a number of parameters, out of which the confining pressure, the shear strain amplitude and the relative density are most important. It is observed that changes in density from medium to dense state have relatively insignificant effect compared to the effect of confining pressure and shear strain amplitude. Since the order of strain level and confining pressure associated with different *in situ* tests are different, tests may be expected to show a large variation, as the strain associated with, say hammer test is very small and that with cyclic plate load test is very large. A rational approach is, therefore, needed to arrive at a suitable design value.

In the range of strains associated with a properly designed machine foundation, the effect of variation in strain on shear modulus is small and the values of G for design purposes may be determined from the *in situ* test values using the relation given below:

$$\frac{G_1}{G} = \left(\frac{\sigma_{01}}{\sigma_0}\right)^m \tag{6.34}$$

where,

G_1 and G = dynamic shear modulus for the prototype and from the field test, respectively

σ_{01}, σ_0 = mean effective confining pressure, associated with prototype foundation and the *in situ* test, respectively

m = constant depending upon the type of soil, shape of grains, etc. Its value has been found to vary from 0.3 to 0.7 and may on the average be taken as 0.5.

In situations where high strain levels are associated as in the case of analysis for earthquake conditions, the effect of strain level shall be considered along with that of the confining pressure.

In such a case, the values of G from different field tests may first be reduced to the same confining pressure (expected below the footing) and their variation with strain levels may be studied to arrive at an appropriate value corresponding to the expected strain level.

The value of damping in soils is also a function of strain level to which the soil is subjected. Damping is less at low strain levels and becomes significantly large at high strain levels.

6.5 SHAKE TABLE TESTING

The shake table is an indispensable testing facility for developing earthquake resistant techniques. It provides experimental data that leads to a better understanding of the behaviour of structures and calibration of various numerical tools used for the analysis and design under earthquake loads. The accurate reproduction of motion as well as fast, synchronous, high resolution measurement of response quantities are essential for the shake table system. The bidirectional, three degree of freedom (i.e., longitudinal, vertical and pitch) shake table is always difficult to control accurately because the shake table with resonating specimen and foundation has properties of strongly coupled multiple input, multiple output (MIMO) system.

The shake table facility is available at a few places in India, especially at IIT Roorkee. At IIT Roorkee, it is housed in a 12 m high test wall, which is served by a 150 ton EOI crane. The facility has 11 kV substation to provide supply of 5000 kVA to the system. Detailed design and development is reported in Basu (2003).

Shake table platform

The shake table platform is 3.5 m by 3.5 m in size and is designed in-house to carry design payload of 200 kN. The table, as shown in Figure 6.42, is a welded plate structure of steel.

Figure 6.42 Shake table platform.

The fundamental frequency of the shake table is 49.5 Hz. The shake table top is fitted with bushes of 30 mm diameter to tie down the test specimen. In a square grid pattern of 400 mm centres, the four bushes are capable of carrying maximum plyload.

Shake table foundation

The foundation block is 8.5 m by 9.35 m in plan, weighs 5450 kN and supports the table and reaction of the actuators. The resultant soil bearing pressure is 77 kPa. The soil foundation system has natural frequency of 12.3 Hz for vertical motion, 14.0 Hz for rocking motion and 9.4 Hz for horizontal motion including the effect of the embedment. This produces coupled natural frequencies as 19.0 Hz and 7.0 Hz. For sinusoidal type of loading at 3 Hz, the calculated vertical, horizontal and rotational amplitudes are 0.059 mm, 0.024 mm and 0.00072 radians, respectively.

Actuators

The driving mechanism of the shake table is servo-hydraulic type and the platform is driven by three double acting actuators (2-vertical and l-horizontal). The important characteristics of actuators are shown in Table 6.5. The actuators were supplied by Silveridge Technology Ltd., UK.

Table 6.5 Characteristics of Servo-hydraulic actuators

Parameter	*Horizontal actuators*	*Vertical actuators*
Static thrust	250 kN	125 kN
Dynamic thrust	200 kN	100 kN
Stroke	300 mm	300 mm
Velocity	1.2 m/s	1.5 m/s
Flow at max. velocity	750 litre/min.	400 litre/min.
Average flow	40 litre/min.	40 litre/min.

Horizontal and vertical motion of shake table

The horizontal and vertical motion of the shake table is geometrically coupled as shown in Figure 6.43. Inertia forces on the specimen during table motion act through centre of gravity and can cause significant pitching, yawing and rolling motions. Each shake table is designed for limited capacities to withstand rolling and yawing by the six hydrostatic bearings of 50 kN capacity each. The specimen is placed and mounted in such a way that minimum yawing and rolling is produced.

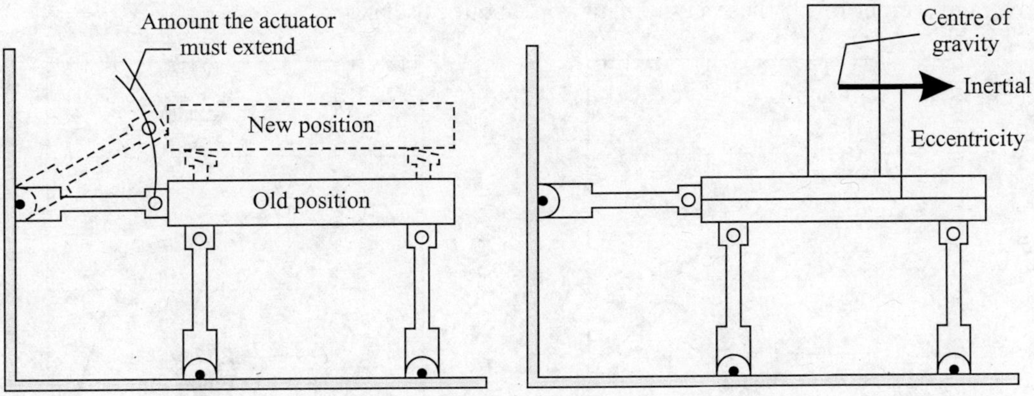

Figure 6.43 Arrangement for horizontal, vertical and pitch motion of the shake table.

The dynamic characteristics of the shake table change with the test specimens mounted on it. The size (mass) and the flexibility of the test specimen will have direct bearing on the frequencies of the coupled system and it should be taken into account in determining the control signal for accurate table motion. Flow management consists of limiting the volume and the force of the flow across the various actuators controlling the table motion. The oil flow at 210 bar pressure from the single power pack has to be regulated in such a way that any instant the total oil demand is less than the capacity of the power pack. At higher frequencies the compressibility of flow becomes significant and causes force limits which need to be taken care of by the controller. The table performance curves in the horizontal direction using the property of horizontal actuators can be derived as shown in Figure 6.44.

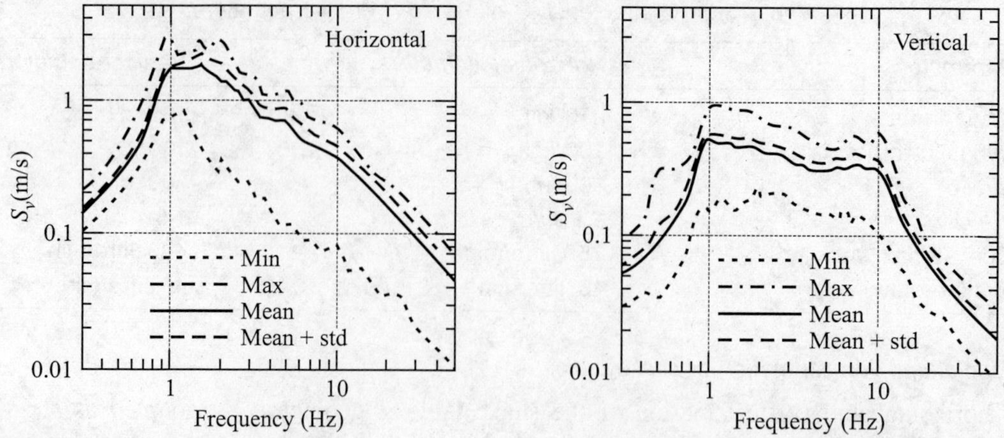

Figure 6.44 Table performance curves from an ensemble of tests (damping 5%).

The shake table since it inception has been used for seismic qualification of equipment and model testing of civil engineering structures. The test spectral velocity curve of shake table motions used for seismic qualification testing of about 24 equipment is shown in Figure 6.45. The average weight of all these equipment was about 50 kN.

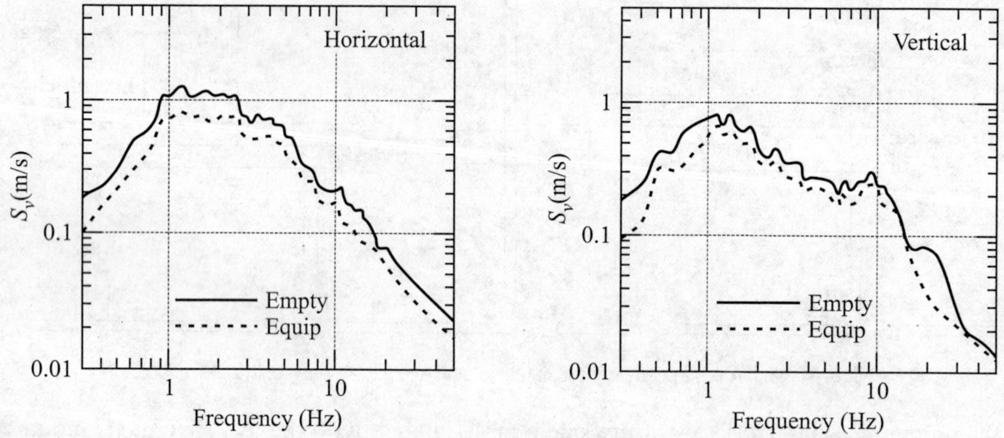

Figure 6.45 Typical response for a 50 kN specimen (damping 5%).

Figure 6.45 shows a typical test response spectra (TRS) with a 50 kN test specimen. It is clear from the data shown in the figure that the near full capacity of the shake table was utilized in testing.

Typically, the shake table testing from command generation point of view can be broadly divided into two categories:

- Periodic wave functions such as sine sweep, square, triangular waves, etc.
- Random motions with either spectrum compatible motion type or time history earthquake type motion. The former kind can be simulated disregarding phase inputs whereas the latter requires phase matching in addition to amplitude meeting.

Basu (2003) reported that at IIT Roorkee, The National Instruments USA supplied a high performance data acquisition system. In this configuration, it is comprised of 64 channels for strain gauges and similar transducers, which need amplification of their voltage output and the remaining 64 are for other output transducers whose voltage does not require amplification. The data acquisition system (DAS) uses oversampling and buffering techniques to provide synchronized sampling of the acquired data. The upgraded system was tested at IIT Roorkee for compliance with the acceptance criterion. The test specimen was mounted on the top of the table and the tests were carried out to test the capabilities of the new system.

Figure 6.46 shows the time history simulation for the Kobe earthquake motion.

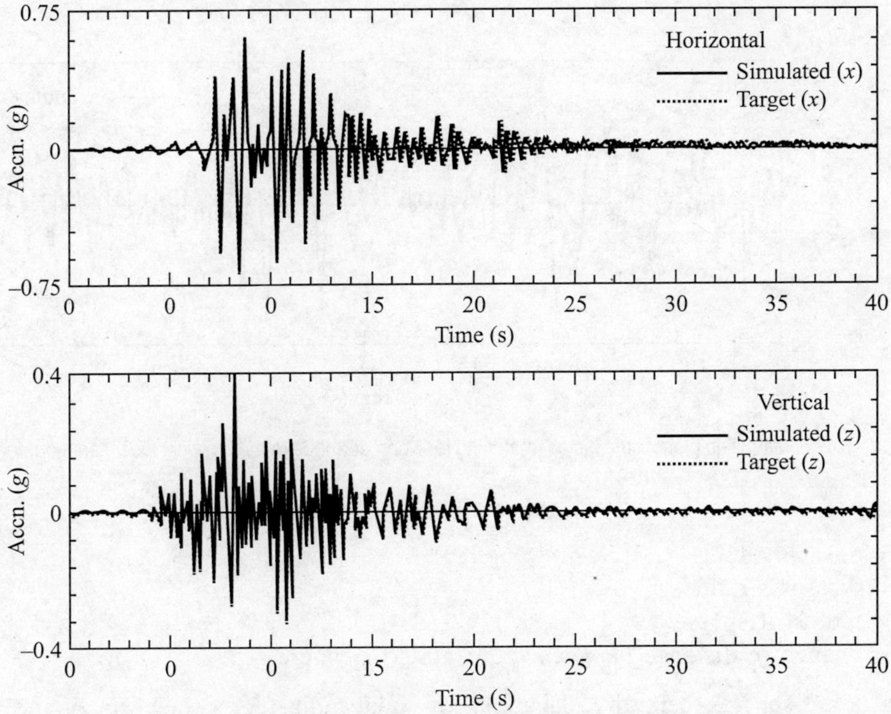

Figure 6.46 Time history matching for Kobe earthquake motion with the specimen mounted on table.

Figure 6.47 shows the enveloping response of the time histories simulated to be compatible with the RRS of mean curves in Figure 6.46. The simulated motions satisfy the limit specified for acceptability.

6.6 SHEAR PHENOMENON OF PARTICULATE MEDIA

Soil is essentially an assemblage of discrete elastic particles (soil skeleton). Hertz studied the group behaviour of identical spherical material having a diameter $2R$. Timoshenko and Goddier studied their behaviour further under a normal force N and showed that

$$N = 2\sqrt{2}\ \frac{G \cdot R^{3/2}}{3(1-\nu)\,\Delta^{3/2}} \tag{6.35}$$

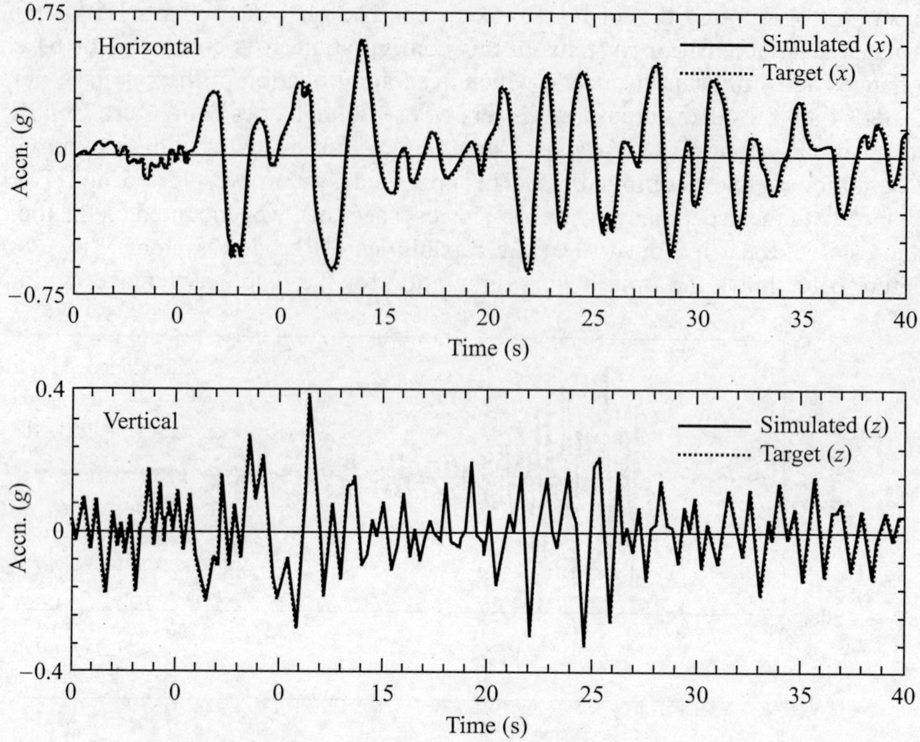

Figure 6.47 Magnified plot of time history matching for Kobe earthquake motion with specimen mounted on table.

where,

G = shear modulus
ν = Poisson's ratio
R = radius of sphere
Δ = change in distance between the centres of spheres.

For 3 × 3 spheres, centrally placed, all of equal radius R as shown in Figure 6.48, at the initiation of gross sliding of the spheres, the shear strain may be expressed as

$$\gamma_{gs} = 2.08\left[\frac{(2\cdot\nu)(1+\nu)f}{(1-\nu^2)^{1/2}}\right]\frac{\sigma^{2/3}}{E^{2/3}} \tag{6.36}$$

where, $f = T/N$ = Coefficient of friction between the spheres.

The shear strain γ_{gs} corresponding to initiation of gross sliding is termed volumetric threshold shear strain. Dobry et al. (1982) studied such micromechanical modelling of soil.

Such basic mechanics of particulate media may provide insight into stress-strain behaviour of soil. At the micro-level itself many studies are in progress wherein special emphasis is on contact interactions of each individual soil particle. However, there is great uncertainty involved in practical use of such studies at micro-level but nevertheless they do provide some fundamental clues to the desired soil behaviour.

The stress-strain (τ- γ) relationship is seldom linear during loading, unloading and reloading cycles. On the contrary, soils exhibit nonlinear τ-γ relationship from the onset of the cyclic loading as shown in Figure 6.4(a).

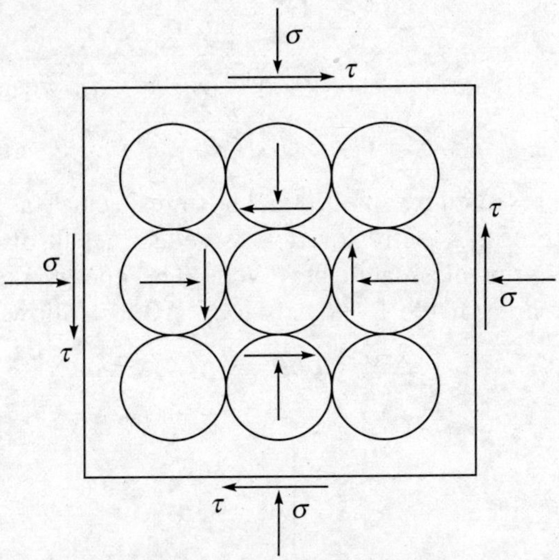

Figure 6.48 Mechanics of particulate media.

6.7 BEHAVIOUR OF SOIL UNDER PULSATING LOAD

A linear model expresses dynamic properties of a material in a manner that is independent of strain amplitude. It is generally useful for small strain levels. Ishihara, JSCE (1997) has presented in detail dynamic deformation of soil:

Let us first discuss the case wherein soil is considered a viscoelastic mass.

If we apply the sinusoidal stress $\bar{\tau}$ having an amplitude τ_a and frequency ω to a viscoelastic mass, the response strain $\bar{\gamma}$ also varies sinusoidally with an amplitude of γ_a and both can be expressed as:

$$\bar{\tau} = \bar{\tau}_a\, e^{i\omega t}, \quad \bar{\gamma} = \gamma_a\, e^{i(\omega t - \delta)} \tag{6.37}$$

where δ is delay in phase of deformation, i.e., strain with respect to stress. It is possible to express the relation between stress and strain using a complex coefficient of elasticity G^* as follows:

$$\bar{\tau} = G^*\, \bar{\gamma} \tag{6.38}$$

where,

$$G^* = G + iG', \quad G = \frac{\tau_a}{\gamma_a} \cos\,\delta, \quad G' = \frac{\tau_a}{\gamma_a} \sin\,\delta \tag{6.39}$$

Here G represents the rigidity of the material and is also called the shear modulus. G' represents energy dissipation and is also called the loss modulus. The ratio of G' to G is termed the loss coefficient η and can be expressed as

$$\eta = \frac{G'}{G} = \tan\,\delta \qquad (6.40)$$

If we use the relations $\tau = \tau_a \cos\,\omega t$ and $\gamma = \gamma_a \cos(\omega t - \delta)$, we get

$$\tau = G\gamma \pm G'\sqrt{(\gamma_a^2 - \gamma^2)} \qquad (6.41)$$

Equation (6.41) can be represented by an elliptical hysteresis curve as shown in Figure 6.49. The first term in Eq. (6.41) represents the restoration force as a result of elastic deformation, while the second term is the energy dissipated per cycle. The energy dissipated per cycle can be expressed in terms of the area of the hysteresis loop ΔW as follows:

$$\Delta W = \oint \tau d\gamma = \pi G' \gamma_a^2 \qquad (6.42)$$

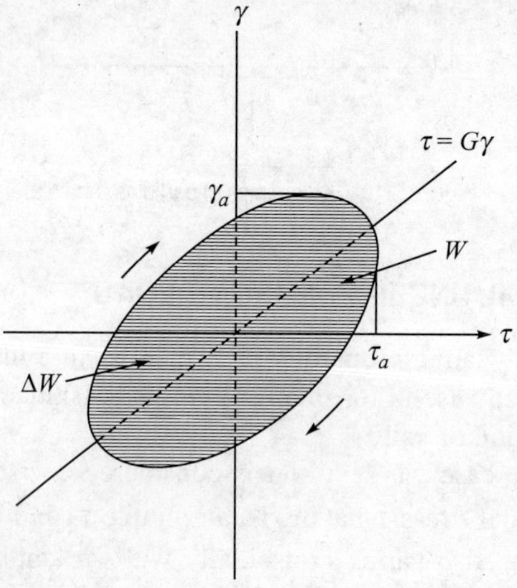

Figure 6.49 Definition of elastic energy W and the energy consumed (dissipated) in one cycle ΔW.

The maximum elastic energy W, stored in a mass, can be obtained as

$$W = \frac{1}{2} G\gamma_a^2 \qquad (6.43)$$

The ratio of energy loss per cycle to total elastic energy $\Delta W/W$ is given by

$$\Delta W/W = \frac{\pi G \gamma_a^2}{G\gamma_a^2/2} = 2\pi G'/G \qquad (6.44)$$

The viscoelastic model is frequently represented as a combination of a spring representing restoration force and a dashpot that gives viscous resistance proportional to velocity. The properties of material can be represented with better accuracy by increasing the number of springs and dashpots. But in doing so the task of defining coefficients for each element becomes difficult, sometimes impractical. Let us first study the Voigt model as a simple example.

Voigt model

As shown in Figure 6.50(a), one spring and one damping system placed in parallel are used in the Voigt model. If we assume that, according to this model, application of stress σ causes strain ε, then the relation between stress and strain can be expressed as

$$\sigma = \mu\varepsilon + \mu'\frac{d\varepsilon}{dt} \tag{6.45}$$

where μ is the modulus of elasticity and μ' is the coefficient of viscosity. The relation between G, G' and η is as follows:

$$G = \mu, \qquad G' = \mu'\omega, \qquad \eta = \tan\ \delta = \mu'\omega/\mu \tag{6.46}$$

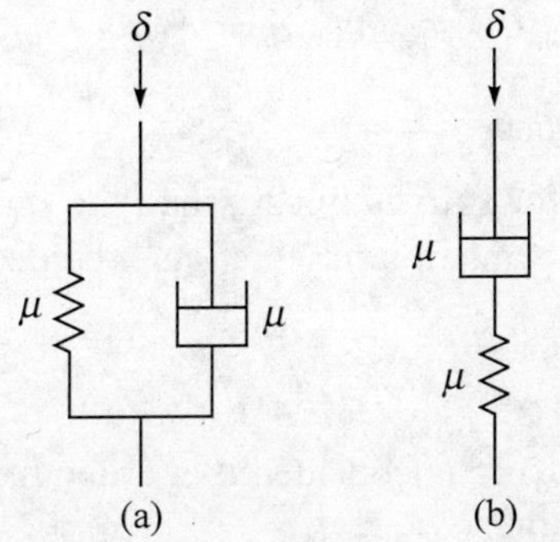

Figure 6.50 Examples of a viscoelastic model (a) Voigt model; (b) Maxwell model.

So, in Eq. (6.4b) according to the Voigt model, G has the same value as the modulus of elasticity μ. However, G' and loss coefficient η are linear functions of ω and both increase with vibration frequency.

Maxwell model

In the Maxwell model the spring and dashpot are connected in series, as shown in Figure 6.50(b) and the relation between stress σ and strain ε in this model is

$$\frac{d\sigma}{dt} + \frac{\mu}{\mu'}\cdot\sigma = \mu\frac{d\varepsilon}{dt} \tag{6.47}$$

If we determine the relation G, G' and η in the same way as in the Voigt model, we get

$$G = \frac{1}{\mu}\frac{1}{[(1/\mu)^2 + (1/\mu'\omega)^2]}$$

$$G' = \frac{1}{\mu'\omega}\frac{1}{[(1/\mu)^2 + (1/\mu'\omega)^2]}$$

$$\eta = \tan\ \sigma = G'/G = \mu/\mu'\omega \tag{6.48}$$

Thus, G, G' and η are functions of vibration frequency ω. The loss coefficient η is inversely proportional to ω and as such decreases with an increase in frequency.

The actual soil behaviour, within the frequency range of practical interest, shows that the modulus of elasticity and damping are both independent of frequency. Hence, it is desirable to build a model in which G and G' are independent of frequency in order to bring it closer to actual soil behaviour. In such a model, loss modulus G' representing damping (energy dissipation) should be independent of frequency, that is, it should be non-viscous damping. It is very difficult to build such a model using just a simple combination of a spring and a dashpot. The following relation between stress and strain, involving complex numbers, can be used to express the desired properties:

$$\sigma = (\mu + i\mu')\varepsilon$$
$$G = \mu,\ G' = \mu''$$
$$\eta = \tan\,\delta = \mu''/\mu \tag{6.49}$$

Dynamic stress–strain characteristics

The shear stress-strain relation for soils is linear when the range of strain is very small. In this linear range, the shear modulus is maximum which may be designated as G_{max}. The shear strain γ_r is defined as

$$\gamma_r = \tau_{max}/G_{max} \tag{6.50}$$

If a soil element dx by dy by dz is considered at a depth z from surface, and if σ_v is taken as the effective vertical stress, then

$$\text{Effective horizontal stress} = k_0\sigma_v$$

where k_0 is the coefficient of earth pressure at rest.
From the Mohr's circle,

$$\tau_{max} = [\{1/2(1 + k_0)\sigma_v \sin\,\phi + c\,\cos\,\phi\}^2 - \{1/2(1 - k_0)\sigma_v\}^2] \tag{6.51}$$

Using hyperbolic stress-strain curve, $\tau = \dfrac{\gamma}{\dfrac{1}{G_{max}} + \dfrac{1}{\gamma_{max}}}$

as $G = \dfrac{\tau}{\gamma}$

Combining these two equations

$$G = \frac{G_{max}}{1 + \dfrac{\gamma}{\gamma_r}} \quad \text{or} \quad \frac{G}{G_{max}} = \frac{1}{1 + \dfrac{\gamma_r}{\gamma_r}}$$

For very small range of strain, $\gamma/\gamma_r = 0$, then $G = G_{max}$

Otherwise for any strain amplitude the value of G may be obtained.

Barkan's formula

Barkan in 1962 proposed the relationship between the coefficient of elastic uniform compression C_u and G as

$$G = \frac{C_u (1-\nu)\sqrt{A}}{2.26} \tag{6.52}$$

in which A = contact area and ν = Poisson's ratio.

Hardin and Black formula

Hardin and Black (1969) proposed that

$$G_{\max} = \frac{1230\,[\text{OCR}]^k\,(2.973 - e^2)}{(1+e)\sqrt{\sigma_0}} \tag{6.53}$$

where,

OCR = over consolidation ratio
σ_0 = effective all round stress, in psi
e = void ratio.
k = a factor that depends upon the P.I. (plasticity index) of clays.

Hardin and Black identified the following soil properties which influence the shear modulus of soil.

(i) Grain size and mineralogy of soil
(ii) Void ratio
(iii) Initial average effective confining pressure
(iv) Degree of saturation
(v) Frequency of vibration and number of cycle of load
(vi) Ambient stress history and vibration history
(vii) Magnitude of dynamic stress
(viii) Time effects

For calculation of $G_{\max}$ by Hardin and Black formula,

$$\sigma_0 = \frac{1}{3}[\sigma_1 + \sigma_2 + \sigma_3]$$

The value of k is given in Table 6.6.

Table 6.6 Values of k

Plastic Index (P.I.)	*k*
0	0
20	0.18
40	0.30
60	0.41
80	0.48
>100	0.50

Prakash and Puri (1980) reported detailed *in situ* data on dynamic constant for several sites from resonance tests on blocks, the shear modulus test and the cyclic plate load test, as per IS recommendations.

Nonlinear expression

The material constituting ground may be safely considered elastic for strain levels below $10^{-5}\%$. As the strain level increases, the inelastic part becomes prominent and can no longer be treated as a linear elastic system. In fact, the soil material cannot be treated in any single manner throughout since it is generally a mixture of three phases: solid, liquid and gases; it has dilatancy due to which its volume varies when it is subjected to stress; the dynamic properties are very sensitive to stress history and they are time-sensitive as well. Soil modeling for nonlinear behaviour under repetitive cyclic loading is one of the major topics of study in the field of soil dynamics. It is necessary to build a model that represents changes in effective stress as a result of dilatancy when a saturated soil is subjected to cyclic loads resulting in softening or liquefaction of sand, etc.

A number of nonlinear models have been proposed from this point of view and are being used in seismic-response analysis. We shall discuss here two representative models, namely: (1) the Hardin-Drenvich and (2) the Ramberg-Osgood. These two models are frequently used in nonlinear response analysis and may be considered basic models for building a more complex model. Here the stress-strain relation is expressed as one of shear stress-shear strain and hence some care is necessary while applying this model to a general stress combination. These two models have been described by Ishihara, JSCE (1997).

Hardin-Drnevich model

In this model the relation between the shear stress τ and the shear strain γ as shown in Figure 6.51 is expressed as follows.

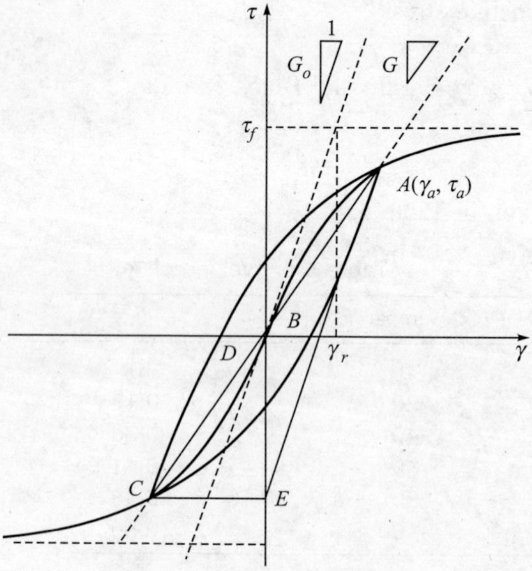

Figure 6.51 Hardin-Drnevich model

$$\tau = \frac{G_0 \gamma}{1 + \dfrac{\gamma}{\gamma_r}}$$

where G_0 is the shear modulus under minimum strain levels and is referred to as initial shear modulus; γ_r is the standard strain parameter defined as τ_f / G_0, where τ_f is shear strength; γ_r indicates the value of γ in the $G/(G_0 - \gamma)$ curve where $G/G_0 = 0.5$. The above equation asymptotically approaches the value of $G_0\, \gamma_r = \tau_f$ as $\gamma \to \infty$. If we replace $G = \tau/\gamma$, then

$$\frac{G}{G_0} = \frac{1}{1 + \dfrac{\gamma}{\gamma_r}}$$

In this model the hysterisis curves corresponding to unloading or reloading are not defined, but specified in terms of h_{max}, the value of damping ratio under infinitely large strain

$$\frac{h}{h_{max}} = 1 - \frac{G}{G_0}$$

This means that nonlinear properties are expressed in terms of three parameters: G_0, h_{max} and γ_r. The values of G/G_0 as calculated from Eqs. (6.50 and 6.51) and the variations in h with the amplitude of shear strain are shown by a broken line in Figure 6.52.

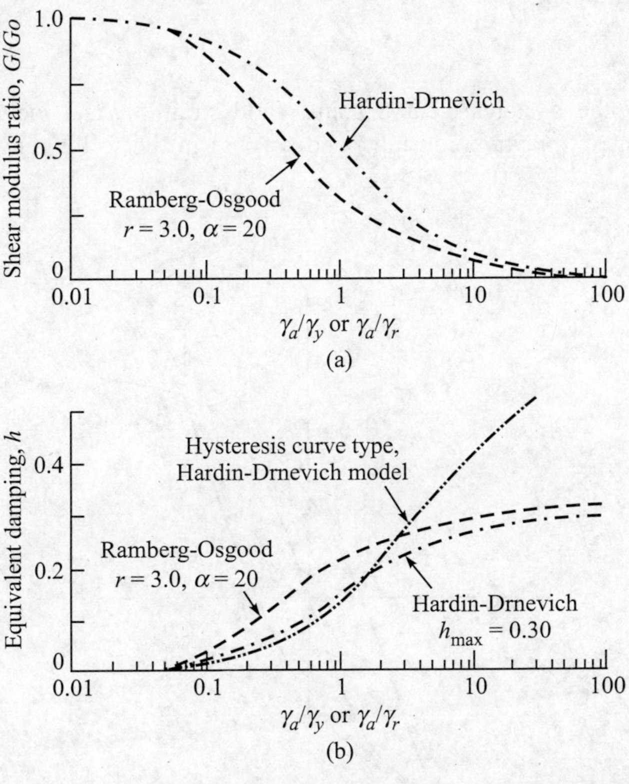

Figure 6.52 Dependence of shear modulus and damping ratio on strain in various models: (a) Dependence of shear modulus ratio on strain as derived from a nonlinear model. (b) Dependence of damping ratio on strain as derived from a nonlinear model.

The hysteresis loop can also be defined on the basis of equation using Masing's law. In this case the ascending and descending curves are expressed as

$$\tau - \tau_a = \frac{G_0(\gamma - \gamma_a)}{\left(1 + \dfrac{|\gamma - \gamma_a|}{2\gamma_r}\right)}$$

$$\tau + \tau_a = \frac{G_0(\gamma + \gamma_a)}{\left(1 + \dfrac{|\gamma + \gamma_a|}{2\gamma_r}\right)}$$

where τ_a and γ_a are shear stress and shear strain, respectively, as shown in Figure 6.53(a).

Using these equations and from the definition of damping ratio, h can be obtained as follows:

$$h = \frac{2}{\pi}\left[\frac{G_0}{G}\left\{\frac{\gamma_r}{\gamma_a} - \left(\frac{\gamma_r}{\gamma_a}\right)^2 \ln\left(1 + \frac{\gamma_a}{\gamma_r}\right)\right\} - 1\right]$$

Accordingly, the peak value of h at $\gamma \to \infty$ is $2/\pi$. The relation between h, as determined from the above equation, and shear strain amplitude is shown by the broken line in Figure 6.52(b). It can be seen that at larger strain levels the value of the damping ratio is even higher than that obtained from above equation.

Ramberg-Osgood model

This model express the hysteresis curve using yield strain γ_y and yield stress τ_y shown in Figure 6.53, constant α (positive constant) and γ (= 1) in the following manner:

$$G_0\gamma = \tau + \frac{\alpha|\tau|^r}{(G_0\gamma_y)^{r-1}}$$

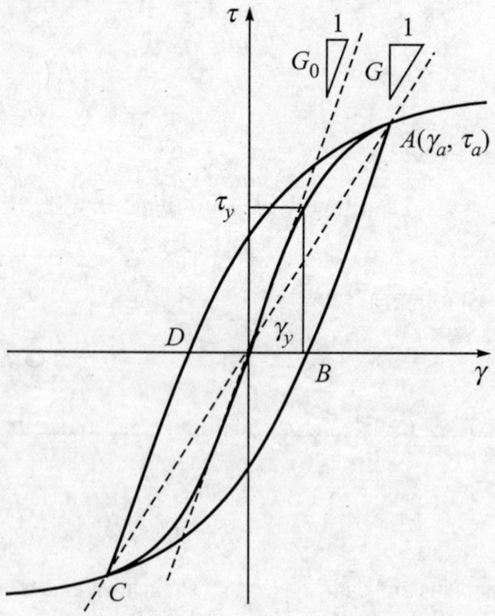

Figure 6.53 Ramberg-Osgood model.

Ascending and descending curves using this equation and Masing's law will be

$$G_0(\gamma - \gamma_a) = (\tau - \tau_a)\left[1+\alpha\left|\frac{\tau - \tau_a}{2G_0\gamma_y}\right|^{r-1}\right]$$

$$G_0(\gamma - \gamma_a) = (\tau + \tau_a)\left[1+\alpha\left|\frac{\tau - \tau_a}{2G_0\gamma_y}\right|^{r-1}\right]$$

The damping ratio h will be:

$$h = \frac{2}{\pi}\,\frac{(r-1)}{(r+1)}\left(1-\frac{G}{G_0}\right)$$

If we use the secant shear modulus $G\ \tau_a/\gamma_a$ shown by the broken line in Figure 6.53, the equation becomes

$$\frac{G}{G_0} = \frac{1}{\left[1+\alpha\left(\frac{\tau_a}{\tau_y}\right)^{r-1}\right]}$$

This gives the relation between G/G_0 and strain amplitude γ_a.

There are four constants involved in this model, namely, τ_y, γ_y, α, γ, which makes it possible to explain the experimental findings. But again, evaluating the values of constants is a difficult task. Various modified models have been proposed for overcoming this problem.

From the models mentioned above, the dynamic properties shown above are used in response analysis based on the equivalent linearization method. All these analyses are based on total stress and additional help is required to calculate directly the changes in effective stress. It is possible, however, to use the results of calculation of dynamic shear stress to indirectly evaluate liquefaction potential and ground settlement.

6.8 DAMPING RATIO

Damping properties play important role in studying dynamic and seismic response of ground material. Damping is basically result of two factors namely

- Viscosity and plasticity properties of soil
- Radiation damping wherein vibration energy is released with propagation of waves

Several viscoelastic model is used to represent damping properties of ground material. In the voigt model damping ratio increases with frequency of vibration, while in the Maxwell model it reduces. In this section the symbol used for damping ratio is h whereas in remaining part of text this has been invariably represented by ξ. The damping ratio of various ground material are determined in laboratory tests as well as by in situ measurements at site. The value of damping ratio (h or ξ) for various type of soils varies in the range of 0.3 to 10.0.

Example 6.1 For seismic design of a machine foundation located in seismic zone II, sa vertical vibration test was done as per IS 5249 recommendations on a 1.5 m × 0.75 m × 0.7 m high concrete block in an open pit of depth 275 cm. The soil parameters are $c = 0$, $\phi = 30°$, $V_{sat} = 18$ kN/m^3, $\nu = 0.30$. The water table is located at a depth of 5.0 m from the existing ground level. For installing the machine, the suitable dimension of foundation is 3.5 m × 2.75 m × 2.5 m. Determine the suitable value of dynamic soil properties C_u, E and G for design. The permissible value of vertical dynamic amplitude of machine is 100 μm and the mass of the machine may be taken as 150 kg. The results of vertical vibration test are shown. [Figure 6.54(b)] Also evaluate the value of C_τ, C_ϕ, C_ψ. The unit weight of the foundation block is 2350 kg/m^3.

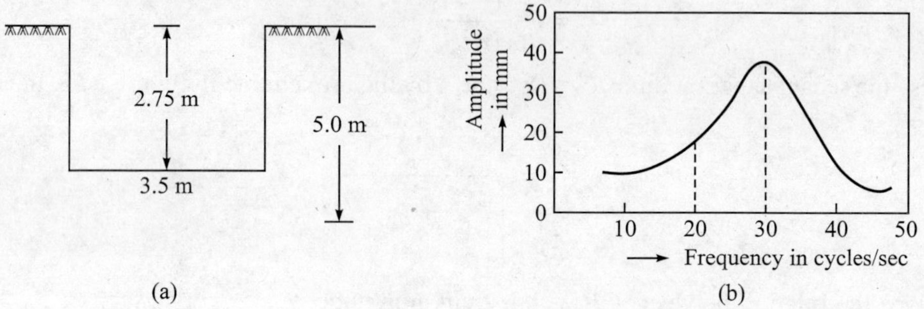

Figure 6.54 Example 7.1.

Solution: The value of C_u may be obtained as

$$C_u = \frac{4\pi^2 f_{nz}^2 m}{A}$$

where,

A = area of block = 1.5 × 0.75 = 1.125 m^2
Mass of block = 1.125 × 0.7 × 2350 = 1850.62 kg where unit wt. of block = 2350 kg/m^3
Mass of machine = 150 kg
Total mass of block and machine, m = 2000.62 kg
From the test result, f_{nz} = 30 cycles/second

$$f_n^2 = 900$$

We have $$C_u = \frac{4\pi^2 (f_{nz}^2)\cdot m}{A} = \frac{4\times\pi^2\times 900\times 2000.62}{1.125\,1000} = 6.3 \times 10^4 \text{ kN/m}^3$$

From $$C_u = \frac{1.13\,E}{(1-\nu^2)}\cdot\frac{1}{\sqrt{A}},\ E = \frac{(1\cdot\nu^2)}{1.13}\sqrt{A}\cdot Cu$$

where,

$$\nu = \text{Poisson's ratio} = 0.3$$

$$E = \frac{(1-\nu^2)\sqrt{A}}{1.13}\cdot C_u \text{ kN/m}^3 = 0.85C_u = 5.37 \times 10^4 \text{ kN/m}^2$$

and, $$G = \frac{E}{2(1+\nu)} = 2.06 \times 10^4 \text{ kN/m}^2$$

Using the relationship,

$$C_\tau = \frac{C_u}{2} = 3.15 \times 10^4 \text{ kN/m}^2$$

$$C_\phi = 3.46C_\tau = 10.8 \times 10^4 \text{ kN/m}^2$$

$$C_\psi = 1.5C_\tau = 4.72 \times 10^4 \text{ kN/m}^2$$

PROBLEMS

6.1 A soil specimen was tested in a resonant column torsional set-up with fixed-free ends. The following details of the specimen are given.

length = 90 mm, diameter = 35 nm, mass = 175 g

In the resonant test the resonant frequency was observed at 750 cycles per second, determine the velocity of shear waves and shear modulus.

6.2 A cyclic plate load test was performed with a plate of 30 cm × 30 cm size. An elastic settlement of 3 mm was observed to a loading intensity of 90 kN/m^2. Determine the coefficient of elastic uniform compression C_u for a foundation block having a size 4 m × 4 m.

6.3 Discuss the factors on which elastic modulus and shear modulus of soil depend. Explain the backbone curve for soils. Describe the provisions and recommendation of the Indian standards for determining dynamic shear modulus.

7

DYNAMIC EARTH PRESSURE

7.1 INTRODUCTION

Earth pressure problems have a special significance in geotechnical engineering for analysis and design of structures like retaining walls, bulkheads, sheet piles, culvert, abutments and coffer-dams. Without the support of structures the soil at higher elevation would tend to move down till its natural, stable configuration is compatible with angle of response. Consequently, true soil that is retained at a slope steeper than it can sustain by virtue of its shearing strength exerts a force on the retaining structure. This force is called the Earth Pressure. The magnitude and classification of the earth pressure is a function of the absolute and relative movement of the soil and the structure. These conditions are very well-known in geotechnical engineering. Depending upon the state of wall movement, there are three possibilities, namely:

(a) State of rest—earth pressure at rest [P_0]
(b) Movement away from the fill—active earth pressure [P_a]
(c) Movement towards the fill—passive earth pressure [P_p]

In static conditions, these states are obtained by different theories mainly by Rankine's theory and Coulomb's theory and their determination depends upon the true coefficient of active pressure (K_a) and the coefficient of passive pressure (K_p).

In classical theory of elasticity the analysis of such long retaining structures compared to the cross-section is regarded as a classical case of plane strain problems in elasticity. However, under dynamic conditions during earthquakes, the retaining structures are subjected to dynamic motion and consequently owing to ground motion the dynamic earth pressure is more than the static earth pressure. So far evaluation of the dynamics earth pressure, there is a need to ascertain the deformation of retaining structures in terms of displacement (sliding) and rotation (overturning). Often it is difficult to ascertain these dynamic soil structure interactions, therefore, pseudostatic approach is employed, wherein the dynamic force is replaced by an equivalent static force. The approach is to determine all static forces along with pseudostatic forces and proceed with conventional stability checks for overturning, sliding and overall stability. Therefore, before taking up the determination of the dynamic earth pressure, the classical earth pressure theory for the static case is discussed first.

7.2 CLASSICAL THEORY FOR STATIC EARTH PRESSURE

The classical earth pressure theory was proposed by Coulomb in 1776 and later on by Rankine in 1857.

7.2.1 Rankine's Earth Pressure Theory

A body of soil is said to be in a state of plastic equilibrium if every part of it is in an incipient failure condition. Plastic equilibrium, which can develop in a semi-infinite mass of cohesionless soils when acted upon by the force of gravity, was investigated by Rankine (1857).

In Figure 7.1(a), AB is a horizontal surface of a semi-infinite mass of cohesionless soil with a unit weight γ. At depth z below AB, the vertical pressure is

$$p_v = \gamma z \tag{7.1}$$

After deposition of this mass of soil, the value of the lateral earth pressure p_h corresponds to the at-rest value; that is

$$p_h = p_0 = K_0\, p_v = K_0\, \gamma z$$

Since this element is symmetrical with respect to a vertical plane, the normal stress may be taken as a principal stress. Consequently, the normal stress on the vertical side also is a principal stress. Thus, the principal stresses in the soil mass at depth z are,

$$\sigma_1 = \gamma z \qquad \text{and} \qquad \sigma_3 = K_0 \gamma z$$

In Figure 7.1(c), circle I corresponds to the at-rest condition. Now, as the soil mass stretches, lateral pressure (a minor principal stress in this case) decreases, and the diameter of the Mohr circle increases. According to Mohr-Coulomb failure criteria, the greatest diameter that

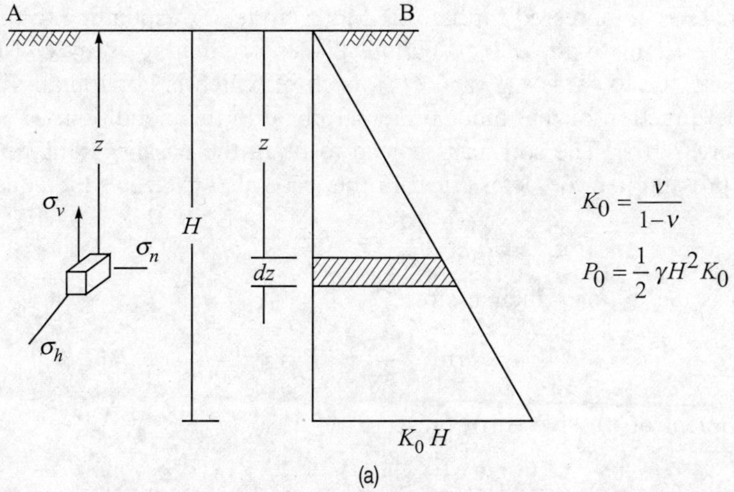

(a)

Figure 7.1(a) Earth pressure at rest condition.

a Mohr circle can have is when the Mohr circle (II) is tangential to the Mohr strength planes inclined at 45° + $\phi/2$, each to the major principal plane. See Figure 7.5(b).

A relationship between major and minor principal stresses at incipient failure is given by

$$\frac{\sigma_1}{\sigma_3} = \frac{1+\sin\phi}{1-\sin\phi} \tag{7.2}$$

or,

$$\frac{\sigma_1}{\sigma_3} = \tan^2\left(45°+\frac{\phi}{2}\right) \tag{7.3}$$

or,

$$\frac{\sigma_1}{\sigma_3} = N\phi, \text{ where } N\phi = \tan^2\left(45°+\frac{\phi}{2}\right) \tag{7.4}$$

or,

$$p_v = \gamma z = p_h\, N\phi \tag{7.5}$$

or,

$$p_h = \frac{\gamma z}{N\phi} = K_a \gamma z \tag{7.6}$$

where K_a, the coefficient of active earth pressure, is given by

$$K_a = \frac{1}{N\phi} = \frac{1-\sin\phi}{1+\sin\phi} \tag{7.7}$$

It should be noted that once the lateral earth pressure is reduced to the active value, further stretching of the mass has no effect on p_h, but sliding occurs along planes in the direction of Of_1 and Of_2, which are horizontally inclined at 45° + $\phi/2$. It should be further noted that failure will be incipient on all planes parallel to Of_1 and Of_2 (Figure 7.1(c)). The vertical traces of such planes in Figure 7.1 constitute the shear pattern. The above concepts of the states of plastic equilibrium in an active condition may be extended to a retaining wall problem if the following assumptions are made: (1) the wall face is smooth and vertical, and (2) the deformation condition for plastic equilibrium is satisfied.

If the soil mass is compressed further, the Mohr circle corresponding to this state of stress is shown by circle III in Figure 7.1(c). Failure planes originating from O' (the origin of the planes) in this case, are towards $O'f_3$ and $O'f_4$, each of which is horizontally inclined at 45° – $\phi/2$, which is the direction of the minor principal plane in this condition. The shear pattern is sketched in Figure 7.1(b). The soil mass is said to be in the passive Rankine state.

The lateral pressure can be determined in this case also by using the equation

$$\frac{\sigma_1}{\sigma_3} = N\phi \tag{7.8}$$

Since σ_3 (= γz) is the minor principal stress,

$$\sigma_1 = \frac{\gamma z}{N\phi} = K_p \gamma z \tag{7.9}$$

where the coefficient of passive earth pressure

$$K_p = N\phi = \frac{1+\sin\phi}{1-\sin\phi} = \tan^2\left(45°+\frac{\phi}{2}\right) \tag{7.10}$$

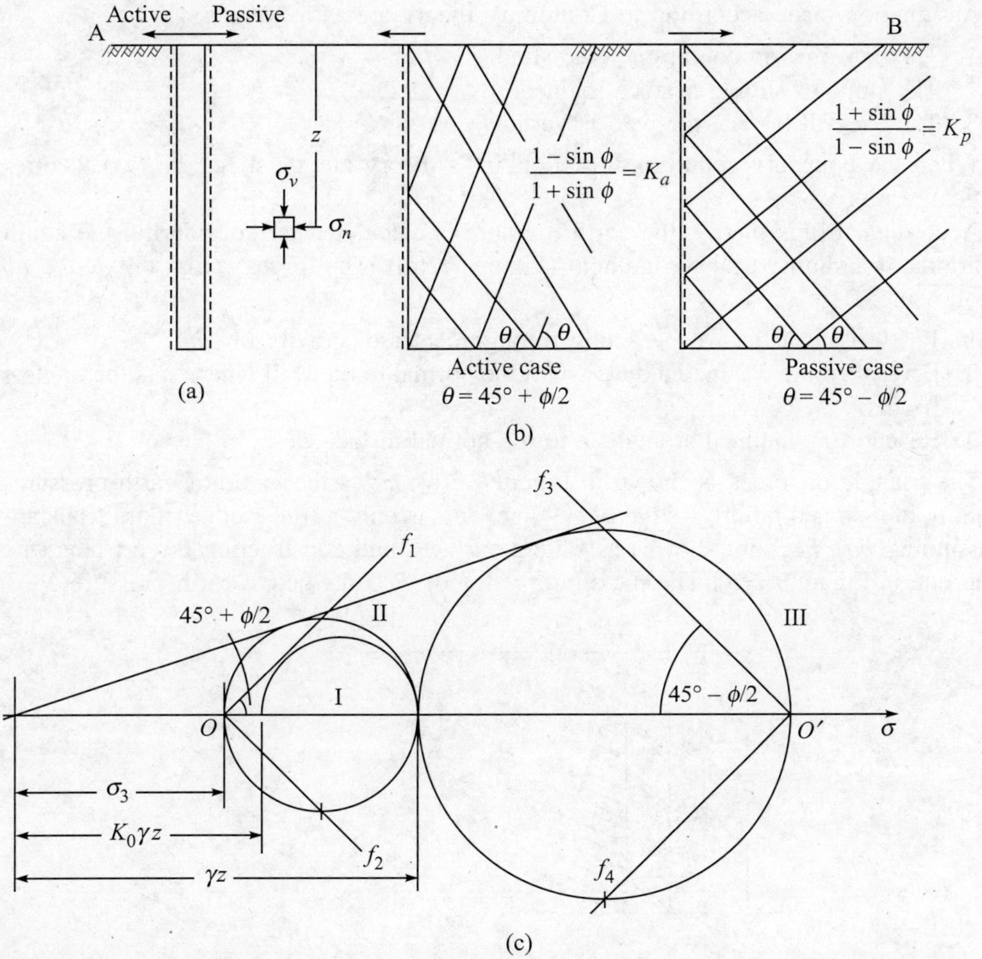

Figure 7.1(b) and (c) (b) Rankine's states of plastic equilibrium illustrating active and passive conditions. (c) Mohr diagrams illustrating and active passive conditions.

It should again be noted that once the Rankine passive resistance has been mobilized, further compression of the soil causes no increase in soil resistance; instead, slippage occurs along the failure planes indicated in the shear pattern.

7.2.2 Coulomb's Earth Pressure Theory

Unlike Rankine's theory, Coulomb's theory of earth pressure does not assume the wall surface to be smooth. In addition, Coulomb's method can be adapted to the any boundary conditions, for example, inclined walls with a break, inclined uniform and non-uniform slopes, and concentrated and distributed surcharge loads. A modified Coulomb method is used to determine the increase in static earth pressure due to a dynamic load.

Assumption made according to Coulomb's theory are as following:

1. The deformation condition is satisfied.
2. The slope of sliding surface is linear.
3. The backfill face of the wall is vertical.

When the boundary conditions for Rankine's theory are satisfied, the two theories yield identical results.

According to this theory, the earth pressure is calculated by considering the equilibrium conditions of a sliding trial wedge abc$_1$ (Figure 7.2(a)). The forces acting on wedge abc_1 are the following:

1. W_1, weight of the wedge acting through centre of gravity of abc_1
2. Earth pressure P_1, inclined at δ with the normal to the wall where δ is the angle of wall friction
3. Reaction R_1 inclined at angle ϕ to the normal to face bc_1

The triangle of forces is shown in Figure 7.2(b). P_1 is the value of earth pressure corresponding to assumed failure wedge abc_1. Since this is only a trial wedge, more trials are made by assuming bc_2, bc_3 (not shown) as failure surfaces and constructing force triangles similar to the one in Figure 7.2(b). The maximum value of P is the active earth pressure P_a.

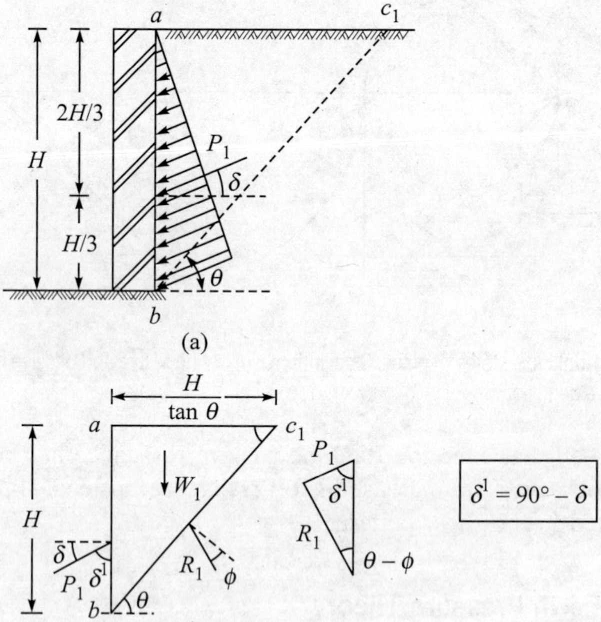

Figure 7.2 Coulomb's theory—active case, $c = 0$ backfill with vertical wall.

Coulomb derived the following analytical expression for the active earth pressure:

$$P_a = \frac{1}{2}\gamma H^2 \frac{\{\cos^2(\phi - \alpha)\}}{\cos^2\alpha \cos(\delta + \alpha)} \cdot \frac{1}{\left\{1 + \left[\dfrac{\sin(\phi + \delta)\sin(\phi - i)}{\cos(\alpha - i)\cos(\delta + \alpha)}\right]^{1/2}\right\}^2} \tag{7.11}$$

For determining earth pressure in cohesive soils, the basic principle remains the same, except that for an active case a force of cohesion $C_1 = c \times bc_1$ acts in the direction bc_1 and a force of adhesion $C_2 = c_a \times ab$ acts in the direction of ba as in Figure 7.2, wherein a force polygon is drawn and P_1 is the earth pressure for assumed failure wedge abc_1. Similarly, for inclined wall with $c = 0$ backfill and $c-\phi$ backfill the force polygon is shown in Figure 7.3 and 7.4, respectively.

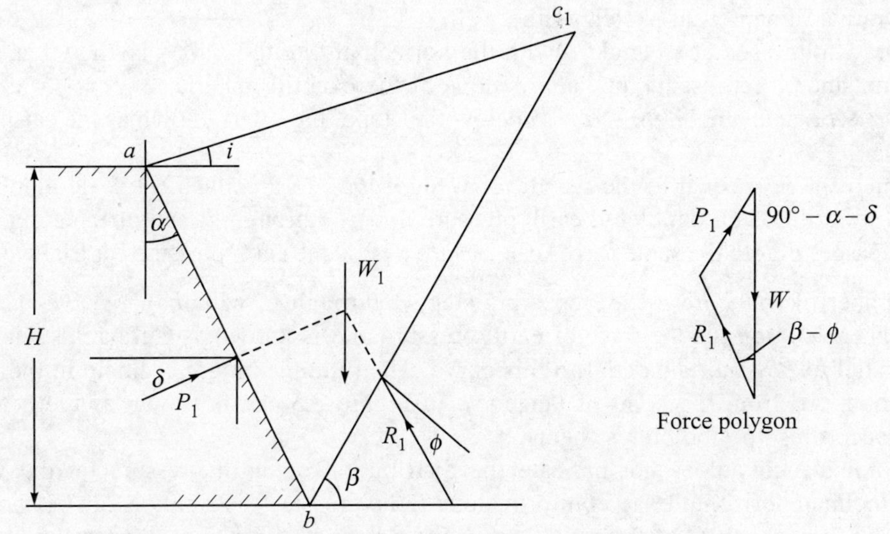

Figure 7.3 Coulomb's theory—active case, $c = 0$ backfill with inclined wall

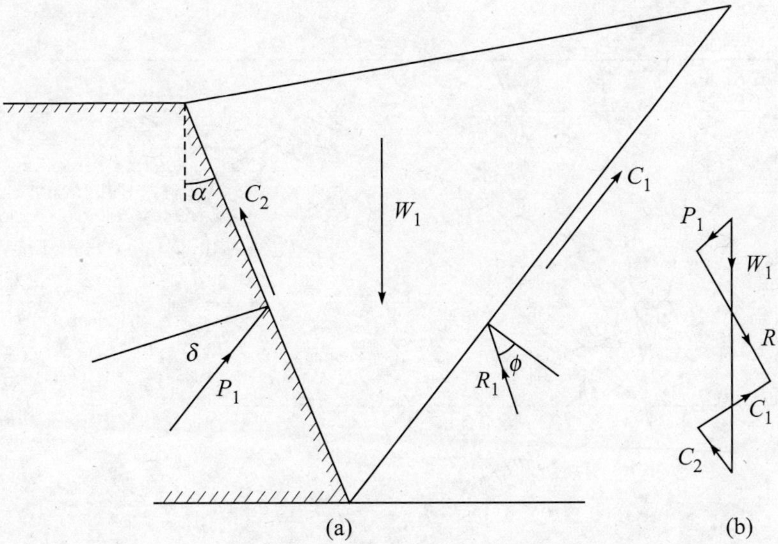

Figure 7.4 Coulomb's theory–active case, $c - \phi$ backfill (inclined wall)

7.2.3 Culmann's Graphical Construction

A graphical construction to determine lateral earth pressure for non-cohesive soils according to Coulomb's theory was suggested by Culmann (1866). Let us consider a retaining wall of height H, vertically inclined at an angle α. The unit weight of soil is γ and its angle of internal friction is ϕ. The angle of wall friction is δ. The steps in Culmann's construction (Figure 7.5) are as follows Prakash (1981):

1. Draw a dimensional sketch of the wall.
2. Draw a line bS_1 at an angle ϕ with the horizontal through b; bS_1 is known as the slope line since it represents the natural slope of the backfill material.
3. Draw bL at an angle $(90 - \alpha - \delta)$ below the slope line; bL is known as the earth pressure line.
4. Intercept bd_1, equal to the weight of wedge abc_1, to a convenient scale along bS_1.
5. Draw a line d_1e_1, parallel to earth pressure line bL, through d_1 and intersecting bc_6 at e_1.
6. Measure d_1e_1 to the same force scale as bd_1; d_1e_1 is the earth pressure for trial wedge abc_1.

A number of trials are made, repeating steps 1 through 6, with bc_2, bc_3, bc_4 as the trial wedges. Then $be_1e_2e_3e_4$ is the trace of earth pressure and is known as Culmann's curve. Draw a line parallel to bS_1 and tangential to this curve. Then the maximum ordinate in the direction of bL is obtained from the point of tangency. In Figure 8.5, de is shown as the active earth pressure according to Coulomb's theory.

Coulomb's theory does not indicate the distribution of earth pressure on the wall. For backfills inclined horizontally at a uniform slope i, the pressure distribution can be shown to be hydrostatic (Terzaghi, 1943); (Prakash, Ranjan and Saran, 1979). Hence, the total earth pressure acts at a height of $H/3$ above the base of the wall and is inclined at an angle δ with the normal to the wall.

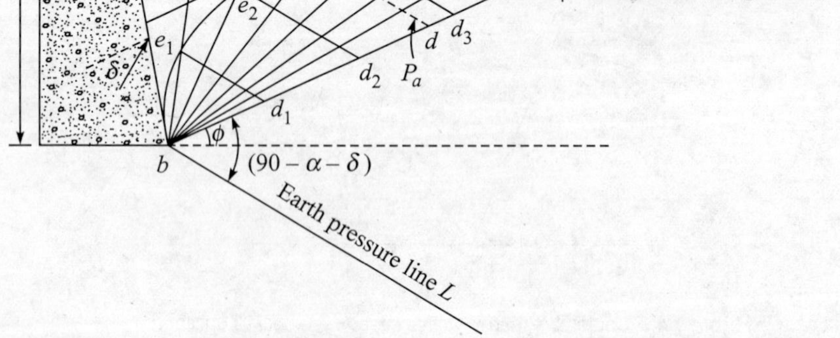

Figure 7.5 Culmann's construction for active earth pressure.

To determine passive pressures, the slope line bS_1 in Figure 7.5 is drawn below the horizontal line and the rest of Culmann's construction is unaltered.

7.3 DYNAMIC EARTH PRESSURE THEORY

A brief review of earth pressure on a retaining wall has been presented. The response of a retaining wall under an earthquake excitation will now be examined. The two coefficients of earth pressure are:

(a) Dynamic Active Earth Pressure coefficient $K_A]_{\text{dyn}}$ expressed as K_{ae}
(b) Dynamic Passive Earth pressure coefficient $K_p]_{\text{dyn}}$ expressed as K_{pe}

The same static formula may be used for evaluating the dynamic earth pressure and the values of dynamic coefficient. The other option is dynamic displacement analysis, wherein the movement (translation and rotation) of the retaining structures under earthquake induced ground motions are considered with the help of a mathematical model. The dynamic displacement and the natural period of motion of the wall–soil system are the important parameters for the dynamic analysis.

During earthquakes, there shall be significant pulses of ground shaking which will try to influence the static equilibrium position of the retaining structures as shown in Figure 7.6(a). For evaluating the dynamic stability of such retaining structures, the important considerations are

- Soil-Structure Displacement Mechanism
- Determination of Dynamic Earth Pressure
- Point of Application of Dynamic Earth Pressure

For a given earthquake the ground motion is as shown in Figure 7.6(b). With each pulse of ground shaking during an earthquake, it will thus be seen in Figure 7.6(c) that the wall keeps on moving from its static equilibrium position. During the earthquake there will be movement of the wall as well as the movement of the failure wedge. The observations may be described as follows as suggested by Prakash (1981).

- If the rate of movement of the wall and the failure wedge is the same, then there is no interaction between the wall and the failure wedge. Therefore, the pressure on the wall remains unaltered.
- If the rate of movement of the wall is greater than that of the failure wedge, then obviously the interaction between the wall and the failure wedge is reduced and the earth pressure may decrease from the true value in static condition.
- If the rate of movement of failure wedge is greater than that of the wall, then the earth pressure on the wall will increase.
- If the earthquake excitation produces an acceleration having the horizontal component as α_h and the vertical component as α_v, then in dynamic case the two inertial forces $W_1\alpha_h/g$ acting horizontally and $W_1\alpha_v/g$ acting vertically come in play where W_1 is the weight of the wedge abc. Thus, the inertial forces will depend upon the horizontal seismic coefficient and the vertical seismic coefficient.
- Based on the records available, the lateral pressure against a retaining wall will increase because of ground motion. The increase may be of the order of ten per cent for the wall of average height. It can be as high as 30 per cent in case of very high walls.

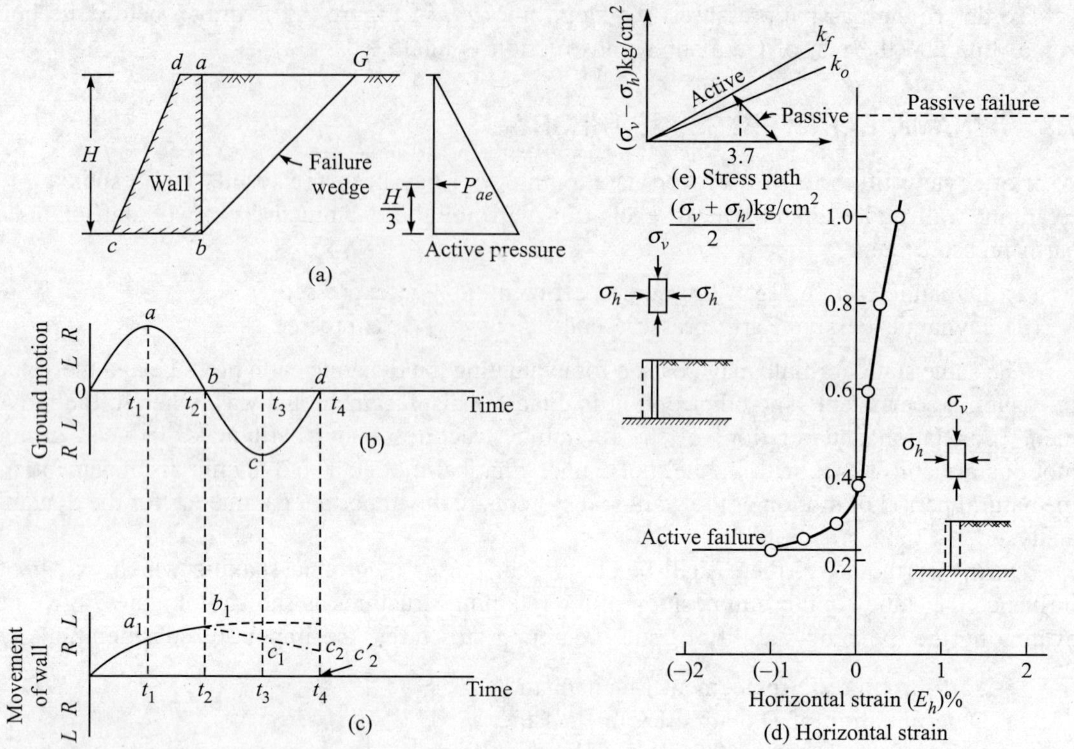

Figure 7.6 (a) Dynamics of retaining wall, (b) ground motion, (c) wall movement, (d) horizontal strain, and (e) stress path.

7.4 MONONOBE-OKABE THEORY FOR DYNAMIC EARTH PRESSURE

In geotechnical engineering for the evaluation of Active Earth Pressure and Passive Earth Pressure, Rankine's theory and Coulomb's classical theory are well-known. Mononobe–Okabe (1929) extended the classical Coulomb's theory for evaluating the dynamic earth pressure by incorporating the effect of inertia force. In the seismic zone the values of the earthquake horizontal and vertical seismic coefficients (α_h and α_v) are known. The Indian Standards IS-1893 have adopted and recommended the values of dynamic earth pressure as given by Mononobe–Okabe. This Code has also adopted the graphical method for evaluation of the dynamic earth pressure.

Consider a wall of height H as shown in Figure 7.7. The wall is inclined at an angle α with the vertical and is retained cohesionless [$c = 0$ soil] having unit weight γ.

The total active pressure P_a according to Mononobe–Okobe method may be expressed as

$$P_{ae} = \frac{1}{2}\gamma H^2(1 - \alpha_v)K_{ae}, \tag{7.12}$$

where the dynamic earth pressure coefficient K_{ae} is given by

$$K_{ae} = \frac{\cos^2(\phi - \theta - \psi)}{\cos\psi \cdot \cos^2\theta \cdot \cos(\delta + \theta + \psi)\left\{1 + \left[\dfrac{\sin(\delta + \phi)\sin(\theta - i - \psi)}{\cos(\delta + \theta + \psi)\cos(i - \theta)}\right]^{1/2}\right\}^2} \tag{7.13}$$

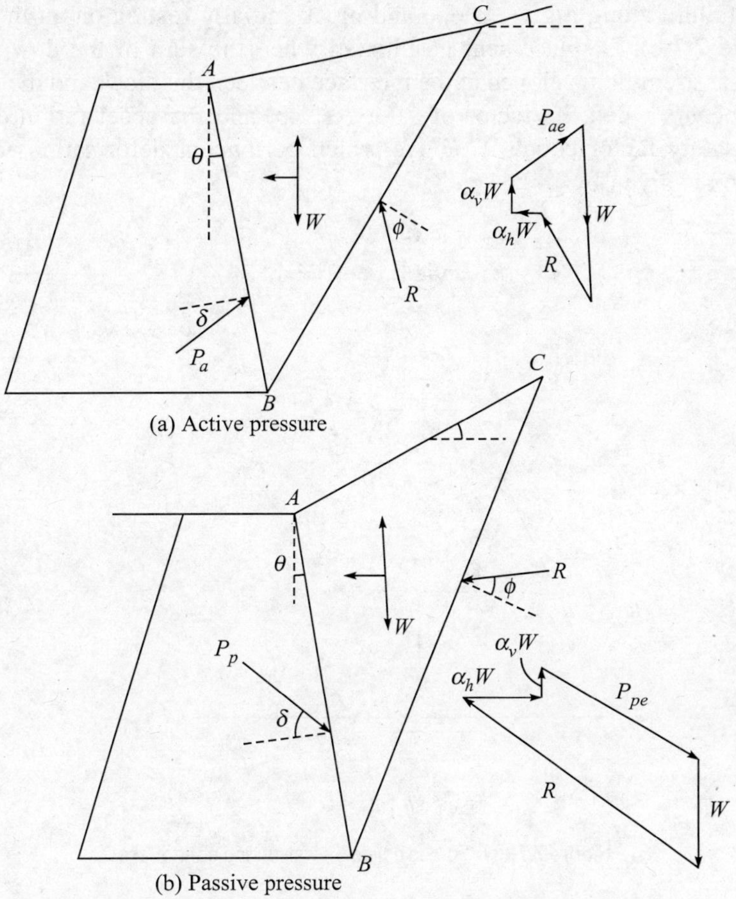

Figure 7.7 Mononobe–Okobe method, forces active on wall: (a) active pressure, and (b) passive pressure.

where $\tan\psi = \dfrac{\alpha_h}{1 \pm \alpha_v} = \tan\lambda$

Similarly, the total passive pressure P_{pe} according to M–O method may be expressed as

$$P_{pe} = \frac{1}{2}\gamma H^2(1 - \alpha_v)k_{pe} \tag{7.14}$$

where the dynamic passive earth pressure coefficient, k_{pe} is given by

$$k_{pe} = \frac{\cos^2(\phi + \theta - \psi)}{\cos\psi \cdot \cos^2\theta \cdot \cos(\delta - \theta + \psi)\left[1 - \sqrt{\dfrac{\sin(\delta + \phi)\sin(\theta + i - \psi)}{\cos(\delta - \theta + \psi)\cos(i - \theta)}}\right]^2} \tag{7.15}$$

7.4.1 Yield Acceleration

Newmark (1965) presented a procedure for estimating displacements associated with time-varying inertial force which is essentially an extension of pseudostatic approach. An analogy is made between failure along a given slope and block initially resting on an inclined surface as shown in Figure 7.7(c). Displacement is initiated when the sum of the downslope static and inertial equals the strength developed at the interface between the block and the undivided plane. The condition occurs when the factor of safety is 1.0, and the acceleration corresponding to a pseudostatic safety factor equals 1, above which permanent deformations accumulate. The yield acceleration is given by

$$a_y = k_y g \tag{7.16}$$

where, k_y = yield coefficient

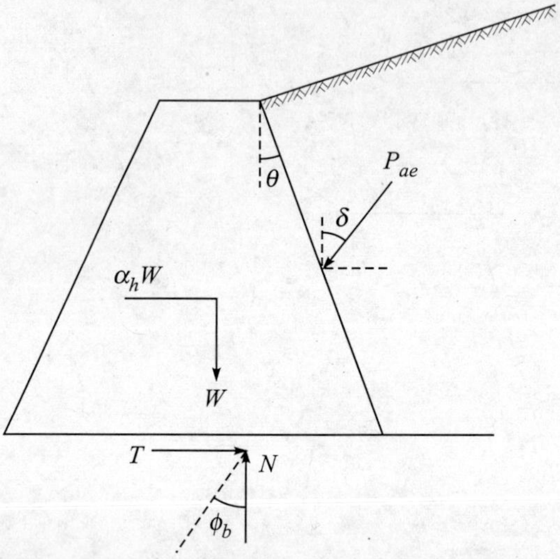

Figure 7.7(c) Forces acting on wall during earthquake.

Thus, the level of acceleration that is just large enough to cause movement or slide is the yield acceleration. Consider a gravity wall acted upon by gravity (g) and pseudostatic acceleration. When the level of acceleration is just large enough to cause the wall to slide on its base, it is called the yield acceleration. The horizontal and vertical equilibrium require that

$$N = W + P_{ae} \sin(\delta + \theta) \tag{7.17}$$

$$T = \alpha_h W + P_{ae} \cos(\delta + \theta) \tag{7.18}$$

$$T = N \cos\phi_b \tag{7.19}$$

where, P_{ae} = dynamic M.O. lateral force during seismic shocking.

ϕ_b = friction angle between to base and the foundation soil.

Makdisi and Seed (1978) also defined yield acceleration and used the concept for obtaining displacement. Displacements will occur along the slide plane whenever the acceleration of the sliding mass exceeds the yield value. Permanent deformation is also generated whenever the acceleration exceeds this value.

These expressions are combined to yield

$$\alpha_h = \tan\phi_b - \frac{P_{ae}\cos(\delta+\theta) - P_{ae}\sin(\delta+\theta)\tan\theta_b}{W} \tag{7.20}$$

Further, $\alpha_h = \dfrac{a_y}{g}$

The term a_y is the yield acceleration beyond which the lateral displacement occurs. Thus, the yield acceleration a_y may be expressed as

$$a_y = \left[\tan\phi_b - \frac{P_{ae}\cos(\delta+\theta) - P_{ae}\sin(\delta+\theta)\tan\phi_b}{W}\right]g \qquad (7.21)$$

7.5 DISPLACEMENT ANALYSIS

There are very few methods available to compute displacements of rigid retaining walls during earthquakes. The most acceptable is the Richard—Elms model based on Newmark's approach.

The displacements of retaining walls under earthquake type ground motion are under persistent investigation and review, but no simple, reliable and satisfactory method has been developed so far to realistically predict the displacement of a retaining wall during a given ground motion. The displacement analysis is based on the Richard—Elms model which relies on Newmarks procedure for evaluating the amount of slide of a soil mass. In this procedure the sliding mass is considered a rigid body and the resistance to sliding is assumed to act on a plane as shown in Figure 7.8(a).

Richard and Elms (1979) recommended that P_{ae} be calculated using the M.O. method. The M.O. method requires that a_y be known hence the solution of Eq. (7.21) for yield acceleration should be obtained. Richard and Elms prepared the following expression for permanent displacement

$$d_{perm} = 0.087\,\frac{v_{max}^2\, a_{max}^3}{a_y^4} \qquad (7.22)$$

where for the wall-backfill system,

v_{max} = peak ground velocity

a_{max} = peak ground acceleration

a_y = yield acceleration.

7.6 DYNAMIC STABILITY ANALYSIS

According to Newmark the slope failure would be initiated and movement would begin to occur if the initial forces on the potential sliding mass were reversed. Newmark evaluated an acceleration at which the inertial forces become sufficiently high to cause yielding to begin and an effective acceleration on sliding mass is obtained in excess of yield acceleration as shown in Figure 7.8(c). From this effective acceleration, the velocity and ultimately the displacements are obtained by integration on time scale.

The displacement analysis based on the Richard–Elms model is as follows. Assume the soil to be a rigid plastic material as shown in Figure 7.8(b).

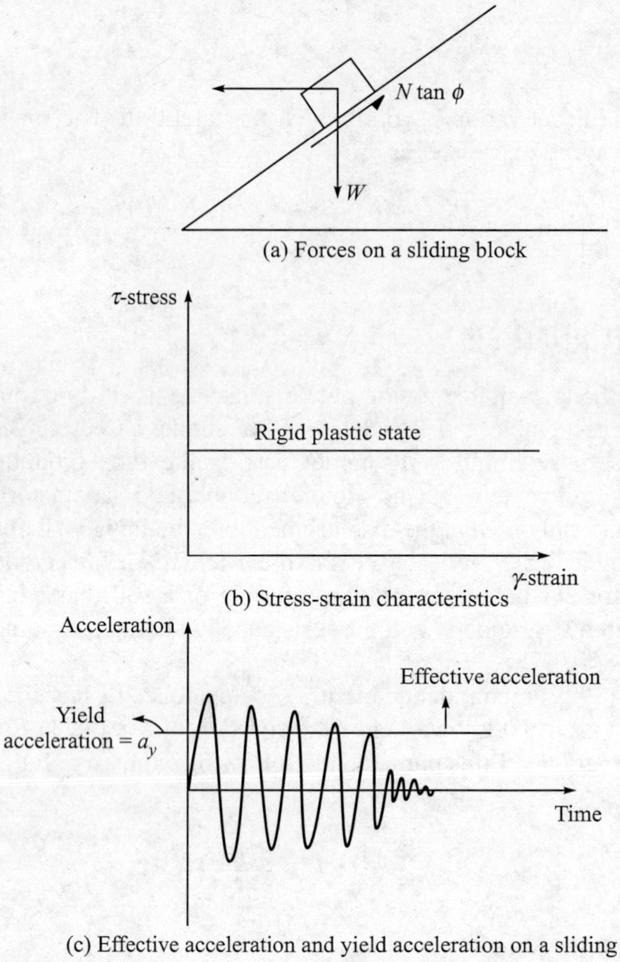

Figure 7.8 (a) Forces acting on sliding block, (b) stress-strain characteristics of soil, and (c) effective acceleration and yield acceleration on a sliding mass.

The earthquake forces have two components in vertical and horizontal directions acting on the c.g. of the retaining wall.

Consider a gravity retaining wall as shown in Figure 7.9 with backfill inclined at an angle i with the horizontal. In this analysis the terms used are as follows:

W_d = weight of the retaining wall
α_h, α_v = horizontal and vertical seismic coefficients
$[P_A]_{\text{dyn}} = P_{ae}$ = dynamic active earth pressure
β = inclination of the wall face with the vertical
δ = angle of the wall friction
ϕ_b = soil friction angle at the base of wall
R_N = vertical component of reaction
R_T = horizontal component of reaction

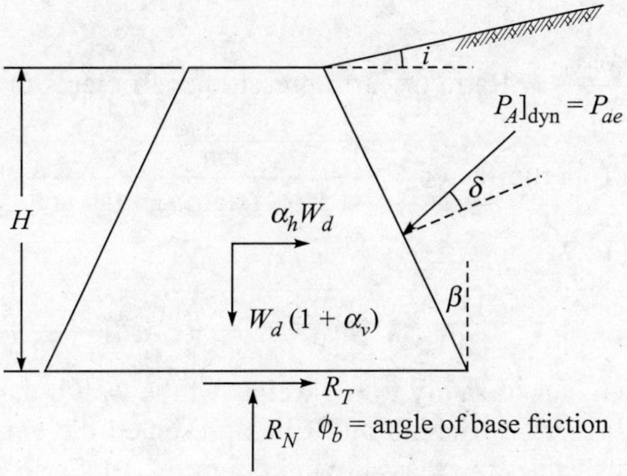

Figure 7.9 Forces on retaining wall.

Using equilibrium of forces,

$R_N = \Sigma$ Forces in vertical direction
$R_T = \Sigma$ Forces in horizontal direction

For sliding at base

$$R_T = R_N \tan \phi_b \tag{7.23}$$

Further considering the forces in vertical directions,

$$R_N = W_d \pm \alpha_v W_d + [P_A]_{\text{dyn}} \sin(\beta + \delta) \tag{7.24}$$

Solving these three equations,

$$W_d = [P_A]_{\text{dyn}} \frac{[\cos(\beta+\delta) - \sin(\beta+\delta)\tan\phi_b]}{(1+\alpha_v)\tan\phi_b - \alpha_h} \tag{7.25}$$

Substituting the value of $[P_A]_{\text{dyn}}$ and $\alpha_h = [1 \pm \alpha_v]\tan\psi$

$$W_d = \frac{1}{2}\gamma H^2 [K_A]_{\text{dyn}} \lambda_d \tag{7.26}$$

where,

$$\lambda_d = \frac{[\cos(\beta+\delta) - \sin(\beta+\delta)\tan\phi_b]}{(1\pm\alpha_v)(\tan\phi_b - \tan\psi)}$$

In static case, the weight of the wall W_s is given by

$$W_s = \frac{1}{2}\gamma H^2 k_a \lambda_s \tag{7.27}$$

where,

$$\lambda_s = \frac{[\cos(\beta+\delta) - \sin(\beta+\delta)\cdot\tan\phi_b]}{\tan\phi_b}$$

Therefore,

$$\frac{[W]_{\text{dyn}}}{[W]_{\text{static}}} = \frac{W_d}{W_s} = \frac{[K_A]_{\text{dyn}}}{K_a} = \frac{k_{ae}}{k_a} = \frac{\tan\phi_b}{[(1\pm\alpha_v)(\tan\phi_b - \tan\psi)]} \tag{7.28}$$

Taking,

$$F_T = \frac{[K_A]_{\text{dyn}}}{K_a} = \text{Ratio of earth pressure coefficients in dynamic and static cases}$$

$$F_1 = \text{wall inertia factor} = \frac{\tan \phi_b}{[(1 \pm \alpha_v)(\tan \phi_b - \tan \psi)]}$$

From Eqs. (7.26) and (7.27)

$$\frac{[W]_{\text{dyn}}}{[W]_{\text{static}}} = \frac{W_d}{W_s} = F_T F_1 = F_w \tag{7.29}$$

Thus, F_w emerges as the factor of safety to the weight of the wall to take into account the effect of soil pressure and wall inertia. Richard and Elms presented the variation of factor of safety F_w, F_T and F_I with various values of α_h as shown in Figure 7.10. From this the critical horizontal acceleration corresponding to $F_w = 1.5$ is equal to 0.105. Thus, for kinematic admissibility as well for realistic design, some lateral displacement of the retaining wall should be considered. Further for such design procedure, Franklind and Chang (1977) proposed that

$$\alpha_{hd} = \alpha_h \left[\frac{5\alpha_h}{d}\right]^{1/4}$$

where,

α_{hd} = design value of horizontal seismic coefficient
α_h = acceleration seismic coefficient from earthquake record
d = displacement of wall in mm.

Further, the design value of the vertical seismic coefficient may be taken as half of design value of the horizontal seismic coefficient.

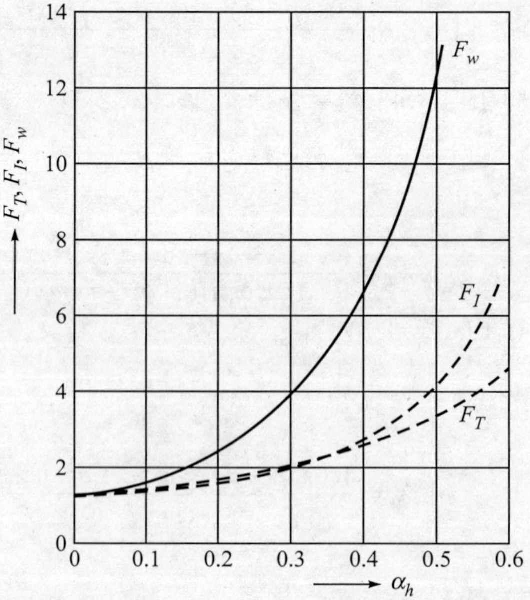

Figure 7.10 Variation of F_T, F_I and F_w with α_h (After Richards and Elms, 1979)

Critical seismicity coefficient

According to Wantane and Baba (1981), the critical seismicity coefficient is defined as that coefficient at which the inertial force of the entire soil mass on the potential sliding face is equal to the resistive moment on the sliding plane. The displacement will occur whenever seismicity coefficient higher than the critical seismicity coefficient acts on the sliding mass. This may be expressed as

$$\alpha_h \text{ or } \alpha_v = \frac{\iint \rho(x,y)\,\alpha(x,y)\,dx\,dy}{g\iint \rho(x,y)\,dx\,dy} \tag{7.30}$$

where,

$\alpha(x, y)$ = horizontal or vertical acceleration at a point (x, y) in the soil mass
$\rho(x, y)$ = density of soil.

Thus, the design procedure for the design of retaining wall based on Richard and Elms is as follows:

- Select an acceptable displacement of retaining wall, d
- Determine the value of α_{hd} and α_{vd}
- Determine the weight of the wall in dynamic condition, W_d
- Adopt a factor of safety, F_w = 1.5

7.6.1 Effect of Saturation on Lateral Earth Pressure

For submerged earth-fill, the dynamic increment (or decrement) in active and passive earth pressure during earthquakes can be found with the following modifications:

- The value of δ shall be taken as half the value of δ for dry backfill.
- The value of λ shall be taken as follows:

$$\lambda = \tan^{-1}\left(\frac{\gamma_s}{\gamma_s - 1}\right) \times \left(\frac{\alpha_h}{1 \pm \alpha_v}\right)$$

where,

γ_s = saturated unit weight of soil in gm/cc,
α_h = horizontal seismic coefficient
α_v = vertical seismic coefficient which is (1/2) α_h.

In addition, the following should also be considered:

- Buoyant unit weight shall be adopted.
- From the value of earth pressure found out as above, subtract the value of earth pressure determined by putting $\alpha_h = \alpha_v = \lambda = 0$ but using buoyant unit weight. The remainder shall be dynamic increment.
- Hydrodynamic pressure on account of water contained in earth-fill shall not be considered separately as the effect of acceleration on water has been considered indirectly.

7.6.2 Partially Submerged Backfill

The ratio of the lateral dynamic increment in active pressures to the vertical pressures at various depths along the height of wall may be taken as shown in Figure 7.11.

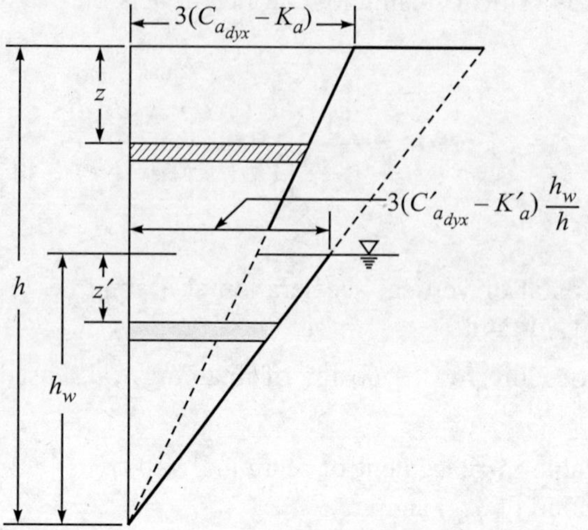

Figure 7.11 Distribution of ratio (lateral dynamic increment/vertical effective pressure) with height of wall (After IS 1893-1984)

The pressure distribution of dynamic increment in active pressures may be obtained by multiplying the vertical effective pressures by the coefficients in Figure 7.11 at the corresponding depths. The procedure may also be used for determining the distribution of dynamic pressure increments. Further,

C_a is computed as in for saturated backfills. (see Eq. (7.32))
C'_a is computed as for submerged backfills.
K_a is the value of C_a when $\alpha_h = \alpha_v = \lambda = 0$.
K'_a is the value of C'_a when $\alpha_h = \alpha_v = \lambda = 0$.

where,

h' is the height of submergence above the base of the wall.
h is the height of the retaining wall.

A similar procedure may be utilized for determining the distribution of dynamic decrement in passive pressures, according to IS 1893–1984.

7.7 RECOMMENDATIONS OF INDIAN STANDARD CODE OF PRACTICE

There is no mention of methods with details in various codes of several countries for computing the dynamic earth pressure under earthquake-type loading or the point of application of the dynamic increments during seismic activity or the displacement of the retaining structures under earthquake. However, Indian Standard 1893–94 does provide some useful information.

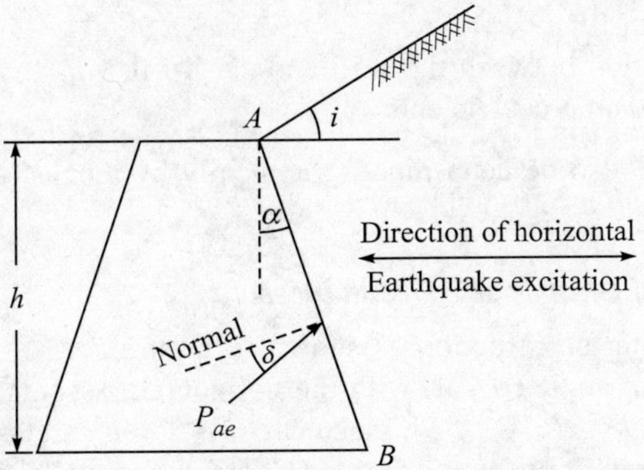

Figure 7.12 Active earth pressure due to earthquake on retaining walls (IS 1893-1984)

The salient features of the IS-1893 are described in the following paragraphs.

7.7.1 Lateral Earth Pressure

The pressure from earth-fill behind retaining walls during an earthquake can be obtained using dynamic earth pressure coefficient given in the IS code. For determination of dynamic co-efficient the cohesion has been neglected in the analysis where on this assumption is on the conservative side.

7.7.2 Dynamic Active Earth Pressure

The general conditions encountered in the design of retaining walls are illustrated in Figure 7.12. The active pressure exerted against the wall is given by

$$P_{ae} = \frac{1}{2} wh^2 C_a \tag{7.31}$$

where,

P_{ae} = dynamic active earth pressure in kg/m length of wall

w = unit weight of soil in kg/m^3

h = height of wall in m

$$C_a = [K_A]_{\text{dyn}} = \frac{(1 \pm \alpha_v)\cos^2(\phi - \lambda - \alpha)}{\cos\lambda \cos^2\alpha \cos(\delta + \alpha + \lambda)} \times \left[\frac{1}{1 + \left\{ \dfrac{\sin(\phi + \delta)\sin(\phi - i - \lambda)}{\cos(\alpha - i)\cos(\delta + \alpha + \lambda)} \right\}^{1/2}} \right]^2 \tag{7.32}$$

the maximum of the two being the value for design,

α_v = vertical seismic coefficient—its direction being taken consistently throughout the stability analysis of wall and equal to $(1/2)\alpha_h$

ϕ = angle of internal friction of soil,

$\lambda = \tan^{-1} \dfrac{\alpha_h}{1 \pm \alpha_v}$

α = angle which the earth face of the wall makes with the vertical

i = slope of earth-fill,
δ = angle of friction between the wall and earth-fill, and
α_h = horizontal seismic coefficient

The active pressure may also be determined graphically by means of the method described in IS-1893–1984 as follows:

Graphical Determination of dynamic active earth pressure

Let us make the following construction as shown in Figure 7.13.

Draw BB' to make an angle $(\phi - \lambda)$ with the horizontal. Assume planes of rupture Ba, Bb, etc., such that $Aa = ab = bc$, etc. On BB', make $Ba' = a'b' = b'c'$, etc., equal to Aa, ab, bc, etc., in length. Draw active pressure vectors from a', b', etc., at an angle $(90° - \delta - \alpha - \lambda)$ with BB' to intersect the corresponding assumed planes of rupture. Draw the locus of the intersection of assumed planes of rupture and the corresponding active pressure vector (modified Culmann's line) and determine the maximum active pressure vector X parallel to BE.

The active earth pressure shall be calculated as follows:

$$P_a = \frac{1}{2}\left\{\frac{(1 \pm \alpha_v)}{\cos\gamma}\right\} wX \cdot BC \tag{7.33}$$

where,

w = unit weight of soil, in kg/m^3

$\gamma = \tan^{-1} \dfrac{\alpha_h}{1 \pm \alpha_v}$

X = active pressure vector,

BC = perpendicular distance from B to AA' as shown in Figure 7.13.

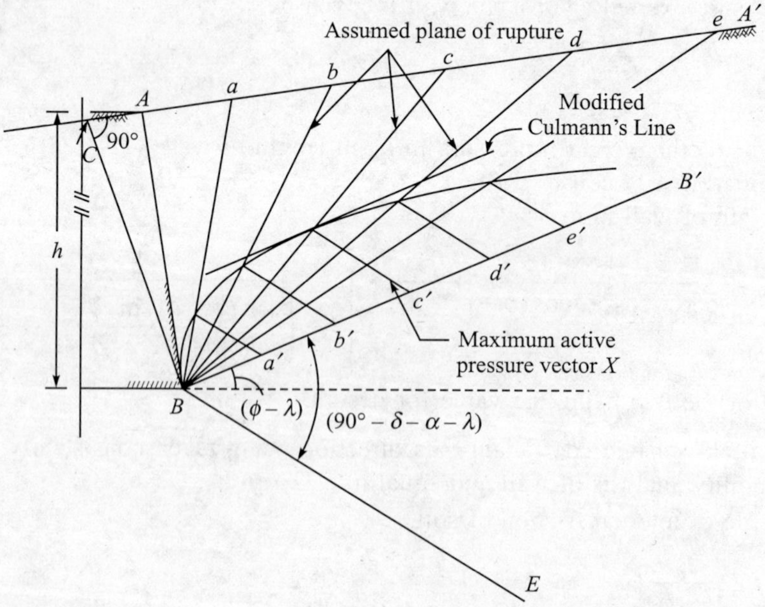

Figure 7.13 Determination of dynamic active earth pressure by graphical method [IS 1893-1984].

Point of application (dynamic active case)

From the total pressure computed as above, subtract the static active pressure obtained by putting $\alpha_h = \alpha_v = \lambda = 0$. The remainder is the dynamic increment. The static component of the total pressure shall be applied at an elevation $h/3$ above the base of the wall. The point of application of the dynamic increment shall be assumed to be at mid-height of the wall.

7.7.3 Dynamic Passive Earth Pressure

The general conditions encountered in the design of retaining walls are illustrated in Figure 7.14. The passive pressure against the walls shall be given by the following formula:

$$P_p]_{\text{dyn}} = \frac{1}{2} wh^2 c_p$$

where,

$P_p]_{\text{dyn}}$ = dynamic passive earth pressure in kg/m length of wall

$$c_p = k_p]_{\text{dyn}} = \frac{(1 \pm \alpha_v)\cos^2(\phi + \alpha - \lambda)}{\cos\lambda \cos^2\alpha \cos(\delta - \alpha + \lambda)} \times \left[\frac{1}{1 - \left\{\dfrac{\sin(\phi + \delta)\sin(\phi + i - \lambda)}{\cos(\alpha - i)\cos(\delta - \alpha + \lambda)}\right\}^{1/2}}\right]^2 \quad (7.34)$$

the minimum of the two being the value for design; w, h, α, ϕ, δ and i are as defined as for active case,

and $\lambda = \tan^{-1} \dfrac{\alpha_h}{1 \pm \alpha_v}$. The dynamic passive earth pressure may also be determined graphically using the modified Culmann's graphical method.

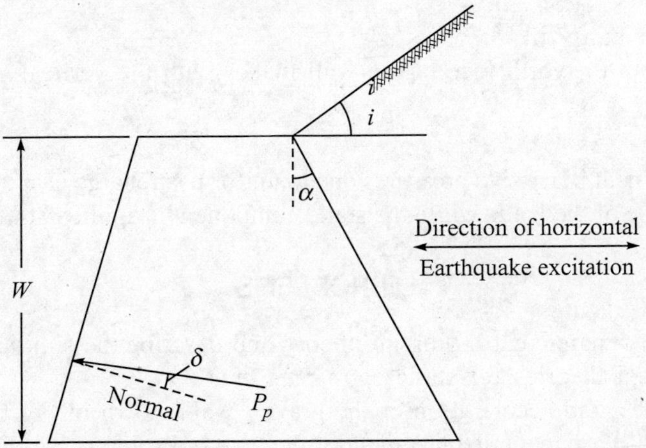

Figure 7.14 Passive earth pressure due to earthquake on retaining wall [After IS 1893-1984]

Point of application (dynamic passive case)

From the static passive pressure obtained by putting $\alpha_h = \alpha_v = \lambda = 0$, subtract the total pressure computed as above. The remainder is the dynamic decrement. The static component of the total pressure is applied at an elevation $h/3$ above the base of the wall. The point of application of the dynamic decrement shall be assumed to be at an elevation $0.66h$ above the base of the wall.

7.7.4 Active Pressure Due to Uniform Surcharge

The active pressure against the wall due to a uniform surcharge of intensity q per unit area of the inclined earth-fill surface shall be:

$$(P_a)_q = \frac{qh \cos \alpha}{\cos (\alpha - i)} \cdot C_a$$

where,

i = slope of earth-fill

α = angle which the earth face of the wall make with the vertical.

Point of application

The dynamic decrement in active pressures due to uniform surcharge is applied at an elevation of $0.66h$ above the base of the walls while the static component is applied at mid-height of the wall.

7.7.5 Passive Pressure to Uniform Surcharge

The passive pressure against the wall due to uniform surcharge of intensity q for unit area of the inclined earth-fill may be obtained as

$$(P_p)_q = \frac{q_h \cos \alpha}{\cos (\alpha - i)} \cdot C_p$$

where,

i = slope of the earth-fill

α = angle which the earth face of the wall makes with the vertical

Point of application

The dynamic decrement in passive pressures due to uniform surcharge is applied at an elevation of $0.66h$ above the base of the walls while the static component is applied at mid-height of the wall.

PROBLEMS

7.1 Discuss the dynamic earth pressure theory and describe the similarities and differences with classical theories for static earth pressure.

7.2 Estimate the yield acceleration of the gravity wall shown in Figure P7.2.

7.3 Estimate the permanent displacement of the wall in Problem 7.2, that would have a 10% probability of exceedance of the following:

(i) Peak acceleration—0.33 g

(ii) Peak velocity—39.2 cm/s

(iii) Peak displacement—13.3 cm

(iv) The wall is located in a seismic zone IV

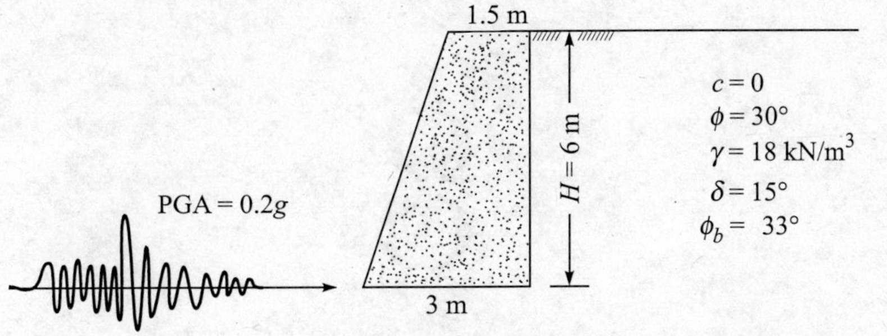

Figure P7.2

7.4 Using the Richart—Elms method, determine the weights sof the retaining wall during static and dynamic conditions. What is the factor of safety? Discuss the dynamics of retaining wall as shown in Figure P7.3?

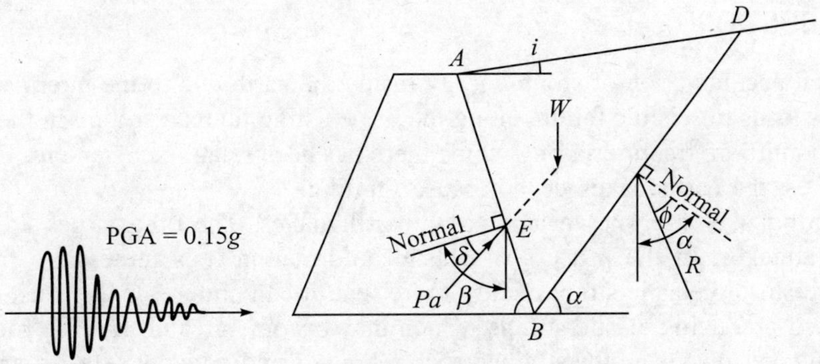

Figure P7.3

7.5 A retaining wall 6.0 m high is inclined at 15° to the vertical but the retaining backfill is horizontal. The wall is located in seismic zone having horizontal seismic coefficient of 0.20. Compute the active earth pressure in static as well as in dynamic state. Determine the percentage increase in dynamic presume over the static pressure. The soil properties are c = 3.0 kN/m^2, ϕ = 33° and γ_t = 18 kN/m^3.

7.6 The backfill of the retaining wall (Problem 7.5) is carrying a surcharge of 2.5 kN/m^2. Determine the % increase in static and dynamic earth pressures.

8

STRONG GROUND MOTION

8.1 INTRODUCTION

The continuual accumulation of strain energy inside the earth due to the interplay of body and surface forces leads to seismic failure along the pre-existing faults resulting in the generation of earthquakes. Faults are fractures along which there has been relative movement. Faulting causes the earthquakes, the earthquakes do not cause faulting.

These earthquakes are known as tectonic earthquakes. The other types are plutonic and volcanic earthquakes. In the process of elastic deformation it is impossible for the rock to rupture before storing elastic strain energy more than it can endure. If the crust of the earth is being displaced producing elastic strains, which may become greater than the rock can endure, a rupture then takes place and the strained rock rebounds under its own elastic stresses until the strain energy is partially or wholly released. The vibrations generated due to the release of strain energy are termed earthquake motion. The great majority of these motions are very weak to the extent that they may not be felt and their recordings need very sophisticated and super sensitive specialized measurements. Such motion or activity is known as micro-seismic activity. The engineers are interested particularly in motion strong enough to damage structures. In qualitative terms, strong ground motions are essentially motions of sufficient strength to affect people and their surroundings. However, the evaluation of the effects of earthquake at a particular site require the quantitative measurement of strong ground motions. These aspects of strong ground motion shall be addressed in this chapter.

Strong ground motion recorded during earthquakes provide the basic information of earthquake engineering. These records are essential for evaluating the earthquake resistant design procedures, for the estimation of attenuation characteristics, assessment of hazards, etc. The analysis of strong ground motion data accordingly leads to a better understanding of the potential effect of strong shaking during earthquakes. The ground motion characteristics determined from strong motion records are studied in various terrain and rock conditions and have been related to various earthquake parameters. The analysis of data has also helped in evaluating the soil-structure interactions, the effects of soil deposits topography and other effects.

There are two basic methods to estimate strong motion in engineering practice. In the first method known as the deterministic estimate, the ground motion is estimated from a given set of seismological parameters, such as earthquake magnitude and the distance from the earthquake rupture zone to the site of interest. In the second method known as the probabilistic method, the ground motion is estimated statistically using all possible earthquake locations and magnitudes together with their expected probabilities of occurring. Both kinds of estimate rely upon a means of estimating strong ground motion from the specified seismological parameters. This estimate is essentially based on an attenuation relation. By attenuation we mean the decrease in amplitude and change in frequency content of the seismic waves with distance.

In addition, the other important aspect is duration of the strong ground motion.

The strong motion information constitutes a vital part of data that enables engineers to understand the response of the foundations and structures to such ground shaking. Economical design of a geotechnical system is possible only when characteristics of input strong motion data as well as those of the structural system to be designed are understood properly. It is a well-recognized fact that only strong earthquake records are useful in disaster management, prevention and mitigation efforts since design guidelines are formulated with this important database in the background. Thus, the records of earthquake strong motion are basic and fundamental to earthquake resistant design of constructed facilities.

The industrialization and urbanization of our country has resulted into a large-scale construction activity, developments of power plants, river valley projects, and other major constructions. As more than 75 per cent area of our country is susceptible to strong earthquakes, it has become absolutely essential to know more about the characteristics of strong ground motion in relation to the future earthquakes in such areas. A correct assessment of the proneness to destructive earthquakes in different regions of the country will lead to a substantial saving in the design of structures and ensure their long-term stability against earthquakes. Ground motion time-histories of past earthquakes provide valuable information on the nature of ground motion during a future earthquake that is expected to occur at a site. Further, the analysis of strong motion data leads to a better assessment of damaging potential of strong shaking during earthquakes to different types of structural systems.

The original purpose of seismograph was neither the location of epicentre nor the investigation of the interior of the earth, but for the detailed study of the strong ground motion. Engineers are interested in motion that is strong enough to damage structures. The idealized ground motion (point source) has been shown in Figure 8.1.

The ground motion produced by the earthquakes can be quite complicated. At a given point, the ground motion can be completely described by three components of translation and three components of rotation. However, in practice, the rotational components are usually neglected, and the three orthogonal components of translational motion are generally measured.

Figure 8.2 shows the acceleration values measured at time increments of seconds. For geotechnical earthquake engineering point of view, the following characteristics of ground motion are of primary significance.

- Amplitude
- Frequency content
- Duration of motion

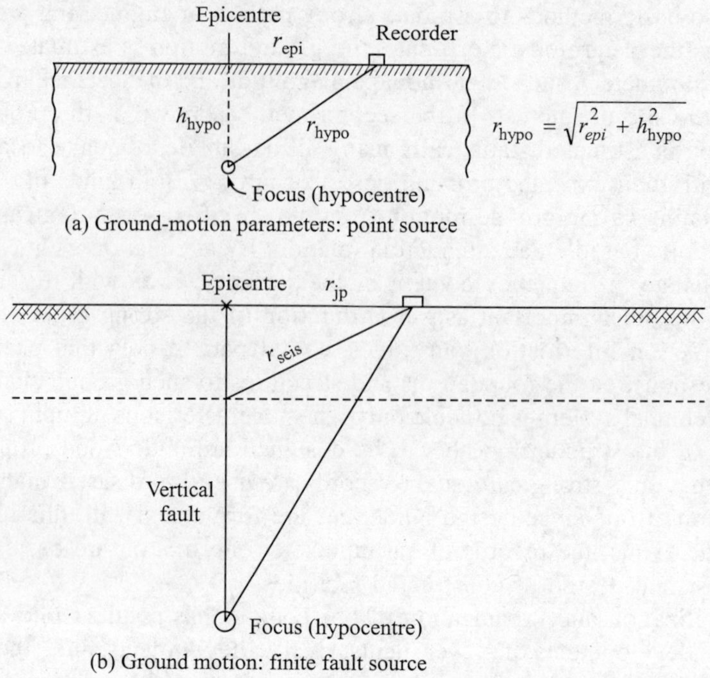

Figure 8.1 (a) Idealized ground motion

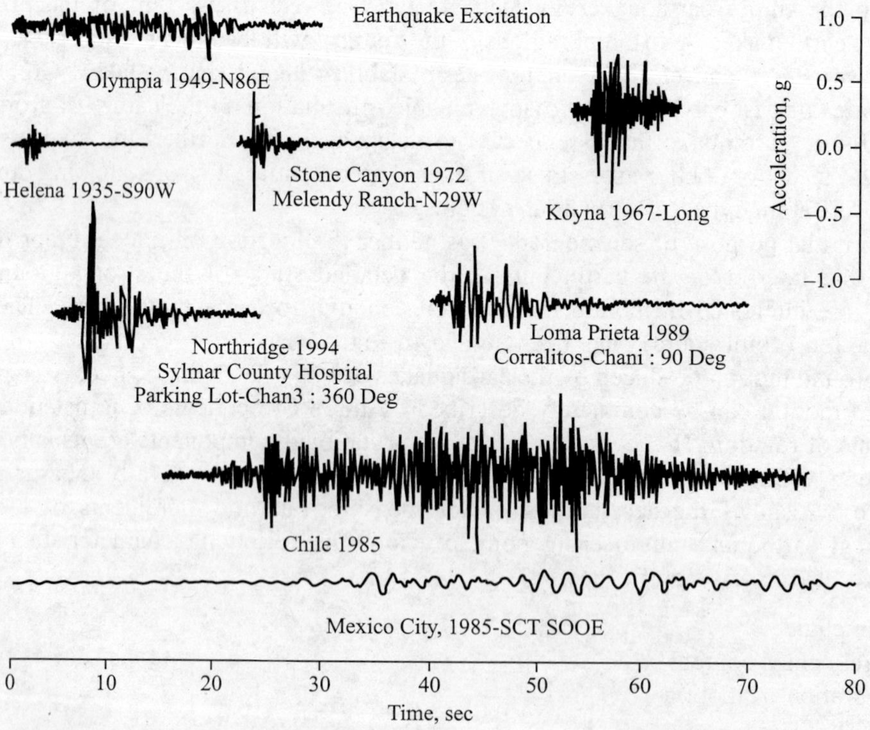

Figure 8.2 Typical ground motions recorded during several earthquakes [After Hudson, 1979].

Proper earthquake resistant design requires the estimation of the level of strong ground motion to which structures will be subjected. This chapter describes the approach and methodology (instruments and techniques) to measure ground motion and the procedure by which the measured motions are corrected. Here, a variety of parameters that can be used to characterize the strong ground motion shall be identified. Relationships that can be used to predict these parameters shall also be presented. The frequency content of a strong ground motion is generally described through several spectra and the duration of such motion shall be adequately addressed.

The earthquake source may be treated as a single point source or as a finite fault rupture. Figure 8.1(b) shows a vertical fault as finite fault rupture. The dotted line below the ground source shows the seismogenic depth, wherein this part of the earth's crust which is capable of generating the ground motion at periods of engineering interest is usually ten seconds.

Figure 8.1(a) shows the point source distance measure which includes the epicentral distance r_{epi} and the hypocentral distance r_{hypo}. The hypocentral distance is the distance from the site to the hypocentre of focus of an earthquake, defined as the point within earth where the earthquake rupture begins. Epicentral distance is the distance from the site to the epicentre defined as the point on the earth surface directly above the hypocentre. The two measures are related to one another by the expression

$$r_{hypo} = \sqrt{r_{epi}^2 + h_{hypo}^2}$$

8.2 STRONG-MOTION OBSERVATIONS STUDIES

Strong-motion observations to explore the behaviour of soil or structures during a strong earthquake are broadly classified into two types. The first is simply called the strong-motion observation. Here, separate strong-motion seismographs are used for the earth's surface and for structures to obtain comparative observations on strong earthquake motions. The second type is called array observations, wherein an array of seismographs is used on the surface as well as on inside of the soil and a temporal correlation is established between the seismic record and such sites.

The strong-motion record obtained by means of these observations is used as input earthquake motion for analyzing the dynamic behaviour of structures or ground or as data for analysis.

The objective of strong-motion observation study is to quantify the level of and the characteristics of ground motion. In general, such a study starts with the detailed examination of available geological, historical and seismological data. The procedures of such studies include the following:

- Evaluating the tectonics and geological settings
- Specifying the faulting sources
- Determining the site soil conditions
- Selecting the motion attenuation relationship
- Determining the maximum source events
- Selecting the site matched spectra

With the above mentioned scope, the seismic instruments are supposed to record the ground motion due to natural and man-made disturbances, and in particular to monitor the seismicity of the given region.

Strong-motion observations in Japan

Strong-motion observations have been carried out in Japan since 1953 at universities, government agencies, industries, etc. However, it was only in 1962 that a Strong-Motion Earthquake Observation Council (SMEOC) was established with a view to providing basic guidelines in obtaining strong-motion records on a national scale and also providing smooth exchange of information between the various agencies on strong-motion records. Various types of strong-motion seismographs are used for these observations. SMEOC has published a Table of National Strong-Motion Observation Points, which includes the observation sites and the status of equipment at these locations with reference to SMAC (Strong-Motion Accelerograph), which is the most widely used in Japan JSCE(1997).

SMEOC publishes the data of strong-motion earthquake records every year and includes such information as sites of installation of strong-motion seismographs, various parameters of earthquakes recorded, maximum acceleration recorded, analogue outputs of important waveforms, etc. Similar information is also published by this council for earthquakes having a magnitude above 5 on the JMA scale, taking into account the records obtained from the SMAC, DC as well as other types of strong-motion seismographs.

Digitization of strong motion records, necessary for the analysis of the behaviour of ground or structures during an earthquake, is carried out separately by the observing institution and some of these agencies publish these digital records.

Strong-motion observations in the USA

Earthquake strong motions are observed in a number of other countries, e.g., the USA, Russia, Italy, Mexico, and so forth. The US records are of the same volume as for Japan. Observations have been ongoing in the USA since 1932 and the maximum acceleration of 353 gal was recorded for the 1940 earthquake EI Centro. About 1300 strong-motion seismographs have been installed in the USA.

Even in the USA strong motions are observed independently by different agencies, the main organizations being the US Geological Survey (USGS), US Army, California State Department of Mining and Geology (CDMG) and the University of Southern California. Even digitization of strong-motion earthquake records is carried out by various agencies, such as the Geological Survey, University of Southern California, and others. The California Institute of Technology has published digitized data for major earthquakes such as the 1971 San Fernando earthquake. Subsequent earthquake records are available from the Geological Survey, California State Department of Mining and Geology and the National Oceanic and Atmospheric Administration (NOAA).

Strong-motion observations in India

Seismic studies in India started in 1898 with the installation of the first seismograph of the country at Alipur, Calcutta (now Kolkata) by the Indian Meteorological Department (IMD). Till 1930, there were only four seismological observatories in operation at Kolkata, Mumbai, Kodaicanal and Agra. However, these were extended to Dehradun and Hyderabad. By the year 1970, the IMD seismological network had grown to 18 permanent observatories.

During 1964–65, some of them were upgraded to conform to the World-Wide Standard Seismological Network (WWSSN) standards under a United States Geological Survey (USGS) collaboration programme. Further, a similar strong-motion programme in India was started mid 1960s when an analogue accelerograph named as RESA (Roorkee Earthquake School Accelerograph) and another low-cost, strong-motion instrument known as Structural Response Recorder (SRR) were developed at the Department of Earthquake Engineering (DEQ), University of Roorkee (now the Indian Institute of Technology, Roorkee).

RESA records the ground acceleration in three mutually perpendicular directions while SRR records the response of classical structural systems during earthquakes. The design modifications in these instruments continued and five models of analogue accelerograph and two models of SRR were developed. Initially these instruments were installed in some river valley projects like Bhakra, Pong, Talwara, Tehri, etc. Later, in 1976, a research project INSMIN (Indian National Strong Motion Instrumentation Network) was funded, on the recommendation of the planning commission, by the Department of Science and Technology (DST) Govt. of India. Further, on the recommendation of the International Association of Earthquake Engineering, funds were sanctioned by the National Science Foundation, USA to DEQ for installation of an array of 50 analogue accelerographs in highly seismically active regions of India. Fifty SMA-1 analogue accelerographs were purchased for this purpose. Subsequently, more funds were sanctioned by DST to procure and install strong-motion instruments at various location in Northern India and Northeast India. These installations have been completed and the instruments have been in operation in these areas. The Department of Earthquake Engineering (DEQ), Indian Institute of Technology Roorkee is now operating an extensive network of strong-motion instruments in the northern, northeastern and western part of the country (Figures 8.3 and 8.4).

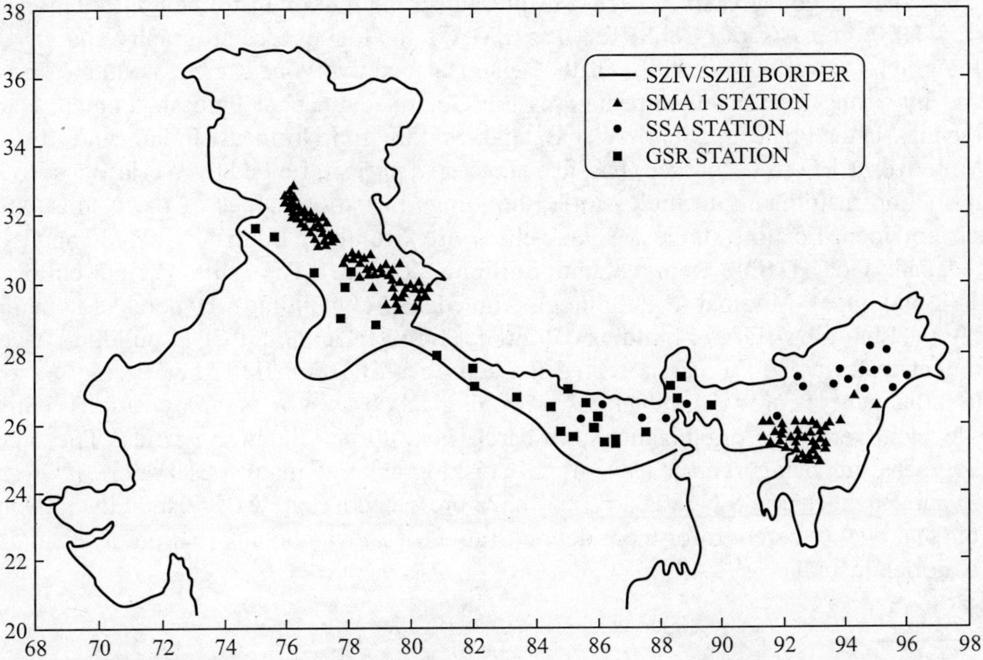

Figure 8.3 Strong-motion network in India showing locations of SMA-1 and GSR stations with SZIV/SZIII border.

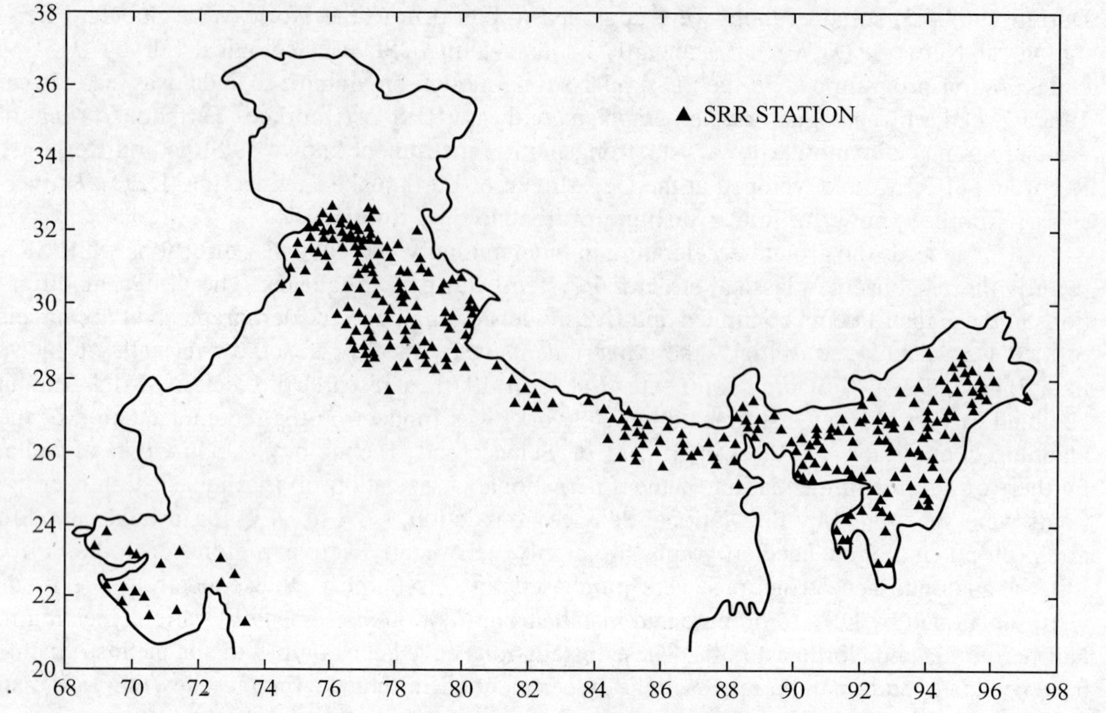

Figure 8.4 Network of structural responses recorder (SRR) In India.

And the activity is going on since 1976. The present strong motion instrument installation looked after by DEQ consists of 338 SRRs, 135 SMA-1 analogue accelerographs and 59 digital accelerographs at various locations in the seismically active zones of the country. The states covered by strong motion instrumentations are Gujarat, Himachal Pradesh, Punjab, Haryana, Uttaranchal, Uttar Pradesh, Bihar, West Bengal, Sikkim and Northeast Indian states.

Moreover, DEQ (IIT Roorkee) has also taken up a project, funded by World Bank, to install strong-motion instruments in multi-storied buildings in various cities of the peninsular India for system identification studies. Select cities are Mumbai, Bangalore, Hyderabad, Pune, Ahmedabad, Goa, Delhi, Roorkee and Shillong. Twelve (12) multi-storied buildings in Delhi (3 buildings), Mumbai (3 buildings), Ahmedabad (1 building), Hyderabad (1 building), Bangalore (1 building), Goa (1 building), Roorkee (1 Tower), and Shilong (1 building) have been instrumented. Basu (2003) has presented the activities of DEQ, IIT, Roorkee.

In India there are now more than 200 seismological observatories run by various central and state level government organizations, research institutions and universities. The analogue seismographs are being replaced by digital seismographs at many observatories. Srivastava (1992) and Bhattacharya S.N. et al., (2005) have presented a review of seismic instrumentation in India and may be referred for more details. Table 8.1 shows the distribution of seismological observatories in India.

Table 8.1 Distribution of seismological observatories in India under different organizations

Organization	*Abbreviation*	*No. of Stations*
Indian Meteorological Department	IMD	17
National Geophysical Research Institute	NGRI	20
Wadia Institute of Himalayan Geology	WIHG	11
Regional Research Laboratory, Jorhat	RRLJ	10
Bhabha Atomic Research Centre	BARG	2
Indian Institute of Geomagnetism	IIG	2
Geological Survey of India	GSI	1
National Institute of Rock Mechanics	NIRM	1
Central Scientific Instruments Organization	CSIO	1
Centre for Earth Science Studies	CESS	1
Maharashtra Engineering Research Institute	MERI	31
Gujarat Engineering Research Institute	GERI	17
Sardar Sarovar Nirman Nigam Ltd.	SSNN	9
Narmada Valley Development Authority	NVDA	10
Kerala State Electricity Board	KSEB	12
IIT Roorkee	IIT/Roorkee	9
Manipur University	M Univ.	4
Delhi University	D.Univ.	3
Guru Nanak Deb University	GNDU Univ	3
Kamaun University	Kun Univ	5
Indian School of Mines	ISM	1
Kurukshetra University	Kur Univ	1

8.3 STRONG-MOTION MEASUREMENT

Housner (1952) has stated that the recordings of strong ground motion provide the basic data for earthquake engineering. Without the knowledge of the ground shaking generated by earthquakes, it is not possible to assess hazard rationally or to develop any appropriate method of seismic design. Earlier the conventional approach of analyzing the seismograms produced by analogue seismographs was mostly confined to the identification of various phases, in an attempt to locate the hypocentral parameters. With the advancements in computer aided digital recording systems, the seismic data analysis not only in time domain but also in the frequency domain are required. Thus, with the availability of digital waveform data, several software packages have been developed in recent years for routine analysis and processing of the data.

8.3.1 Seismographs

A seismograph consists of a seismometer which senses the ground motion and the motion is recorded. The strong-motion record obtained by means of these observations is used as input earthquake motion for analyzing the dynamic behaviour of structures or ground or as data for analysis.

Wave propagation especially the body wave studies emphasized the determination of the arrival times of phases; the analysis of surface waves demanded the registration of dispersed trains. Thus, any instrument or a combination of instruments, for the recording of surface waves must possess a frequency response of adequate width.

A seismograph is any instrument, which provides a visual record of some characteristic of ground motion during the arrival of seismic waves. The heart of the instrument is the seismometer or detector, which responds to one of the following: ground displacement, velocity or acceleration; it must produce a signal, usually electric, which may be recorded. Normally, a seismometer responds to a particular component of ground disturbance, and for a complete description of the arrival, three instruments, corresponding to three coordinate directions are required at each station.

It is evident that a general requirement for seismological instruments is adequate for coupling to the earth. This is usually accomplished by mounting the seismometers on piers, which are preferably isolated from the observatory building, and if possible, constructed on rock outcrop. A second requirement is the placing of accurately controlled time marks on the record. While this appears straightforward, it is in fact only very recently that absolute timing accuracy to about 0.1 second has been achieved at a majority of seismograph stations of the world.

The early history of global seismometry has been described by Dewey and Byerly (1969). Bhattacharya, S.N. et al., (2001) has presented a review of seismic instrumentation in India.

Regardless of the type of transducer employed in a seismometer, it must respond to ground displacement or its time derivatives, and this appears to require that there be a fixed reference point. No point of the seismometer remains truly fixed during the arrival of a seismic wave, but a mass of large inertia, loosely coupled to the frame of the instrument, remains nearly so. The various type may be listed as

- Inertial pendulum seismometer
- Electromagnetic seismometer
- Optomechanical seismometer
- Broadband seismometer
- Visible recording seismometer (Modern analogue seismometer)
- Digital seismometer

Inertial pendulum seismometer

We begin with a simple type of horizontal seismometer, consisting of a pendulum free to rotate about an axis, which is slightly inclined to the vertical as shown in Figure 8.5.

The ground motion causes the mass of the seismometer to undergo forced motion of damped pendulum. A pendulum of total mass M, as shown in Figure 8.5 is free to rotate about an inclined axis. The plane containing the pendulum, when it is at equilibrium position, is known as the neutral plane. We take θ as the rotation of the boom from the neutral plane and u as the displacement of the frame perpendicular to the neutral plane. The forces exerted by the frame on the pendulum can be represented by a force F_1, perpendicular to the neutral plane, and F_2, in the neutral plane and perpendicular to the axis; a force parallel to the axis does not contribute to the rotation. Garland (1971) has presented a detailed analysis of inertial pendulum seismometer.

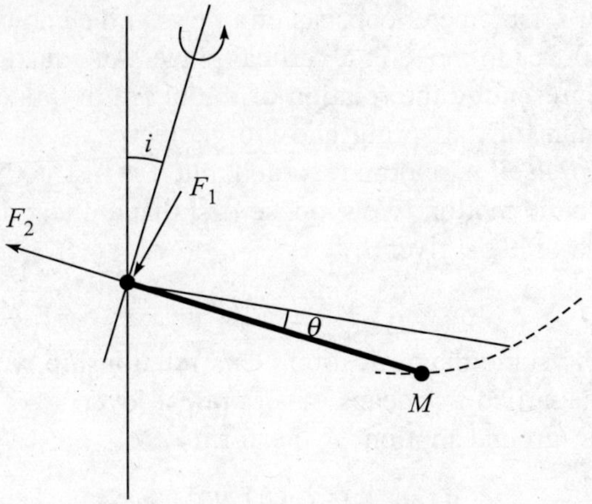

Figure 8.5 The horizontal seismograph. (A pendulum of total mass M is free to rotate about an inclined axis).

Then, in the standard notation for time differentiation,

$$\begin{aligned} F_1 &= M(\ddot{u} + h\ddot{\theta}) \\ F_2 &= Mg \sin i, \\ &= Mgi \text{ (if } i \text{ is small)} \end{aligned} \tag{8.1}$$

If we consider the rotation of the frame about an axis through the mass centre parallel to the axis of rotation, we have

$$Mk^2\ddot{\theta} = -F_1h - F_2h\theta \tag{8.2}$$

where k is the radius of gyration of the suspension about this axis.

Substitution for F_1 and F_2 from Eq. (8.1) leads to

$$(k^2 + h^2)\ddot{\theta} + ghi\theta = -h\ddot{u}$$

or,

$$\ddot{\theta} + \omega^2\theta = -\frac{\ddot{u}}{L} \tag{8.3}$$

where,

$$\omega^2 = g\frac{i}{L} \quad \text{and} \quad L = \frac{(k^2 + h^2)}{h}$$

Equation (8.3) is evidently the equation of (undamped) forced oscillations with ω, the natural angular frequency of the pendulum, being determined by the inclination i of the axis. The quantity L, which becomes equal to h in the case of a point mass on a mass-less boom, is known as the reduced pendular length.

In practice there will always be some damping, so that a term proportional to θ must be added to Eq. (8.3); a standard form is then

$$\ddot{\theta} + 2\zeta\omega\dot{\theta} + \omega^2\theta = -\frac{\ddot{u}}{L} \tag{8.4}$$

where ζ is the damping coefficient.

A vertical component seismometer consists of a mass at the end of a light boom, suspended by an inclined spring, and free to move in a vertical plane. An equation identical to Eq. (8.4) is obtained for it, with θ representing the rotation of the boom in its vertical plane, and ω determined by the force constant of the spring and the geometry.

In practical the angle θ itself is seldom recorded, but it is instructive to consider the solution of Eq (8.4) for a given ground motion. We suppose first that a linear displacement x is recorded, where x a simple multiple of θ, is given by

$$x = L\theta M_s \tag{8.5}$$

Where M_s is known as the static magnification. This relationship would arise, for example, if small rotations θ were magnified by means of an optical lever.

Then, for a harmonic ground motion of the form

$$u = a \cos pt \tag{8.6}$$

we have

$$\ddot{x} + 2\xi\omega\dot{x} + \omega^2 x = M_s p^2 a \cos pt \tag{8.7}$$

of which the solution is

$$x = a M_s p^2\{(\omega^2 - p^2)^2 + 4\xi^2\omega^2 p^2\}^{-\frac{1}{2}} \cos(pt - \delta) \tag{8.8}$$

where, $\tan \delta = \dfrac{2\xi\omega p}{\omega^2 - p^2}$

Equation 8.8 is the familiar solution for forced oscillations, displaying a resonant peak at $p = \omega$. In the absence of damping, the peak becomes infinite, but with increasing ξ, the response is smoothened out.

Two important special cases follow from Eq. (8.8). If the natural period of the seismometer is large compared to the period of the seismic waves, $\omega \ll p$, and the equation reduces to

$$\ddot{x} = -M_s\ddot{u} \tag{8.9}$$

So that x behaves as u, and the instrument is a displacement meter. If, however, the natural period is small compared to the period of the seismic waves, $\omega \gg p$, and we have

$$\omega^2 x = -M_s\ddot{u} \tag{8.10}$$

So that x records essentially the ground acceleration, and the instrument is an accelerometer.

Regardless of the type of response, the ratio of the amplitude of trace displacement to that of the ground displacement at any frequency is known as the dynamic magnification.

Virtually all seismographs in use today record the seismometer motion electromagnetically with a technique invented by Galitzin (1914). The seismometer pendulum carries a permanent magnet, free to move relative to coils mounted on the frame, or vice versa; the induced emf causes a current to flow through a galvanometer connected to the coils and the deflections of the galvanometer are recorded photographically.

Accelerographs

The instrument required for ground motion measurement, i.e., the seisometer can measure displacement, velocity or accelerations of the ground motion. It is more convenient to measure the acceleration and the velocity, and displacement can be obtained with help of integrating circuits. Strong 'ground motion' are usually measured by accelerographs and expressed in the form of accelerograms. The simplest type of seismogram to measure acceleration can be illustrated by a mass-spring-damper (SDF) system as shown in Figure 8.6.

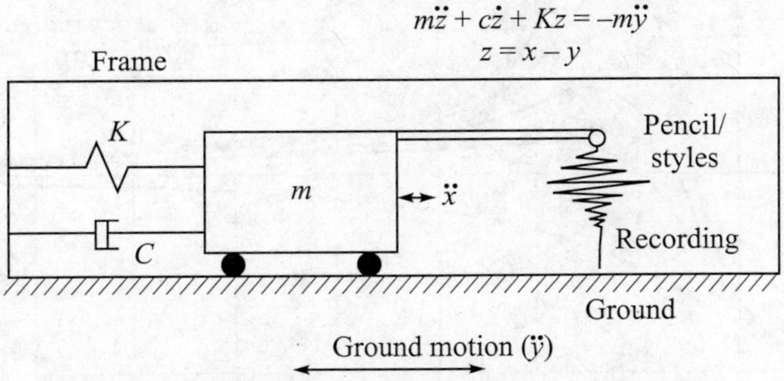

Figure 8.6 Mass-spring dashpot (SDF) type of seismograph.

A rotating drum is connected to such a seismogram frame with a pencil or stylus attached to a mass. The mass is connected to the seismogram housing (frame) by a linear elastic Hookean spring and a newtonian dashpot arranged in parallel and the housing (frame) is connected to the ground. Since the spring and dashpot are not rigid, the motion of the mass will not be identical to the motion of the ground during an earthquake. The relative movement of the mass and the ground will be indicated by the trace made by the stylus on the rotating drum. A typical seismogram station may have three seismographs, oriented to record motion in the vertical and two perpendicular horizontal directions. Therefore, the transactions in the three mutually perpendicular directions may be measured. The details of the governing equation of motion and solution for such a SDF system subjected to ground motion are already explained in Section 3.7. The variation of dynamic magnification acceleration and phase angle with frequency ratio has been shown in Figure 8.7. The curves are drawn for several coefficients fixed values of the damping coefficients from 0.05 to 1.0.

8.3.2 Other Types of Seismograms

Electromagnetic seismometer

The essential part of electromagnetic seismometer is same as that of an inertial pendulum seismometer.

In addition, a coil is attached to the mass which is kept in a magnetic field. As the ground moves, the coil attached to the mass also moves in the magnetic field and generates a voltage across the coil terminals, which is proportional to the velocity of the mass of the seismometer. The constant of proportionality is referred to as the electromagnetic constant of the seismometer and expressed in [mV/(mm/sec)]. The output voltage of the coil terminal gives the measure of the

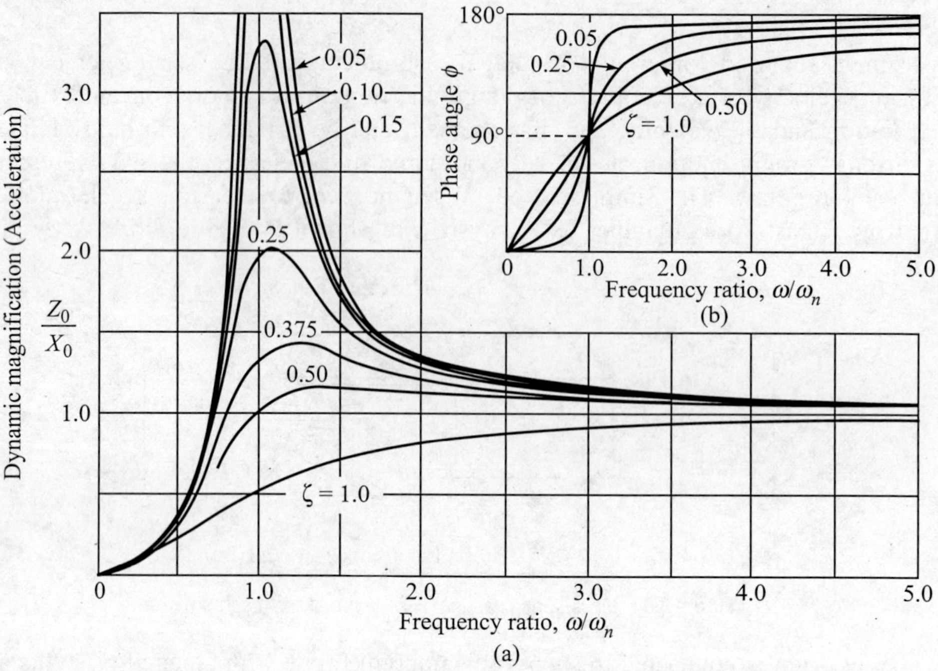

Figure 8.7 Dynamic Magnification (Acc^n.) and phase angle versus frequency ratio (damped seismogram).

ground motion. An external resistance provides the required electromagnetic damping of the seismometer. Further the output of the electromagnetic seismometer is connected to a sensitive mirror galvanometer. The mirror reflects a light beam towards a photographic paper on a recorder. In an electromagnetic seismogram, the seismometer and the galvanometer are connected by a combination of resistances to give proper damping to the seismometer and the galvanometer, as well as to have a suitable magnification. Electromagnetic seismograms offer much higher magnification in comparison to the optomechanical seismometers.

Optomechanical seismogram (Wood-Anderson seismogram)

In order to record small ground motions, seismograms are designed to magnify the ground motion before it is actually recorded. The magnification of a seismogram gives the factor by which the ground motion is magnified. Higher magnification can be obtained or achieved by employing an optical method. In an optomechanical seismograph, a ray of light is made to fall on a mirror fixed to the mass of the seismometer and the reflected light is recorded on a photographic film. An example of one of the earliest seismographs of this type which is widely in use, is Wood-Anderson seismogram. This seismogram also records the horizontal component of the ground motion and was developed in 1920 by two scientists Wood and Anderson. In this type, the pendulum consists of a cylindrical mass (generally, made of copper), which is attached to a vertical suspension wire. During ground motion the cylinder reflects a light beam on the photographic paper wrapped on a recorder drum. A horse magnet surrounding the copper cylinder acts as a damping device.

Broadband seismometer

In strong-motion measurement, it is often required to increase the frequency range of recording. As such it is necessary to increase the free period of the seismometer. But increase of free period of a pendulum seismometer causes instability and nonlinearity. In recent times, broadband seismometers with electronic feedback mechanism can provide increased period without compromising stability and linearity.

It may be recalled that in a pendulum-type seismometer, the relative motion between the frame and the mass produces the signal. The force on the mass due to ground motion is equal to the product of the mass of the seismometer with the sum or acceleration of the frame and the relative acceleration between the frame and the mass (Bhattacharya & Dattatrayam, 2000). However, if a force is applied by the frame to the swinging mass in such a way that the relative displacement (and consequently, the relative velocity as well as acceleration) between the frame and the mass is reduced to years, the applied force would then be a direct measure of the acceleration of the mass. This can be done in principle by detecting the relative displacement between the frame and the mass, generating a current signal corresponding to it and feeding the current back to the soil moving with the mass and the magnet fixed on the frame. In such a system, the mass–spring arrangement becomes much less critical so far as the response characteristic is concerned. However, the feedback amplifier has to be of a wide bandwidth and a high dc gain. By making the gain of the feedback loop dependant on frequency, a very wide variety of response characteristics may be obtained.

Modern analogue seismograph

The conventional photographic recording is getting obsolete because of high recurring costs involved in photographic charts. Further, it is sometimes necessary to observe records continuously. In a modem analogue seismograph, the output from the electromagnetic seismometer is fed into an amplifier, the amplified voltage is fed to a galvanometer attached to a pen for recording on a paper. The recording is usually made through the following:

- Ink recording on a plain paper
- Scratching on smoked paper/heat sensitive paper

In the heat sensitive recording mode, the pen remains hot and it removes the white chemical cover of the recording paper by scratching, to bring out the black background.

Digital recording seismograph

In a digital recording seismograph, the amplifier output is fed to a digital recorder which records the ground motion in terms of digital counts. The output from the amplifier when goes to a digital recorder, it converts the voltage to counts (say, L counts/mV). In a digital recorder, the analogue-to-digital converter (ADC), also called a digitizer, samples the input signal at regular intervals, defined by N samples per second (sps).

For a unique representation, any harmonic must have at least three samples per wavelength. Thus, the frequency defined by $f_N = 0.5$ represents the harmonic with the lowest frequency and is called Nyquist frequency. The minimum digitization step is called a digit or count. The smallest unit of a digital value or word is called byte (= 8 bits). A 16-bit ADC can count values from –32768 (-2^{15}) to 32768 (2^{15}), giving a dynamic range of 90 dB. Higher dynamic range in digital recording, compared to analogue recording, gives the advantage to record the ground motions from very small magnitude earthquakes as well as large magnitude earthquakes without saturation. The true parameters crucial in the design of a digital seismograph are bandwidth, dynamic range and the bit resolution of the digitizer. In present days, several types of large storage devices (of the order of a few gigabytes) and data compression techniques are available to store the data so generated. The present-day system offers recording in different streams in continuous mode or in trigger modes, i.e., when the signal amplitude exceeds a predefined threshold value. In the trigger mode, normally recording is based on the ratio of short-term average (STA) to long-term average (LTA) of the recorded signal. The time at which this ratio exceeds a predefined threshold value is called the trigger time. Figure 8.8 shows digital telemetry field stations and central receiving stations of digital telemetry system.

In the field of seismological observations, recently a 16-element VSAT-based seismic telemetry system has been deployed in and around Delhi by IMD (Shukla et al., 2002).

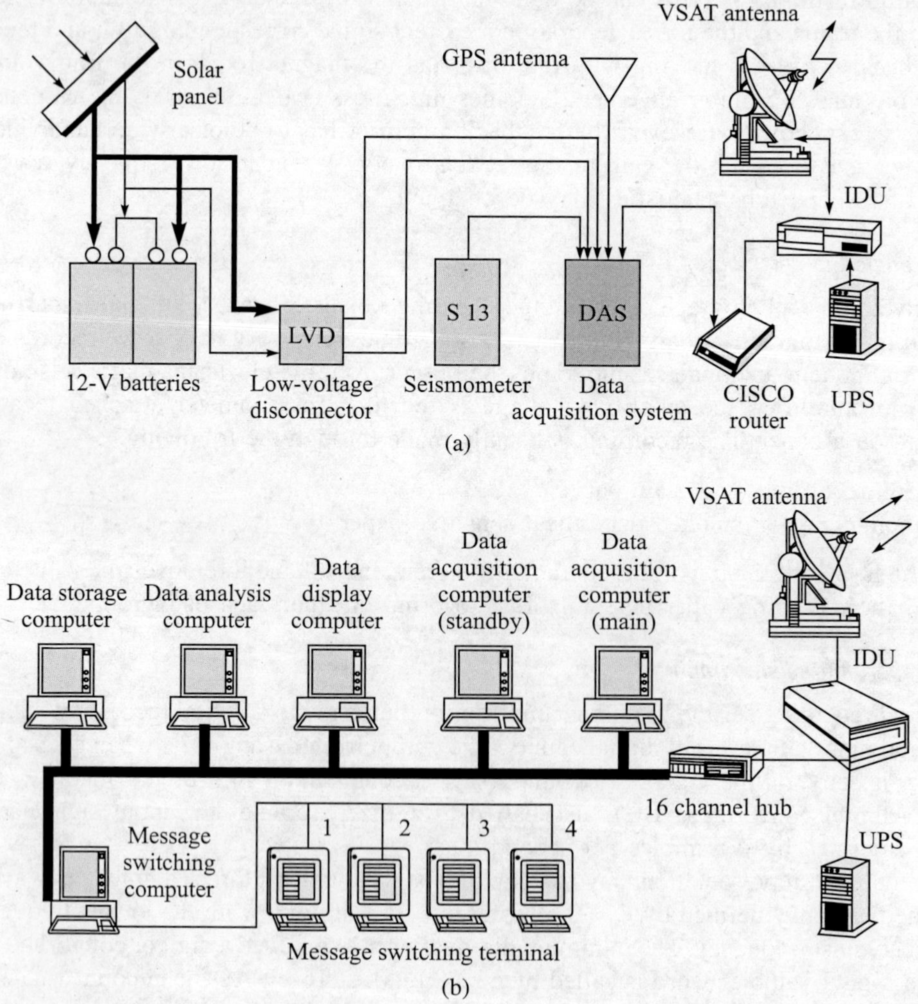

Figure 8.8 (a) Digital telemetry field stations, and (b) central receiving stations of digital telemetry system.

Standard seismographs and seismograms

The requirement of a seismograph station is essentially for the recording of both body and surface waves. What has become known as a "standard station" contains three components of short-period instruments (0.5 to 1.0 second natural period), with a magnification, in a seismic areas, as great as 500,000. The rapid response of short-period instruments provides records on which the onset of body-wave phases may be accurately read. A standard station also contains three components of long-period instruments of the Press-Ewing type, whose magnification is of the order of 10,000. Time marks are placed on all records by means of clocks whose error is not greater than 0.05 seconds. The growth of a worldwide network of stations of this type has come about chiefly as a result of the desire to detect nuclear explosions and to distinguish them from earthquakes.

Seismology owes much to international cooperation in the exchange of data. Original seismograms are normally kept at the observatories where they were produced, but microfilm copies are filed at World Data Centers in Washington and Moscow. Routine determination of epicentres, and the publication of bulletins, is carried out on the basis of arrival times of body-wave phases, read at cooperating stations throughout the world and reported to a central agency. At the present time this work is conducted by the United States Coast and Geodetic Survey, and by centres set up under the International Association of Seismology and Physics of the Earth's Interior at Strasbourg and Edinburgh.

8.3.3 Data and Digitization

For engineering computations the recorded ground motion and analogue records must be digitized. Although analogue transducers are used, digital instruments convert the analogue signal to digital form in the field. Because digital system are more complex, more expensive and more difficult to maintain in the field, they have not replaced analogue system.

The temporal variation of the acceleration component of earthquake motion (horizontal and vertical component) is recorded by the strong-motion seismograph as the analogue wave. To digitize the same for dynamic analysis purposes, it is sampled at the time interval of about 0.01 second and then the time series is subjected to various corrections.

The strong-motion record contains the various types of errors due to the properties of the seismograph, recording equipment and the digitization process. According to studies carried out to evaluate and eliminate these errors, it is necessary to consider the following points while using these records.

The ground strong-motion seismograph measures the ground acceleration and naturally, in principle, is not very sensitive to high-frequency components. Thus, if we assume that it has a sensitivity of 1 at 1 Hz, it will be 0.5 at 7 Hz and 0.2 at 15 Hz. Accordingly, some correction is applied to compensate this loss of sensitivity. However, such correction also increases the high-frequency noise component and hence is not useful beyond a certain frequency. The errors due to the recording equipment and subsequent digitization can be attributed to the errors in paper feeding, writing or friction resistance, reading error, etc. Among these, zigzag paper feeding or frictional resistance of the pen, zero adjustment error, etc. is prominent in the case of longer period components. As such, during digitization it is necessary to eliminate the components below a certain frequency. The extent of reading error is generally the same over a wide range.

Error correction of the high-frequency component considerably affects the short-period components of maximum acceleration of the seismic wave or response spectra, while the error correction of the low-frequency component affects the velocity or displacement (quantities) obtained by integration of the acceleration waveform. Various correction filters have been proposed which take these points into consideration so as to obtain accurate digital records.

The digital strong-motion records published are generally subjected to circular arc correction only. It is necessary to note while using strong-motion records that the seismic properties are affected not just according to whether the correction is applied or not, but also by the properties of the (correction) filter used.

8.3.4 Strong-Motion Records

The Internet has become an excellent resource for obtaining all sorts of information regarding seismology and earthquake engineering ranging from reports on the source of parameters of earthquake that occurred within the last few hours to collections of photographs of earthquake damage. Many agencies operating strong–motion recording networks maintain websites and some of these allow users to access and download their accelerograms. These networks make digital accelerograms available online within a few hours of them being recorded, as well as providing remarkably detailed geotechnical information. Probably the most useful sites for general engineering applications are those that provide access to data from many countries or regions.

Some of the popopular websites are listed below:

http:/www.geophys.washington.edu./seismosurfing.html
http:/www/http://db.cosmos-eq.org
http:/www/ isesd. cv.ic.ac.,uk
http:/www/seismolinks.com
http:/www/seismosoft.com
http:/www.quake.ca.gov

The Cosmos website (http://db.cosmos-eq.org) contains a database of more than 18,000 freely available acceleration traces from about 400 events and 2500 stations around the world, 35 per cent which are from the western US, 23 per cent from Japan, 14 per cent from New Zealand and the remainder from other part of the world, Stepp. (2000)

8.4 ARRAY OBSERVATIONS

For array observations, an array of seismographs is used on the surface as well as inside the soil and a temporal correlation is established between the seismic record and such sites. The properties of strong motion at any given point are affected by factors such as

- earthquake mechanism
- propagation properties of the path between the hypocentre and the observation site
- the ground properties at and near the point of observation

To explore the relation between these factors and the resultant seismic properties, observations at individual sites may not be sufficient; it is necessary to carry out array observations by placing a number of seismographs systematically over the desired area. The location or number of seismographs to be used in array observation varies according to the area, which in turn is decided by the purpose of observations. For example, we can mention observations involving a wide area, such as those for studying the fault rupture process or for studying the path of propagation, or observations for studying the effect of topography or ground, in which the seismographs are concentrated in a localized manner, or observations for studying the response of structures, etc. The seismographs may be installed on the surface as well as underground depending on the purpose. A pore-water pressure meter and subsurface strain meter are used additionally.

The presence of noise has been observed on seismograms. Where body waves from small events at great distances are to be recorded, the signal on a single seismograph may be lost in this noise. In order to improve the signal-to-noise ratio, arrays of seismometers have been designed in such a way that their outputs can be combined with chosen time delays. The result bears some similarity to a grating as used in spectroscopy, or to the receivers used in radio astronomy.

The noise on seismograms may be of various types. It may be produced near the station by man or by wind, or over the sea by pressure variations associated with storms. Seismic surface waves from these sources are known as microseisms. On the other hand, as the sensitivity of a seismograph is increased, it is found that there is a background of noise, which probably represents true body waves from many small earthquakes. The principle of the array is to remove the first type of noise, by virtue of its surface wave characteristics. A body wave, arriving at near vertical incidence from a distant source, has an apparent velocity along the earth's surface, which is large compared to the velocity of surface waves, and it is this difference, which permits the discrimination.

8.4.1 Array Observations in Japan and USA

Array observation is carried out in Japan by universities, government and public agencies, or industries. These can be broadly divided into observations by the Public Works Research Institute under the Ministry of Construction, Post and Harbor Research Institute under the Ministry of Transport, the Institute of Industrial Science under Tokyo University and Building Research Institute under the Ministry of Construction, and others—all of which seek to study the effect of local topography and ground conditions on an earthquake, and observations by the National Research Center for Disaster Prevention, Science and Technology Agency or Earthquake Research Institute under Tokyo University, and others—which seek to study the mechanism of an earthquake or the properties of the path of propagation. A number of industries are also involved in carrying out these observations with a view to studying ground conditions or the properties of structures. (JSCE, 1997)

Aside from Japan, array observations are also being carried out or planned by the USA, Taiwan, Turkey and other countries. In the USA, commencing with EI Centro array which obtained strong-motion records near the fault area of the 1979 Imperial Valley earthquake, a number of array observations have been and still are carried out to study the source mechanism or propagation properties. Local array installations have recently been planned to study the effect of topography or ground-surface geology.

The seismographs used in array observations need to give better performance as enumerated in Table 8.2 than conventional seismographs.

Table 8.2 Main specifications required in new models of strong-motion seismographs

Frequency range	0.05 ~ 25 Hz (sampling 10 Hz min)
Measurement range	0.1 ~ 2,000 gal (dynamic range 86 dB min) (AD conversion corresponding to 12 bit binary gain amp. Or 16 bit)
Precision of absolute time	1/100 s (with automatic timing correction)
Signal delay	5 s (min)
Recording method	Digital recording

8.5 CHARACTERISTIC OF STRONG GROUND MOTION

The important characteristics of strong ground motion like amplitude, frequency, duration, peak ground acceleration (PGA), peak ground velocity (PGV), peak ground displacement (PGD) and time history may be expressed in a compact, quantitative form by suitable ground motion parameters.

8.5.1 Earthquake Magnitude

Earthquake magnitude is generally used to define the size of the earthquake. The magnitude scales have been described in Section 2.8. Accordingly to Bolt (2003), there is an increasing tendency to adopt M_w (moment magnitude) as the worldwide standard for qualifying magnitude because of its physical and seismological basis. It may be recalled that by definition, M_w is related to seismic moment M_o, a measure of the seismic energy radiated by an earthquake.

Hanks and Kanamori (1979) proposed that

$$M_w = \frac{2}{3} \log M_o - 10.7 \tag{8.11}$$

where,

$M_o = GA_fD = 2GEs/\Delta\sigma$

G = shear modulus of the crust (source region)

A_f = rupture area

D = average displacement on a rupture plane

$\Delta\sigma$ = stress drop

E_s = radiated seismic energy

The $M_o = GA_fD$ may be obtained from the geological faulting parameter whereas $M_o = 2GE_s/\Delta\sigma$ may be estimated from seismological measurements.

8.5.2 Peak Ground Acceleration (PGA), Peak Ground Velocity (PGV), Peak Ground Displacement (PGD)

PGA, PGV and PGD are the most common and easily recognizable time domain parameters of the strong ground motion. Typically, the ground motion records termed seismograph or time histories have recorded acceleration (these records are termed accelerograms) for many years in analogue form and more recently, digitally. Time histories theoretically contain complete information about the motion at the instrument location, recording time traces or orthogonal records (two horizontal and one vertical). The maximum amplitude of recorded acceleration is termed the peak ground acceleration (PGA)—peak ground velocity (PGV) and peak ground displacement (PGD) are the maximum respective amplitudes of velocity and displacement. Acceleration is normally recorded and expressed in units of cm/s/s (termed gals), but it is also expressed in terms of fraction of or per cent of the acceleration of the gravity (980.66 gals, termed $1g$). The focus is usually on peak horizontal acceleration (PHA) due to its role in determining the lateral inertial forces in structures. An acceleration time trace series can be integrated numerically to obtain the corresponding velocity and displacement time series as shown in Figure 8.9. Peak horizontal velocity (PHV) and peak horizontal displacement (PHD) are sometimes reported. They reflect different sensitivities to the frequency component of the motion. It has been a common practice in the past to estimate a design response spectrum from PGA or a combination of PGA or PGV and the peak ground displacement (PGD).

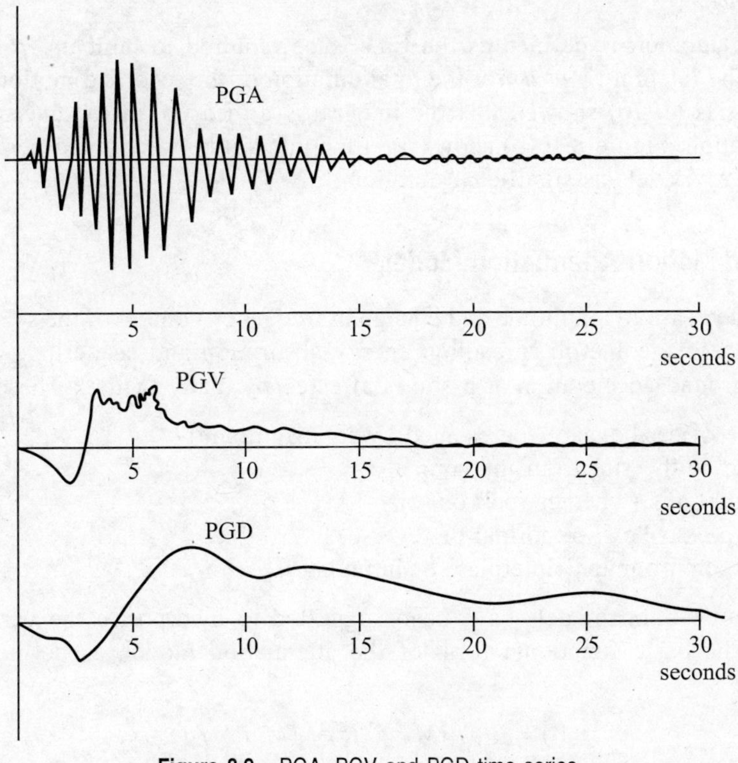

Figure 8.9 PGA, PGV and PGD time series.

The relationship of these parameters may be expressed as

$$\text{PGA} = \left(\frac{2\pi}{T_n}\right) \text{PGV} = \left(\frac{2\pi}{T_n}\right)^2 \text{PGD} \tag{8.12}$$

where T_n = undamped time period.

8.5.3 Duration of the Strong Ground Motion

Duration of strong ground motion is an important parameter and is directly responsible for damages by an earthquake. The most widely used definitions of durations for strong ground motion are:

- Bracketed duration
- Significant duration

Bracketed duration

The bracketed duration is the interval between the points in time where the acceleration amplitude first and last exceeds a prescribed level such as 0.05 g (Bolt, 1993). Figure 8.10(a) shows the time interval between the first-exceedance and last-exceedance as bracketed duration.

Significant duration

The significant duration is defined as the time trace required to built up from 5 per cent to 95 per cent of the integral ($\int a\, dt$) for the total duration of the recorded motion, where a is the acceleration. Arias (1970) showed that the integral is a measure of the energy in the ground motion acceleration. Figure 8.10(b) shows the time interval between the 5 per cent energy and 95 per cent energy levels as significant duration.

8.5.4 Ground Motion Attenuation Model

Attenuation is decrease in amplitude and change in frequency content of the seismic waves with distance, because of geometric spreading, energy absorption and scattering.

The peak ground acceleration at a site is affected by many factors. These include:

(i) The size of earthquake (represented by its magnitude)
(ii) Distance of the site from the source
(iii) Site conditions (whether rock or soil)
(iv) Fault type (strike slip, normal or reverse)
(v) Tectonic environment (interplate or intraplate).

A number of attenuation models have been suggested to incorporate the various parameters noted above. The basic functional form of the attenuation model, as defined by Campbell (1985) is:

$$f(Y) = b_1 f_1(M)\, f_2(R)\, f_3(M, R)\, f_4(S)\varepsilon \tag{8.13}$$

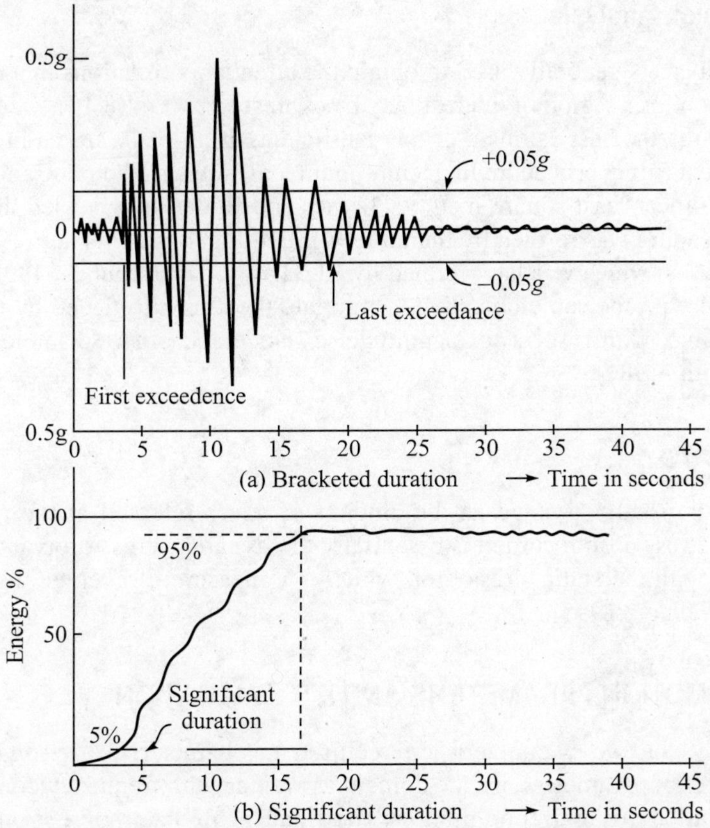

Figure 8.10 Bracketed duration and significant duration illustrated from a typical ground motion produced by an earthquake.

where Y is the response variable (dependent variable), b_1 is a constant scaling factor, $f_1(M)$ is a function of magnitude M, $f_2(R)$ is a function of source-to-side distance R, $f_3(M, R)$ is a joint function of M and R, $f_4(S)$ is a function representing the parameters of the earthquake path, site or structure and ε is a random variable representing the uncertainty in Y.

The estimated transformation of distance and PGA show a logarithmic behaviour, whereas the transformation of magnitude is seen to be approximately linear. The functions of Eq. (8.13) were, therefore, formulated as

$$f(Y) = \log A; \quad f_1(M) = b_2, m; \quad f_2(R) = b_3 \log R \tag{8.14}$$

After applying the relation of Eq. (9.14) to the relation of Eq. (8.12), we get the following generalized form

$$\log A = b_1 + b_2 M + b_3 \log R + b_4 S + \varepsilon \tag{8.15}$$

To develop an attenuation relation of the type of Eq. (8.15) for a region of interest, coefficients b_1, b_2, etc. are evaluated by regression analysis of the strong-motion data available. However, due to strong dependence on regional geology, use of an attenuation relation for another area with different geological and tectonic may lead to PGA values which are usually unacceptable for earthquakes resistant design of amplitudes at close distances.

8.5.5 Regression Analysis

A regression analysis is generally used to obtain the attenuation relations in a given region. The attenuation relation for a region of interest may be expressed by Eq. (8.15). A regression analysis is used to determine the best estimate of the coefficients b_1, b_2, b_3.... b_n as in Eq. (8.15), using a suitable statistical fitting procedure like a minimum least square method. Campbell (1981) used the weighted nonlinear least square method. Joyner and Boore (1981) used the step regression and later the procedure was refined by them (1994). Brillinger and Preisler (1984) used random effects regression and which was later refined by Abrahamson and Youngs (1992). All these three methods are used with the same objective to mitigate the bias introduced by the uneven distribution of recordings with respect to magnitude, distance and other seismological parameters. (Draper and Smith, 1981).

8.5.6 Stress Drop

The stress drop is usually defined as the amount of stress released at the rupture during an earthquake. The stress drop recorded is essentially the dynamic stress drop and in fact it should not be confused with the static stress drop which is a measure of average displacement in the fault (Lay and Wallace, 1995).

8.6 STRONG-MOTION PARAMETERS AND ITS EVALUATION

Generally, more than two parameters are required to characterize a strong-ground motion. However, a variety of parameters in time domain as well as with frequency content are available for the description of strong-ground motion. The frequency content of a strong-ground motion is generally described through the different types of spectra, namely:

- Power spectra
- Response spectra
- Fourier spectra
- Ground motion spectra

The frequency content describes how the amplitude of a ground motion is distributed among different frequencies. The effect of earthquake motion on structures, foundations, bridges, slope or soil deposits are very sensitive to the frequency of the motion, therefore, the characterization of the motion cannot be complete without consideration of the frequency content.

8.6.1 Frequency Content Parameters

The amplitude of the ground motion is distributed among different frequencies. This can be easily observed with frequency content description of the strong-ground motion. The characterization of the ground motion should be considered with its frequency content. Using Fourier analysis, a periodic function $x(t)$ may be expressed as

$$x(t) = a_0 + \sum_{n=1}^{\alpha} (a_n \cos \omega_n t + b_n \sin \omega_n t) \tag{8.16}$$

where,

$$a_0 = \frac{1}{T}\int_0^T x(t)\ dt$$

$$a_n = \frac{2}{T}\int_0^T x(t)\cos\ \omega_n t\ dt$$

$$b_n = \frac{2}{T}\int_0^T x(t)\sin\ \omega_n t\ dt$$

Further if $a(t)$ represents the acceleration—time history, then a complete frequency-domain description of the ground motion may be expressed as

$$F(\omega) = \int_0^T a(t)\ e^{-i\omega t}\ dt \tag{8.17}$$

A plot of the Fourier amplitude versus frequency from the above equation is known as Fourier Amplitude Spectrum (FAS). The amplitude of Fourier amplitude may be expressed as

$$|F(\omega)| = \left\{\left[\int_0^T a(t)\cos\omega t\ dt\right]^2 + \left[\int_0^T a(t)\sin\omega t\ dt\right]^2\right\}^{1/2} \tag{8.18}$$

where,

T = duration of ground motion

ω = circular frequency

In frequency domain, the Fourier amplitude, the phase spectrum and several definitions of response spectra are used in the quantification of strong ground motion.

8.6.2 Power Spectra

The frequency content of a ground motion may also be depicted by power spectra. The total intensity of a ground motion of duration S is given by the time domain by the area under the time-history of squared acceleration and may be expressed as

$$I_0 = \int_0^S [a(t)]^2\ dt \tag{8.19}$$

The same equation may be expressed in frequency domain as

$$I_0 = \frac{1}{\pi}\int_0^{\omega_n} C_n^2\ d\omega \tag{8.20}$$

where $\omega_n = \frac{\pi}{\Delta t}$ is the Nyquist frequency (the highest-frequency in the fourier series).

The average intensity λ_0 in time domain or frequency domain may be expressed as

$$\lambda_0 = \frac{1}{S}\int_0^S [a(t)]^2\ dt = \frac{1}{\pi S}\int_0^{\omega_n} C_n^2\ d\omega \tag{8.21}$$

Thus, the power spectral density $G(\omega)$ is defined such that

$$\lambda_0 = \int_0^{\omega_n} G(\omega)\, d\omega \tag{8.22}$$

By comparing Eq. (8.22) with Eq. (8.21), we have

$$G(\omega) = \frac{1}{\pi S} C_n^2 \tag{8.23}$$

The power spectral density function is useful in characterizing the earthquake as a random process. The power spectral density function by itself can describe a stationary random process, i.e., where the statistical parameters do not vary with time.

8.6.3 Bandwidth and Predominant Period

Bandwidth

The range of frequency over which some level of Fourier amplitude is exceeded represents the bandwidth. As explained in Section 3.7, the bandwidth is usually measured at the level where power of the spectrum is half its minimum value; in other words it corresponds to a level of $\frac{1}{\sqrt{2}}$ times the maximum Fourier amplitude.

Predominant period

In Fourier amplitude spectrum, the period of vibration corresponding to the maximum value as shown in Figure 8.11 is called the predominant period. Figure 8.11 shows two Fourier amplitude spectra, wherein case I describes a wideband motion and the lower (Case II) a narrow band motion.

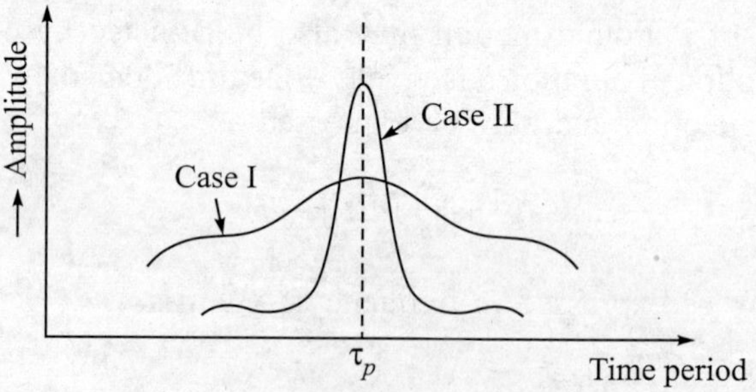

Figure 8.11 Typical Fourier amplitude spectra (broadband and narrowband).

8.6.4 Spectral Parameters

Spectral parameters have been proposed by different authors to describe the dynamics and other information from each spectrum, namely:

- Central frequency
- Shape factor

- Bandwidth
- Predominant period
- V_{max}/a_{max} ratio

Central frequency

Vanmarcke (1976) introduced the central frequency, Ω as a measure of frequency where power spectral density is concentrated. The theoretical medium peak acceleration may be expressed as

$$\ddot{u}_{max} = \sqrt{2\lambda_0 \log\left(\frac{2.8\,\Omega S}{2\pi}\right)} \tag{8.24}$$

where, λ_0 = mean-squared acceleration
S = time period
Ω = Central frequency

Shape factor

Vanmarcke (1976) proposed a shape factor δ, which indicates the dispersion of power spectral density function about the central frequency and may be expressed as

$$\delta = \sqrt{1 - \frac{\lambda_1^2}{\lambda_0 \lambda_2}} \tag{8.25}$$

where, λ_0 = mean-squared acceleration.

V_{max}/a_{max}

Newmark (1973) observed that the ratio V_{max}/a_{max} should be related to the frequency content of the motion, as both are associated with motions of different frequency. For SHM of period τ, for example, $V_{max}/a_{max} = \tau/2\pi$. As such the quantity $2\pi(V_{max}/a_{max})$ should be considered as the period of vibration of an equivalent harmonic wave for the earthquake motion under consideration.

8.6.5 Other Ground-Motion Parameters

The other ground-motion parameters in addition to those described in the previous sections may include the following

- RMS acceleration
- Arias intensity
- Corner frequency and cut-off frequency

RMS acceleration

RMS acceleration as a single parameter, to include the effects of amplitude as well as frequency content of a strong motion record, may be expressed as

$$a_{rms} = \sqrt{\frac{1}{S}\int_0^S [a(t)]^2\, dt} = \sqrt{\lambda_0} \tag{8.26}$$

where,

S = duration of the strong ground motion
$a(t)$ = acceleration at any time t
λ_0 = mean squared acceleration
a_{rms} = root mean square value of acceleration.

Predicted ground motion studies have shown that the rms acceleration is very useful for engineering purposes.

Arias intersity

Arias (1970) defined a parameter as

$$I_A = \frac{\pi}{2g} \int_0^{\infty} [a(t)]^2 \, dt \tag{8.27}$$

This parameter has units of velocity and is usually expressed in metre per second. Arias intensity is thus essentially a ground-motion parameter derived from an accelerogram and proportional to the integral overtime of the acceleration squared. This parameter is obtained by integration over the entire duration rather then the duration of strong motion; its value is independent of the method used to define the duration of the strong-ground motion.

8.6.6 Corner Frequency and Cut-off Frequency

At times the Fourier amplitude spectra (FAS) of the actual earthquake are smoothed and for better representation they are plotted on logarithmic scales. As such their characteristics shapes can be observed more easily. See Figure 8.12.

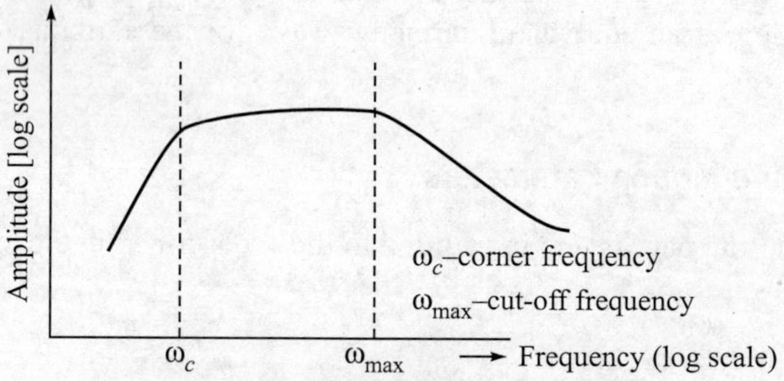

Figure 8.12 An idealized Fourier amplitude spectrum.

shows FAS on logarithmic scale, where FAS tends to be the largest over an intermediate range of frequencies bounded by the corner frequency (f_c) on the low side and cut-off frequency (f_{max}) on the higher side. Brune (1970) showed that the corner frequency (f_c) is inversely proportional to the cube root of seismic moment, M_o. As such large earthquake produce greater low-frequency motions.

8.7 EVALUATION OF STRONG-MOTION PARAMETERS

As discussed in the previous sections, the important characteristics of a strong ground motion are amplitude, frequency and duration. Various parameters can describe the strong ground motion. Some parameters describe only the amplitude of the motion, some rely heavily on the characterization of the strong ground motion. Characterization of a single parameter is rare, the use of several parameters is required to describe the important characteristics of a particular ground motion.

Predictive relationships usually express ground motion parameters as functions of magnitude, distance and in some cases, other variable, for example

$$Y = \phi\ (M,\ R,\ Y_j) \tag{8.28}$$

where,

Y = ground-motion parameter of interest
M = magnitude of the earthquake
R = measure of the distance from source to the site under consideration
Y_j = any other parameter

The predictive relationships are further developed by regression analysis of the recorded strong-motion databases. The functional form of the predictive relationship is usually selected to reflect the mechanics of the ground motion as closely as possible. The earthquake resistant design of foundation, substructures and soil retaining structures require estimation of the level of ground shaking to which they will be subjected. Since the level of shaking is most conveniently described in terms of ground-motion parameters, methods for evaluating the ground-motion parameters are required.

With the involvement and subsequent development of the fault model, hopefully it is now possible to predict the earthquake motion based on this model. With rapid advances in seismology, an earthquake is now attributed to fault movement within the crust. Hence, it is possible to express the seismic motion in terms of parameters of fault movement Heaton and Harzell (1986). The seismic motion $U(t)$ at the observation point can be expressed as

$$U(t) = \int_0^L \int_0^W \dot{D}(x,\ y,\ t) * G(x,\ y,\ t)\ dy\ dx \tag{8.29}$$

where,

D = shear slip
L = length of the fault
W = width of the fault
$*$ = a time convolution
$\dot{D}$ = slip velocity
$G(x,\ y,\ t)$ = green function

The Figure 8.13 shows the fault model in which the dislocation on the fault plane is specified as a kinematic model. In this figure, the fault model shown describes:

- Shear slip D as the source process of an earthquake wherein a displacement discontinuity vector across a finite fault plane propagates over the plane at a finite rupture velocity v.
- The rupture front propagates with a constant velocity.
- The slipline function $D(t)$ at a point on the fault is a ramp function.

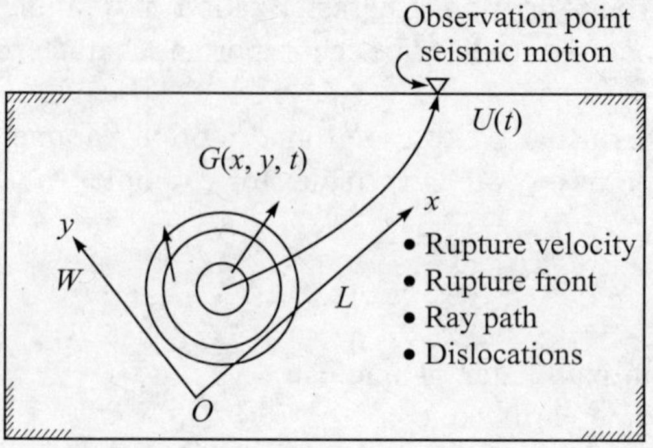

Figure 8.13 Schematic kinematic fault.

As such a total of some dozen physical parameters exist for describing the source process, including three coordinates of one corner of the fault plane, six static parameters (L, W, D, v, δ and λ), two coordinates of the point where the rupture nucleates, mode of rupture front propagation besides the two kinematic parameters. A comparison can be made with the conventional point source model prepared by Guntenberg and Richter (1942), wherein only a few parameters were considered, namely, the three coordinates of the hypocentral location and the empirical magnitude M. However, the following relationship exists between the fault parameters and the seismic moment M_o representing the physical strength of an earthquake.

$$M_o = G\,L\,D_f\,W \tag{8.30}$$

where,

G = shear modules near the focal area
D_f = average final slip over the fault plane
W = width of the fault
L = length of the fault

With the recent advances in seismology, an earthquake is now attributed to fault movement within the crust. Seismic motion may be expressed in terms of parameters of the fault movement. The term seismic moment, M_o is used to express the moment of a point double couple equivalent to fault slip in terms of the theory of elasticity. This parameter is proportional to the amplitude of the long-period seismic waves. Once the fault dimensions and the final offset are determined, the stress drop as the initial tectonic stress minus the final stress over the fault plane can be estimated by

$$\Delta\sigma = c\,GD_f/L \tag{8.31}$$

where c is the non-dimensional constant.

Using the definition of seismic moment, Eq. (8.31) may be expressed as

$$M_0 = C_1\, \Delta\sigma\, L_{c1} \tag{8.32}$$

where c_1 is another constant that depends upon the rupture surface geometry.

Kanamori and Anderson (1975) have shown that for a circular fault of radius a, the stress drop and the average dislocation (fault offset) may be further expressed as

$$\Delta\sigma = \left(\frac{7\pi}{16}\right)\cdot G \cdot \left(\frac{D_f}{a}\right) \tag{8.33}$$

where D_f is the average dislocation.

$$\Delta\sigma = \frac{7}{16}\, M_o a^{-3} \tag{8.34}$$

For a circular fault, the dislocation radius may be related to seismic moment and stress drop as

$$a = \left(\frac{7}{16}\right)^{1/3}\left(\frac{M_o}{\Delta\sigma}\right)^{1/3} \tag{8.35}$$

Aki, K (1987) states that for a wide range of rupture models the relationship between the stress drop and the slip velocity is given by

$$\Delta\sigma = K\left(\frac{G}{V_s}\right)\left(\frac{D}{\tau_r}\right) \tag{8.36}$$

where,

K = scaling factor that varies between 0.5 and 1.0
D = average slip on fault plane
τ_r = rise time
V_s = shear wave propagation velocity
V = slip velocity = $\dfrac{D}{\tau_r}$.

Brune (1970), by the use of physical arguments obtained the following far-field average shear wave displacement function:

$$u(r, t) = \left(\frac{\Delta\sigma V_s}{G}\right)\left(\frac{a}{r}\right) t^1\, e^{-\omega_c \cdot t'} \tag{8.37}$$

where,

r = epicentral distance
a = radius of the circular fault
a/r = geometric spreading
ω_c = corner frequency
$t' = t - r/V_s$
V_s = velocity of shear waves

Figure 8.14 shows a plot of the Brune's far-field Fourier amplitude spectrum in log-log plot. As can be seen for frequencies less than ω_c, the spectral amplitudes decay by ω^2. The spectral amplitude asymptote to a constant level equals to $K(\Delta\sigma V_s/G)(a/r)$. The effects of high frequency diminution, not accounted for by Eq. (8.37) are indicated by a dashed line in the high frequency regions.

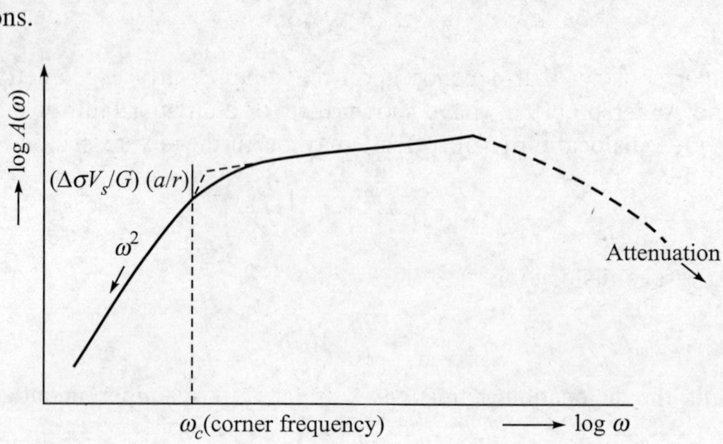

Figure 8.14 Brune far-field S-wave acceleration Fourier amplitude spectrum (After Brune, 1970)

8.8 METHOD FOR SIMULATING STRONG GROUND MOTION

The method for simulating strong-motion is called the theoretical method when the Green function G in Eq. 8.29 is calculated for the relatively simple model of the medium, such as parallel layer structure. The slip on the fault is specified kinematically in this model in Figure 8.13 and in the semi-empirical model as well. However, in the case of seismic motion with a wavelength of a few hundred metres or periods of about 1 second, the engineers are interested on account of the following:

- There are a few places only where the detailed subsurface structures (between the fault and the site) are known.
- In such places, the calculation of seismic motion might take time and involve more labour.
- The calculation of the Green function would require some knowledge of mathematics.

Thus kinematic modeling and simulation of the strong ground motion with Green's function techniques may help performance based seismic resistent design of earth retaining structures.

PROBLEMS

8.1 Describe the methods for strong ground motion recordings. How will you process digitization of the observed records?

8.2 How will you determine the bracketed durations of a given ground motion? Figure 8.2 shows the general ground motion (accelerograms), show by a suitable sketch how will you obtain the bracketed duration and significant duration of these ground motions. If the Arias intensity is expressed as

$$I_a = \frac{\pi}{2g} \int_0^{\infty} [a(t)]^2 dt.$$

Show that its value is independent of the method used to define the duration of the strong motion.

8.3 What are the methods used to measure the earthquake magnitude? Hanks and Kanamori (1979) proposed that earthquake magnitude is better determined by

$$M_w = 2/3 \log M_o - 10.7$$

where M_w = moment magnitude and M_o = a measure of seismic energy. Discuss the relationship.

8.4 Describe the Campbell attenuation model and compare with other models. How can attenuation relationships be developed with the models using regression analysis?

8.5 What do you mean by power spectra and spectral parameters in strong ground motion studies? How will you obtain the following:

(a) Central frequency
(b) Predominant period
(c) Shape factor

8.6 Describe the procedure for strong-motion observations. Describe the activities of such observations in Japan, USA and India.

9

SEISMIC HAZARD ANALYSIS

9.1 INTRODUCTION

The term 'earthquake-hazard analysis' is generally used to mean predicting the various parameters and properties of an earthquake that is likely to take place within a certain period in the future at a given site. The basic consideration in seismic hazard analysis is the uncertainty in decision-making process. In predicting earthquakes and estimating strong motions, the uncertainties are largely due to

- Earthquake occurrence
- Source process
- Propagation
- Soil condition near the surface.

The detailed explanations are as under.

Earthquake occurrence

The time and place of a future earthquake are both uncertain. The period of a few tens or hundreds of years, which is considered in earthquake-resistant design, is too long for earthquake prediction except for a few sites. On the other hand, it is too short for recurrence of steady seismic activity.

Source process

The direction and shape of the seismic fault, the direction and amount of dislocation of the fault movement, stress drop, seismic moment, starting point of fault rupture and its direction as well as velocity of propagation are the parameters for expressing the source process in physical terms. All these factors affect the intensity of an earthquake at a given site. However, it is not possible to accurately predict the value of these parameters for a future earthquake from the engineering point of view. In conventional earthquake-hazard analysis the source properties are expressed in terms of parameters indicating the size of the earthquake, such as magnitude (JMA magnitude, surface wave magnitude, or seismic moment, based on the information about earthquakes in the past. Even today, when the source process is better understood, the situation has not changed much because of the limited information available. Predicting the size of a future earthquake is uncertain in the foregoing sense and there is no direct correlation between such a magnitude parameter and the resultant earthquake motion.

Propagation

During the propagation of a seismic wave from its origin to the site of observation, the waveform changes as a result of reflection or refraction. Thus, change in seismic waveform due to its propagation is generally considered to be controlled by the distance of propagation from its origin except when the ground structure of the path can also be estimated. The effect of a complex layer structure can be seen as an uncertainty in correlation between ground motion and distance.

Soil condition near the surface

Properties of earthquake motion are considerably affected by ground conditions over a depth of about 100 m from the surface. This is the issue of earthquake response of the ground, however, because of the limitations of information about the ground and some imperfections in the evaluation of the effect of these limitations, the prediction of ground motion contains some elements of uncertainty.

Seismic hazard analysis may be analyzed deterministically (DSHA) or probabilistically (PSHA) in which uncertainties in earthquake size, location, frequency of occurrence are explicitly considered. In general, a natural hazard associated with earthquakes includes ground shaking, fault rupture, landslide, liquefaction, etc. However, the major interest is in the probabilistic estimation of ground shaking hazard since it causes loss of life and property. In this context the seismic zones are based on expected damage. It is universally accepted that countries around the world have employed different hazard computation methods for the development of seismic zoning maps. While carrying out the hazard analysis deterministically or probabilistically, the area is divided into seismic zones and the seismic activity in each zone is studied and reclassified according to the changes in the frequency of earthquake occurrence.

This chapter discusses mainly the considerations in seismic hazard analysis developed for decision-making under some uncertainty. Thus, one can think of numerous ways in which seismic hazard analysis can be used for dynamic analysis. In each method the uncertainty of future seismic activity is taken into consideration and the seismic properties are determined by using a term index representing the possibility of occurrences.

9.2 MEANING OF EARTHQUAKE-HAZARD ANALYSIS

The term 'earthquake-hazard analysis' has been assigned various meanings. In general, this term used is to mean "predicting the properties of an earthquake that is likely to take place within a certain period in future at a given site". Considerable knowledge of the source process of an earthquake has recently been acquired. The fault model of an earthquake has also been developed and it is now possible to estimate earthquake motion based on this model.

The factors of uncertainty during seismic-hazard analysis have been discussed by a number of researchers. Efforts have been made to reduce this uncertainty by trying to understand the physical background that supports (preexists) the different phenomena and by collecting data, which could form the basis of quantitative analysis. The source spectrum and its scaling or they methods of evaluation of nonlinearity in earthquake response properties of ground, etc. are examples of this. However, it is not possible to understand all the elements of uncertainty due to the fact that an earthquake is a failure phenomenon under non-uniform geological conditions, which cause the rupture of a fault.

The ultimate objective of evaluation of strong motion considered in earthquake engineering is not only to explain earthquakes of the past, but also to take appropriate measures against future earthquakes. As such, it is necessary to establish some system of evaluation considering the uncertainties involved. Accordingly, the fundamental requirement in earthquake-hazard analysis is to consider such uncertainties.

9.3 PARAMETERS FOR SEISMIC HAZARD ASSESSMENT

A lot of complete scientific perception and analytical modelling is involved in the seismic hazard estimation. The main parameters are

- Evaluating of seismic source
- Ground motion attenuation
- Earthquake recurrence relations
- Fault activity
- Local site and soil conditions

9.3.1 Evaluation of Seismic Source

The objective of seismic hazard is to evaluate ground motion characteristics at a particular site, so that earthquake resistant design is carried out. As such the structures so designed may withstand a certain level of shaking with excessive damage. Seismic hazard analysis can be realistically carried out provided the potential source of seismic activity is idealized on the basis of the following:

- Geological event
- Fault activity
- Magnitude and intensity of past earthquakes
- Tectonic evidence and status of strain energy storage
- Historical seismicity

9.3.2 Ground Motion Attenuations

The ground motion attenuation relationships are generally regionally dependent. This relationship can provide the meaning of assessing a strong motion parameter of interest from the seismic data such as magnitude source to site distance, faulting mechanism and local site conditions. (see Section 8.5.4). The choice of appropriate relationship is governed by the regional tectonic features of the concerned site, whether it is located within a stable continental region or an active tectonic region or whether the site is in proximity to a subduction zone tectonic environment.

9.3.3 Earthquake Recurrence Analysis

The average number of earthquakes N with magnitude greater than or equal to M is given by the equation (Richter, 1958).

$$\text{Log } N = a - bM \tag{9.1}$$

The parameters a and b are estimated from the catalogue of historical seismicity. These parameters play an important role in assessing the seismic hazard. Statistical uncertainty in these estimates can have a dominant role on the seismic hazard (McGuire, 1977). Because of the moderate seismicity and the incomplete reporting of earthquakes in the catalogue, reliable estimates of the a and b parameters may not be easily obtainable from the data. On the hypothesis that all source zones in the Indian peninsula would exhibit similar seismic tectonic behaviour, the earthquake catalogues for all zones were merged to form a viable data set for analysis. *Stepp* (1972) procedure may be adopted to minimize the effects of incomplete reporting before the values of the a and b statistics are estimated. Figure 9.1 shows the derived log N versus M plot. The a value, which denotes the number of events with $M \geq 0$ per 40 years was reapportioned among the constituent source zones. Fitting lines obtained the interim estimates of a value for the constituent source zones with the estimated b slope of the reported earthquakes in a set of magnitude classes. The corresponding a values were averaged and formed the weights for redistributing a value, obtained from combined analysis, to various constituent zones.

Gutenberg–Richter law is illustrated schematically in Figure 9.1, where N is the mean annual rate of exceedance of magnitude M, 10^a is the mean yearly number of earthquakes of magnitude greater than or equal to zero, and the b value describes the relative likelihood of large and small earthquakes.

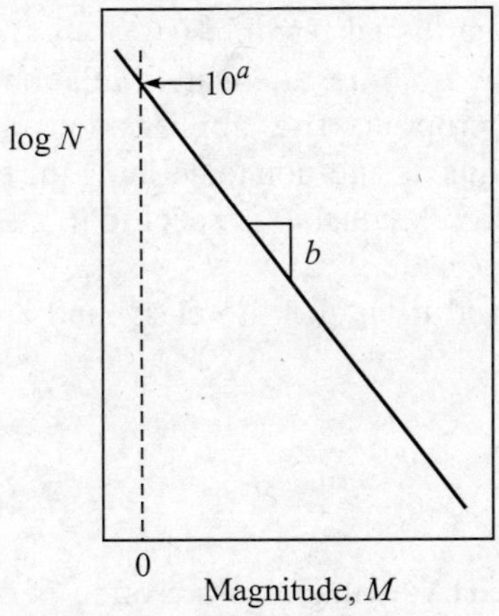

Figure 9.1 Gutenberg–Richter recurrence law.

The worldwide recurrence data as proposed by Esteva (1970) are shown in Figure 9.2. The a and b parameters are generally obtainable by least square method or regression on a database of seismicity from the source zone of interest. The typical values of b may be taken as 1 ± 0.3.

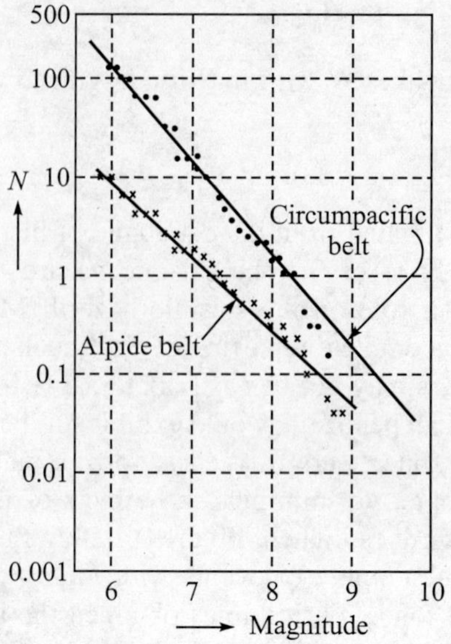

Figure 9.2 Worldwide seismicity data [After Esteva, 1970]

9.3.4 Local Site and Soil Conditions

The influence of local geologic and soil condition on the intensity of ground shaking and earthquake damage has been known for many years. For each source zone, uncertainty in earthquake location is characterized by a probability density function of source-to-site distance. Evaluation of the probability density function requires estimation of the geometry of the source zone and of the distribution of earthquake within it.

In case the estimates of PGA from the attenuation relationship are for hard rocks, it is then necessary to multiply by an appropriate site amplification factor in order to account for the existing soil conditions at site. Thus, in addition to defining the earthquake source, the application of the ground motion attenuation relationship is specific to a soil or rock type on which PSHA ground motion estimate is to be made.

The ground types are referred to as the site class, and are defined as:

- Hard Rock
- Soft Rock
- Firm Soil
- Soft Soil

Site-specific engineering PSHA evaluations are often performed to obtain a more precise measure of ground motion amplitude.

9.4 RISK INDEX AND EVALUATION OF EARTHQUAKE MOTION

In the consideration of various uncertainties it is important to represent the properties of earthquake motion along with a "risk index", a parameter describing the possibility of their occurrence.

Thus, earthquake-hazard analysis can also mean evaluation of various properties of earthquake motion likely to occur at a given point within the specified period in the future in terms of the risk index.

The probability of earthquake occurrence in a year, or recurrence time, is frequently used as the risk index. For this purpose, the earthquake occurrence or properties of earthquake motion are expressed in terms of a probability model. This does not mean that the earthquake phenomenon is a statistical (probability) phenomenon, but the element of uncertainty present in the quantitative evaluation of related parameters can be considered a model in the form of relative frequency (probability distribution). The problem can then be superimposed on the process of determining the risk index.

Earthquake-hazard analysis based on the probability model exhibits clarity in the steps involved and the result obtained. As such, it is very useful from the engineering point of view. Hence, we shall explain here the methods of earthquake-hazard analysis based on the probability model.

The probability model for earthquake-hazard analysis naturally reflects the physical properties of the region concerned but due to lack of adequate data the model cannot be made highly rational. Calculation of model parameters is also difficult and quite often the reliability of the model itself is questionable. Let us discuss the uncertainties involved in an analytical model.

9.4.1 Historical Earthquake Data

The earthquakes mentioned in catalogues of past earthquakes are called historical earthquakes and are widely used as data that directly represents the seismic activity of the region. Earthquakes catalogues and databases make the first essential input for the delineation of seismic source zones and their characterization. The preparation of a uniform or homogeneous catalogue for a region under consideration is an important task. Globally, considering the data from historic time to recent time can broadly be divided into three temporal categories.

- Since 1964 till recent time for which modern instrumentation based data is reliable
- 1900–1963—early instrumental data
- Pre 1900–pre instrumental data, based primarily on historic information

Historical accounts of ground-shaking effects can be used to confirm the occurrence of past earthquakes and to estimate their geographic distribution of intensity. As historical data are dated, they can also be used to evaluate the rate of recurrence of earthquakes in particular or the seismicity of the region in general.

9.4.2 Aleratory and Epistemic Variability

Aleratory variability is uncertainty in the data used in the analysis and generally accounts for randomness associated with the prediction of a parameter from a specific model, assuming that the model is correct.

Epistemic variability or modelling uncertainty accounts for incomplete knowledge in the predictive models and the variability in the interpretations of the data used to develop the models.

However, due to large measure of uncertainty in each of the above parameters, a probabilistic approach provides a more realistic basis for estimating the seismic risk provided the data is available (Algermissen, 1982). From an engineering point of view seismic risk may be defined as the probability of the occurrence of a generalized intensity (for example, peak ground acceleration) greater than a specified value at a site in a specified interval of time (service life). Assessment of seismic risk from the basis of the probabilistic approach is possible (Lomnitz, 1974). The assessment of seismic risk at a site involves the following:

- Identification of the source of earthquake, which may be idealized as point, area or volume source.
- Construction of stochastic models of the strong motion activity in each source incorporating the temporal, spatial and magnitude characteristics.
- The nature of the attenuation in the region, including the effect of scatter in the data.

9.4.3 Logic Tree

Simple logic trees are framed for incorporation of model uncertainty (Kulkarni et al., 1984). They provide a convenient framework for the explicit treatment of model uncertainty. The logic tree approach allows the use of alternative models, each of which is assigned a weighting factor that is interpreted as the relative likelihood of that model being correct.

The yearly frequency of occurrence of earthquakes having a magnitude greater than m_0 may be expressed as

$$V_h = 10^a \tag{9.2}$$

where m_0 is the minimum meaningful earthquake magnitude from engineering consideration. (See Figure 9.1)

9.4.4 Active-Fault Data

Since an earthquake was earlier considered fault movement, studies were conducted jointly by experts in the field of geology and seismology to investigate the faults (active lineaments), which could possibly have caused the earthquake. While the active-fault data indicates the average fault activity over a period of one million years, the period of historical data is only about 1000 years. Reliable data is therefore restricted. As such, the active-fault data can be considered to supplement historical records. While historical data is available irrespective of the place of occurrence, many active faults may not have been detected so far, which means there is some imperfection in spatial distribution.

In addition to these imperfections, we also have to consider the difference in basic properties of the two types of data. Thus, V_h obtained from historical data represents the recent seismic activity whereas V_f obtained from active fault data represents the average seismic activity over a period of a million years. Naturally, these two values will not be the same.

9.4.5 Evaluation of Probability of Earthquake Occurrence Based on Historical Earthquake Data

The accuracy of historical earthquake data diminishes as one goes backward in time. This is particularly important for the quantitative evaluation of probability (incidence) of earthquake occurrence. Moving backward in time, we find records of only major earthquakes. If only earthquakes of large magnitude are of interest, useful data even for very old periods can be found. On the other hand, reliable data for minor (smaller) earthquakes became available only after recording instruments were developed.

9.4.6 Calculation of Earthquake Occurrence Based on Active-Fault Data

If we assume that all fault movement is caused by an earthquake and the average yearly seismic moment accumulated in a fault is the same as average yearly seismic moment released by that fault causing an earthquake, then the frequency of earthquake incidence by that fault is given by the following expression:

$$V_f = \frac{\mu W L \lambda}{\int_{m_0}^{m_u} G(m) f_M(m)\, dm} \tag{9.3}$$

where L is the length of fault; W its width, μ the shear modules; λ the average dislocation rate of fault; m_0 the lower limit of meaningful magnitude from the engineering point of view (for example, $m_0 = 5.5$); m_u is the maximum magnitude generated by the fault; $G(m)$ represents the seismic moment M_0 as the function of magnitude m; and $f_M(m)$ is the probability density function of earthquake magnitude generated by fault.

9.4.7 Considerations of Combined Historical Earthquake Data and Active-Fault Data

While active-fault data indicates the average fault activity over a period of one million years, the period of historical data is only about 1,000 years. Reliable data is therefore restricted. As such, the active-fault data can be considered to supplement historical records. While historical data is available irrespective of the place of occurrence, many active faults may not have been detected so far, which means there is some imperfection in spatial distribution.

In addition to these imperfections, we also have to consider the difference in basic properties of the two types of data. Thus, V_h obtained from historical data represents recent seismic activity whereas V_f obtained from active-fault data represents the average seismic activity over a period of a million years. Naturally, these two values will not be the same.

9.5 METHOD OF ANALYSIS

The methodology for assessing the probability of seismic hazards grew out of an engineering need for better design in the context of structural reliability. The paper entitled, "Engineering

Seismic Risk Analysis" by Cornell, C.A. (1968) created a great awarness worldwide for seismic hazard analysis. The recent trends in reliability-based structural design include design models which consider dispersal of ground properties by deterministic method as well as by probabilistic method. The geotechnical consultant is supposed to provide the structural designer with the estimated probability of experiencing ground shaking more severe than the given accelerogram. It is no longer acceptable to provide a ground motion without identifying its expected frequency of occurrence. The estimates of frequency of occurrence are commonly given in one of the two ways. The probability of an earthquake in any one year for an event with a return period of R is $1/R$, this can be used to calculate the probability of occurrence in a longer period of time. The estimation of the probability of exceeding some amplitude of shaking at a site in some period of interest requires that a probability distribution of the ground motion amplitudes be assumed. The Poisson model (see Section 9.7.1) serves as a reasonable assumption in non-engineering applications except the rare cases where a single earthquake source may dominate the hazard at a site and the earthquake occurrence model for the source can be considered time-dependent or non-Poissonian (see Section 9.7.2). Poisson model has generally and traditionally been used throughout seismic hazard assessment.

However, scientific knowledge for the accurate quantification of these hazards is always limited. The balance of the hazard assessment finally hinges on technical judgement.

The seismic hazard assessment can be carried out by two methods, namely

- Deterministic Seismic Hazard Analysis (DSHA)
- Probabilistic Seismic Hazard Analysis (PSHA)

The deterministic method considers the effect at a site of either a single scenario earthquake, or a relatively small number of individual earthquakes. The probabilistic methodology quantifies the hazard at a site from all earthquakes of all possible magnitudes, at all significant distances from the site of interest, as a probability by taking into account this frequency of occurrence. The deterministic earthquake scenario therefore, is a subset of the probabilistic methodology (Chen, W.F., 2002).

PSHA can address any natural hazard associated with earthquakes, including ground shaking, fault rupture, landslide or liquefaction, etc. However, in this section the presentation is restricted to the estimation of earthquake ground motion hazard. Figure 9.3 shows and illustrates the elements of the probabilistic ground motion hazard methodology in the context of seismic design for a site of significant engineering importance. The process begins with the characterization of the earthquake occurrence using two sources of data: observed seismicity (historical and instrumental) and geologic. The occurrence information is combined with data on the propagation of seismic shaking.

By this process the seismotectonic model is obtained. Since uncertainty is inherent in the earthquake process, the parameters of the seismotectonic model are systematically varied by logic tree, Monto Carlo simulation and other techniques, to provide the probabilistic seismic hazard model's results. The results may be disaggregated (also termed deaggregation) to identify specific contributory parameters to the overall results. These results should also consider the site-specific soil results (Chen, W.F., 2002). The final results, presented in many different ways depending upon the user's needs, are termed seismic design criteria.

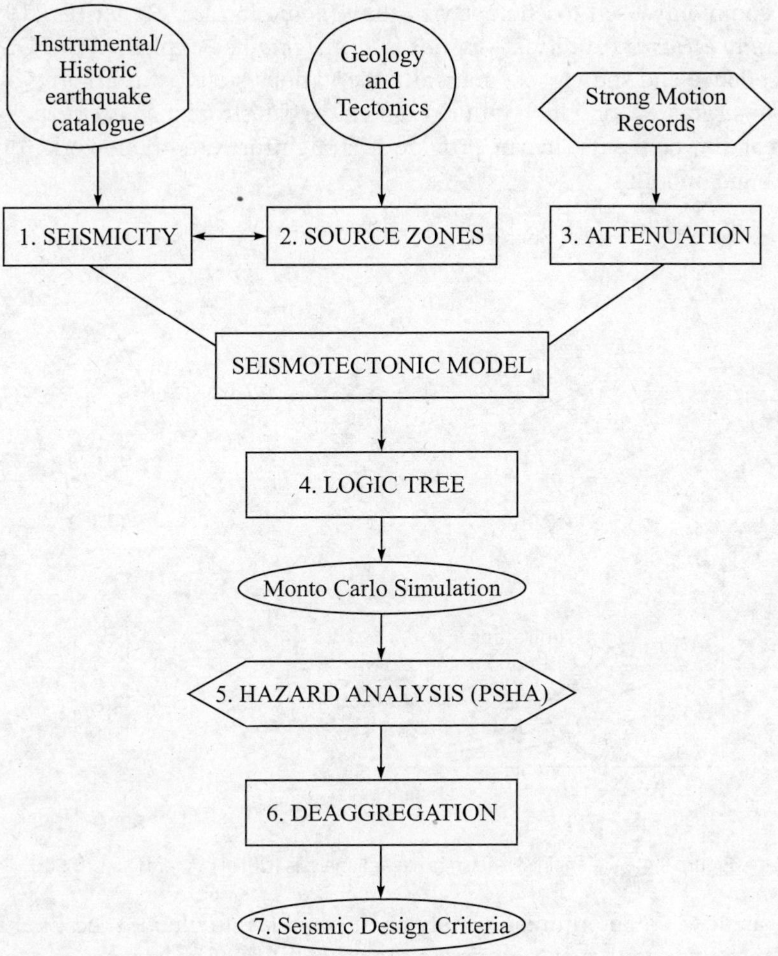

Figure 9.3 Probabilistic hazard methodology in the context of a seismic design criteria [After Chen, W.F. 2002]

9.5.1 Deterministic Seismic Hazard Analysis (DSHA)

The deterministic seismic hazard analysis (DSHA) is based on postulated occurrence of an earthquake of a specified size occuring at a specified location. For carrying out DSHA, the following are to be ascertained:

- Characterization of all earthquake sources
- Solution of a source to site distance parameter
- Selection of controlling earthquake which produces the maximum level of shaking
- Hazard is finally defined in terms of ground motion.

As such the hazard is computed in terms of ground motion at the site by the controlling earthquakes. Thus, DSHA provides evaluation of the worst-case ground motion at the site by controlling earthquakes. Peak ground acceleration (PGA), peak velocity and response spectrum ordinates are commonly used to characterize the seismic hazard. Reiter, L. (1990) in the text entitled *Earthquake-Hazard Analysis—issues and insights* has explained the DSHA procedure with great excellence and shown schematically the various steps as in Figure 9.4. Thus, DSHA when applied to structures for which failure could have catastrophic *consequences,* such as large dams or nuclear power plants, it can provide a straightforward framework for evaluation of worst-case ground motions.

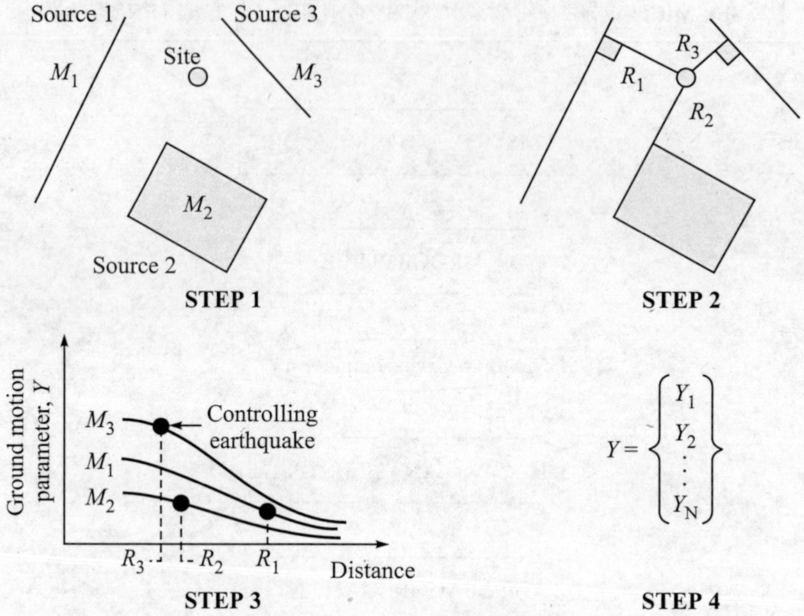

Figure 9.4 Deterministic seismic hazard analysis (DSHA) (After Reiter, 1990).

However, as this is a deterministic analysis, it provides no clue on the likelihood of occurrence of the controlling earthquake. As far as the probability of occurrence is concerned, DSHA implicitly assumes that the probability of occurrence is at a point in each source zone closest to the site and zero elsewhere. In this context several terms are used to describe earthquake potential

- MCE (maximum credible earthquakes)
- DBE (design basis earthquakes)
- SSE (safe shutdown earthquakes)
- MPE (maximum probable earthquakes)
- OBE (operating basis earthquakes)

They are defined as the maximum earthquake that appears capable of occuring *under* the known tectonic features.

9.5.2 Probabilistic Seismic Hazard Analysis (PSHA)

Probabilistic Seismic Hazard Analysis (PSHA) takes into account the ground motions from the full range of earthquake magnitudes that can occur on each fault or source zone that can affect the site. (Cornell, 1968: Kulkarni, et al., 1979). The PSHA numerically integrates the information using the probability theory to produce the annual frequency of exceedance of each different ground-motion level for each ground-motion parameter of interest, reflecting the effects of all the postulated seismic sources in the region. PSHA thus incorporates for each fault or source zone the following three factors.

- Mean frequency per year of occurrence of each different magnitude level of earthquakes
- Mean frequency per event of each possible source to site distance
- Mean frequency per event of each different level of ground motion from each possible magnitude distance pair

The PSHA yields the annual frequency of exceedance of each different ground-motion level for each ground-motion parameter of interest. This relationship between ground-motion level and annual frequency of exceedance is called a ground motion hazard curve. The products of a PSHA are ideally suited for performance-based design, because they specify the ground motions that are expected to occur for a range of different annual probabilities (or return periods) [Somerville and Moriwaki (2003)].

The seismic hazard analysis in the context of engineering design is generally defined as the predicted level of ground acceleration which would be exceeded with 10% probability at the site under consideration due to occurrence of an earthquake in the region in the next 50 years. PSHA may be summarized as the solution of the following expression (Chen, W.F. et al., 2002).

$$\lambda[X > x] = \sum_{\text{Source } i} V_i \int_{M_0}^{M_{\max}} \int_{R/M} P[X \geq x | M, R]\, f_M(m) \cdot f_{R/M}(r/m)\, \text{dr dm} \tag{9.4}$$

where,

$\lambda[X > x]$ = annual frequency that ground motion at a site exceeds the choosen level $X = x$

V_i = annual rate of occurrence of earthquakes on seismic source i, having magnitude M_0 and $M_{\max}$

M_0 = minimum magnitude of engineering significance

$M_{\max}$ = maximum magnitude assumed to occur on the source

$P[X \geq x, | M, R]$ = conditional probability that the chosen ground motion level is exceeded for a given magnitude and distance

$f_M(m)$ = probability density function of earthquake magnitude

$f_{R/M}(r/m)$ = probability density of distance from earthquake source to the site of interest.

Once the annual exceedance rate $\lambda[X \geq x]$ is known, the probability that an observed ground motion parameter X will be greater than or equal to the value x in the next 't' years (the exposure period) is easily computed by the equation

$$P[X \geq x] = 1 - \exp(-\, t\, \lambda\, [X > x]) \tag{9.5}$$

where the return period of x is defined as

$$R = \frac{1}{\lambda[X \geq x]} = \frac{-t}{\log(1 - P[X \geq x])} \tag{9.6}$$

However, the probability of an earthquake exceeding a certain value in specified time t may also be simply expressed as

$$P_t = 1 - (P_1)^t \tag{9.7}$$

where,

P_t = probability of exceedance of magnitude x or greater in t years
P_1 = annual probability

The return period R may be expressed as

$$R = \frac{1}{1 - P_1} \tag{9.8}$$

EXAMPLE 9.1 Using the PSHA method find the return period for a seismic event that has a 10% probability of being exceeded in a 50 year period. What will be the return period if there is a 50% probability of being exceeded in 50 years? How many such events are likely to occur in the next 100 years?

Solution: Using Eq. (9.6)

$$R = \frac{-50}{\log(1 - 0.1)} = 475 \text{ years}$$

Again using Eq. (9.7)

$$0.10 = 1 - (P_1)^{50}$$

$$P_1 = 0.997895$$

$$R = \frac{1}{1 - P_1} = 475 \text{ years}$$

Further, for a 50% chance of exceedance in 50 years

$$0.50 = 1 - (P_1)^{50}$$

$$P_1 = 0.98624$$

$$\text{Annual probability} = 1 - P_1 = 0.01376$$

$$R = \frac{1}{1 - P_1} = \frac{1}{0.01376} = 72 \text{ years}$$

$$\frac{100}{R} = 1.38 \text{ events/100 years}$$

9.6 CLASSIFICATION OF SEISMIC ZONES

The zone (region) in which there is a likelihood of occurrence of an earthquake of the magnitudes considered in earthquake engineering is termed the seismic zone. It is necessary to quantitatively evaluate the seismic activity in and around this region as the basic information needed in earthquake hazard analysis.

In a broad sense, seismic zoning can be described as a process of demarcating areas of equal seismicity, or of equal hazard related to a characteristic of strong ground shaking.

This evaluation comprises the following:

1. Seismic zoning according to the differences in seismic activity.
2. Calculation of the probability of earthquake occurrence in each such zone.
3. Identification and modelling of temporal and spatial correlation of earthquake occurrence in each region.
4. Modelling of distribution properties of parameters, which express earthquake magnitude.

Among the above, 1 and 2 express the overall phenomenon of seismic activity while 3 and 4 express the time-space distribution of earthquake occurrence.

In constructing the model of seismic activity the following basic data is used. While carrying out earthquake-hazard analysis, the target area is divided into a number of seismic zones and the seismic activity in each zone is studied. Seismic zoning maps indicate a quantitative assessment of the seismic risk in different zones of a region. Thus, in a broad sense, the seismic zoning can be defined as a process of demarcating or mapping areas of equal seismicity or of equal hazard related to a characteristic of strong shaking and of site or structural response. Further seismic zoning can be done on a macro or on a micro scale depending on the size of the area.

The preparation of such maps involves a synthesis of the available historical and seismological data, an understanding of the geological and geophysical features and a balance between cost and risk. Early seismic zoning maps were prepared in different countries Hodgson (1956), Medvedev (1958), Richter (1958), Guha (1962) using the available historical and seismological data, on the assumption that earthquakes recur at places where they were previously recorded and with the same intensity. It was further assumed that at places where earthquakes have not been recorded, strong earthquakes will not occur in the future. Such maps of many regions proved wrong because in many cases strong shocks occurred in locations where shocks were not recorded before (Gubin, 1960). It was thus established that recorded seismic manifestation may not fully represent the seismogenic process of many zones, especially where strong shocks occur rarely. This suggested a seismotectonic approach, based on the confrontation of geological, geophysical and seismological data. This approach supplements the information contained in the historical and seismological data with geological and geophysical information, such as, faults, their sizes, and competency, grade of seismotectonic, mechanisms and rate of tectonic differential movements, and the correlations between the seismic and the tectonic features (Gubin 1967, 1969).

In seismic zoning, seismological data, including magnitude, time of occurrence, epicentre, focal depths of recorded earthquakes and the attenuation laws provide the quantitative basis for estimating seismic risk. The risk analysis may be carried out within a deterministic or a probabilistic framework. Due to a large measure of uncertainty, in each of the above parameters, a probabilistic approach provides a more realistic basis for estimating the seismic risk, provided adequate data is available (Algermissen, 1982). A probabilistic approach may be used to estimate the seismic risk, in terms of 100-year return period, peak-ground-acceleration. The zone boundaries given in IS 1893–1975 are broadly accepted, except for a few modifications in the peninsular region following Gubin's (1969) multi-element-map, and minor changes in other regions based on the risk analysis. A simple procedure is proposed to determine the peak ground acceleration for any return period or for a specified exceedance probability over a given service life (Basu and Nigam, 1978).

From an engineering point of view seismic risk may be defined as the probability of the occurrence of a generalized intensity (for example, peak ground acceleration) greater than a specified value at a site, in a specified interval of time (service life). Assessment of seismic risk forms the basis of the probabilistic approach (Lomnitz, 1974). The assessment of the seismic risk at a site involves the following.

- Identification of the sources of earthquake, which may be idealized as point, area or volume sources,
- Construction of stochastic model of the strong-motion earthquake activity, in each source, incorporating the temporal, spatial and magnitude characteristics, and
- The nature of the attenuation laws in the region, including the effect of scatter in the data.

9.6.1 Parameters for Seismic Zoning

The parameters used to map the seismicity may be the *a* and *b* values in the Gutenberg–Richter's magnitude–frequency relationship (see Eq. (9.1)), the maximum magnitude expected to occur during a specified exposure, and the total seismic energy released during a specified duration, to name a few.

The ground-motion parameter that was used in seismic zoning was the intensity of shaking. Even though the intensity scale is not based on instrumental data in many parts of the world, where the instrumental data is lacking or is insufficient, zoning in terms of site intensity is the only choice. In many studies, the site intensity has been also correlated to instrumentally measured parameters of strong ground motions, such as peak acceleration, for example. However, due to the subjective nature of defining the site intensity, such correlations are generally associated with large uncertainties. The next most widely used ground-motion parameter for seismic zoning has been the peak ground acceleration (Algermissen and Perkins, 1982; Khattri et al. 1984). Both the site intensity and the peak ground acceleration are simple, single parameter characterizations of earthquake shaking at a site. Consequently, the spatial distribution of seismic hazards can be represented with only a few zoning maps, corresponding to two or more probabilities of exceedence and a few exposure periods (e.g., 50 years or 100 years).

9.6.2 Seismic Zoning of India

Just as there cannot be any statistical approach to the problem of earthquake intensity, an entirely scientific basis for zoning is also not possible in view of the scanty data available. While the problem of the paucity of data is serious for a probabilistic approach, adequate data is available in some regions and techniques have now become available (Cornell 1968, Esteva 1968, Algermissen 1982) for an acceptable probabilistic estimate of seismic risk. A probabilistic analysis provides a broad basis for relative assessment of risk as well as a probabilistic interpretation of the maps prepared using a deterministic approach. The maps prepared through a probabilistic analysis make it possible to incorporate the effect of the service life, and acceptable risk in arriving at the design loads. Kaila et al., (1972) has carried out statistical analysis of seismic data to draw *a* and *b* value contours indicating seismicity of different regions. Basu and Nigam (1977) have carried out a systematic probabilistic analysis of available seismological data to incorporate the effects of uncertainties in various parameters affecting the seismic risk and has interpreted the present zoning map in probabilistic terms, Prasad (2003).

The contour map shown in Figure 9.5 provides a probabilistic estimate of seismic risk at different locations. It is based on a stochastic analysis of available seismological data and to some extent reflects the geological features through the identified faults. The small period of time (55 years) and paucity of data, particularly in the peninsular region, needs to be recognized in using the contour map for preparing the zoning map. For this purpose it is necessary to supplement this analysis with the available geological and seismotectonic information. The seismic zoning map, shown in Figure (9.5), has been prepared using the following stepwise procedure:

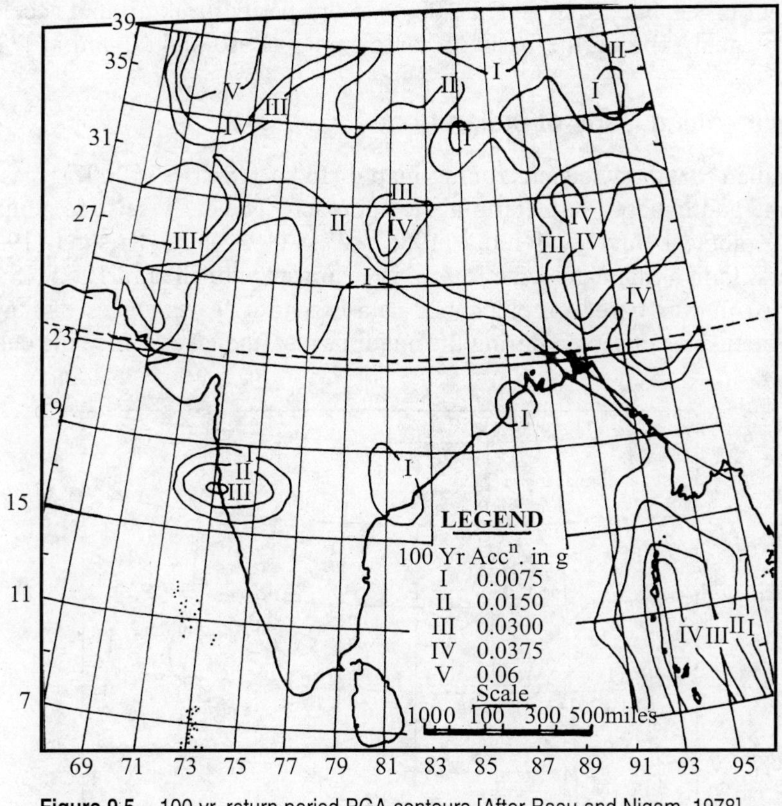

Figure 9.5 100 yr. return period PGA contours [After Basu and Nigam, 1978]

1. The Indian Standard seismic zoning map contained in IS: 1893 (1975) is treated as the first approximation of the proposed map. The Lateral force coefficient given in the code is divided by 2.6 to give peak ground acceleration at 5% damping.
2. For the peninsular region, the multi-element map due to Gubin (1969) is used as a guide to modify the zone boundaries of Indian Standard map for the peninsular region. In particular, the following premises are accepted:

- Narmoda-Son and Tapti fault system is treated as a homogeneous seismogenic region with a ceiling of 8 on M.M.I.
- Western Ghat, with faults extending down to Lat. 12, is accepted as seismogenic with a ceiling of 8 on M.M.I.
- Nilgiri, Shevaroy and Cardamon fault systems are accepted as homogeneous seismogenic regions with a ceiling of 7 on M.M.I.
- Godwana rift zone and adjacent parts of the shield including boundary faults are accepted as a seismogenic region with a ceiling of 7 on M.M.I.

Further, it is accepted that if an earthquake of certain maximum magnitude is recorded somewhere in a homogeneous active fault zone, then an earthquake of the same magnitude can occur anywhere in the zone. The contour map as shown in Figure 9.5 may be used as a guide for improvement in Seismic Zoning. The 100-yr return period peak ground acceleration values are assigned to equal zone on the basis of contour map. Basu and Nigam (1978).

9.6.3 Seismic Zoning Maps of Indian Code

The present Indian Standard seismic zoning map of India {IS: 1893 (2002)} has evolved over a period of time and is based on deterministic approach. The early seismic zoning maps were prepared by Geological Survey of India (1935), West (1937), Jai Krishna (1959) and Guha (1962). The first Indian Standard seismic zoning map was published in 1962 (IS: 1893 (1962)) (See Figure 9.6). It was based on epicentral data of known earthquakes of magnitude 5 and above and the intensities observed during the major past earthquakes. Since the data was limited,

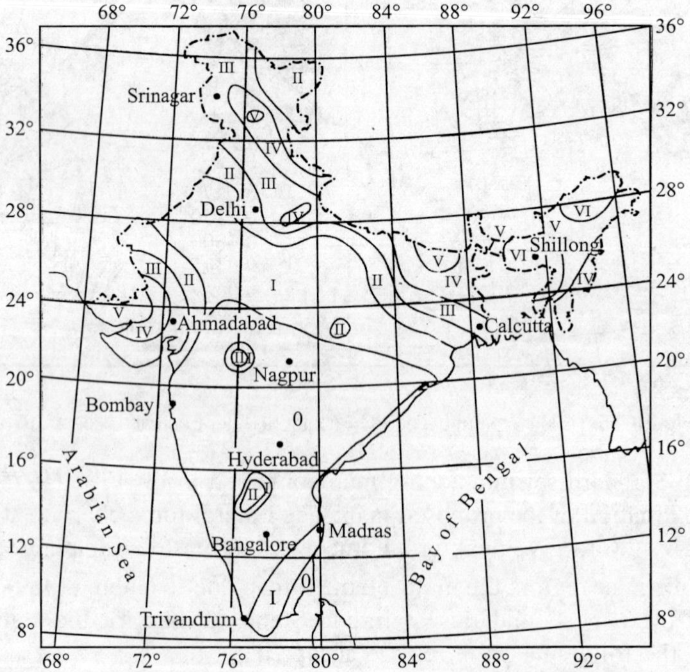

Figure 9.6 Seismic zoning map of IS : 1893(1962).

the seismic zoning map reflected a dominant influence of known earthquakes and not necessarily the "seismotectonic set-up of the region, in which major earthquakes could occur due to operative geotectonic processes (Srivastava, 1969)". The 1966 revision IS: 1893 reflected this view and was based on additional data on geology, tectonics and earthquake occurrence as shown in Figure 9.7. The occurrence of Koyna (1967) earthquake in the peninsular shield, hitherto assumed to be a seismogenic, highlighted the need for greater emphasis on the

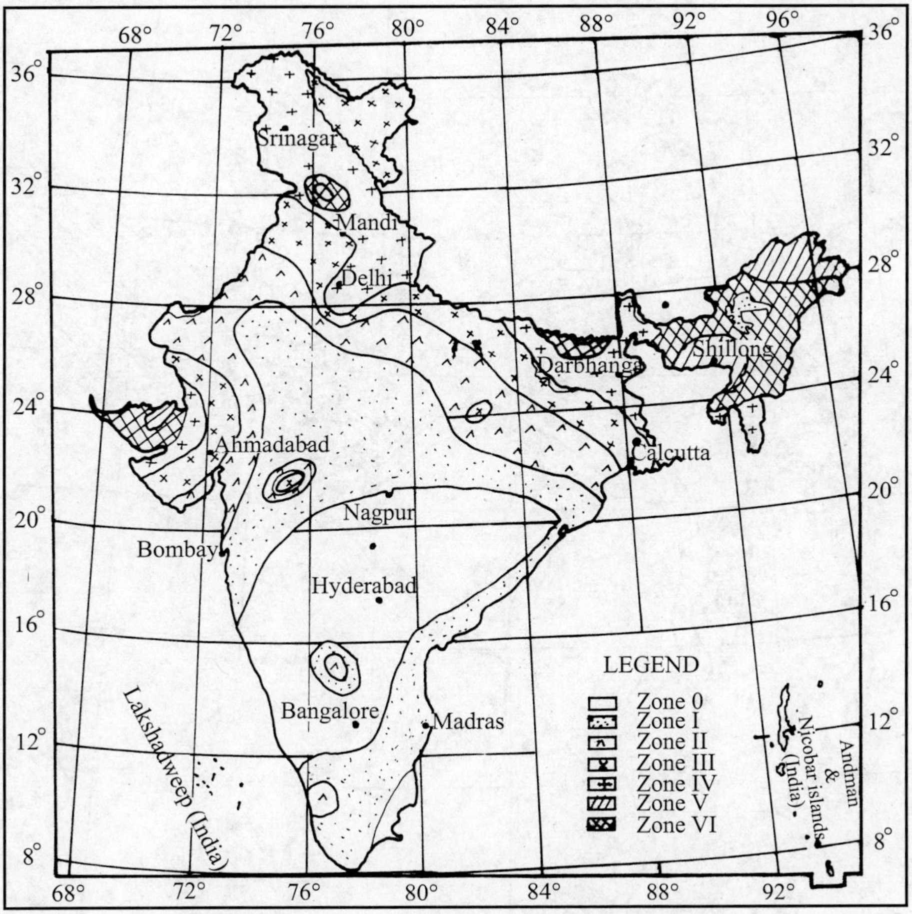

Figure 9.7 Seismic zoming map of India (IS : 1893(1966)).

tectogenesis and geological history of the country and understanding of the operative process, which are responsible for development of the various structural and related geomorphologic features and could lead to occurrence of earthquakes in future (Srivastava, 1969). The 1970 revision of the seismic zoning map, which is included as such in IS: 1873 (1975), gives greater emphasis to the seismotectonic framework of the country. Srivastava (1969) has given an excellent review of the historical development of seismic zoning in India and the basis for the zoning map contained in IS: 1893 (1975). Gubin (1969) has prepared a multi-element seismic zoning map for the peninsular region using the seismotectonic method. This map is of special significance because it covers an area where the rate of seismic activity is low.

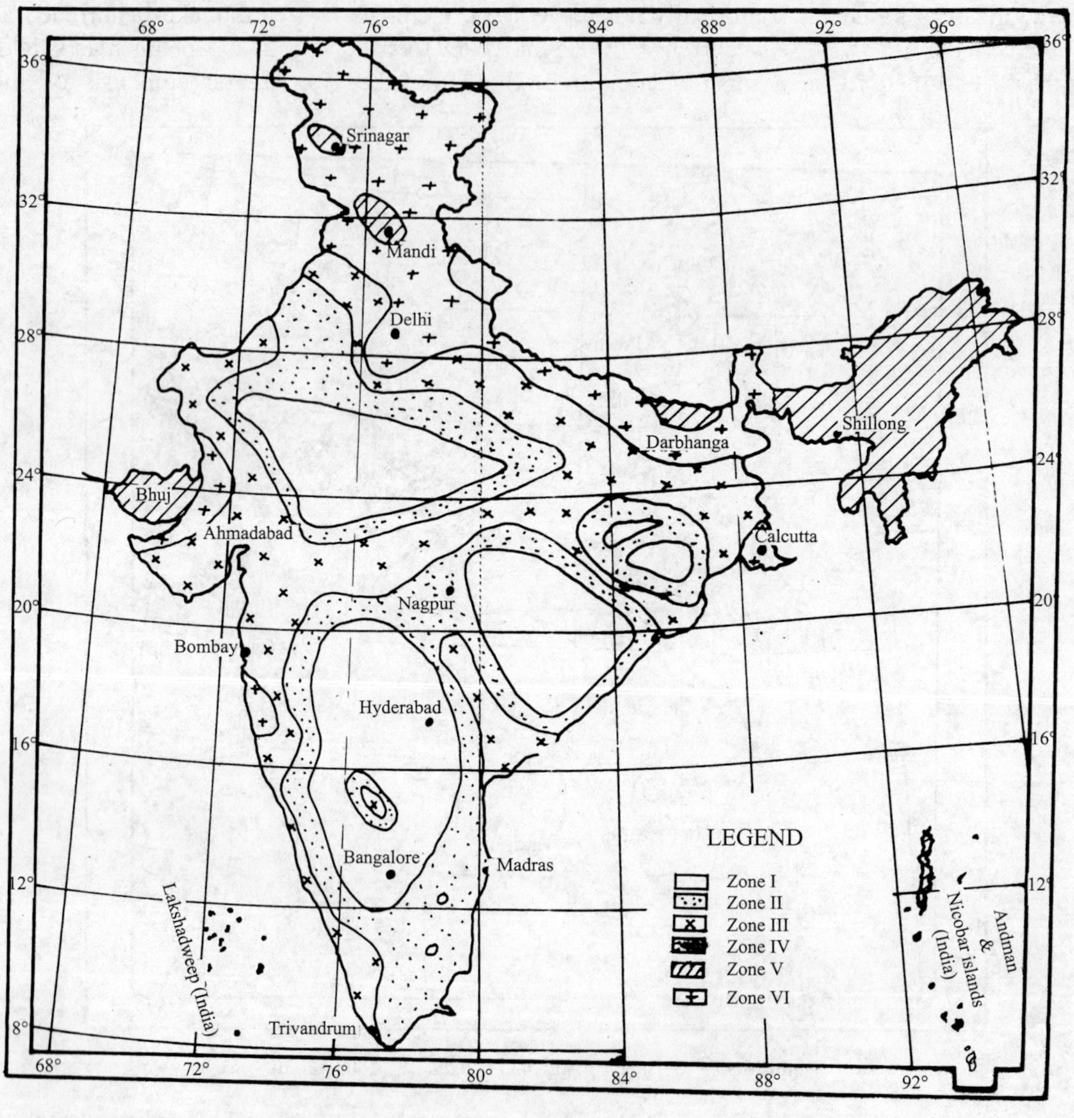

Figure 9.8 Seismic zoning map of India (IS : (1893(1975)).

The most significant feature of the seismic zoning maps described above is the fact that they are based on a deterministic analysis of seismic data. In fact, Section 0.6.1 of IS: 1893(1975) states that "The Sectional Committee has appreciated that there cannot be any statistical approach to the problem of earthquake intensity and an entirely earthquake intensity and an entirely scientific basis for zoning is also not possible in view of the scanty data available".

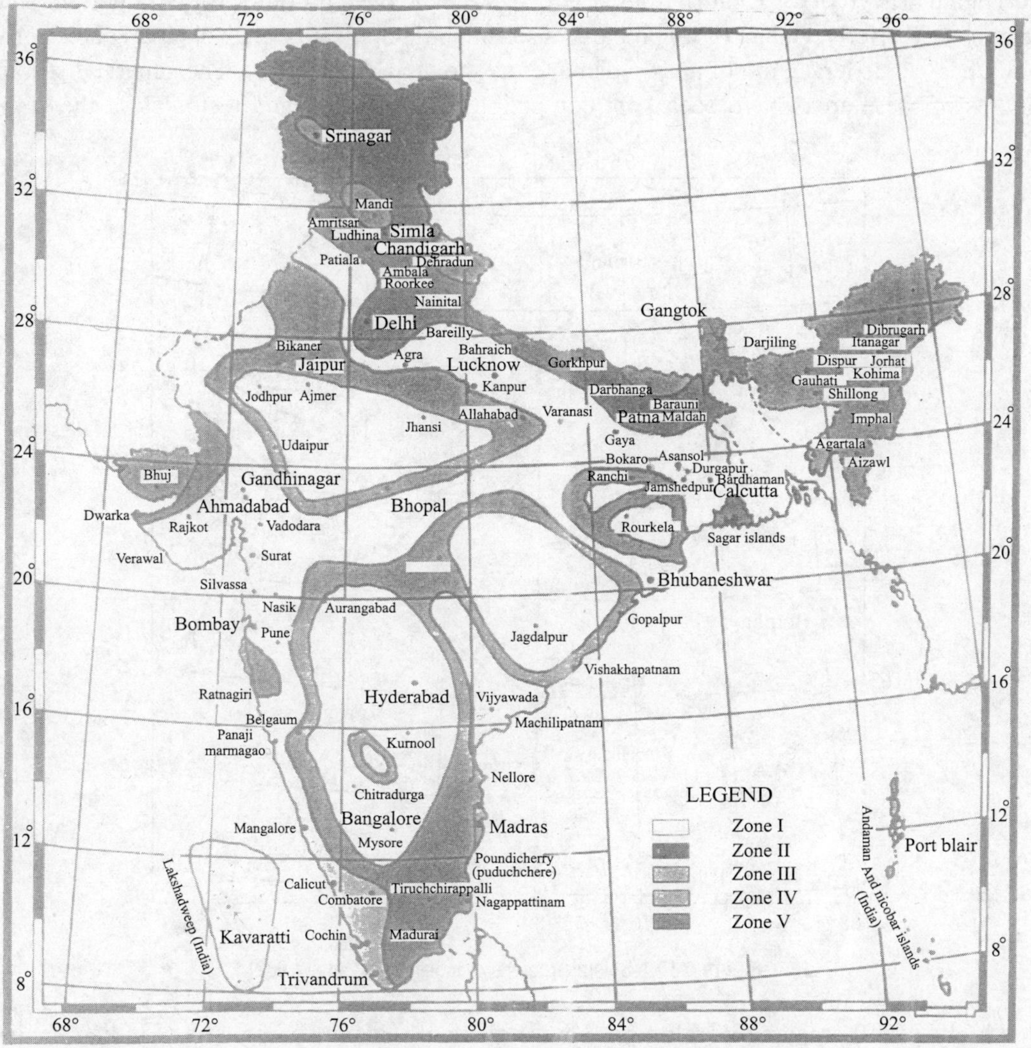

Figure 9.9 Seismic zoning map of India (IS : 1893(1984)).

9.6.4 Seismic Zoning Maps by Individual Studies

Several individuals have carried out studies for seismic zoning of India before the publication of the first zoning map by Bureau of Indian Standards in 1962. The early seismic maps were qualitative in nature, and demarcated the areas of severe, moderate, light, etc. damages Tandon, (1956) and Krishna (1959). The later studies quantified the seismic zones on the basis of Modified Mercalli Intensity levels, similar to the IS code zoning (Guha, 1962). To overcome the shortcomings associated with the intensity-based zoning, some studies have also been attempted for seismic zoning on the basis of probabilisitic hazard analysis (Kaila and Rao, 1979, 1980; Basu and Nigam, 1977, 1978; Khattri et al., 1978, 1984). The description of some important studies are described briefly. Guha (1962) prepared a seismic regionalization map depicting areas of very heavy, heavy, moderate, minor and no damages as shown in Figure 9.10. The general distribution patterns of these areas have been kept compatible with geology and tectonics of the country.

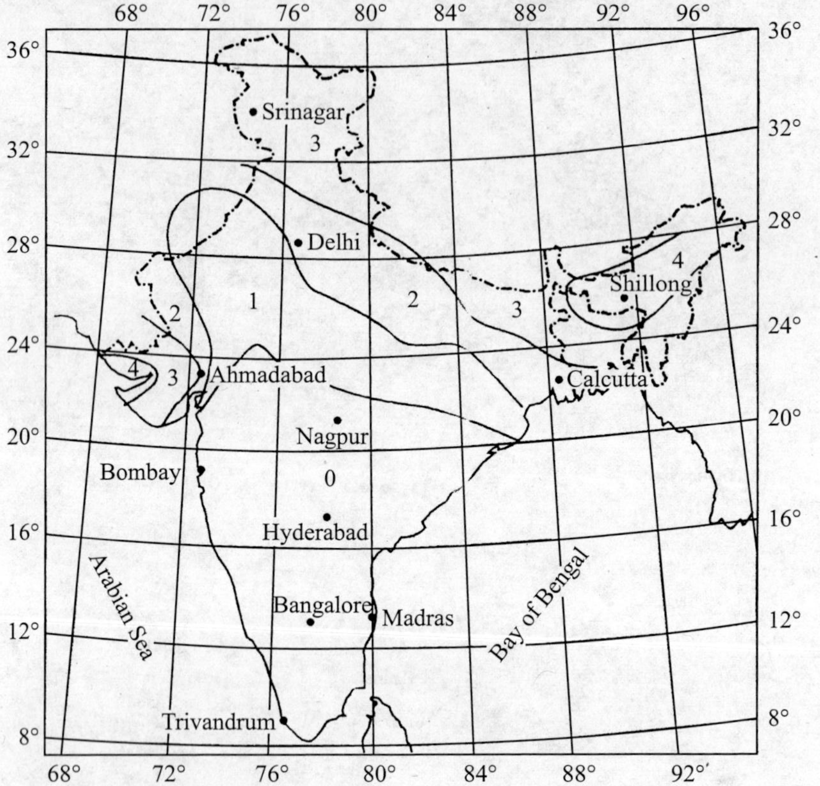

Figure 9.10 Seismic regions of India [After Guha, 1962]

Gubin (1969) prepared a preliminary multi-element seismic zoning map of Indian Peninsula by studying the available data on geology, tectonic features and past earthquake occurrences. This map is shown in Figure 9.11.

Unlike single element maps based only on the occurrences of past earthquakes, this map represents the zones of very recent differential motions where strong tremors will also occur in those areas where no tremors are known to have occured in the past. Thus, this map is essentially a multi-element zoning map, in other words, is a map of seismogenic zones.

Kaila et al., (1972) and Kaila and Rao (1979) presented the seismic risk map of India as shown in Figure 9.12. This shows the probability of occurrence of earthquakes capable of producing accelerations exceeding 10% of gravity in a design of 50 years.

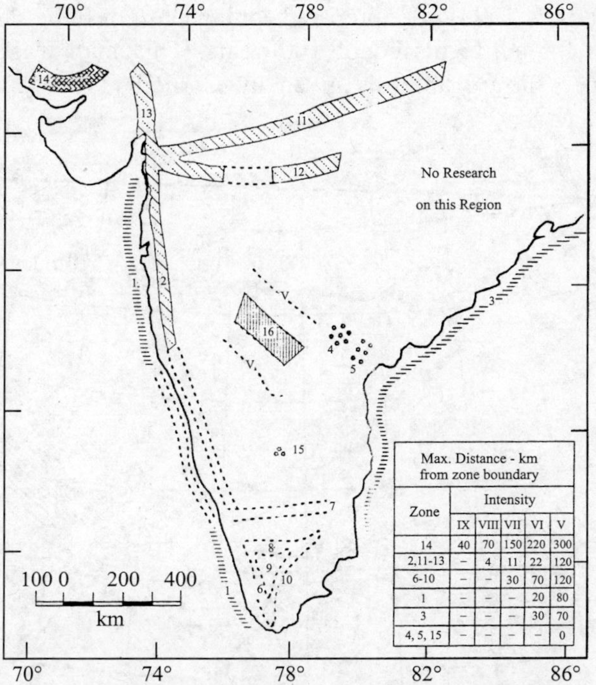

Figure 9.11 Various seismogenic zones in Peninsular India (After Gubin, 1969)

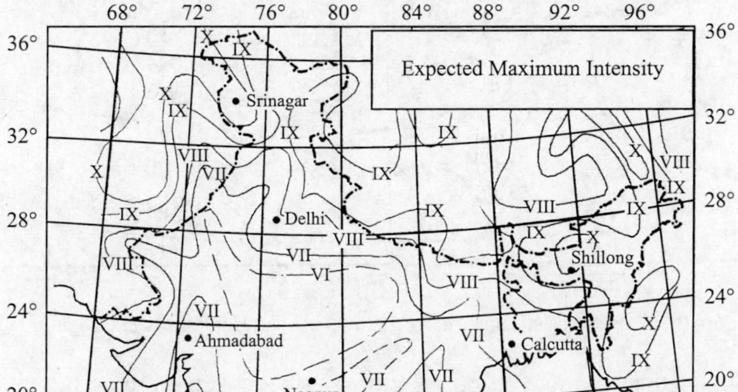

Figure 9.12 Seismic zoning map of Indian mainland showing the boundaries of areas with different maximum intensity values (After Kaila and Rao, 1979)

Basu and Nigam (1977, 1978) have prepared zoning maps of India showing the peak ground acceleration contours for a life period of 100 years. This map was based on the method of probabilistic seismic hazard analysis given by Cornell (1968) and has been shown in Figrue (9.13).

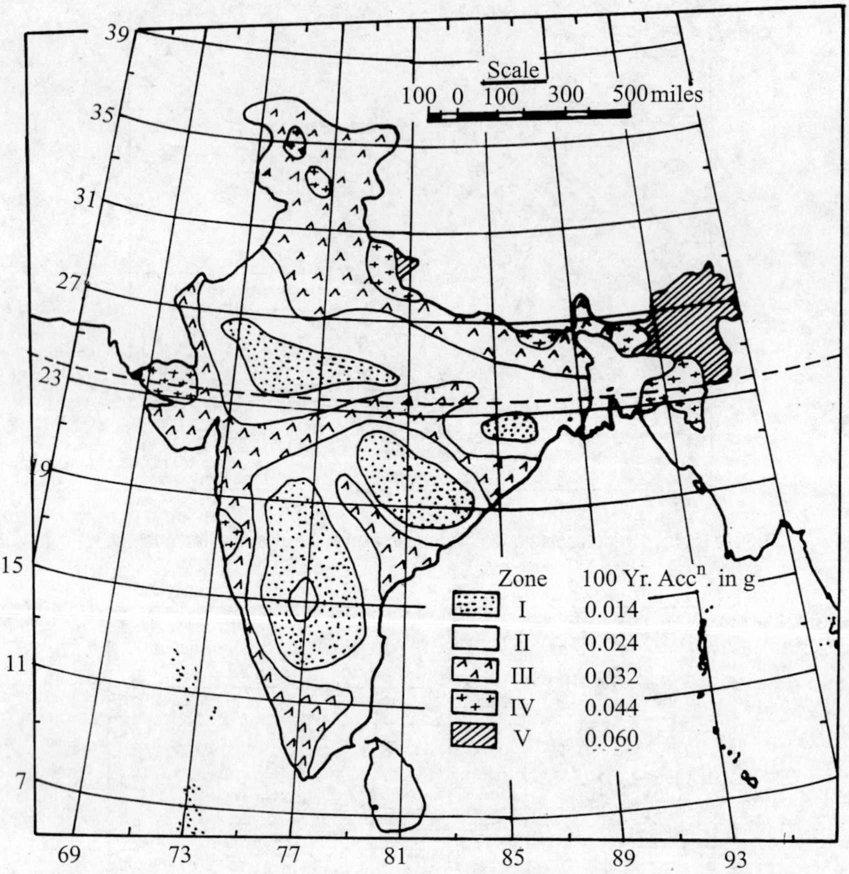

Figure 9.13 Probabilistic seismic zoning map of India (After Basu and Nigam, 1977)

Khattri et al., (1984) using the correlation between seismicity and geological features, divided the Indian subcontinent into 24 seismogenic sources and then carried out the hazard calculation to prepare the 50-year peak accelaration zoning map with a probability of exceedance equal to 0.10, and this has been shown in Figure 9.14. If we compare the zoning of Basu and Nigam with that by Khattri (1984), it may be observed that the probability of exceedance taken by Basu and Nigam equals to 0.63 in 100 years and that is why the acceleration values are unrealistically small, even in the highly seismic Himalayan region. The general methodology as used by Khattri et al. (1989), to prepare the zoning map is very simillar to that used by Basu and Nigam. The peak acceleration contour map thus obtained by Khattri et al. (1984), is shown in Figure 9.15.

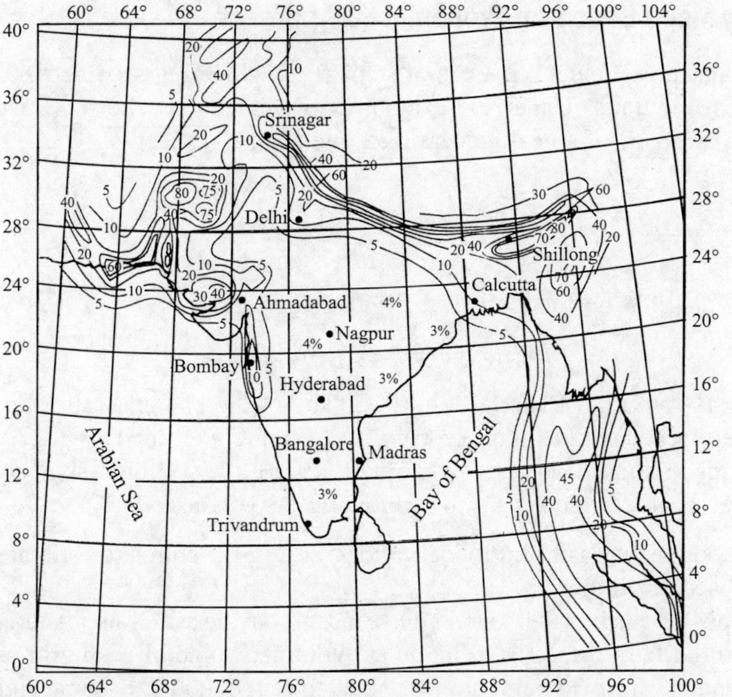

Figure 9.14 Map showing peak acceleration contours for India and adjacent areas (After Khattri et al., 1984)

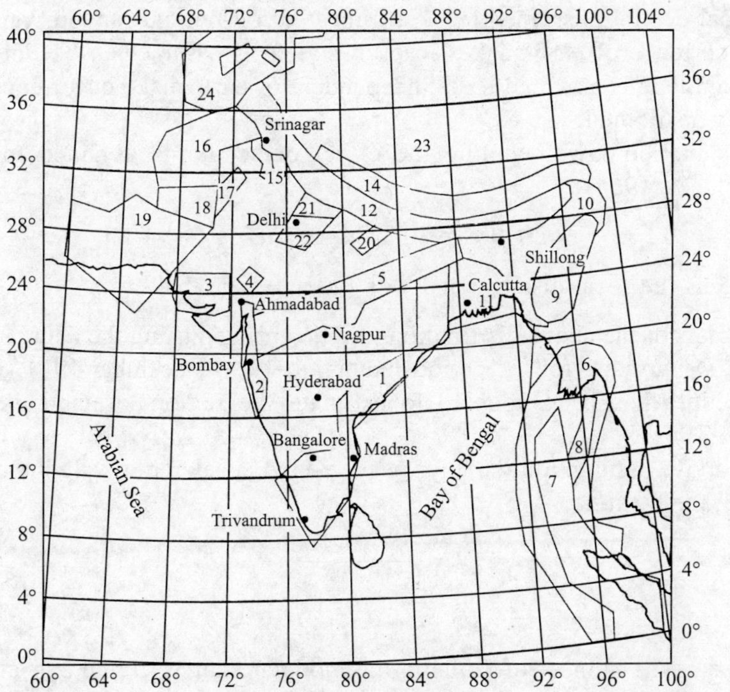

Figure 9.15 Various seismogenic soures identified by Khattri et al., (1989)

9.6.5 Zoning Maps Based on Probabilistic Approach

Cornell (1968) has presented a sequence of steps for engineering seismic risk analysis at a site, under certain assumptions. Under these assumptions it can be shown (Basu, 1977) that the average return period, T, of a peak-generalized intensity, Y, exceeding y, due to a single source can be expressed as

$$T = \frac{1}{\mu P[Y > y]} \tag{9.9}$$

The probability distribution function $F_{\text{Ymax}}(y, t)$ is given by

$$F_{\text{Ymax}}(y, t) = \exp\{-\mu t P[Y > y]\} \tag{9.10}$$

Substituting $t = T$ from Eq. (9.11) into Eq. (9.12) shows that the generalized intensity predicted for a T-year, return period, has 0.63 probability of being exceeded.

Following the method developed by Cornell (1968), the seismic risk analysis of the Indian subcontinent has been carried out in the following steps (Basu, 1977).

1. The subcontinent has been divided into 2° × 2° grid points to estimate the 100-yr peak ground acceleration.
2. Two types of earthquake sources have been considered (i) area sources corresponding to identified faults, and (ii) a 'modular' volume source of each grid point. The sources of earthquake at a grid point have been taken as the modular source and the faults within it have been assumed to be statistically independent mutually, and with respect to the volume source.
3. The focal depth is assumed to be a truncated log-normal random variable.
4. The occurrence of earthquakes in each source is assumed to be Poisson.
5. The magnitude is assumed to be independent of earthquake occurrence rates and exponentially distributed.
6. The attenuation law including the effect of scatter is assumed to be (Esteva and Rosenblueth, 1964)

$$Y = 2000 \exp[0.8M - 2\ \ln(R + 25) + \varepsilon] \tag{9.11}$$

where R is the focal distance and ε is the unobservable error.

The parameters of the above distributions are determined using the data compiled by Indian Meteorological Department for the period January 1917 to December 1971 (Basu, 1977). The Bayesian procedure has been adopted to estimate the regional seismicity (Newmark and Rosenblueth, 1971).

Under the above assumption the T-yr peak ground acceleration value is computed at each grid point using the formula:

$$T = \frac{1}{\sum_{i=1}^{m+1} \mu_i P[Y_i > y]} \tag{9.12}$$

where m represents the number of faults in the modular source, μ_i the expected arrival rate in the ith source and $P[Y_i > y]$ is the exceedance probability of peak generalized intensity, Y_i, due to the ith source. The peak ground acceleration is obtained at each grid point by numerical solution of Eq. (9.12) as 100-yr return period intensity. Figure 9.8 shows the contours of 100-yr peak ground acceleration for five levels.

In many applications it becomes necessary to establish the intensity for a specific return period or for a specified exceedance probability over a given period. Since earthquake sources at each grid point include faults, which vary from grid-points to grid-point, fresh computations are needed for such cases. If, however, the earthquake source is of identical geometry, as the modular source, simple extrapolation procedures are justified because a comparison of the contour maps obtained by neglecting the faults shows only a marginal local change in contours near the faults (Basu, 1977).

9.7 MODEL FOR EVALUATION OF SEISMIC HAZARD

9.7.1 Poisson Model

This is the basic model of earthquake-hazard analysis with the assumption that the possibility of earthquake occurrence is random and independent on the time axis, and even the place of earthquake is random and independent within the seismic zone. The process of earthquake is random and independent within the seismic zone. The process of earthquake generation in such a model forms the Poisson process on the time axis. Earthquake-hazard analysis using the Poisson model was proposed by Cornell (1968), developed by Der-Kiureghian (1977), and is currently used as the standard method in such analysis.

Since the stresses within the seismic zone accumulate and are released repetitively in an actual earthquake, the foregoing assumptions are quite rational. However, the utility of the Poisson model lies in the fact that it considers even the rarest phenomena, that is, its precision increases more when we consider seismic motion in regions where the probability of earthquake is less JSCE (1997).

Thus, the simplest reference model for the distribution of earthquakes in time is the stationary Poisson process, in which all events occur independently and uniformly in time. This process is characterized by one parameter, the rate of occurrence, μ. If N events occur during a sufficiently long time interval T, the rate μ is given by $\mu = N/T$. For stationary Poisson process, the frequency of events n in an interval of length Δt has a Poisson distribution

$$f(n) = (\mu\ \Delta t)^n \frac{e^{-\mu \Delta t}}{n!} \tag{9.13}$$

The time interval τ between successive events has an exponential distribution

$$\omega(\tau) = \mu\ e^{-\mu\tau} \tag{9.14}$$

The power spectrum of a point process t_k (occurrence time) of the kth event, $k = 1, 2, \ldots,$ N, is given by

$$S(\omega) = \frac{\left| \sum_{k=1}^{N} e^{-i\omega t_k} \right|^2}{N} \tag{9.15}$$

or,
$$S(\omega) = \frac{\left\{\left(\sum_{k=1}^{N} \cos \omega t_k\right)^2 + \left(\sum_{k=1}^{N} \sin \omega t_k\right)^2\right\}}{N} \tag{9.16}$$

where ω denotes the angular frequency. For a Poisson process, $S(\omega)$ is independent of ω one has an exponential distribution of parameter 1. Therefore, the probability that $S(\omega)$ for a randomly chosen ω exceeds a certain value α is $e^{-\alpha}$.

9.7.2 Non-Poisson Model

If a large earthquake occurs in a certain region, the stress fields near that region are released and the next earthquake can probably occur only after a considerably long period of calm has elapsed. As a result, the earthquake occurrence probability becomes a function of time. The seismic-activity model based on this consideration is called the non-Poisson model.

Non-Poisson models include those based on the regeneration process, or those based on Shimazaki's time-predictable or slip-predictable models, Shimazaki and Nakata (1980). It is possible to study the effect of irregularity of earthquake occurrence, ignored in the Poisson model, using these models. However, since the number of parameters increases as we go deeper, it becomes more difficult to collect sufficient data for quantitative evaluation and the applicable region becomes smaller. As a result, it is not practical to use the non-Poisson model as a general method for earthquake-hazard analysis, but it is still very useful for detailed investigation of earthquake hazards in a region for which proper data is available or for studying the range of application of the Poisson model JSCE (1997).

9.7.3 Other Models

The magnitude of an earthquake occurring in a region or in a specific fault is assumed to possess the distribution properties. Thus, as is known from experience, the greater the magnitude of an earthquake, the lower its frequency of incidence. Since this property is defined by b and b_f, it is called the b-value model after Wesnousky et al., (1983).

Maximum-moment model

Wesnousky et al. (1983) suggessted that the major earthquake generated by various active faults is one in which the faults are completely destroyed. The magnitude of the earthquake in a zone is decided by the length of the fault, which varies considerably. This is called the maximum-moment model. According to this model, an earthquake generated by a specific fault has the characteristic maximum moment

M_0^{max} and its frequency of occurrence V'_f is given by the following equation:

$$V'_f = M_0^g / M_0^{max} \tag{9.17}$$

where M_0^g is the seismic moment release ratio and is given by

$$M_0^g = \mu W L \lambda \tag{9.18}$$

While calculating $M_0^{\max}$, we can obtain the following empirical relation between rupture length of a fault 'l' (km) and M_0 for large inland earthquakes ($M_0 = 2 \times 10^{25} - 5 \times 10^{27}$ dyn.cm) occurring in Japan:

$$\log M_0 = 23.5 + 1.94 \log 1 \tag{9.19}$$

Since the maximum-moment model mainly targets earthquakes of large magnitude, the b-value model is useful for obtaining the intensity of earthquake motion wherever the probability of occurrence is low. The maximum-moment model is useful while carrying out the earthquake-hazard analysis from the engineering point of view using the active-fault data.

9.7.4 Seismic Hazard Analysis Based on Poisson Model

We can assume that the area, with the point under consideration at its centre, can be divided into seismic zones S_1, S_2, S_ι, as shown in Figure 9.16, and the frequency of earthquake occurrence in each zone is denoted by ν_1, ν_2, ν_ι. We can also consider the Poisson model for time-space distribution of earthquake occurrence.

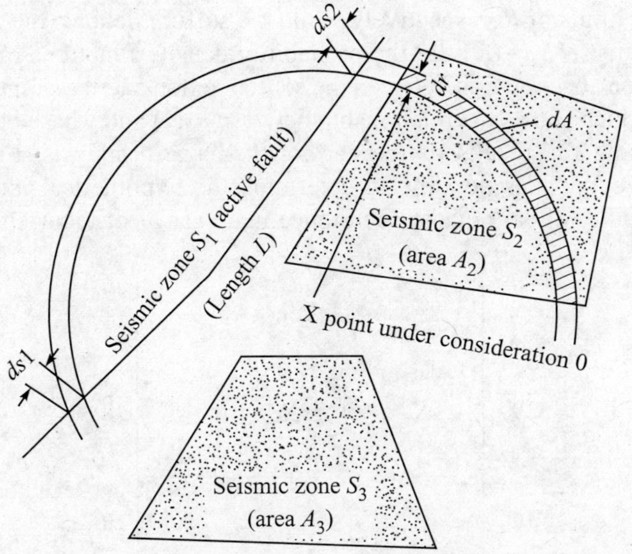

Figure 9.16 Site and seismic zones.

In a Poisson model, the time series, in which the intensity of earthquake motion Y at a given point O exceeds the specified value Y, is also a Poisson process. If its frequency of occurrence is denoted by $\nu_O(y)$, then the probability of intensity of earthquake motion Y at point O exceeding y in a year is given by the following equation:

$$p_O(y) = 1 - \exp[-\nu_O(y)] = \nu_O(y) \tag{9.20}$$

The curve showing the relation between $p_O(y)$ and Y is called the 'hazard curve'. If we denote the return period of seismic motion of intensity exceeding y_0 as $T_R(y)$, then using Eq. (9.20) we get

$$T_R(y) = \frac{1}{\nu_0(y)} = \frac{1}{p_0(y)} \tag{9.21}$$

If $p_0(y)$ is determined, the probability $p_1(y, t)$ of earthquake motion at point O exceeding intensity y in t years from now, can be obtained from the following equation:

$$p_1(y, t) = 1 - [1 - p_0(y)]^t \tag{9.22}$$

The frequency of occurrence $\nu_0(y)$ can be calculated from the following equation:

$$\nu_0(y) = \sum \nu_i\, q_i(y) \tag{9.23}$$

where $q_i(y)$ is the ratio of earthquakes occurring in seismic zone S_i exceeding the intensity of seismic motion y at point O. It is related to the magnitude of earthquake M and the distance of point O from the hypocentre R as given below JSCE (1997):

$$q_i(y) = \int_{m_0}^{m_{ui}} \int_{r_{li}}^{r_{ui}} [P(Y > y \mid m, r)\, f_R(r)\, dr]\, f_M(m)\, dm \tag{9.24}$$

where m_u is the upper limit of M, m_0 the lower limit of M considered during analysis; r_u and r_l the upper and lower limits of R, respectively while the suffix i denotes the values corresponding to the seismic zone i; $f_R(r)$, $f_M(m)$ are the probability density functions of R and M, and $f_R(r)$ is decided by the geometric relation between the source region and the point under consideration. The hazard curve, which shows the relationship between $p_0(y)$ and y based on Eq. 9.16 is a direct method for expressing the results of the earthquake-hazard analysis. Figure 9.17 shows an example. With this curve it is possible to determine the hypothetical intensity of earthquake motion if the probability of earthquake occurrence in a year p_0 or recurrence time T_R is known.

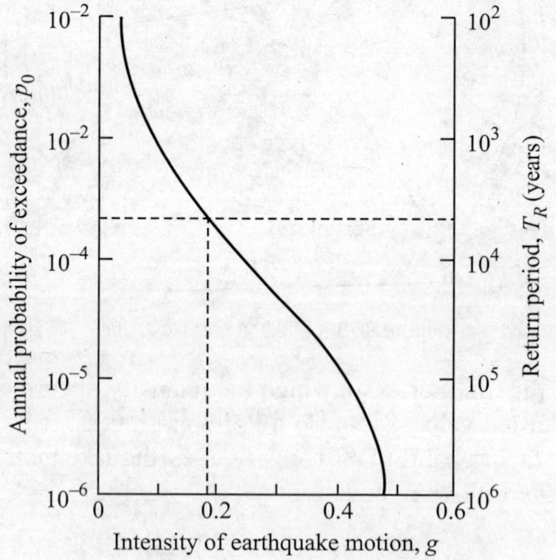

Figure 9.17 Schematic diagram of hazard curve.

Thus, the hazard curve is meaningful from the engineering viewpoint as we can consider y a function of p_0 or T_R.

The Hazard Map-A Case Study

Figure 9.18 shows the seismic hazard map of Bombay and the adjoining area. The contours show the peak acceleration with a probability of ten per cent of being exceeded in 50 years. The acceleration values contoured range from 0.04*g* to 0.09*g*. The contours follow the source zones in a N-S direction, except that where the source zones 2 and 5 meet. The contour for 0.04*g* bends to the east, owing to the influence of source zone 5. The highest acceleration contour encloses two areas—the bigger one to the south of Bombay and the smaller one in the north near Ahmedabad. The northern high reflects the joint influence of hazards from source zones 2, 3 and 4.

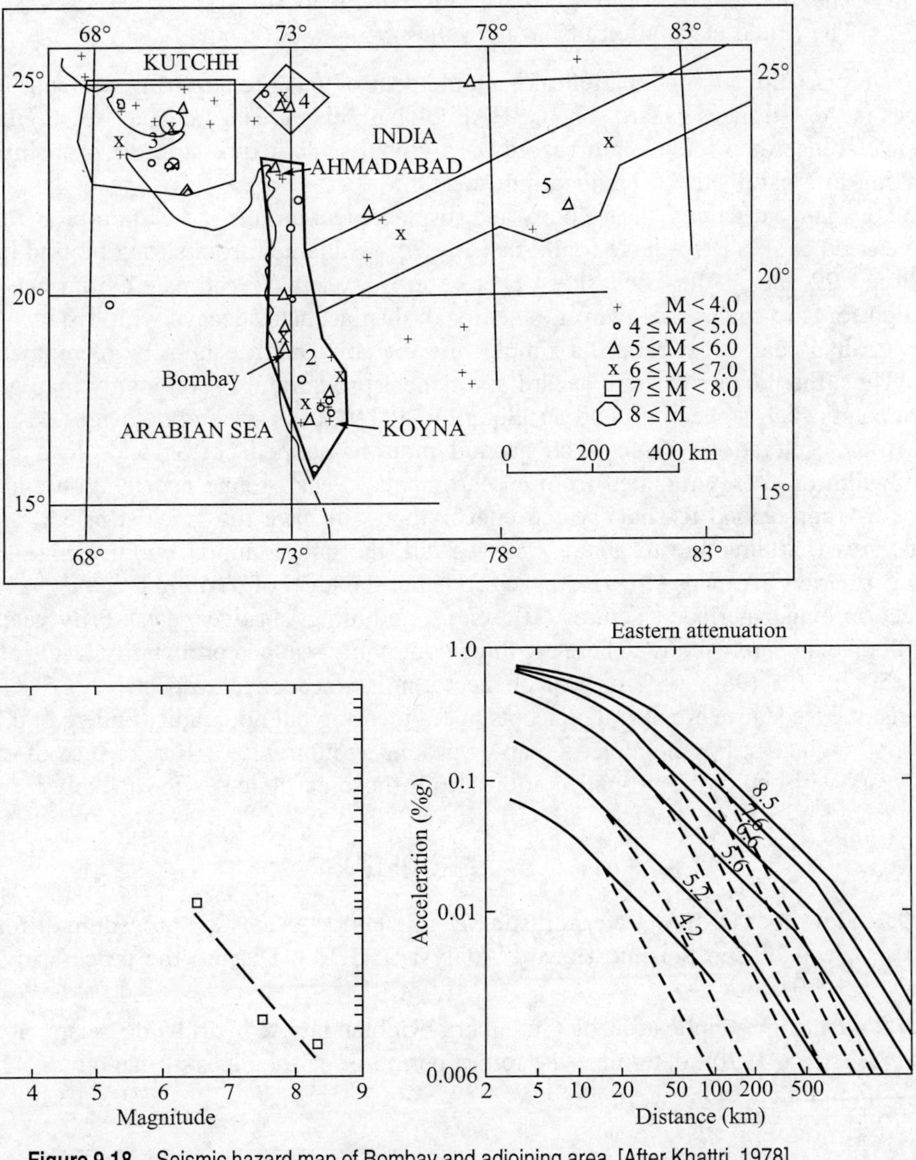

Figure 9.18 Seismic hazard map of Bombay and adjoining area. [After Khattri, 1978]

The Koyna dam area is not enclosed within the highest peak acceleration contour. However, the fact that an area has low expected acceleration does not mean that damaging events cannot occur there; it only means that on an average the likelihood that acceleration will exceed the estimated value in 50 years is only 10 per cent (Khattri et al., 1978).

The accelerations reported here are estimates for hard rock and appropriate scaling factors should be used to take into account the local geological conditions.

Finally, there are three major elements of modern probabilistic seismic hazard assessment namely

- The characterization of seismic source
- The characterization of ground motion attenuations
- The actual calculation of hazard values.

The variations in application of each element of seismic hazard assessment lead to differences in the estimates hazard. The GSHAP Global Seismic Hazard Map (see Figure 9.17) is the first reference map for seismic hazard on a global scale, expressing the probability of ground shaking in a parameter of engineering interest.

For many structures, the velocity and displacement are the more appropriate design parameters. Seed et al., (1976) have found that A_{max}/A_{min} obtained from strong motion records ranges between 22 and 29 for rock sites. Therefore, V_{max} can be estimated from acceleration in a straightforward manner. However, since peak displacement decays with distance considerably slower than peak acceleration, a simple rule for conversion cannot be recommended.

Thus, the modern seismic hazard assessment programmes scommonly map ground motion parameters such as peak ground acceleration (PGA), peak ground velocity (PGV) and several spectral accelerations (SA). Each ground motion mapped corresponds to a portion of the bandwidth of energy radiated from an earthquake. Peak ground acceleration and 0.2–0.5 seconds SA correspond to short-period energy that will have the greatest effect on short-period structures (buildings up to about 7 stories tall, the non-common building size in the world). Longer period SA maps (1.0 s, 2.0 s, etc.) depict the level of shaking that will have the greatest effect on longer-period structures (10 + storey buildings, bridges, etc.). Fifty years is the most common exposure window. There are three commonly mapped probability levels of exceedance; 2%, 5% or 10% (98%, 95% or 90% chance of non-exceedance, respectively). These probability levels of exceedance are useful concepts in engineering but not readily understood by those who are not engineers. For example, a map of ground motion with a 10% chance of exceedance in 50 years will depict the ground motions from those earthquakes most likely to occur.

PROBLEMS

9.1 How does DSHA (Deterministic Seismic Hazard Analysis) procedure differ from PSHA (Probabilistic Seismic Hazard Analysis (PSHA). Discuss the process given by Reiter (1990).

9.2 Using the application of Gutenberg–Richter law to worldwide seismicity data (After Esteva, 1970), determine the return period of earthquakes of magnitude $M = 8.5$ on the circumpacific and Alpide belts.

9.3 The general form of the attenuation relation considered in the present study is taken as follows:

$$\log(a) = f_1(M) + f_2(r,\ E) + f_3(r,\ M,\ E) + f_4(F) + \varepsilon$$

where, $f_1(M)$ = function of earthquake magnitude

$f_2(r, E)$ = function of earthquake-to-recording site distance and the tectonic environment

$f_3(r, M, E)$ = function of magnitude, distance and tectonic environment

$f_4(F)$ = function of fault type

ε = a random variable

and in this context develop the attenuation relation for the Himalayan region.

10

LIQUEFACTION OF SOILS

10.1 INTRODUCTION

The destructive power of liquefaction dramatically attracted world's attention during the following earthquakes:

- 1934 Bihar Earthquake, India ($M_w = 8.3$)
- 1964 Great Alaskan Earthquake, U.S.A. ($M_w = 9.2$)
- 1964 Niigata Earthquake, Japan ($M_w = 7.5$)

During an earthquake the ground structure may fail owing to various reasons, namely:

- Fissures
- Differential movements
- Faults
- Loss of strength

The reason pertaining to loss of strength means loss of shear strength of soil. The shear strength of soil is mainly due to cohesion and frictional resistance. The intermolecular attraction and the frictional resistance contribute to shear strength of soil.

The shear strength is expressed as

$$\tau = c + \sigma_n \tan \phi \tag{10.1}$$

For $c = 0$, cohesionless soil, i.e., sands

$$\tau = \sigma_n \tan \phi \tag{10.2}$$

The soil is a polyphase material consisting of water, air in pores and the solid soil skeleton. The pore water pressure u does make reduction in effective normal stress, so that

$$\tau = (\sigma_n - u) \tan \phi \tag{10.3}$$

During an earthquake owing to ground motion, there is instantaneous increase in pore water pressure and so reduction in shear strength. In other words during an earthquake, the propagation of shear waves causes the loose soil to contract, resulting in and increase in pore water pressure. The loss of strength is more pronounced in sandy soil due to increase in pore pressure. This phenomenon of loss of strength owing to rise in pore pressure is called liquefaction. As the term suggests the sand no longer behaves like a solid, rather it acts like a viscous fluid. Loose saturated sand deposits may lose a part of their shear strength when subjected to sudden earthquake excitation. It is evident that owing to increase in pore water pressure, the effective stress reduces, resulting in loss of strength. After sudden rise in pore pressure and thereafter, stress transfer take place and the resulting effective stress controls the shear strength. If this stress transfer is complete, there is total liquefaction. However, when only partial stress transfer takes place,

$$\lim(\sigma_n - u) \to 0$$

There is a partial loss of strength resulting in partial liquefaction. Apart from earthquake, strength may be reduced due to sudden shock or dynamic load due to pile driving, explosive blasting, bomb blast loading, vibration in machinery or even rapid draw-down in dams. Thus, owing to total loss of strength, the soil is said to have liquefied. Indeed, liquefaction is an external manifestation of decrease in shear strength, due to the cyclic pore-pressure generation mechanism. In other words, if a loose saturated sand is rapidly deformed due to an earthquake, the grains tend to become closely spaced and compact. For this to happen, water must flow out of the voids. If the loading is so rapid that there is no time for drainage of water, the entire load on the soil must be carried instantaneously by pore water. Pore water is more compressible than the soil skeleton. Thus, the intergranular stress is largely reduced due to increase in pore water pressure. The increase in pore water pressure causes a reduction in shear strength. This loss of strength due to transfer of intergranular stress from soil grain to pore water due to dynamic load is known as liquefaction. When loss of strength occurs, the soil behaves like a viscous fluid. As the bearing capacity of soil to sustain foundation loads is directly related to strength, liquefaction poses a serious hazard to structures and must be assessed in areas where liquefaction prone deposits exist.

In this chapter, the phenomenon of liquefaction shall be addressed. As far as geotechnical earthquake engineering is concerned, this phenomenon of liquefaction may be divided into three main groups, namely:

(a) Flow liquefaction
(b) Cyclic mobility
(c) Ground level liquefaction

Flow Liquefaction

The flow liquefaction has already been explained and this can occur when the static shear stress in a soil deposit during earthquake excitation is greater than the steady-state strength of the soil. It can produce devastating flow slide failures during or even after earthquake shaking. However, flow liquefaction can occur only in loose soil.

Cyclic mobility

Cyclic mobility can occur when the static shear stress is less than the steady state strength and the cyclic shear stress is large enough, then the steady state strength is exceeded momentarily. Deformations produced by cyclic mobility develop incrementally but can become substantial by the end of a strong and/or long-duration earthquake. Cyclic mobility can occur in both loose and dense soils but deformations decrease markedly with increased density.

Ground-level liquefaction

Ground-level liquefaction can occur when cyclic loading is sufficient to produce high excess pore pressure even when static driving stresses are absent. Its occurrence is generally manifested by ground oscillation, post-earthquake settlement or development of sand boils. Level-ground liquefaction can occur in loose and dense soils. This level-ground liquefaction is a special case of cyclic mobility. Level-ground liquefaction failures are caused by the upward flow of water that occurs when seismically induced excess pore presssure dissipates.

The different faces of liquefaction have been observed and reported in different earthquakes. In Bihar earthquake of January 15, 1934 having a magnitude more than 8.0 on Richter scale, liquefaction appeared as sand fountain. As the soil lost almost all of its strength, the structures resting on it simply sank into it. This liquefaction phenomenon was of exceptional violence. This was largely due to the fact that alluvial deposits in Bihar are 1000 m (approx) in thickness, and the groundwater table is only 2 m below the ground surface.

The Niigata earthquake of June 16, 1964 had a magnitude of 7.5 wherein the earthquake phenomenon of liquefaction was observed on a large scale.

The destruction was observed to be largely limited to buildings that were founded on top of loose, saturated soil deposits. The majority of the buildings settled or tilted rigidly without appreciable damage to the superstructure. The main reason for these failures was ground failure, particularly due to liquefaction.

More recently on the morning of January 26, 2001, a devasting earthquake occurred near Bhuj in the Kutchh district of Gujarat. Liquefaction was the most widespread phenomenon resulting from this earthquake (Figures 10.2(a) and 10.2(b)). Along with the liquefaction, dark coloured mudflows were also observed at several places especially in the Rann of Kutchh, which the local people described as "Lava Flows".

(a)

(b)

Figure 10.1 Liquefaction during Bhuj Earthquake 2001

10.2 THEORY OF LIQUEFACTION

The shear strength of sand in saturated condition may be expressed as

$$\tau = (\sigma_n - u) \tan \phi \tag{10.4}$$

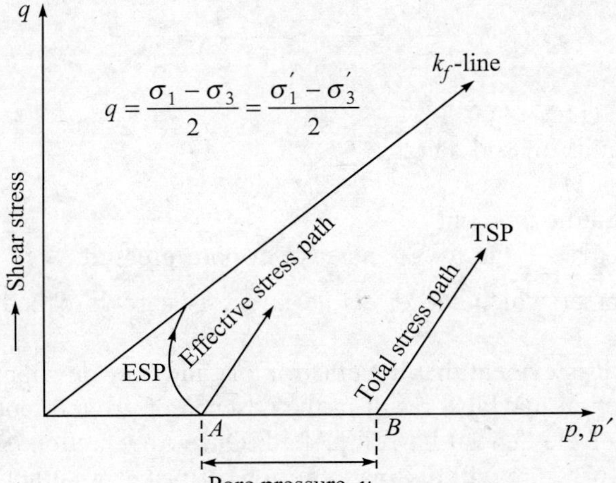

Figure 10.2 Shear stress and pore pressure in a soil element.

where,

τ = shear strength
σ_n = normal stress on a soil element at depth z
u = pore pressure
ϕ = angle of internal friction

The vertical stress on a horizontal plane of elemental soil under consideration at a depth z is given by

$$\sigma_n = \gamma_{sat}\, z \tag{10.5}$$

where, γ_{sat} = unit weight of saturated soil.
Thus,

$$\sigma_{effective} = (\sigma_n - u) = \gamma_{sat}\, z - \gamma_w\, z = (\gamma_{sat} - \gamma_w)\, z \tag{10.6}$$

During the ground motion due to earthquake, the static pore pressure may increase by an amount $+\nabla_n$, then

$$\nabla_n = \gamma_w \cdot hw' \tag{10.7}$$

$$\sigma_{effective} = (\gamma_b\, z - \nabla_n)$$

$$= (\gamma_b\, z - \gamma_w\, hw') \tag{10.8}$$

$$\tau = (\gamma_b\, z - \gamma_w\, hw') \tan\phi \tag{10.9}$$

or in other form it can be written as

$$\tau_{dyn} = (\sigma_n - \nabla_{u\ dyn}) \tan\phi_{dyn} \tag{10.10}$$

For complete loss of strength,

$$\sigma_n - \nabla_{u\ dyn} = 0$$

or,

$$\gamma_b\, z - \gamma_w\, hw' = 0$$

$$\frac{hw'}{z} = \frac{\gamma_b}{\gamma_w} = \frac{G-1}{1+e} = i_{cr} \tag{10.11}$$

where,

G = specific gravity of soil solids
e = void ratio
i_{cr} = critical hydraulic gradient
hw' = rise in water head due to ∇_n increase in pore pressure.

Thus, the gradient at which the effective stress is zero is called the critical hydraulic gradient.

From Eq. (10.11) it is evident that liquefaction of sand may develop at any zone of deposit at any depth. Liquefaction may also result in the absence of ground motion or such motions if the underlying zones of the deposit have liquefied. Once liquefaction develops at some depth, the excess pore water pressure will dissipate by flow of water in an upward direction. For this the hydraulic gradient may be large enough to induce a quick sand condition. When a natural surface or soil deposit is in quick sand condition, it cannot support the weight of a person or an animal.

10.3 LIQUEFACTION ANALYSIS

The basic objective of the liquefaction analysis is to ascertain if the soil has the ability or potential to liquefy during an earthquake. Let us consider a soil column of unit width and length as shown in Figure 10.3. It is assumed that the soil column will move horizontally as a rigid body in response to the maximum horizontal acceleration a_{max} exerted by the earthquake at ground surface.

If $\vec{P}$ denotes the horizontal seismic force acting on a soil column of unit width and unit length, then

$$\vec{P} = \text{mass} \times \text{acceleration} \tag{10.12}$$

where mass $= \dfrac{W}{g} = \dfrac{\gamma_t \cdot z}{g}$

Substituting the value of mass in Eq. (10.12),

$$\vec{P} = \frac{\gamma_t \cdot z}{g} \cdot a_{max} = \sigma_{v_o} \cdot \left(\frac{a_{max}}{g}\right) \tag{10.13}$$

where,

W = total weight of soil column

γ_t = total unit weight of soil $\left(\dfrac{\text{kN}}{m^3}\right)$

z = depth below ground level as shown in Figure 10.3

a = acceleration which is equal to maximum horizontal acceleration at ground surface, a_{max}

a_{max} = maximum horizontal acceleration at ground surface due to earthquake $\left(\dfrac{\text{m}}{\text{s}^2}\right)$

σ_{v_o} = total vertical stress at bottom of soil column (kPa)

Considering the force equilibrium of the soil column the force $\vec{P}$ acting on the rigid soil column is equal to the maximum shear force at the base of the soil column.

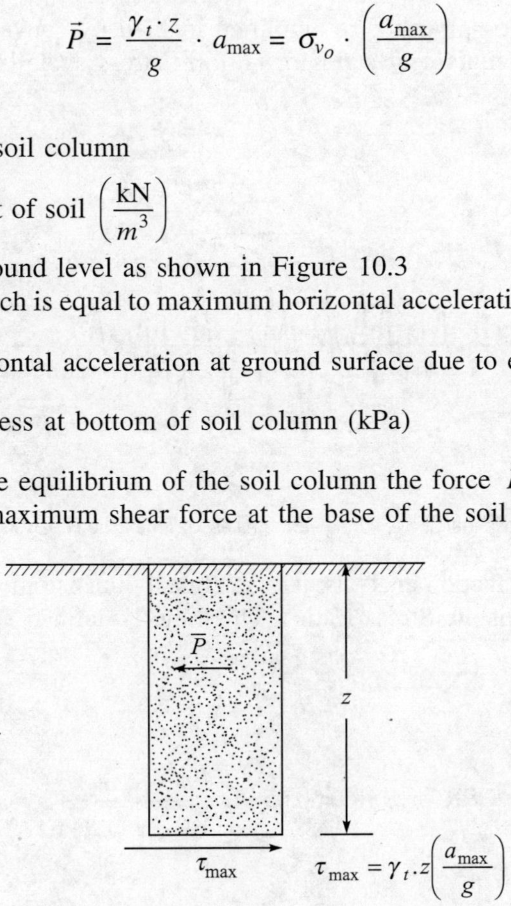

Figure 10.3 Soil column for liquefaction analysis.

Since the soil element is assumed to have a unit base width and length, the maximum shear force $\vec{P}$ is equal to the maximum shear stress τ_{max}. The equation of the force equilibrium may then be written as

$$\tau_{max} = \vec{P} = \sigma_{v_o}\left(\frac{a_{max}}{g}\right) \tag{10.14}$$

Let σ'_{v_o} be the vertical effective stress, then dividing both sides of Eq. (10.14) by σ'_{v_o},

$$\frac{\tau_{\max}}{\sigma'_{v_o}} = \left(\frac{\sigma_{v_o}}{\sigma'_{v_o}}\right)\frac{a_{\max}}{g} \tag{10.15}$$

Since the soil column does not act as a rigid body in reality during the earthquake, Seed and Idriss (1971) proposed a depth reduction factor r_d into the right-side as the soil is deformable, so

$$\frac{\tau_{\max}}{\sigma'_{v_o}} = r_d \cdot \left(\frac{\sigma_{v_o}}{\sigma'_{v_o}}\right)\frac{a_{\max}}{g} \tag{10.16}$$

Further (Seed et al., 1975) proposed a simplified method by converting the typical irregular earthquake record to an equivalent series of uniform stress cycle by assuming the following (see Figure 10.4).

$$\tau_{av} = \tau_{cyc} = 0.65\,\tau_{\max} \tag{10.17}$$

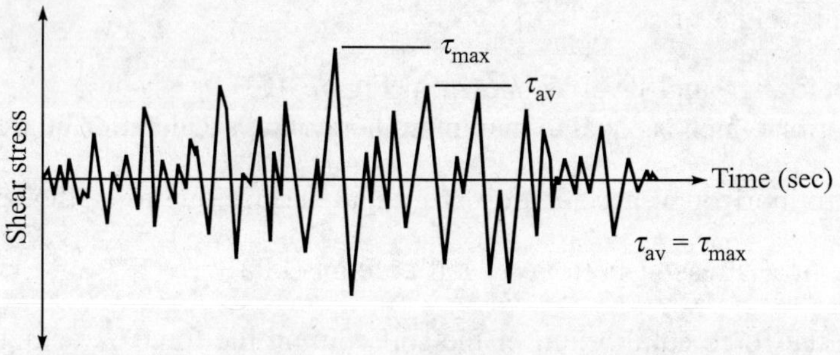

Figure 10.4 Time history of shear stress during earthquakes for liquefaction analysis (After Seed and Idriss, 1971)

To felicitate the liquefaction analysis, let us define a dimensionless parameter CSR (Cyclic Stress Ratio) or SSR (Seismic Stress Ratio). This ratio is defined as

$$\text{CSR or SSR} = \frac{\tau_{cyc}}{\sigma_{v_o}{}'} \tag{10.18}$$

Thus,

$$\text{CSR or SSR} = 0.65\ r_d \cdot \left(\frac{\sigma_{v_o}}{\sigma'_{v_o}}\right) \cdot \frac{a_{\max}}{g} \tag{10.19}$$

where,

$a_{\max}$ = maximum horizontal acceleration at ground surface or PGA (m/s^2)
g = acceleration due to gravity
σ_{v_o} = total vertical stress at a particular depth where liquefaction analysis is being carried out
σ'_{v_o} = vertical effective stress at the same depth
r_d = depth reduction factor or stress reduction coefficient

Depth reduction factor r_d

As already explained the depth reduction factor has been introduced for the fact that the soil column in Figure 10.5 does not behave as rigid body during earthquake shaking. Computations of the value of r_d for a wide variety of earthquakes motion and soil conditions having sand in upper 15 m falls within the range of values shown in Figure 10.5. (Seed and Idriss, 1971)

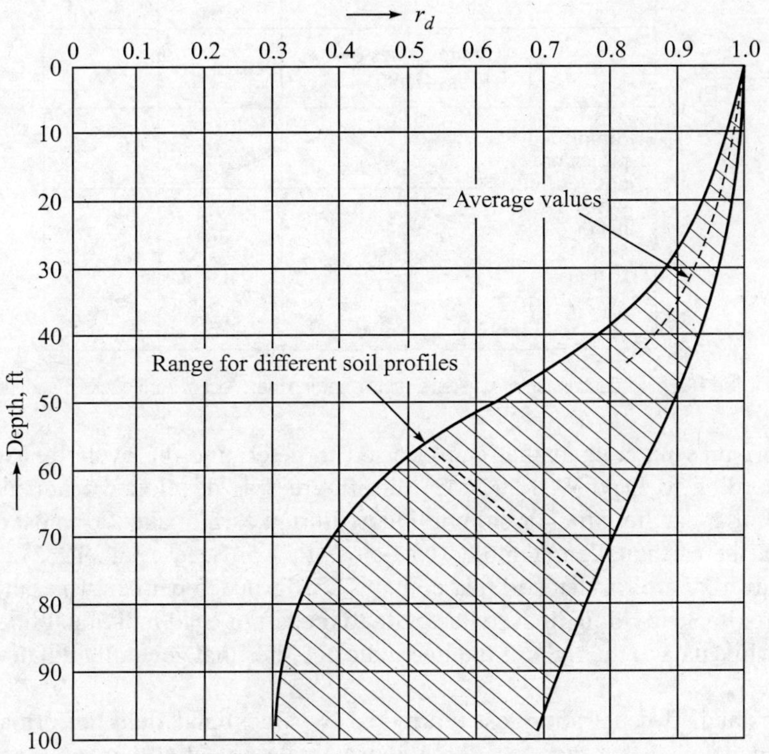

Figure 10.5 Reduction factor to estimate the variation of cyclic shear stress with depth below ground level (After Seed and Idriss, 1971)

Figure 10.6 shows that the value of r_d depends upon the magnitude of the earthquake. Another option is to assume a linear relationship of r_d versus depth and use the following equation (Kayen et al., 1992)

$$r_d = 1 - 0.012z \tag{10.20}$$

where z = depth in metres below the ground surface where the liquefaction analysis is being performed.

10.3.1 Cyclic Resistance Ratio

The cyclic resistance ratio represents the liquefaction resistance of the soil. The most commonly used method for determining the liquefaction resistance is to use the data obtained from the standard penetration test. In general, the factors that increase the liquefaction resistance of a soil will also increase the corrected N values from the standard penetration test.

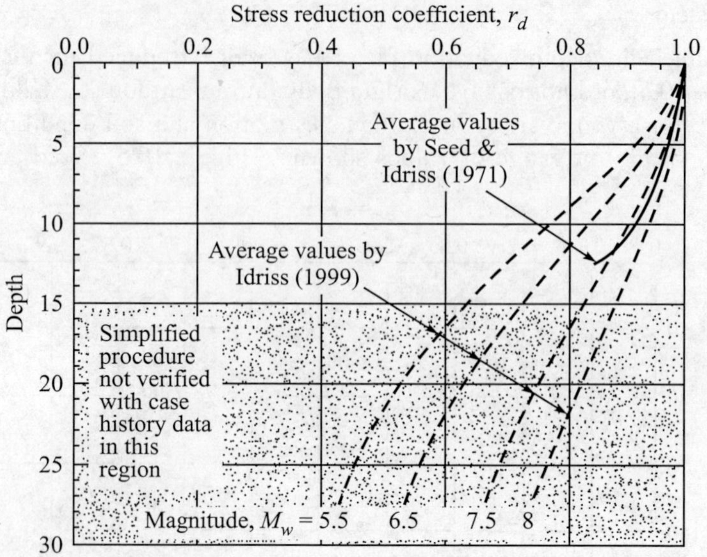

Figure 10.6 Reduction factor r_d versus depth showing the effect of earthquake magnitude

Figure 10.7 presents a chart that can be used to determine the cyclic resistance of the *in situ* soil. According to Day, W. R. (2002) this figure was developed from investigations of numerous sites that had liquefied or did not liquefy during earthquake. For most of the data used in preparation, the earthquake magnitude was close to 7.5 (Seed et al., 1985). The three lines shown in Figure 10.7 are for that soil that contains 35, 15 or ≤ 5 per cent fines. The lines shown represent approximate dividing lines, where data to the left of each individual line indicates field liquefaction, while data to the right of the line indicates sites that generally did not liquefy during the earthquake.

Based on standard penetration test (corrected N values) and field performance data, Seed et al., (1985) concluded that there are three approximate potential damage ranges than can be identified and they are listed in Table 10.1.

As indicated in Table 10.1, a corrected N value of 20 is the approximate boundary between the medium and dense states of sand. Above a corrected N value of 30, the sand is generally in a dense state. For this condition, initial liquefaction does not produce large deformation because of the dilation tendency of the sand upon reversal of the cyclic shear stress. This is the reason that such soils produce no significant damage as indicated in the table.

Table 11.1 Liquefaction potential damage

S. no.	*Range of SPT $N_{corrected}$*	*Potential Damage*
1	0–20	High
2	20–30	Medium
3	> 30	No significant damage

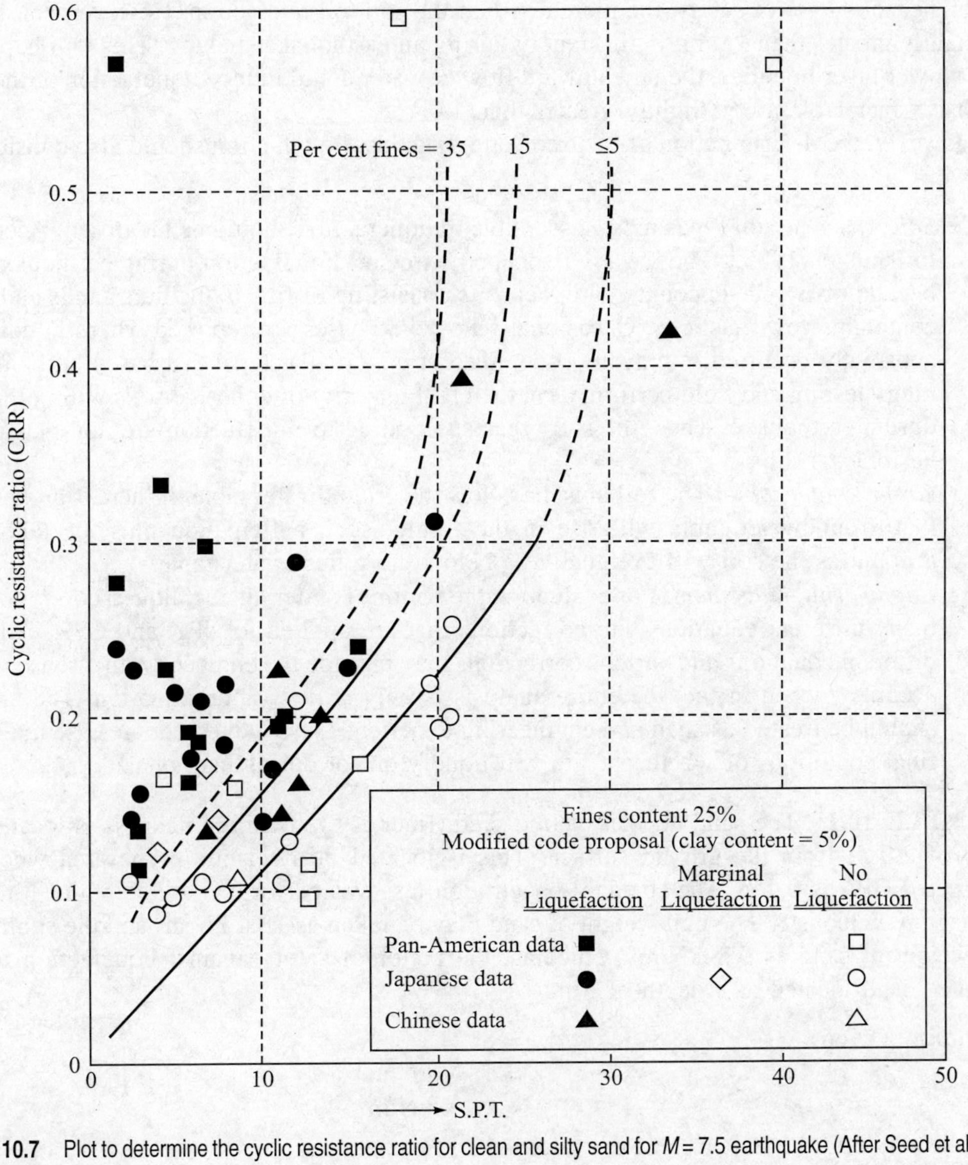

Figure 10.7 Plot to determine the cyclic resistance ratio for clean and silty sand for $M = 7.5$ earthquake (After Seed et al. 1975)

10.4 FACTOR OF SAFETY AGAINST LIQUEFACTION

The liquefaction analysis culminates in determining the factor of safety against liquefaction. If the cyclic stress ratio (CSR) caused by the anticipated earthquake as in Eq. (10.19) is greater than the cyclic resistance ratio (CRR) of the *in situ* soil (Figure 10.7) the liquefaction could occur during the earthquake. Thus, the factor of safety (FS) against liquefaction may be defined as

$$FS = \frac{CRR}{CSR} \tag{10.21}$$

The higher the factor of safety, the more resistant the soil to liquefaction. However, soil having FS slightly greater than 1.0 may still liquefy during an earthquake shaking. This will happen in case a lower layer liquefies, then the upward flow of water could induce liquefaction of the layer that has a factor of safety slightly greater than 1.0.

However, the determination of factor of safety against liquefaction should also consider the following.

- *Soil type*: The soil types most susceptible to liquefaction should be identified. According to Ishihara, 1985, "The hazard associated with soil liquefaction during earthquakes has been known to be encountered in deposits consisting of fine to medium sands and sands containing low plasticity. Occasionally however, cases are reported where liquefaction apparently occurred in gravelly soil". Seed et al., (1983) stated that based on both laboratory testing and field performance, the great majority of cohesive soils will not liquefy during earthquake. Thus, the soil types susceptible to liquefaction are non-plastic (cohesionless) soils.
- *Groundwater table*: The soil must be below the groundwater table. If there is a likelihood that groundwater table will rise in the future, such anticipation must be taken into account as the soil will eventually be below the groundwater table.
- *Engineering judgement*: For evaluating the factor of safter against liquefaction, there are many different equations and corrections that are applied to CRR and CSR. All these different equations and various corrections may provide the engineer with a sense of high accuracy, when in fact the entire analysis is only a gross approximation. The analysis should be treated as such and engineering experience and judgement are essential in the final conditions of whether a site will liquefy or not during earthquake.

EXAMPLE 10.1 The sand deposit of fine sand (finer ≤ 5) of finite thickness is located at a depth of 3.0 m from the ground surface. This is located in a seismic prone area where the anticipated GPA is $0.40g$. The standard penetration test was performed at a depth of 3 m. The corrected N-value is 8. The unit weight of sand may be taken as 18.4 kN/m^3 and the submerged unit weight of sand as 9.2 kN/m^3. Calculate the factor of safety against liquefaction for the saturated sand located at a depth of 3 m.

Solution: Given

$$N_{\text{cor}} = 8$$

$$\gamma_t = 18.4 \text{ kN/m}^3$$

$$\gamma_o = 9.24 \text{ kN/m}^3$$

$$\text{Taking the unit weight of water} \approx 10 \text{ kN/m}^3$$

$$z = 3 \text{ m}$$

$$\text{PGA} = 0.4g$$

$$\text{Effective stress} = \sigma'_{v_o} = \sigma_{v_o} - u = \gamma_t(1.5) + r_b(1.5)$$

$$= 18.4(1.5) + 9.24(1.5)$$

$$= 27.6 + 13.86 = 41.46$$

$$\text{Total stress} = \sigma_{v_o} = 1.5(18.4) + 1.5(9.24 + 10)$$

$$= 27.6 + 13.86 + 15 = 56.46$$

Using the linear relationship for the stress reduction factor (depth reduction factor)

$$r_d = 1 - 0.012z = 1 - 0.036 = 0.964$$

Thus, $$\text{CSR} = 0.65 r_d \left(\frac{\sigma_{v_o}}{\sigma'_{v_o}}\right)\left(\frac{a_{max}}{g}\right)$$

$$= 0.65 \times 0.0964 \times \frac{56.46}{41.46} \times 0.40$$

$$= 0.341$$

Using Figure 10.7 with $N_{cor} = 8$

$$\text{CCR} = 0.10$$

Hence, for Factor of Safety (FS)

$$\text{FS} = \frac{\text{CRR}}{\text{CSR}} = \frac{0.10}{0.341} = 0.293$$

Thus, based on the factor of safety of 0.293, the sand deposit at a depth of 3.0 m will liquefy.

10.5 FACTORS RESPONSIBLE FOR LIQUEFACTION

There are many factors that control the liquefaction of soil. They may be summarized as follows:

Particle size distribution of soil

Liquefaction occurs in cohesionless soil (sands) and does not occur in cohesive soils. However, highly sensitive clays lose the strength substantially during earthquake excitation. As far as sands are concerned, fine and uniform sands are more prone to liquefaction. As the pore pressure is dissipated more quickly in coarse-sand, hence the chances of liquefaction are reduced in coarse-sand deposits.

Further, uniformly graded non-plastic soil tends to form more unstable particle arrangements and is more susceptible to liquefaction than the well-graded soils. Well-graded soils will also have small particles that fill in the void space between the large particles. This tends to reduce the potential contraction of the soil, resulting in less excess pore pressure generation during an earthquake. Majority of the field evidence indicates that most liquefaction failures have involved uniformly graded granular soil.

Relative density

The most important index aggregate property of a cohesionless soil is its relative density D_r. The relative density is defined as

$$D_r = \frac{e_{max} - e}{e_{max} - e_{min}} \times 100 \tag{10.22}$$

The relative density is a measure of denseness of sand deposit and so it is one of the most important factors controlling liquefaction. The chance of liquefaction is much reduced if the relative density is high.

Based on field studies, cohesionless soils in loose relative density state are susceptible to liquefaction. Loose non-plastic soils will contract during the seismic shaking which will cause the development of excess pore water pressure. After reaching initial liquefaction, there is always a sudden and dramatic increase in shear displacement for loose sands.

For dense sands, the state of initial liquefaction does not produce large deformation because of the dilatation tendency of the sand upon reversal of the cyclic shear stress. Poulos et al. (1985) stated that if the *in situ* soil can be shown to be dilative, then it is need not be evaluated because it will not be susceptible to liquefaction. As such dilative soils are not susceptible to liquefaction became their undrained shear strength is greater than the drained shear strength.

Dynamic characteristics of soil

The dynamic characteristics are: (i) Frequency, (ii) Amplitude, (iii) Acceleration, and (iv) Velocity.

The frequency of the dynamic load plays a vital role specially if it is close to natural frequency of the system. Florin, V.A. and Ivanov, P.L. (1961) reported that liquefaction depends on nature, magnitude and type of dynamic loading. The whole stratum may be liquefied at the same time under shock and transient loading while liquefaction may start from the top and proceed downward under steady state vibrations. Further, Seed and Lee (1966) observed that under steady state vibrations the maximum pore pressure develops only after a certain number of cycles have been imparted to the desopit. Prakash and Gupta (1970) studied this problem under site conditions and concluded that the horizontal vibrations in dry sand lead to a larger settlement than when in a state of vertical vibrations.

Seed and Idriss (1982) reported that landslides during *Alaska* earthquake were triggered about 90 seconds after the start of the ground motion. They opined that if the ground motion had lasted less than 90 seconds, say 60 seconds, liquefaction or instability would not have developed.

Further, Seed in his research paper published in 1976 and 1979 states that multidirectional shaking as in earthquake excitation is more severe than that in one-directional vibrations. Under multi-directional shaking, pore pressure builds up faster and consequently the loss of the strength of the soil is much faster and the stress ratio required is about ten per cent less than that required for unidirectional shaking.

Strain history

The studies on liquefaction characteristics of fresh deposits or of recent origin and their comparison with similar soil deposits previously subjected to some strain history reveal that there is no significant change in relative density owing to the previous strain history. However, Seed (1995) showed that although the prior strain history caused no significant change in the density of the sand, but it increased the stress that causes liquefaction by a factor of about 1.5.

This older soil deposits that have already been subjected to seismic shaking in past have increased liquefaction resistance compared to a newly formed specimen of the soil having an identical density [Finn et al., 1970]

Liquefaction resistance also increases with an increase in the over consideration ratio (OCR) and the coefficient of lateral earth pressure at rest k_0 [Seed and Peacock, 1971, Ishihara et al., 1978]. As such a soil that has been preloaded will be more resistant to liquefaction than the same soil that has not been preloaded.

Influence of superimposed load and overburden pressure

In soil deposits at any depth the effective stress depends upon the magnitude of superimposed load and the overburden pressure. In the field the initial stress conditions are not isotropic, i.e., the lateral stress is not equal to the normal stress, rather the lateral stress depends upon the coefficient of earth pressure at rest K_0 which is usually defined as

$$K_0 = \frac{\mu}{1-\mu}$$

where μ is the Poisson's ratio.

The stress conditions causing liquefaction depend upon the value of K_0, the coefficient of earth pressure at rest.

Entrapped air

The soil is a polyphase material consisting of solid soil grains, water and air entrapped in voids and the pores. If the air is trapped, during the rise of pore pressure due to earthquake excitation, part of it is consumed and is dissipated due to compression of air. Hence, the trapped air helps in reducing the chances of liquefaction.

Drainage facility

If the excess water pressure can quickly dissipate, the soil may not liquefy. This highly permeable gravel drains or gravel layers can reduce the liquefaction potential of adjacent soil.

Confining pressures

The greater the confining pressure, the less susceptible the soil is to liquefaction. The condition that can create a higher confining pressure on a deeper groundwater table is the soil that is located at a depth below ground surface and a surface charge pressure applied at the ground surface. Day, R.W. (2002) states that case studies have shown that the possible zone of liquefaction usually extends from ground surface to a maximum depth of about 15 m. Deeper soils generally do not liquefy because of the higher confining pressure.

Particle shape

The soil particle shape can also influence liquefaction potential. For example, soils having rounded particles tend to densify more easily than angular-shape soil particles.

Building load

The construction of a heavy building on top of a sand deposit can decrease the liquefaction resistance of the soil. The reason is that a smaller additional shear stress will be required from the earthquake in order to cause contraction and hence liquefaction of the soil.

Groundwater table

The condition most conducive to liquefaction is a near surface groundwater table. Unsaturated soil located above the groundwater table will not liquefy. At sites where the groundwater table significantly fluctuates, the liquefaction potential will also fluctuate.

10.6 CRITERION FOR ASSESSING LIQUEFACTION

The criterion for deciding in the field whether the soil would liquify or not, has been under investigation and the following parameters have been suggested for a possible criterion of liquefaction.

- Critical void ratio [After Casagrande, 1936, 1976]
- Critical acceleration [After Maslov, s1975]
- Stress condition and hydraulic gradient [After Florin, V.A. and Lvanov, P.L. (1961)]
- Standard Penetration Value (*N*) [After Trifunac, 1995]
- Energy based liquefaction criterion [After Todoroska, 1996]

The various studies have shown that a large number of parameter and factors influence the liquefaction. At this stage it is not possible to select one single parameter for an acceptable criterion for assessing liquefaction.

10.6.1 Criteria Based on Grain Size

The criteria for fine-grained soil to liquefy can be evaluated with modified Chinese criteria [Finn, et al., 1994] shown in Figure 10.8. According to these criteria, soil may liquefy if the clay fraction is less than 15% (using the Chinese definition of clay size as less than 0.005 mm), the liquid limit is less than 35% and the water content is larger than 0.9 times the liquid limit. Thus, three criteria, must be met for soil to be liquefiable, namely:

- Clay fraction (% finer than 0.005 mm) < 15%
- Liquid limit (LL) < 35%
- Moisture content (MC) > 0.9LL

However, these criteria are somewhat controversial and they do not represent a global consensus. Establishing more precise and reliable measures for identifying which finer grained soils are potentially susceptible to liquefaction is an area of ongoing research [Seed et al., 2001]

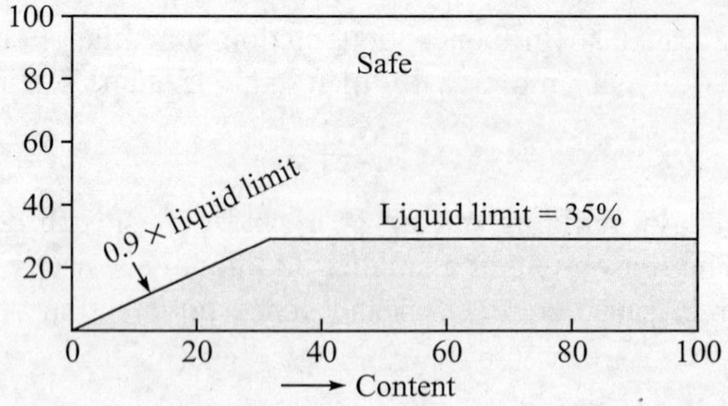

Figure 10.8 Modified Chinese criteria for liquefaction assessment [After Finn et al. 1994]

10.6.2 ENERGY BASED LIQUEFACTION CRITERION

Initiation of liquefaction in water-saturated cohesionless sand is assumed to occur when the effective stress in the ground approaches zero. Starting from this hypothesis, Nemat-Nassor and Sokooh (1979) introduced an energy concept for producing the liquefaction. Davis and Berril (1982) formulated energy-based empirical criteria for liquefaction which was later defined by Berril and Davis (1985) and Trifunac (1995). Further, evaluation methodology was provided by Todoroska (1996).

Trifunac (1995) studied observations worldwide of occurrence as well as of no occurrence of liquefaction after an earthquake. He proposed five empirical models to compute the corrected standard penetration tests (SPT) value, $\vec{N}$ (corrected for overburden stress) that separates the observed cases, when liquefaction occurred from these, and when it did not occur.
The corrected $\overline{N}$ values Peck, Hanson and Thonburn (1974) are obtained as

$$\overline{N}_{\text{cor}} = 0.77 \log_{10} \frac{2000}{\sigma_0} N_{\text{obs}}$$

where, σ_0 = effective overburden pressure, in kN/m^2

$\overline{N}$ = corrected N values

N_{obs} = observed N values

The models are based on seismic energy and are functions of the following:

- overburden pressure, σ_0
- magnitude of earthquake
- epicentral distance
- local site conditions

The value of $\overline{N}$ given by the model can be viewed as a critical value of N, and estimating $\overline{N}$ – N_{crit} for liquefaction to occur under the specified conditions. In other words liquefaction will not occur at the site if $\overline{N}$ at the site is smaller than N_{crit}. As there are many uncertainties in the ground motion description and in the site characterization, the data points are scattered about the prediction of the model. From the points in the database that violates prediction, a distribution of the N_{crit} can be constructed, Todoroska (1996). Probability of liquefaction at site characterized by some $\overline{N}$ is equal to the probability that $\overline{N}_{\text{crit}} > \overline{N}$. Trifunac (1995) assumed that N_{crit} has Gaussian distribution and then evaluated the standard deviation.
Then the probability density of N_{crit} may be expressed as

$$\text{PDF}(N_{\text{crit}}) = \frac{1}{\sqrt{2\pi}\,\sigma\overline{N}} \exp\left\{-\frac{1}{2}\left(\frac{\overline{N}-\mu}{\sigma}\right)\right\}^2 \tag{10.23}$$

where, μ = mean value given by the model

σ = standard deviation

Further, the probability of N_{crit} exceeding $\overline{N}$ may be obtained as

$$\text{Prob.}\left\{\overline{N}_{\text{crit}} > \overline{N}\right\} = \frac{1}{\sqrt{2\pi}\,\sigma\overline{N}} \int_{\overline{N}}^{\infty} -\frac{1}{2}\frac{(x-\mu)^2}{\sigma}\,dx \tag{10.24}$$

Thus, based on above Eq. (10.24), the following may be obtained

- Conditional probability q_{ik} for an earthquake of type (l, k) (an earthquake of magnitude μ_k and distance R)
- Average return period of occurrence of liquefaction for a site with the given value of N and σ_0.

Finally, according to Todoroska (1996) the probabilistic seismic hazard methodology may be applied to the probability of occurrence that may happen during the specified exposure.

10.7 EVALUATION OF LIQUEFACTION POTENTIAL

In Section 10.5, all the factors influencing the liquefaction characteristics of soil have been described. In order to evaluate the liquefaction potential of a deposit it is thus necessary to determine the shear stress induced at any depth by an earthquake excitation. If it is sufficiently large enough for liquefaction at that depth, Seed and Idriss (1971), proposed a method of analysis consisting of the following steps, based on cyclic stress approach as explained in Section 10.3.

Computation of τ_{max} induced by earthquake

The shear stresses τ developed in soil deposit during earthquake depends largely on the upward propagation of shear wave in the deposit. Thus, the maximum shear stress $\tau_{\max}$ would be

$$\tau_{\max}\Big|_{\text{surface}} = \frac{\gamma h}{g}\alpha_{\max} \tag{10.25}$$

where,

$\alpha_{\max}$ = maximum ground surface acceleration

γ = unit weight of soil

h = depth from ground level.

However, the actual $\tau_{\max}$ at any depth will be less than that on surface. So,

$$\tau_{\max}\Big|_{\text{at any depth } d} = r_d\,\tau_{\max}\Big|_{\text{surface}} \tag{10.26}$$

where r_d = a stress reduction coefficient < 1.

Seed and Idriss presented a range of values of r_d for different soil profiles in liquefaction analysis and has been shown in Figure 10.5.

Thus, the τ_{max} for evaluation of liquefaction potential at any depth h may be expressed as

$$\tau_{max} = \frac{\gamma h}{g} \cdot \tau_{max} \cdot r_d \tag{10.27}$$

Determination of average stress τ_{av}

Seed and Idriss (1971) studied the time history of shear stresses during earthquake and the variation of shear stress with time is shown in Figure (10.4). Based on experimental evidence, it has been shown that the average shear stress τ_{av} is about 65% of τ_{max}. Therefore,

$$\tau_{av} \cong 0.65 \frac{\gamma h}{g} \alpha_{max} r_d \tag{10.28}$$

Determination of significant stress cycles N_c

Experimental evidences have shown that the appropriate number of significant stress cycles N_c will depend upon the duration of ground shaking and finally on magnitude of the earthquake. The representative number of stress cycles are shown in Table 10.2.

Table 10.2

Magnitude on Richter scale	*Number of significant stress cycles, N_c*
7	10
7.5	20
8.0	30

Determination of stress causing liquefaction

Experimental results by a cyclic triaxial shear test or simple shear test is used to determine stresses that cause liquefaction within a given number of cycles, depending upon the magnitude of the earthquake. The laboratory test results have been correlated with stress causing liquefaction under field conditions. Seed and Idriss have proposed that

$$\left.\frac{\tau}{\sigma'_0}\right|_{\text{field condition}} \cong \frac{\sigma_{dc}}{2\sigma_a} \cdot \frac{C_r \cdot D_r}{50} \tag{10.29}$$

where,

τ = shear stress at any depth on horizontal plane in the field
σ'_0 = initial effective overburden pressure
σ_{dc} = cyclic deviator stress in laboratory conditions
σ_a = initial cell pressure in triaxial test
C_r = correction factor
D_r = relative density.

Seed and Idriss have adopted the following values of the correction factor (Table 10.3).

Table. 10.3 Correction factor

Relative density D_r in %	*Correction factor, C_r*
0 – 50	0.57
60	0.60
80	0.68

Evaluation of liquefaction potential

To evaluate the liquefaction potential of a deposit it will be sufficient to determine whether the shear stress determined by Eq. (10.7) induced at any depth by the earthquake is sufficiently larger than the value determined by Eq. (10.16).

Liquefaction potential may be evaluated in the cyclic strain approach in a manner similar to that used in cyclic stress approach. In the 1960s and 1970s, many advances in the state of knowledge of liquefaction phenomenon resulted from the pioneering work of H.B. Seed and his colleagues at the university of California at Berkeley. This research was directed towards evaluation of the loading conditions required to initiate liquefaction. This loading was described in terms of cyclic shear stress and liquefaction potential was evaluated on the basis of the amplitude and the number of cycles of earthquake induced shear stress. This approach is conceptually very simple. The seismic induced loading expressed in terms of cyclic shear stress is compared with the liquefaction resistance of the soil, also expressed in terms of cyclic shear stresses. At locations where the loading exceeds the resistance, liquefaction is expected to occur.

The evaluation of liquefaction potential is easily performed graphically as shown in Figure 10.9. The variation of equivalent cyclic shear stress with depth is plotted. The variation of the cyclic shear stress required to cause liquefaction with depth is then plotted on the same graph. Liquefaction can be expected at depths where the loading exceeds the resistance. In Figure 10.9 where loading is exceeding the strength, that zone is marked (identified) as zone of liquefaction. As an alternative way (method) a factor of safety against liquefaction can be expressed as

$$FS]_L = \frac{\tau_{cycl,L}}{\tau_{cyl}} = \frac{\text{cyclic shear stress to cause liquefaction}}{\text{equivalent cyclic shear stress induced by earthquake}} = \frac{\text{CRR}}{\text{CSR}}$$

Thus, the simplified approach required the calculation of two primary variables the seismic demand placed on a soil layer, expressed in terms of CSR (Cyclic Stress Ratio) and the capacity of the soil to resist liquefaction, expressed in terms of cyclic resistance ratio (CRR). This procedure has been updated periodically since that time with a landmark paper by Seed (1979), Seed and Idriss (1982), Seed et al., (1985), and Youd and Gilstrap (1999).

EXAMPLE 10.2 The soil deposit in an alluvial terrain is 80 ft deep from the ground level. The groundwater table is located at 5 ft from the ground level. During an earthquake of magnitude 8.0 on Richter scale, damage to building and other structures occurred owing to liquefaction of the sand. The maximum intensity of ground shaking is 0.12g. The unit weight of sand is

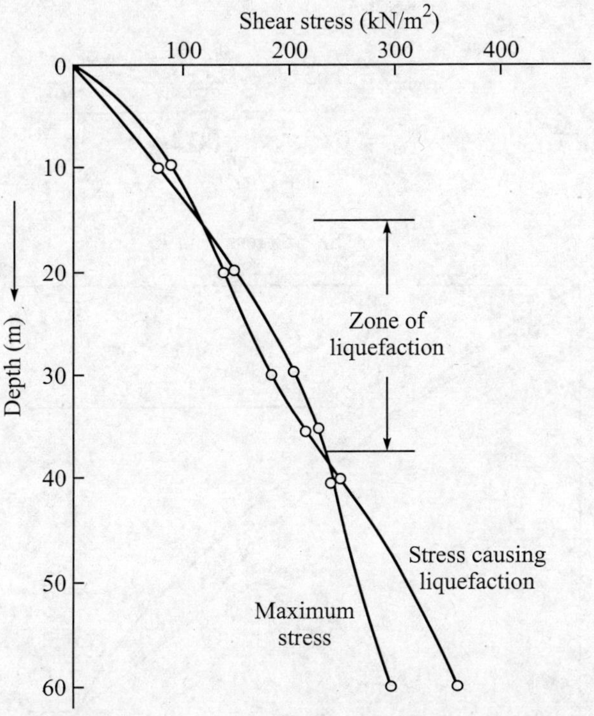

Figure 10.9 Process by which zone of liquefaction is identified.

118 lb/ft^3 and submased wt. of sand is 55 lb/ft^3. D_{50} of the sand is 0.25 mm and its relative density D_r is 50%. Using the Seed and Idriss method of analysis, determine the zone of liquefaction.

Solution:

1. Determination of τ_{max}

   Using the relationship $\tau_{max} = \dfrac{\gamma_h}{g} \alpha_{max} \cdot r_d$, where h = depth from the ground level

   then $\tau_{av} = 0.65\ \tau_h \dfrac{a_{max}}{g} \cdot r_d$

Depth in feet	γh (lb/ft^2)	$\dfrac{\alpha_{max}}{g}$	r_d	τ_{av} (lb/ft^2)	*Remarks*
10	1177.50	0.12	0.98	90.00	γ = 118 (lb/ft^3)
20	2352.50	0.12	0.96	176.15	up to 5′–0″ depth and
30	3527.50	0.12	0.94	258.63	(55 + 62.5 below 5 feet.
40	4702.50	0.12	0.90	330.11	= 117.5 lb/ft^3)
50	5877.50	0.12	0.85	389.67	
60	7052.50	0.12	0.70	385.0	
70	8227.50	0.12	0.60	385.047	
80	9402.50	0.12	0.50	366.69	

2. Determination of τ_{av}

$$\tau_{max} = 0.65 \frac{\gamma h}{g} \alpha_{max} \cdot r_d$$

$$\tau_{av} = 0.65. \ \gamma h \ . \ 0.12 r_d$$
$$= 0.715 \ \gamma h r_d$$

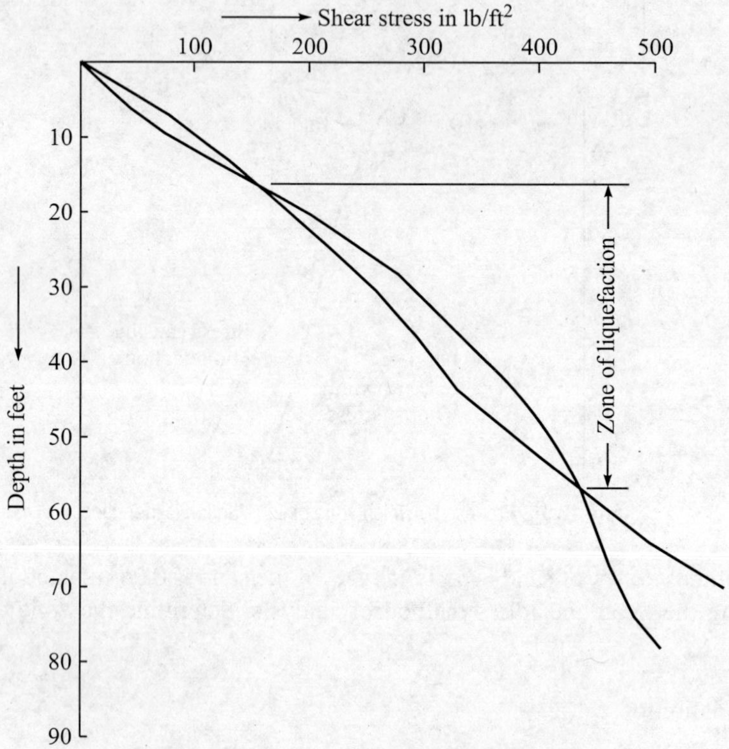

Figure 10.10 Example 10.2: (evaluation of zone of liquefaction).

10.8 LABORATORY INVESTIGATIONS OF SOIL LIQUEFACTION

Generally, following two types of laboratory tests have been used for studies of liquefaction of sand, namely:

- Triaxial or simple shear test
- Vibration table studies

Liquefaction of soils has been extensively studied in the laboratory. There is a considerable amount of published data covering laboratory liquefaction testing. However, in this section examples and salient features of laboratory liquefaction data are included, namely:

- Laboratory test data from Seed and Lee (1965)
- Laboratory test data from Ishihara (1985)

Laboratory test data from Seed and Lee (1965)

In the early 1960s, a detailed testing programme on liquefaction of sand was initiated at the university of California at Berkeley under the leadership of H.B. Seed. Seed and Lee reported the first set of comprehensive data on sand.

Figure 10.11 (From Seed and Lee, 1965) shows a summary of laboratory tests from cyclic triaxial tests performed on saturated specimen of Sacraments River sand. Cylindrical sand specimen were first saturated and subjected in the triaxial apparatus to an isotropic effective confining pressure of 100 kPa (2000 lb/ft^3). The saturated sand specimen were then subjected to undrained conditions during the application of the cyclic deviator stress in the triaxial apparatus.

Numerous sand specimen were prepared at different initial void ratio, e_i. They were subjected to different values of cyclic deviator stress, σ_{de} and the number of cycles of deviator stress required to produce initial liquefaction and 20 per cent axial strain were recorded. Figure 10.11 shows the cyclic deviator stress, σ_{dc} for different number of cycles of deviator stress on log scale. The laboratory data shown in Figure 10.11 indicates the following:

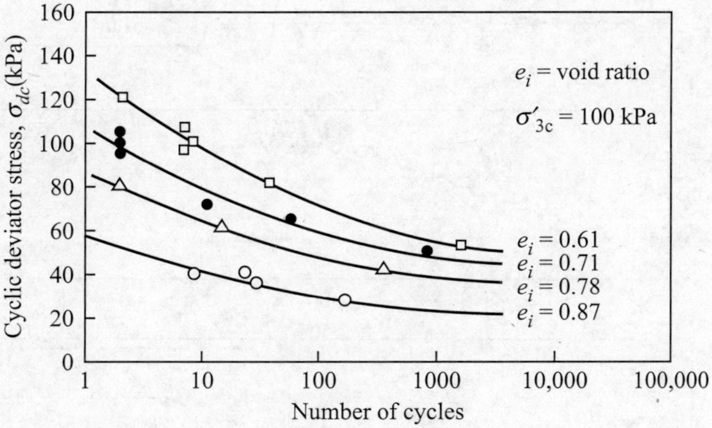

Figure 10.11 Laboratory test data from cyclic triaxial tests performed on Sacraments River sand (After Seed and Lee, 1965).

- For sand having the same initial void ratio e_i and the same effective confining pressure, the higher the cyclic deviator stress σ_{dc}, the lower the number of cycles of deviator stress required to cause initial liquefaction.
- Sand having the same initial void ratio e_i and the same effective confining pressure, the cyclic deviator stress σ_{dc} required to cause initial liquefaction will decrease as the number of cycles of deviator stress is increased.
- Sand in dense state will require a higher cyclic deviator stress σ_{dc} or more cycles of the deviator stress in order to cause initial liquefaction compared to the same soil in a loose state.
- Sand in a loose state will require a lower cycle deviator stress σ_{dc} or fewer cycles of the deviator stress in order to cause initial liquefaction compared to the same soil in a dense state.

10.8.1 Laboratory Test Data

Laboratory tests show that the number of loading cycles required to produce liquefaction failure, decreases with increasing shear stress amplitude and with decreasing density (Figures 10.12 and 10.13). While liquefaction failure can occur in only a few cycles in a loose specimen subjected to larger cyclic shear stresses, thousand of cycles of low amplitude shear stress may be required to cause liquefaction of a dense specimen.

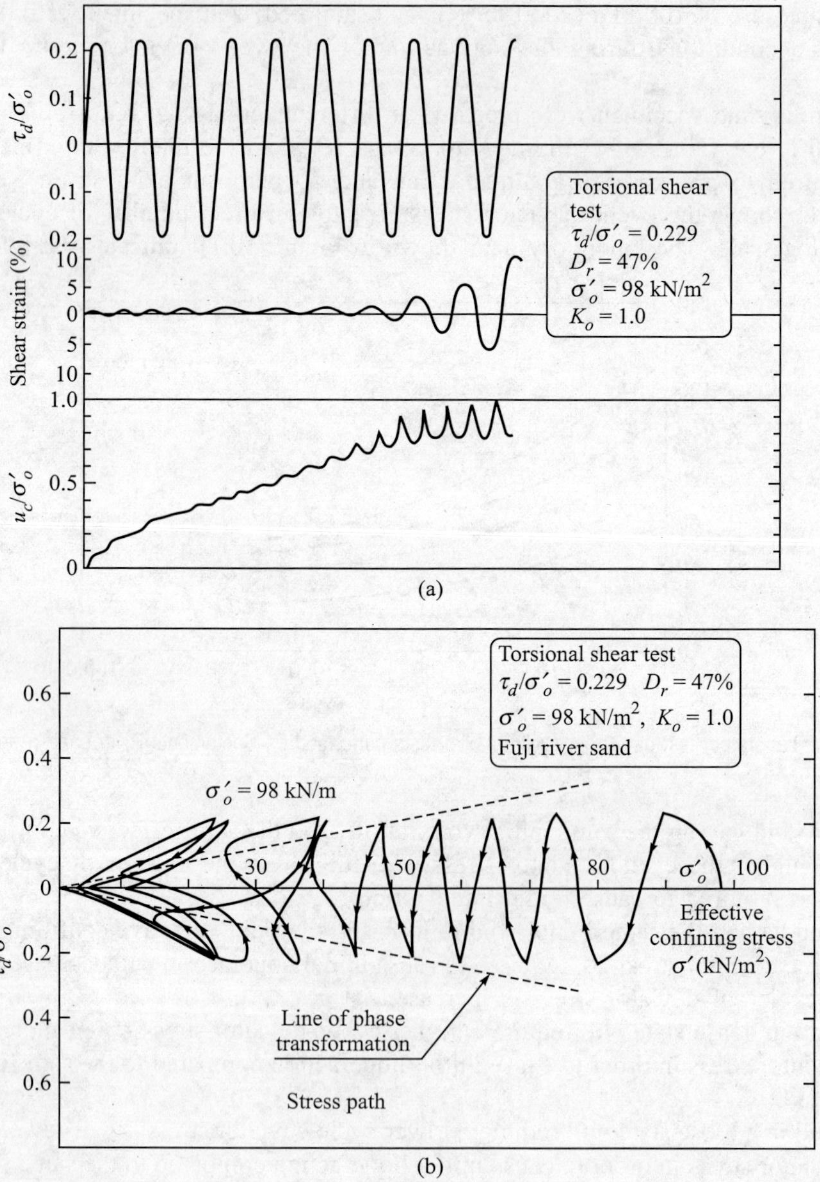

Figure 10.12 Laboratory test data from cyclic torsional shear tests performed on Fuji river sand having a relative density (D_r = 47%) (After Ishihara, 1985).

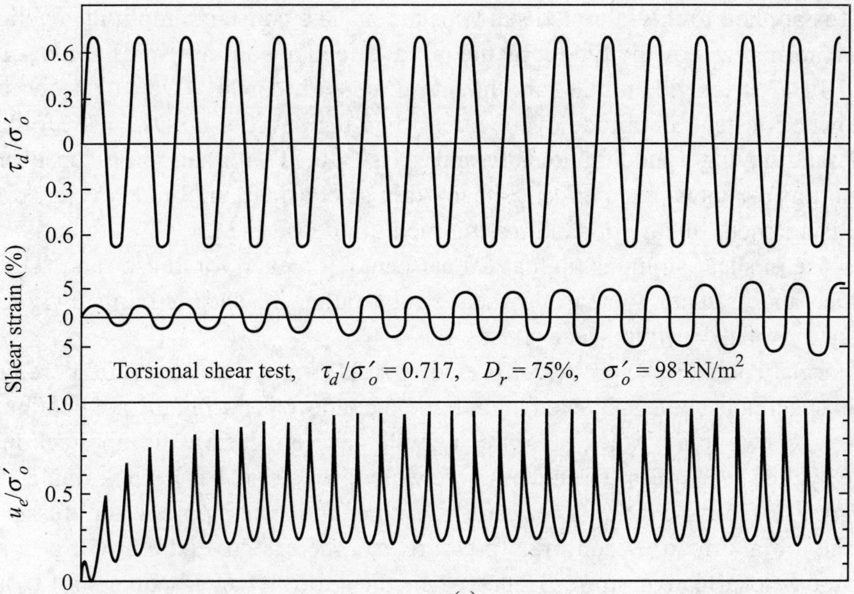

(a)

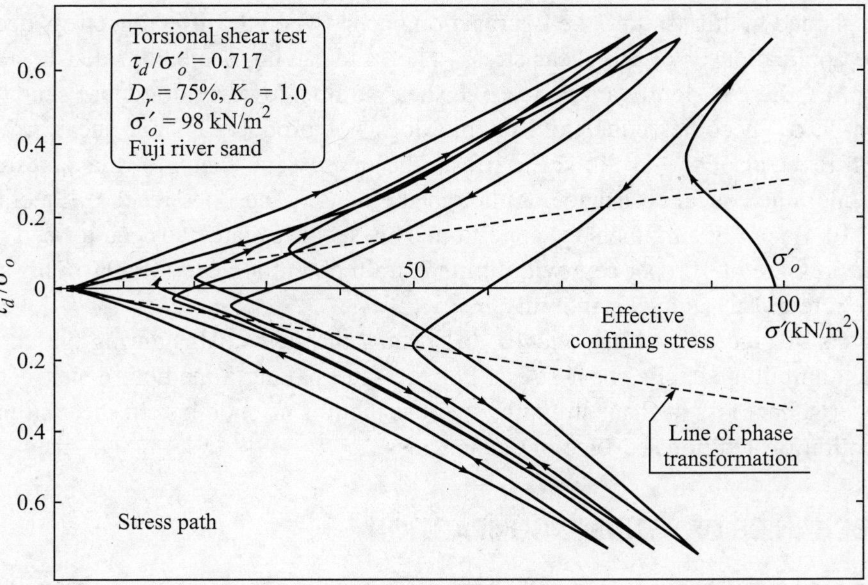

(b)

Figure 10.13 Laboratory test data from cyclic torsional shear tests performed on Fuji river having a relative density ($D_r = 75\%$) (After Ishihara, 1985).

Figures 10.12 and 10.13 present the result of laboratory tests performed on a hollow cylindrical specimen of saturated Fuji River sand tested in a torsional shear test apparatus. The uppermost part (Figures 10.12(a) and 10.13(a)) shows the constant amplitude cyclic shear stress that is applied to the saturated sand specimen. The constant amplitude cyclic shear stress r_d has been normalized by dividing it by the initial effective confining pressure σ_o'. From Figures 10.12 and 10.13 it is evident that sand having a medium density (D_r = 47%) was subjected to a much lower constant amplitude cyclic stress than the dense sand (D_r = 75%): that is τ_d/σ_0' = 0.229 for sand having a medium density and τ_d/σ_0' = 0.717 for sand in a dense state. The plot of shear strain (%) shows the per cent shear strain as constant amplitude cyclic stress is applied to the soil specimen. In case of sand having medium density (D_r = 47%) there is a sudden and rapid increase in shear strain as high as 20 per cent. However, for dense sand (D_r = 75%) there is no sudden and dramatic increase in shear strain, rather the shear strain slowly increases with the application of the cyclic shear stress.

The normalized excess pore water pressure is also known as the cyclic pore pressure ratio. As the soil specimens were subjected to undrained conditions during the application of the cyclic shear stress, excess pore water pressure v_e will develop during the application of constant amplitude cyclic shear stress. The excess pore water pressure v_e has been normalized by dividing it by the initial effective confining pressure σ_o'. When the excess pore water pressure v_e becomes equal to the initial effective confining pressure σ_o', the effective stress will become zero. The condition of zero effective stress occurs when the ratio v_e/σ_o' is equal to 1.0. Figure 10.12 shows that the shear strain dramatically increases when the effective stress is equal to zero. As already explained in this section, the liquefaction occurs when the effective stress becomes zero during the application of cyclic shear stress. The sand having a medium density (D_r = 47%) when liquefies, there is significant increase in shear strain. However, for dense sand (D_r = 75%) even when v_e/σ_o' becomes equal to 1.0, this does not produce a large shear strain. This is because on reversal of the cyclic shear stress, the dense sand tends to dilate, resulting in an increased undrained shear resistance. Although the dense sand does reach a liquefaction state (v_e/σ_o' = 1.0), it is only a momentary condition. This specific state has been termed peak cyclic pore water pressure of 100 per cent with limited strain potential Seed (1979a). This state is also commonly referred to as cyclic mobility.

The plots of Figures 10.12(b) and 10.13(b) show the stress paths during the application of the constant amplitude cyclic shear stress. For the sand having a medium density (D_r = 47%), there is a permanent loss in shear strength, whereas there is no such loss in shear strength during the application of additional cycles of shear stress.

10.9 MECHANICS OF DYNAMIC COMPACTION

Liquefaction may also be identified with increase in dynamic compaction. As already explained, the propagation of shear wave causes the loose soil to contract resulting in an increase in pore water pressure. As such there is compaction and subsequent change in voids and porosity of soil. In this perspective Maslov (1987) introduced a new quantity as the coefficient of dynamic compaction v_n. This is defined as an index characteristic that is indicative of the rate of compaction of sand with a specified density related to porosity n acted on.

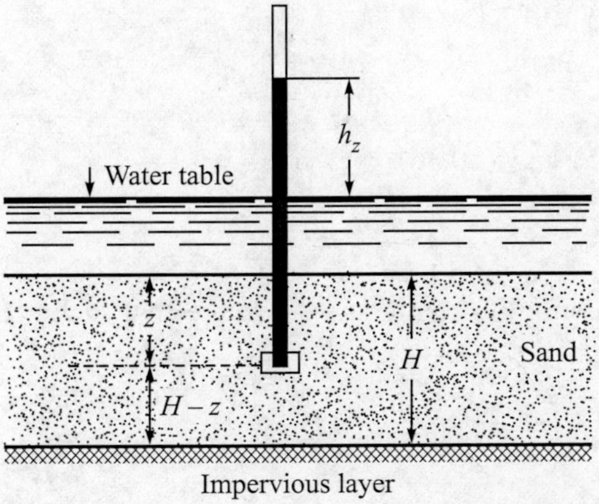

Figure 10.14 Parameters of the dynamic regime of a submerged sand mass.

By virtue of the above definition according to Maslov (1987), v_n may be expressed as

$$v_n = \frac{dn}{dt} \tag{10.30}$$

Thus, if all the voids are filled by water, the coefficient v_n will be indicative of the rate of compression of sand (referred to porosity n) and of the amount of water squeezed out of the voids per unit of time following a shock acting on the sand. The coefficient of compression is measured in (t^{-1}), which is a reciprocal of time.

Let us select an elementary layer of sand, which is dz in thickness in the mass of sand Fig. (10.14). The increment of the discharge of water in the total upward flow dq_z due to the shock-induced compression of sand in the particular layer can be found from this relationship:

$$dq_z = -\frac{dn}{dt}\,dz \tag{10.31}$$

The minus sign on the right-hand side of this equation implies that with reduction in porosity of the sand the seepage of water through the sand mass increases. If we assume the cross-sectional area of passing water ω, to be unity, we can write $\omega = 1.0$. Let us define the coefficient of permeability as K_p. Then

$$q_z = K_p \frac{dh_z}{dz} \tag{10.32}$$

In the particular case the discharge of water will be determined in conformity with the Darcy's law, i.e.,

$$q_z = \omega K_p I_z \tag{10.33}$$

The value of the hydraulic gradient at the horizon z will be found from the relationship:

$$I_z = \frac{dh_z}{dz} \tag{10.34}$$

By differentiating Eq. (10.32) with respect to z, we obtain

$$\frac{dq_z}{dz} = K_p \frac{d^2 h_z}{dz^2} \tag{10.35}$$

Let us transform Eq. (10.31) to

$$\frac{dq_z}{dz} = -\frac{dn}{dt} \tag{10.36}$$

By comparing Eq. (10.35) and (10.36), we have

$$\frac{d^2 h_z}{dz^2} = -\frac{I}{K_P}\frac{dn}{dt} \tag{10.37}$$

In the particular case, to simplify matters, we assume that in Eq. (10.37)

$$\frac{dn}{dt} = v_n = \text{const.}$$

whence,

$$\frac{d^2 h_z}{dz^2} = -\frac{I}{K_P} v_n \tag{10.38}$$

Clearly,

$$h_z = -\frac{v_n}{K_P} \iint dz^2 \tag{10.39}$$

At the same time,

$$dh_z = -\frac{v_n}{K_P} \int dz \tag{10.40}$$

or,

$$I_z = \frac{dh_z}{dz} = -\frac{v_n z}{K_P} + C_1 \tag{10.41}$$

Let us determine the constant of integration C_1 proceeding from the condition that for $z = H$, that is, at the contact with an aquiclude in the absence of discharge ($q_H = 0$), the gradient is $I_H = 0$. As a result, we have

$$I_z = -\frac{v_n}{K_P}(H - z) \tag{10.42}$$

Hence, it is possible to obtain a differential equation for determining the dynamic pressure head h_z, i.e.,

$$dh_z = \frac{v_n}{K_P}(H - z)dz \tag{10.43}$$

Consequently,

$$h_z = \frac{v_n}{K_P}\int (H - z)dz \tag{10.44}$$

Having integrated and determined the constant of integration C_2, given that for $z = 0$, that is on the surface of the sand layer, the pressure head is $h_z = 0$, we finally obtain

$$h_z = \frac{v_n}{K_P}(Hz - z^2/2) \tag{10.45}$$

Thus, it follows that the relationship between the dynamic head and the depth $h_z = f(z)$ is of parabolic character. On the surface layer for $z = 0$ the dynamic head $h_z = 0$. The dynamic head h_z attains its maximum value at the foot of the layer at the contact with the aquiclude for $z = H$.

The relationship between J_z and the depth z is linear. Given $z = 0$, at the upper surface of the layer, this gradient is of maximum value, which indicates that the sand condition there is the most complicated. The value of the dynamic gradient is $J_H = 0$ at the contact with the aquiclude, which is quite natural.

If the sand mass is underlain by a layer of material exhibiting good draining properties (for example, gravel) the water removed from the sand mass as this layer is compressed may percolate upwards and downwards. The above formulae hold good for the given case as well, provided the values of H substituted there are halved.

It is often necessary to allow for a reduction in the value of v_n with increasing depth in the sand mass, especially, if the latter is of conspicuous thickness (several tens of metres). The first version of the solution of the problem assumes a linear relationship between the coefficient v_z and the depth z, i.e.,

$$v_z = f(z); \qquad v_z = v_0 \frac{(L - z)}{L}$$

where v_0 is the coefficient of dynamic compaction at the surface of the sand mass; L is the depth at which $v_z = 0$ where $\alpha_z = \alpha_{cr}$, viz. where the critical acceleration decreasing with depth becomes equal to α_{cr}.

The solution of the problem for such conditions involves the following relationships for determining the dynamic heads h_z and gradients J_z acting at a depth z

$$h_z = \frac{1}{2}\frac{v_0}{K_p}\left(Lz - z^2 + \frac{z^2}{3L}\right) \tag{10.46}$$

$$J_z = \frac{dh_z}{dz} = \frac{1}{2}\frac{v_0}{K_p}\left(L - 2z + \frac{z^2}{L}\right) \tag{10.47}$$

The reduction in the value of the coefficient of dynamic compaction accordingly decreases the values of h_z, yet the mechanism of the event remains the same.

For the determination of the coefficient of dynamic compaction, the relationship may be written as

$$v_n = \frac{\Delta n}{\Delta t} \tag{10.48}$$

where Δn is the decrease in sand porosity within a discrete minor interval of time Δt.

With increasing sand porosity n, smoothness of its grains and intensity of the dynamic load, the coefficient v_n increases from zero to a value corresponding to the conditions that obtain after a dynamic load. The pattern of the alteration of v for a definite sand variety, depending on the above factors, is presented in Figure 10.15. The magnitude of this coefficient for the sand of interest to us must invariably be established by laboratory tests.

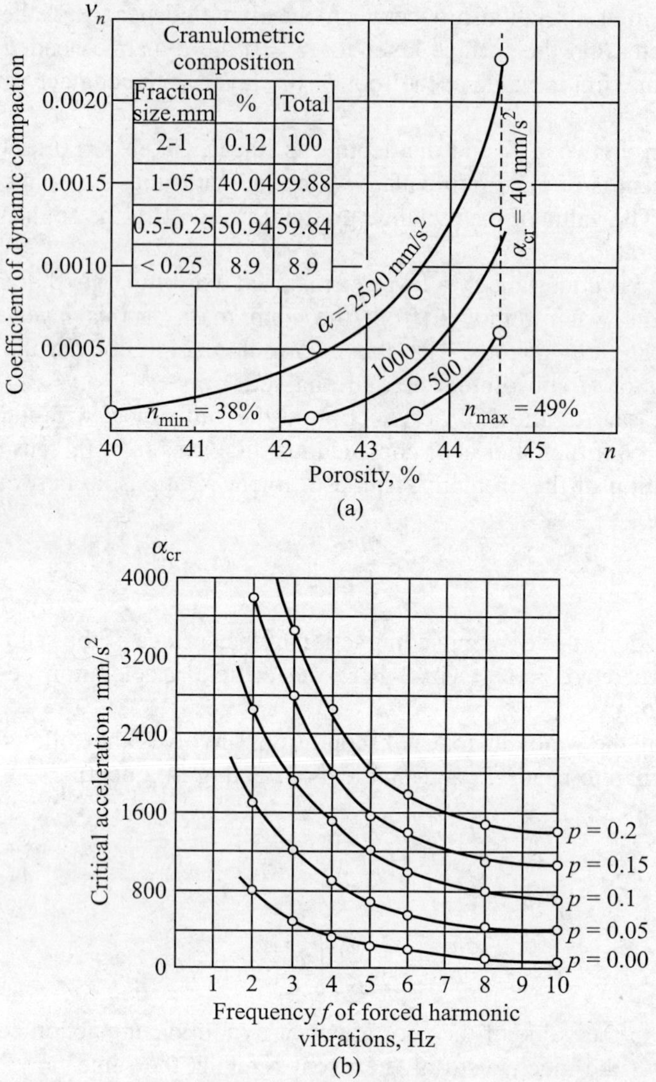

Figure 10.15 Diagram showing a relationship between the coefficient of dynamic compaction v_n and the sand porosity n and acceleration α of vibrational motion.

Thus, the evaluation of the dynamic stability, including seismic, of a mass of saturated cohesionless soils requires their index characteristics, primarily, the value of the critical acceleration α_{cr}, and the coefficient of dynamic compaction v_n. Apart from this, it is necessary to know the value of the coefficient of the permeability of the soil K_p.

The constants α_{cr} and v_n must be determined by taking into account the specified design seismic characteristics (acceleration α, amplitude A, vibrational frequency f or period T) and the particular density of soil (porosity n). These parameters are determined by using special vibration machines that simulate soil vibrations under a specified dynamic regime. Such tests generally employ vibrations at a frequency close to that of the soil (15–30 Hz).

Figure (10.16) is a diagram of an elementary machine for laboratory tests. Similar installations for *in situ* horizontal vibrational and dynamic load test are also available. Maslov (1987).

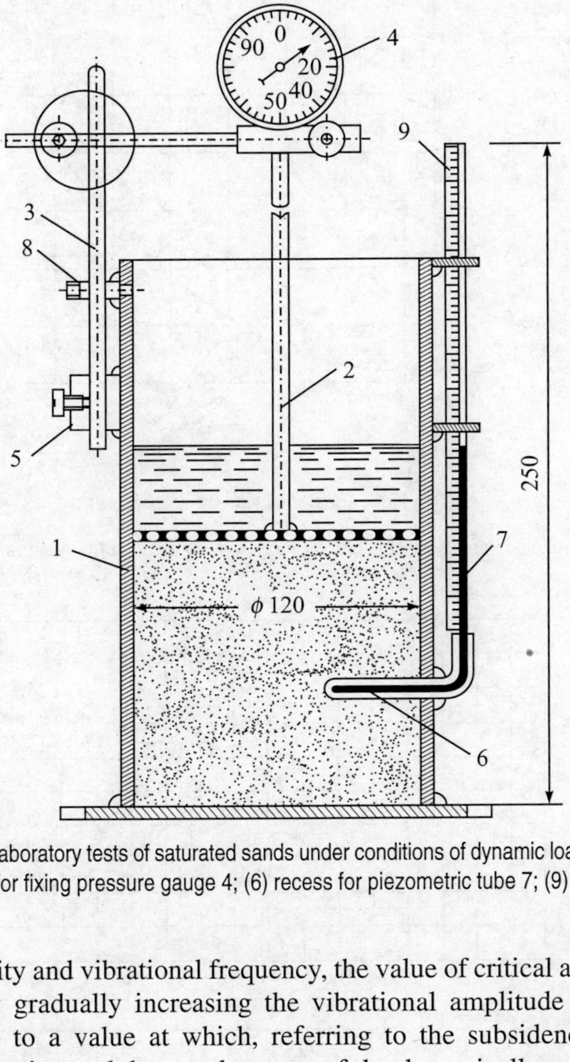

Figure 10.16 Apparatus for laboratory tests of saturated sands under conditions of dynamic load: (1) testing pot; (2) pressure gauge rod; (3, 5, 8) elements for fixing pressure gauge 4; (6) recess for piezometric tube 7; (9) scale for measuring pressure head.

Given the soil density and vibrational frequency, the value of critical acceleration α_{cr} is found for a saturated soil by gradually increasing the vibrational amplitude and, consequently, its acceleration from zero to a value at which, referring to the subsidence of the soil surface, we determine its compaction and, hence, the onset of the dynamically excited state of the sand. The relationship between α_{cr} and density for a definite sand variety is graphically shown in Figures (10.17) and (10.18).

The same installation is used to determine the value of the coefficient of dynamic compaction for the same conditions, allowing both for the porosity of the sand and seismic parameters (α, f or T). Maslov (1987).

Evidently, seismic loads on a saturated sand mass at an intensity referred to the acceleration $\alpha_p > \alpha_{cr}$ will reduce the stability of the sand.

The shearing resistance of loose sand in static conditions is found from the familiar relation

$$\tau_{st} = \sigma_{st} \tan \varphi \qquad (10.49)$$

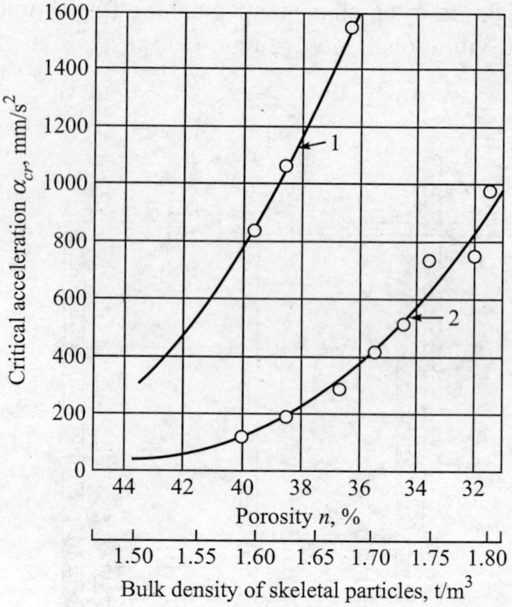

Figure 10.17 Relationship of critical acceleration α_{cr} with porosity (content of grains 20.25 mm in size)

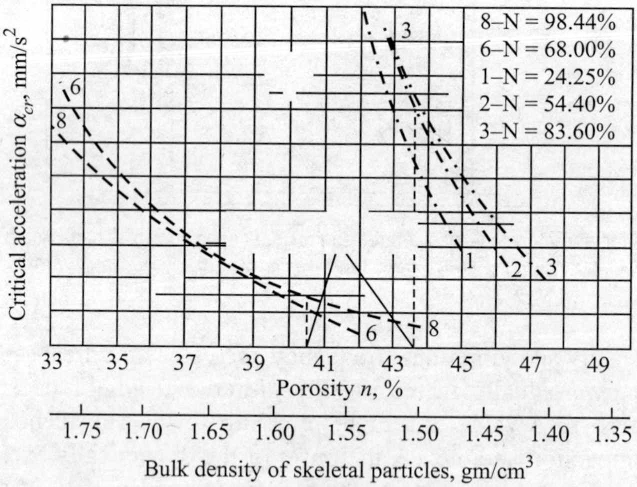

Figure 10.18 Critical acceleration α_{cr} depending upon porosity n

where σ_{st} is the normal stress at a horizon z induced by the weight of the overburden or structure (dead load) and φ is the angle of internal friction of sand.

If the load acting on loose sand is of seismic origin, conditions change. Then, we have

$$\tau_{dyn} = \sigma_{dyn} \tan \varphi \tag{10.50}$$

where τ_{dyn} and σ_{dyn} have the previous value, yet as applied to new dynamic conditions.

The variation of the normal stress under dynamic conditions is accounted for by the buoyancy effect of the dynamic pressure head h_z appearing in the sand mass as it is compacted by vibrations, i.e.,

$$\sigma_{\text{dyn}} = \sigma_{\text{st}} - \rho_w h_z \tag{10.51}$$

where ρ_w is the density of water.

If there is no dead load acting on the soil the value of σ_{st} at the horizon z is determined by the weight of the overburden, i.e.,

$$\sigma_{\text{st}} = \rho_s z \tag{10.52}$$

where ρ_s is the bulk density of sand, determined by taking into account, when necessary, the hydrostatic uplift pressure acting on it (in the submerged state $\rho_w \approx 1$ t/m^3). By virtue of these two relations, we can rewrite Eq. (10.50) as

$$\tau_{\text{dyn}} = (\rho_s z - \rho_w h_z) \tan \varphi \tag{10.53}$$

As follows from this equation, for $z = h_z$ the shearing resistance of sand is $\tau_{\text{dyn}} = 0$. In this case, sand completely loses its stability and transforms to a heavy fluid (liquefaction of sand).

Analysis of equations (10.51) and (10.45) demonstrates that, other conditions being equal, the hazard of liquefaction of sand increases with following:

- porosity, n
- thickness of sand layer, H
- decrease in coefficient of permeability, K_p
- duration of seismic action, t

The 1934 Bihar earthquake of magnitude 8.3, in Indo-Gangetic alluvial belt is an glaring example of this phenomenon and reaffirms the above conclusions.

That is why under seismic conditions the greatest hazard is presented by saturated fine and small-grained sands exhibiting reduced permeability induced by the small value of K_p and occurring as deposits of conspicuous thickness.

When conducting dynamic tests of sands, it is often sufficient to determine only the value of the critical acceleration α_{cr} corresponding to the particular conditions. In doing so the degree of aseismic stability of sand is established by direct correlation of the values of α_{cr} and design acceleration.

If a saturated sand mass is overlain by structures or a fairly thick layer of dry sand, when determining the coefficient of dynamic compaction v_n or critical acceleration α_{cr}, this condition must accordingly be taken into account. If the values of normal stresses are relatively small, the relationship $\alpha_{cr} = f(\sigma_n)$ is of a linear character.

It should be made clear that the stability and deformations of structures resting on any soil mass (dry or, particularly, submerged sand mass) in seismic regions in generally important only if during the construction the dry sand soil is submerged (say, when excavating a pond or reservoir). It is assumed that the sand occurring below the table of surface or groundwater was subjected in the geologic past to dynamic action and after being compacted is already found in an inert state.

It was pointed out in the proceedings of the 3rd World conference on Earthquake Engineering that the Niigata earthquake of 1964 provided classical proof of the seepage theory given by Maslov (1987). This recognition is another piece of evidence suggesting the necessity of taking into account not only the inertia seismic forces but also the foundation soils of structures to be designed in seismic areas.

10.10 ADVANCES IN THE ANALYSIS OF SOIL LIQUEFACTION

The advances in the analysis of soil liquefaction have been based on the following:

- Mathematical analysis
- Case Histories
- Physical modelling of liquefaction using sine specimens in the laboratory
- Physical modelling at reduced scales in Shake Tables and centrifuges
- Empirical modelling of liquefaction
- Numerical modelling of liquefaction
- Generalized modelling–Effective stress approach.
- Material Instability

10.10.1 Effective Stress Method for Liquefaction Analysis

The stress in saturated soils may be expressed as

$$\sigma' = \sigma - u \tag{10.54}$$

where, σ' = effective stress
σ = total stress
u = pore water pressure (fluid pressure)

In the 3 D generalized form the same equation may be expressed as

$$\sigma'_{ij} = \sigma_{ij} - p\delta_{ij} \tag{10.55}$$

where σ'_{ij} = effective stress
σ_{ij} = total stress
p = interstitial fluid pressure

Using the non-linear effective stress model, the constitutive equations are

$$d\sigma'_{ij} = C_{ijkl}\, d\epsilon_{kl} \tag{10.56}$$

where, C_{ijkl} = fourth order tensor that characterizes the tangential material moduli and varies with stress and strain
$d\sigma'_{ij}$ = increment of effective stress
$d\epsilon_{kl}$ = increment of strain

In literature many constitutive models have been proposed to simulate soil liquefaction. Most models are only applicable to represent the cyclic generation of pore pressure and strain. Very few models are capable of describing the large strain post-liquefied stress-strain response of saturated sands.

As described in Zienkiewicz and Shiomi (1984), soil liquefaction analyses can be derived from several different approximations of the governing differential equation for two-phase materials (soil and water) and may be expressed as

$$\sigma'_{ji,j} - \rho\frac{d^2u_i}{dt^2} + \rho b_i + \frac{2p}{2x_i} = 0,\ \mathrm{i} = 1,2,3 \tag{10.57}$$

where, b_i = earth gravity,

ρ = unit mass of saturated soil

t = time, and

x_i(i = 1,2,3) are the spatial coordinates

The fluid pressure p obeys the flow conservation law, and the governing equation (Zienkiewicz et al., 1990) may be expressed as

$$\frac{\partial}{\partial x_i}\left(-K\frac{-\delta p}{-\delta x_i} + K\rho_w b_i\right) + \frac{\delta}{-\delta x_i}\left(\frac{du_i}{dt}\right) - \frac{1}{K_w}\frac{dp}{dt} = 0 \tag{10.58}$$

where, ρ_w = fluid unit mass,

K = hydraulic permeability

K_w = bulk compressibility (two phase material)

and,
$$K_w = \frac{n}{K_f} + \frac{1-n}{K_s}$$

where n = porosity,

K_s = solid grain bulk modulus,

K_f = fluid bulk modulus.

Specific problems and sites may be discretized using a finite element mesh and the material stress-strain responses may be simulated with appropriate models (e.g., Iai et al., 2001). Computer programs like FLIP may be used for detailed analysis (Iai, et al., 2001).

10.10.2 Liquefaction Analysis Based on Material Instability

Two-phase materials consisting of non-linear pervious solids and interstitial fluids like saturated soil can become mechanically instable. Iai and Bardet (2001) discovered new type of instability for two phase material resulting from coupling between non-linear soils and interstitial fluid flow. This study revealed the following:

- Non-homogenous field of solid deformation with vortex-like pattern growing exponentially with time
- Non-uniform flow of interstitial water exhibiting source-sink flow patterns.

It may be pointed out that the two-phase instability is different from strain localization, strain softening, and constitutive singularities and has practical implications on the numerical modelling of soil liquefaction Bardet et al. (2003).

10.11 REMEDIAL MEASURES FOR LIQUEFACTION

When the liquefaction potential has been assessed for a particular site, mitigative remedial measures may be suggested to reduce the liquefaction hazard to an acceptable level depending upon the importance of the site. The remedial measures may described as

- Identification based on evaluation of liquefaction hazard so that its consequences (risk) to life are minimal
- Avoidance of liquefaction prone areas through zoning restrictions
- Rehabilitation and relocation to comparatively safer sites
- Strengthening of the ground to prevent liquefaction.
- Use of Pile or other deep foundations that transfer structural loads to competent layers beneath the liquefaction prone sediments
- Insurance for protection against large financial loss

PROBLEMS

10.1 Discuss the criteria for liquefaction. Describe the modified Chinese criteria for fine grained soil.

10.2 Describe the factors on which liquefaction of sand depends. How will you estimate the settlement after liquefaction?

10.3 Describe the mechanics of liquefaction. What are the residual shear strengths of sand after liquefaction and discuss the methods for its determination?

10.4 A 9 m thick deposit of loose sand (D_r = 42%, finer ≤ 5%) is saturated below a depth of 3 m. The stratum and this region is highly prone to liquefaction. Estimate the ground surface acceleration that would be required to produce sand soils in a M = 7.7 earthquake.

10.5 Loose sand deposits are located at a depth of 3.5 m from the level ground surface. These are located in seismic active area where the expected QPA is 0.35g. The corrected S.P.T. value at a depth of 3.5 m is 9. Calculate the factor of safety against liquefaction, the other given data is

$$\gamma_t = 18.5 \text{ kN/m}^3,\ G = 2.4 \text{ gm/cc and}$$

$$\gamma_b = 9.5 \text{ kN/m}^3, \text{ with weight of water} = 10 \text{ kN/m}^3.$$

10.6 How will you use shear wave velocity in sand deposit to determine the factor of safety against liquefaction? Explain the variation of CRR with shear wave velocity as shown in Figure E10.6.

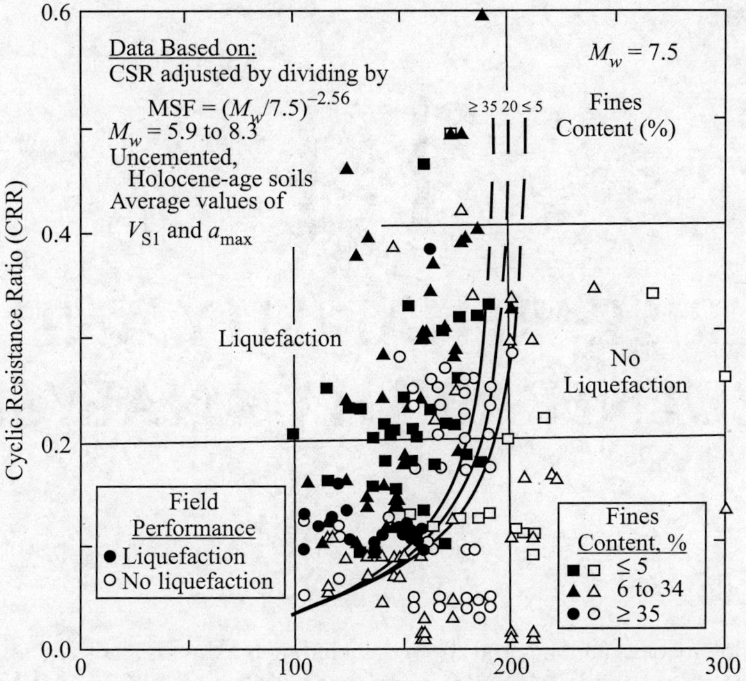

Figure E10.6 Corrected shear wave velocity, V_{s1} (m/s)

11

RISK, RELIABILITY AND VULNERABILITY ANALYSIS

11.1 INTRODUCTION

When data relating to loads and material strength including maximum loads and minimum strengths are not completely predictable, absolute safety cannot be assured. When such a state of uncertainty exists, and the load and strength variables can be described in a probabilistic sense, the safety of a structure can truthfully be specified only with an associated probability. The assurance of structural safety and reliability is an important objective in engineering design. Indeed, the evaluation of safety and reliability of a given or proposed system is one of the purposes of structural analysis and is necessary for the development of proper criteria for design. Engineering designs, however, are normally formulated under conditions of uncertainty. The loading processes and material properties are usually random, or the information and relationship used in the development of design are imperfect and thus are potentially subject to error; in either case, there could be significant uncertainty underlying the formulation of the design. In the face of such uncertainty, absolute assurance of safety and reliability would ordinarily be difficult to achieve; realistically, safety band reliability may be assured only with a tolerable risk or probability of failure.

Due to computational difficulties of computing the reliability, safety factors are used to provide structural safety. A safety factor, however, relates to the failure of a single member under a single load condition and a constant safety factor cannot be used to ensure that structures will have the same overall reliability. The entire structure must be examined to predict its probability of survival and the safety factors of the individual members should be adjusted so that the structure as a whole attains a prescribed reliability.

A common type of reliability prediction assumes that by using a design load which may be encountered with a frequency of only 1/100 and a design strength which may be realized with a frequency of only 1/100, of a probability of failure 1/10,000 can be obtained. On the contrary, the reliability of a structure depends on the frequency distributions used to characterize loads and strengths.

Much of the reported work in reliability analysis has considered the case in which all the loads can be lumped as a single load and all the strengths of the individual members can be jumped into a single resistance for the entire structure. Results have been reported for loads and resistances with normal, lognormal and external distributions for various ratios of mean resistance to mean load and different values of the standard deviations.

Several authors have presented reliability analysis for full structures. Freudenthal, Garrets, and Shinozuka (1966) considered several examples including a tower subjected to wind loading. Hilton (1972) showed how the weight of a structure could be minimized with a maximum probability of failure being the only constraint. However, in computing the reliability, Hilton assumed that all failure modes were statically independent. This assumption leads to values of reliability, which are too conservative. A significant degree of statistical dependence exists between failure modes because they are subjected to the same environmental loading. Tichy and Vorlicek (1964) included dependence between failure modes but only the strength had a frequency distribution and the load was assumed to be a deterministic quantity. Ditlevsen (1966) analyzed structures in which the loading had a frequency distribution with fixed strength.

In other words, reliability specifies the failure-free performance of the foundation and the soil retaining structure for a designed life of the structure. As such unreliability means failure, total or partial, of the structure. Basically reliability is a random phenomenon. Thus the randomness of the failure occurrence is necessary for the reliability of the structure. By randomness, all that we mean is that the failure cannot be predicted accurately.

Let X be the random variables that represent the designed life of the structure. The failure probability $F(t)$ of the structure is defined as the structure that will fail by time t, that is the life of the structure, X, is less than the time t,

$$F(t) = P(X < t) \tag{11.1}$$

As $F(t)$ specifies the failure probability up to a given time t, which changes with time t, one can specify a function for $F(t)$. Such a function is called the failure distribution function. Obviously, these functions have a value of 0 at time $t = 0$ (a structure cannot fail before time, $t = 0$) and a value 1 at time, $t = \infty$ (all structure must fail before infinity). As already explained the reliability of the structure, $R(t)$, is the probability that the system has not failed by time t, then

$$R(t) = 1 - F(t) \tag{11.2}$$

If $F(t)$ is differentiable, its first derivative $dF(t)/dt = f(t)$ is called failure density function. The failure density function represents the instantaneous failure probability at time t and $(t + \Delta t)$ is given by $f(t)$.

Hazard Rate

These definitions as stated above provide the failure probability and reliability, failure density as a function of time at the initial time. All this is being predicted in a sense that the probability that the structure will fail by some time t is $F(t)$. What if we find that by time t the system has not failed (after all $F(t)$ is only a probability?). That is as time passes we may find that the system has not failed by some time t. In that case, at time t, we would like to know the future failure probabilities from that time onwards. In other words, we would like to know the failure probability for a system, given that the system has not failed by time t. This is specified by hazard rate, $Z(t)$, that is the conditional failure density at time t, given that no failure has occurred between 0 and t. Thus by definition,

$$Z(t) = \frac{f(t)}{R(t)} \tag{11.3}$$

11.2 RELIABILITY AND PROBABILITY OF FAILURE

An important objective in geotechnical engineering design is the determination of the resistance that must be developed by a soil structure so that an adequate level of safety can be ensured. Conventionally, this is achieved through the "Factor of Safety (FS)" which is defined as the ratio of the resistance available over that required for failure to occur. In geotechnical practice, however, FS is often determined as the result of a worst-case analysis in which estimated lower values of the parameters influencing soil resistance are assumed to occur simultaneously and at the critical sections of a given structure. For example, in determining the factor of safety of the soil slope, constant and generally minimum values of the two strength parameters c and ϕ are assumed to hold simultaneously and along the critical failure surface, as a result, geotechnical designs tend to be over-conservative, a situation that often entails more waste than is realized and leads to a less satisfactory solution than would be achieved by accepting reasonable risks.

A more recent approach to the determination of the safety of soil structures, one that makes use of tools available in the areas of probability theory and statistical analysis, has provided an alternative to the factor of safety, i.e. "the probability of failure P_f". Based on a more rational analysis, the probability of failure accounts not only for the uncertainties present in soil resistance but also for those in evidence in the applied loads and the analytical modelling of soil and soil–structure behaviour.

A probabilistic description of the safety of a soil structure is based on the recognition that both its loading (S) and resistance (R) are random variables and that failure may occur when the difference between R and S (called the safety margin, SM) receives a negative value, as shown in Fig. 11.1, i.e.

$$\text{Failure} = [SM = R - S < 0]$$

The probability of the occurrence of this event is then equal to the probability of failure P_f. Thus,

$$P_f = P\{\text{Failure}\} = P[SM = R - S < 0] \tag{11.4}$$

If $P(R)$ and $P(S)$ denote the probability density functions of the resistance R and loading S, respectively, the expression for the probability of failure P_f becomes (Harr, 1976).

$$P_f = \int_{-\infty}^{\infty} F_R(S)\, P_s(S)\, ds \tag{11.5}$$

where $F_R(S)$ is the cumulative distribution of the resistance R.

In the case where the density functions of resistance and loading receive simple analytical expressions (i.e. uniform, exponential, normal, lognormal etc.) the probability of failure may be determined by performing the integration indicated in Eq. (11.5). If, however, the expressions for the density functions of R and S are complex (as is often the case in actual geotechnical

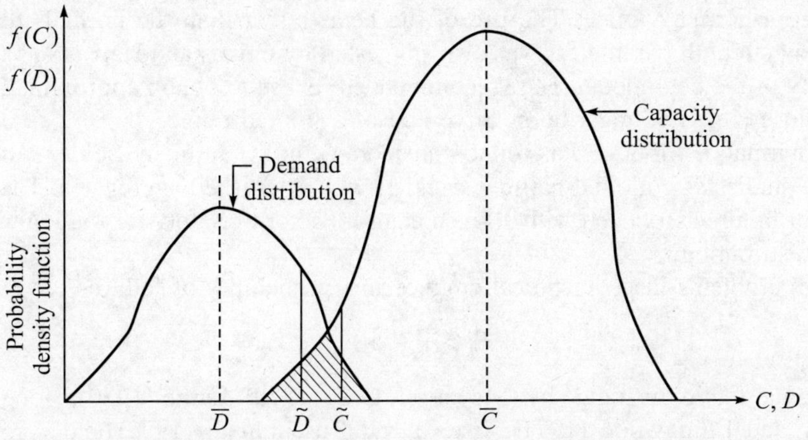

Figure 11.1 Presentation of Loadings (Demand) {*S*} and Resistance (Capacity) {*R*} As Random Variables.

situations) in spite of the simplicity offered by Eq. (11.5), the indicated integration is not easy to a accomplish analytically and numerical solutions to the problem have to be obtained.

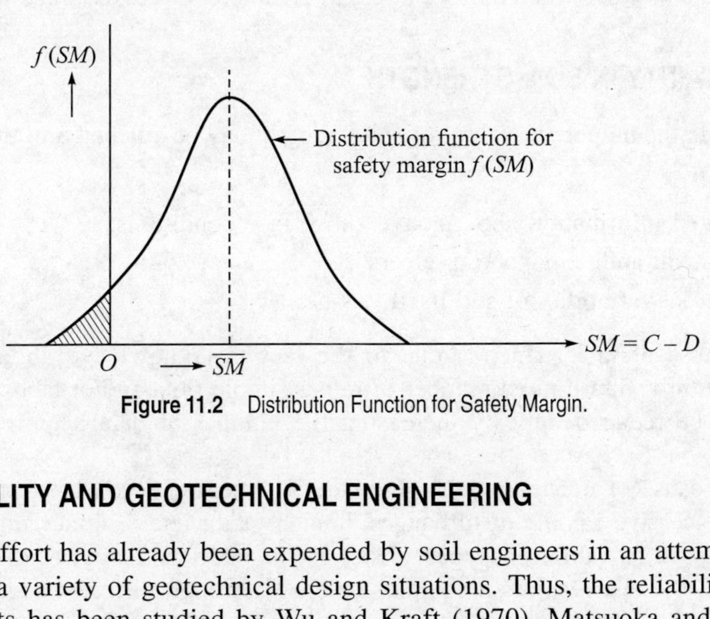

Figure 11.2 Distribution Function for Safety Margin.

11.3 RELIABILITY AND GEOTECHNICAL ENGINEERING

A considerable effort has already been expended by soil engineers in an attempt to obtain risk assessments for a variety of geotechnical design situations. Thus, the reliability of soil slopes and embankments has been studied by Wu and Kraft (1970), Matsuoka and Kuroda (1974), Vanmarcke (1977), the reliability of foundations by Wu and Kraft (1970), Grivas and Harr (1977), the safety of braced excavation by Tang et.al (1999), and the reliability of earth retaining structures by Hoeg and Murarka (1974). Lumb (1966) in his pioneer work found that random variables c and ϕ followed a normal distribution. In additional studies of frequency distributions of soil properties, Schultze (1972) Singh (1972) support Lumb's conclusion that strength parameters are normal variates. In a more recent work, Lumb (1970) found that the c parameter of strength followed more closely a beta distribution and that only its central portion could be approximated a normal variate. The use of the beta (rather than the normal) distribution for modelling soil strength parameters was also suggested by Grivas and Harr (1977). Recognizing the versatility of the beta model, Harr recommended its use to obtain approximations for many data sets whose measures must be positive and of a limited range.

The lognormal distribution has found considerable utility in many fields. Among these are such diverse studies as colloidal particle sizes, as well as crushed aggregates. There is a higher probability of failure associated with a given central Factor of Safety for the lognormal than for the normal distribution.

For most problems in geotechnical engineering, probability of failures,

$$P_f > 1 \times 10^{-3}$$

have been suggested in this range by (Mayerhof, G.G. (1970), Lumb (1970), Hoeg and Murarka (1974). Ang and Ellingwood (1971), Esteva and Rosenblueth (1972) and Ang and Cornell (1974), found similar results in this range for other types of probability distribution functions.

D. A. Grivas and M.E. Harr (1977), analyzed the slope stability with the beta distribution function for soil strength parameter assuming rupture surface as a log spiral curve. They found the probability of failure in the range of 8.8% for the coefficient of correlation between c and ϕ, and in the range of 4.8% when the two strength parameters were assumed to be independent.

11.4 UNCERTAINTY IN SOIL STRENGTH

The uncertainty in the numerical values of soil strength may be attributed to the following three reasons:

1. The limited information about actual subsurface conditions.
2. Measurement and "engineering errors".
3. The inherent variability of soil itself.

The first cause of uncertainty is due to the fact that knowledge about soil profiles is usually obtained through a limited number of rather simple field and/or laboratories tests. This uncertainty can be reduced only by increasing the number of data acquired during the site exploration.

The second cause of uncertainty involves measurement errors, which may be due to test imperfections, excessive sample disturbances, human factor, etc. Engineering errors may also arise if, for example, two different types of soil are treated identically; or if no differentiation is made between the test results on samples obtained through different sampling and testing procedures, etc.

The last cause of uncertainty, i.e. the inherent variability of soil itself is by far the most important one. On the basis of results obtained from a large number of tests on natural soils, it was found that the effects of test imprecision may be overwhelmed. Research studies have been undertaken recently with an objective to quantify this uncertainty and describe the spatial variation of soil strength and strength parameter Lumb (1970), Alonso (1976) and Vanmarcke (1974).

11.4.1 Variation of Strength Parameters of Soil

The soil strength test results in a wide range of values for the two strength parameters c and ϕ. If N different series of tests are performed in the same or different laboratories and on nearly identical samples, in general, N pairs of values of c and ϕ are obtained. To account for this variability, strength parameters have been taken to be random variables and probability models have been proposed for their determination.

11.5 GENERAL PRINCIPLES OF RELIABILITY

The performances of an engineering system obviously, depends on the characteristics of the system as well as the environmental conditions to which it is subjected. A system and the relevant environment are usually characterized or defined by certain variables that may be called the design variables; accordingly, the safety and reliability of a given system would be a function of these design variables. If the design variables are $X_1, X_2, \ldots, X_n$, we may define a performance function as

$$Z = g(X_1, X_2, \ldots, X_n) \tag{11.6}$$

Invariably, one or more of the design variables will contain uncertainty and thus should be considered random variables. It follows, therefore, that Z is also a random variable whose values Z represent specific levels of performance.

If the required minimum level of performance is Z_0, then satisfactory performance would mean $Z \geq Z_0$ whereas failure to perform would be $Z < Z_0$. Accordingly, we obtain the probability of successful performance, which is the measure of safety or reliability, as

$$P_s = P[Z \geq Z_0] \tag{11.7}$$

Conversely, the measure of unreliability is the probability of failure to perform:

$$P_f = P[Z < Z_0] = 1 - P_s \tag{11.8}$$

Equations (11.7) and (11.8) may be evaluated readily if the probability distribution of Z is given; that is

$$P_s = \int_{Z_0}^{\infty} P_z(z)dz = 1 - F_z(Z_0) \tag{11.9}$$

and thus,

$$P_f = F_z(Z_0) \tag{11.10}$$

As expected $f_z(Z)$ or $F_z(Z)$ is related [through Eq. (11.4)] to the distributions of the design variables conceivably, therefore, if the distributions of the basic design variables $X_1, X_2, \ldots, X_n$ are given or assumed, the distribution of z may (theoretically) be derived from those of $X_1, X_2, \ldots, X_n$, i.e. given the density functions $(fx_1, fx_2, \ldots, fx_n)$ we would have (for uncorrelated $X_1, X_2, \ldots, X_n$)

$$F_z(Z_0) = \iint\limits_{[g(x_1 \cdot x_2 \ldots x_n) \leq Z_0]} \ldots \int fx_1\, fx_2 \cdots fx_n\, dx_1\, dx_2 \cdots dx_n \tag{11.11}$$

The evaluation of the multiple integral in Eq. (11.11), however, is a formidable task; except for very specialized and simple cases, direct integration of Eq. (11.11) would be quite impractical to implement. In general, therefore, efforts to derive the exact distribution of z would not be warranted. Often the distributions $fx_1,\ldots, fx_n$ of the basic variables are not well defined, the information for these variables are not well defined, the information for these variables may be limited only to the respective means and variances.

A practical alternative, therefore, is to prescribe the required distribution of z and evaluate the necessary reliability measure on the basis of this distribution and the estimated means and variances of the design variables. The required distribution, however, should be judiciously chosen, taking into account all relevant physical conditions that are pertinent to the problem and that may have bearing on the form of the distribution function. In this regard, it is important to point out that when the failure probability P_r is not too small, say $P_r > 10^{-3}$, the choice of the form of $F_z(Z)$ is not too critical meaning that the calculated P_f will not be very sensitive to the form of the distribution. In such cases, therefore, a mathematically convenient form of $F_z(Z)$ may be prescribed for practical purposes. However, for very small failure probabilities, i.e. $P_f < 10^{-5}$ the correct choice of the form of $F_z(Z)$ becomes more critical the calculated P_f could be vastly different for different forms of $F_z(Z)$, and thus a much more careful and accurate determination of the pertinent distribution would be required. Even in the latter cases, however, there are occasions when a relative measure of reliability is largely all that is necessary for this comparative purpose or a convenient distribution form of Z may also suffice.

Once the form of $F_z(Z)$ is prescribed, the remaining problem is the determination of the mean and variance of Z. By first order approximations, there are

$$\mu_z \cong g(\mu x_1 \cdot \mu x_2 . \cdots \mu x_n) \tag{11.12}$$

and (for uncorrelated $X_1, \ldots, X_n$)

$$\sigma_z \cong \sum_{i=1}^{h} C_i^2 \, \sigma x_1^2 \tag{11.13}$$

in which the constant $C_i = \dfrac{\delta g}{\delta X_i}$

evaluated at $(\mu x_1 \cdot \mu x_2 . \cdots \mu x_n)$

Therefore, for a prescribed distribution $F_z(Z)$, the failure probability can be evaluated as

$$P_f = F_z(Z_0, \mu_z, \sigma_z) \tag{11.14}$$

where μ_z and σ_z are given by Eq. (11.12) and Eq. (11.13), respectively.

It is quite common to consider structural safety from the standpoint of a resistance R (or capacity C) relative to a load S (or demand D) in these terms. If the respective probability density functions are $f_R(r)$ and $f_S(s)$, the probability of failure can be expressed as

$$P_f = P[R \leq S]$$

$$= \int_0^{\infty} f_R(s) f_S(s)\, ds \tag{11.15}$$

or

$$P_f = \int_0^{\infty} [1 - f_S(r)] f_R(r)\, dr \tag{11.16}$$

Equation (11.15) or (11.16) would be useful when numerical evaluation of P_f becomes necessary.

11.6 RELIABILITY AND DISTRIBUTION FUNCTION

It is the experience that provides the allowable factor of safety for a particular structure—a measure that is discovered often with great difficulty and only after many trial and error excursions. The factor of safety that is necessary to prevent failure, would be very much different for each problem situation—ranging from about 1.3 for the slope stability to 3 for bearing capacity. Clearly not having the advantage of experience with a particular soil structure under the action of unknown ambient and induced loading, the conventional design procedure should not be expected to be able to assess its reliability for experiments yet to be performed. The natural variability of the material parameters is largely overlooked in the conventional practice. Basically the conventional approach seeks to minimize the maximum tolerable risk. Estimated lower limits of capacity (strength) and upper limit of demand (loading) are often taken to occur simultaneously and at critical locations. The ability of a structure to perform adequately, i.e. incapability is a random variable. This is illustrated schematically in Fig. 11.1. The actual distribution may depart considerably from the normal. The single calculated value of the capacity that would be used for the conventional approach is denoted on the figure as $\tilde{C}$; $\overline{C}$ indicates the mean value. A simplified distribution of the demand is also shown in Fig. 11.1. $\tilde{D}$ denotes the single anticipated value that would be used in the conventional approach and $\overline{D}$ is the mean value of the distribution.

In general, the demand function will be the resultant of many component factors, such as vehicle loading, wind loading, earthquake acceleration, location of water table, stress history, construction procedure, etc. Needless to say its distribution may also depart widely from normality.

The conventional definition of the factor of safety for the conditions depicted in Fig. 11.1 is

$$\text{FS} = \frac{\tilde{C}}{\tilde{D}} \tag{11.17}$$

The ratio of mean values is called the central factor of safety (CFS). It is defined as

$$\text{CFS} = \frac{\text{expected capacity}}{\text{expected demand}} = \frac{\overline{C}}{\overline{D}} \tag{11.18}$$

In general as indicated on Fig. 11.1, the central factor of safely exceeds the conventional factor of safety or

$$\text{CFS} > \text{FS} \tag{11.19}$$

The difference between the random variables C and D is called the safety margin (SM), that is

$$SM = C - D \tag{11.20}$$

Obviously, the safety margin is a random variable. Failure is associated with that portion of its probability density function wherein it becomes negative; that is SM = $C - D < 0$, shown shaded in Fig. 11.2. As the shaded area is the probability of failure P_f, we have

$$P_f = P[C - D \leq 0] \tag{11.21}$$

The component of the probability of failure is the probability of success, which is more commonly known as reliability R. Hence we have

$$R + P_f = 1 \tag{11.22}$$

11.6.1 Normal Distribution Function

The most common type of function is the normal distribution in which the mean value is the most commonly used value. In general the density of the normal distribution is defined by

$$f(x) = \frac{1}{S\sqrt{2\pi}} \exp\left[-\frac{1}{2}\left(\frac{X - \overline{X}}{SD}\right)^2\right]$$

where $\overline{X}$ is the mean value given by

$$\overline{X} = \sum_{X=1}^{X=n} x$$

and S is the standard deviation given by

$$S = \left[\frac{\sum_{X=1}^{X=n} (X - \overline{X})^2}{n}\right]^{\frac{1}{2}}$$

Similarly the probability distribution function for capacity C, domain D and safety margin SM may be obtained as follows.

If C and D are normal variates, then their probability density functions $f_C(C)$ and $f_D(D)$ are

$$f_C(C) = \frac{1}{S_C\sqrt{2\pi}} \cdot \exp\left[\frac{-1}{2}\left\{\frac{(C - \overline{C})}{S_C}\right\}^2\right]$$

$$f_D(D) = \frac{1}{S_D\sqrt{2\pi}} \cdot \exp\left[\frac{-1}{2}\left\{\frac{(D - \overline{D})}{S_D}\right\}^2\right] \tag{11.23}$$

where S designates, the standard deviation as $\overline{C}$, $\overline{D}$ are their mean values. Recalling that the safety margin is $SM = C - D$ and accepting (Haugen, 1968) the mathematical fact that the difference between normal variates is a normal variate, the safety margin will have the probability density function for this case of

$$f_{SM}(SM) = \frac{1}{S_{SM}\sqrt{2\pi}} \cdot \exp\left[\frac{-1}{2}\left\{\frac{(SM - \overline{SM})}{S_{SM}}\right\}^2\right] \tag{11.24}$$

where $\overline{SM} = \overline{C} - \overline{D}$ and $S_{SM} = \sqrt{S_D^2 + S_C^2}$

As failure occurs when $SM \leq 0$, the probability of failure becomes

$$P_f = \frac{1}{2} - \psi\left[\frac{C - D}{\sqrt{S_D^2 + S_C^2}}\right] \tag{11.25}$$

where $\psi(z)$ is the cumulative-probability function of the standard normal variate as given in Table 11.1.

Table 11.1 Normally distributed comulative probability function

z	0	2	4	6	8
0.0	0.000000	0.007978	0.015953	0.023922	0.031881
0.2	0.079260	0.087064	0.094835	0.102568	0.110251
0.4	0.155422	0.162757	0.170031	0.177242	0.184386
0.6	0.225747	0.232371	0.234914	0.245373	0.251748
0.8	0.288145	0.293892	0.299546	0.305105	0.310570
1.0	0.341345	0.346136	0.350830	0.355428	0.359929
1.2	0.384930	0.388768	0.392512	0.396165	0.399727
1.4	0.419243	0.422196	0.425006	0.427855	0.430563
1.6	0.445201	0.447384	0.449497	0.451543	0.453521
1.8	0.464070	0.465620	0.467116	0.468557	0.469946
2.0	0.477250	0.478308	0.479325	0.480301	0.481237
2.2	0.486097	0.486791	0.487455	0.488089	0.488696
2.4	0.491802	0.492240	0.492657	0.493053	0.493431
2.6	0.495339	0.495604	0.495855	0.496093	0.496319
2.8	0.497445	0.497599	0.497744	0.497882	0.498012
3.0	0.498650	0.498736	0.498817	0.498893	0.498965
3.2	0.499313	0.499359	0.499402	0.499443	0.499481
3.4	0.499663	0.499687	0.499709	0.499730	0.499749
3.6	0.499841	0.499853	0.499864	0.499874	0.499883
3.8	0.499928	0.499933	0.499938	0.499943	0.499948

$$\psi(z) = \frac{1}{\sqrt{2\pi}} \int_0^z \exp\left|-\frac{U^2}{2}\right| dU$$

for $z > 2.2\ \psi(z) \approx \frac{1}{2} - \frac{1}{2}(2\pi)^{1/2} \exp\left|-\frac{z^2}{2}\right|$

Equation for P_f can also be written as given by Ang and Cornell (1974) and Garrets and Shinozuka (1966).

$$P_f = \frac{1}{2} - \psi\left[\frac{CFS - 1}{\sqrt{(CFS)^2\,(V_D^2 + V_C^2)}}\right] \tag{11.26}$$

where V_C and V_D are coefficients of variation of Capacity and Demand respectively.

Figure 11.3 represents the overlapped region of Fig. 11.1. The probability associated with a level of demand, say D_1, is

$$P\left[D_1 - \frac{dD}{2} < D < D_1 + \frac{dD}{2}\right] = f_D(D_1)\,dD$$

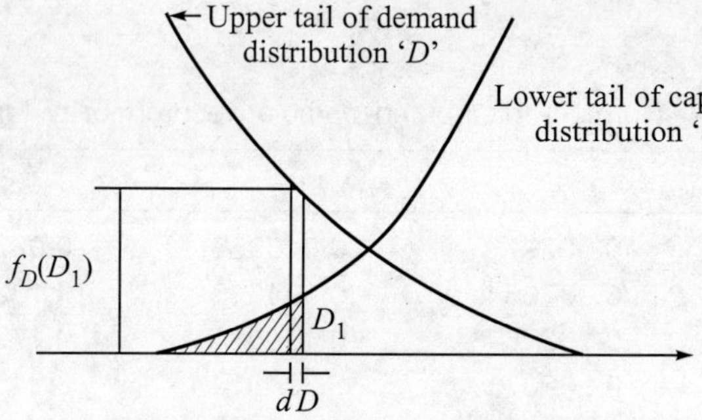

Figure 11.3 Overlapped region of capacity/demand distribution curve as in Figure 11.1.

where $f_D(D)$ is the probability density function of the demand. The probability that the capacity is less than D, $C < D_1$, is represented by the shaded area shown on the figure and is given by

$$P[C < D_1] = \int_{-\infty}^{D_1} f_C(C)\,dC$$

The probability of failure at the demand level D_1 is the product of these two probabilities, i.e.

$$dP_f = f_D(D)\,dD \int_{-\infty}^{D_1} f_C(C)\,dC \tag{11.27}$$

To account for all possible values of demand, we integrate this expression to obtain

$$P_f = \int_{-\infty}^{\infty}\left[\int_{-\infty}^{D_1} f_C(C)\,dC\right] f_D(D)\,dD \tag{11.28}$$

or the equivalent form (Freudenthal, Garrelts and Shinozuka)

$$P_f = \int_{-\infty}^{\infty} F_C(D)\,f_D(D)\,dD \tag{11.29}$$

where $F_C(D)$ is the cumulative-distribution function of the capacity C.

11.6.2 Lognormal Distribution

Some special problems in geotectonical earthquake engineering involving ground motion are formulated in terms of logarithm of a parameter rather than the parameter itself. For example, x is a random variable then $y = \ln x$ (i.e. logy) is also a random variable. If y is normally distributed, then x is log normally distributed. In other words, a random variable is log normally distributed, if its logarithm is normally distributed.

Thus if $y = \ln x$ is a normal variate, the random variable x is said to have a lognormal distribution. Given that a variable $y = \ln x$ is normally distributed, its probability density function will be

$$f_x(x) = \frac{1}{S_x\sqrt{2\pi}} \cdot \exp\left[\frac{-1}{2}\left\{\frac{(x-\mu_x)}{S_x}\right\}^2\right] \quad -\infty < x < \infty \tag{11.30}$$

Substituting $x = \ln x$, we obtain the lognormal distribution function for the random variable y.

$$f_y(y) = \frac{1}{y\,S_x\sqrt{2\pi}} \cdot \exp\left[\frac{-1}{2}\left\{\frac{(\ln y-\mu_x)}{S_x}\right\}^2\right] \quad y \geq 0 \tag{11.31}$$

$$f_y(y) = \frac{1}{y\,S_{\ln y}\sqrt{2x}} \cdot \exp\left[\frac{-1}{2}\left\{\frac{(\ln y-\mu_{\ln y})}{S_{\ln y}}\right\}^2\right] \tag{11.32}$$

If the capacity and demand distribution are lognormal variates with means and coefficients of variation of $\overline{C}$, V_C and V_D, $\overline{D}$, respectively, the probability of failure is given by

$$P_f = \frac{1}{2} - \psi\left[\frac{\ln\left\{\left(\frac{\overline{C}}{\overline{D}}\right)\sqrt{\left\{\frac{(1+V_C^2)}{(1+V_D^2)}\right\}}\right\}}{\sqrt{\ln(1+V_D^2)\,(1+V_C^2)}}\right] \tag{11.33}$$

The shape of lognormal distribution is shown in Fig. 11.4 and the probability distribution function is shown in Fig. 11.5.

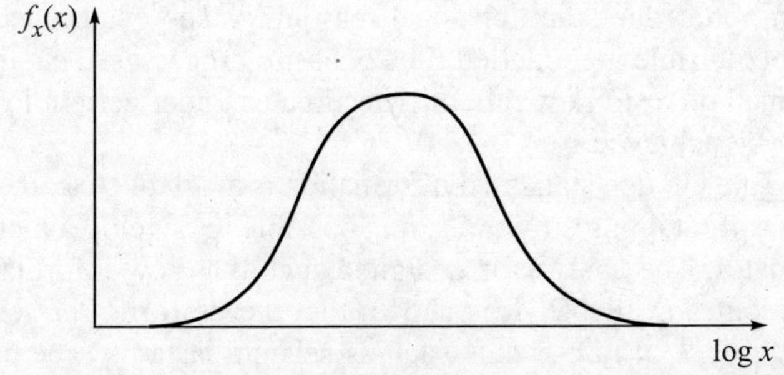

Figure 11.4 Lognormally distribution of random variable.

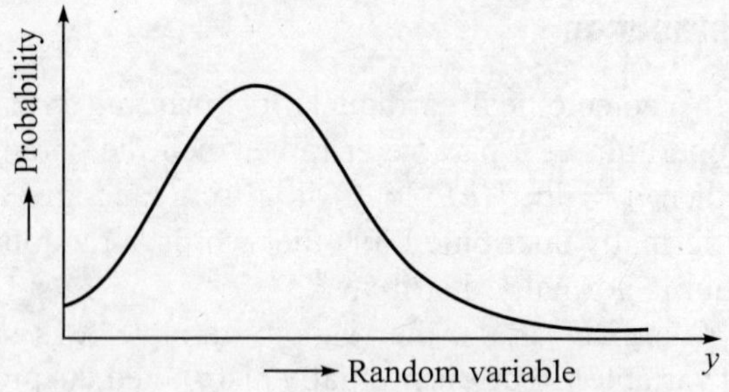

Figure 11.5 Probability distribution function.

11.6.3 Beta Distribution Function

When the random variable x is bounded, say between limits a and b, the beta distribution has been found to be of much use. The density function for this distribution may be expressed as

$$f(x) = \frac{1}{B(\alpha, \beta)} (x - a)^{\alpha} (b - x)^{\beta} (b - a)^{-1-\alpha-\beta} \tag{11.34}$$

where

a = minimum value of x
b = maximum value of x
α, β = parameters of distribution
$B(\alpha, \beta)$ = beta function

The probability of failure can be obtained by integrating Eqs. (11.28) and (11.29). But the integration is very complex and can be solved only with the help of numerical methods. Approximate solution can also be obtained graphically.

11.7 RISK AND RELIABILITY

The performance of a risk-related design requires the consideration of the loading conditions, which the structure is expected to experience during its design life, as well as its structural resistance as random variables. Integral part of the analysis is the selection of appropriate types of probability distributions describing these variables. This is mainly done on the basis of statistical curve fitting procedures and physical reasoning. The occurrence of extreme loading conditions with respect to time is modelled by stochastic processes. The analysis shown in this section is to show a method as to how reliability estimates under certain loading conditions and its combinations can be achieved.

Although at times not enough statistical information is available to carry a rigorous structural reliability analysis of soil retaining structures, it is possible to develop a method that can provide good reliability estimates. The analysis is designed such that new information can be incorporated readily as it becomes available. It is shown that the uniform Poisson process can model satisfactorily the occurrence of rare events such as seismic hazards. The parameter estimate of the process (mean rate occurrence) is obtained from statistical data. The occurrence of extreme conditions due to internal load is taken from failure free analysis developed elsewhere. The basic equation of the reliability analysis may be expressed as

$$L_T = \frac{\sum_{k=0}^{k=\infty} (vt)^k}{k!e^{-vt}(1-p_f)^k} \tag{11.35}$$

where L_T is the reliability, v the mean rate of occurrence of the load, k the number of load occurrences and p_f the probability of failure under a single load application. p_f is calculated in terms of upper and lower bounds using theory of plasticity including the stochastic properties of the strength variations. The result of the analysis indicates clearly the risk involved in choosing certain values of design parameters such as loads and resistance. Only within a probabilistic framework, rational decisions concerning these choices can be made.

11.7.1 Risk Analysis

Acceptable risk is defined in terms of its five basic parameters:

(1) The hazard or problem
(2) The probability of occurrence
(3) The consequence
(4) The possible alternative actions
(5) The value system of the community or the society.

Techniques for consistent deterministic and probabilistic setting limits and design standards may be considered. The influence of level of consequence and general methodology should be evolved. The concept of acceptable risk is put in a quantitative format that can be used by engineers and planners. Bayesian statistical methods are used to develop the methodologies. Recent developments in the field of acceptable risk in reactor safety closely parallel other studies in the overall natural and man-made hazard area. The key words are acceptable risk. What is risk, acceptable or unacceptable, and how can it be consistently quantified so that decision makers can label the risk as to being acceptable or unacceptable?

Generally the decision process is to include the risk assessment or definition of the problem, a scale of values on a value space, and four alternative actions that can be taken by the decision maker. The decision maker can label the risk as acceptable or unacceptable, but he can also reject the assessment or take no action. The latter two decision alternatives are all too often found as a consequence of the result of the risk assessment being either off the scale of values or more often not even being included in the decision maker's value space.

Fortunately, the development of a rational methodology for dealing with such problems is being attacked on many fronts.

Thus, while comparative loss rates are a valuable tool, they do not by themselves define a region of acceptability and a region of rejection. Engineers are lured by numbers but few decision makers are engineers.

11.7.2 The Role of Acceptable Risk

Traditionally, risk assessment and the judgment of acceptability have not played a recognized formal role in professional engineering practice. As long as problems were more or less routine, conventional standards of practice appeared adequate and acceptability was achieved in terms of experience, successes and failures.

The importances of the structure and environmental hazards, etc. have forced a new look at criteria, codes, standards for design, and construction practice. Acceptable risk is built into all criteria. Note, however, that the consequence is not specifically identified in any design standard or criteria. We do not find a numerical mean annual rate of fatalities per year although it exists.

In addition, if we accept that all real loads, capacities, strengths, rigidities, etc. are fundamentally probabilistic in nature but that designs, drawings, construction, etc. must be deterministic, the meeting ground for establishment of criteria is that of acceptable risk.

11.7.3 Risk

There is a growing trend to define risk as mean annual loss. Risk then includes both the occurrence of damage and the consequence of that damage. Risk is then not a probability measure. It is a mean or average annual loss measured over a sufficiently long time span to allow a reasonably reliable estimate to be made. Risk can be defined as mean annual deaths, injuries, economic loss, impact loss, or in a variety of useful inferential forms such as loss of function in which the consequence is included but not specifically identified.

One of the problems associated with risk assessment is the use of very small numbers. Human decision makers find it very difficult to differentiate between 10^{-4} and 10^{-5}.

11.7.4 Decision Rules

Some study weighs consequences by probabilities to obtain extremely small estimates of loss. The object of the decision rule is to prescribe a particular action as a function of the characteristics of the problem.

The decision rule is the key to assessing acceptability and three rules dominate:

- Reasonable rule
- Minimax rule
- Expected value rule.

The reasonable rule says to take the reasonable action, or it states that any reasonable decision maker under the same circumstances and with the same background would take the same action. The minimax rule says that the worst possible future is certain to be found so that we should take the action that minimizes the maximum possible loss. Finally, the expected value decision rule says to take the action with the largest expected value, the sum of the products of values by probabilities being the expected value.

Studies have been made to evaluate the expected value rule in assessing losses and risks employ the expected value rule. Human experience has shown that the rare catastrophe can really occur, so that the weighing of consequences by probabilities found with expected value concepts is questioned when the consequences are catastrophic.

Having recognized that decision rules vary and include value considerations, it is useful to discuss value briefly. Three different value spaces exist: economic, human, and impact. Thus far, despite many efforts, utility functions combining the three or even two value sources into one common value measure have not been particularly successful. The most common approach has been to assign economic values to human losses, death and injury, and then to consider all values as economic. It is interesting, however, to note that impact values dominate both economic and human values in the long run.

As an example of the problem involved in comparison of values, assume that the risk of acute fatalities is 2.5×10^{-4} per year, the previous estimated expected value. The proponent of one more reactor with a useful life of 30 years might estimate the economic gain to be $\$10^6$ per year or some other value. The decision maker is then faced with balancing 2.5×10^{-4} fatalities per year against $\$10^6$ per year. Note that if the decision maker is minimax oriented, the tradeoff is between 2300 humans fatally endangered and $\$10^6$. If human life has infinite value, the problem is obvious.

One answer is not to compare human and economic values but to compare human losses with human gains, economic losses with economic gains, and impact losses with impact gains. For example, if additional power will save 1 to 100 lives a year, a comparison is possible. Note, however, that opponents and proponents can, in full honesty, supply totally different sets of data. The argument is then not about facts but about decision rules and values.

11.7.5 Risk Assessment

Risk assessment includes a definition of the hazard, the likelihood, the nature, and the consequence of damage, the decision alternatives to mitigate the hazard, the effectiveness of such measures, and the presentation of the study in appropriate form for consideration by the decision maker. Note that all aspects of the assessment can be considered under the general classification of criteria.

Engineering studies have too often failed to realize that this is a chain situation and each step or level is of near equal importance. For example, if core melt is the hazard of concern and core melt can be initiated by an accident or by a damaging earthquake, both must be given a complete treatment regardless of probabilities of occurrence. A minimax decision maker may make the decision as to acceptable risk, and probabilities may not be considered important if consequences can be extreme.

The key to the problem of defining acceptable risk appears to lie in the loop: communication – assessment – decision maker – decision maker's scale of values. This is the loop that receives little attention. The other three elements, the hazard, damage, and mitigation involve good solid science and engineering while the communication loop involves the human.

The very practical problem of how the decision maker can use the risk, fatalities per reactor per year, for example, is found here. If such an estimate cannot be accepted or used in decision making for reasons that have been discussed briefly, it must be obvious that alternative techniques of conveying the results of the study must be found. The problem is not one of optimizing an engineering technique, but one of recognizing that the decision maker is the ultimate user of the study and the results must fit into the world in which he functions.

11.7.6 Common Consequence Analysis

When several hazards involve common consequences, the risk assessment can be greatly simplified. Note that this consequence can be quite general. The requirement is that all elements of the analysis have some reasonable common denominator. It is then possible to consider all hazards together in a common context, evaluate design in that same context, and relate design to the hazards, thereby identifying weak links, mitigation measures and effectiveness, and alternative decision actions. There is a general assumption that the basic problem is one of choices of alternative comparable actions.

The basic concept is to consider each hazard a loading or demand dimension and to probabilistically model all possible loading combinations in a time context such that the probability of every loading combination is defined at any instant of time for the life of the facility of concern. The capacity space of the facility is then projected on the loading space. The comparison of loading space and capacity space affords a means of forecasting the future for the trial design, its weak points or design consistency, possible mitigation measures and effectiveness, and action alternatives. Furthermore, the function and optimization of design criteria can be studied.

It is possible to consider probabilistic uncertainties in capacity calculations from the start, but it is strongly recommended that these judgments be added using Bayesian techniques after the deterministic studies are completed. The object is to avoid a multitude of unsolved problems in probabilistic mechanics.

As a simple example, assume that the cooling system of a nuclear power plant can be damaged by an earthquake or a flood. The problem of concern is to first assess the properties of the trial design and then accept or reject it based on acceptable risk. The consequence is loss of function, and if both earthquake and flood occurred together, the effect is cumulative, making loss of function more likely than if either acted separately. It is not reasonable to combine the maximum credible earthquake and the PMF, probable maximum flood, as the design criteria, but it is possible for a lesser earthquake to react with a smaller river flow.

There have been a number of studies of probabilistic capacities and loading combinations with the goal of estimating reliability bounds for various types of systems, including Shinozuka and Shao [1973], Ang and Cornell [1974], Borges and Castanheta [1972], and Tichy [1974]. The technique that follows is based on the use of simple probabilistic models that are selected to fit the demands on the system at any instant of time, rather than extreme values and their combinations. The rigorous probabilistic modeling of step-by-step problems in time is most complex but a simple approximate method has been developed.

11.8 VULNERABILITY ANALYSIS

The consequences of seismic shaking are loss of life and property or facility. The relationship between damage or loss to structure or facility and seismic shaking are addressed by vulnerability functions. The probability of occurrence of future earthquakes falls in the domain of seismic hazard analysis as described in Chapter 10. The damage or loss conditioned on the effects of the earthquakes at the site of all assets are assumed by seismic vulnerability analysis.

The damage of a structure is essentially degradation of the structure to earthquake effects. This degradation may be to the structural system or to the non-structural components. The damage factor (DF) is defined as repair cost as a fraction of the replacement cost. The degree of damage may be ascertained by qualitative description of the facility's physical appearance or functionality, or quantitatively in terms of repair cost. However loss is a measure of the severity of an undesirable outcome in terms of repair cost, insurance clean amount, number of fatalities etc.

Earthquakes are essentially high consequence-low probability (HCLP) events. Thus probabilistic riser analysis (PRA) may establish the relationship between loss and probability of exceeding that loss in a particular future time period. In a site specific PRA, the risk analysis seeks to estimate the relationship between frequency of earthquakes and severity of earthquake losses. The general care of a seismic PRA expresses loss as an uncertain function of intensity (Chen, 2002). Such a probabilistic seismic vulnerability function is shown in Figure 11.6 where the vertical axis measures loss as a function of seismic intensity.

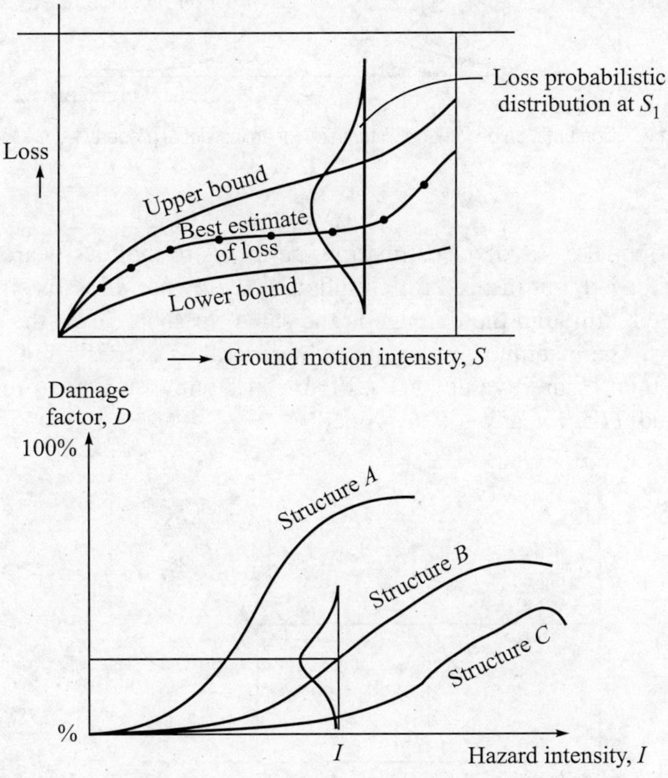

Figure 11.6 Probability Vulnerability Functions.

11.9 DAMAGE AND LOSS ESTIMATION

The damage factor represents the relative cost of repair that the structure is likely to need, if it is hit by a hazard of a given intensity. The hazard yields a probability density function (pdf) of the hazard intensity as shown on the horizontal axis of Figure 11.7. For a given value of the hazard intensity I, a mean damage factor, $\mu_D(I)$, is obtained from the vulnerability function, with its corresponding standard deviation, $\sigma_D(I)$, derived from the associated COV (coefficient of variation i.e. standard deviation) curve. Combining the vulnerability assessment over all values of hazard intensity results in a probability distribution for the damage factor described by the PDF on the vertical axis in Figure 11.7. From this distribution the mean value, μ_D, and standard deviation, σ_D, for the damage to a structure can be determined [Chen, 2002].

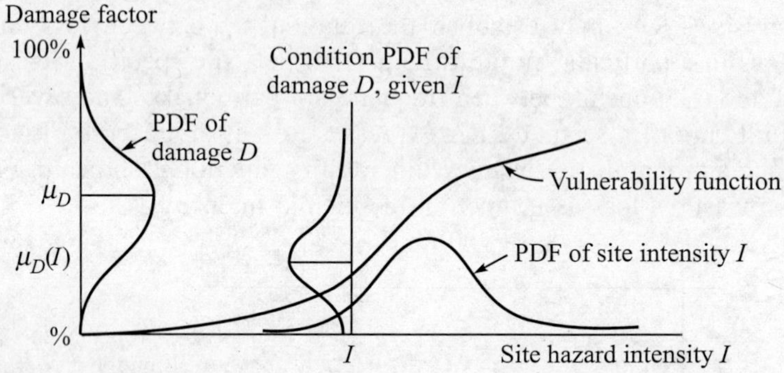

Figure 11.7 Computation of damage distribution for an individual property (After Chen, 2002).

Loss

Damage during earthquakes involves estimating the losses to the loss-bearer. For example, if a house is heavily damaged, but insured for earthquake, the owner's loss is significantly less than if it were not insured, although the damage is the same. In such cases, the insurance company may also run in loss. The insurance information, in general, is expressed in terms of deductibles (d), limits (I), and total insured value (TIV). Figure 11.8 shows a typical relation between loss and damage d, I and TIV, for any given property.

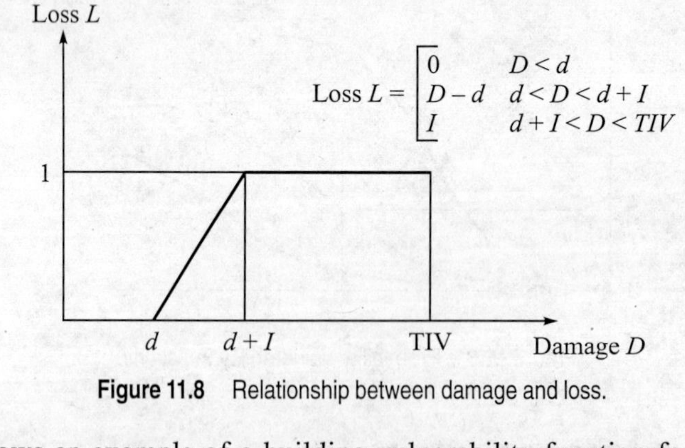

Figure 11.8 Relationship between damage and loss.

Figure 11.9 shows an example of a building vulnerability function for a particular type of reinforced concrete structure. Given the hazard intensity for a site, the vulnerability function provides an estimate of the mean damage factor for the structure [Chen, 2002].

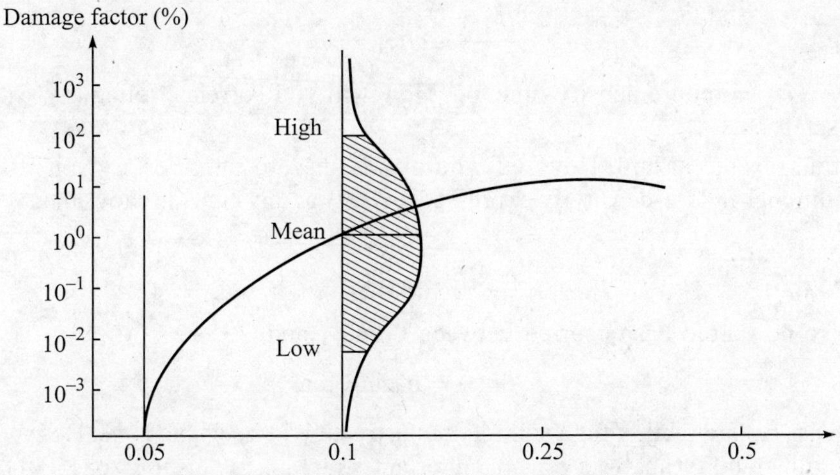

Figure 11.9 Typical vulnerability function and variation of damage factor.

The estimate of damage or loss may be expressed as

$$E(D) = \int E(D|H)\, p(H) \tag{11.36}$$

$$E(L) = \int E(L|D) E(D|H)\, p(H) \tag{11.37}$$

where

$E(\cdot)$ = expected value
L = loss to the assets of interest
D = damage
$E(L|D)$ = the expected loss, given the damage
$p(\cdot)$ = the probability density function (PDF)
$p(H)$ = the probability of the hazard: that is, the combined effect of the probability of the initiating earthquake occurring on the seismogenic fault, and the attenuation of the hazard in terms of shaking from the fault to the site of interest
$E(D|H)$ = the damage, given the hazard; that is, the seismic vulnerability

Each element of Eq. (11.36) or (11.37) is uncertain to a varying degree and can be treated as random variable.

PROBLEMS

11.1 Discuss the following relationship and explain with suitable examples for a foundation structure

$$R + P_f = 1$$

where $R(t)$ = reliability of the structure

and
$$\frac{dF(t)}{at} = f(t)$$

where $f(t)$ = failure density function. How will you correlate reliability with the factor of safety?

11.2 Define seismic hazard. How will you express hazard rate. For a given structure if the conditional failure density is expressed by $f(t)$ at any time t, show that

$$z(t) = \frac{f(t)}{R(t)}$$

where no failure has occurred between 0 and t and

$$z(t) = \text{hazard rate}$$

11.3 Discuss risk, reliability for a soil retaining structure in a seismic zone. How will you carry out vulnerability analysis for an important structure in seismic zone IV. Describe the methods for damage and loss estimation after an earthquake of $M \geq 8.0$.
Show that

$$\text{Hazard} \times \text{Vulnerability} - \text{Capacity} = \text{Risk}$$

Appendix

VIBRATION MEASUREMENTS

A.1 INTRODUCTION

In the study of soil dynamics and earthquake engineering the measurement of dynamic soil properties and the measurement of the effect of dynamic forces require a high standard of vibration measurement devices. The vibration phenomenon may be described by displacement, velocity or acceleration of the vibrating system and accordingly the instrument, which reproduces signals proportional to these, must be designed with precision.

In Chapters 3 and 4, analytical tools have been developed to study the free and forced vibrations of discrete and continuous systems. In developing these methods, several assumptions are made essentially to simplify the physical system to a mathematical model, and therefore the accuracy with which the mathematical model can predict the behaviour of the physical system would depend a lot on these assumptions.

The system properties can be identified experimentally to build a mathematical model in order to study its dynamic behaviour analytically or to analyze its vibration signatures (vibration records in time domain) experimentally. These tests can be conducted in the laboratory so that its supporting structure model has insignificant influence on the behaviour of the component itself. If such is not the case, the tests have to be conducted in the field itself. In this chapter, the types of instrumentation that are used for conducting vibration tests and other tests are briefly described.

The major developments in systematic form of vibration testing have taken place in the last century; the most modern vibration testing techniques got developed during the Second World War. Electronic equipment have used widely for the generation of vibration signals and for the provision of essential control functions. Fundamental features of reliability, stability and reproducibility are obtained. Further, more voltage analogies of frequency and vibration variables are being used to provide hard copies of test results.

The earthquakes seismic recording systems monitor the ground movements continuously, and major developments in this field have taken place since 1930. Institutions around the globe have been active in making such recordings. Ground motion studies have recorded acceleration in analog form on smoked paper, photographic paper, common ink recordings on standard paper and more recently in digital format (on computers, tapes and CD ROMs). As far as measurements relevant to earthquakes are concerned, seismic intensity is a measure of the effect of an earthquake hazard at a specific location. While their terms can be well meaning for engineering measures (quantitatively), the terms such as peak ground acceleration (PGA) are usually meant for qualitative measure of location-specific earthquake effects, based on observed human behaviour and structural damage.

The study of vibration measurements is presented in this Appendix to address the following:

- Understanding the basic principles and procedures of vibration measurement.
- How are vibration measurements useful for soil dynamics?
- Measurement procedures for random vibrations.
- Vibration measurements relevant to earthquake engineering.

A.2 GENERAL CONSIDERATIONS FOR MEASUREMENTS

As presented in Chapter 3, there are three quantities, which are of interest in vibration studies—the vibratory displacement, velocity and acceleration (peak or rms values). These three quantities are related by a simple relation. If the phase relationships between these three parameters are neglected, as is always the case when making time-average measurements, then the velocity at a given frequency can be obtained by dividing the acceleration by a factor proportional to the frequency, and the displacement can be obtained by dividing the acceleration level by a factor proportional to the square of the frequency. This operation is performed in electronic measuring instruments by an integration process. The relationship between the three parameters as a function of frequency is shown in Figure A.1 in which the axes are logarithmic.

The earlier vibration pickups producing an electrical output were rather bulky, velocity sensitive devices. During the last few years there has been a marked move towards the use of

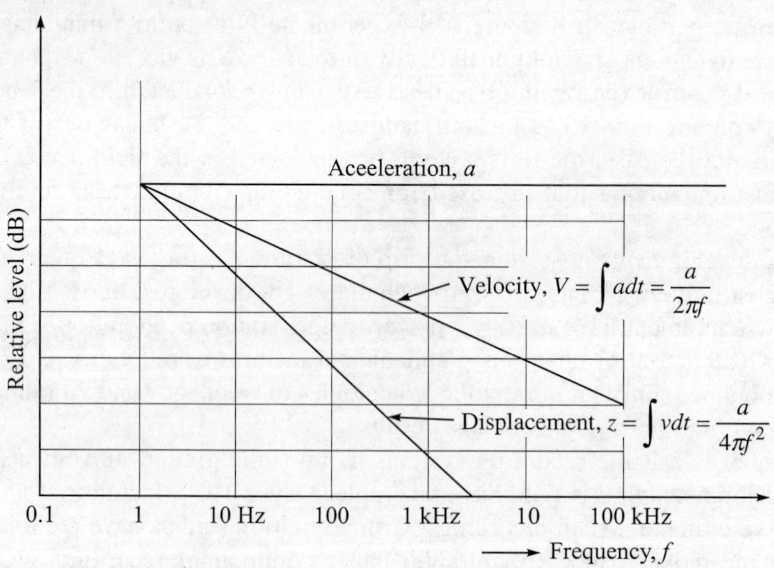

Figure A.1 The integration and double integration of acceleration to obtain velocity and displacement respectively.

acceleration-sensitive transducers, called *accelerometers*. The accelerometers are generally much smaller physically than the velocity pick-ups and their frequency and dynamic ranges are significantly wider even after integration to velocity. A wider dynamic and frequency range is a prime requirement of a modern vibration pickup, particularly to cater for the growing interest in high frequency vibration as a carrier of information on the running condition of machinery and the corresponding wide range of vibration levels that need to be detected. An additional factor which underlines the benefits of accelerometers, is the fact that an acceleration signal can be easily and validly integrated electronically to obtain velocity and displacement whereas electronic differentiation used with velocity and displacement transducers is a more complex and dubious affair.

In theory it is irrelevant as to which of the three parameter—acceleration, velocity or displacement—is chosen to measure vibration. If one plotted a narrowband frequency analysis of a vibration signal in terms of the three parameters, they would all show the same frequency components but would have different average slopes as seen in Figure A.1. It can be seen that displacement measurements give most weight to low frequency components and conversely the acceleration measurements give most weight to the high frequency components. This leads to a practical consideration that can influence the choice of a parameter. It is advantageous to select that parameter which gives the flattest frequency spectrum in order to best utilize the dynamic range of the measuring instrumentation.

The nature of mechanical systems is such that appreciable displacements only occur at low frequencies; therefore the displacement measurements are of limited value. The rms velocity measurements are widely used for the measurements of severity of vibrations. This is due to the fact that the vibratory velocity is simply related to vibratory energy and is therefore a measure of the destructive effect of vibrations. A given velocity level also signifies constant stress for geometrically similar constructions vibrating in the same mode.

Because acceleration measurements are weighted towards high frequency vibration components, this parameter is preferred where the frequency range of interest covers high frequencies.

Vibration pickups of small size and weight are desirable because the pickup should not load the structural member on which it is mounted; also it is often required to measure vibrations at a point on a structure rather than over an area.

The frequency range of interest in vibration measurements has been increasing steadily over the past two or three decades. Today many vibration measurements are carried out up to 10 kHz, and often higher. The increased interest in higher frequencies has been prompted by the development of high speed machinery and the recognition that high frequency vibrations carry valuable information. These measurements are of special significance in the design of machine foundations.

The vibration associated with seismic motion is essentially random in nature and must often be measured either alone or together with the periodic vibration components. This again calls for more complicated measurement techniques than those available in the earlier days.

To be able to predict the effects of vibration on mechanical structures and man, it is not normally sufficient to measure the overall vibration level over the frequency range of interest. Frequency analysis is necessary to reveal the individual frequency components making up the wideband signal. For this purpose, a filter is contained in or attached to the vibration-measuring instrument, thus making it a frequency analyzer. The filter allows only the frequency components to be measured, which are contained in a specific frequency band. The pass band of the filter is moved sequentially over the whole frequency range of interest so that a separate vibration level reading can be obtained at each frequency.

The filter section can consist of a number of individual contiguous fixed frequency filters, which are scanned sequentially or a tunable filter which can be tuned continuously over the frequency range. A third alternative, which is becoming rapidly widespread due to advances in digital signal handling technology, is the use of real-time analyzers, which present a continuously updated complete frequency spectrum on a display screen. Another fundamental difference between the various filter and analyzer types is in the filter bandwidth, narrow or wide, and whether it is a fixed percentage of the tuned frequency or is a constant number of Hz absolute bandwidth independent of tuned frequency. It is very often necessary in practical vibration work to use analyzers with an exceedingly narrow bandwidth in order to separate the closely-spaced sinusoidal components.

Which type of instrumentation to use in a particular measurement situation must be decided by the ultimate use of the data obtained and the measuring equipment available. This appendix will demonstrate the facilities provided by a wide range of instruments and discuss their application to practical problems.

Figure A.2 illustrates the fundamental approach for a vibration measurement scheme. In this scheme the physical motion (or dynamic input) of the vibrating foundation is converted into an electrical signal by the vibration transducer of the pick-up.

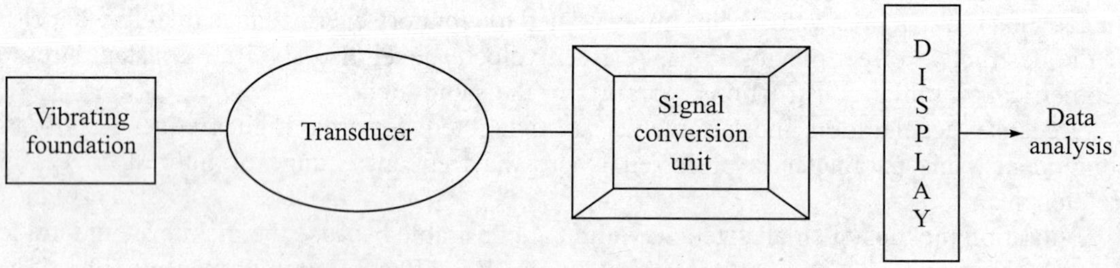

Figure A.2 Basic vibration measurement scheme.

A.3 PRINCIPLE OF VIBRATION MEASUREMENT

The basic elements of vibration measuring instruments are again mass, spring and a dashpot, as shown in Figure A.3. The parameters required to be determined are displacement, velocity and acceleration.

If the base of the instrument is subjected to an external excitation of the form with coordinate y, then

$$y = Y \sin \omega t \tag{A.1}$$

where ω is the forcing frequency of the vibrating body. The equation of motion with representative coordinate x, may be expressed as

$$m\ddot{x} + c(\dot{x} - \dot{y}) + k(x - y) = 0 \tag{A.2}$$

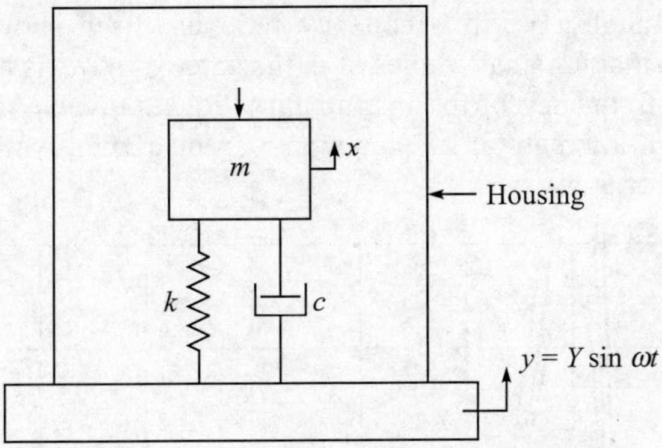

Figure A.3 Seismic instrument (vibration pick-ups)

Introducing $$u = x - y, \tag{A.3}$$

the equation of motion of the mass m may be expressed as

$$m\ddot{u} + c\dot{u} + ku \; + = \; -m\ddot{y} \tag{A.4}$$

From Eq. (A.1) $$\ddot{y} = -\omega^2 Y \sin \omega t$$

Substituting for $\ddot{y}$ in the right-hand side of Eq. (A.4),

$$m\ddot{u} + c\dot{u} + ku = m\,\omega^2 Y \sin \omega t \tag{A.5}$$

Assuming a solution of the form

$$u(t) = U \sin(\omega t - \psi) \tag{A.6}$$

$$U = \frac{\omega^2 Y}{\sqrt{[(k - m\omega^2)^2 + (c\omega)^2]}} \quad \text{and} \quad \tan \psi = \frac{\omega c}{k - m\omega^2}$$

Substituting $\beta = \omega/\omega_n$ and $\zeta = \dfrac{c}{2\sqrt{km}} = \dfrac{c}{2m\omega_n}$, we have

$$U = \frac{Y\beta^2}{\sqrt{[(1-\beta^2)^2 + (2\zeta\beta)^2]}} \tag{A.7}$$

or $$H(\omega) = \frac{U}{Y} = \frac{\beta^2}{\sqrt{[(1-\beta^2)^2 + (2\zeta\beta)^2]}}$$

and $$\tan \psi = \frac{2\zeta\beta}{1 - \beta^2} \tag{A.8}$$

From Eq. (A.7) it is evident that the parameters involved for measurement are the frequency ratio, β, and damping factor ζ. The variation of $H(\omega)$ with β has been shown in Figure A.4 and the variation of phase angle ψ with frequency ratio has been shown in Figure A.5. From Figure A.4 it is obvious that for small values of β, the value of $H(\omega)$ is high and for higher values of β where the natural frequency of the instrument is low, the value of $H(\omega)$ is close to unity. Here an instrument with low natural frequency and an instrument with high natural frequency should be designed accordingly.

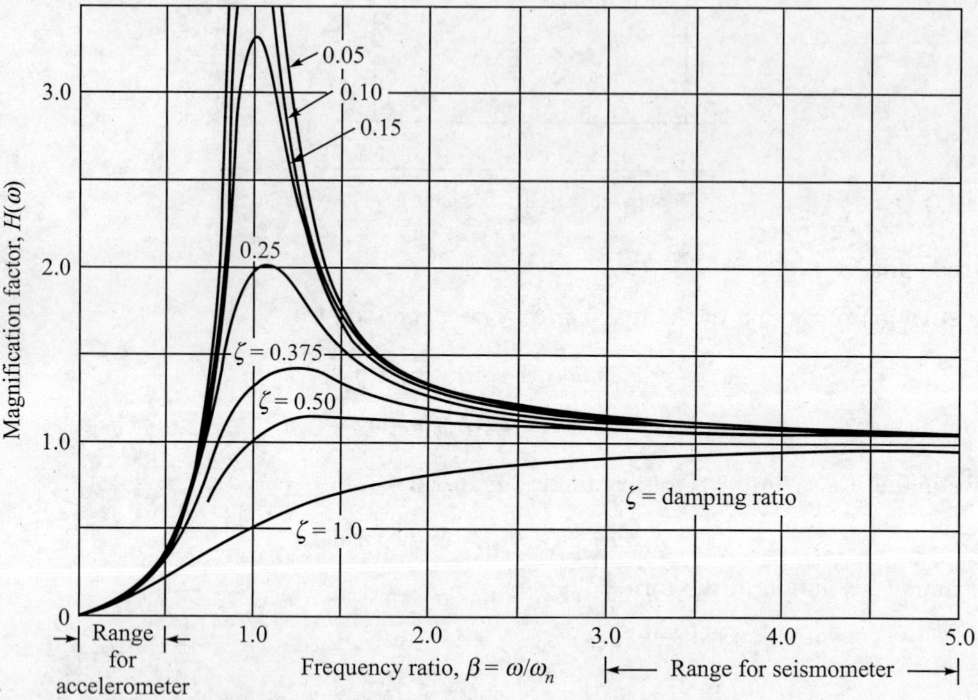

Figure A.4 Response of vibration measuring instruments.

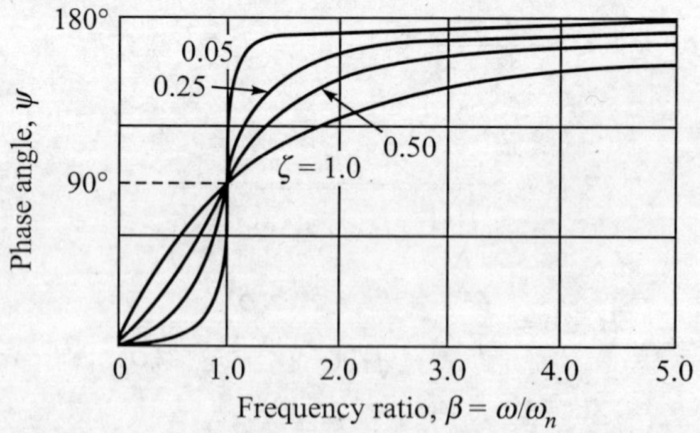

Figure A.5 Variaton of phase angle ψ with frequency ratio ω/ω_n

The instruments with low natural frequency are popularity known as *seismometers*. From Figure A.4 it can be observed that when the natural frequency ω_n of the instrument is low in comparison to the vibration frequency ω to be measured, then the ratio ($\beta = \omega/\omega_n$) approaches a high value to the extent that the relative displacement approaches Y regardless of the damping ratio. As such the mass m then remains stationary while the supporting base moves with the vibrating body. The greatest drawback is its large size in order to keep the natural frequency low.

A seismometer is an instrument that measure the displacement of a vibrating body. It can be noted in Figure A.4, that the magnification factor approach unity when $\beta(= \omega/\omega_n) \geq 3$. As such in this range the relative displacement between the mass and the base is essentially the same as the displacement of the base.

Further, the mechanical signal of the instrument is converted into an electrical pulse in terms of electric voltage and a lot more sophistication/improvement has been incorporated recently. A device which converts a mechanical signal into an electrical pulse, is called a *transducer*. The other aspects of such instrumentation are not presented here owing to the limited scope of this text; only the linear vibration aspects are discussed. In a modern seismometer, both displacement and acceleration are available from the velocity transducer by means of the integrating circuits of the differentiator provided in the signal conditioner units.

A.4 VIBRATION MEASUREMENT FOR EARTHQUAKES

The fundamental observations used in seismology (the study of earthquakes) are seismograms which are a record of the ground motion at a specific location. Seismograms come in many forms, on "smoked" paper, photographic paper, common ink recordings on standard paper, and in digital format (on computers, tapes, CD ROMs). Careful observations of ground vibrations during the last 80 years or so have led to our understanding these vibrations, which are caused by seismic waves. Seismic waves and their propagation have been discussed in Chapter 5, however, a few basics are again addressed here so that we can understand the variety of seismic signals and instruments.

A wave is a disturbance that transfers energy through a medium. Waves are very common in nature: light is a wave, sound is a wave, ocean surf is generated by waves, and even matter has wave-like properties. The "disturbance" can be an alternating electromagnetic field strength (light), a variation in water height (ocean waves), a variation in material density (sound waves), or a distortion of the shape of the ground (seismic waves). If you have felt Earth shake during an earthquake or explosion then you have felt seismic waves. These vibrations travel outward in all directions from their source. Waves generated by large earthquakes can be detected throughout the world and are routinely recorded and analyzed by seismologists.

Seismic waves are generated by many different processes:

- Earthquakes
- Volcanoes
- Explosions (especially nuclear bombs)
- Wind
- Airplanes (supersonic)
- People
- Vehicles

The range of ground motion amplitudes that are of interest in earthquake studies is very large and the seismometers used are very sensitive. They can detect motions that are much smaller than the thickness of a sheet of paper or as tall as a room. Waves are energy transmitting phenomena that have amplitude and a wavelength.

The period and the wavelength are related by a simple expression connecting the two with the speed of the wave

$$\text{Wavelength (km)} = \text{speed (km/s)} \times \text{period (s)}.$$

In many ways seismic vibrations are analogous to light and sound waves, so let us discuss these more familiar waves first.

Light is actually just one of a range of electromagnetic waves that we use each day. Other related waves, which differ primarily from visible light in their wavelength, include microwaves, radar, AM and FM radio, X-rays, etc. Different colours are actually light waves with different frequencies. For example, red light has a longer wavelength than violet light. The wavelength for red light is about 650 nanometres (billionth of a metre), violet is about 400 nanometres, an X-ray is about 1/3 of a nanometre. FM radio and TV waves have wavelengths around 2–12 metres, AM radio signals have a wavelength of about 30 metres. These terms derive from the wavelength of the colours. TV, VCR, stereo, etc. remote controls transmit signals in the infrared, and high-energy signals like X-rays and ultra-violet radiation are in the part of the spectrum beyond violet.

White light is composed of all visible colours and sunlight contains a large range of the visible spectrum. We could use prism to split light into its different colour components. The same thing happens when water vapour in the atmosphere is aligned with the rays of light arriving from the sun in just the right geometry, the water vapor acts like a prism and splits the sunlight into a rainbow.

Another common example of multiple-frequency signals are the everyday sounds. Humans can hear frequencies in the range of 20 to 20,000 hertz and the musical note "middle A" is around 440 hertz. We can calculate the wavelength of sound waves that we can hear using the formula wavelength = velocity/frequency. In air sound travels at about 340 metres per second (the precise value depends on temperature, humidity, and pressure). We hear (measured in air) the range from about two-thirds of an inch to 50 feet per second.

Seismic Signals

Just like sunlight contains different colours that can be split apart with a prism, seismic waves contain many different "frequencies" that we can record with specially "tuned" seismometers. The idea is completely analoguous with light and sound

Seismic	Light	Sound
Short-Period	Blue	Treble
Long-Period	Red	Bass

Seismographs

Seismographs, which generally consist of two parts, comprise a sensor of ground motion which we call a **seismometer**, and a **seismic recording system**. Modem seismometers are sensitive electromechanical devices but the basic idea behind measuring ground movement can be illustrated using a simpler physical system that is actually quite similar to some of the earliest seismograph systems. A simple system for seismic recording is shown in Figure A.6.

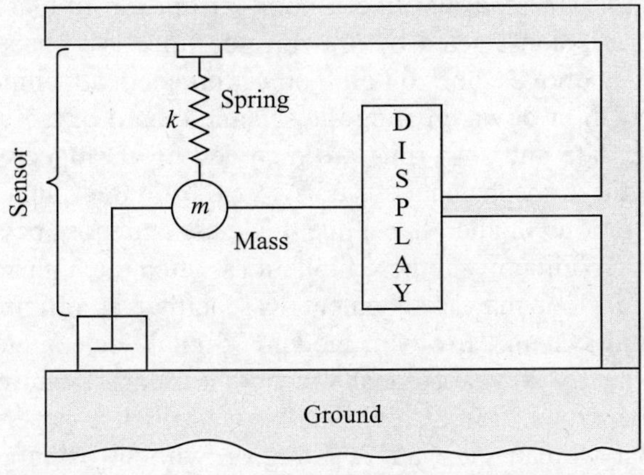

Figure A.6 A simple mechanical system for seismic recording.

The physics behind the sensor is Newton's Law of Inertia:

> "A body in motion tends to stay in motion unless acted upon by a force, and a body at rest tends to remain at rest unless acted upon by another force."

The sensor and the recorder do not have to be located next to one another. Seismometers are spread throughout the world, but are usually concentrated in regions of intense earthquake activity or research. These days, the recording system is invariably a computer, custom designed for seismic data collection in harsh weather. Often they are also connected to a satellite communication system. Such systems enable us to receive seismic signals from all over the world, soon after an earthquake. Figure A.7 shown the arrangement for digital storage and associated satellite communication.

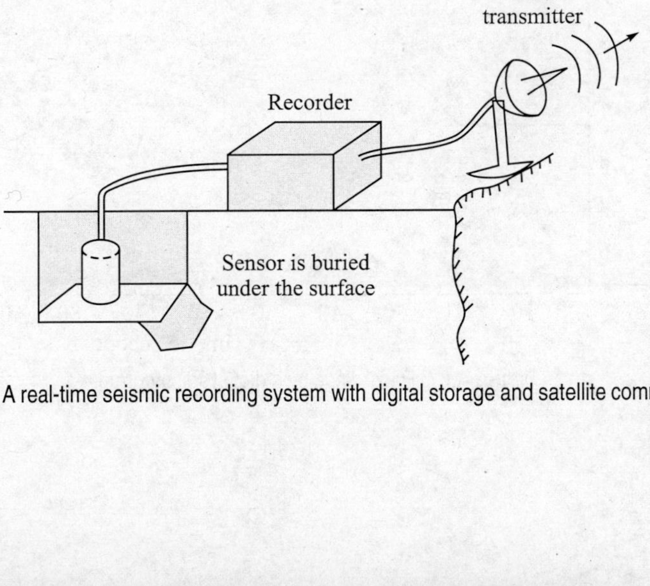

Figure A.7 A real-time seismic recording system with digital storage and satellite communications.

Classic Seismograms

For most of the last century, seismograms were recorded on a sheet of paper, either with ink or photographically. We call such records "analog" records to distinguish them from digital recordings. These records are read just like a book—from top-to-bottom and left-to-right.

A continuous record is constructed by drawing the line, as a sheet of paper fastened to a rotating drum constantly moves horizontally on a threaded attachment. When the ground vibrates, the pen moves up or down creating the seismic record of the vibrations. Seismograph station and component, date and start time are recorded on the upper left of this paper.

One problem with these mechanical systems was the limited range of ground motion that could be recorded—vibrations smaller than a line thickness and those beyond the physical ranges of the ink pen were lost. To circumvent these limitations, often the high and low-gain instruments are operated side-by-side, but that arrangement was neither as efficient nor effective as the modem digital electronic system. However, modem "digital" or computerized instruments are relatively new, only about 15–20 years old, and most of our data regarding large earthquakes are actually recorded on paper (or film). Additionally, we still use paper recording systems for display purposes so that we can see what is going on without using a computer.

Digital Seismograms

Today, most seismic data are recorded digitally, which facilitates quick interpretations of the signals using computers. Digital seismograms are "sampled" at an even time interval that depends on the type of seismic instrument and the interest of the people who deploy the seismometer. The same principle is used to provide "digital" sound on compact disks. The motion of the ground is continuous, but we can pick only certain positions and reconstruct the motion (within certain limits).

Further, a digital seismogram is a record of the ground movement stored as an array of numbers which indicate the time and the movement of the ground for a range of times and are easily analyzed using computers. The principle is the same as that used for digital audio signals that are stored on Music CD's. A typical digital seismogram record is shown in Figure A.8.

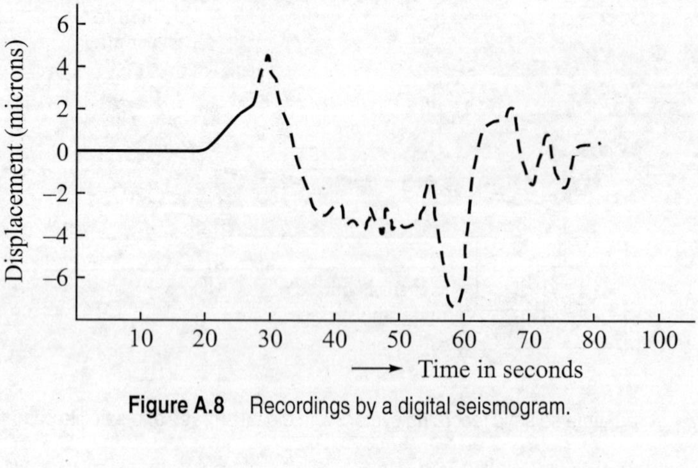

Figure A.8 Recordings by a digital seismogram.

Also, since we live in a three-dimensional space, to record the complete ground motion, we must record the motion in three directions. Usually, we choose:

- Up-down (Vertical)
- North-south (Horizontal)
- East-west (Horizontal)

With three records of ground motion in three directions, a single seismic station that records about 20 samples per second must store or transmit about

$$3 \times 1.7 \times 10^6 \text{ samples/day} \sim 5 \times 10^6 \text{ samples/day}$$

which equals about 5 megabytes per day? For each 200 seismometers of this type around the world, we are recording about 1,000 Mbytes/day ~ 1 Gigabyte/day.

Although we can do better using compression algorithms to save storage, keeping up with this flux of data is a challenge.

Seismometers

Before the advent of technological advances in the last few decades, many different kinds of seismometers were developed and deployed to record seismic signals. In the 1960s, as part of an effort to verify underground nuclear test threshold treaties, two seismometer models became standard for global earthquake analyses, i.e., the world-wide network *long-period* and *short-period* instruments. These instruments have been replaced by "broad-band" instruments which can detect ground motions over large ranges, or "bands", of periods. However, since much of our historical data are recorded on the older "narrow-band" short-period and long-period instruments, they remain important sources of data. Different seismometers record different frequencies (or periods) of ground motion.

Seismometer Response Curves

Seismometers are usually designed to record signals over a specified range of frequencies (or periods) so it is convenient to discuss instruments based on the range of vibration frequencies that they can detect. Thus, one way to characterize seismometers is to describe the range of vibration frequencies that they can detect. A plot of the amplification versus frequency is called an instrument response. An instrument is sensitive to vibrations at frequencies for which the "response" curve is relatively large. Five sample instrument response curves are shown in Figure A.9. The frequency is shown along the bottom horizontal axis, the equivalent period (period = 1/frequency) is shown along the top horizontal axis. The vertical axis shows the ground–motion amplification factor.

To characterize an instrument, what is really important is the range of amplitudes, not the specific amplification, which is usually adjusted depending on the location of the seismometer.

The broadband instrument senses most frequencies equally well; the long-period and short-period instruments are called "narrowband", because they preferentially sense frequencies near 1/(15 s) and 1 Hz respectively.

Figure A.9a is a comparison of a modern broadband seismometer response and the classic World-Wide Standard Seismic Network (WWSSN) long- and short-period instruments. The same broadband response is shown in Figure A.9b to compare the response with a special short-period instrument, the Wood-Anderson, and an accelerometer. The Wood-Anderson short-period instrument was the one that Charles Richter used to develop his magnitude scale for southern California. The accelerometer is an instrument designed to record large amplitude and high-frequency shaking near large earthquakes. Those are the vibrations that are important to consider in the design of buildings, highways, etc.

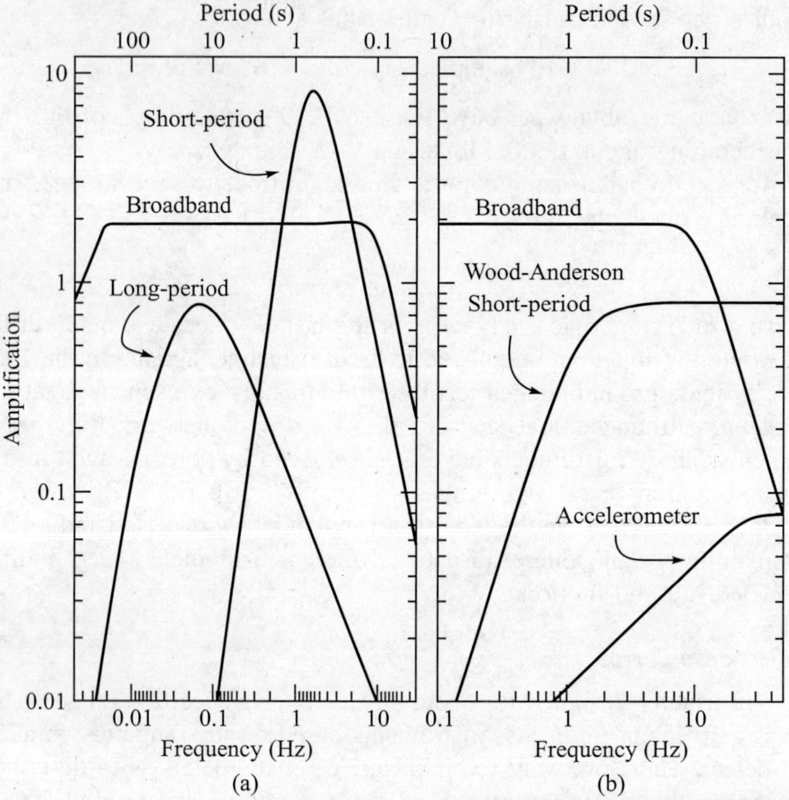

Figure A.9 Display of recordings from broadband instruments and narrowband instruments.

Accelerometers

Another important class of seismometers was developed for recording large amplitude vibrations that are common within a few tens of kilometres of large earthquakes—these are called *strong-motion* seismometers. Strong-motion instruments were designed to record the high accelerations that are particularly important for designing buildings and other structures. One place with which we are all familiar with acceleration changes is an elevator. The acceleration that you experience in an elevator is about 2 m/s^2 or about 0.2g. However, in an elevator the transition from 0.0g (not moving) to 0.2g is smooth and comfortable. During an earthquake, you can feel that the ground accelerations vary between –0.1g to +0.1g several times each second, for at least 10–15 seconds. That is not very gentle shaking.

Seismometer Networks

Seismic recording systems monitor the ground movements continuously and we have records of the motion of the ground at some places dating back nearly 100 years. Strong ground motion records during earthquakes provides the basic information for earthquake engineering. These records are essential for evaluating earthquake resistant deign procedures, estimation of attenuation characteristics, assessment of seismic hazard, etc. Analysis of strong ground motion data accordingly lead to a better understanding of the potential effect of strong shaking during earthquakes. The ground motion characteristics determined from strong motion records are studied in various terrain and rock conditions and have been related to various earthquake parameters. The analysis of data has also helped in understanding the soil–structure interactions, effect of soil deposits, topography and other effects. Ground motion time histories recorded during past earthquakes provide valuable information on the nature of ground motion during a future earthquake that may be expected to occur at site. Further, analysis of strong motion data lead to a better assessment of damaging potential of strong shaking during earthquakes for different types of foundations and sub-structures.

In general all the countries in the world where earthquakes are of any concern have their strong-motion instrumentation programmes. However, countries like the USA, Japan and Taiwan have extensive deployment of strong-motion instruments. Mexico also has an impressive network of strong-motion instruments. Other countries where strong-motion instruments network exist are Albania, Algeria, Austria, Bosnia, Bulgaria, Canada, Croatia, Greece, Iceland, India, Iran, Italy, Korea, Macedonia, New Zealand, Pakistan, Portugal, Romania, Serbia, Span, Slovenia, Switzerland and Turkey.

In the USA there are large numbers of organizations, which are managing the strong-motion network in the country. However, the major players in this field are, California Institute of Technology, University of California Berkeley, California Division of Mines and Geology and US Geological Survey (USGS).

Many institutions have been monitoring earthquake activity since the development of seismometers, for example, Caltech and UC Berkeley have relatively long histories of monitoring California seismic activity. That work continues to upgrade the 1970s instruments with modem, broader-band digital equipment.

International monitoring of earthquakes also has a long history. The WWSSN was deployed by the US Government in the 1960s to monitor nuclear explosions and help verify nuclear test treaties since the 1960s. Several smaller networks were deployed before that and today's global seismic network is a cooperative, international effort that consists of more than 100 seismic recording stations.

The USGS are operating 840 accel-erograph at 640 stations. A national committee, called US Committee for advancement of Strong-Motion Programme, coordinates the activities of different organizations. Japanese strong-motion instrumentation programme is called K-Net. There are 1000 strong-motion digital instruments deployed all over Japan at an average station-to-station distance of 25 km. Taiwan has four different strong-motion programmes. There is a network of 100 analog type instruments in which records are obtained on photographic film. Most of these are installed at true field sites and some are installed on structures. Strong motion in Mexico was started in 1960 and at present has 497 accelerograph stations.

After an earthquake we often descend on the epicentre region with portable seismic instruments to carefully and closely monitor aftershock sequences that follow most large earthquakes. Portable seismic recording systems have been designed for this purpose and they are similar to the permanent stations but often run on battery and solar power. The data are retrieved by a scientist who visits the site and downloads the digital signals from a computer hard disk that is part of the portable seismic recording system.

In addition to their value in monitoring and studying earthquakes and aftershocks, seismograms can be used to probe the internal structure of Earth. To facilitate such studies we often deploy temporary seismometer networks for several months or years in geographic regions that we want to study more closely.

Since 1930 seismometers have been used to record actual ground motion specific to their locations. In such ground motion records, time histories have recorded acceleration and as such these records are termed acclerograms. Recently such recordings are being taken digitally. Acceleration is normally recorded directly, and velocity, displacements are being obtained by integration. Acceleration can be expressed in units of cm/s^2 (terms gals), but is often also expressed in terms of fraction or per cent of the acceleration of gravity (980.66 gals, termed 1g).

Velocity is expressed in cm/s (termed kine) and displacement in cm. Modern instruments use sets of accelerometers to measure components of acceleration in three orthogonal directions, two in horizontal directions and one in the vertical direction.

Figures A.10a, b and c show these three recordings for a typical earthquake ($5 < M < 7.5$). Time histories, such as these in Figure A.10 represent complete record of motion at the instrument site. From geotechnical point of view, the duration, the frequency content and amplitude of motion are important. A number of parameters have been in use to express the ground motion, the most commonly used is peak ground acceleration (PGA).

Figure A.10a shows a peak horizontal acceleration of 0.35 g, and a peak vertical acceleration of 0.25 g. Using the method of integration, velocity–time trace may be obtained.

Figure A.11a shows the time history for variation of velocity while Figure A.11(b) shows the time history for variation of displacement. Geotechnical engineers traditionally use acceleration, rather than velocity or displacement, because acceleration is directly related to the dynamic forces that an earthquake induces on the soil mass.

For geotechnical analyses, the measure of the cyclic ground motion is represented by the maximum horizontal acceleration at the ground surface. The maximum horizontal acceleration at the ground surface is also known as the peak horizontal ground acceleration. For most earthquakes, the horizontal acceleration is greater than the vertical acceleration and thus the peak horizontal ground acceleration also turns out to be the peak ground acceleration (PGA). For earthquake engineering analysis the peak ground acceleration (PGA) is one of the most difficult parameters to determine. It represents an acceleration that will be reduced sometime in the future by an earthquake. At present it is not possible to predict earthquakes, so the choice of assigning the value of the peak ground acceleration must be based on prior earthquakes, geotechnical history, fault studies and sound engineering judgement.

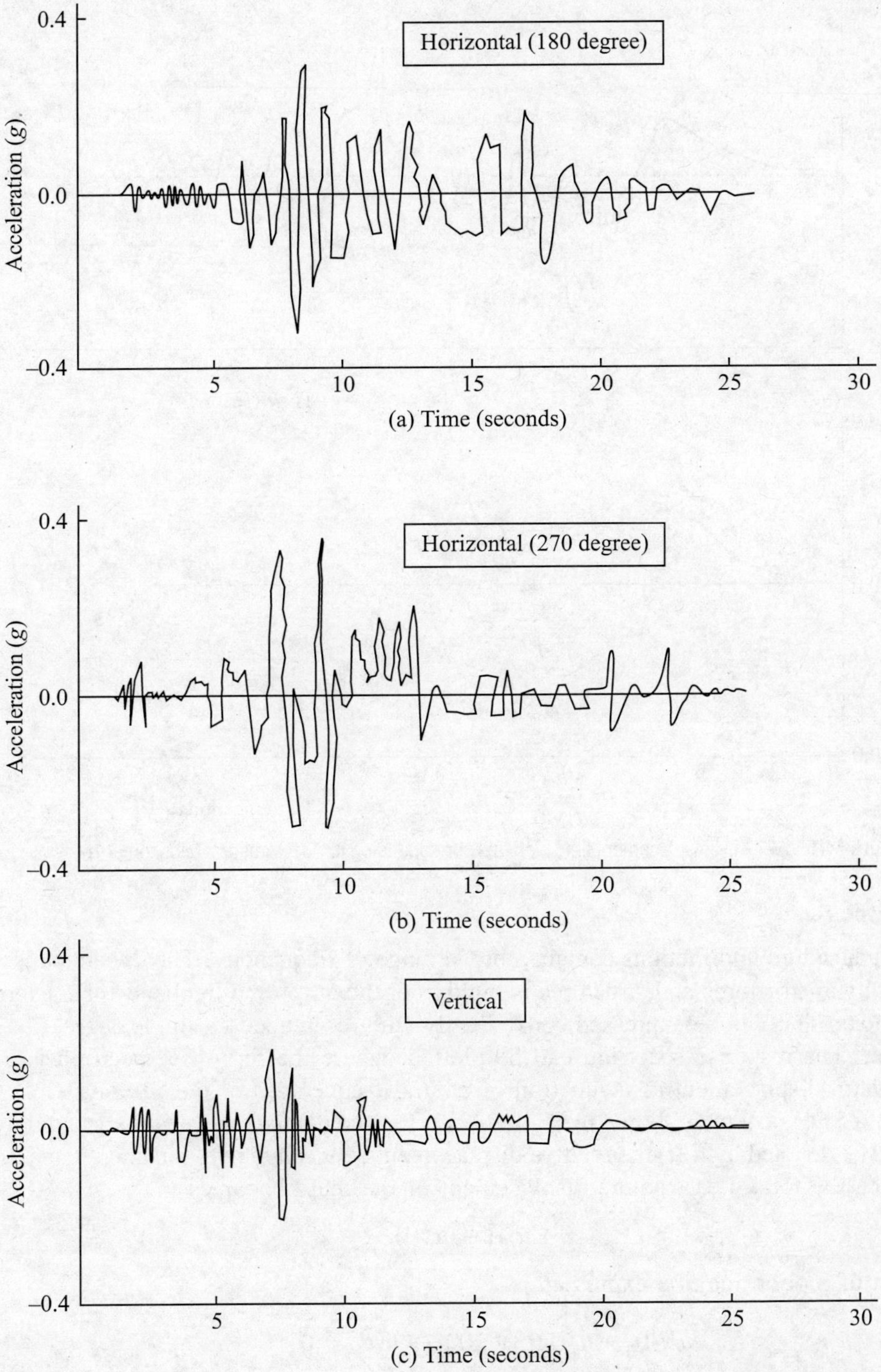

Figure A.10 Accelerogram of recorded motions along three mutually perpendicular directions for a typical earthquake (5 < M < 7.5).

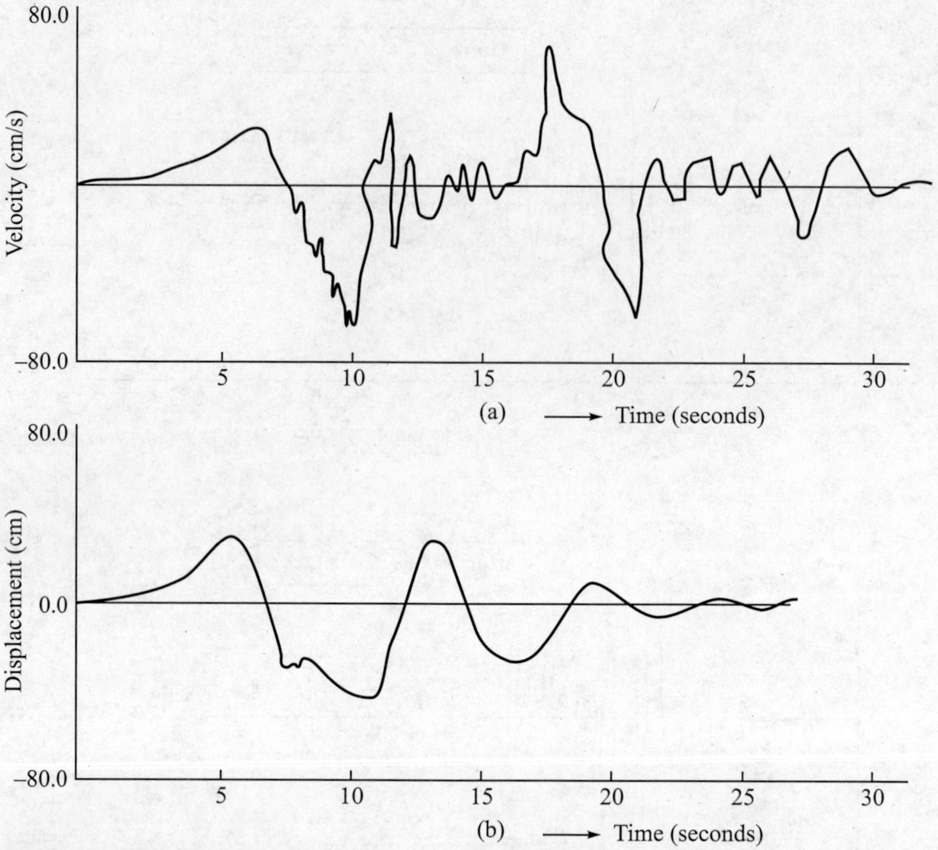

Figure A.11 Velocity, displacement time histories corresponding to Accelerogram (Horizontal 180).

Response Spectra

Earthquake induced ground motions contain a broad range of frequencies. The dynamic response of foundation, sub-structures and structures is highly dependent on the frequency characteristics of these motions. This can be expressed conveniently with reference to a single degree freedom (SDF) system. The response spectrum can be plotted, where the response spectrum is a plot of the maximum displacement as a function of natural frequencies of the SFS system.

Consider a SDF system as shown in Figure A.12. Let the displacement of the ground owing to earthquake be denoted by $u_g(t)$ the relative displacement between the ground and the maximum by $u(t)$. Hence the total, i.e. absolute displacement of the mass becomes

$$u(t) + u_g(t)$$

The equation of motion may be expressed as

$$m[\ddot{u}(t) + \ddot{u}_g(t)] + c\dot{u}(t) + ku(t) = 0 \tag{A.9}$$

where

m = mass

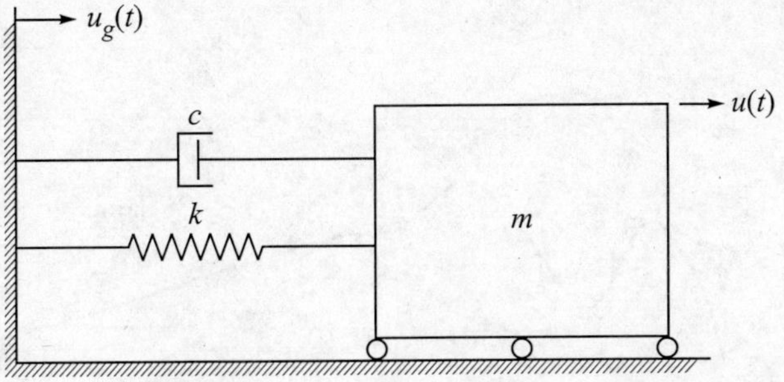

Figure A.12 A single degree freedom system subjected to ground motion $u_g(t)$.

c = viscous damping coefficient of the dashpot
k = spring stiffness.

The first term on the left-hand side of the Eq. (A.9) represents the inertia force, the second term the damping force, and the third term the elastic force. Equation (A.9) can be written as

$$z\zeta\omega\dot{u}(t) + \omega^2 u(t) = -[\ddot{u}(t) + \ddot{u}_g(t)] \qquad \text{(A.10)}$$

where

ω = resonant natural frequency = $\sqrt{k/m} = \dfrac{2\pi}{T}$

T = resonant period

ζ = damping ratio = $\dfrac{c}{2m\omega}$.

The ground acceleration $u_g(t)$ would be given by site specific acceleration time histories, such as the ones in Figure A.10. Solution of Eq. (A.10), which for the case of complex ground motions $ug(t)$ must be accomplished numerically, fields the relative displacement history $u(t)$.

The response spectrum is a plot of maximum displacement and may be expressed as

$$S_d(\omega, \zeta) = \max\ |u(t, \omega, \zeta)| \qquad \text{(A.11)}$$

where

S_d = maximum displacement
ζ = damping ratio

Spectra are determined for a given level of damping, typically 5%. Similarly the velocity $x(t)$ and acceleration $x(t)$ spectra can be determined by numerical differential of the displacement $x(t)$. The velocity spectra and acceleration spectra may computed directly from S_d using the following relationship.

$$S_v = \omega S \qquad \text{(A.12)}$$

$$S_a = \omega^2 S \qquad \text{(A.13)}$$

S_d, S_v and S_a spectra are shown in Figure A.13. For the accelerogram as shown in Figure A.l0a, the three spectra illustrate how displacement, velocity and acceleration are associated with different frequency ranges. S_v and S_a, which are different in magnitude from the spectra based

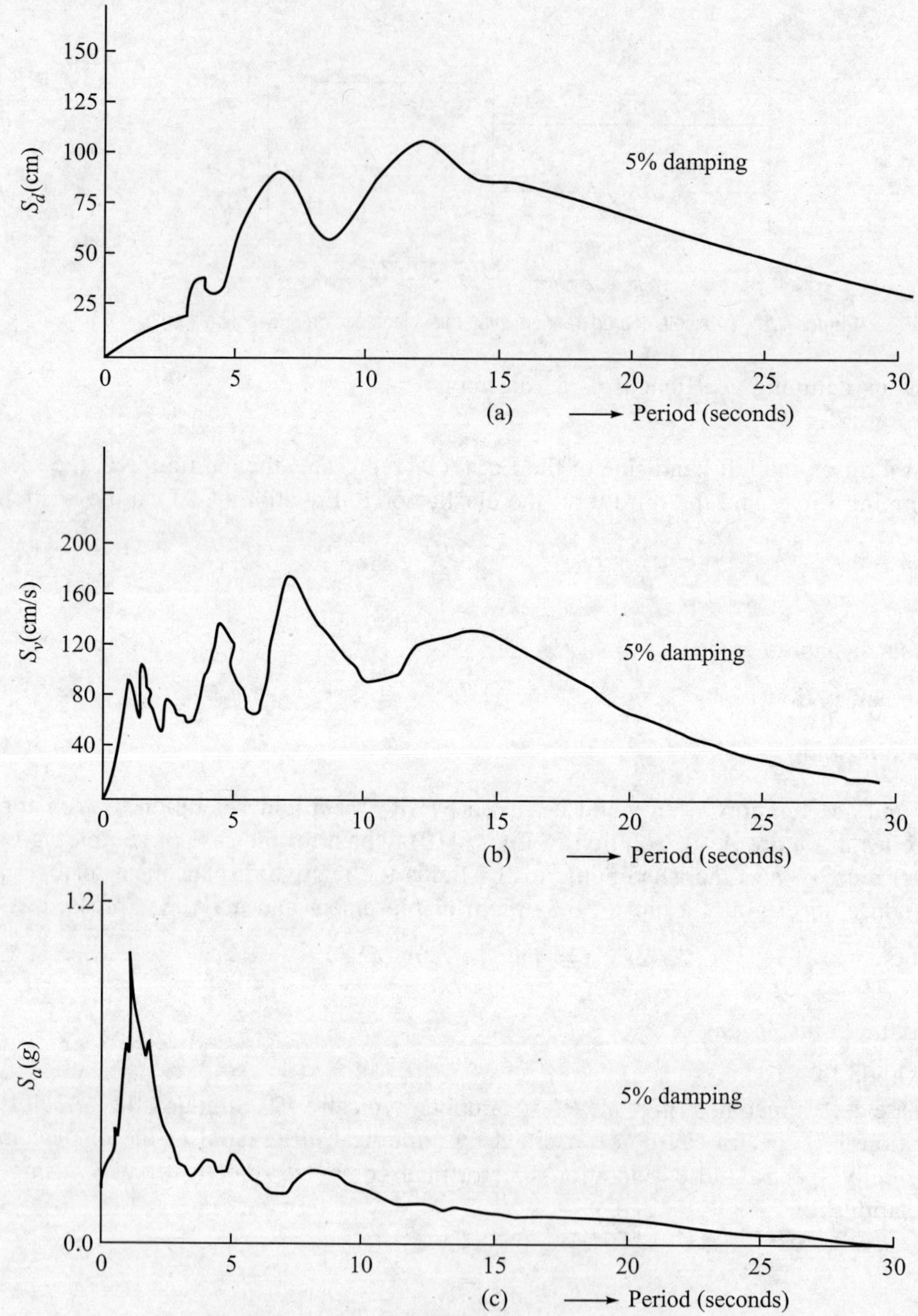

Figure A.13 Response spectra to accelerograph in Figure A.10a.

on $u(t)$ and $u_g(t)$, are more meaningful quantities. S_v can be related to the peak value of strain energy stored in the system during a given earthquake, and S_d can be related to the base shear coefficient used in building codes to estimate the base shear forces in structures.

Peak displacement occurs at comparatively high periods (or low frequencies), whereas peak accelerations are found at low periods (or high frequencies). Stiff or rigid foundations with low resonant period are most sensitive to the low period components of the ground motion, therefore, consideration of the acceleration spectrum is of prime importance. On the other hand, flexible foundations with a high resonant period are more sensitive to the high-period components of the motion, which are better reflected in the displacement spectrum. As S_d, S_v and S_a are related, they are also presented in a tripartite display plot that uses four logarithmic scales to display velocity, acceleration and displacement—either period or frequency. Such a tripartite plot for the spectra is shown in Figure A.14.

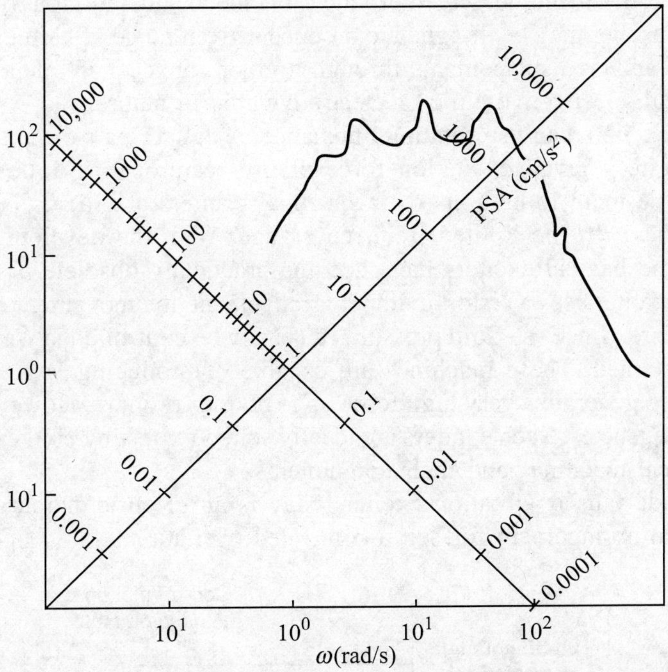

Figure A.14 Tripartite plot of response spectral displacement, velocity and acceleration.

Thus the duration of the strong ground motion in general and in particular the number of load reversals has a strong influence on the damage that a particular earthquake can cause. A moderate earthquake of a longer duration can result in equal or larger destruction than a shorter, intense one. Structures that are loaded beyond a cyclically stable range may fail after a given number of load reversals. The same is true for foundation elements that may lose their support upon repeated load reversals due to reduction in soil stiffeners and strength. Also, pore pressure in saturated, loose granular soils can accumulate with load cycles and lead to liquefaction after a given number of load reversals. The duration of the strong-motion component of a seismic event can be determined from an accelerogram using one of a number of criteria. According to Bolt (1969), the most straightforward approach is to measure the time between the first and the last exceedance of some threshold acceleration, typically 0.05 g. This is referred to as the bracketed duration.

A.5 VIBRATION INSTRUMENTS

The vibration instruments can be broadly classified into following groups:

1. Vibration exciters
2. Vibration measurement devices
3. Analysers.

A.5.1 Vibration Exciters

A vibration exciter is a machine, which produces the mechanical motion to which the test object is subjected. The exciter may be designed to produce a given range of harmonic or time dependent excitation forces and/or displacements through a given range of frequencies. These machines can be mechanical, electro-hydraulic or electrodynamic in nature.

Mechanical exciters can use rotating unbalance, usually a pair of contra-rotating eccentric masses, to generate a given excitation force or use reciprocating follower motion from an eccentric cam. The main limitations of mechanical exciters are (a) no control over the force or displacement and (b) the limited frequency range. With the development of sophisticated electronics, the mechanical exciters have become practically obsolete.

A hydraulic exciter uses a piston-cylinder arrangement, the movement of which is controlled by the fluid pressure. Since the fluid pressure can easily be controlled, a wide range of excitation forces can be obtained. These machines are capable of producing large displacements up to 50 cm and can also generate a very high force up to 50 tons or more. However, they can generate only very low frequencies. Such exciters are ideally suited for testing civil engineering structures, but not mechanical machines and their components.

The most widely used vibration exciter today is an electrodynamic exciter. Figure A.15 shows the general arrangement for servo-controlled excitation.

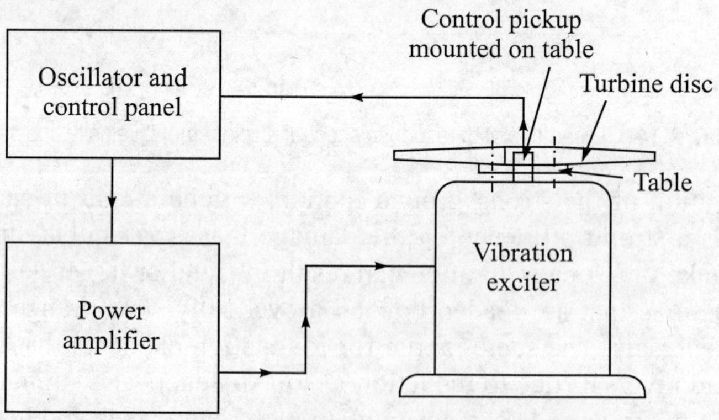

Figure A.15 Instrumentation for servo-controlled excitation.

Since the moving element and the coil should have a one-dimensional motion, they need to be suspended by a flexible support system with as low a natural frequency as possible. This gives rise to suspension resonance at a low frequency as shown in Figure A.16. The moving element itself has a resonance frequency, which can be made fairly high. Between these two

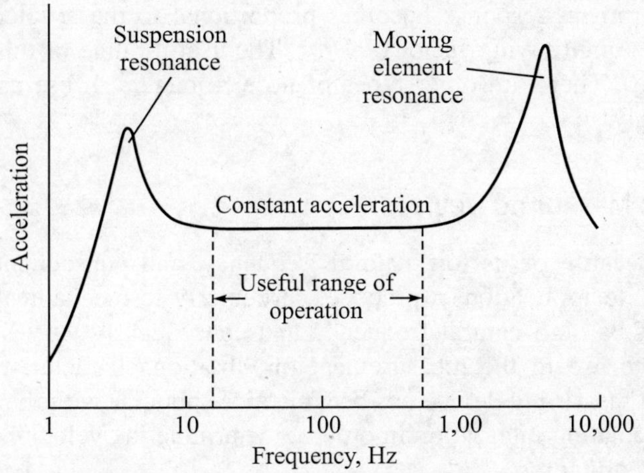

Figure A.16 Typical resonance characteristics of an electrodynamic vibration exciter

resonances we have the useful range of the exciter giving a constant acceleration for a given input power.

To minimize the influence of these resonances a servo-control system can be used as shown in Figure A.15. With the help of such a control system, the useful frequency range of an exciter can be extended considerably.

The exciters discussed above need to be physically connected to the member under test. The excitation is through this connection and can be assumed to be acting at a single point. A non-contact type of electromagnetic exciter can be used where it is not desirable to load the test member. When the member under test is small, the exciter spring and mass characteristics may affect the test procedure because the physical connection can alter the dynamic characteristics. An electromagnetic shaker consists of a coil wound on a strong permanent or dc electromagnet. The coil is fed with ac current generated by an oscillator. The test specimen should be of magnetic material and is kept at a suitable distance away from the shaker so that no contact is made between them. The force exerted by the electromagnetic exciter is small, and is harmonic only if the vibratory amplitude of the test specimen is small compared to the gap between the exciter and the test specimen.

A.5.2 Instruments with High Natural Frequency

If the natural frequency ω_n of the instrument is high compared to that of the forcing frequency ω, the frequency of vibration to be measured, the relative response $H(\omega)$ indicates

$$H(\omega) = \frac{\beta^2}{\sqrt{[(1-\beta^2)^2 + (2\zeta\beta)^2]}} \tag{A.14}$$

that for small values of β, the denominator approaches unity. In such a situation, then

$$H(\omega) = \beta^2 \frac{\omega^2}{\omega_n^2} = \frac{\text{acceleration}}{\omega_n^2} \quad \text{as } \omega^2 \text{ is proportional to acceleration}$$

Thus, the instrument response becomes proportional to the acceleration of the motion to be measured with a multiplying factor of $1/\omega_n^2$. The useful range of this type of accelerometer is $0 < \omega/\omega_n < 0.20$. There are other types of accelerometers in use as well and they will be discussed separately.

A.5.3 Vibration Measuring Devices

The vibrometer, because of its low natural frequency, and consequently because of its high mass, rarely finds its application in practice, particularly to mechanical systems. The accelerometer because of its high natural frequency and consequently very light in its construction, finds an immediate use in the measurement of vibration characteristics of a machine. The accelerometer as a measuring device has become more popular with the advent of sophisticated electronics for calculating integration in order to determine the velocity as well as the displacement of vibrating structures.

Since very small levels of vibration have to be measured in practice, a direct recording scheme is impossible and some kind of an electromechanical principle is necessary in the design of a vibration measuring instrument. In the past, several principles have been used, electrodynamic, differential transformer (inductive type), capacitance and strain gauge, however, the piezoelectric principle is found to be most suitable in designing a small size transducer, having least influence on the mass of the system whose vibrations are to be measured. Thus, compression type piezoelectric accelerometers are widely used, whose construction is schematically shown in Figure A.17. A relatively heavy mass is suspended from the base of the instrument through a stiff spring. This assembly is mounted in the instrument housing and exerts a variable force on a pair of piezoelectric crystals (lead zirconate, barium titanate, etc.). When this accelerometer is subjected to vibratory motion, the force exerted by the mass on the piezoelectric crystals is exactly proportional to its acceleration. Due to the piezoelectric effect a variable voltage is developed across the two discs, which will be proportional to the acceleration of the mass. If the spring is very stiff, the voltage developed will be proportional to acceleration except in the vicinity of resonance of the accelerometer.

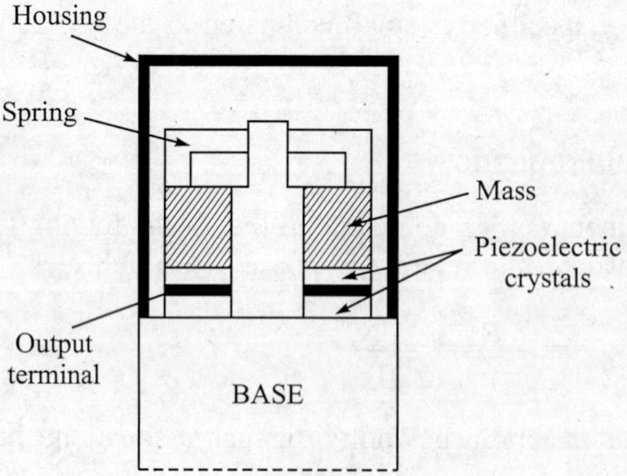

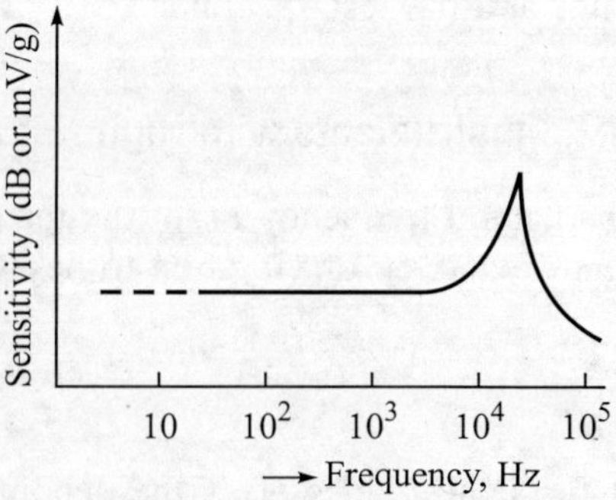

Figure A.17 (a) A piezoelectric accelerometer and (b) its typical frequency response.

The electrical voltage signals thus obtained from an accelerometer are usually very low and for the purpose of readout, a preamplifier is generally necessary. The signal thus amplified can be observed on an oscilloscope. If the signal is harmonic, the amplitude can be directly read out on a voltmeter. If the velocity and displacement amplitudes are required, a double integrator is to be included which is a standard feature in present-day analyzers.

The vibration measuring transducers/devices can be broadly classified into two categories: (1) contact type and (2) non-contact (non-intrusive or proximity) type. The accelerometer, the LVDT (Linear Variable Differential Transformer) and a class of electromagnetic/inductive transducers based on the movement of a coil or permanent magnet or the core, as well as the strain gauge are of the contact type and have to be attached to the test specimen. These transducers have the effect of loading the test specimen and also present difficulties when measurements are to be made on rotating members.

The non-contact type transducers work on the principle of change of capacitance or reluctance or the loss of energy in eddy current field in the measurement circuit due to a change in the gap between the transducer and the test member. The possible physical configurations of transducers based on the above principles are too numerous to be described here. The details of these are available in texts on Measurements and Instrumentation as well as in the catalogues of the manufacturers.

A.6 ROLE OF TRANSDUCERS IN INSTRUMENTATION

Transducer is the name given to any device, which converts a physical quantity, e.g. a vibration of small amplitude, into another more-easily usable (say, an electrical) quantity either for measurement, control or for operation of processes. For measurement purposes it is usual to transduce into electrical voltage or current and to record what is happening, e.g. amplified as a visual image on the screen of a cathode-ray oscilloscope, which is especially useful for the display of qualitative results. Electrical signals can also be processed, e.g. integrated, differentiated, selectively filtered, immediately recorded and presented in analogue or digital form for use in a computer. Electrical filters can suppress those frequencies in which engineers are not interested; or, alternatively, the graphical recording of a total vibration on a magnetic tape can be filtered. One of the effects most used for changing a mechanical quantity into a corresponding electrical one is the change in electrical resistance when a wire is stretched; another is the piezo-electric effect, namely when certain crystals are deformed, a voltage is produced between their surfaces. The other effects known to physics are used for the transduction of mechanical effects into their electrical analogies or equivalents, e.g. thermo-electric voltages, produced when a junction between two dissimilar metals is heated, are used for temperature measurement; and changes in capacitance or inductance when coils carrying alternating current, or plates fed with direct current, move relative to each other, are used for measuring displacement and pressure. Thus, the range of electrical transducers available is extensive for the measurement of physical quantities including displacement, velocity, pressure, flow rate, viscosity, temperature, etc. In particular, as the resistance strain-gauge (which may form one arm of a Wheatstone bridge) is a *pickup* in which the electrical resistance of a *zig-zag fine wire*, foil or filament of semiconductor material changes as it is stretched when glued or cemented to the test specimen.

To record what is happening is, therefore, relatively easy since amplification is electrically straightforward, and recording apparatus can be placed away from the vibrating system. For automatic control purposes, however, it may be preferable to transduce a measurement into such terms capable of being applied directly to the operation of a mechanism being controlled. Transducers are, of course, only part of a complete instrumentation system, the integrity of which is very important since unless the initial transfer from mechanical to electrical effect is accurate, all else is useless for quantitative and evaluation purposes. Hence, it is important to make the choice of a transducer in accordance with the type of problem being investigated, and the range of vibrations involved. Care is needed with calibration, e.g. the inclusion of connecting wires in the calibration of pickups. Also, unless the mains and other leads are screened, the magnetic fields of currents flowing may cause interference and distortion of a genuine signal being transmitted.

The voltage waveform which emanates from vibration pickups can be amplified for display on a cathode-ray screen across which the spot can be adjusted to trace one complete cycle, and then continually repeat the cycle. A predominating waveform envelope associated with the period of rotation of a mechanism may be revealed containing several higher-order frequency components caused by the vibrations of certain parts composing the mechanism. By increasing the time base frequency, i.e. the rate at which the spot traverses the screen, a picture of these higher-order components can be obtained.

A.6.1 Seismic Pickups

The earliest displacement pickups (referred to as 'seismic') were mechanical and designed for recording earth tremors and transient shocks. Certain natural crystals such as quartz, Rochelle salt and some semiconductors show a piezoelectric effect, which produces a voltage on opposite faces when such crystals are deformed or strained and, in effect, act mainly as condensers. Lead zirconate has a greater sensitivity than quartz for the same frequency, but its temperature rating is limited to 250°C as against 350°C for quartz when attached, say, to hot engine parts. These types of pickups have a desirably high resonant frequency, e.g. from 30 to 100 kHz, and can be compacted within a 1 cm^3. In principle, pickups which are piezoelectric in function consist of a fairly small mass held firmly or cemented to a crystal as shown diagrammatically in Figure A.18.

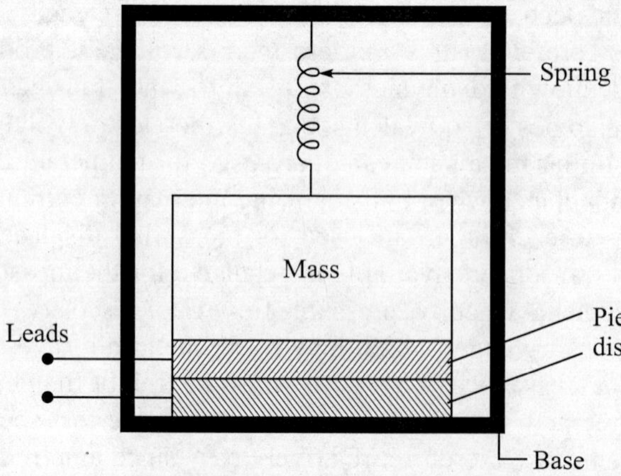

Figure A.18 Piezoelectric pickup.

Another type is the seismic moving coil pickup shown in Figure A.19, basically comprised of a mass so mounted such that its deflection is linear with the imposed displacement, and the electrical output depends on the rate of cutting of lines of force from the magnets, i.e. the output is velocity responsive which requires a single integration to give displacement or a single differentiation to give acceleration. Moving coil types are used at frequencies from 25 Hz up to 2 kHz. Seismic pickups comprising a mass mounted on springs may be used, when required to measure displacement; its design must be such that its resonant frequency $\omega_n = \sqrt{(k/m)}$ is lower than the frequency to be measured, i.e. soft springs and relatively large mass, and when required to measure acceleration its resonant frequency must be higher than that of the frequency to be measured. Seismic pickups read movements relative to the earth; hence to obtain movement or acceleration relative to any other part of the system it is necessary to combine the outputs of two such pickups vectorially. Also, all seismic pickups are sensitive to side effects, e.g. to movements transverse to their axes, and from torsional oscillations which may *swivel* a moving coil in the magnetic field.

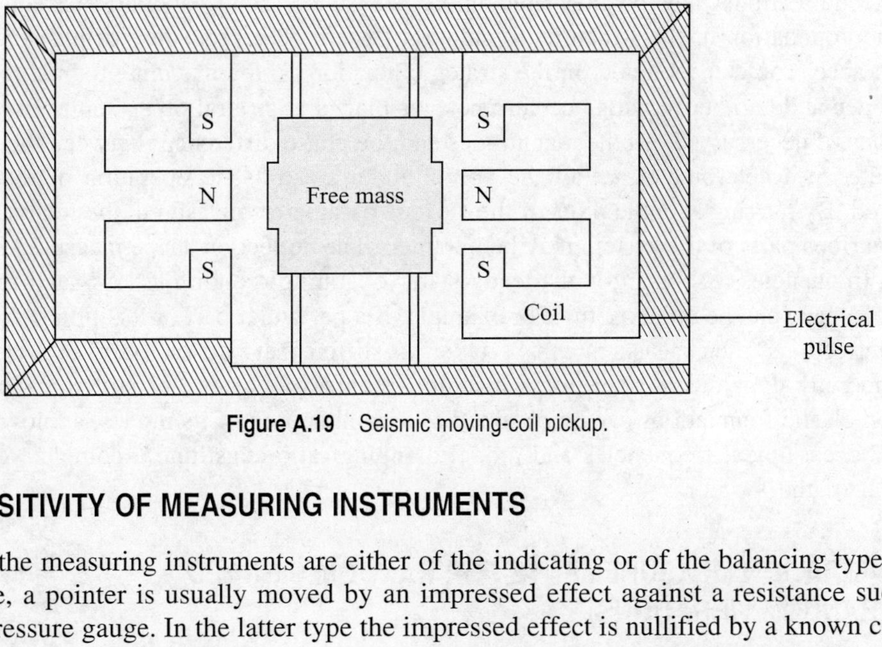

Figure A.19 Seismic moving-coil pickup.

A.7 SENSITIVITY OF MEASURING INSTRUMENTS

In general, the measuring instruments are either of the indicating or of the balancing type. In the former type, a pointer is usually moved by an impressed effect against a resistance such as a spring or pressure gauge. In the latter type the impressed effect is nullified by a known counter-effect to give zero deflection of the indicating element such as an electrical bridge. If x_0 is the correct value of a measurement record such as x, then $\delta x = (x_0 - x)$ is the absolute error, and $\delta x/x$ is the relative error. The sensitivity of an instrument is the change in reading per unit change in the measured quantity, e.g. the sensitivity of a spring balance may be quoted in mm/kg and that of an electrical strain gauge in fractional change of resistance per unit of strain.

If an instrument is being used to measure a quantity, which suddenly changes its value, e.g. a galvanometer when current is switched on, the indicating pointer oscillates at the natural frequency. The same thing happens if the instrument is given a jolt. An instrument having a light spring control (i.e. k is small) will have a low undamped natural frequency ω_n but is incapable of faithfully recording a rapidly-changing quantity; and a low-frequency instrument is susceptible to extraneous disturbances. If, however, the instrument is correctly damped it will record the mean value of a fluctuating quantity.

Since $\omega_n = \sqrt{\dfrac{k}{m}}$, it follows that to obtain a sensitive instrument for measuring the fluctuating quantities (including the effect of jolts in, say, moving vehicles), it is necessary that its natural frequency should be high, which requires m to be relatively small. This is most effectively done by the use of an electrical pickup or transducer in conjunction with an amplifier and indicating meter such as an oscilloscope whose electron beam has negligible inertia and, hence has an extremely rapid response.

A.8 DYNAMIC TESTING OF FOUNDATIONS AND STRUCTURES

The dynamic testing foundations and structures involves obtaining experimentally the deformations at a critical frequency. Bruel & Kjaer (1987) have suggested that the forced dynamic deflections may be measured under the steady state (operating) frequency of the machine–foundation or the structure–foundation system. The procedure of measurement starts with an accelerometer mounted on the strategic location on the machine–foundation system as a reference and another moving accelerometer is placed at several other points and in different directions, if necessary. Then the magnitudes and the phase differences between the moving and the reference accelerometers at all the points under steady-state operation of the system are measured. By making suitable plots of the various parameters measured, the relative movement of the various parts of the system may be obtained. The deflection shape measured is valid only for the frequencies associated with steady-state (operating) conditions. As such the measured deflation shape can be quite useful. For example, if a particular part or location is found to have large deflections, then measures can be taken to stiffen that part or location.

Since any dynamic response of a machine–foundation system or structure–foundation or beam on elastic foundations can be obtained as a combination of its modes, a knowledge of the mode shapes, modal frequencies and modal damping ratio constitute a complete dynamic description of the system.

A.9 VIBRATION MEASUREMENTS FOR RANDOM SIGNALS (RANDOM VIBRATIONS)

A building subjected to ground acceleration due to an earthquake, a water tank under a wind loading, a car running on a rough road and an aircraft flying in air represent random processes. Such random excitation is represented by random signals. Thus random signal analysis becomes necessary in vibration testing. If $x(t)$ is a random signal, as shown in Figure A.20, its average or mean, denoted by $\bar{x}$ is defined as

$$\bar{x}(t) = \lim_{T\to\infty} \frac{1}{N} \int_0^T x(t)\, dt$$

which, for a digital signal, may be written as

$$\overline{x} = \lim_{N\to\infty} \frac{1}{N} \sum_{i=1}^{N} x(t_i)$$

With reference to any random signal $y(t)$, a new variable as

$$x(t) = y(t) - \overline{y}(t)$$

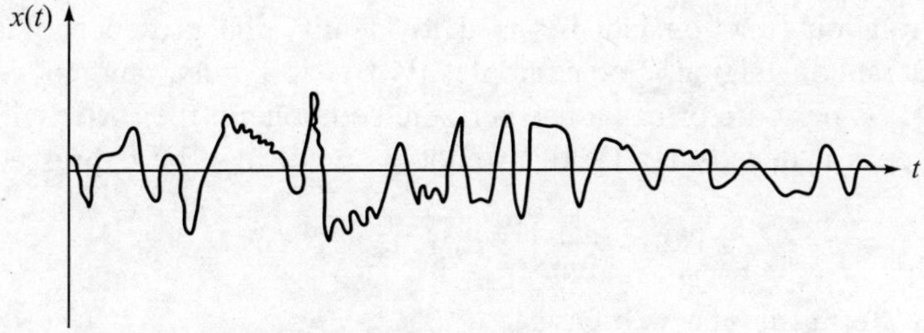

Figure A.20 A random signal, $x(t)$.

may be assumed such that the mean value of $x(t)$ is zero. Thus the signal represented by $x(t)$ may be assumed to have zero mean value. Then the mean square value or variance of $x(t)$, denoted as $x^2(t)$, may be described by

$$x^2(t) = \lim_{N\to\infty} \frac{1}{T} \int_0^T x^2(t)\, dt \tag{A.15}$$

Such parameter for digital signal takes the form

$$= \lim_{N\to\infty} \frac{1}{N} \sum_{i=1}^{N} x^2(t_i)$$

The root mean square (rms) value of $x(t)$ is given by

$$X_{\text{rms}} = \sqrt{\overline{x}_2} \tag{A.16}$$

The power spectral density (PSD) of a random signal $x(t)$, denoted by $S(\omega)$, represents the speed with which the signal changes in frequency domain.

Correlation is a measure of how well two functions relate with each other. Consider two functions $x(t)$ and $f(t)$, the cross-correlation between them is given by the average of the product $x(t)\, f(t + \tau)$

$$R_{fx}(\tau) = \lim_{T\to\infty} (1/T) \int_{-T/2}^{T/2} x(t)\, f(t+\tau)\, dt \tag{A.17}$$

The autocorrelation function of a random signal, denoted by $R_{ff}(\tau)$ gives a measure of the speed with which the signal changes in the time domain and is defined by

$$R_{ff}(\tau) = \lim_{T\to\infty} (1/T) \int_{-T/2}^{T/2} f(t)\, f(t+\tau)\, dt \tag{A.18}$$

The autocorrelation function $R_{ff}(\tau)$ for a digital signal may be expressed as

$$R(n,\ \Delta t) = \frac{1}{N-n}\sum_{i=0}^{N-n} x_i \cdot x_{i+n} \tag{A.19}$$

where

N = number of samples
Δt = sampling interval.

The autocorrelation function can be used to identity the presence of (hidden) periodic components in a random signal. For example, if $x(t)$ is purely random then $R(t) \to 0$ as $T \to \infty$. But if $x(t)$ is periodic or has a has periodic component, then $R(t)$ will also be periodic

The power spectral density (PSD) of a random signal may be expressed as

$$S(\omega) = \frac{1}{2\pi}\int_{-\infty}^{\infty} R_{ff}(\tau)\cdot e^{-i\omega\tau}\cdot d\tau \tag{A.20}$$

This is digital form may be written as

$$S(\Delta\omega) = \frac{|x(\omega)|^2}{N\,\Delta t} \tag{A.21}$$

where $|x(\omega)^2|$ represents the magnitude of the Fourier transform of the sampled data of $x(t)$.

The definition of autocorrelation and PSD functions can be extended for two different signals, such as a displacement signal $x(t)$ and an applied signal $f(t)$. This can lead to cross-correction function $R_{xf}(t)$ as

$$R_{xf}(t) = \lim_{T\to\infty}\frac{1}{T}\int_0^T x(\tau)\, f(t+\tau)\, d\tau \tag{A.22}$$

The cross-PSD $S_{xf}(\omega)$ may be written as

$$S_{xf}(\omega) = \frac{1}{2\pi}\int_{-\infty}^{\infty} R_{xf}(\tau)\cdot e^{-i\omega\tau}\cdot d\tau \tag{A.23}$$

If $f(t+\ \tau)$ is replaced by $x(t+\ \tau)$, then

$$R_{xx}(t) = \lim_{T\to\infty}\frac{1}{T}\int_0^T x(\tau)\, x(t+\tau)\cdot d\tau$$

and

$$S_{xx}(\omega) = \frac{1}{2\pi}\int_{-\infty}^{\infty} R_{xx}(\tau)\cdot e^{-i\omega z}\cdot d\tau$$

The frequency response function $H(\omega)$ may be expressed as

$$S_{xx}(\omega) = |H(\omega)|^2\ S_{ff}(\omega) \tag{A.24}$$

The $S_{xx}(\omega)$ contains information about the magnitude of the transfer function of the system (structures of foundations). Thus the frequency response function, $H(\omega)$ may be used to find the natural frequencies, damping ratios and mode shapes corresponding to all resonant peaks observed in the trace of $H(\omega)$. Figure A.21 shows the frequency-response function with its three peaks or resonances suggesting that system undergoing testing has three degrees of freedom.

However at times it is difficult to assign the number of degrees of freedom as the peaks are often closely spaced.

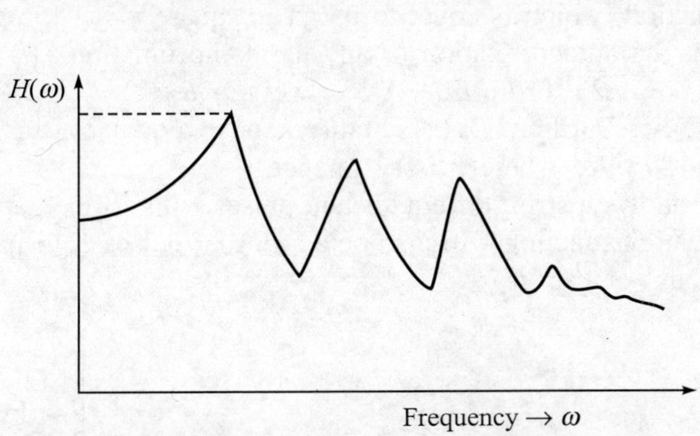

Figure A.21 A typical display of frequency-response function.

It is obvious that the cross-correlation between two functions depends on time delay τ. Once this value of τ is determined, further information about the possible transmission path can be obtained from the knowledge of velocity of propagation of signal. It may be noted that even if the two signals are buried in noise, the cross-correlation function will still remain the same, because the cross-correlation between two truly random functions will be zero.

Next, let us examine some properties of the cross-correlation function. It has two important applications.

1. To determine the time delay τ_d, which provides information about the transmission path of a signal
2. To detect common/similar signals from two noises contaminated signals.

Let us examine further properties of the autocorrelation function given above.

- Figure A.22 shows a function $f(t)$ delayed with itself by a time τ. As the delay period τ is increased, the probability of the product at some instant being positive and at another instant being negative increases. The net result is that the value of $R_{ff}(\tau)$ decreases as τ increases. The autocorrelation $f(t)\, f(t + \tau)$ reduces to $f^2(t)$ and $R_{ff}(0)$ reduces to the

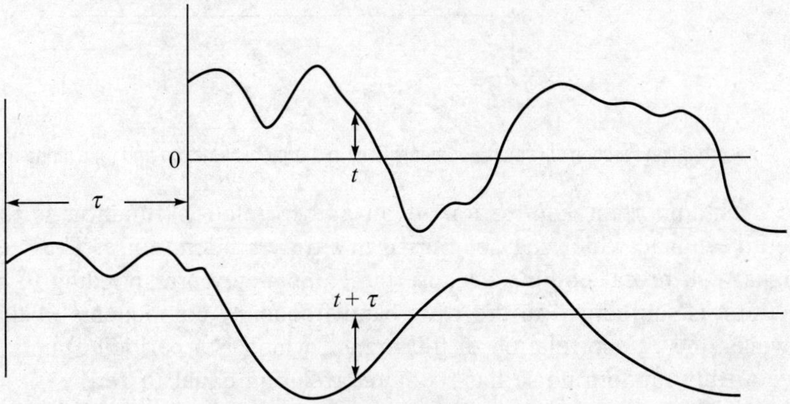

Figure A.22 Original and delayed time functions

mean square value of the function. Figure A.23a shows the autocorrelation function of a random function, which is equal to its mean square value for $\tau = 0$.

- Next consider the autocorrelation of any harmonic function $f(t) = F \sin(\omega t + \theta)$ as shown in Figure A.23. From Eq. (A.18), we get, $R_{ff}(\tau) = (1/2)F^2 \cos \omega t$. For $\tau = 0$. $R_{ff}(\tau) = (1/2)F^2$. Further, $R_{ff}(\tau)$ attains a peak equal to this mean square value whenever $\tau = 27\tau n/\omega$, where n is an integer.
- Figure A.23c shows a string of equidistant pulses with a time gap T. It is obvious that the autocorrelation function will also be a string of pulses equispaced on the τ-axis by the same amount of gap T.

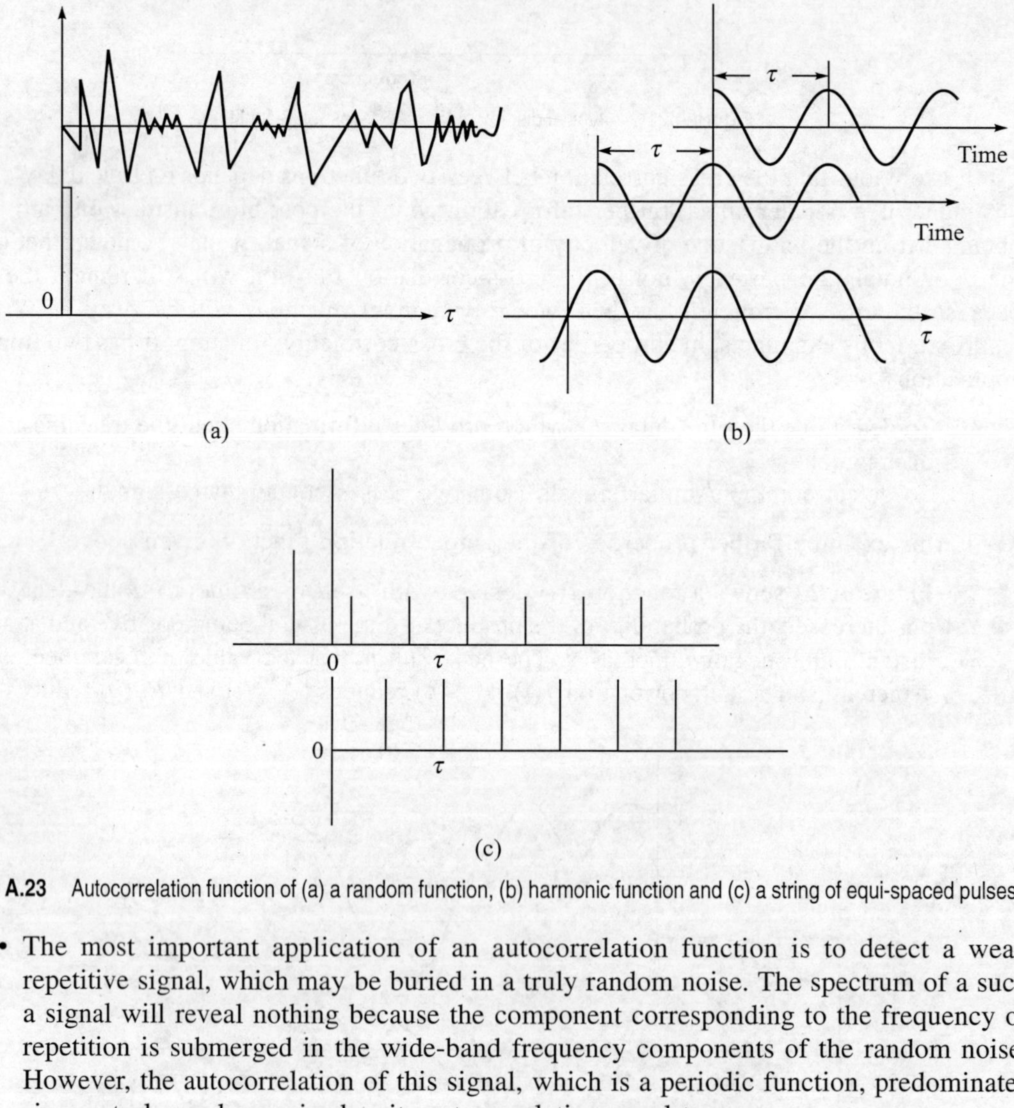

Figure A.23 Autocorrelation function of (a) a random function, (b) harmonic function and (c) a string of equi-spaced pulses.

- The most important application of an autocorrelation function is to detect a weak repetitive signal, which may be buried in a truly random noise. The spectrum of a such a signal will reveal nothing because the component corresponding to the frequency of repetition is submerged in the wide-band frequency components of the random noise. However, the autocorrelation of this signal, which is a periodic function, predominates since a truly random noise has its autocorrelation equal to zero.

A.9.1 Signal Analysis Techniques

The raw data from a vibration transducer mounted on a test structure is obtained in time domain. Therefore, the time domain analysis of the signal using strip-chart recorders or oscilloscopes is quite simple. While the time domain analysis is direct and does provide useful information, it has also severe limitations that it does not provide clear information about all the frequencies present in the response from the structure. To overcome this difficulty, the signal analysis in the frequency domain is performed. With the advent of Fast Fourier Transform (FFT) and faster microprocessors, it has now become easy to perform quick (almost instantaneously) transformations of functions from time domain to frequency domain and vice versa, so much so, that we call such an analysis to be in real time. Because of this fast transformation technique, a correlation function, which is essentially in time domain, is now evaluated via the frequency domain using FFT. Similarly, modulations in other systems which can be identified through time domain signals are better studied through spectral analysis.

A.9.2 Time Domain Analysis

When the signal is periodic, its time period can be measured as shown in Figure A.24. The frequency of repetition is $\omega = 2\pi/T$ rad/s. From the theory of Fourier analysis, we know that higher harmonics 2ω, 3ω, ... are present in a periodic signal. However, a limitation of this analysis is that the relative strengths of each of these harmonics cannot be determined.

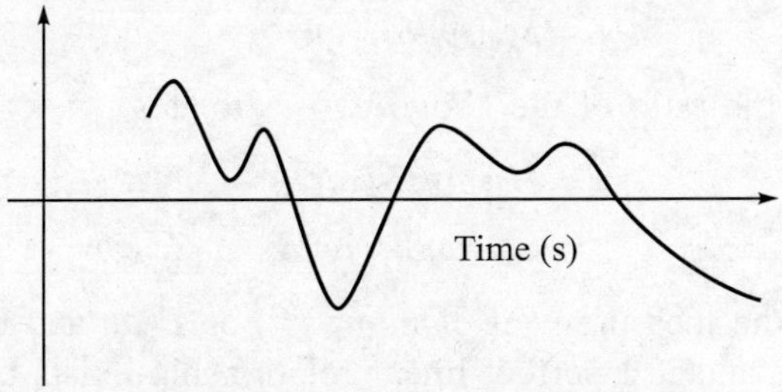

Figure A.24 A signal in time domain.

A.9.3 Frequency Domain Analysis

Frequency spectrum analysis or simply spectral analysis is a powerful technique of signal analysis. Frequency analysis essentially means determination of the spectrum of a vibratory signal. This can be carried out through hardware or through software based on either Fourier series, or by the Fourier transform. The transform equation may be expressed as

$$F(\omega) = \psi f(t) = \int_{-\infty}^{+\infty} f(t)\, e^{-i\omega t}\, dt \tag{A.25}$$

$$f(t) = \psi^{-1}\, F(\omega) = (1/2\pi) \int_{-\infty}^{+\infty} F(\omega)\, e^{i\omega t}\, dt \tag{A.26}$$

The integrals in the above equations are replaced by summations taken over selected intervals of a vibration record in time domain of time duration T to give the Discrete Fourier Transform (DFT).

A.9.4 Transfer Function

Coherence is a measure of degree of linear dependence between two signals, e.g. between the excitation force (input) and vibration response (output). The frequency response function (transfer function) $H(\omega)$ is given by

$$H(\omega) = X(\omega)/F(\omega) \tag{A.27}$$

Let us first multiply the numerator and denominator of the above equation by $F^*(\omega)$ to give an estimate $H_1(\omega)$ of the frequency response function as

$$\begin{aligned} H_1(\omega) &= X(\omega)/F(\omega) \\ &= X(\omega)F^*(\omega)/F(\omega)F^*(\omega) && \text{(A.28)} \\ &= S_{fx}(\omega)/S_{ff}(\omega) && \text{(A.29)} \end{aligned}$$

In a similar manner, we can write

$$\begin{aligned} H_2(\omega) &= X(\omega)/F(\omega) \\ &= X(\omega)X^*(\omega)/F(\omega)X^*(\omega) \\ &= S_{xx}(\omega)/S_{xf}(\omega) && \text{(A.30)} \end{aligned}$$

Coherence is defined as the ratio of the above two estimates

$$\begin{aligned} \gamma^2 &= H_1(\omega)/H_2(\omega) \\ &= S_{xf}(\omega)S_{fx}(\omega)/S_{ff}(\omega)S_{xx}(\omega) && \text{(A.31)} \end{aligned}$$

It is obvious that if the measurement of input $f(t)$ or $F(\omega)$ and the response $x(t)$ or $X(\omega)$ is free of any noise/error and if a perfect linear relationship exists between the input and the output as shown in Figure A.25a, the coherence should be unity. This, however, is rare. To ensure the reliability of any measurement, its coherence level should be sufficiently high, approaching unity.

Since $S_{xf}(\omega) = S^*_{fx}(\omega)$, coherence given by Eq. (A.31) can also be written as

$$\begin{aligned} \gamma^2 &= S^*_{xf}(\omega)S_{fx}(\omega)/S_{ff}(\omega)S_{xx}(\omega) \\ &= | S_{fx}(\omega) |^2/S_{ff}(\omega)\, S_{xx}(\omega) && \text{(A.32)} \end{aligned}$$

It may be noted that $S_{ff}(\omega)$, $S_{xx}(\omega)$ and the product $S^*_{fx}(\omega)\, S_{fx}(\omega)$ are all scalar quantities, hence coherence is a real number. Figure A.25b shows the presence of a linear relationship, however, there is random noise and coherence will be less than unity. In Figure A.25c, a nonlinear relationship can be clearly observed, where coherence will be less than unity. A case where there is no correlation at all between the input and output is shown in Figure A.25d and its coherence will be almost zero.

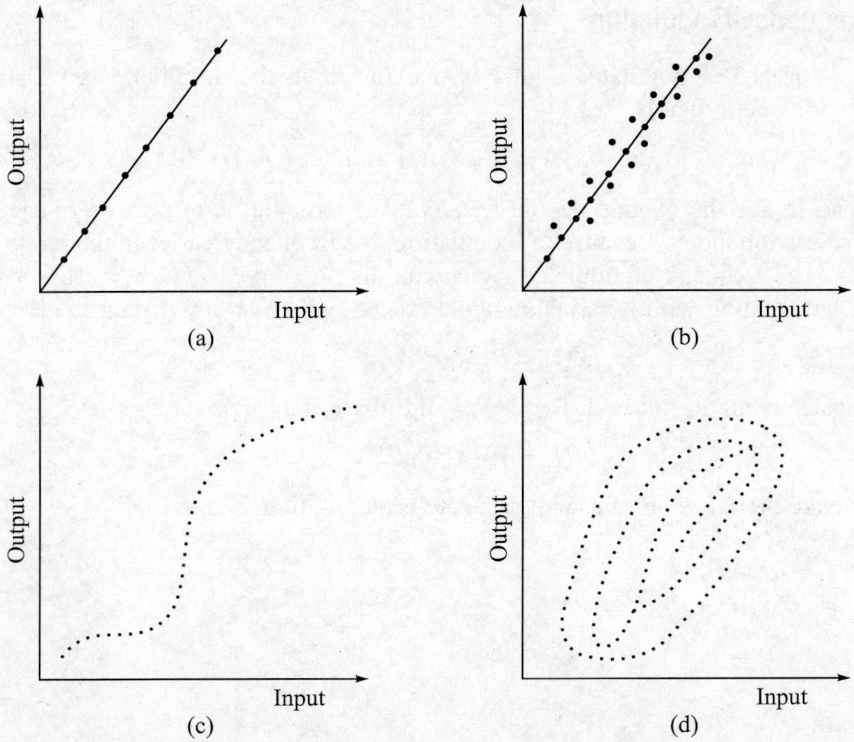

Figure A.25 (a) Good coherence; linear relationship between input and output, (b) linear relationship but noise is present, (c) non-linear relationship without noise, and (d) input and output are uncorrelated

A.9.5 Amplitude Modulation

Modulation is defined as a regular variation in amplitude or frequency or both in a periodic signal. An amplitude modulated signal can be defined as

$$x(t) = [x + x_a \cos \omega_a t] \cos \omega t \tag{A.33}$$

where x_a is the modulation amplitude and ω_a is the frequency of modulation. The amplitude varies in the range $(x - x_a)$ to $(x + x_a)$ periodically with modulation frequency ω_a. In Eq. (A.34), ω is the carrier frequency, usually much greater than the modulation frequency ω_a. It may be noted that both the functions related to the carrier frequency and modulation frequency need not be harmonic but are periodic in reality. In such a case, we can apply Fourier series expansion to both the carrier and modulation frequencies and get a series of carrier frequencies $n\omega$ ($n = 1, 2,$ ---------) with each frequency having an infinite number of modulating frequencies $n\omega_a$ ($n = 1, 2, ...$). In practical terms, however, first few harmonics will prevail.

An amplitude modulated signal given by Eq. (A.33) can also be denoted by a phasor of variable magnitude but rotating at a uniform rate of ω rad/s.

$$x(t) = ReX(t)\ e^{i\omega t} \tag{A.34}$$

A.9.6 Frequency Modulation

A frequency or phase modulated signal is one in which the amplitude is constant but the frequency varies periodically, i.e.

$$x(t) = X \cos (\omega + \omega' \cos \omega_a t)t \tag{A.35}$$

where ω' and ω_a are the magnitude and frequency of modulation respectively. The ratio ω'/ω_a is called modulation index. Because of modulation, the frequency varies in the range of $(\omega - \omega')$ to $(\omega + \omega')$. The frequency modulated signal can also be represented by a phasor of constant magnitude but rotating with a variable angular speed, which at any instant is given by $d\phi/dt$.

$$X(t) = \mathrm{Re}\, Xe^{i\phi(t)} \tag{A.36}$$

Thus a signal modulating in both frequency and phase may be expressed as

$$X(t) = \mathrm{Re}\, X(t)\, e^{i\phi(t)} \tag{A.37}$$

where $X(t)$ and $\phi(t)$ are periodic with respective modulation frequencies.

REFERENCES

Abramowitz, M. and I. Stegun (1964), *Tables of Mathematical Functions*, Dover, New York.

Abrahamson, N.A. and R.R. Young (1992), A Stable Algorithm for Regression Analyses Using the Random Effects Model, *Bulletin, Seismological Society of America*, Vol. 82, pp. 505–510.

Acharyya, S.K. and K.K. Ray (1977), Geology of Darjeeling Sikkim Himalaya: Guide to Excursion, No. 4, *4th International Gondwana Symposium*, New Delhi, pp. 1–25.

Agarwal, B.S. (1999), *Oscillations and Waves*, KNR, Nath Publishers, New Delhi.

Agrawal, S.K. and C. Ajay (2004), Estimation of Seismic Vulnerability of Buildings in Delhi, *World Congress on Natural Disaster Mitigation*, Institution of Engineers (I), Kolkata.

Airey, D.W. and D.M. Wood (1987), An Evaluation of Direct Simple Shear Tests on Clay, *Geotechnique*, Vol. 37, pp. 25–35.

Aki, K. (1993), Local Site Effects on Weak and Strong Ground Motion, *Tectonophysics*, Vol. 218, pp. 93–111.

Aki, K. and B. Chouet (1975), Origin of Coda Waves: Source, Attenuation and Scattering Effects, *Journal of Geophysics Research*, Vol. 80, pp. 3322–3342.

Aki, K. and K. Irikura (1991), Characterization and Mapping of Earthquake Shaking for Seismic Zonation, *Proc. 4th Intern. Conf. Seismic Zonation*, EERI, Oakland, Vol. 1, pp. 61–110.

Aki, K. and P.G. Richards, (2002), *Quantitative Seismology*, 2nd ed., University Science Books, Sausalito.

Algermissen, S.T., D.M. Perkins, P.C. Thenhaus, S.L. Hanson, and B.L. Bender (1982), A Probabilistic Estimate of Maximum Acceleration in Rock in the Contiguous United States, *United States, Geological Survey* (USGS), Open-file Report, Vol. 82–1033, p. 99.

Alonso, E.E. (1976), Risk Analysis of Slopes and Its Application to Slopes in Canadian Sensitive clays, *Geotechnique*, Vol. 26, No. 3, pp. 453–472.

Alvarez, L., F. Vaccari, and G.F. Panza (1999), Deterministic Seismic Zoning of Eastern Cuba, *Journal of Pure Appied Geophysics*, Vol. 156, pp. 469–486.

Ambraseys, N. and R. Bilham (2000), A Note on the Kangra Ms=7.8 Earthquake of 4 April 1905, *Current Science*, Vol. 79(4), pp. 45–48.

Ambraseys, N.N. (1995), The Prediction of Earthquake Peak Ground Acceleration in Europe, *Journal of Earthquake Engineering and Structural Dynamics*, Vol. 24, pp. 467–490.

Ambraseys, N.N. (2001), The Earthquake of 10 July 1894 in the Gulf of Izmit (Turkey) and Its Relation to the Earthquake of 17 August 1999, *Journal of Seismology*, Vol. 5, pp. 117–128.

Ambraseys, N.N. and C.F. Finkel (1995), The Seismicity of Turkey and Adjacent Areas, *EREN*, Istanbul, p. 240.

Ambraseys, N.N. and C.P. Melville (1982), *A History of Persian Earthquakes*, Cambridge Earth Science Series, Cambridge, UK.

Ambraseys, N.N., K.A. Simpson, and J.J. Bommer (1996), The Prediction of Horizontal Response Spectra in Europe, *Earthquake Engineering and Structural Dynamics*, Vol. 25, pp. 371–400.

Anderson, D.G., C. Espana, and V.R. Mclamore (1978), Estimating In-Situ Shear Moduli at Competent Sites, *Proc. ASCE Specialty Conference on Earthquake Engineering and Soil Dynamics*, Pasadena, Vol. 1, pp. 181–197.

Anderson, D.L., A. Ben-Menahem, and C.B. Archambeau (1965), Attenuation of Seismic Energy in the Upper Mantle, *Journal of Geophical Research*, Vol. 70, pp. 1441–1448.

Anderson, J.G. and J.E. Luco (1983), Parametric Study of Near-field Ground Motions or Oblique-slip and Dip-slip Dislocation models, *Bulletin Seismological Society of America*, Vol. 73, pp. 45–57.

Anderson, J.G., P. Bodin, J.N. Brune, J. Prince, S.K. Singh, R. Quaas, and M. Onate (1986), Strong Ground Motion from the Michoacan, Mexico, *Earthquake Science*, Vol. 233, pp. 1043–1049.

Anderson, R.A. (1967), *Fundamentals of Vibration*, Macmillan Co., New York.

Andrews, D.J. (1986), Objective Determination of Source Parameters and Similarity of Earthquakes of Different Size, *Earthquake Source Mechanics*, S. Das, J. Boatwright and C.H. Scholz (Eds.), AGU, Washington D.C., pp. 259–268.

Ang, A.H.S. (1973), Structural Risk Analysis and Reliability-based Design, *Journal of the Structural Division*, ASCE, Vol. 99, No. ST9, pp. 1891–1909.

Ang, A.H.S. and B. Ellingwood, (1971), Critical Analysis of Reliability Principles Relative to Design, *Proceedings International Conference on Application of Statistics and Probability to Soil and Structural Engineering*, Hongkong University Press, Hong Kong.

Ang, A.H.S. and C.A. Cornell (1974), Reliability Bases of Structural Safety and Design, *Journal Structural Division*, ASCE, Vol. 100, No. ST 9, pp. 1755.

Ang, A.H.S. and W.H. Tang, (1975), Probability Concepts in Engineering Planning and Design, *Basic Principles*, Vol. 1, Wiley, New York.

Ang, A.H.S. and W.H. Tang (1984), *Probability Concepts in Engineering Planning and Design: Decision, Risk, and Reliability*, Vol. II, John Wiley, New York.

Ansary, M.A., E. Yamazaki, and T. Katayama (1995), Statistical Analysis of Peak and Directivity of Earthquake Ground Motion, *Earthquake Engineering and Structural Dynamics*, Vol. 24, pp. 1527–1539.

Aoudia, A., F. Vaccari, P. Suhadolc, and M. Meghraoui (2000), Seismogenic Potential and Earthquake Hazard Assessment in the Tell Atlas of Algeria, *Journal of Seismology*, Vol. 4, pp. 79–98.

Arias, A. (1970), *A Measure of Earthquake Intensity in Seismic Design for Nuclear Power Plants*, R.J. Hansen (Ed.), MIT Cambridge, Massachusetts, pp. 438–483.

Arnold, R.N., G.N. Bycroft, and G.B. Warburton (1955), Forced Vibrations of a Body of an Infinite Elastic Solid, *J. of Appl. Mech.*, Trans. ASME, pp. 391–400.

ATC-21 (1988), Rapid Visual Screening of Buildings for Potential Seismic Hazards: A Handbook, *Applied Technology Council*, Redwood City, CA, USA.

ATC-40 (1996), Seismic Evaluation and Retrofit of Concrete Buildings, Vol. 1, *Applied Technology Council*, Redwood City, CA, USA.

Athanasopoulos, G.A., P.C. Pelekis, and E.A. Leonidou (1999), Effects of Surface Topography on Seismic Ground Response in the Egion (Greece) 15 June 1995 Earthquake, *Journal of Soil Dynamics and Earthquake Engineering*, Vol. 18, pp. 135–149.

Atkinson, G.M. and D.M. Boore (1995), New Ground Motion Relations for Eastern North America, *Bulletin Seismological of Society of America*, Vol. 85, pp. 17–30.

Auden, J.B. (1959), Earthquakes in Relation to the Damodar Valley Project, *Proc. 1st Symp. Earthq. Eng*., University of Roorkee, Roorkee.

Aviles, J. and L.E. Perez-Rocha (1998), Site Effects and Soil-Structure Interaction in the Valley of Mexico, *Journal of Soil Dynamics and Earthquake Engineering*, Vol. 17, pp. 29–39.

Ayyub, B.M. (Ed.) (1997), *Uncertainty Modeling and Analysis in Civil Engineering*, CRC Press, 528 pp.

Ayyub, B.M. and R.H. McCuen (1997), *Probability, Statistics, and Reliability for Engineers*, CRC Press, Boca Raton, 514 pp.

Baecher, G.B. and J.T. Christian (2003), *Reliability and Statistics in Geotechnical Engineering*, John Wiley & Sons, London and New York.

Baldi, G., R. Bellotti, V. Ghionna, M. Jamiolkowsky, S. Marchetti, and E. Pasquelini (1986). Flat Dilatometer Tests in Calibration Chambers, Proceedings, Situ,'86, Geotechnical Special Publication 6, ASCE, New York, pp. 431–446.

Bardet, J.P, H.K. Lee, H. Kanamori, P.C. Jennings, and C. Kisslinger (2003), Advances in Analysis of Soil Liquefaction during Earthquakes, *International Handbook of Earthquake and Engineering Seismology*, Vol. 2, IASPEI, pp. 1175–1201.

Bardet, J.P. et al. (2000), Soil Liquefaction, Landslides and Subsidence, Chapter 7 in 1999 Kokaeli. Turkey, *Earthquake*, Reconnaissance Report, pp. 141–162.

Bardet, J.P. and C.A. Davis (1996), Performance of San Fernando dam during 1994 Northbridge Earthquake, *Journal Geotechnical Engineering*, ASCE, Vol. 122(7), pp. 554–564.

Barkan, D.D. (1962), *Dynamics of Bases and Foundations*, McGraw-Hill, New York.

Basu, S. (1977), Statistical Analysis of Seismic Data and Seismic Risk Analysis of Indian Peninsula, *Ph.D Thesis*, Deptt. of Civil Engineering, IIT Kanpur, India.

Basu, S. (2003), Infrastructure Facility at DEQ, IIT Roorkee, Presentation at National Workshop on Earthquake Engineering Education by NPEEE, IIT Delhi.

Basu, S. and N.C. Nigam (1977), Seismic Risk Analysis of Indian Peninsula, *Proc. 6th World Conf. on Earthquake Engineering*, New Delhi, Vol. 2, pp. 425–433.

Basu, S. and N.C. Nigam (1978), On Seismic Zoning Map of India, *Proc. 6th Symposium, Earthquake Engineering*, Roorkee, Vol. 1, pp. 83–90.

Bath, M. (1966), *Earthquake Energy and Magnitude, Physics and Chemistry of Earth*, Ahren, L.H. Press, pp. 115–165.

Bath, M. and S.J. Duda (1964), Earthquake Volume, Fault Plane Area, Seismic Energy, Strain, Deformation and Related Quantities, *Ann. Geofis*., Vol. 17, pp. 353–368.

Baziar, M.H. and R. Dobry (1995), Residential Strength and Large Deformations Potential of Loose Silty Sand, *Journal Geotechnical Engineering*, ASCE, Vol. 121, pp. 896–906.

Beasley, M. (1981), *Reliability for Engineers: An Introduction*, Macmillan, London.

Beatty, J.K. and Chaikin, A. (Eds.) (1990), *The New Solar System*, 3rd ed., Sky Publishing Massachusetts.

Benioff, Hugo (1934), The Physical Evaluation of Seismic Destructiveness, *ibid*., Vol. 24, pp. 398–403.

Benjamin, J.R. (1974), Probabilistic Decision Analysis Applied to Earthquake Damage Surveys, Earthquake Engineering Research Institute, Open Report.

Benjamin, J.R. and C.A. Cornell (1970), *Probability Statistics and Decision for Civil Engineers*, McGraw-Hill, New York.

Ben-Menahem, A. and S.J. Singh (1981), *Seismic Waves and Sources*, Springer–Verlag, New York.

Ben-Menahem, A., E. Aboodi, and R. Schild (1974), Source of the Great Assam, Earthquake and Interplate Wedge Motion, *Phys. Earth and Planet. Institute*., Vol. 9, pp. 265–289.

Berg, G.V. (1961), Response of Multi-storied Structure to Earthquake, *Journal of Engineering Mechanics Division*, American Society of Civil Engineers, Vol. 87, EM2, pp. 1–16.

Berlin, G.L. (1980), *Earthquakes and Urban Environment*, Vol. 1, CRC, Florida.

Berrill, J.B. and R.O. Davis (1985), Energy Dissipation and Seismic Liquefaction of Sands, Revised Model. Soils and Foundations, *Japanese Society Soil Mechanics and Foundation Engineering*, Vol. 25(2), pp. 106–118.

Bhatia, S.C., H.K. Guta, and M. Ravi Kumar (2002), Probabilistic Seismic Hazard Map of India and Adjoining Areas, National Geophysical Research Institute, Hyderabad, India.

Bhattacharya, S.N., A.K. Shukla, R. Prakash, R. Shukla, and R.K. Verma (2001), Seismicity of National Capital Territory, Delhi, *Proc. International Seminar on Earthquake Strategies* with Particular Reference to India pp. 1–10.

Bhattacharya, S.N. and R.S. Duttatrayan (2003), Some Characteristics of Recent Earthquake Sequences in Peninsular India, *Gondwana Geol. Magazine*, Vol. 5, pp. 67–85.

Bhattacharya, S.N. and R.S. Dattatrayam (2000), Recent Advances in Seismic Instrumentation and Data Interpretations in India, *Current Science*, Vol. 79, No. 9.

Biggs, J.M. (1964), *Introduction to Structural Dynamics*, McGraw-Hill, New York.

Bilham, R. and V.K. Gaur (2000), The Geodetic Contribution to the Study of Seismotectonics in India, *Current Science*, Vol. 79, 1259–1269.

Biot, M.A. (1941), General Theory of Three-dimentional Consolidation, *Journal of Applied Physics*, Vol. 12, pp. 155–164.

Biot, M.A. (1962), The Mechanics of Deformable and Acoustic Propagation in Porous Media, *Journal of Applied Physics*, Vol. 33, No. 4, pp. 1482–1498.

Biot, M.A. and D.G. Willis (1957), The Elastic Coefficients of the Theory of Consolidation, *Journal Applied Mechanics*, Vol. 24, pp. 594–601.

Bjerrum, L. and A. Landra (1966), Direct Simple Shear Tests on Norwegian Quick Clay, *Geotechnique*, Vol. 26, No. 1, pp. 1–20.

Bolt, B.A. (1969), Duration of Strong Motion, *Proceedings of the 4th World Conference on Earthquake Engineering*, Santiago, Chile, pp. 1304–1315.

Bolt, B.A. (1975), *Nuclear Exlosion and Earthquakes*, W.H. Freeman, San Fransisco.

Bolt, B.A. (1988), *Earthquakes*, W.H. Freeman, New York, p. 282.

Bolt, B.A. (1989), The Nature of Earthquake Ground Motion, in F. Naeim (Ed.) *The Seismic Design Handbook*, Van Nostrand Reinhold, New York.

Bolt, B.A. (2003), *Earthquakes*, 5th ed., W.H. Freeman, New York.

Borges, J.F. and M. Castanheta (1972), Statistical Definition and Combination of Loads, Probabilistic Design of Reinforced Concrete Buildings, ACI, SP 31, 43.

Bott, M.H.P. (1982), *The Interior of the Earth, Its Structure, Constitution and Evolution*, 2nd ed., Edward Arnold, London.

Boussinesq, J.V. (1885), Application des potentials a 1 etude de 1' equilibre et du mouvement des solides elastiques, Paris, Gauthier Villars.

Bozorgnia, Y. and V.V. Bertero (2001), Improved Shaking and Damage Parameters for Post-earthquake Applications, *Proceedings of the SMIP01 Seminar on Utilization Strong-Motion Data*, Los Angeles, California, pp. 1–22.

Brillinger, D.R. and H.K. Preisler (1984), An Exploratory Analysis of the Joyner –Boore Attenuation Data, Bulletin Seismological Society of America, Vol. 74, pp. 1441–1450.

Broach, J.T. (1980), *Mechanical Vibration and Shock Measurements*, Bruel and Kjear, Naurem, Denmark.

Brune, J.N. (1970), Tectonic Stress and the Spectra of Seismic Shear Waves from Earthquakes, *Journal of Geophysical Research*, Vol. 75, pp. 4497–5009.

Burland, J.B. (1989), Small is Beautiful: the Stiffness of Soils at Small Strains, *Canadian Geotechincal Journal*, Vol. 26(4), pp. 449–516.

Bycroft, G.N. (1956), Forced Vibrations of a Rigid Circular Palate on a Semi-infinite Elastic Space and on an Elastic Stratum, Phil. Transactions Royal Society of London, Series A, Vol. 248, pp. 327–368.

Cagniard, L. (1935), *Reflexinet Refration des aOndessSeismiques Progresssives*, Gauthiers-Villars, Paris.

Campbell, K.W. (1981), Near Source Attenuation of Peak Horizontal Acceleration, *Bulletin Seismological Society of America*, Vol. 71, pp. 2039–2070.

Campbell, K.W. (1985), Strong Motion Attenuation Relations: A Ten-Year Perspective, Earthquake Spectra, Vol. I, pp. 759–804.

Cancani, A. (1904), Sur I'emploi d'une double echelleseismique des intensites, empirique et absolue, G. Beite., *Erganzungsband* 2, pp. 281–283. (Scale in terms of acceleration; modification of Mercalli scale to 12 degrees).

Carder, D.S. (1945), Seismic Investigations on the 5000 Foot Level, Homes Take Mine, S.D. Lead, *Earthquake Notes*, Vol. 21, pp. 13–14.

Casagrande, A. (1936), Characteristics of Cohensionless Soils Affecting the Stability of Slopes and Earth Fills, *Journal of the Boston Society of Civil Engineers*, reprinted in contributions, to Soil Mechanics Boston Society of Civil Engineers, 1940, pp. 257–276.

Casagrande, A. (1976), Liquefaction and Cyclic Mobility of Sands: A Critical Review, *Harvard Soil Mechanics Series*, 28, Harvard University, Cambridge, Massachusetts.

Casagrande, A. and W.L. Shannon (1948a), Stress Deformation and Strength Characteristics of Soils under Dynamic Loads, *Proc. Second International Conference Soil Mech Foundation Engineering*, Vol. 5, pp. 29–34.

Casagrande, A. and W.L. Shananon (1948b), Research on Stress Deformation and Strength Characteristics of Soils and Soft Rocks under Transient Loading, *Harvard Soil Mechanics Series*, 31.

Casciati, F. (1996), *Mathematical Models for Structural Reliability Analysis*, CRC Press, Boca Raton, p. 384.

Catuneanu, V.M. (1989), *Reliability Fundamentals*, Elsevier, New York, 1989.

Chatterjee, S. (2000), The Drift of India, A Conflict in Plate Tectonics in Paper on Earthquake Hazards and Mitigation in India with special reference to North Eastern Region, Bulletin (EN VIS), *Himalayan Ecology and Development*, Vol. 8, No. 2.

Chen, W.F. and C. Scawthorn (2002), *Earthquake Engineering Handbook*, CRC Press, Florida, USA.

Chi, Y.Y., L.T. Ou, and Y.R. Chen (2003), The Liquefaction Probability Model via Seismic Energy Dissipation Theory, The 10th Conference on Current Research in Geotechnical Engineering in Taiwan, Sanshia.

Cho, Y., P.C. Rizzo, and W.K. Humphries (1976), Saturated Sand and Cyclic Dynamic Tests, ASCE, *Ann. Conv. Expo*., Philadelphia, PA, Meet. Prepr. 2752, pp. 285–312.

Chopra, A.K. (2001), *Dynamics of Structures*, Prentice-Hall of India, New Delhi.

Christian, J.T., C.C. Ladd, and G.B. Baecher (1994). Reliability Applied to Slope Stability Analysis, *Journal of Geotechnical Engineering*, ASCE, Vol. 120, No. 12, pp. 2180–2207.

Christian, J.T. (2004), Geotechnical Engineering Reliability: How Well Do We Know What We Are Doing?, *Geotechnical and Geoenvironmental Engineering*, ASCE, Vol. 130, No. 10, pp. 985–1003.

Christian, J.T. and A. Urzua (1998), Probabilistic Evaluation of Earthquake-Induced Slope Failure, *Journal of Geotechnical and Geoenvironmental Engineering*, ASCE, Vol. 124, No. 11, pp. 1140–1143.

Church, A.H. (1963), *Mechanical Vibrations*, John Wiley & Sons, New York.

Clough, R.W. and J. Penzien (1993), *Dynamics of Structures*, 2nd ed., McGraw-Hill, New York.

Cornell, C.A. (1968), Engineering Seismic Risk Analysis, *Bulletin Seismological Society of America*, Vol. 58, pp. 1583–1606.

Cornell, C.A. (1971), Probabilistic Analysis of Damage to Structure under Seismic Loads, in *Dynamic Waves in Civil Engineering*, Howells D.A., I.P. Haigh and C. Taylor (Eds.), Wiley-Interscience.

Cornell, C.A. and H.A. Merz (1974), Earthquake Risk Analysis of Boston, ASCE National Structural Engineering Meeting, Cincimati, Ohio.

Couloumb (1776), Essai sur une application des regles de maximis et minimis a quelquePs problemes de statique relatifs a 1' architecture. Memoires de la mathematique et de physique, presentes *a 1' Academie Royale des sciences*, Paris, Vol. 7, pp. 343–382.

Cox, S.J. (1991), *Reliability, Safety, and Risk Management*: *An Integrated Approach*, Butterworth-Heinemann.

Craig, R.R. (1981), *Structural Dynamics*, John Wiley & Sons, New York.

Crandall, S.H. and N.C. Dahl (1959), *An Introduction to Mechanics of Solids*, McGraw-Hill, New York.

Criner, H.R. and G.B. Mecann (1953), Rail on Elastic Foundation under the Influence of High Speed Travelling Loads, *Journal of Applied Mechanics*, ASME, Vol. 20, pp. 13–22.

Culmann, K. (1866), Die graphische statik, Mayer and Zeller, Zurich.

Dai, S-H. (1992), *Reliability Analysis in Engineering Applications*, Van Nostrand Reinhold, New York, June, 448 pp.

Darrabh, R.B., M.J. Huang, and A.F. Shakal (1994), *Earthquake Engineering Aspects of Strong Motion.* Data from Recent California Earthquake, Proceedings, Fifth US National Conference Earthquake, *Engineering*, Vol. 3, Earthquake Engineering Research Institute, Oakland, CA, pp. 99–108.

Das, B.M. (1992), *Principles of Soil Dynamics*, McGraw Hill.

Das, S., I.D. Gupta, and V.K. Gupta (2006), A Probabilistic Seismic Hazard Analysis of North–East India, *Earthquake Spectra*, Vol. 22, No. 1, pp. 1–27.

Davis, R.O. and Berrill (1982), Energy Dissipation and Seismic Liquefaction in Sands, *Earthquake Engineering and Structural Dynamics*, Vol. 10, pp. 59–68.

Davison, C. (1933), On Scales of Seismic Intensity and on the Construction and Use of Isoseismic Lines, *B.S.S.A.*, Vol. 11(1921), pp. 95–129.

Day, W.R. (2002), *Geotechnical Earthquake Engineering Handbook*. McGraw Hill, New York,

Den Hartog, J.P. (1956), *Mechanical Vibrations*, 4th ed., McGraw-Hill.

Der Kiureghian, A. (1981), A Response Spectrum Method for Random Vibration Analysis of MDF Systems, *Journal of Earthquake Engineering Structural Dynamics*, Vol. 9, pp. 419–435.

Desai, C.S. and H.J. Siriwardane (1984), *Constitutive Laws for Engineering Materials with Emphasis on Geologic Materials*, Prentice Hall Inc., New Jersey.

Desai, C.S. and J.T. Christian (1977), *Numerical Methods in Geomechanics*, McGraw-Hill, New York.

Dewey, J. and P. Byerly (1969), The Early History of Seismometry [to 1900], *Bulletin, Seismological Society of America*, Vol. 59, p. 183.

Dier-Kiureghen, A. and Ang. (1977), A Fault Rupture Model for Seismic Risk Analysis, *Bulletin Seismological Society of America*, Vol. 67(4), pp. 1173–1194.

Dimarogonas, A.D. and S. Haddad (1992), *Vibrations for Engineers*, Prentice-Hall, Englewood Cliffs, New Jersey.

Ditlevsen, O. (1962), On the Probability of Failure of a Structure Subjected to a Countinuous and Random Varying Load Intensity, *Bygningsststiske Meddlelsee*, Copenhagen, Vol. 33, No. 3.

Ditlevson, O. (1996), *Structural Reliability Methods*, John Wiley & Sons, New York.

Dix, C.H., Reflection and Refraction of Seismic Waves, II, Discussion of Physics of Refraction Prospecting, *Geophysics*, Vol. 4, pp. 238–241.

Dobrin, M.B. (1957), Dispersion in Seismic Waves, *Geophysics*, Vol. 16, pp. 63–80.

Dobrin, M.B. (1973), *Introduction to Geophysical Prospecting*, 3rd ed., McGraw-Hill.

Dobry, R., A. Vasquez-Herrera, R. Mohammad, and M. Vucetic (1985), Liquefaction Flow Failure of Silty Sand by Torsional Cyclic Tests, *Advances in the Art of Testing of Soil Under Cyclic Loadings Condition*, ASCE, New York, pp. 29–50.

Dobry, R., R.S. Ladd, F.Y. Yokel, R.M. Chung, and D. Powell (1982), Prediction of Pore Water Pressure Build up, and Liquefaction of Sands during Earthquake by the Cyclic Stain Method, *NBS Building Science Series* 138, National Bureau of Standards, Gaithersburg, Maryland, p. 150.

Draper, N.R, and H. Smith (1981), *Applied Regression Analysis*, 2nd ed., John Wiley and Sons, New York.

Drnevich, V.P. (1967), Effect of Strain History on the Dynamic Properties of Sand, *Ph.D. Dissertation*, University of Michigan and Ann. Arbor.

Drnevich, V.P. (1972), Undrained Cyclic Shear of Saturated Sand, *Journal of Soil Mechanics and Foundation Engineering Division*, ASCE, SM 8, 807–825.

Duncan, J.M. (2000), Factors of Safety and Reliability in Geotechnical Engineering, *Journal of Geotechnical and Geoenvironmental Engineering*, ASCE, Vol. 126, No. 4, pp. 307–316.

Duncun, J.M. and S.G. Wright (2005), *Soil Strength and Slope Stability*, McGraw-Hill, New Jersey.

Elasser, W.M. (1969), Convection and Stress Propagation in the Upper Mantle. *The Application of Modern Physics to the Earth and Planetary Interiors*, S.K. Runcorn (Ed.), Wiley Interscience, London, pp. 223–246.

Esteva, L. (1970), Seismic Risk and Seismic Design Decisions, *Seismic Design of Nuclear Power Plants*, in R.J. Hansen (Ed.), MIT Press, Cambridge.

Esteva, L. and E. Rosenblueth (1964), Espectros de Tremblores a Distencies Moderodes Grendes, *Bulletin Seismological Society of Mexico*, Ing. Sismica, Vol. 2(1), pp. 1–18.

Esteva, L. and E. Rosenblueth (1972), *Use of Reliability Theory in Buildings Codes, Statistics and Probability in Civil Engineering*, Hong Kong University Press.

Esteva, L. and Villarverde, R. (1973), Seismic Risk, Design Spectra, and Structural Reliability, Proceedings, *5th World Conference, Earthquake Engineering*, Rome, pp. 2586–2596.

Esteva, L., C. Lomnitz, and E. Rosenblueth (1958), Seismic Risk and Engineering estimated from Acceleration Data and 1D Soil Profiles, *Bull. Seism. Soc. Am.,* Vol. 89.

Evans, P. (1964), The Tectonic Framework of Assam, *Journal, Geological Society*, India, Vol. 5, pp. 80–96.

Ewing, W.M., W.S. Jardetzky, and F. Press (1957), *Elastic Waves in Layered Media*, McGraw-Hill, New York.

Fellenius, W. (1948), *Erdstatische Berechungen*, 4th ed., W. Ernst Und Sohn, Berlin.

Fellenius, W. (1979), Calculation of the Stability of Earth Dams, Trasactions. Second Congress on Large Dams, Washington.

Fertis, D.G. (1973), *Dynamics and Vibrations of Structures*, John Wiley & Sons, New York.

Finn, W.D.L, P.M. Byrne, and R. Martui, (1976), Seismic Response and Liquefaction of Sands, *Journal of the Geotechnical Engineering Division*, ASCE, Vol. 102, GT8.

Finn, W.D.L. (1984), Dynamic Response Analysis of Soils in Engineering Practice, *Mechanics of Engineering Material*, John Wiley & Sons, Chapter 13.

Finn, W.D.L., R.H. Ledbetter, and G. Wu (1994), Liquefaction in Silty Soil: Design and Analysis, Ground Failures under Seismic Conditions, *Geotechnical Special Publications* 44, ASCE, New York, pp. 51–76.

Finn, W.D.L., K.W. Lee, and G.R. Martin (1977), An Effective Stress Model for Liquefaction, *Journal of Geotechnical Engineering Division*, ASCE, Vol. 193(6), pp. 517–533.

Finn, W.D.L., P.L. Bransby, and D.J. Pickering (1970), Effect of Strain History on Liquefaction of Sands, *Journal of Soil Mechanics and Foundations Division*, ASCE, Vol. 96, No. SM6, pp. 1917–1934.

Florin, V.A. and P.L. Ivanov (1961), Liquefaction of Saturated Sandy Soils, Proceedings *Fifth International Conference on Soil Mechanics and Foundation Engineering*, Paris, Vol. 1, pp. 107–111.

Foppal, A. (1922), *Verlesungen uber technische Mechanik*, Leipzig, p. 258.

Ford, G. and J.B. Haddow (1960), Determinging Machine Foundation Natural Frequencies by Analysis, *The Engineering Journal*, Vol. 43, No. 12, pp. 76–80.

Forel, F.A. (1884), Les tremblements de terre etudies par la commission sismologique suisse pendant I'annee 1881; 2me rapport, *Arch. Sciences phys. et nat.*, Vol. 11, pp. 147–182.

Forsyth, D.W. and S. Uyeda (1975), On the Relative Importance of the Driving Forces of Plate Motion, Geophysical, *Journal Royal Astr.* Soc., Vol. 43, pp.163–200.

Foscoe, K.H. (1953), An Apparatus for the Application of Simple Shear to Soil Samples, Proceedings, *International Conference Soil Mechanics and Foundation Engineering 3rd*, Azurich, Vol. 1, pp. 186–191.

Foster, R.J. (1971), *Physical Geology*, Charles E. Merrill, Columbus, Ohio.

Fowler, C.M.R. (1990), *The Solid Earth, An Introduction to Global Geophysics*, Cambridge University Press Cambridge, England, p. 472.

Franklin, A.G. and F.K. Chang (1977), Permanent Displacements of Earth Embankments, by Newmark Sliding Block Analysis, Report 5, Miscellaneous Paper, S-71-17, US, Army Corps of Engineers Waterway Experiment Station, Vicksburg, Mississippi.

Freudenthal, A.M. (1949), Safety of Structures, ASCE, *Transactions*, Vol. 112, pp. 125–180.

Freudenthal, A.M., J.M. Garretts, and M. Shinozuka (1966), The Analysis of Structural Safety *Journal of the Structural Division*, ASCE, Vol. 92, No. ST1.

Frohlick, D.R. (1953), The Factor of Safety with respect to Sliding of a Mass of a Soil along the Arc of a Logarithmic Spiral, Proceedings, *Third International Conference on Soil Mechanics and Foundation Engineering*, Vol. 2, Zurich, Switzerland.

Fung, Y.C. (1960), *Foundations of Solid Mechanics*, Prentice-Hall of India, New Delhi.

Galitzin, B. (1914), Vorlesungen, uber Seismometrie, Teubner, Leipzig.

Gannser, A. (1966), The Indian Ocean and the Himalayans: A Geological Interpretation, *Eclog. Geol. Helv.*, Vol. 39, p. 831.

Gannser, A. (1980), The Significance of the Himalayan Suture Zone, *Technophysics*, Vol. 62, pp. 37–52.

Gansser, A. (1964)., *Geology of the Himalayas*, Interscience Publications.

Garland, George D. (1971), *Introduction to Geo-Physics*, W.B. Saunders Company, Philadelphia.

Gaur, V.K. and R.K.S. Chouhan (1968), Quantitative Measures of Seismicity Applied to Indian Regions, *Bulletin Indian Society of Earthquake Technology*. Vol. 5, pp. 63–78.

Goto, N., H. Kagami, K. Shiono, and Y. Ohta (1977), An Easy-capable and High Precise Shear Wave Measurement by Means of the Standard Penetration Test. Proceedings, *Sixth World Conference on Earthquake Engineering*, pp. 171–176.

Griffiths, D.H. and R.F. King. (1965), *Applied Geophysics for Engineers and Geologists*, Pergamon Press, New York.

Grivas, D.A. and M.E. Harr (1977), Reliability with respect of Bearing Capacity Failures of Structures on Ground, Ninth International Conference on Soil Mechanics and Foundation Engineering, Tokyo, Japan.

Grivas, D.A. and M.E. Harr (1979), A Reliability Approach to Design of Soil Slope, Design Parameters in Geotechnical Engineering, *Seventh International Conference Soil Mechanics Foundation Engineering*, Vol. 1, British Geotechnical Society, London.

GSI (1939), The Bihar–Nepal Earthquake of 1934, Geological Survey of India, *Memoirs*, Vol. 7, p. 280.

GSI (1993), Bihar–Nepal Earthquake, August 20, 1988, Geological Survey of India, Special. Publications.

GSI (2000), *Seismotectonic Atlas of India and Its Environs*, P.L. Narula, S.K. Acharyya and J. Banerjee (Eds.), Geological Survey of India, Special Publications, Kolkata.

Gubbins, D. (1990), *Seismology and Plate Techtonics*, Cambridge University Press, Cambridge.

Gubin, I.E. (1960), *Pattern of Seismic Manifestation in Tajikistan*, U.S.S.R., Academy Press, Moscow.

Gubin, I.E. (1967), Basic Problems Seismotechnics, Lecture Notes No. 3, *International Institute of Seismology and Earthquake Engineering*, Press, Tokyo.

Gubin, I.E. (1968), Seismic Zoning of the Indian Peninsula, *Bulletin International Institute Seismology and Earthquake Engineering*.

Gubin, I.E. (1969), *Earthquakes and Seismic Zoning of Indian Peninsula*, C.W.P.C. 188, CWPW, Press, New Delhi.

Gubin, I.E. (1971), Multi-element Seismic Zoning Considered on the Example of Indian Peninsula, Izu, *Earth Physics*, Vol. 12, pp. 10–23.

Gucunski, N. and R.D. Wood (1991), Use of Rayleigh Modes in Interpretation of SASW Tests, Proceedings, *International Conference on Recent Advances in Geotechnical Engineering* and Soil Dynamics, St. Louis. Missouri, Vol. 2, pp. 1399–1408.

Guha, S.K. (1962), *Seismic Regionalization of India*, 2nd Symposium on Earthquake Engineering, Roorkee, 191–207.

Guntenberg, B. and C.F. Richter (1945), *Seismicity of Earth and Related Phenomenon*, Princeton University Press, Princeton, New Jersey.

Gupta, H.K. (1992), *Reservoir Induced Earthquakes*, Elsevier, Amserdam.

Gutenberg, B. (1945), Magnitude Determination for Deep Focus Earthquakes, *Bulletin of the Seismological Society of America*, Vol. 35, pp. 117–130.

Gutenberg, B. (1956), Magnitude Determination for Deep Focus Earthquakes, *Bulletin of Seimological Society of America*, Vol. 35, pp. 117–130.

Gutenberg, B. (1957), *Effects of Ground on Earthquake Motion*, BSSA, Vol. 47, pp. 221–250.

Gutenberg, B. and C.F. Richter (1954), *Seismicity of the Earth*, Princeton University Press.

Gutenberg, B. and C.F. Richter (1944), Frequency of Earthquakes in California, *Bullettin Seismological Society of America*, Vol. 34, pp. 1985–1988.

Gutenberg, B. and C.F. Richter (1954), *Seismicity of Earth and Associated Phenomena*, Princeton University Press, Princeton, New Jersey.

Gutenberg, B. and C.F. Richter (1956), Earthquake Magnitude, Intensity, Energy and Acceleration, *Bulletin of the Seismological Society of America*, Vol. 46, pp. 143–145.

Haldar, A. and S. Mahadevan (2000), *Probability, Reliability and Statistical Methods in Engineering Design*, John Wiley & Sons, p. 304.

Haldar, A. and W.H. Tang, (1979), Probabilistic Evaluation of Liquefaction Potential, *Journal of Geotechnical Engineering Division*, ASCE, Vol. 105, No. GT2, pp. 145–162.

Hall, J.R. Jr. and F.E. Richart, Jr. (1963), Dissipation of Elastic Wave Energy in Granular Soils, *Journal of Soil Mechanics and Foundations Division*, ASCE, Vol. 89 (SM-6), pp. 27–56.

Handy, Richart, L. (1995), The Day the House Fell, ASCE.

Hanks, T.C. and H. Kanamori (1979), A Moment-Magnitude Scale, *Journal of Geophysical Research*, Vol. 84, pp. 2348–2350.

Hanks, T.C. and R.K. McGuire (1981), The Character of High-frequency Strong Ground Motion, *Bulletin of the Seismological Society of America*, Vol. 71, pp. 2071–2095.

Hardin, B.C. (1971), Program of Simple Shear Testing of Soils, Univ. Ky., *Soil Mech.*, Ser. No. 8, pp. 1–14.

Hardin, B.O. and F.E. Richart, Jr. (1963), Elastic Wave Velocities in Granular Soils, *Journal of the Soil Mechanics and Foundations Division*, ASCE, Vol. 89(SM1), pp. 33–65.

Hardin, B.O. and J. Music (1965), Apparatus for Vibration of Soil Specimens during the Tri-axial Test, Instruments and Apparatus for Soil and Rock Mechanics, ASTM, STP 392, Vol. 55–74.

Hardin, B.O. and V.P. Drnevich (1972), Shear Modulus and Damping in Soils, Measurement and Parameter Effects, Proceedings, ASCE, Vol. 98, No. SM6, pp. 603–624.

Hardin, B.O. and W.L. Black, (1968), Vibration Modulus of Normally Consolidated Clay, *Journal of the Soil Mechanics and Foundations Division*, ASCE, Vol. 94, No. SM2, pp. 352–369.

Hardin, B.O., The Nature of Stress Strain Behaviour of Soils, State of The Art Report, Proc. *ASCE Specialty Conference on Earthquake Engineering and Soil Dynamics*, Pasadena.

Harr, M.E. (1966), *Foundations of Theoretical Soil Mechanics*, McGraw-Hill Inc., New York.

Harr, M.E. (1987), *Reliability-Based Design in Civil Engineering*, McGraw-Hill, New York.

Hart, Gary C. (1982), *Uncertainty Analysis of Loads and Safety in Structural Engineering*, Prentice Hall, Englewood Cliffs, p. 240.

Hateyni (1946), *Beams on Elastic Foundations*, The University Michigan Press, USA.

Haugen, E.B. (1968), *Probabilistic Approach to Design*, John Wiley & Sons, London.

Hazrell, S.H. (1978), Earthquake Aftershocks as Gereen's Function, *Geophysical Research Letters*, Vol. 5(1), pp. 1–4.

Heaton, T.H. and S.H. Hazrell (1986), Estimation of Strong Motions from Hypothetical Earthquakes on the Cascadia Subduction Zone, Pacific Northwest USGS Open File Report, 86–328.

Heaton, T.H., F. Tajimaand, and A.W. Mori (1986), Estimating Ground Motions Using Recorded Accelerograms, *Surv. Geophys*, Vol. 8, pp. 25–83.

Henley, E.J. (1981), *Reliability Engineering and Risk Assessment*, Prentice-Hall, Englewood Cliffs, New Jersey.

Hilton, H.H and M. Feigen (1972), Minimum Weight Analysis Based on Structural Reliability, *Journal of the Aerospace Science*, Vol. 27, No. 9.

Hodgson, J.H. (1956), A Seismic Probability Map of Canada, Canadian Underwriter, 23.

Housner, G.W. (1947), Characteristics of Strong-motion Earthquakes, *Journal of the Engineering Mechanics Division*, ASCE, Vol. 37, No. EM1, pp. 19–31.

Housner, G.W. (1952), Intensity of Ground Motion during Strong Earthquakes, California Institute of Technology, Pasadena.

Housner, G.W. (1959), Behaviour of Structures during Earthquakes, *Journal Engineering Mechanics Division*, ASCE, Vol. 85, No. EM 14, pp. 109–129.

Housner, G.W. (1965), Intensity of Earthquake Ground Shaking Near the Causative Fault, Proceedings, *Third World Conference on Earthquake Engineering*, Auckland, Vol. 1.

Housner, G.W. (1970), Measurement of Severity of Earthquake Ground Shaking, in Proceedings of the *US National Conference on Earthquake Engineering*, E.E.R.I. Ann Arbor, Michigan, pp. 25–33.

Housner, G.W. (1970), Strong Ground Motion, Chapter 4 of *Earthquake Engineering*, Robert L. Wiegel (Ed.), Prentice Hall, Englewood Cliffs, NJ, pp. 75–91.

Hubbert, M.K. and W.W. Rubey (1959), Role of Fluid Pressure in Mechanics Overthust Faulting, *Bulletin Seismological Society of America*, Vol. 70, pp. 115–166.

Hudson, D.E. (1970), *Reading and Interpreting Strong Motion Accelerograms*, Earthquake Engineering Research Institute, Berkeley, California, pp. 22–70.

Iai, S. (2001), Recent Advances in Seismic Analysis and Design of Retaining Structures, State of Art, Proceedings, *Fourth International Conference on Recent Advances in Geotechnical Earthquake Engineering and Soil Dynamics*, SOAP 4, pp. 1–28.

Iai, S. and J.P. Bardet (2001), Plain Strain Instability of Saturated Elasto-plastic Soil, *Geotechnique*, Vol. 51, No. 5, pp. 389–398.

Idriss, I.M. (1985), Evaluating Seismic Risk in Engineering Practice, Proceedings, *Eleventh International Conference on Soil Mechanics and Foundation Engineering*, San Francisco, Vol. 1, pp. 255–320.

Idriss, I.M. and J.I. Sun (1992), SHAKE91: A Computer Program for Conducting Equivalent Linear Seismic Response Analyses of Horizontally Layered Soil Deposits. *User's Guide*, University of California, Davis, California.

Idriss, K. (1984), Post Earthquake Failure of a Tailings Dam due to Liquefaction of the Pond Deposit, Proceedings, *International Conference on Case Histories in Geotechnical Engineering*, Rolla, Missouri, Vol. 3, pp. 1129–1143.

Iida, K. (1938), The Velocity of Elastic Waves in Sand, *Bulletin*, *Earthquake Research Institute*, Tokyo, Imp. University, Vol. 16, pp. 131–144.

Iida, K. (1940), On the Elastic Properties of Soil particularly, in relation to its water content, *Bulletin, Earthquake Research Institute*, Tokyo, Imp. University, Vol. 18, pp. 675–690.

Imai, T. (1977), P and S Wave Velocities of the Ground in Japan, Proceedings, *Ninth International Conference on Soil Mechanics and Foundation Engineering* Tokyo, Vol. 2, pp. 257–260.

Indian Standard Criteria for Earthquake Resistant Design of Structures (1962), IS:1893–1962, Indian Standard Institutions, New Delhi.

Indian Standard Criteria for Earthquake Resistant Design of Structures (1966), IS:1893–1966, 1st Rev., Indian Standard Institutions, New Delhi.

Indian Standard Criteria for Earthquake Resistant Design of Structures (1970), IS:1893–1970, 2nd Rev., Indian Standard Institutions, New Delhi.

Indian Standard Criteria for Earthquake Resistant Design of Structures (1975), IS:1893–1975, Indian Standard Institutions, New Delhi.

Indian Standard Criteria for Earthquake Resistant Design of Structures (1984), IS:1893–1984, 4th Rev., Indian Standard Institutions, New Delhi.

Indian Standard Criteria for Earthquake Resistant Design of Structures (2002), IS:1893–2002, Part 1, Indian Standard Institutions, New Delhi.

Indian Standard Criteria for Earthquake Resistant Design of Structures (2005), IS:1893–2002, Part 4, Indian Standard Institutions, New Delhi.

Indian Standard on Classification of Soil for General Engineering Purpose: IS (1971) 1498–1971, First Revision, Indian Standard Institutions, New Delhi.

Indian Standard Recommendations for Method of Load Test (1982), IS: 1888-1982, Indian Standard Institutions, New Delhi.

Indian Standard Recommendations for Method for Standard Penetration Test for Soils (1981), IS: 2131–1981, Indian Standard Institutions, New Delhi.

Indian Standard Recommendations for Determination of Dynamic Properties & Soil (1992), IS: 5244–(1992), Indian Standard Institutions, New Delhi.

Indian Standard Method of Test for Determination of Dynamic Properties of Soil, IS 5249–1978, First Revision, Indian Standard Institution, New Delhi.

Ishihara, K. (1982), Evaluation of Soil Properties for Use in Earthquake Response Analysis, *International Symposium on Numerical Model in Geomechanics*, Zurich, pp. 237–259.

Ishihara, K. (1985), Stability of Natural Deposits During Earthquakes, Proceedings of the Eleventh International on Soil Mechanics and Foundation Engineering, Vol. 1, San Fransisco, pp. 321–376.

Ishihara, K. (1971), Factors Affecting Dynamic Properties of Soil, Proceeding, *Fourth Asian Regional*, Conference in Soil Mechanics and Foundation Engineering Bangkok, Vol. 2.

Ishihara, K. (1993), Liquefaction and Flow Failures during Earthquakes, *Geotechnique*, Vol. 43, No. 3, London, England, pp. 351–415.

Ishihara, K. (1996), *Soil Behaviour in Earthquake*, Geotechnical Clarendon Press, Oxford.

Ishihara, K. and S. Li (1972), Liquefaction of Saturated Sand In Tri-axial Torsion Shear Test, *Soils Foundation* (Japan), Vol. 12(2), pp. 19–39.

Ishihara, K. and S. Yasuda (1975), Sand Liquefaction in Hollow Cylinder Torsion under Irregular Excitation, *Soils Found* (Japan), Vol. 15(1), pp. 45–59.

Ishihara, K., H. Tsuchiya, Y. Huang, and K. Kamada (2001), Recent Studies on Liquefaction Resistance of Sand-effect of Saturation, (CD-ROM), Keynote Lecture, *4th Intl. Conf. on Recent Advances in Geotechnical Earthquake Engineering and Soil Dynamics*, A. Prakash, S. (Ed.), March 26–31, San Diego.

Ishihara. K. (1993), Liquefaction and Flow Failure during Earthquake, *Geotechnique*, Vol. 43, No. 3, pp. 351–415.

Ishimoto, M. and K. Iida, (1937), Determination of Elastic Constants of Soil by Vibration Methods, *Bulletin Earthquake Research Institute*, Tokyo Imp. University, Vol. 15, p. 67.

Iwasaki, T., F. Tatsuokak, and Y. Takagi (1977), Shear Moduli of Sands under Cyclic Torsional Shear Loading. *Tech. Memo*., No. 1264, Public Works Res. Inst., Ministry of Construction, Chiba-Shi, Japan.

Iwasaki, T., T. Arakawa, and K. Tokida (1982), Simplified Procedures for Assessing Soil Liquefaction During Earthquakes, *Soil Dynamics and Earthquake Engineering Conference Southampton*, pp. 925–939.

Iyengar, R.N. (2000), Seismic Status of Delhi Megacity, *Curr. Sci*., Vol. 78, pp. 568–574.

Jacobean, L.S. and R.S. Ayer (1958), *Engineering Vibrations*, McGraw-Hill Inc.

Johnson, Richard A. (2002), *Probability and Statistics for Engineers*, Prentice-Hall of India, New Delhi.

Joyner, H.B. and D.M. Boore (1981), Peak Horizontal Acceleration and Velocity from Strong Motion Records including Records from 1979 Imperial Valley, California Earthquake, *Bulletin Seismological Society of America*, Vol. 71, pp. 2011–2038.

Joyner, H.B. and D.M. Boore (1988), Measurement, Characterization and Prediction of Strong Ground Motion, Proceedings, *ASCE Conference on Earthquake Engineering and Soil Dynamics II*, pp. 103–155.

JSCE (1997), *Dynamic Analysis and Earthquake Resistant Design*, Vols. 1–3, Oxford & IBH, New Delhi.

Juang, C.H., C.J. Chen, T. Jiang, and D. Andrus (2000), Risk-based Liquefaction Potential Evaluation Using Standard Penetration Test, *Journal of Canadian Geotechnical*, Vol. 37, pp. 1195–1208.

Juang, C.H., D.V. Rosowsky, and W.H. Tang (1999), Reliability-based Method for Assessing Liquefaction Potential of Soils, *Journal of Geotechnical and Geoenvironmental Engineering*, pp. 684–689.

Kaila, K.L. and N.M. Rao (1979), Seismic Zoning Maps of the Indian Subcontinent, *Geophysics Research Bulletin*, NGRI, Hyderabad, Vol. 17, pp. 293–301.

Kaila, K.L. and N.M. Rao (1980), Expected Maximum Magnitude Intensity and Peak Horizontal Ground Accelerations Maps of Himalayan Region, Proceedings *Seventh World Conference Earthquake Engineering*, Istanbul, Turkey, Vol. I, pp. 233–236.

Kaila, K.L., H.A. Narain, and V.K. Gaur (1972), Quantitative Seismology Map of India, *Bulletin Seismological Society of America*, Vol. 62, pp. 1119.

Kameswara Rao, N.S.V. (1998), *Vibration Analysis and Foundation Dynamics*, Wheeler Publishing, New Delhi.

Kanai, K. (1952), Relation between the Natural of Surface Layer and the Amplitude of Earthquake Motions, *Bulletin, E.R.R.I*, Tokyo, Vol. 30, pp. 31–37.

Kanai, K. (1957), Semi-empirical Formula for the Seismic Characteristics of the Ground, *Bulletin of the Earthquake Resarch Institute*, Tokyo University, Vol. 35, pp. 308–325.

Kanai, K. and T. Tanaka (1951), Observations of the Earthquake Motion at the Different Depths of the Earth, *Bulletin Earthquake Research Institute*, Tokyo, Vol. 29, pp. 107–113.

Kanamori, H. (1977), The Energy Release in Great Earthquakes, *Journal of Geophysical Research*, Vol. 82, pp. 2981–2987.

Kanamori, H. (1978), Quantification of Earthquakes, *Nature*, Vol. 271, pp. 411–414.

Kanamori, H. (1983), Magnitude and Qualification of Earthquakes, *Technophysics*, Vol. 93, pp. 185–199.

Kanamori, H., Anderson (1975), Theoretical Basis of Some Empirical Relation in Seismology, *Bulletin Seismic*, *Soc. Am.*, Vol. 65, pp. 1073–1095.

Kapur, K.C. (1977), *Reliability in Engineering Design*, John Wiley, New York, 1977.

Kasahara, K. (1981), *Earthquake Mechanics*, Cambridge University Press, Cambridge.

Kayal, J.R. (1987), Microseismicity and Source Mechanism Study, Shilong massif, Northeast India, *Bulletin Seismological Society of America*, Vol. 77, pp. 1513–1527.

Kayal, J.K. (1998), Seismicity of North East India and Surroundings—Development over 100 years, *Journal AaGeo*., Vol. XIX (1), pp. 9–34.

Kayal, J.R. (2001), Microearthquake Activity in Some Parts of the Himalaya and the Tectonic Model, *Tectonophys*., Vol. 339, pp. 331–351.

Kayen R.E., and J.K. Mitchell (1997), Assessment of Liquefaction Potential during Earthquakes by Arias Intensity, ASCE, *Geotechnical and Geoenvironmental Engineering*, Vol. 123(12), pp. 1162–1174.

Kayen, R.E., J.K. Mitchell, R.B. Seed, S. Nishio, and R. Continho (1992), Evaluation of SPT_CPT and Shear Wave Based Methods for Liquefaction Potential Assessment Using Loma Prieta Data, Proceedings, *Fourth U.S. and Japan Workshop on Earthquake Resistant Design of Lifeline Facilities and Countermeasure for Soil Liquefaction*, Vol. 1, pp. 177–204, NCEER 92-0019, Buffalo.

Keaey, P. and F.J. Vine (1990), *Global Tectonics*, Blackwell Scientific, London.

Kelly, S.G. (1993), *Fundamentals of Mechanical Vibrations*, McGraw-Hill Inc., New York.

Kerr, A. D. (1964), Elastic and Viscoelastic Foundations Models, JAM, ASME, Vol. 31, pp. 491–498.

Khattri, K., M. Wyss, V.K. Gaur, S.N. Saha and V.K. Bansal (1985), Seismic Gap. in Assam Area, *Indian Geology*, Vol. 6, pp. 615.

Khattri, K.N., A.M. Rogers, and D.M. Perkins (1978), Seismic Risk Map of Koyana, Bombay, Ahmedabad areas, India, *Sixth Symposium on Earthquake Engineering*, University of Roorkee, Roorkee, Vol. 1, pp. 91–96.

Khattri, K.N., A.M. Rogers, S.T. Algermissen, and D.M. Perkins (1984), A Seismic Hazard Map of India and Adjoining Areas, *Technophysics*, Vol. 108, pp. 93–134.

Kolsky, H. (1963), *Stress Waves in Solids*, Dover Publications, Inc., New York.

Kramer, S.L. (1966), *Geotechnical Earthquake Engineering*, Prentice Hall, International. Upper Saddle River, New Jersey.

Kreyszig, E. (1967), *Advanced Engineering Mathematics*, John Wiley & Sons, Inc., New York, USA.

Krishna, J. (1958), Earthquake Engineering Problems of India, *Journal of Institution of Engineers*, India.

Krishna, J. (1959), Seismic Zoning of India, Earthquake Engineering Seminar, University of Roorkee, India, pp. 24–38.

Krishna, J. (1954), Seismic Zoning of India, First Symposium, Earthquake Engineering, University of Roorkee, Roorkee.

Krishna, J., A.R. Chandrasekaran, and Brijesh Chandra (1994), *Elements of Earthquake Engineering*, South Asian Publishers, New Delhi, 1994.

Krishnan, M.S. (1953), The Structure and Tectonics of India, *Memoirs*, Geological Survey, India, Vol. 81, pp. 1–109.

Krishnan, M.S. (1956), *Geology of India and Burma*, Higginbothams, Madras.

Kulkarni, R.B., K. Sadigh, and I.M. Idriss (1979), Probabilistic Evaluation of Seismic Exposure, Proceedings, *Second US National Conference on Earthquake Engineering*, pp. 90–99.

Kulkarni, R.B., R.R Youngs, K. Sadgeh, and I.M. Idriss (1984), Assessment of Confidence Intervals for Results of Seismic Hazard Analysis, Proceedings of *Eighth World Conference on Earthquake Engineering*, Vol. 1, pp. 263–270, San Francisco.

L. Prandtl's: Uber die Harle plastischer Korper, Nachrichten von der Koni-glichen Gesellschaft der Wissenschaften zu Gottingen (Mathematisch-physikalische Klasse aus dem Jahre 1d920, Berlin, 1920).

Lacasse, S. and F. Nadim (1998), Risk and Reliability in Geotechnical Engineering, Proceedings, *4th International Conference on Case Histories in Geotechnical Engineering*, St. Louis, MO.

Lai, S., T. Morita, T. Kameoka, Y. Matsunaga, and K. Abiko (1995), Response of Dense Sand Deposit During 1993 Kushiro, Oki Earthquakes, *Soils and Foundation*, Vol. 35, pp. 115–131.

Lai, S., Y. Matsunaga, and T. Kameoka (1992), Strain Space Plasticity Model for Cyclic Mobility, Soils and Foundation, *Japanese Society of Soil and Foundation*, Vol. 32(2), pp. 1–15.

Lais, R., Die Wirkungen (1914), des Erdbebens vom 20. Juli 1913 in der Stadt Freiburg i. Br., *Mitt. Badische Geol. Landesanstalt*, Vol. 7, pp. 671–698.

Lamb, H. (1904), On the Propagation of Tremors Over the Surface of an Elastic Solid, *Philosophical Transactions*, London, Vol. 203, pp. 1–42.

Lawrence, F.V. Jr. (1963), Propagation Velocity of Ultrasonic Waves through Sand, MIT Research Report, R63–8.

Lay and Wallace, T.C. (1995), Modern Global Seismology, Academic Press, New York.

Lazer, A.C. and P.J. McKenna (1990), Large Amplitude Oscillations in Suspension Bridges: Some new Connections with Nonlinear Analysis, *SIAM Review*, Vol. 32, pp. 537–575.

Le Pichon, X. (1968), Sea Floor Spreading and Continental Drift, *Journal of Geophysical Research*, Vol. 73, pp. 3661.

Leemis, L.M. (1995), *Reliability: Probabilistic Models and Statistical Methods*, Prentice Hall, Englewood Cliffs, New Jersey.

Liao, S.S., D. Veneziano, and R.V. Whitman (1998), Regression Model for Evaluating Liquefaction Probability, *Journal of Geotechnical Engineering*, ASCE, Vol. 114, No. 4, pp. 389–411.

Lomnitz, C. (1974), *Global Tectonics and Earthquake Risks*, Elsevier Scientific Publishing Co., Amsterdam.

Love, A.E.H. (1911), *Some Problems of Geodynamics*, Cambridge University Press.

Love, A.E.H. (1944), *A Treatise on Mathematical Theory of Elasticaity*, Dovery Publications, New York.

Lumb, P. (1966), Variability of Natural Soils, *Canadian Geotechnical Journal*, Vol. 3, No. 2, pp. 74–97.

Lumb, P. (1970), Safety Factors and the Probability Distribution of Soil Strength, *Canadian Geotechnical Journal*, Vol. 7, No. 3, pp. 225–242.

Lumb, P. (1974), Application of Statistics in Soil Mechanics, J.K. Lee (Ed.), *Soil Mechanics – New Horizon*, American Elsevier Publishing Company, Inc., New York, pp. 44–111.

Mair, R.J. and D.M. Woods (1987), *Pressuremeter Testing: Method and Interpretations*, Butterworth, London, p. 160.

Major, A. (1980), *Dynamics in Civil Engineering*, Vol. 1, Akademical, Kiado, Budapest.

Makdsi, F.I. and H.B. Seed (1978), Simplified Procedures for Estimating Dam and Embankment Earthquake–induced Deformations, *Journal of the Geotechnical Engineering Division*, ASCE, Vol. 104, No. GT7, pp. 849–867.

Mal, A.K. and S.J. Singh (1991), *Deformations of Elastic Solids*, Prentice Hall, Englewood Cliffs, New Jersey.

Mallet, Robert (1862), *Effects of Neapolitan Earthquake of 1857*: The First Principle of Observational Seismometry, Chapmann & Hall, London, Vol. 1, pp. 26–29.

Marchetti, Silvano (1980), In-situ Tests by Flat Dilatometer, *Journal of the Geotechnical Engineering Division*, Vol. 106, No. GT3, pp. 299–321.

Maria, Paz (1983), *Structural Dynamics Theory and Computations*, 2nd ed., CBS Publication and Distribution, New Delhi.

Maslov, N.N. (1987), *Basic Engineering Geology and Soil Mechanics*, Mir Publishers, Moscow.

Matsuoka, M. and S. Midorikawa (1994), The Digital National Land Information and Maximum Acceleration in Bulgaria for Seismic Risk Reduction Purposes.

Matsuoka, M., and K. Kuroda (1974), Probabilistic Approach to Design of Embankments, Soils and Foundations.

McGuire, R.K. (1977), Seismic Design Spectra and Mapping Procedures using Hazard Analysis Directly on Oscillator Response, *Journal of Earthquake Engineering and Soil Dynamics*, Vol. 69, No. 4, pp. 1253–1265.

McKenna, P.J. and W. Walter (1987), Nonlinear Oscillations in a Suspension Bridge, *Archives for Rational Mechanics and Analysis*, Vol. 98, pp. 167–177.

Medvedev, S.V. (1953), Novaya seysmicheskaya shkala, Trudy Geofiz. Inst. Akad. Nauk SSSR, No. 21 (148).

Medvedev, S.V. (1956), Seysmometr dlya opredeleniya ball'nosti zemletryaseniy, *ibid.*, No. 36 (163), pp. 127–133.

Medvedev, S.V. (1958), A Seismic Zoning Map of USSR, in 1957, Trans. Inst. Phys. Earth. Acad., *Science*, USSR, 1 (168).

Medvedev, S.V. and V. Sponheur (1969), Scale of Seismic Intensity, Proceedings of *4th World Conference on Earthquake Engineering*, Santiago, Chile, pp. 143–153.

Meirovitch, L. (1971), *Analytical Methods in Vibration*, Macmillan Co., London.

Meirovitch, L. (1975), *Elements of Vibration Analysis*, McGraw-Hill, New York.

Mercalli, G. (1902), *Bull. Soc. Sismologica italiana*, Vol. 8, pp. 184–191, (Improved Mercalli scale, still with 10 degrees).

Meyerhof, G.G. (1970), Safety Factors in Soil Mechanics, *Canadian Geotechnical Journal*, Vol. 7, No. 4.

Miklowitz, J. (1953), Flexural Wave Solutions of Coupled Equations, Representing the More Exact of Bending, *Journal of Applied Mechanics*, ASME, Vol. 20, pp. 511–514.

Miklowitz, J. (1960), *Applied Mechanics Review*, 13, pp. 865–878.

Miller, R.P., J.H. Troncosco, and F.R. Brown (1975), In Situ Impulse Test For Dynamic Shear Modulus of Soils, Proceedings of the *Conference on In Situ Measurement of Soil Properties*, Geo-technical Engineering Division, ASCE. Specialty Conference, Rayleigh, North Carolina, Vol. 1, pp. 319–335.

Milne, W.A. and A.G. Davenport (1969), *Distribution of Earthquake Risk in Canada*, Missouri, pp. 1643–1648.

Mohorovicic, A. (1909), Das Beben Vom 8 x 1909, Jb, Met, Obs. Zagreb, Vol. 9, pp. 1–63.

Mononobe, N. (1929), Earthquake Proof Construction of Masonary Dams, Proceedings, *World Engineering Congress*, Vol. 9, p. 275, Tokyo.

Mononobe, N. and H. Matsuo (1929), On the Determination of Earth Pressure During Earthquakes, Proceedings, *World Engineering Congress*, p. 9.

Moon, F.C. (1987), *Chaotic Vibrations, an Introduction for Applied Scientists and Engineers*, Wiley, New York.

Mukhopadhyay, M. (2000), *Vibrations, Dynamics and Structural System*, Oxford & IBH, New Delhi.

Mukhopadhyay, S., Y. Pandey, R. Dharmaraju, R.K.S. Chauhan, P. Singh, and A. Dev (2002), Seisimic Microzonation of Delhi for Ground-shaking Site Effects, *Current Science*, Vol. 82, No. 7, p. 877.

Myslivec, Alois and Kysela (1978), *Developments in Geo-technical Engineering 21—The Bearing Capacity of Building Foundations*, Elsevier Scientific Publishing Company, Amsterdam.

Narayanan, S. and K. Jayaraman (1988), Chaotic Vibration in a Nonlinear Oscillator with Coulomb Damping, *Journal of Sound and Vibration*, Vol. 146, pp. 17–31.

Nasser, S.N. and A. Shokooh (1979), A Unified Approach to Densification and Liquefaction of Cohesionless Sand in Cyclic Shearing, *Journal of Canadian Geotech*, Vol. 16, pp. 659–678.

Nath, S.K. (2004), Seismic Hazard Mapping and Microzonation in the Sikkim Himalaya.

Nemat-Nasser's (1980), Liquefaction of Soil during Earthquakes, Dynamic Response of Structures, Experimentation, Observation, Prediction and Control, Proceedings of the *ASCE-EMD Second Speciality Conference*, Atlanta, Georgia, G. Hart. (Ed.), ASCE, NY, pp. 376–385.

Nemat-Nasser, S. and Sokooh (1979), A Unified Approach to Densification and Liquefaction of Cohesionless Sands in Cyclic Shearing, *Canadian Geotechnical Journal*, Vol. 16, pp. 659–678.

Neumann, F. (1954), *Earthquake Intensity and Related Ground Motion*, University of Washington Press, Seattle.

Newmark, N.M. (1965), Effects of Earthquakes on Dams and Embankments, *Geotechnique*, Vol. 15, No. 2. pp. 139–160.

Newmark, N.M. (1973), Ninth International Conference on Soil Mechanics and Foundation Engineering, Tokyo, pp. 605–50, July 1977.

Newmark, N.M. (1974), Comments of Conservatism in Earthquake Resistance Design, Unpublished report, presented to the US Atomic Energy Commission.

Newmark, N.M. and E. Rosenblueth (1971), *Fundamentals of Earthquake Engineering*, Prentice Hall, Englewood Cliffs, New Jersey, pp. 640.

Nordquist, J.M. (1945), Theory of Largest Values Applied to Earthquake Magnitudes, North America, *Bull. Seis. Soc. Am.*, Vol. 85, pp. 17–30.

Noson, L., A. Qamar, and G.W. Thorsen (1988), Washington State Earthquake Hazards, Information Circular 85, *Washington Division of Geology and Earth Resources*, Olympia, Washington.

Nowak, A.S. and K.R. Collins (2000), *Reliability of Structures*, McGraw-Hill, New York.

Nowacki, W. (1963), *Dynamics of Elastic Systems*, Chapman & Hall, London.

O. Mohr (1906). *Technische Mechanic*, Berlin, William Ernst und Sohn.

Okabe, S. (1924), General Theory on Earth Pressure and Seismic Stability of Retaining Wall and Dams during Earthquakes, *Journal, JSCE*, Vol. 6.

Okada, N. and S. Nemat-Nasser (1994), Energy Dissipation in Inelastic Flow of Saturated Cohesionless Granular Media. *Geotechnique*, Vol. 44(1), pp. 1–19.

Oldham, R.D. (1882–1883), A Catalogue of Indian Earthquakes from the Earliest Time to the End of A.D. 1869, *Memoirs*, X., Geophysical Survey of India.

Oldham, R.D. (1899), Report on the Great Earthquake of 12th June 1987, *Memoirs*., Geol. Survey India, Vol. 29, pp. 1–379.

Oldham, R.D. (1906), The Constitution of the Interior of the Earth, as Revealed by Earthquakes, Quarterly *Journal of Geological Society of London*, Vol. 62, pp. 456–475.

Omachi, T., Y. Nakamura, and T. Toshinawa (1991), Ground Motion Characteristics on Deterministic Hazard Computation, *Pure Appl. Geophys.*, Vol. 157, p. 171.

Orowan, E. (1965), Convection in a Non-Newtonian Fluid, Continental Drift and Mountain Building, *Philosphical Transactions*, Royal Society of London, Vol. 258A, pp. 284–313.

Pande, G.N. and O.C. Zienkiewicz (Eds.) (1982), *Soil Mechanics Transient and Cyclic Loads*, Wiley Chichester, U.K.

Papoulis, A. (1965), *Probability, Random Variables, and Stochastic Processes*, McGraw-Hill, New York.

Parzen, E. (1960), *Modern Probability Theory and Its Applications*, John Wiley & Sons, Inc., New York.

Peacock, W.H. and H.B. Seed (1968), Sand Liquefaction under Cyclic Loading Simple Shear Conditions, *J. Soil Mech. Found. Div.*, Am. Soc. Civ. Engg., Vol. 94(SM-3), pp. 689–708.

Peirs, T.A. and N. Yoshida (1996), Modelling of Volume Change Characteristics of Sand Cyclic under Loading, Proceedings *Eleventh World Conference on Earthquake Engineering*, Acapulco, Mexico, Paper No. 1087.

Phillips, W.S. and K. Aki (1986), Site Amplification of Coda Waves from Local, *Planet. Inter.*, Vol. 137, pp. 13–29.

Plaut, R.H. and J.C. Hsieh (1987), Chaos in a Mechanism with Time Delays Under Parametric and External Excitation, *Journal of Sound and Vibration*, Vol. 114, pp. 73–90.

Porter, L.D. and D.J. Leeds (2000), Correlation of Strong Ground Motion for the 1994 Possibilities and Limitations of Nakamura's Technique, *J. Phy. Earth*, Vol. 42, pp. 377–397.

Poulos, S.J., G. Castro, and J.W. France (1985), Liquefaction Evaluation Procedures, *Journal of Geotechnical Engineering*, ASCE. New York, Vol. 111, No. 6, pp. 772–792.

Prakash, S. (1981), *Soil Dynamics*, McGraw-Hill, New York.

Prakash, S. and V.K. Puri (1981), Dynamic Properties of Soils From In-Situ Tests, *J. Geotech. Engineering*. Division, American. Society of Civil. Engineers, Vol. 107 (GT-7), pp. 943–963.

Prakash, S., G. Ranjan, and S. Saran (1979), *Analysis and Design of Foundations and Retaining Structures*, Sarita Prakashan, Meerut, India.

Prakash, S., P. Nandkumaran, and V.H. Joshi (1973), Design and Performance of an Oscillatory Shear Box, *J. Indian Geotech. Soc.*, Vol. 3, No. 2, pp. 101–112.

Prasad, Bharat Bhushan (1978), Propagation of Flexural Wave in Beams on Elastic Foundations, Proceedings *6th Symporium on Earthquake Engineering*, University of Roorkee, pp. 395–400.

Prasad, Bharat Bhushan (1983), Soil-structure in Nuclear Structures, Proc. *International Workshop on Soil-structure Interactions*, University of Roorkee, Roorkee, pp. 194–197.

Prasad, Bharat Bhushan and N.S.V. Kameshwar Rao (1979), Beam Foundation Interaction due to Impact Loads, *Indian Geotechincal Journal*, New Delhi, Vol. 9, pp. 257–278.

Prasad, Bharat Bhushan and N.S.V. Kameswar Rao (1980), Response of Beams on Elastic Foundation to Dynamic Loading, *Journal of the Engineering Mechanics Division*, Proceedings of the *American Society of Civil Engineers*, Vol. 106, No. EM1, USA, pp. 179–185.

Prasad, Bharat Bhushan and N.S.V. Kameswar Rao (1975), Impact Loads on Beam on Elastic Foundation, 3rd SmiRT Conference on Structure Mechanics in Reactor Technology, London, 1975, pp. J5/8.

Prasad, Bharat Bhushan (1984), Dynamic Stability of Beam on Elastic Foundation. Proc. *5th ASCE Specialty Conference*, Wyoming, USA, pp. 1476–1479.

Prasad, Bharat Bhushan (2003), Seismic Hazard Analysis and Vulnerability Studies of Seismic Prone Areas of Bihar, Workshop on Retrofitting of Structures, Department of Earthquake Engineering and Indian Society of Earthquake Engineering, IIT Roorkee, pp. 45–65.

Press, E. and R. Siever (1994), *Understanding Earth*, Freeman & Company, p. 593.

Prevost, J.H. (1985), A Simple Plasticity Theory for Frictionless Cohesionless Soils, *Journal Soil Dynamics and Earthquake Engineering*, Vol. 4, No. 1, pp. 9–17.

Rai, S.N., D.M. Ramana, and A. Manglik (Eds.) (2002), *Dynamics of Earth's Fluid System*, Oxford & IBH, New Delhi.

Rankine, W.J.M. (1857), On the Stability of Loose Earth, *Philosophical Transactions*, Royal Society, London.

Rao, J.S. and K. Gupta (1999), *Theory and Practice of Mechanical Vibrations*, New Age International Publishers, New Delhi.

Rao, S.S. (1990), *Mechanical Vibrations*, 2nd ed., Addison-Wesley, Reading, Mass.

Rao, S.S. (1992), *Reliability Based Design*, McGraw Hill.

Rastogi, B.K. (2000), Chamoli Earthquake of Magnitude 6.6 on 29th March 1999, *J. Rates, Geophys. J. Int.*, Vol. 152, pp. 94–112.

Rastogi, B.K., R.K. Chadha, C.S.P. Sarma, P. Mandal, H.V.S. Satanarayana, I.P. Raju, N. Kumar, Satyamurty, and A. Nageshwar Rao (1997), Seismicity at Warna Reservoir (near Koyana), through 1995, *Bulletin, Seismological Society of America*, pp. 1184–1194.

Rayleigh, Lord (1885), On Waves Propagated along the Plane Surface of an Elastic Solid, Proceedings of the *London Mathematical Society*, Vol. 17, pp. 4–11.

Reedy, J.N. (1993), *An Introduction to Finite Element Method*, 2nd ed., McGraw-Hill, New York.

Reid, H.F. (1910), Mechanics of the Earthquake, the California Earthquake of April 18, 1906, Report Earthquake Investigation Commission, Vol. 2, Carnegie Inst. Washington.

Reissner, E. and H.F. Sagoci (1944), Forced Torsional Oscillations of an Elastic Half-space, *Journal of Appl. Phys.*, Vol. 15, pp. 652–662.

Reiter, L. (1990), *Earthquake Hazard Analysis—Issues and Insights*, Colombia University Press, New York.

Richard, A. (2000), *Probability and Statistics for Engineers*, Prentice-Hall of India, New Delhi.

Richard, R., Jr. and D.G. Elms (1979), Seismic Behavior of Gravity Retaining Wall, *Journal Geotechnical Engineering Division*, ASCE, Vol. 105, No. GT4, pp. 449–464.

Richart, F.E. (1962), *Foundation Vibrations*, Trans. ASCE, Vol. 127, Part I, pp. 863–898.

Richart, F.E. and R.D. Woods (1972), In-situ Shear Wave Velocity by Cross-Hole Method, *J. Soil Mech. Found. Div.*, ASCE, Vol. 98, p. SM 5, pp. 443–460.

Richart, F.E. and R.J. Hoar (1978), Variables Affecting In-Situ Seismic Measurements, Proc. *ASCE Specialty Conference on Earthquake Engineering and Soil Dynamics*, Pasadenma, Vol. 2, pp. 919–938.

Richart, F.E., Jr. (1997), Dynamic Stress Strain Relations For Soils, State of the Art Report, Proc. *Ninth International Conference on Soil Mechanics and Foundation Engineering*, Tokyo, Vol. 2, pp. 605–612.

Richter, C.F. (1935), An Instrumental Earthquake Scale, *Bulletin Seismological Society of America*, Vol. 25, pp. 1–32.

Richter, C.F. (1957), *Elementary Seismology*, Eurasia Publishing House, New Delhi.

Richter, C.F. (1958), *Elementary Seismology*, W.H. Freeman, San Francisco.

Richter, C.F. (1959), Seismic Regionalisation, *Bull., Scism. Soc. America*, Vol. 49, pp. 123–162.

Robertson, P.K. and C.E. Wride (1998), Evaluation Cyclic Liquefaction Potential Using the Cone Penetration Test, *Canadian Geotechnical Journal*, Vol. 35, pp. 442–459.

Robertson, P.K. and R.G. Campanella (1983), *Interpretations of Cone Penetration Tests*, Parts 1 and 2, *Canadian Geotechnical Journal*, Vol. 20, pp. 718–745.

Robertson, P.K. and R.G. Campanella (1989), *Guidelines for Geotechnical Design Using the Cone Penetration Test and CPT with Pore Pressure Measurement*, 4th ed., Hogentogler and Co., Columbia.

Robertson, P.K. and R.G. Campanella (1983), Interpretation of Cone Penetration Tests: Part 1 and 2, *Canadian Geotechnical Journal*, Vol. 20, pp. 718–745.

Robertson, P.K., and C.E. Wride (1998), Evaluation Cyclic Liquefaction Potential Using the Cone Penetration Test, *Canadian Geotechnical Journal*, Vol. 35, pp. 442-459.

Robertson, P.K., R.G. Campanella, D. Gillespie, and J. Geig (1986), Use of Piezometer Cone Data, Use of In-situ Tests in Geotechnical Engineering, In-situ' 86, ASCE, *GSPG*, pp. 1263–1280.

Rokini, M.A. and B.S. Berger (1988), Chaotic Motion of a Two Link Mechanism, *Journal of Sound and Vibration*, Vol. 147, pp. 349–357.

Romani, F., C.W. Lovell Jr., and M.E. Harr (1972), Influence of Progressive Failure on Slope Stability, *Journal of American Society of Civil Engineers*, Soil Mech. and Foundation Division.

Rossi, M.S. (1883), Programma dell'osservatorio ed afrchivo centrale geodinamico", *Boll. Del vulcanismo italiano*, Vol. 10, pp. 3–124.

Roy R. Craig, Jr. (1981), *Structural Dynamics*, John Wiley & Sons.

Saxena, K.R. (1994), *Geo-technical Engineering—Emerging Trends in Design and Practice*, Oxford & IBH Publishing Co., New Delhi.

Schlaifer, R. (1969), *Analysis of Decisions Under Uncertainty*, McGraw-Hill, New York.

Schnabel, P.B., J. Lysmer, H.B. Seed (1972), SHAKE, A Computer Program for Earthquake Response Analysis of Horizontally Layered Sites. Report No. EERC 72-12, Earthquake Engineering Research Centre, University of California, Berkeley, California.

Scholz, C.H. (2002), *The Mechanics of Earthquake Faulting*, Cambridge University Press.

Schultze, E. (1972), Frequency Distribution and Correlation of Soil Properties, Statistics and Probability in Civil Enginreering, Hong Kong University Press.

Schwarz, K., *Bulletin of the Earthquake Research Institute*, Tokyo, Japan, Vol. 6, 1929, pp. 1–18.

Schwarz, S.D. and J.M. Musser, Jr. (1972), Various Techniques for Making In-Situ Test Shear Wave Velocity Measurements—A Description and Evaluation, Proceedings *International Conference on Microzonation*, Seattle, Washington, Vol. 2, pp. 594–608.

Seed, H.B. (1960), Soil Strength during Earthquakes, Proc. *Second World Conf. Earthquake Engineering*, Vol. 1, pp. 183–194.

Seed, H.B. (1970), Soil Problems and Soil Behaviour, Chapter 10 of *Earthquake Engineering*, Robert L. Wiegel (Ed.), Prentice-Hall, Englewood Cliffs, New Jersey, pp. 227–251.

Seed, H.B. (1979a), Soil Liquefaction and Cyclic Mobility Evaluation for Level Ground During Earthquakes, *Journal of the Geotechnical Engineering Division*, ASCE, Vol. 105, No. GT2, pp. 201–255.

Seed, H.B. (1979b), Considerations in the Earthquake-resistant Design of Earth and Rockfill Dams, Geotechnique, Vol. 29, No. 3, pp. 215–263.

Seed, H.B. (1987), Design Problems in Soil Liquefaction, *Journal of Geotechnical Engineering*, ASCE, Vol. 113, No. 8, pp. 827–845.

Seed, H.B. (1987), Design Problems in Soil Liquefaction, *Journal Geotechnical Engineering*, ASCE, 113(8), pp. 1129–1143.

Seed, H.B. and C.K. Chan (1966), Clay Strength Under Earthquake Loading Conditions, *Journal of the Soil Mechanics and Foundation Division*, ASCE, Vol. 92, No. SM 2, pp. 53–78.

Seed, H.B. and I.M. Idriss (1970), Soil Moduli and Damping Factors for Dynamic Response Analyses, Report No. EERC-70/10, Earthquake Engineering Research Centre, University of California, Berkeley, California.

Seed, H.B. and I.M. Idriss (1971), Simplified Procedure for Evaluating Soil Liquefaction Potential, *Journal of the Soil Mechanics and Foundation Division*, ASCE, Vol. 107, No. SM 9, pp. 1249–1274.

Seed, H.B. and I.M. Idriss (1982), Ground Motion and Soil Liquefaction During Earthquakes, Earthquake Engineering Research Institute, Berkeley, California, p. 134.

Seed, H.B. and J.R. Booker, (1977), Stabilisation of Potentially Liquefiable Sand Deposits, ASCE, *Journal of Geotechnical Engineering*, Proceedings of ASCE, Vol. 103, No. GT7, pp. 757–768.

Seed, H.B. and J.W.N. Fead (1959), *Apparatus for Repeated Load Tests on Soils*, Special Technical Publication No. L, 204, ASTM, Philadelphia.

Seed, H.B. and K.L. Lee (1965), Studies of Liquefaction of Sand Under Cyclic Loading Conditions Report 65, Department of Civil Engineering, University of California, Berkeley.

Seed, H.B. and K.L. Lee (1966), Liquefaction of Saturated Sand during Cyclic Loading, *Journal of Soil Mechanics and Foundation Engineering*, ASCE, Vol. 92, No. SM 6, pp. 105–134.

Seed, H.B. and Theirs (1957), The Behaviour of Soils under Transient Loadings, Proceedings *Fourth International Conference on Soil Mechanics and Foundation Engineering*, Vol. 1, pp. 207–210, 1957b.

Seed, H.B. and W.H. Peacock (1971), Test Procedures for Measuring Soil Liquefaction, Characteristics, *Journal of Soil Mechanics and Foundations Division*, ASCE, Vol. 97, SM 8, pp. 1099–1119.

Seed, H.B., C. Ugas, and J. Lysmer (1976), Site Dependant Spectra for Earthquake–resistant Design, *Bulletin of Seismological Society of America*, Vol. 66, pp. 221–243.

Seed, H.B., I. Arango, and C.K. Khan (1975), Evaluation of Soil Liquefaction Potential during Earthquake, Report on EERC 75–28, Earthquake Engineering Reaseach Centre, University of California, Berkeley.

Seed, H.B., I.M. Idriss, and I. Arango (1983), Evaluation of Liquefaction Potential Using Field Performance Data, *J. Geotechnical Engineering*, Div., ASCE, 109 No. 3, pp. 458–482.

Seed, H.B., K. Tokimatsu, L.F. Harder, and R.M. Chung (1985), Influences of SPT Procedures in Soil Liquefaction Resistance Evaluation, *Journal of Geotechnical Engineering*, ASCE, Vol. 112, No. 12, pp. 1425–1445.

Seed, H.B., P.P. Martin, and J. Lysmer (1975), The Generation and Dissipation of Pore Water Pressures during Soil Liquefaction. Rep. No. UCB/EERC 75-26, Earthq. Eng. Research Centre, University of California, Berkeley, California.

Seed, H.B., R. Murrakka, J. Lysmer, and I.M. Idriss (1976), Relationship of Maximum Acceleration, Maximum Velocity, Distance from Source and Local Site Conditions for Moderately Strong Earthquakes, Bulletin of Seismological Society of America, Vol. 66, No 4, pp. 1323–1342.

Seed. H.B. (1979), Soil Liquefaction and Cyclic Mobility Evaluation for Level Ground during Earthquakes, *Journal of the Geotechnical Engineering Division*, ASCE, New York, Vol. 105, No. GT2, pp. 201–255.

Seed. H.B., K.O. Cetin, R.E.S. Moss, A.M. Kemmerer, T. Wu, J.M. Pestana, and M.F. Riemer (2001), Recent Advances in Soil Liquefaction Engineering and Seismic Site Response Evaluation, in Proceedings; *Fourth International Conference in Recent Advances in Geotechnical Earthquake Engineering and Soil Dynamics*, Paper No. SLP-2.

Shinozuka, M. and L.C. Shao (1973), Basic Probabilistic Considerations on the Safety of Prestressed Concrete Pressure Vessels, Second International Conference on Structural Mechanics in Reactor Technology, Paper No. H 1/4.

Shocklley, W.G. and R.G. Ahlvin (1960), Non-uniform Conditions in Triaxial Test Specimens, Proc. *Res. Conference on Shear Strength of Cohesive Soils*, ASCE, Boulder, Colorado, pp. 341–357.

Shukla, A.K., R. Prakash, H.P. Shukla, H. V. Gupta, S.N. Bhattacharya, and S.K. Srivastava, (2002), Delhi Seismic Telemetry Network and Analysis of Digital Waveform Data Collected Through this Network, 12th Symposium Earthquake Engineering, IIT Roorkee, pp. 105–113.

Shukla, A.K., R. Prakash, R.K. Singh, P.S. Mishra, and A.K. Bhatnagar (2007), Seismotectonic Implication of Delhi Region Through Fault Plane Solution of Some Recent Earthquake, *Current Science*, Vol. 93, No. 12, pp. 1848–1853.

Sieberg, A. (1930), Die Erdbeben, Part V in Handbuch der Geophysik, Vol. 4, Bornatraeger, Berlin, pp. 527–686. (Sieberg's scale, pp. 552–554).

Silver, M.L., C.K. Chan, et al. (1976), Cyclic Triaxial Strength of Standard Test Sand, *Journal Geot. Engg. Div.* ASCE, Vol. 102, No. GT 5, pp. 511–523.

Singh, A. (1972), How Reliable is the Factor of Safety in Foundation Engineering, Proceedings of the International Conference on Statistics and Probability in Civil Engineering, Hong Kong University Press, pp. 389–424.

Singh, V.P. (2001), *Mechanical Vibrations*, Dhanpat Rai & Co., New Delhi.

Smith, D.J. (1993), *Reliability, Maintainability, and Risk: Practical Methods for Engineers*, 4th ed., Butterworth-Heinemann, Boston.

Somerville, Paul and Moriwaki (2003), Seismic Hazards and Risk Assessment in Engineering Practice, *Handbook of Earthquake Engineering and Engineering Seismology*, Part B, Academic Press, London, pp. 1065–1080.

Srinivasan, P. (1982), *Mechanical Vibration Analysis*, Tata McGraw-Hill, New Delhi.

Srinivasan, P. (1995), *Non-Linear Mechanical Vibrations*, John Wiley & Sons, New York.

Srivastava, H.N. (1992), *Current Science*, 62, pp. 34–39.

Srivastava, L.S. (1969), A Note on the Seismic Zoning Map of India, *Bulletin, Indian Society of Earthquake Technikogy*, Vol. 6(4), pp. 185–194.

Srivastava, L.S. (1974), *Seismic Zoning of India, Earthquake Engineering*, Arya, A.S. (Ed.), Sarita Prakashan, Meerut, pp. 49–65.

Stephenesen, R.W. (1978), Ultrasonic Testing for Determining Soil Modulli, Denver, *Dynamical Geotechnical Testing*, ASTM, STP, pp. 179–195.

Stephens, R.W.P. and A.E. Bate (1950), *Wave Motion and Sound*, Arnold, London.

Stepp, J.C. (1972), Analysis of Completeness of the Earthquake Sample in the Puget Sound Area, and its Effect on Statistical Estimate of Earthquake Hazard, Proceedings International Conference on Microzonation, Seatle, Washington, Vol, 2, pp. 897–910.

Stepp, J.C. (2000), Co-ordination of Strong–motion Programs and Strong Motion Data Dissemination, Proceedings of *12th World Conference on Earthquake Engineering*, Auckland, and Paper No. 2600.

Stokoe, K.H. and F.E. Richart (1973), In-Situ And Laboratory Shear Wave Velocities, Proceedings, *Seventh International Conference on Soil Mechanics and Foundation Engineering*, Vol. 1, pp. 403–409.

Sung, T.Y. (1953), Vibrations in Semi-infinite Solids due to Periodic Surface Loading, ASTM Special Technical Publication, No. 156, *Symposium on Soil Dynamics*, pp. 35–64.

Swami Saran (1999), *Soil Dynamics and Machine Foundations*, Galgotia Publications, New Delhi.

Talwani, P. (1997), Seismotechnotonics of the Koyana-Warna, *Pure and Applied Geophysics*, Vol. 150, pp. 511–550.

Talwani, P. and S. Acree (1984), Pore Pressure Diffusion and the Mechanism of Reservoir Induced Seismicity, *Pure and Applied Geophysics*, Vol. 122, pp. 947–965.

Talwani, P., S.V. Kumarswamy, and C.B. Sawalwade (1996), The Revaluation of Seismicity Data in the Koyana-Warna Area (report), South Carolina, Columbia, South Carolina, USA.

Tandon, A.N. (1954), Study of the Great Assam Earthquake of August, 1950 and its After Shocks, *Indian Journal Meteorology Geophysics*, Vol. 10, pp. 137–146.

Tandon, A.N. (1956), Zones of India Liable to Earthquake Damage, *Indian Journal Meteorology Geophysics*, Vol. 10, pp. 179–191.

Tang, W.H. (1979), Probabilistic Evaluation of Penetration Resistances, *Journal of the Geotechnical Engineering Division*, ASCE, Vol. 105, No. 10, pp. 1173–1191.

Taylor, F.B. (1910), Bearing of the Tertiary Mountain Belt on the Origin of Earth's Plan, *Bulletin Geological Society of America*. Vol. 21, p. 179.

Terzaghi, K. (1925), Erdbaumechanik auf boden physikalischer Grundlage, Leipzig und Wien, Franz Deuticke Vienna.

Terzaghi, K. (1943), *Theoretical Soil Mehanics*, John Wiley & Sons, Inc., New York.

Thatcher, A.R. (1964), Relationships between Bayesian and Confidence Limits for Prediction, *J. Royal Statist. Soc.*, B, Vol. 26, pp. 176–210.

Thevanayagam, S. and G.R. Martin (2002), Liquefaction in Silty Soils-screening and Remediation Issues. *Soil Dynamics and Earthquake Engineering*, Vol. 22, pp. 1035–1042.

Thiers, G.R. and H.B. Seed (1968), Cyclic Stress Strain Characteristics of Clay, *J. Soil Mech. Found. Div*., ASCE, Vol. 94, No. SM 6, pp. 555–569.

Thomsom, W.T. (1965), *Vibration Theory and Applications*, Prentice-Hall Inc., Englewood Cliffs, NJ, USA.

Thomson, W.T. and Kobori, T. (1957), Dynamical Compliance of Rectagular Foundations on Elastic Half Space, *Journal of Applied Mechanics*, *Transactions*, ASME, March 1957, pp. 579–584.

Tichy, M. (1974), A Probabilistic Model for Structural Actions, ACTA Techica CSAV, No. 5, p. 562.

Tichy, M. (1993), *Applied Methods of Structural Reliability*, Kluwer Academic Publishers, Norwell, 416 pp.

Tichy, M. and Vorlicek, M. (1964), Safety of Reforced Concrete Framed Structure, Proceedings of the *International Conference on Flexural Mechanics of Reinforced Concrete*, Miami, Florida, USA.

Timoshenko, S. and J.N. Goodier (1951), *Theory of Elasticity*, McGraw Hill Inc.

Todorovska, M.I. (1996), Liquefaction Hazard Assessment via seismic Wave Energy and SPT Values, *Earthquake Engineering*, Vol. X-N2, pp. 24–37.

Tokimastu, K. and Y. Yoshimi (1983), Empirical Correlation of Soil Liquefaction Based on SPT-N Value and Fines Content, *Soil and Foundations*, Vol. 23, No. 4, pp. 56–74.

Tokimatsu, A.M. and H.B. Seed (1987), Evaluation of Settlements in Sands Due to Earthquake Shaking, *Journal of the Geotechnical Division*, ASCE, Vol. 113, No. 8, pp. 861–878.

Trifunac, M. (1995), Empirical Criteria for Liquefaction in Sands via Standard Penetration Tests and Seismic Wave Energy, *Journal of Soil Dynamics and Earthquake Engineering*, 14(4), pp. 419–426.

Trifunac, M.D. (1988), Seismic Microzonation Mapping via Uniform Risk Spectra, Proceedings *Ninth World Conference Earthquake Engineering*, Tokyo, VII, pp. 187–219.

Trifunac, M.D. (1993), Long Period Fourier Amplitude Spectra of Strong Motion Acceleration, *Soil Dynamics and Earthquake Engineering*, Vol. 12(6), pp. 363–382.

Tschebotarioff, G.P. (1949), Discussion on Strength of Soils Under Dynamic Loads by A. Casagrande and W.L. Shannon, Trans. ASCE, p. 825.

Tse, F.S., I.E. Morse and R.T. Hinkle (1983), *Mechanical Vibrations—Theory and Applications*, 2nd ed., Allyn & Bascon Inc., London, UK.

Tung, P.C. and S.W. Shaw (1988), The Dynamics of an Impact Print Hammer, *Journal of Vibrations*, Acoustics, Stress and Reliability in Design, Vol. 110, pp. 193–200.

Tung, T.P. and N.M. Newmark, (1955) Numerical Analysis of Earthquake Response of a Tall Structure, *Bulletin Seismological Society of America*, Vol. 45, No. 4, pp. 269–278.

Turcotte, D.L. and G., Schubert (1982), *Geodynamics*. John Wiley, New York.

Vanmarcke, E.H. (1972), Properties of Spectral Moments with Application to Random Vibrations, *Journal of EM Division*, ASCE, Vol. 98, No. 3, pp. 425–464.

Vanmarcke, E.H. (1974), Decision Analysis in Dam Safety Monotoring, Proceedings of *Conference on Safety of Small Dams*, Henniku, New Hampshire.

Vanmarcke, E.H. (1976), Structural Response to Earthquakes, Chapter 8 in C. Lomnitz and E. Rosenbleuth, (Eds.), *Seismic Risk and Engineering Decisions*, Elsevier, Amsterdam, pp. 287–338.

Vanmarcke, E.H. (1977), Probabilistic Modeling of Soil Profiles, *Journal of the Geotechnical Engineering Division*, ASCE, Vol. 103, No. GT11, pp. 1227–1246.

Vanmarcke, E.H. (1983), *Random Fields: Analysis and Synthesis*. MIT Press, Cambridge.

Vardoulakis, I. and J. Sulem (1995), *Bifurcation Analysis in Geomechanics*, Blackie Academic and Professional, Glasgow, U.K.

Vlasov, V.Z. and U.N. Leontev (1966), *Beams, Plates and Shells on Elastic Foundations*, National Technical Information Service, Springfield, USA.

Voigt, Dorothy S. and P. Byerly (1949), The Intensity of Earthquakes as Rated from Questionnaires", *ibid.*, Vol. 39, pp. 21–26.

Volterra, E. and E.C. Zachmanoglou (1965), *Dynamics of Vibrations*, Charles E. Merrill Books Inc., Columbus, Ohio, USA.

Vyas, M., S.K. Nath, I. Pal, P. Sengupta and W.K. Mohanty (2004), GSHAP Revisited.

Wadia, D.N. (1957), *Geology of India*, Macmillan, London.

Watanbe, H. and K. Baba (1979), Consideration for an Earthquake Resistant Design Method for Rock-fill Dam, 13th *ICOLD*, Q.51, R.15, pp. 1048–1074.

Weaver, W. Jr., S.P. Timoshenko, and D.H. Young, (1990), *Vibration Problems in Engineering*, John Wiley & Sons.

Wegener, A. (1924), *The Origin of Continents and Occeans*, Metheun.

Wegener, A. (1966), *The Origin of Continents and Oceans*, Dover New York.

Weighardt, K. (1922), Under den ballllken suf Nauchgiebiger unter lage,zeistschrift fur angwandte Mathematic and physik, Vol. 2, pp. 165–184.

Wesnousky, S.G., C.H. Scholz, Shimazaki, and T. Matsuda (1983), Earthquake Frequency Distribution and the Mechanics of Faulting, *J. Geophysical Research*, 88 (BH), pp. 9331–9340.

West, W.D. (1937), *Earthquakes in India*, Presidential Address, 24th Indian Science Congress, pp. 189–227.

Wiegel, Robert L. (Ed.) (1970), *Earthquake Engineering*, Prentice Hall, Upper Saddle River, New Jersey.

Whitman, R.V. (1984), Evaluating Calculated Risk in Geotechnical Engineering, The Seventeenth Terzaghi Lecture, *Journal of Geotechnical Engineering*, ASCE, Vol. 110, No. 2, pp. 145–188.

Whitman, R.V. (2000), Organizing and Evaluating Uncertainty in Geotechnical Engineering, *Journal of Geotechnical and Geoenvironmental Engineering*, ASCE, Vol. 126, No. 7, pp. 583–593.

Wittmann, F.H. (1987), Structural Reliability and Probabilistic Safety Assessment, Ashgate Publishing Co., Brookfield, 498 pp.

Wilson, S.D. and R.J. Dietrich (1960), Effect of Consolidation Pressure on Elastic and Strength Properties of Clay, Proceedings, American Society of Civil Engineers, Research Conference Shear Strength Cohesive Soil, Boulder, Colorado, pp. 419–435.

Winkler (1867), Die lehre von der elastizitat and festigketi, Prague. p. 182.

Wolf, John P. (1985), *Dynamic Soil-Structure Interaction*, Prentice-Hall, Englewood, New Jersey.

Wood, Harry O. and Neumann, Frank, Modified Mercalli Intensity Scale of 1931, B.S.S.A., Vol. 21 (1931), pp. 277–283.

Woods, R.D. (1968), Screening of Surface Waves in Soils, *Journal Soil Mechanics and Foundation Engineering Division*, ASCE, Vol. 94, No. SM 4, pp. 951–979.

Woods, R.D. (1978), Measurement of Dynamics Soil Properties–State of the Art, Proc. ASCE Specialty Conference on Earthquake Engineering and Soil Dynamics, Pasadena, California, Vol. 2.

Wu, T.H. (1974), Uncertainty, Safety, and Decision in Soil Engineering, *Journal of Geotechnical Engineering Division*, ASCE, Vol. 100, No. 3, pp. 329–348.

Wu, T.H. and L.M. Kraft (1967), Safety Analysis of Slopes, *Journal of Soil Mechanics and Foundation Division*, ASCE, Vol. 96, No. SN 2.

Wu, T.H. and L.M. Kraft (1967), The Probability of Foundation Safety, *Journal of Soil Mechanics and Foundation Division*, ASCE, Vol. 93, No. SN 5.

Wysession, M. (1995), *The Inner Working of the Earth*, American Science, March–April, p. 10.

Yilmaztürk, A. and P.W. Burton (1999), An Evaluation of Seismic Hazard Parameters, New York, USA.

Yoshimi, Y., F. E. Richart, S. Prakash, and Barkan, D.D (1977), General Report on Soil Dynamics, Proc. *Ninth International Conference on Soil Mechanics and Foundation Engineering*, Tokyo, Vol. II, pp. 605–650.

Youd, T.L. and D.M. Perkins (1998), Mapping of Liquefaction Severity Index, *Journal of the Geotechnical Engineering Division*, ASCE, Vol. 118, No. 11, pp. 1374–1392.

Youd, T.L. and D.M. Perkins (1978). Mapping of Liquefaction Induced Ground Failure Potential, *Journal of the Geotechnical Engineering Division*, ASCE, Vol. 104, No. 4, pp. 433–446.

Youd, T.L and AM. Idriss (2001), Liquefaction Resistance of Soils, Summary Report from 1996 NCEER and 1998 NCEER/NSF Workshop on Evaluation of Liquefaction Resistance of Soils, *Journal of Geotechnical Engineering*, ASCE, Vol. 127(4), pp. 297–313.

Youd, T.L. (1955), Liquefaction Induced Lateral Ground Displacements, in Proceedings of the *Third International Conference in Recent Advances in Geotechnical Earthquake Engineering and Soil Dynamics*, St. Louis, Missouri, Vol. II, pp. 911–925.

Youd, T.L. and S.D. Gilstrap (1999), Liquefaction and Associated Ground Failures, Proceedings, *Geologic and Hydrologic Hazards Training Program*, Open File Report 84–760, US Geological Survey, Meulo Park, CA, pp. 210–232.

Ziegler, P.A. (1993), Plate Moving Mechanism: their Relative Importance, *Journal Geological Science*, London, pp. 927–940.

Zienkiewicz, O.C. (1991), *The Finite Element Method*, Vol. 2, Solid and Fluid Mechanics Dynamics and Non-Linearity, 4th ed., McGraw Hill, London.

Zienkiewicz, O.C. and T. Shiomi (1984), Dynamic Behaviour of Saturated Porous Media: The Generalized Biot Formation and Its Numerical Solution, *International Journal for Numerical Analytical Method in Geomechanics*, Vol. 18, pp. 71–96.

Zienkiewicz, O.C., A.H.C. Chen, M. Paul, and T. Shiomi (1990a), Static and Dynamic Behaviour of Soil: A Rational Approach to Quantitative Solutions, I, Fully Saturated Problems, Proc. Royal Society, London, A 429, pp. 285–309.

Zienkiewicz, O.C., Y.M. Xie, B.A. Schrefler, A. Ledsesma, and N. Bicannie (1990b), Static and Dynamic Behaviour of Soils, *A Rational Approach to Quantitative Solutions*, Part II: Semi Saturated Problems, Proc. Royal Society, London, A 429, pp. 311–321.

INDEX